FUNDAMENTAL FINITE ELEMENT ANALYSIS AND APPLICATIONS

FUNDAMENTAL FINITE ELEMENT ANALYSIS AND APPLICATIONS

With *Mathematica*® and MATLAB® Computations

M. ASGHAR BHATTI

WILEY

JOHN WILEY & SONS, INC.

This book is printed on acid-free paper. ♾

Published by John Wiley & Sons, Inc., Hoboken, New Jersey
Published simultaneously in Canada.

Library of Congress Cataloging-in-Publication Data

Bhatti, M. Asghar
Fundamental finite element analysis and applications: with Mathematica and Matlab computations / M. Asghar Bhatti.
p. cm.
Includes index.
ISBN 0-471-64808-6
1. Structural analysis (Engineering) 2. Finite element method. I. Title.
TA646.B56 2005
620′.001′51825—dc22 2004002270

Printed in the United States of America
10 9 8 7 6 5 4 3

CONTENTS

CONTENTS OF THE BOOK WEB SITE (www.wiley.com/go/bhatti)

ABAQUS Applications

AbaqusUse\AbaqusExecutionProcedure.pdf
AbaqusUse\HeatFlow
AbaqusUse\PlaneStress
AbaqusUse\TrussAnalysis

ANSYS Applications

AnsysUse\AppendixA
AnsysUse\Chap5
AnsysUse\Chap7
AnsysUse\Chap8
AnsysUse\GeneralProcedure.pdf

Full Detail Text Examples

Full Detail Text Examples\Chap1Examples.pdf
Full Detail Text Examples\Chap2Examples.pdf
Full Detail Text Examples\Chap3Examples.pdf
Full Detail Text Examples\Chap4Examples.pdf
Full Detail Text Examples\Chap5Examples.pdf
Full Detail Text Examples\Chap6Examples.pdf
Full Detail Text Examples\Chap7Examples.pdf

Full Detail Text Examples\Chap8Examples.pdf
Full Detail Text Examples\Chap9Examples.pdf

Mathematica Applications

MathematicaUse\MathChap1.nb
MathematicaUse\MathChap2.nb
MathematicaUse\MathChap3.nb
MathematicaUse\MathChap4.nb
MathematicaUse\MathChap5.nb
MathematicaUse\MathChap6.nb
MathematicaUse\MathChap7.nb
MathematicaUse\MathChap8.nb
MathematicaUse\Mathematica Introduction.nb

MATLAB Applications

MatlabFiles\Chap1
MatlabFiles\Chap2
MatlabFiles\Chap3
MatlabFiles\Chap4
MatlabFiles\Chap5
MatlabFiles\Chap6
MatlabFiles\Chap7
MatlabFiles\Chap8
MatlabFiles\Common

Sample Course Outlines, Lectures, and Examinations

Supplementary Material and Corrections

PREFACE

Large numbers of books have been written on the finite element method. However, effective teaching of the method using most existing books is a difficult task. The vast majority of current books present the finite element method as an extension of the conventional matrix structural analysis methods. Using this approach, one can teach the mechanical aspects of the finite element method fairly well, but there are no satisfactory explanations for even the simplest theoretical questions. Why are rotational degrees of freedom defined for the beam and plate elements but not for the plane stress and truss elements? What is wrong with connecting corner nodes of a planar four-node element to the midside nodes of an eight-node element? The application of the method to nonstructural problems is possible only if one can interpret problem parameters in terms of their structural counterparts. For example, one can solve heat transfer problems because temperature can be interpreted as displacement in a structural problem.

More recently, several new textbooks on finite elements have appeared that emphasize the mathematical basis of the finite element method. Using some of these books, the finite element method can be presented as a method for obtaining approximate solution of ordinary and partial differential equations. The choice of appropriate degrees of freedom, boundary conditions, trial solutions, etc., can now be fully explained with this theoretical background. However, the vast majority of these books tend to be too theoretical and do not present enough computational details and examples to be of value, especially to undergraduate and first-year graduate students in engineering.

The finite element courses face one more hurdle. One needs to perform computations in order to effectively learn the finite element techniques. However, typical finite element calculations are very long and tedious, especially those involving mapped elements. In fact, some of these calculations are essentially impossible to perform by hand. To alleviate this situation, instructors generally rely on programs written in FORTRAN or some other conventional programming language. In fact, there are several books available that include these types of programs with them. However, realistically, in a typical one-semester course, most students cannot be expected to fully understand these programs. At best they use them as *black boxes*, which obviously does not help in learning the concepts.

In addition to traditional research-oriented students, effective finite element courses must also cater to the needs and expectations of practicing engineers and others interested only in the finite element applications. Knowing the theoretical details alone does not help in creating appropriate models for practical, and often complex, engineering systems.

This book is intended to strike an appropriate balance among the theory, generality, and practical applications of the finite element method. The method is presented as a fairly straightforward extension of the classical weighted residual and the Rayleigh-Ritz methods for approximate solution of differential equations. The theoretical details are presented in an informal style appealing to the reader's intuition rather than mathematical rigor. To make the concepts clear, all computational details are fully explained and numerous examples are included showing all calculations. To overcome the tedious nature of calculations associated with finite elements, extensive use of MATLAB® and *Mathematica*® is made in the book. All finite element procedures are implemented in the form of interactive *Mathematica* notebooks and easy-to-follow MATLAB code. All necessary computations are readily apparent from these implementations. Finally, to address the practical applications of the finite element method, the book integrates a series of computer laboratories and projects that involve modeling and solution using commercial finite element software. Short tutorials and carefully chosen sample applications of ANSYS and ABAQUS are contained in the book.

The book is organized in such a way that it can be used very effectively in a lecture/computer laboratory (lab) format. In over 20 years of teaching finite elements, using a variety of approaches, the author has found that presenting the material in a two-hour lecture and one-hour lab per week is ideally suited for the first finite element course. The lecture part develops suitable theoretical background while the lab portion gives students experience in finite element modeling and actual applications. Both parts should be taught in parallel. Of course, it takes time to develop the appropriate theoretical background in the lecture part. The lab part, therefore, is ahead of the lectures and, in the initial stages, students are using the finite element software essentially as a black box. However, this approach has two main advantages. The first is that students have some time to get familiar with the particular computer system and the finite element package being utilized. The second, and more significant, advantage is that it raises students' curiosity in learning more about why things must be done in a certain way. During early labs students often encounter errors such as "negative pivot found" or "zero or negative Jacobian for element." When, during the lecture part, they find out mathematical reasons for such errors, it makes them appreciate the importance of learning theory in order to become better users of the finite element technology.

The author also feels strongly that the labs must utilize one of the several commercially available packages, instead of relying on simple home-grown programs. Use of commercial programs exposes students to at least one state-of-the-art finite element package with its built-in or associated pre- and postprocessors. Since the general procedures are very similar among different programs, it is relatively easy to learn a different package after this exposure. Most commercial programs also include analysis modules for linear and nonlinear static and dynamic analysis, buckling, fluid flow, optimization, and fatigue. Thus with these packages students can be exposed to a variety of finite element applications, even though there generally is not enough time to develop theoretical details of all these topics in one finite element course. With more applications, students also perceive the course as more *practical* and seem to put more effort into learning.

TOPICS COVERED

The book covers the fundamental concepts and is designed for a first course on finite elements suitable for upper division undergraduate students and first-year graduate students. It presents the finite element method as a tool to find approximate solution of differential equations and thus can be used by students from a variety of disciplines. Applications covered include heat flow, stress analysis, fluid flow, and analysis of structural frameworks. The material is presented in nine chapters and two appendixes as follows.

1. *Finite Element Method: The Big Picture.* This chapter presents an overview of the finite element method. To give a clear idea of the solution process, the finite element equations for a few simple elements (plane truss, heat flow, and plane stress) are presented in this chapter. A few general remarks on modeling and discretization are also included. Important steps of assembly, handling boundary conditions, and solutions for nodal unknowns and element quantities are explained in detail in this chapter. These steps are fairly mechanical in nature and do not require complex theoretical development. They are, however, central to actually obtaining a finite element solution for a given problem. The chapter includes brief descriptions of both direct and iterative methods for solution of linear systems of equations. Treatment of linear constraints through Lagrange multipliers and penalty functions is also included.

This chapter gives enough background to students so that they can quickly start using available commercial finite element packages effectively. It plays an important role in the lecture/lab format advocated for the first finite element course.

2. *Mathematical Foundations of the Finite Element Method.* From a mathematical point of view the finite element method is a special form of the well-known Galerkin and Rayleigh-Ritz methods for finding approximate solutions of differential equations. The basic concepts are explained in this chapter with reference to the problem of axial deformation of bars. The derivation of the governing differential equation is included for completeness. Approximate solutions using the classical form of Galerkin and Rayleigh-Ritz methods are presented. Finally, the methods are cast into the form that is suitable for developing finite element equations. Lagrange and Hermitian interpolation functions, commonly employed in derivation of finite element equations, are presented in this chapter.

3. *One-Dimensional Boundary Value Problem.* A large number of practical problems are governed by a one-dimensional boundary value problem of the form

$$\frac{d}{dx}\left(k(x)\frac{du(x)}{dx}\right) + p(x)\,u(x) + q(x) = 0$$

Finite element formulation and solutions of selected applications that are governed by the differential equation of this form are presented in this chapter.

4. *Trusses, Beams, and Frames.* Many structural systems used in practice consist of long slender members of various shapes used in trusses, beams, and frames. This chapter presents finite element equations for these elements. The chapter is important for civil and mechanical engineering students interested in structures. It also covers typical modeling techniques employed in framed structures, such as rigid end zones and rigid floor diaphragms. Those not interested in these applications can skip this chapter without any loss in continuity.

5. *Two-Dimensional Elements.* In this chapter the basic finite element concepts are illustrated with reference to the following partial differential equation defined over an arbitrary two-dimensional region:

$$\frac{\partial}{\partial x}\left(k_x \frac{\partial u}{\partial x}\right) + \frac{\partial}{\partial y}\left(k_y \frac{\partial u}{\partial y}\right) + pu + q = 0$$

The equation can easily be recognized as a generalization of the one-dimensional boundary value problem considered in Chapter 3. Steady-state heat flow, a variety of fluid flow, and the torsion of planar sections are some of the common engineering applications that are governed by the differential equations that are special cases of this general boundary value problem. Solutions of these problems using rectangular and triangular elements are presented in this chapter.

6. *Mapped Elements.* Quadrilateral elements and other elements that can have curved sides are much more useful in accurately modeling arbitrary shapes. Successful development of these elements is based on the key concept of mapping. These concepts are discussed in this chapter. Derivation of the Gaussian quadrature used to evaluate equations for mapped elements is presented. Four- and eight-node quadrilateral elements are presented for solution of two-dimensional boundary value problems. The chapter also includes procedures for forming triangles by collapsing quadrilaterals and for developing the so-called *infinite* elements to handle far-field boundary conditions.

7. *Analysis of Elastic Solids.* The problem of determining stresses and strains in elastic solids subjected to loading and temperature changes is considered in this chapter. The fundamental concepts from elasticity are reviewed. Using these concepts, the governing differential equations in terms of stresses and displacements are derived followed by the general form of finite element equations for analysis of elastic solids. Specific elements for analysis of plane stress and plane strain problems are presented in this chapter. The so-called *singularity* elements, designed to capture a singular stress field near a crack tip, are discussed. This chapter is important for those interested in stress analysis. Those not interested in these applications can skip this chapter without any loss in continuity.

8. *Transient Problems.* This chapter considers analysis of transient problems using finite elements. Formulations for both the transient field problems and the structural dynamics problems are presented in this chapter.

9. *p-Formulation.* In conventional finite element formulation, each element is based on a specific set of interpolation functions. After choosing an element type, the only way to

obtain a better solution is to refine the model. This formulation is called *h-formulation*, where *h* indicates the generic size of an element. An alternative formulation, called the *p-formulation*, is presented in this chapter. In this formulation, the elements are based on interpolation functions that may involve very high order terms. The initial finite element model is fairly coarse and is based primarily on geometric considerations. Refined solutions are obtained by increasing the order of the interpolation functions used in the formulation. Efficient interpolation functions have been developed so that higher order solutions can be obtained in a *hierarchical* manner from the lower order solutions.

10. *Appendix A: Use of Commercial FEA Software.* This appendix introduces students to two commonly used commercial finite element programs, ANSYS and ABAQUS. Concise instructions for solution of structural frameworks, heat flow, and stress analysis problems are given for both programs.

11. *Appendix B: Variational Form for Boundary Value Problems.* The main body of the text employs the Galerkin approach for solution of general boundary value problems and the variational approach (using potential energy) for structural problems. The derivation of the variational functional requires familiarity with the calculus of variations. In the author's experience, given that only limited time is available, most undergraduate students have difficulty fully comprehending this topic. For this reason, and since the derivation is not central to the finite element development, the material on developing variational functionals is moved to this appendix. If desired, this material can be covered with the discussion of the Rayleigh-Ritz method in Chapter 2.

To keep the book to a reasonable length and to make it suitable for a wider audience, important structural oriented topics, such as axisymmetric and three-dimensional elasticity, plates and shells, material and geometric nonlinearity, mixed and hybrid formulations, and contact problems are not covered in this book. These topics are covered in detail in a companion textbook by the author entitled *Advanced Topics in Finite Element Analysis of Structures: With Mathematica® and MATLAB® Computations*, John Wiley, 2006.

UNIQUE FEATURES

(i) All key ideas are introduced in chapters that emphasize the method as a way to find approximate solution of boundary value problems. Thus the book can be used effectively for students from a variety of disciplines.

(ii) The "big picture" chapter gives readers an overview of all the mechanical details of the finite element method very quickly. This enables instructors to start using commercial finite element software early in the semester, thus allowing plenty of opportunity to bring practical modeling issues into the classroom. The author is not aware of any other book that starts out in this manner. Few books that actually try to do this do so by taking discrete spring and bar elements. In my experience this does not work very well because students do not see actual finite element applications. Also, this approach does not make sense to those who are not interested in structural applications.

(iii) Chapters 2 and 3 introduce fundamental finite element concepts through one-dimensional examples. The axial deformation problem is used for a gentle introduction to the subject. This allows for parameters to be interpreted in physical terms. The derivation of the governing equations and simple techniques for obtaining exact solutions are included to help those who may not be familiar with the structural terminology. Chapter 3 also includes solution of one-dimensional boundary value problems without reference to any physical application for non-structural readers.

(iv) Chapter 4, on structural frameworks, is quite unique for books on finite elements. No current textbook that approaches finite elements from a differential equation point of view also has a complete coverage of structural frames, especially in three dimensions. In fact, even most books specifically devoted to structural analysis do not have as satisfactory a coverage of the subject as provided in this chapter.

(v) Chapters 5 and 6 are two important chapters that introduce key finite element concepts in the context of two-dimensional boundary value problems. To keep the integration and differentiation issues from clouding the basic ideas, Chapter 5 starts with rectangular elements and presents complete examples using such elements. The triangular elements are presented next. By the time the mapped elements are presented in Chapter 6, there are no real finite element–related concepts left. It is all just calculus. This clear distinction between the fundamental concepts and calculus-related issues gives instructors flexibility in presenting the material to students with a wide variety of mathematics background.

(vi) Chapter 9, on *p*-formulation, is unique. No other book geared toward the first finite element course even mentions this important formulation. Several ideas presented in this chapter are used in recent development of the so-called *meshless* methods.

(vii) *Mathematica* and MATLAB implementations are included to show how calculations can be organized using a computer algebra system. These implementations require only the very basic understanding of these systems. Detailed examples are presented in Chapter 1 showing how to generate and assemble element equations, reorganize matrices to account for boundary conditions, and then solve for primary and secondary unknowns. These steps remain exactly the same for all implementations. Most of the other implementations are nothing more than element matrices written using *Mathematica* or MATLAB syntax.

(viii) Numerous numerical examples are included to clearly show all computations involved.

(ix) All chapters contain problems for homework assignment. Most chapters also contain problems suitable for computer labs and projects. The accompanying web site (www.wiley.com/go/bhatti) contains all text examples, MATLAB and *Mathematica* functions, and ANSYS and ABAQUS files in electronic form. To keep the printed book to a reasonable length most examples skip some computations. The web site contains full computational details of these examples. Also the book generally alternates between showing examples done with *Mathematica* and MATLAB. The web site contains implementations of all examples in both *Mathematica* and MATLAB.

TYPICAL COURSES

The book can be used to develop a number of courses suitable for different audiences.

First Finite Element Course for Engineering Students About 32 hours of lectures and 12 hours of labs (selected materials from indicated chapters):

Chapter 1: Finite element procedure, discretization, element equations, assembly, boundary conditions, solution of primary unknowns and element quantities, reactions, solution validity (4 hr)

Chapter 2: Weak form for approximate solution of differential equations, Galerkin method, approximate solutions using Rayleigh-Ritz method, comparison of Galerkin and Rayleigh-Ritz methods, Lagrange and Hermite interpolation, axial deformation element using Rayleigh-Ritz and Galerkin methods (6 hr)

Chapter 3: 1D BVP, FEA solution of BVP, 1D BVP applications (3 hr)

Chapter 4: Finite element for beam bending, beam applications, structural frames (3 hr)

Chapter 5: Finite elements for 2D and 3D problems, linear triangular element for second-order 2D BVP, 2D fluid flow and torsion problems (4 hr)

Chapter 6: 2D Lagrange and serendipity shape functions, mapped elements, evaluation of area integrals for 2D mapped elements, evaluation of line integrals for 2D mapped elements (4 hr)

Chapter 7: Stresses and strains in solids, finite element analysis of elastic solids, CST and isoparametric elements for plane elasticity (4 hr)

Chapter 8: Transient problems (2 hr)

Review, exams (2 hr)

About 12 hours of labs (some sections from the indicated chapters supplemented by documentation of the chosen commercial software):

Appendix: Introduction to *Mathematica* and/or MATLAB (2 hr)

Chapters 1 and 4: Software documentation, basic finite element procedure using commercial software, truss and frame problems (2 hr)

Chapters 1 and 5: Software documentation, 2D mesh generation, heat flow problems (2 hr)

Chapters 1 and 7: 2D, axisymmetric, and 3D stress analysis problems (2 hr)

Chapter 8: Transient problems (2 hr)

Software documentation: Constraints, design optimization (2 hr)

First Finite Element Course for Students Not Interested in Structural Applications Skip Chapters 4 and 7. Spend more time on applications in Chapters 5 and 6. Introduce Chapter 9: *p*-Formulation. In the labs replace truss, frame, and stress analysis problems with appropriate applications.

Finite Element Course for Practicing Engineers From the current book: Chapters 1, 2, 6, and 7. From the companion advanced book: Chapters 1, 2, and 5 and selected material from Chapters 6, 7, and 8.

Finite Element Modeling and Applications For a short course on finite element modeling or self-study, it is suggested to cover the first chapter in detail and then move on to Appendix A for specific examples of using commercial finite element packages for solution of practical problems.

ACKNOWLEDGMENTS

Most of the material presented in the book has become part of the standard finite element literature, and hence it is difficult to acknowledge contributions of specific individuals. I am indebted to the pioneers in the field and the authors of all existing books and journal papers on the subject. I have obviously benefited from their contributions and have used a good number of them in my over 20 years of teaching the subject.

I wrote the first draft of the book in early 1990. However, the printed version has practically nothing in common with that first draft. Primarily as a result of questions from my students, I have had to make extensive revisions almost every year. Over the last couple of years the process began to show signs of convergence and the result is what you see now. Thus I would like to acknowledge all direct and indirect contributions of my former students. Their questions hopefully led me to explain things in ways that make sense to most readers. (A note to future students and readers: Please keep the questions coming.)

I want to thank my former graduate student Ryan Vignes, who read through several drafts of the book and provided valuable feedback. Professors Jia Liu and Xiao Shaoping used early versions of the book when they taught finite elements. Their suggestions have helped a great deal in improving the book. My colleagues Professors Ray P.S. Han, Hosin David Lee, and Ralph Stephens have helped by sharing their teaching philosophy and by keeping me in shape through heated games of badminton and tennis.

Finally, I would like to acknowledge the editorial staff of John Wiley for doing a great job in the production of the book. I am especially indebted to Jim Harper, who, from our first meeting in Seattle in 2003, has been in constant communication and has kept the process going smoothly. Contributions of senior production editor Bob Hilbert and editorial assistant Naomi Rothwell are gratefully acknowledged.

CHAPTER ONE

FINITE ELEMENT METHOD: THE BIG PICTURE

Application of physical principles, such as mass balance, energy conservation, and equilibrium, naturally leads many engineering analysis situations into differential equations. Methods have been developed for obtaining exact solutions for various classes of differential equations. However, these methods do not apply to many practical problems because either their governing differential equations do not fall into these classes or they involve complex geometries. Finding analytical solutions that also satisfy boundary conditions specified over arbitrary two- and three-dimensional regions becomes a very difficult task. Numerical methods are therefore widely used for solution of practical problems in all branches of engineering.

The *finite element method* is one of the numerical methods for obtaining approximate solution of ordinary and partial differential equations. It is especially powerful when dealing with boundary conditions defined over complex geometries that are common in practical applications. Other numerical methods such as *finite difference* and *boundary element* methods may be competitive or even superior to the finite element method for certain classes of problems. However, because of its versatility in handling arbitrary domains and availability of sophisticated commercial finite element software, over the last few decades, the finite element method has become the preferred method for solution of many practical problems. Only the finite element method is considered in detail in this book. Readers interested in other methods should consult appropriate references. Books by Zienkiewicz and Morgan [45], Celia and Gray [32], and Lapidus and Pinder [37] are particularly useful for those interested in a comparison of different methods.

The application of the finite element method to a given problem involves the following six steps:

1. Development of element equations
2. Discretization of solution domain into a finite element mesh
3. Assembly of element equations
4. Introduction of boundary conditions
5. Solution for nodal unknowns
6. Computation of solution and related quantities over each element

The key idea of the finite element method is to discretize the solution domain into a number of simpler domains called *elements*. An approximate solution is assumed over an element in terms of solutions at selected points called *nodes*. To give a clear idea of the overall finite element solution process, the finite element equations for a few simple elements are presented in Section 1.1. Obviously at this stage it is not possible to give derivations of these equations. The derivations must wait until later chapters after we have developed enough theoretical background. Few general remarks on discretization are also made in Section 1.1. More specific comments on modeling are presented in later chapters when discussing various applications. Important steps of assembly, handling boundary conditions, and solutions for nodal unknowns and element quantities remain essentially unchanged for any finite element analysis. Thus these procedures are explained in detail in Sections 1.2, 1.3, and 1.4. These steps are fairly mechanical in nature and do not require complex theoretical development. They are, however, central to actually obtaining a finite element solution for a given problem. Therefore, it is important to fully master these steps before proceeding to the remaining chapters in the book.

The finite element process results in a large system of equations that must be solved for determining nodal unknowns. Several methods are available for efficient solution of these large and relatively sparse systems of equations. A brief introduction to two commonly employed methods is given in Section 1.5. In some finite element modeling situations it becomes necessary to introduce constraints in the finite element equations. Section 1.6 presents examples of few such situations and discusses two different methods for handling these so-called *multipoint constraints*. A brief section on appropriate use of units in numerical calculations concludes this chapter.

1.1 DISCRETIZATION AND ELEMENT EQUATIONS

Each analysis situation that is described in terms of one or more differential equations requires an appropriate set of element equations. Even for the same system of governing equations, several elements with different shapes and characteristics may be available. It is crucial to choose an appropriate element type for the application being considered. A proper choice requires knowledge of all details of element formulation and a thorough understanding of approximations introduced during its development.

A key step in the derivation of element equations is an assumption regarding the solution of the governing differential equation over an element. Several practical elements are available that assume a simple linear solution. Other elements use more sophisticated functions to describe solution over elements. The assumed element solutions are written in terms of unknown solutions at selected points called *nodes*. The unknown solutions at the nodes are generally referred to as the *nodal degrees of freedom*, a terminology that dates back to the early development of the method by structural engineers. The appropriate choice of nodal degrees of freedom depends on the governing differential equation and will be discussed in the following chapters.

The geometry of an element depends on the type of the governing differential equation. For problems defined by one-dimensional ordinary differential equations, the elements are straight or curved line elements. For problems governed by two-dimensional partial differential equations the elements are usually of triangular or quadrilateral shape. The element sides may be straight or curved. Elements with curved sides are useful for accurately modeling complex geometries common in applications such as shell structures and automobile bodies. Three-dimensional problems require tetrahedral or solid *brick*-shaped elements. Typical element shapes for one-, two-, and three-dimensional (1D, 2D, and 3D) problems are shown in Figure 1.1. The nodes on the elements are shown as dark circles.

Element equations express a relationship between the physical parameters in the governing differential equations and the nodal degrees of freedom. Since the number of equations for some of the elements can be very large, the element equations are almost always written using a matrix notation. The computations are organized in two phases. In the first phase (the element derivation phase), the element matrices are developed for a typical element that is representative of all elements in the problem. Computations are performed in a symbolic form without using actual numerical values for a specific element. The goal is to develop general formulas for element matrices that can later be used for solution of any numerical problem belonging to that class. In the second phase, the general formulas are used to write specific numerical matrices for each element.

One of the main reasons for the popularity of the finite element method is the wide availability of general-purpose finite element analysis software. This software development is possible because general element equations can be programmed in such a way that, given nodal coordinates and other physical parameters for an element, the program returns numerical equations for that element. Commercial finite element programs contain a large library of elements suitable for solution of a wide variety of practical problems.

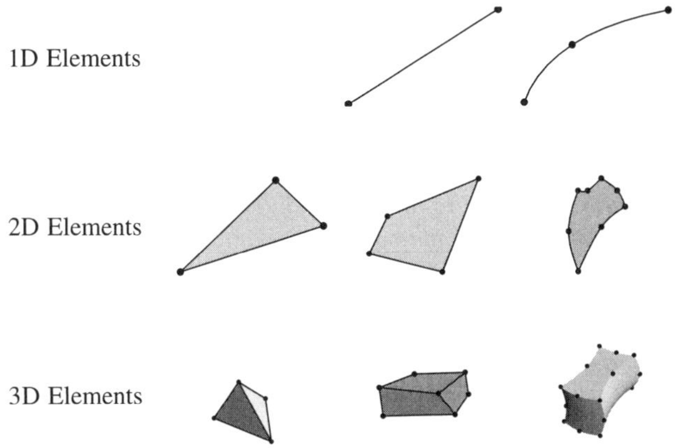

Figure 1.1. Typical finite element shapes

To give a clear picture of the overall finite element solution procedure, the general finite element equations for few commonly used elements are given below. The detailed derivations of these equations are presented in later chapters.

1.1.1 Plane Truss Element

Many structural systems used in practice consist of long slender shapes of various cross sections. Systems in which the shapes are arranged so that each member primarily resists axial forces are usually known as *trusses*. Common examples are roof trusses, bridge supports, crane booms, and antenna towers. Figure 1.2 shows a transmission tower that can be modeled effectively as a plane truss. For modeling purposes all members are considered pin jointed. The loads are applied at the joints. The analysis problem is to find joint displacements, axial forces, and axial stresses in different members of the truss.

Clearly the basic element to analyze any plane truss structure is a two-node straight-line element oriented arbitrarily in a two-dimensional x-y plane, as shown in the Figure 1.3. The element end nodal coordinates are indicated by (x_1, y_1) and (x_2, y_2). The element axis s runs from the first node of the element to the second node. The angle α defines the orientation of the element with respect to a global x-y coordinate system. Each node has two displacement degrees of freedom, u indicating displacement in the x direction and v indicating displacement in the y direction. The element can be subjected to loads only at its ends.

Using these elements, the finite element model of the transmission tower is as shown in Figure 1.4. The model consists of 16 nodes and 29 plane truss elements. The element numbers and node numbers are assigned arbitrarily for identification purposes.

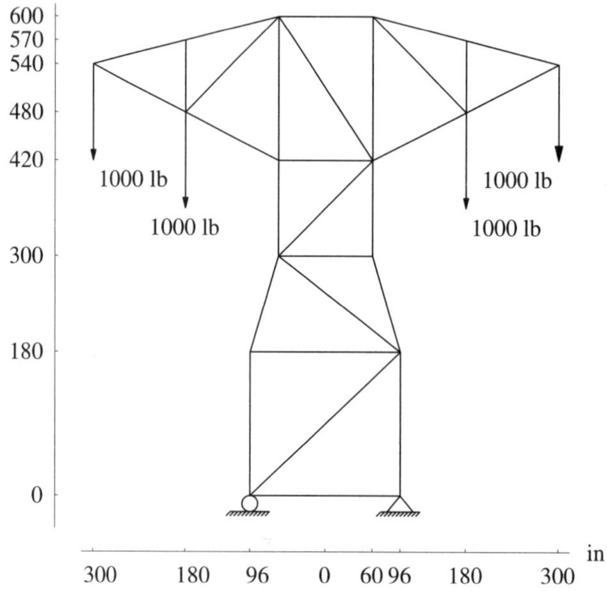

Figure 1.2. Transmission tower

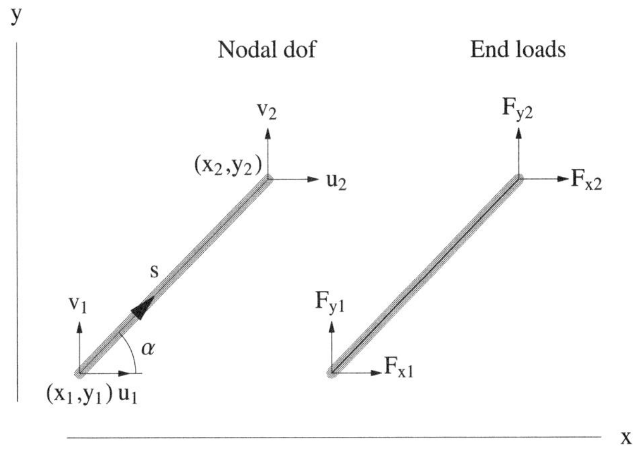

Figure 1.3. Plane truss element

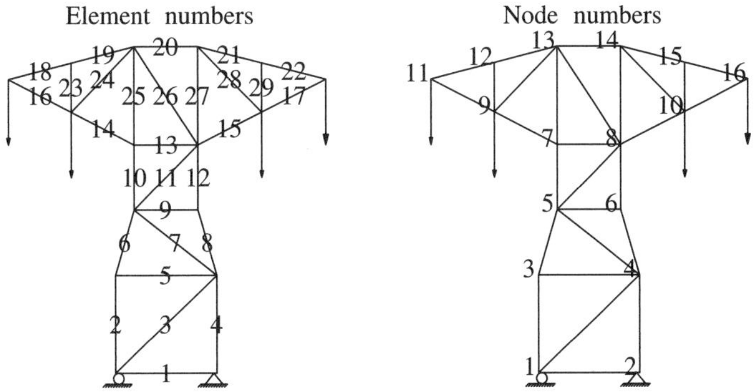

Figure 1.4. Plane truss element model of the transmission tower

Using procedures discussed in later chapters, it can be shown that the finite element equations for a plane truss element are as follows:

$$\frac{EA}{L}\begin{pmatrix} l_s^2 & l_s m_s & -l_s^2 & -l_s m_s \\ l_s m_s & m_s^2 & -l_s m_s & -m_s^2 \\ -l_s^2 & -l_s m_s & l_s^2 & l_s m_s \\ -l_s m_s & -m_s^2 & l_s m_s & m_s^2 \end{pmatrix}\begin{pmatrix} u_1 \\ v_1 \\ u_2 \\ v_2 \end{pmatrix} = \begin{pmatrix} F_{1x} \\ F_{1y} \\ F_{2x} \\ F_{2y} \end{pmatrix}$$

where E = elastic modulus of the material (Young's modulus), A = area of cross section of the element, L = length of the element, and l_s, m_s are the direction cosines of the element axis (line from element node 1 to 2). Here, l_s is the cosine of angle α between the element axis and the x axis (measured counterclockwise) and m_s is the cosine of angle between the element axis and the y axis. In terms of element nodal coordinates,

$$L = \sqrt{(x_2 - x_1)^2 + (y_2 - y_1)^2}; \quad l_s = \cos\alpha = \frac{x_2 - x_1}{L}; \quad m_s = \cos(90-\alpha) = \sin\alpha = \frac{y_2 - y_1}{L}$$

In the element equations the left-hand-side coefficient matrix is usually called the *stiffness matrix* and the right-hand-side vector as the nodal load vector. Note that once the element end coordinates, material property, cross-sectional area, and element loading are specified, the only unknowns in the element equations are the nodal displacements.

It is important to recognize that the element equations refer to an isolated element. We cannot solve for the nodal degrees of freedom for the entire structure by simply solving the equations for one element. We must consider contributions of all elements, loads, and support conditions before solving for the nodal unknowns. These procedures are discussed in detail in later sections of this chapter.

Example 1.1 Write finite element equations for element number 14 in the finite element model of the transmission tower shown in Figure 1.4. The tower is made of steel ($E = 29 \times 10^6 \text{lb/in}^2$) angle sections. The area of cross section of element 14 is 1.73 in^2.

The element is connected between nodes 7 and 9. We can choose either node as the first node of the element. Choosing node 7 as the first node establishes the element s axis as going from node 7 toward 9. The origin of the global x-y coordinate system can be placed at any convenient location. Choosing the centerline of the tower as the origin, the nodal coordinates for the element 14 are as follows:

$$\text{First node (node 7)} = \{-60, 420\} \text{ in;} \qquad x_1 = -60; \qquad y_1 = 420$$
$$\text{Second node (node 9)} = \{-180, 480\} \text{ in;} \qquad x_2 = -180; \qquad y_2 = 480$$

Using these coordinates, the element length and the direction cosines can easily be calculated as follows:

$$\text{Element length:} \qquad L = \sqrt{(x_2 - x_1)^2 + (y_2 - y_1)^2} = 60\sqrt{5} \text{ in}$$

$$\text{Element direction cosines:} \qquad l_s = \frac{x_2 - x_1}{L} = -\frac{2}{\sqrt{5}}; \qquad m_s = \frac{y_2 - y_1}{L} = \frac{1}{\sqrt{5}}$$

From the given material and section properties,

$$E = 29000000 \text{ lb/in}^2; \qquad A = 1.73 \text{ in}^2; \qquad \frac{EA}{L} = 373945. \text{ lb/in}$$

Using these values, the element stiffness matrix (the left-hand side of the element equations) can easily be written as follows:

$$k = \frac{EA}{L}\begin{pmatrix} l_s^2 & l_s m_s & -l_s^2 & -l_s m_s \\ l_s m_s & m_s^2 & -l_s m_s & -m_s^2 \\ -l_s^2 & -l_s m_s & l_s^2 & l_s m_s \\ -l_s m_s & -m_s^2 & l_s m_s & m_s^2 \end{pmatrix} = \begin{pmatrix} 299156. & -149578. & -299156. & 149578. \\ -149578. & 74789. & 149578. & -74789. \\ -299156. & 149578. & 299156. & -149578. \\ 149578. & -74789. & -149578. & 74789. \end{pmatrix}$$

The right-hand-side vector of element equations represents applied loads at the element ends. There are no loads applied at node 7. The applied load of 1000 lb at node 9 is shared by elements 14, 16, 23, and 24. The portion taken by element 14 cannot be determined without detailed analysis of the tower, which is exactly what we are attempting to do in the first place. Fortunately, to proceed with the analysis, it is not necessary to know the portion of the load resisted by different elements meeting at a common node. As will become clear in the next section, in which we consider the assembly of element equations, our goal is to generate a global system of equations applicable to the entire structure. As far as the entire structure is concerned, node 9 has an applied load of 1000 lb in the $-y$ direction. Thus, it is immaterial how we assign nodal loads to the elements as long as the total load at the node is equal to the applied load. Keeping this in mind, when computing element equations, we can simply ignore concentrated loads applied at the nodes and apply them directly to the global equations at the start of the assembly process. Details of this process are presented in a following section.

Assuming nodal loads are to be added directly to the global equations, the finite element equations for element 14 are as follows:

$$\begin{pmatrix} 299156. & -149578. & -299156. & 149578. \\ -149578. & 74789. & 149578. & -74789. \\ -299156. & 149578. & 299156. & -149578. \\ 149578. & -74789. & -149578. & 74789. \end{pmatrix} \begin{pmatrix} u_7 \\ v_7 \\ u_9 \\ v_9 \end{pmatrix}, = \begin{pmatrix} 0 \\ 0 \\ 0 \\ 0 \end{pmatrix},$$

◆ *Mathematica*/MATLAB Implementation 1.1 on the Book Web Site: *Plane truss element equations*

1.1.2 Triangular Element for Two-Dimensional Heat Flow

Consider the problem of finding steady-state temperature distribution in long chimneylike structures. Assuming no temperature gradient in the longitudinal direction, we can take a unit slice of such a structure and model it as a two-dimensional problem to determine the temperature $T(x, y)$. Using conservation of energy on a differential volume, the following governing differential equation can easily be established.:

$$\frac{\partial}{\partial x}\left(k_x \frac{\partial T}{\partial x}\right) + \frac{\partial}{\partial y}\left(k_y \frac{\partial T}{\partial y}\right) + Q = 0$$

where k_x and k_y are thermal conductivities in the x and y directions and $Q(x, y)$ is specified heat generation per unit volume. Typical units for k are W/m · °C or Btu/hr · ft · °F and those for Q are W/m^3 or Btu/hr · ft^3. The possible boundary conditions are as follows:

(i) Known temperature along a boundary:

$$T = T_0 \text{ specified}$$

(ii) Specified heat flux along a boundary:

$$-k\frac{\partial T}{\partial n} \equiv -\left(k_x \frac{\partial T}{\partial x} n_x + k_y \frac{\partial T}{\partial y} n_y\right) = q_0 \text{ specified}$$

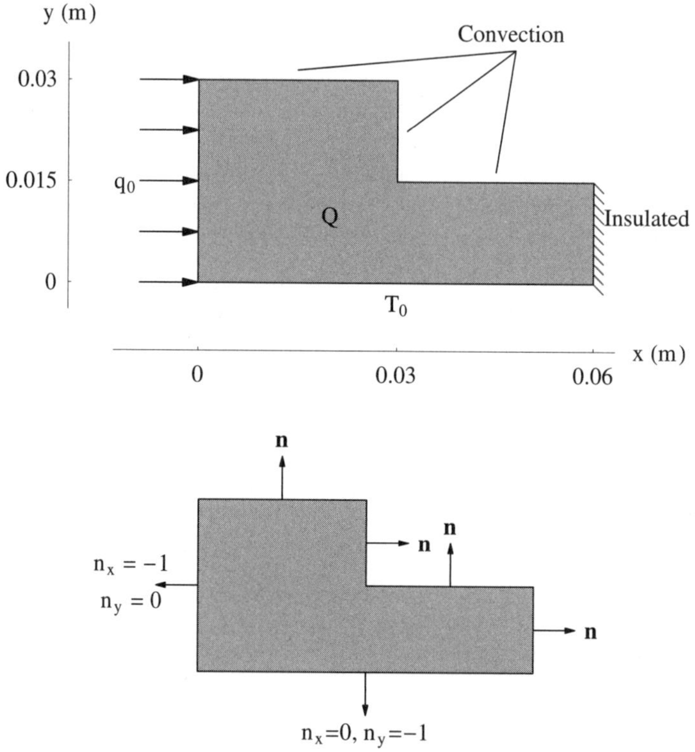

Figure 1.5. Heat flow through an L-shaped solid: solution domain and unit normals

where n_x and n_y are the x and y components of the outer unit normal vector to the boundary (see Figure 1.5 for an example):

$$n = \begin{pmatrix} n_x \\ n_y \end{pmatrix}; \qquad |n| = \sqrt{n_x^2 + n_y^2} = 1$$

On an insulated boundary or across a line of symmetry there is no heat flow and thus $q_0 = 0$. The sign convention for heat flow is that heat flowing into a body is positive and that flowing out of the body is negative.

(iii) Heat loss due to convection along a boundary:

$$-k\frac{\partial T}{\partial n} \equiv -\left(k_x\frac{\partial T}{\partial x}n_x + k_y\frac{\partial T}{\partial y}n_y\right) = h(T - T_\infty)$$

where h is the convection coefficient, T is the unknown temperature at the boundary, and T_∞ is the known temperature of the surrounding fluid. Typical units for h are $W/m^2 \cdot {}^\circ C$ and $Btu/hr \cdot ft^2 \cdot {}^\circ F$.

As a specific example, consider two-dimensional heat flow over an L-shaped body shown in Figure 1.5. The thermal conductivity in both directions is the same, $k_x = k_y =$

45 W/m · °C. The bottom is maintained at a temperature of $T_0 = 110$°C. Convection heat loss takes place on the top where the ambient air temperature is 20°C and the convection heat transfer coefficient is $h = 55$ W/m^2 · °C. The right side is insulated. The left side is subjected to heat flux at a uniform rate of $q_0 = 8000$ W/m^2. Heat is generated in the body at a rate of $Q = 5 \times 10^6$ W/m^3.

Substituting the given data into the governing differential equation and the boundary conditions, we see that the temperature distribution over this body must satisfy the following conditions:

Over the entire L-shaped region	$45\left(\dfrac{\partial^2 T}{\partial x^2} + \dfrac{\partial^2 T}{\partial y^2}\right) + 5 \times 10^6 = 0$
On the left side ($n_x = -1, n_y = 0$)	$-\left(45\dfrac{\partial T}{\partial x}(-1)\right) = 8000 \Longrightarrow \dfrac{\partial T}{\partial x} = \dfrac{8000}{45}$ along $x = 0$
On the bottom of the region	$T = 110$ along $y = 0$
On the right side ($n_x = 1, n_y = 0$)	$\dfrac{\partial T}{\partial x} = 0$ along $x = 0.06$
On the horizontal portions of the top side ($n_x = 0, n_y = 1$)	$-\left(45\dfrac{\partial T}{\partial y}(1)\right) = 55(T - 20) \Longrightarrow \dfrac{\partial T}{\partial y} = -\dfrac{55}{45}(T - 20)$
On the vertical portion of the top side ($n_x = 1, n_y = 0$)	$-\left(45\dfrac{\partial T}{\partial x}(1)\right) = 55(T - 20) \Longrightarrow \dfrac{\partial T}{\partial x} = -\dfrac{55}{45}(T - 20)$

Clearly there is little hope of finding a simple function $T(x, y)$ that satisfies all these requirements. We must resort to various numerical techniques. In the finite element method, the domain is discretized into a collection of elements, each one of them being of a simple geometry, such as a triangle, a rectangle, or a quadrilateral.

A triangular element for solution of steady-state heat flow over two-dimensional bodies is shown in Figure 1.6. The element can be used for finding temperature distribution

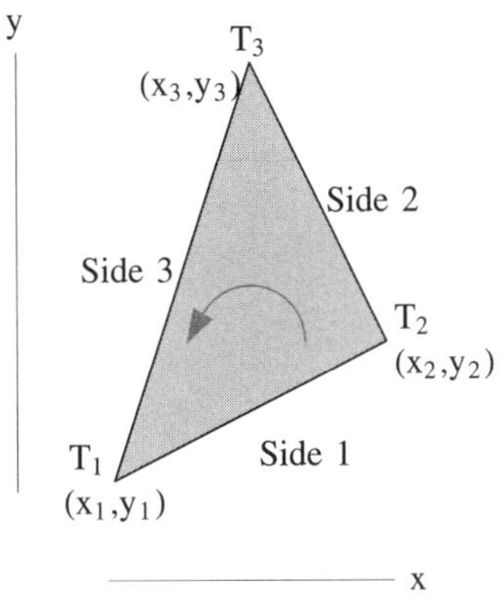

Figure 1.6. Triangular element for heat flow

over any two-dimensional body subjected to conduction and convection. The element is defined by three nodes with nodal coordinates indicated by (x_1, y_1), (x_2, y_2), and (x_3, y_3). The starting node of the triangle is arbitrary, but we must move counterclockwise around the triangle to define the other two nodes. The nodal degrees of freedom are the unknown temperatures at each node T_1, T_2, and T_3.

For the truss model considered in the previous section, the structure was discrete to start with, and thus there was only one possibility for a finite element model. This is not the case for the two-dimensional regions. There are many possibilities in which a two-dimensional domain can be discretized using triangular elements. One must decide on the number of elements and their arrangement. In general, the accuracy of the solution improves as the number of elements is increased. The computational effort, however, increases rapidly as well. Concentrating more elements in regions where rapid changes in solution are expected produces finite element discretizations that give excellent results with reasonable computational effort. Some general remarks on constructing good finite element meshes are presented in a following section. For the L-shaped solid a very coarse finite element discretization is as shown in Figure 1.7 for illustration. To get results that are meaningful from an actual design point of view, a much finer mesh, one with perhaps 100 to 200 elements, would be required.

The finite element equations for a triangular element for two-dimensional steady-state heat flow are derived in Chapter 5. The equations are based on the assumption of linear

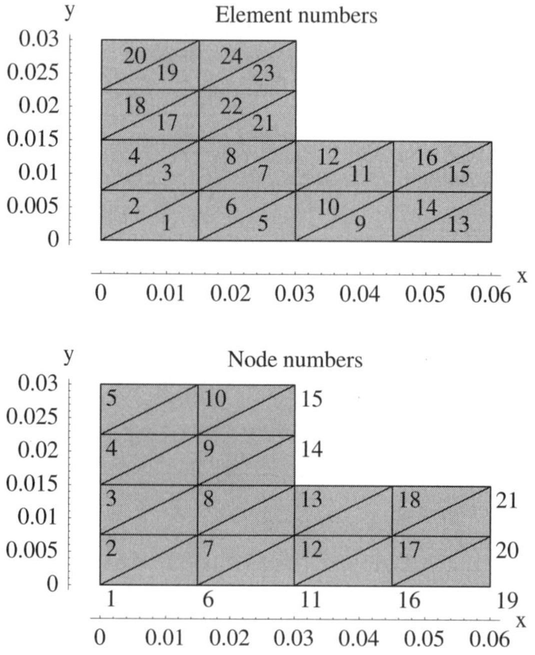

Figure 1.7. Triangular element mesh for heat flow through an L-shaped solid

temperature distribution over the element. In terms of nodal temperatures, the temperature distribution over a typical element is written as follows:

$$T(x, y) = N_1 T_1 + N_2 T_2 + N_3 T_3 = \begin{pmatrix} N_1 & N_2 & N_3 \end{pmatrix} \begin{pmatrix} T_1 \\ T_2 \\ T_3 \end{pmatrix} \equiv \boldsymbol{N}^T \boldsymbol{d}$$

where

$$\boldsymbol{N} = \begin{pmatrix} N_1 \\ N_2 \\ N_3 \end{pmatrix} = \frac{1}{2A} \begin{pmatrix} xb_1 + yc_1 + f_1 \\ xb_2 + yc_2 + f_2 \\ xb_3 + yc_3 + f_3 \end{pmatrix}$$

The quantities N_i, $i = 1, 2, 3$, are known as interpolation or shape functions. The superscript T over $\boldsymbol{N}$ indicates matrix transpose. The vector $\boldsymbol{d}$ is the vector of nodal unknowns. The terms $b_1, c_1, \ldots$ depend on element coordinates and are defined as follows:

$$\begin{array}{lll} b_1 = y_2 - y_3; & b_2 = y_3 - y_1; & b_3 = y_1 - y_2 \\ c_1 = x_3 - x_2; & c_2 = x_1 - x_3; & c_3 = x_2 - x_1 \\ f_1 = x_2 y_3 - x_3 y_2; & f_2 = x_3 y_1 - x_1 y_3; & f_3 = x_1 y_2 - x_2 y_1 \end{array}$$

The area of the triangle A can be computed from the following equation:

$$A = \frac{1}{2} \det \begin{pmatrix} 1 & x_1 & y_1 \\ 1 & x_2 & y_2 \\ 1 & x_3 & y_3 \end{pmatrix} \equiv \frac{f_1 + f_2 + f_3}{2}$$

where det indicates determinant of the matrix.

A note on the notation employed for vectors and matrices in this book is in order here. As an easy-to-remember convention, all vectors are considered *column vectors* and are denoted by boldface italic characters. When an expression needs a row vector, a superscript T is used to indicate that it is the transpose of a column vector. Matrices are also denoted by boldface italic characters. The numbers of rows and columns in a matrix should be carefully noted in the initial definition. Remember that, for matrix multiplication to make sense, the number of columns in the first matrix should be equal to the number of rows in the second matrix. Since large column vectors occupy lot of space on a page, occasionally vector elements may be displayed in a row to save space. However, for matrix operations, they are still treated as column vectors.

As shown in Chapter 5, the finite element equations for this element are as follows:

$$[\boldsymbol{k}_k + \boldsymbol{k}_h]\boldsymbol{d} = \boldsymbol{r}_Q + \boldsymbol{r}_h + \boldsymbol{r}_q$$

$$\boldsymbol{k}_k = \frac{1}{4A}\begin{pmatrix} k_x b_1^2 + k_y c_1^2 & k_x b_1 b_2 + k_y c_1 c_2 & k_x b_1 b_3 + k_y c_1 c_3 \\ k_x b_1 b_2 + k_y c_1 c_2 & k_x b_2^2 + k_y c_2^2 & k_x b_2 b_3 + k_y c_2 c_3 \\ k_x b_1 b_3 + k_y c_1 c_3 & k_x b_2 b_3 + k_y c_2 c_3 & k_x b_3^2 + k_y c_3^2 \end{pmatrix}$$

$$\boldsymbol{r}_Q = \frac{QA}{3}\begin{pmatrix} 1 \\ 1 \\ 1 \end{pmatrix}$$

where k_x = heat conduction coefficient in the x direction, k_y = heat conduction coefficient in the y direction, and Q = heat generated per unit volume over the element. The matrix $\boldsymbol{k}_h$ and the vector $\boldsymbol{r}_h$ take into account any specified heat loss due to convection along one or more sides of the element. If the convection heat loss is specified along side 1 of the element, then we have

Convection along side 1: $\boldsymbol{k}_h = \frac{hL_{12}}{6}\begin{pmatrix} 2 & 1 & 0 \\ 1 & 2 & 0 \\ 0 & 0 & 0 \end{pmatrix}$; $\boldsymbol{r}_h = \frac{hT_\infty L_{12}}{2}\begin{pmatrix} 1 \\ 1 \\ 0 \end{pmatrix}$

where h = convection heat flow coefficient, T_∞ = temperature of the air or other fluid surrounding the body, and L_{12} = length of side 1 of the element. For convection heat flow along sides 2 or 3, the matrices are as follows:

Convection along side 2: $\boldsymbol{k}_h = \frac{hL_{23}}{6}\begin{pmatrix} 0 & 0 & 0 \\ 0 & 2 & 1 \\ 0 & 1 & 2 \end{pmatrix}$; $\boldsymbol{r}_h = \frac{hT_\infty L_{23}}{2}\begin{pmatrix} 0 \\ 1 \\ 1 \end{pmatrix}$

Convection along side 3: $\boldsymbol{k}_h = \frac{hL_{31}}{6}\begin{pmatrix} 2 & 0 & 1 \\ 0 & 0 & 0 \\ 1 & 0 & 2 \end{pmatrix}$; $\boldsymbol{r}_h = \frac{hT_\infty L_{31}}{2}\begin{pmatrix} 1 \\ 0 \\ 1 \end{pmatrix}$

where L_{23} and L_{31} are lengths of sides 2 and 3 of the element. The vector $\boldsymbol{r}_q$ is due to possible heat flux q applied along one or more sides of the element:

Applied flux along side 1: $\boldsymbol{r}_q = \frac{qL_{12}}{2}\begin{pmatrix} 1 \\ 1 \\ 0 \end{pmatrix}$

Applied flux along side 2: $\boldsymbol{r}_q = \frac{qL_{23}}{2}\begin{pmatrix} 0 \\ 1 \\ 1 \end{pmatrix}$

Applied flux along side 3: $\boldsymbol{r}_q = \frac{qL_{31}}{2}\begin{pmatrix} 1 \\ 0 \\ 1 \end{pmatrix}$

If convection or heat flux is specified on more than one side of an element, appropriate matrices are written for each side and then added together. For an insulated boundary $q = 0$, and hence insulated boundaries do not contribute anything to the element equations.

As mentioned in the previous section, we cannot solve for nodal temperatures by simply solving the equations for one element. We must consider contributions of all elements and specified boundary conditions before solving for the nodal unknowns. These procedures are discussed in detail in later sections in this chapter.

Example 1.2 Write finite element equations for element number 20 in the finite element model of the heat flow through the L-shaped solid shown in Figure 1.7.

The element is situated between nodes 4, 10, and 5. We can choose any of the three nodes as the first node of the element and define the other two by moving counterclockwise around the element. Choosing node 4 as the first node establishes line 4–10 as the first side of the element, line 10–5 as the second side, and line 5–4 as the third side. The origin of the global *x*-*y* coordinate system can be placed at any convenient location. Choosing node 1 as the origin, the coordinates of the element end nodes are as follows:

$$\begin{array}{lll} \text{Node 1 (global node 4)} = \{0., 0.0225\}\text{ m}; & x_1 = 0.; & y_1 = 0.0225 \\ \text{Node 2 (global node 10)} = \{0.015, 0.03\}\text{ m}; & x_2 = 0.015; & y_2 = 0.03 \\ \text{Node 3 (global node 5)} = \{0., 0.03\}\text{ m}; & x_3 = 0.; & y_3 = 0.03 \end{array}$$

Using these coordinates, the constants b_i, c_i, and f_i and the element area can easily be computed as follows:

$$\begin{array}{lll} b_1 = 0.; & b_2 = 0.0075; & b_3 = -0.0075 \\ c_1 = -0.015; & c_2 = 0.; & c_3 = 0.015 \\ f_1 = 0.00045; & f_2 = 0.; & f_3 = -0.0003375 \end{array}$$

$$\text{Element Area} = 0.00005625$$

From the given data the thermal conductivities and heat generated over the solid are as follows:

$$k_x = 45; \qquad k_y = 45; \qquad Q = 5000000$$

Substituting these numerical values into the element equation expressions, the matrices $\boldsymbol{k}_k$ and $\boldsymbol{r}_Q$ can easily be written as follows:

$$\boldsymbol{k}_k = \begin{pmatrix} 45. & 0. & -45. \\ 0. & 11.25 & -11.25 \\ -45. & -11.25 & 56.25 \end{pmatrix}; \qquad \boldsymbol{r}_Q = \begin{pmatrix} 93.75 \\ 93.75 \\ 93.75 \end{pmatrix}$$

There is an applied heat flux on side 3 (line 5–4) of the element. The length of this side of the element is 0.0075 m and with $q = 8000$ (a positive value since heat is flowing into the body) the $\boldsymbol{r}_q$ vector for the element is as follows:

Heat flux on side 3 with coordinates ({0., 0.0225} {0., 0.03}),
$L = 0.0075$; $q = 8000$

$$r_q = \begin{pmatrix} 30. \\ 0 \\ 30. \end{pmatrix},$$

The side 2 of the element is subjected to heat loss by convection. The convection term generates a matrix k_h and a vector r_h. Substituting the numerical values into the formulas, these contributions are as follows:

Convection on side 2 with coordinates ({0.015, 0.03} {0., 0.03}),
$L = 0.015$; $h = 55$; $T_\infty = 20$

$$k_h = \begin{pmatrix} 0 & 0 & 0 \\ 0 & 0.275 & 0.1375 \\ 0 & 0.1375 & 0.275 \end{pmatrix}; r_h = \begin{pmatrix} 0 \\ 8.25 \\ 8.25 \end{pmatrix},$$

Adding matrices k_k and k_h and vectors r_Q, r_q, and r_h, the complete element equations are as follows:

$$\begin{pmatrix} 45. & 0. & -45. \\ 0. & 11.525 & -11.1125 \\ -45. & -11.1125 & 56.525 \end{pmatrix} \begin{pmatrix} T_4 \\ T_{10} \\ T_5 \end{pmatrix} = \begin{pmatrix} 123.75 \\ 102. \\ 132. \end{pmatrix}$$

◆ *Mathematica*/MATLAB Implementation 1.2 on the Book Web Site: *Triangular element for heat flow*

1.1.3 General Remarks on Finite Element Discretization

The accuracy of a finite element analysis depends on the number of elements used in the model and the arrangement of elements. In general, the accuracy of the solution improves as the number of elements is increased. The computational effort, however, increases rapidly as well. Concentrating more elements in regions where rapid changes in solution are expected produces finite element discretizations that give excellent results with reasonable computational effort. Some general remarks on constructing good finite element meshes follow.

1. *Physical Geometry of the Domain.* Enough elements must be used to model the physical domain as accurately as possible. For example, when a curved domain is to be discretized by using elements with straight edges, one must use a reasonably large number of elements; otherwise there will be a large discrepancy in the actual geometry and the discretized geometry used in the model. Figure 1.8 illustrates error in the approximation of a curved boundary for a two-dimensional domain discretized using triangular elements. Using more elements along the boundary will obviously reduce this discrepancy. If available, a better option is to use elements that allow curved sides.

2. *Desired Accuracy.* Generally, using more elements produces more accurate results.

3. *Element Formulation.* Some element formulations produce more accurate results than others, and thus formulation employed in a particular element influences the number of elements needed in the model for a desired accuracy.

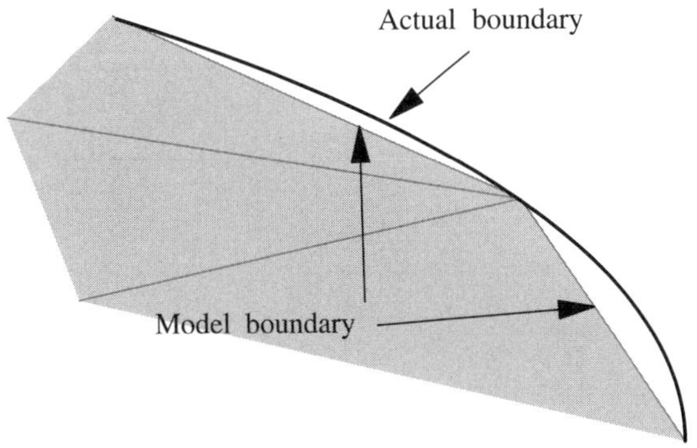

Figure 1.8. Discrepancy in the actual physical boundary and the triangular element model geometry

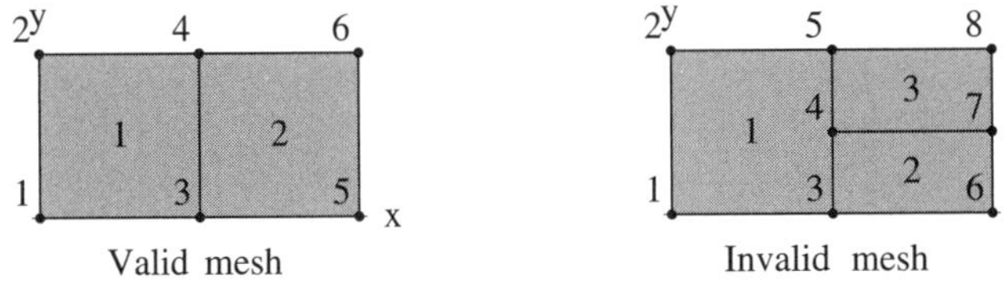

Figure 1.9. Valid and invalid mesh for four-node elements

4. *Special Solution Characteristics.* Regions over which the solution changes rapidly generally require a large number of elements to accurately capture high solution gradients. A good modeling practice is to start with a relatively coarse mesh to get an idea of the solution and then proceed with more refined models. The results from the coarse model are used to guide the mesh refinement process.

5. *Available Computational Resources.* Models with more elements require more computational resources in terms of memory, disk space, and computer processor.

6. *Element Interfaces.* Elements are joined together at nodes (typically shown as dark circles on the finite element meshes). The solutions at these nodes are the primary variables in the finite element procedure. For reasons that will become clear after studying the next few chapters, it is important to create meshes in which the adjacent elements are always connected from corner to corner. Figure 1.9 shows an example of a valid and an invalid mesh when employing four-node quadrilateral elements. The reason why the three-element mesh on the right is invalid is because node 4 that forms a corner of elements 2 and 3 is not attached to one of the four corners of element 1.

7. *Symmetry.* For many practical problems, solution domains and boundary conditions are symmetric, and hence one can expect symmetry in the solution as well. It is important to recognize such symmetry and to model only the symmetric portion of the solution domain that gives information for the entire model. One common situation is illustrated in the modeling of a notched-beam problem in the following section. Besides the obvious advantage of reducing the model size, by taking advantage of symmetry, one is guaranteed to obtain a symmetric solution for the problem. Due to the numerical nature of the

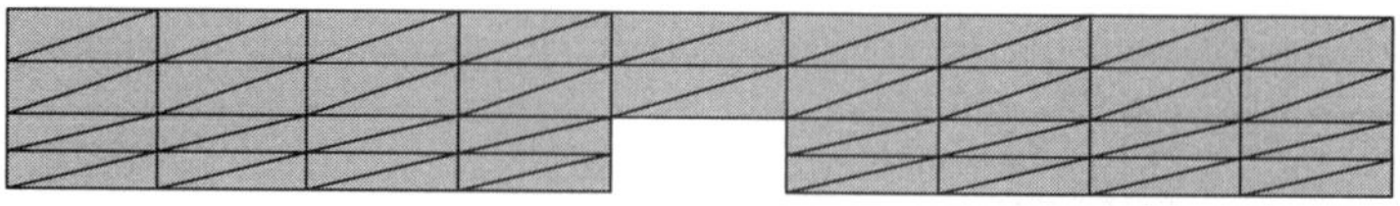

Figure 1.10. Unsymmetrical finite element mesh for a symmetric notched beam

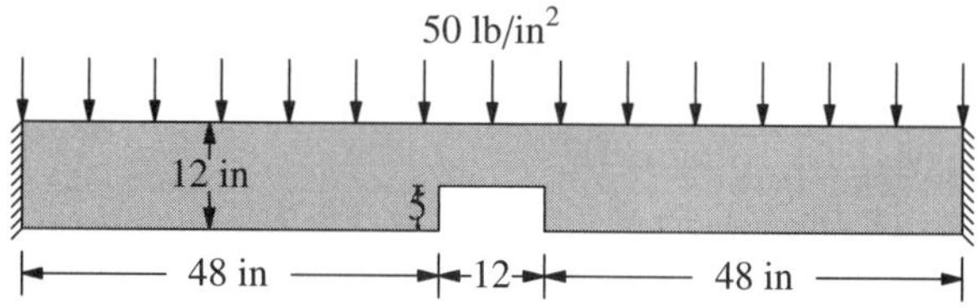

Figure 1.11. Notched beam

finite element method and the unique characteristics of elements employed, modeling the entire symmetric region may in fact produce results that are not symmetric. As a simple illustration, consider the triangular element mesh shown in Figure 1.10 that models the entire notched beam of Figure 1.11. The actual solution should be symmetric with respect to the centerline of the beam. However, the computed finite element solution will not be entirely symmetric because the arrangement of the triangular elements in the model is not symmetric with respect to the midplane.

A general rule of thumb to follow in a finite element analysis is to start with a fairly coarse mesh. The number and arrangement of elements should be just enough to get a good approximation of the geometry, loading, and other physical characteristics of the problem. From the results of this coarse model, select regions in which the solution is changing rapidly for further refinement. To see solution convergence, select one or more critical points in the model and monitor the solution at these points as the number of elements (or the total number of degrees of freedom) in the model is increased. Initially, when the meshes are relatively coarse, there should be significant change in the solution at these points from one mesh to the other. The solution should begin to stabilize after the number of elements used in the model has reached a reasonable level.

1.1.4 Triangular Element for Two-Dimensional Stress Analysis

As a final example of the element equations, consider the problem of finding stresses in the notched beam of rectangular cross section shown in Figure 1.11. The beam is 4 in thick in the direction perpendicular to the plane of paper and is made of concrete with modulus of elasticity $E = 3 \times 10^6$ lb/in^2 and Poisson's ratio $\nu = 0.2$.

Since the beam thickness is small as compared to the other dimensions, it is reasonable to consider the analysis as a plane stress situation in which the stress changes in the thickness direction are ignored. Furthermore, we recognize that the loading and the geometry are symmetric with respect to the plane passing through the midspan. Thus the displacements must be symmetric and the points on the plane passing through the midspan do not experience any displacement in the horizontal direction. Taking advantage of these simpli-

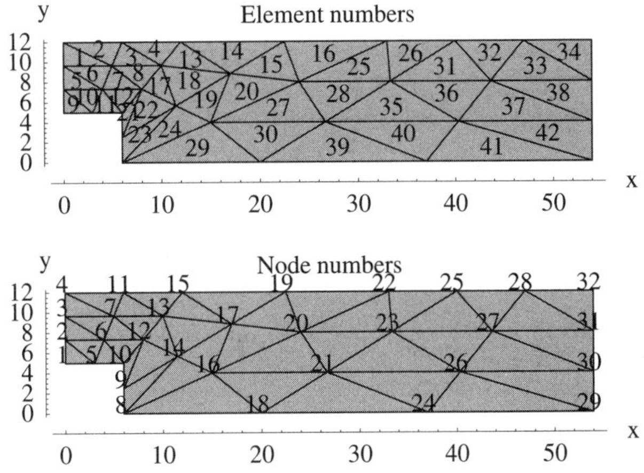

Figure 1.12. Finite element model of the notched beam

fications, we need to construct a two-dimensional plane stress finite element model of only half of the beam. As an illustration, a coarse finite element model of the right half of the beam using triangular elements is shown in Figure 1.12. All nodes on the right end are fixed against displacement because of the given boundary condition. The left end of the model is on the symmetry plane, and thus nodes on the left end cannot displace in the horizontal direction. Once again, in an actual stress analysis a much finer finite element mesh will be needed to get accurate values of stresses and displacements. Even in the coarse model notice that relatively small elements are employed in the notched region where high stress gradients are expected.

A typical triangular element for the solution of the two-dimensional stress analysis problem is shown in Figure 1.13. The element is defined by three nodes with nodal coordinates indicated by (x_1, y_1), (x_2, y_2), and (x_3, y_3). The starting node of the triangle is arbitrary, but we must move counterclockwise around the triangle to define the other two nodes. The nodal degrees of freedom are the displacements in the x and y directions, indicated by u and v. On one or more sides of the element, uniformly distributed load in the normal direction q_n and that in the tangential direction q_t can be specified.

The element is based on the assumption of linear displacements over the element. In terms of nodal degrees of freedom, the displacements over an element can be written as follows:

$$u(x, y) = N_1 u_1 + N_2 u_2 + N_3 u_3$$

$$v(x, y) = N_1 v_1 + N_2 v_2 + N_3 v_3$$

$$\begin{pmatrix} u(x, y) \\ v(x, y) \end{pmatrix} = \begin{pmatrix} N_1 & 0 & N_2 & 0 & N_3 & 0 \\ 0 & N_1 & 0 & N_2 & 0 & N_3 \end{pmatrix} \begin{pmatrix} u_1 \\ v_1 \\ u_2 \\ v_2 \\ u_3 \\ v_3 \end{pmatrix} \equiv \boldsymbol{N}^T \boldsymbol{d}$$

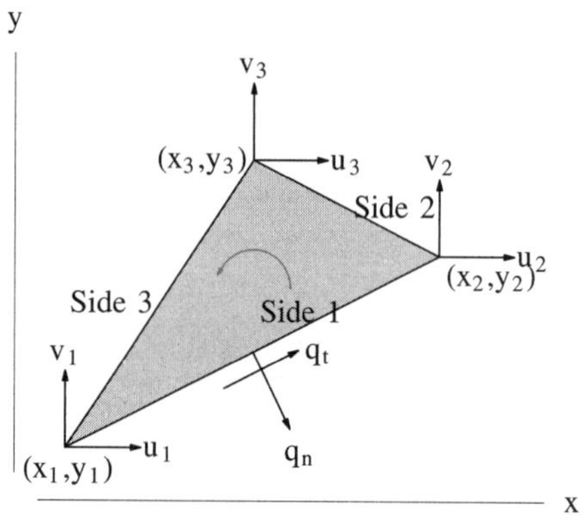

Figure 1.13. Plane stress triangular element

where the N_i, $i = 1, 2, 3$, are the same linear triangle interpolation functions as those used for the heat flow element:

$$\begin{pmatrix} N_1 \\ N_2 \\ N_3 \end{pmatrix} = \frac{1}{2A} \begin{pmatrix} xb_1 + yc_1 + f_1 \\ xb_2 + yc_2 + f_2 \\ xb_3 + yc_3 + f_3 \end{pmatrix}$$

$$\begin{array}{lll} b_1 = y_2 - y_3; & b_2 = y_3 - y_1; & b_3 = y_1 - y_2 \\ c_1 = x_3 - x_2; & c_2 = x_1 - x_3; & c_3 = x_2 - x_1 \\ f_1 = x_2y_3 - x_3y_2; & f_2 = x_3y_1 - x_1y_3; & f_3 = x_1y_2 - x_2y_1 \end{array}$$

The element area A can be computed as follows:

$$A = \frac{1}{2} \det \begin{pmatrix} 1 & x_1 & y_1 \\ 1 & x_2 & y_2 \\ 1 & x_3 & y_3 \end{pmatrix} \equiv \frac{f_1 + f_2 + f_3}{2}$$

Using these assumed displacements, the element strains can be written as follows:

$$\epsilon = \begin{pmatrix} \epsilon_x \\ \epsilon_y \\ \gamma_{xy} \end{pmatrix} = \begin{pmatrix} \frac{\partial u}{\partial x} \\ \frac{\partial v}{\partial y} \\ \frac{\partial u}{\partial y} + \frac{\partial v}{\partial x} \end{pmatrix} = \frac{1}{2A} \begin{pmatrix} b_1 & 0 & b_2 & 0 & b_3 & 0 \\ 0 & c_1 & 0 & c_2 & 0 & c_3 \\ c_1 & b_1 & c_2 & b_2 & c_3 & b_3 \end{pmatrix} \begin{pmatrix} u_1 \\ v_1 \\ u_2 \\ v_2 \\ \vdots \end{pmatrix} \equiv \boldsymbol{B}^T \boldsymbol{d}$$

where ϵ_x and ϵ_y are the normal strains in the x and y directions and γ_{xy} is the shear strain. The corresponding stresses are denoted by σ_x, σ_y, and τ_{xy}, respectively. The vector of stresses is denoted by $\boldsymbol{\sigma}$ and is related to the strain vector through Hooke's law as follows:

$$\boldsymbol{\sigma} = \boldsymbol{C\epsilon} = \boldsymbol{CB}^T \boldsymbol{d}$$

where $\boldsymbol{C}$ is the appropriate constitutive matrix. For homogeneous, isotropic, and elastic materials under plane stress conditions,

$$\boldsymbol{C} = \frac{E}{1-\nu^2}\begin{pmatrix} 1 & \nu & 0 \\ \nu & 1 & 0 \\ 0 & 0 & \frac{1-\nu}{2} \end{pmatrix}$$

where E = Young's modulus and ν = Poisson's ratio.

The finite element equations for this element are derived in Chapter 7 and are as follows:

$$\boldsymbol{kd} = \boldsymbol{r}_q$$

where $\boldsymbol{k}$ is the element stiffness matrix given by

$$\boldsymbol{k} = hA\boldsymbol{BCB}^T$$

where h = element thickness. The vector $\boldsymbol{r}_q$ represents equivalent nodal load due to any applied distributed loads along one or more sides of an element. For a uniformly distributed load on side 1 of the element with components q_n and q_t in the normal and tangential directions of the surface, the equivalent load vector is as follows:

$$\boldsymbol{r}_q^T = \frac{hL_{12}}{2}(\,n_x q_n - n_y q_t \quad n_y q_n + n_x q_t \quad n_x q_n - n_y q_t \quad n_y q_n + n_x q_t \quad 0 \quad 0)$$

where $L_{12} = \sqrt{(x_2-x_1)^2 + (y_2-y_1)^2}$ is the length of side 1 and n_x and n_y are the components of the unit normal to side. Note that $\boldsymbol{r}_q$ is a 6×1 column vector. It is written as a row to save space. The components of the unit normal to the side can be computed as follows:

$$n_x = \frac{y_2 - y_1}{L_{12}}; \qquad n_y = -\frac{x_2 - x_1}{L_{12}}$$

A pressure component is considered positive if it is along the positive direction of the normal or tangent to the side. As shown in Figure 1.13, while moving counterclockwise around the element, the positive normal vector points in the outward direction. The positive tangent vector is 90° counterclockwise from the positive normal vector.

For a uniformly distributed load on side 2,

$$\boldsymbol{r}_q^T = \frac{h\,L_{23}}{2}(0 \quad 0 \quad n_x q_n - n_y q_t \quad n_y q_n + n_x q_t \quad n_x q_n - n_y q_t \quad n_y q_n + n_x q_t)$$

where $L_{23} = \sqrt{(x_3-x_2)^2 + (y_3-y_2)^2}$ is the length of side 2 and

$$n_x = \frac{y_3 - y_2}{L_{23}}; \qquad n_y = -\frac{x_3 - x_2}{L_{23}}$$

For a uniformly distributed load on side 3,

$$\boldsymbol{r}_q^T = \frac{h\,L_{31}}{2}(\,n_x q_n - n_y q_t \quad n_y q_n + n_x q_t \quad 0 \quad 0 \quad n_x q_n - n_y q_t \quad n_y q_n + n_x q_t)$$

where $L_{31} = \sqrt{(x_3 - x_1)^2 + (y_3 - y_1)^2}$ is the length of side 3 and

$$n_x = \frac{y_1 - y_3}{L_{31}}; \qquad n_y = -\frac{x_1 - x_3}{L_{31}}$$

If loads are specified on more than one side of an element, appropriate vectors are written for each side and then added together. As mentioned with the plane truss element, any concentrated applied load at a node is added directly to the global equations during assembly. This will be illustrated in a later section. Furthermore, we cannot solve for nodal displacements by simply solving the equations for one element. We must consider contributions of all elements and specified boundary conditions before solving for the nodal unknowns. These procedures are discussed in detail in later sections in this chapter.

Example 1.3 Write finite element equations for element number 2 in the finite element model of the notched beam shown in Figure 1.12.

The element is connected between nodes 4, 7, and 11. We can choose any of the three nodes as the first node of the element and define the other two by moving counterclockwise around the element. Choosing node 4 as the first node establishes line 4–7 as the first side of the element, line 7–11 as the second side, and line 11–4 as the third side. The origin of the global x-y coordinate system can be placed at any convenient location. Choosing the origin as shown in Figure 1.12, the coordinates of the element end nodes are as follows:

Node 1 (global node 4) = {0., 12.} in; $x_1 = 0.$; $y_1 = 12.$

Node 2 (global node 7) = {5., 9.66667} in; $x_2 = 5.$; $y_2 = 9.66667$

Node 3 (global node 11) = {6., 12.} in; $x_3 = 6.$; $y_3 = 12.$

Using these coordinates, the constants in the $\boldsymbol{B}$ matrix can easily be computed as follows:

$$b_1 = -2.33333; \quad b_2 = 0.; \quad b_3 = 2.33333$$
$$c_1 = 1.; \quad c_2 = -6.; \quad c_3 = 5.$$

Substituting the given data, we have

$A = 7.$; $h = 4$; $E = 3000000$; $\nu = 0.2$

$$\boldsymbol{B}^T = \begin{pmatrix} -0.166667 & 0 & 0. & 0 & 0.166667 & 0 \\ 0 & 0.0714286 & 0 & -0.428571 & 0 & 0.357143 \\ 0.0714286 & -0.166667 & -0.428571 & 0. & 0.357143 & 0.166667 \end{pmatrix}$$

$$\text{Plane stress } \boldsymbol{C} = \begin{pmatrix} 3.125 \times 10^6 & 625000. & 0 \\ 625000. & 3.125 \times 10^6 & 0 \\ 0 & 0 & 1.25 \times 10^6 \end{pmatrix}$$

Thus the element stiffness matrix is

$$\boldsymbol{k} = hABCB^T = 10^6\begin{pmatrix} 2.60913 & -0.625 & -1.07143 & 1.25 & -1.5377 & -0.625 \\ -0.625 & 1.41865 & 2.5 & -2.67857 & -1.875 & 1.25992 \\ -1.07143 & 2.5 & 6.42857 & 0 & -5.35714 & -2.5 \\ 1.25 & -2.67857 & 0 & 16.0714 & -1.25 & -13.3929 \\ -1.5377 & -1.875 & -5.35714 & -1.25 & 6.89484 & 3.125 \\ -0.625 & 1.25992 & -2.5 & -13.3929 & 3.125 & 12.1329 \end{pmatrix}$$

There is an applied load in the negative outer normal direction on side 3 (nodes {11, 4}) of the element. The equivalent nodal load vector $\boldsymbol{r}_q$ for the element is computed as follows:

Specified load components: $q_n = -50$; $q_t = 0$

End nodal coordinates: ({6., 12.} {0., 12.}), giving side length $L = 6$

Components of unit normal to the side: $n_x = 0.$; $n_y = 1.$

Using these values, we get $\boldsymbol{r}_q^T = (0. \quad -600. \quad 0 \quad 0 \quad 0. \quad -600.)$

Thus the complete element equations are as follows:

$$10^6\begin{pmatrix} 2.60913 & -0.625 & -1.07143 & 1.25 & -1.5377 & -0.625 \\ -0.625 & 1.41865 & 2.5 & -2.67857 & -1.875 & 1.25992 \\ -1.07143 & 2.5 & 6.42857 & 0 & -5.35714 & -2.5 \\ 1.25 & -2.67857 & 0 & 16.0714 & -1.25 & -13.3929 \\ -1.5377 & -1.875 & -5.35714 & -1.25 & 6.89484 & 3.125 \\ -0.625 & 1.25992 & -2.5 & -13.3929 & 3.125 & 12.1329 \end{pmatrix}\begin{pmatrix} u_4 \\ v_4 \\ u_7 \\ v_7 \\ u_{11} \\ v_{11} \end{pmatrix} = \begin{pmatrix} 0. \\ -600. \\ 0. \\ 0. \\ 0. \\ -600. \end{pmatrix}$$

◆ ***Mathematica*/MATLAB Implementation 1.3 on the Book Web Site:**
Triangular element for plane stress

1.2 ASSEMBLY OF ELEMENT EQUATIONS

The finite element discretization divides a solution domain or structure into simple elements. For each element the finite element equations can be written by substituting numerical values into the formulas for appropriate element type. In the assembly process, we must put the split-up solution domain back together before proceeding with the solution. The key concept in the assembly process is that at a node common between several elements the nodal solution is the same for all elements sharing this node. Thus contributions to that degree of freedom from all adjacent elements must be added together.

To illustrate the assembly process, consider the 10-node and 8-element finite element mesh for the heat flow problem shown in Figure 1.14. The nodal degrees of freedom are the temperatures at the nodes. Since there are 10 nodes, the global system of equations is 10×10. Thus we start the assembly process by initializing a 10×10 system of equations

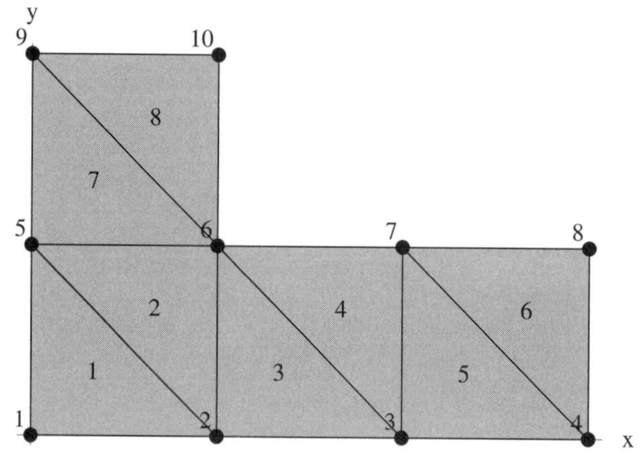

Figure 1.14. Eight-element finite element mesh

as follows:

$$\begin{pmatrix} 0&0&0&0&0&0&0&0&0&0\\ 0&0&0&0&0&0&0&0&0&0\\ 0&0&0&0&0&0&0&0&0&0\\ 0&0&0&0&0&0&0&0&0&0\\ 0&0&0&0&0&0&0&0&0&0\\ 0&0&0&0&0&0&0&0&0&0\\ 0&0&0&0&0&0&0&0&0&0\\ 0&0&0&0&0&0&0&0&0&0\\ 0&0&0&0&0&0&0&0&0&0\\ 0&0&0&0&0&0&0&0&0&0 \end{pmatrix}\begin{pmatrix} T_1\\T_2\\T_3\\T_4\\T_5\\T_6\\T_7\\T_8\\T_9\\T_{10}\end{pmatrix} = \begin{pmatrix}0\\0\\0\\0\\0\\0\\0\\0\\0\\0\end{pmatrix}$$

The element equations can be written using the formulas given in the previous section. However, for simplicity of presentation we assume arbitrary values for the element equations. For element 1, assume that the element equations are as follows:

$$\begin{pmatrix} 111 & 201 & 301\\ 201 & 222 & 232\\ 301 & 232 & 333 \end{pmatrix}\begin{pmatrix} T_1\\T_2\\T_5\end{pmatrix} = \begin{pmatrix}11\\12\\13\end{pmatrix}$$

Now we consider the assembly of element 1 into the global system. This element contributes to the nodal degrees of freedom (1, 2, 5). Since the element involves degrees of freedom T_1, T_2, and T_5, these equations will be added to global equations 1, 2, and 5, respectively. Written in expanded form, the first equation of this element is

$$111T_1 + 201T_2 + 301T_5 = 11$$

Expanding it to include all 10 degrees of freedom in the model, we have

$$111T_1 + 201T_2 + 0T_3 + 0T_4 + 301T_5 + 0T_6 + 0T_7 + 0T_8 + 0T_9 + 0T_{10} = 11$$

In the global system this is the first equation, and therefore the equation can be inserted into the global system as follows:

$$\begin{pmatrix} 111 & 201 & 0 & 0 & 301 & 0 & 0 & 0 & 0 & 0 \\ 0 & 0 & 0 & 0 & 0 & 0 & 0 & 0 & 0 & 0 \\ 0 & 0 & 0 & 0 & 0 & 0 & 0 & 0 & 0 & 0 \\ 0 & 0 & 0 & 0 & 0 & 0 & 0 & 0 & 0 & 0 \\ 0 & 0 & 0 & 0 & 0 & 0 & 0 & 0 & 0 & 0 \\ 0 & 0 & 0 & 0 & 0 & 0 & 0 & 0 & 0 & 0 \\ 0 & 0 & 0 & 0 & 0 & 0 & 0 & 0 & 0 & 0 \\ 0 & 0 & 0 & 0 & 0 & 0 & 0 & 0 & 0 & 0 \\ 0 & 0 & 0 & 0 & 0 & 0 & 0 & 0 & 0 & 0 \\ 0 & 0 & 0 & 0 & 0 & 0 & 0 & 0 & 0 & 0 \end{pmatrix} \begin{pmatrix} T_1 \\ T_2 \\ T_3 \\ T_4 \\ T_5 \\ T_6 \\ T_7 \\ T_8 \\ T_9 \\ T_{10} \end{pmatrix} = \begin{pmatrix} 11 \\ 0 \\ 0 \\ 0 \\ 0 \\ 0 \\ 0 \\ 0 \\ 0 \\ 0 \end{pmatrix}$$

The other two equations for the element are expanded in a similar manner.

Element Equation Number	Global Equation Number	Equation
2	2	$201T_1 + 222T_2 + 0T_3 + 0T_4 + 232T_5 + 0T_6 + 0T_7 + 0T_8 + 0T_9 + 0T_{10} = 12$
3	5	$301T_1 + 232T_2 + 0T_3 + 0T_4 + 333T_5 + 0T_6 + 0T_7 + 0T_8 + 0T_9 + 0T_{10} = 13$

Placing these equations in the second and fifth rows, the global equations after assembly of element 1 are as follows:

$$\begin{pmatrix} 111 & 201 & 0 & 0 & 301 & 0 & 0 & 0 & 0 & 0 \\ 201 & 222 & 0 & 0 & 232 & 0 & 0 & 0 & 0 & 0 \\ 0 & 0 & 0 & 0 & 0 & 0 & 0 & 0 & 0 & 0 \\ 0 & 0 & 0 & 0 & 0 & 0 & 0 & 0 & 0 & 0 \\ 301 & 232 & 0 & 0 & 333 & 0 & 0 & 0 & 0 & 0 \\ 0 & 0 & 0 & 0 & 0 & 0 & 0 & 0 & 0 & 0 \\ 0 & 0 & 0 & 0 & 0 & 0 & 0 & 0 & 0 & 0 \\ 0 & 0 & 0 & 0 & 0 & 0 & 0 & 0 & 0 & 0 \\ 0 & 0 & 0 & 0 & 0 & 0 & 0 & 0 & 0 & 0 \\ 0 & 0 & 0 & 0 & 0 & 0 & 0 & 0 & 0 & 0 \end{pmatrix} \begin{pmatrix} T_1 \\ T_2 \\ T_3 \\ T_4 \\ T_5 \\ T_6 \\ T_7 \\ T_8 \\ T_9 \\ T_{10} \end{pmatrix} = \begin{pmatrix} 11 \\ 12 \\ 0 \\ 0 \\ 13 \\ 0 \\ 0 \\ 0 \\ 0 \\ 0 \end{pmatrix}$$

The above procedure of reordering and expanding element equations is quite tedious. Fortunately, it is not necessary to formally carry out these steps in detail. The appropriate locations of the entries in the global equations can be determined simply by taking the list of degrees of freedom to which the element is contributing. This list is called the *location vector*. For element 1 the location vector is as follows:

$$\begin{pmatrix} 1 \\ 2 \\ 5 \end{pmatrix}$$

For assembling into a global vector (right-hand side), the entries in the location vector directly indicate the locations where the corresponding element quantity will contribute:

$$\text{Locations for element 1 contributions to a global vector:} \quad \begin{pmatrix} 1 \\ 2 \\ 5 \end{pmatrix}$$

To determine the global locations of the entries in an element matrix (left-hand-side coefficient matrix), we simply take all combinations of the indices in the location vector. For the first row, the locations have the row index of 1 and column indices are 1, 2, and 5. For the second row, the row index is 2 and the column indices are 1, 2, and 5. For the third row, the row index is 5 and the column indices are 1, 2, and 5. Thus the locations in the global matrix where the corresponding element quantities are added are as follows:

$$\text{Locations for element 1 contributions to the global matrix:} \quad \begin{pmatrix} [1,1] & [1,2] & [1,5] \\ [2,1] & [2,2] & [2,5] \\ [5,1] & [5,2] & [5,5] \end{pmatrix}$$

This indicates that 111 from the element matrix goes to the location [1, 1] in the global matrix, 201 into the [1, 2], etc. Clearly, this gives us exactly the same global matrix after assembly of this element as before.

Each element in a finite element model is processed in exactly the same manner. As a further illustration, consider assembly of element 2. Assume that the equations for element 2 are as follows:

$$\begin{pmatrix} 77 & 80 & 90 \\ 80 & 88 & 100 \\ 90 & 100 & 99 \end{pmatrix} \begin{pmatrix} T_2 \\ T_6 \\ T_5 \end{pmatrix} = \begin{pmatrix} 21 \\ 22 \\ 23 \end{pmatrix}$$

The location vector and the locations to which this element contributes in the global matrix are as follows:

$$\text{Element 2 location vector:} \quad \begin{pmatrix} 2 \\ 6 \\ 5 \end{pmatrix}$$

$$\text{Locations for element 2 contributions to a global vector:} \quad \begin{pmatrix} 2 \\ 6 \\ 5 \end{pmatrix}$$

$$\text{Locations for element 2 contributions to a global matrix:} \quad \begin{pmatrix} [2,2] & [2,6] & [2,5] \\ [6,2] & [6,6] & [6,5] \\ [5,2] & [5,6] & [5,5] \end{pmatrix}$$

This indicates that 77 from the element matrix is added to the location [2, 2] in the global matrix, 80 into the [2, 6], etc. Thus the global equations after assembly of this element are as follows.

$$\begin{pmatrix} 111 & 201 & 0 & 0 & 301 & 0 & 0 & 0 & 0 & 0 \\ 201 & 222+77 & 0 & 0 & 232+90 & 80 & 0 & 0 & 0 & 0 \\ 0 & 0 & 0 & 0 & 0 & 0 & 0 & 0 & 0 & 0 \\ 0 & 0 & 0 & 0 & 0 & 0 & 0 & 0 & 0 & 0 \\ 301 & 232+90 & 0 & 0 & 333+99 & 100 & 0 & 0 & 0 & 0 \\ 0 & 80 & 0 & 0 & 100 & 88 & 0 & 0 & 0 & 0 \\ 0 & 0 & 0 & 0 & 0 & 0 & 0 & 0 & 0 & 0 \\ 0 & 0 & 0 & 0 & 0 & 0 & 0 & 0 & 0 & 0 \\ 0 & 0 & 0 & 0 & 0 & 0 & 0 & 0 & 0 & 0 \\ 0 & 0 & 0 & 0 & 0 & 0 & 0 & 0 & 0 & 0 \end{pmatrix} \begin{pmatrix} T_1 \\ T_2 \\ T_3 \\ T_4 \\ T_5 \\ T_6 \\ T_7 \\ T_8 \\ T_9 \\ T_{10} \end{pmatrix} = \begin{pmatrix} 11 \\ 12+21 \\ 0 \\ 0 \\ 13+23 \\ 22 \\ 0 \\ 0 \\ 0 \\ 0 \end{pmatrix}$$

The equations for the remaining six elements can be assembled in exactly the same manner. The following examples, involving plane truss, heat flow, and plane stress elements, further illustrate the assembly procedure.

◆ *Mathematica*/MATLAB Implementation 1.4 on the Book Web Site: *Finite element assembly procedure*

Example 1.4 Five-Bar Truss Write element equations and assemble them to form global equations for the five-bar plane truss shown in Figure 1.15. The area of cross section for elements 1 and 2 is 40 cm^2, for elements 3 and 4 is 30 cm^2, and for element 5 is 20 cm^2. The first four elements are made of a material with $E = 200$ GPa and the last one with $E = 70$ GPa. The applied load $P = 150$ kN.

Each node in the model has two displacement degrees of freedom. They are identified by the letters u and v with a subscript indicating the corresponding node number and are shown in Figure 1.16. Without considering the specified zero displacements at the supports, the model has a total of eight degrees of freedom. Thus the global equations will be a system of eight equations in eight unknowns.

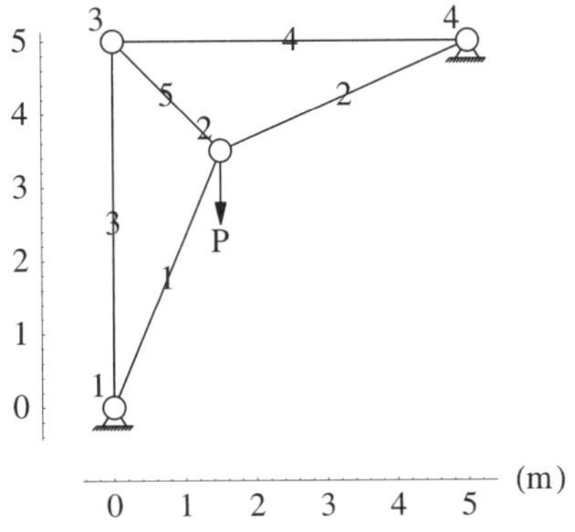

Figure 1.15. Five-bar plane truss

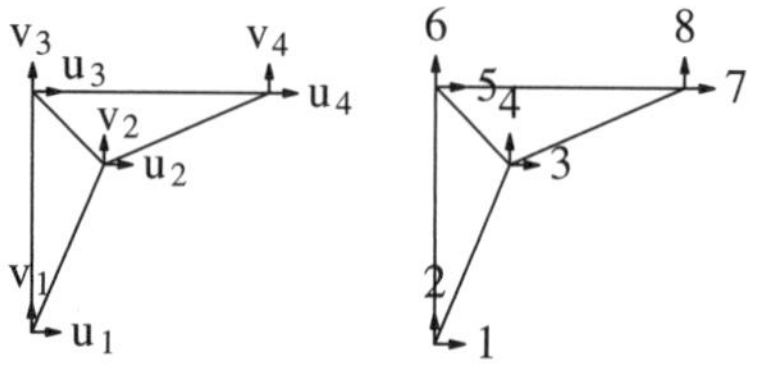

Figure 1.16. Five-bar plane truss finite element model

The next step is to get finite element equations for each element in the model. We simply need to substitute the appropriate numerical values into the plane truss element equations. The concentrated nodal load is added directly into the global equations at the start of assembly. Since the load is acting downward and the displacements are assumed positive along the positive coordinates, the load at node 2 is (0, −150 kN). Since the displacements are usually small, it is convenient to use newton-millimeters. The displacements will come out in millimeters and the stresses in megapascals.

For each element we substitute the numerical data into the plane truss stiffness matrix and assemble them into the global equations using the assembly procedure discussed earlier. The complete computations are as follows. All numerical values are in newtons and millimeters.

The specified nodal loads are as follows:

Node	dof	Value
2	u_2	0
	v_2	−150000

The global equations at the start of the element assembly process are

$$\begin{pmatrix} 0 & 0 & 0 & 0 & 0 & 0 & 0 & 0 \\ 0 & 0 & 0 & 0 & 0 & 0 & 0 & 0 \\ 0 & 0 & 0 & 0 & 0 & 0 & 0 & 0 \\ 0 & 0 & 0 & 0 & 0 & 0 & 0 & 0 \\ 0 & 0 & 0 & 0 & 0 & 0 & 0 & 0 \\ 0 & 0 & 0 & 0 & 0 & 0 & 0 & 0 \\ 0 & 0 & 0 & 0 & 0 & 0 & 0 & 0 \\ 0 & 0 & 0 & 0 & 0 & 0 & 0 & 0 \end{pmatrix} \begin{pmatrix} u_1 \\ v_1 \\ u_2 \\ v_2 \\ u_3 \\ v_3 \\ u_4 \\ v_4 \end{pmatrix} = \begin{pmatrix} 0 \\ 0 \\ 0 \\ -150000 \\ 0 \\ 0 \\ 0 \\ 0 \end{pmatrix}$$

The equations for element 1 are as follows:

$E = 200000; A = 4000$

Element Node	Global Node Number	x	y
1	1	0	0
2	2	1500.	3500.

$x_1 = 0; y_1 = 0; x_2 = 1500.; y_2 = 3500.$

$L = \sqrt{(x_2 - x_1)^2 + (y_2 - y_1)^2} = 3807.89$

Direction cosines: $l_s = \dfrac{x_2 - x_1}{L} = 0.393919$; $m_s = \dfrac{y_2 - y_1}{L} = 0.919145$

Substituting into the truss element equations, we get

$$\begin{pmatrix} 32600.2 & 76067.2 & -32600.2 & -76067.2 \\ 76067.2 & 177490. & -76067.2 & -177490. \\ -32600.2 & -76067.2 & 32600.2 & 76067.2 \\ -76067.2 & -177490. & 76067.2 & 177490. \end{pmatrix} \begin{pmatrix} u_1 \\ v_1 \\ u_2 \\ v_2 \end{pmatrix} = \begin{pmatrix} 0. \\ 0. \\ 0. \\ 0. \end{pmatrix}$$

The element contributes to {1, 2, 3, 4} global degrees of freedom:

Locations for element contributions to a global vector: $\begin{pmatrix} 1 \\ 2 \\ 3 \\ 4 \end{pmatrix}$

and to a global matrix: $\begin{pmatrix} [1,1] & [1,2] & [1,3] & [1,4] \\ [2,1] & [2,2] & [2,3] & [2,4] \\ [3,1] & [3,2] & [3,3] & [3,4] \\ [4,1] & [4,2] & [4,3] & [4,4] \end{pmatrix}$

Adding element equations into appropriate locations, we have

$$\begin{pmatrix} 32600.2 & 76067.2 & -32600.2 & -76067.2 & 0 & 0 & 0 & 0 \\ 76067.2 & 177490. & -76067.2 & -177490. & 0 & 0 & 0 & 0 \\ -32600.2 & -76067.2 & 32600.2 & 76067.2 & 0 & 0 & 0 & 0 \\ -76067.2 & -177490. & 76067.2 & 177490. & 0 & 0 & 0 & 0 \\ 0 & 0 & 0 & 0 & 0 & 0 & 0 & 0 \\ 0 & 0 & 0 & 0 & 0 & 0 & 0 & 0 \\ 0 & 0 & 0 & 0 & 0 & 0 & 0 & 0 \\ 0 & 0 & 0 & 0 & 0 & 0 & 0 & 0 \end{pmatrix} \begin{pmatrix} u_1 \\ v_1 \\ u_2 \\ v_2 \\ u_3 \\ v_3 \\ u_4 \\ v_4 \end{pmatrix} = \begin{pmatrix} 0 \\ 0 \\ 0 \\ -150000. \\ 0 \\ 0 \\ 0 \\ 0 \end{pmatrix}$$

The equations for element 2 are as follows:

$E = 200000$; $A = 4000$

Element Node	Global Node Number	x	y
1	2	1500.	3500.
2	4	5000	5000

$x_1 = 1500.$; $y_1 = 3500.$; $x_2 = 5000$; $y_2 = 5000$

$L = \sqrt{(x_2 - x_1)^2 + (y_2 - y_1)^2} = 3807.89$

Direction cosines: $l_s = \dfrac{x_2 - x_1}{L} = 0.919145$; $m_s = \dfrac{y_2 - y_1}{L} = 0.393919$

Substituting into the truss element equations, we get

$$\begin{pmatrix} 177490. & 76067.2 & -177490. & -76067.2 \\ 76067.2 & 32600.2 & -76067.2 & -32600.2 \\ -177490. & -76067.2 & 177490. & 76067.2 \\ -76067.2 & -32600.2 & 76067.2 & 32600.2 \end{pmatrix} \begin{pmatrix} u_2 \\ v_2 \\ u_4 \\ v_4 \end{pmatrix} = \begin{pmatrix} 0. \\ 0. \\ 0. \\ 0. \end{pmatrix}$$

The element contributes to {3, 4, 7, 8} degrees of freedom:

$$\text{Locations for element contributions to a global vector: } \begin{pmatrix} 3 \\ 4 \\ 7 \\ 8 \end{pmatrix}$$

$$\text{and to a global matrix: } \begin{pmatrix} [3,3] & [3,4] & [3,7] & [3,8] \\ [4,3] & [4,4] & [4,7] & [4,8] \\ [7,3] & [7,4] & [7,7] & [7,8] \\ [8,3] & [8,4] & [8,7] & [8,8] \end{pmatrix}$$

Adding element equations into appropriate locations, we have

$$\begin{pmatrix} 32600.2 & 76067.2 & -32600.2 & -76067.2 & 0 & 0 & 0 & 0 \\ 76067.2 & 177490. & -76067.2 & -177490. & 0 & 0 & 0 & 0 \\ -32600.2 & -76067.2 & 210090. & 152134. & 0 & 0 & -177490. & -76067.2 \\ -76067.2 & -177490. & 152134. & 210090. & 0 & 0 & -76067.2 & -32600.2 \\ 0 & 0 & 0 & 0 & 0 & 0 & 0 & 0 \\ 0 & 0 & 0 & 0 & 0 & 0 & 0 & 0 \\ 0 & 0 & -177490. & -76067.2 & 0 & 0 & 177490. & 76067.2 \\ 0 & 0 & -76067.2 & -32600.2 & 0 & 0 & 76067.2 & 32600.2 \end{pmatrix} \begin{pmatrix} u_1 \\ v_1 \\ u_2 \\ v_2 \\ u_3 \\ v_3 \\ u_4 \\ v_4 \end{pmatrix} = \begin{pmatrix} 0 \\ 0 \\ 0 \\ -150000. \\ 0 \\ 0 \\ 0 \\ 0 \end{pmatrix}$$

The remaining elements can be processed in exactly the same manner. After assembling all elements, the global equations for the model are as follows:

$$\begin{pmatrix} 32600.2 & 76067.2 & -32600.2 & -76067.2 & 0 & 0 & 0 & 0 \\ 76067.2 & 297490. & -76067.2 & -177490. & 0 & -120000 & 0 & 0 \\ -32600.2 & -76067.2 & 243089. & 119136. & -32998.3 & 32998.3 & -177490. & -76067.2 \\ -76067.2 & -177490. & 119136. & 243089. & 32998.3 & -32998.3 & -76067.2 & -32600.2 \\ 0 & 0 & -32998.3 & 32998.3 & 152998. & -32998.3 & -120000 & 0 \\ 0 & -120000 & 32998.3 & -32998.3 & -32998.3 & 152998. & 0 & 0 \\ 0 & 0 & -177490. & -76067.2 & -120000 & 0 & 297490. & 76067.2 \\ 0 & 0 & -76067.2 & -32600.2 & 0 & 0 & 76067.2 & 32600.2 \end{pmatrix} \begin{pmatrix} u_1 \\ v_1 \\ u_2 \\ v_2 \\ u_3 \\ v_3 \\ u_4 \\ v_4 \end{pmatrix} = \begin{pmatrix} 0 \\ 0 \\ 0 \\ -150000. \\ 0 \\ 0 \\ 0 \\ 0 \end{pmatrix}$$

◆ *Mathematica*/MATLAB Implementation 1.5 on the Book Web Site: *Five-bar plane truss assembly*

Example 1.5 Heat Flow through a Square Duct The cross section of a 20 × 20-cm duct made of concrete walls 20 cm thick is shown in Figure 1.17. The inside surface of the duct is maintained at a temperature of 300°C due to hot gases flowing from a furnace. On the outside the duct is exposed to air with an ambient temperature of 20°C. The heat conduction coefficient of concrete is 1.4 W/m · °C. The average convection heat transfer coefficient on the outside of the duct is 27 W/m · °C.

Because of symmetry, we can model only one-eighth of the duct as shown in Figure 1.18. There cannot be any heat flow across a line of symmetry, and hence a symmetry line is equivalent to a fully insulated boundary. The solution domain is discretized into four triangular elements. This model is very coarse and therefore we do not expect very accurate results. However, showing all calculations for a larger model will be too tedious.

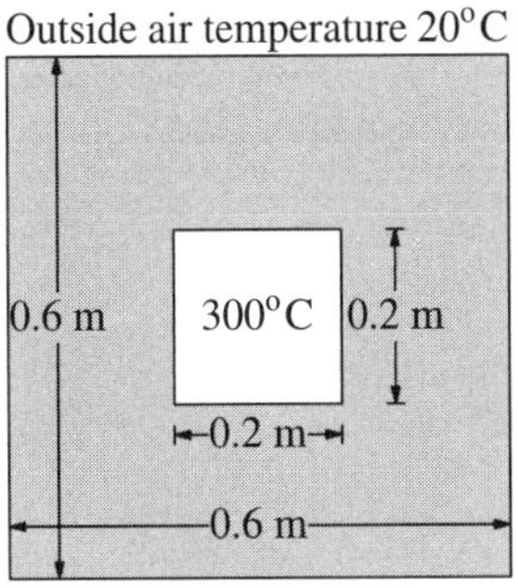

Figure 1.17. Cross section of a square duct

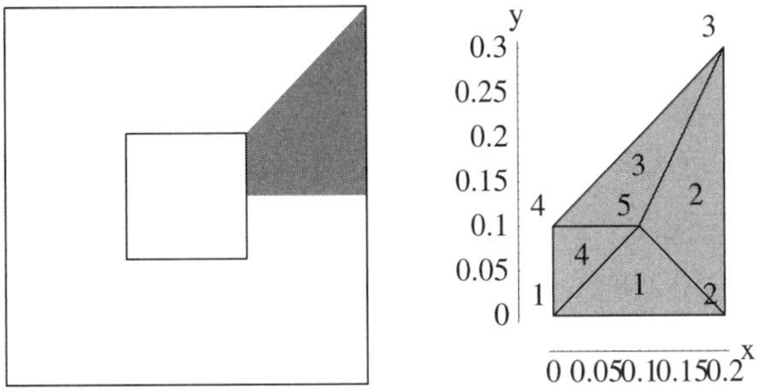

Figure 1.18. Model of eighth of a square duct

For each element we substitute the numerical data into the triangular element equations for heat flow and assemble them into the global equations using the assembly procedure discussed earlier. There is no heat generation, thus the Q term is 0. We need to compute the $\boldsymbol{k}_k$ matrix for each element. Convection boundary conditions are specified on the outside surface. Thus the convection terms in the element equations will only affect element 2. Assuming this element is defined by starting with node 2, the convection term is on side 1 of the element. On the top and bottom sides, due to symmetry, there is no heat flow. They behave as insulated sides and do not require any special consideration during the finite element solution. The complete computations of element equations and their assembly follow.

The global equations at the start of the assembly of the element equations are

$$\begin{pmatrix} 0 & 0 & 0 & 0 & 0 \\ 0 & 0 & 0 & 0 & 0 \\ 0 & 0 & 0 & 0 & 0 \\ 0 & 0 & 0 & 0 & 0 \\ 0 & 0 & 0 & 0 & 0 \end{pmatrix} \begin{pmatrix} T_1 \\ T_2 \\ T_3 \\ T_4 \\ T_5 \end{pmatrix} = \begin{pmatrix} 0 \\ 0 \\ 0 \\ 0 \\ 0 \end{pmatrix}$$

The equations for element 1 are as follows:

$k_x = 1.4;\ k_y = 1.4;\ Q = 0$

Nodal coordinates:

Element Node	Global Node Number	x	y
1	1	0.	0.
2	2	0.2	0.
3	5	0.1	0.1

$x_1 = 0.; x_2 = 0.2; x_3 = 0.1$
$y_1 = 0.; y_2 = 0.; y_3 = 0.1$
Using these values, we get

$$b_1 = -0.1; \quad b_2 = 0.1; \quad b_3 = 0.$$
$$c_1 = -0.1; \quad c_2 = -0.1; \quad c_3 = 0.2$$

Element area, $A = 0.01$
Substituting these values, we get

$$\boldsymbol{k}_k = \begin{pmatrix} 0.7 & 0. & -0.7 \\ 0. & 0.7 & -0.7 \\ -0.7 & -0.7 & 1.4 \end{pmatrix}; \qquad \boldsymbol{r}_Q = \begin{pmatrix} 0 \\ 0 \\ 0 \end{pmatrix}$$

Complete element equations:

$$\begin{pmatrix} 0.7 & 0. & -0.7 \\ 0. & 0.7 & -0.7 \\ -0.7 & -0.7 & 1.4 \end{pmatrix} \begin{pmatrix} T_1 \\ T_2 \\ T_5 \end{pmatrix} = \begin{pmatrix} 0 \\ 0 \\ 0 \end{pmatrix}$$

The element contributes to {1, 2, 5} degrees of freedom:

Locations for element contributions to a global vector: $\begin{pmatrix} 1 \\ 2 \\ 5 \end{pmatrix}$

and to a global matrix: $\begin{pmatrix} [1,1] & [1,2] & [1,5] \\ [2,1] & [2,2] & [2,5] \\ [5,1] & [5,2] & [5,5] \end{pmatrix}$

Adding element equations into appropriate locations, we have

$$\begin{pmatrix} 0.7 & 0 & 0 & 0 & -0.7 \\ 0 & 0.7 & 0 & 0 & -0.7 \\ 0 & 0 & 0 & 0 & 0 \\ 0 & 0 & 0 & 0 & 0 \\ -0.7 & -0.7 & 0 & 0 & 1.4 \end{pmatrix} \begin{pmatrix} T_1 \\ T_2 \\ T_3 \\ T_4 \\ T_5 \end{pmatrix} = \begin{pmatrix} 0 \\ 0 \\ 0 \\ 0 \\ 0 \end{pmatrix}$$

The equations for element 2 are as follows:

$k_x = 1.4; k_y = 1.4; Q = 0$

Nodal coordinates:

Element Node	Global Node Number	x	y
1	2	0.2	0.
2	3	0.2	0.3
3	5	0.1	0.1

$x_1 = 0.2; x_2 = 0.2; x_3 = 0.1$
$y_1 = 0.; y_2 = 0.3; y_3 = 0.1$
Using these values, we get

$$b_1 = 0.2; \qquad b_2 = 0.1; \qquad b_3 = -0.3$$
$$c_1 = -0.1; \qquad c_2 = 0.1; \qquad c_3 = 0.$$

Element area, $A = 0.015$
Substituting these values, we get

$$\boldsymbol{k}_k = \begin{pmatrix} 1.16667 & 0.233333 & -1.4 \\ 0.233333 & 0.466667 & -0.7 \\ -1.4 & -0.7 & 2.1 \end{pmatrix}; \qquad \boldsymbol{r}_Q = \begin{pmatrix} 0 \\ 0 \\ 0 \end{pmatrix}$$

Convection on side 1 (nodes {2, 3}) with $h = 27$ and $T_\infty = 20$
End nodal coordinates: ({0.2, 0.} {0.2, 0.3}), giving side length $L = 0.3$
Using these values, we get

$$\boldsymbol{k}_h = \begin{pmatrix} 2.7 & 1.35 & 0 \\ 1.35 & 2.7 & 0 \\ 0 & 0 & 0 \end{pmatrix}; \qquad \boldsymbol{r}_h = \begin{pmatrix} 81. \\ 81. \\ 0 \end{pmatrix}$$

Complete element equations:

$$\begin{pmatrix} 3.86667 & 1.58333 & -1.4 \\ 1.58333 & 3.16667 & -0.7 \\ -1.4 & -0.7 & 2.1 \end{pmatrix} \begin{pmatrix} T_2 \\ T_3 \\ T_5 \end{pmatrix} = \begin{pmatrix} 81. \\ 81. \\ 0 \end{pmatrix}$$

The element contributes to {2, 3, 5} degrees of freedom:

Locations for element contributions to a global vector: $\begin{pmatrix} 2 \\ 3 \\ 5 \end{pmatrix}$

and to a global matrix: $\begin{pmatrix} [2,2] & [2,3] & [2,5] \\ [3,2] & [3,3] & [3,5] \\ [5,2] & [5,3] & [5,5] \end{pmatrix}$

Adding element equations into appropriate locations, we have

$$\begin{pmatrix} 0.7 & 0 & 0 & 0 & -0.7 \\ 0 & 4.56667 & 1.58333 & 0 & -2.1 \\ 0 & 1.58333 & 3.16667 & 0 & -0.7 \\ 0 & 0 & 0 & 0 & 0 \\ -0.7 & -2.1 & -0.7 & 0 & 3.5 \end{pmatrix} \begin{pmatrix} T_1 \\ T_2 \\ T_3 \\ T_4 \\ T_5 \end{pmatrix} = \begin{pmatrix} 0 \\ 81. \\ 81. \\ 0 \\ 0 \end{pmatrix}$$

The remaining elements can be processed in exactly the same manner. After assembling all elements the global equations for the model are as follows:

$$\begin{pmatrix} 1.4 & 0 & 0 & -0.7 & -0.7 \\ 0 & 4.56667 & 1.58333 & 0 & -2.1 \\ 0 & 1.58333 & 3.51667 & 0.35 & -1.4 \\ -0.7 & 0 & 0.35 & 3.15 & -2.8 \\ -0.7 & -2.1 & -1.4 & -2.8 & 7. \end{pmatrix} \begin{pmatrix} T_1 \\ T_2 \\ T_3 \\ T_4 \\ T_5 \end{pmatrix} = \begin{pmatrix} 0 \\ 81. \\ 81. \\ 0 \\ 0 \end{pmatrix}$$

◆ *Mathematica*/MATLAB Implementation 1.6 on the Book Web Site: *Heat flow example assembly*

Example 1.6 Stress Analysis of a Bracket The top surface of a thin cantilever bracket is subjected to normal pressure $q = 20\ \text{lb/in}^2$, as shown in Figure 1.19. The bracket is 4 in long and is 2 in wide at the base and 1 in wide at the free end. The thickness of the bracket perpendicular to the plane of paper is $\frac{1}{4}$ in. The material properties are $E = 10^4\ \text{lb/in}^2$ and $\nu = 0.2$. A very coarse four-element finite element model using plane stress triangular elements is shown in Figure 1.20. For each element we substitute the numerical data into the element equations given in the previous section and then assemble them into the global equations using the assembly procedure discussed earlier.

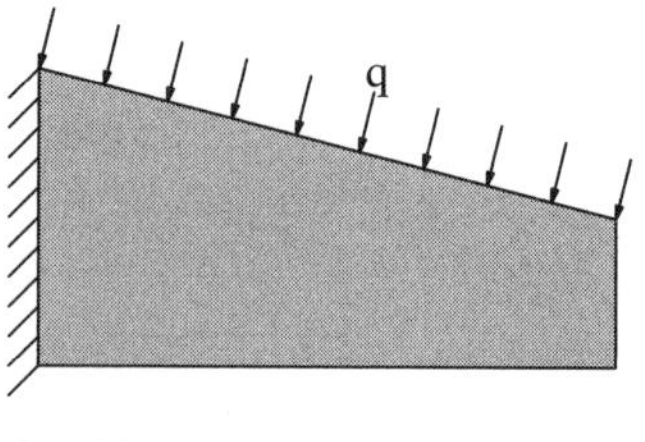

Figure 1.19. Cantilever bracket subjected to normal pressure

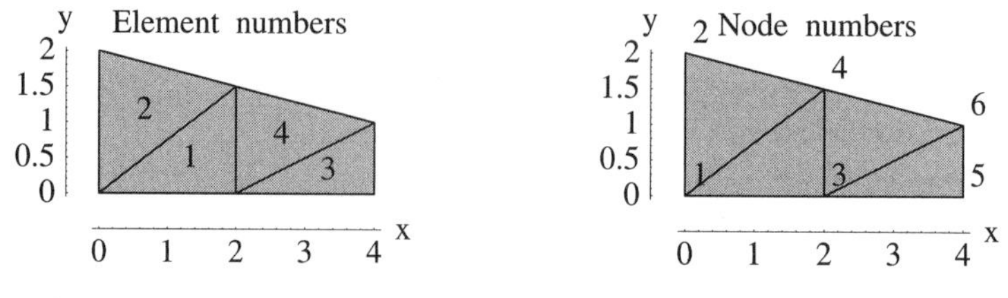

Figure 1.20. Finite element model of the cantilever bracket

The pressure will be accounted for in elements 2 and 4. Choosing node 4 as the first node for element 2, the pressure term will be applied to side 1 of the element. The normal pressure is in the direction opposite to the outer normal to the surface and therefore it must be assigned a negative sign. For element 4, node 6 is chosen as the first node and thus the pressure on this element is also on side 1. The complete computations of element equations and their assembly are as follows. All numerical quantities are in pounds and inches.

The equations for element 1 are as follows:

$h = 0.25;\ E = 10000;\ \nu = 0.2$

Plane stress constitutive matrix, $\boldsymbol{C} = \begin{pmatrix} 10416.7 & 2083.33 & 0 \\ 2083.33 & 10416.7 & 0 \\ 0 & 0 & 4166.67 \end{pmatrix}$

Nodal coordinates:

Element Node	Global Node Number	x	y
1	1	0.	0.
2	3	2.	0.
3	4	2.	1.5

$x_1 = 0.;\ x_2 = 2.;\ x_3 = 2.$
$y_1 = 0.;\ y_2 = 0.;\ y_3 = 1.5$
Using these values, we get

$$b_1 = -1.5; \quad b_2 = 1.5; \quad b_3 = 0.$$
$$c_1 = 0.; \quad c_2 = -2.; \quad c_3 = 2.$$

Element area, $A = 1.5$

$$\boldsymbol{B}^T = \begin{pmatrix} -0.5 & 0 & 0.5 & 0 & 0. & 0 \\ 0 & 0. & 0 & -0.666667 & 0 & 0.666667 \\ 0. & -0.5 & -0.666667 & 0.5 & 0.666667 & 0. \end{pmatrix}$$

Thus the element stiffness matrix is:

$$\boldsymbol{k} = hA\boldsymbol{B}\boldsymbol{C}\boldsymbol{B}^T = \begin{pmatrix} 976.563 & 0 & -976.563 & 260.417 & 0 & -260.417 \\ 0 & 390.625 & 520.833 & -390.625 & -520.833 & 0 \\ -976.563 & 520.833 & 1671.01 & -781.25 & -694.444 & 260.417 \\ 260.417 & -390.625 & -781.25 & 2126.74 & 520.833 & -1736.11 \\ 0 & -520.833 & -694.444 & 520.833 & 694.444 & 0 \\ -260.417 & 0 & 260.417 & -1736.11 & 0 & 1736.11 \end{pmatrix}$$

Complete element equations:

$$\begin{pmatrix} 976.563 & 0 & -976.563 & 260.417 & 0 & -260.417 \\ 0 & 390.625 & 520.833 & -390.625 & -520.833 & 0 \\ -976.563 & 520.833 & 1671.01 & -781.25 & -694.444 & 260.417 \\ 260.417 & -390.625 & -781.25 & 2126.74 & 520.833 & -1736.11 \\ 0 & -520.833 & -694.444 & 520.833 & 694.444 & 0 \\ -260.417 & 0 & 260.417 & -1736.11 & 0 & 1736.11 \end{pmatrix} \begin{pmatrix} u_1 \\ v_1 \\ u_3 \\ v_3 \\ u_4 \\ v_4 \end{pmatrix} = \begin{pmatrix} 0. \\ 0. \\ 0. \\ 0. \\ 0. \\ 0. \end{pmatrix}$$

The element contributes to {1, 2, 5, 6, 7, 8} degrees of freedom.

Locations for element contributions to a global vector: $\begin{pmatrix} 1 \\ 2 \\ 5 \\ 6 \\ 7 \\ 8 \end{pmatrix}$

and to a global matrix: $\begin{pmatrix} [1,1] & [1,2] & [1,5] & [1,6] & [1,7] & [1,8] \\ [2,1] & [2,2] & [2,5] & [2,6] & [2,7] & [2,8] \\ [5,1] & [5,2] & [5,5] & [5,6] & [5,7] & [5,8] \\ [6,1] & [6,2] & [6,5] & [6,6] & [6,7] & [6,8] \\ [7,1] & [7,2] & [7,5] & [7,6] & [7,7] & [7,8] \\ [8,1] & [8,2] & [8,5] & [8,6] & [8,7] & [8,8] \end{pmatrix}$

Adding element equations into appropriate locations, we have

$$\begin{pmatrix} 976.563 & 0 & 0 & 0 & -976.563 & 260.417 & 0 & -260.417 & 0 & 0 & 0 & 0 \\ 0 & 390.625 & 0 & 0 & 520.833 & -390.625 & -520.833 & 0 & 0 & 0 & 0 & 0 \\ 0 & 0 & 0 & 0 & 0 & 0 & 0 & 0 & 0 & 0 & 0 & 0 \\ 0 & 0 & 0 & 0 & 0 & 0 & 0 & 0 & 0 & 0 & 0 & 0 \\ -976.563 & 520.833 & 0 & 0 & 1671.01 & -781.25 & -694.444 & 260.417 & 0 & 0 & 0 & 0 \\ 260.417 & -390.625 & 0 & 0 & -781.25 & 2126.74 & 520.833 & -1736.11 & 0 & 0 & 0 & 0 \\ 0 & -520.833 & 0 & 0 & -694.444 & 520.833 & 694.444 & 0 & 0 & 0 & 0 & 0 \\ -260.417 & 0 & 0 & 0 & 260.417 & -1736.11 & 0 & 1736.11 & 0 & 0 & 0 & 0 \\ 0 & 0 & 0 & 0 & 0 & 0 & 0 & 0 & 0 & 0 & 0 & 0 \\ 0 & 0 & 0 & 0 & 0 & 0 & 0 & 0 & 0 & 0 & 0 & 0 \\ 0 & 0 & 0 & 0 & 0 & 0 & 0 & 0 & 0 & 0 & 0 & 0 \\ 0 & 0 & 0 & 0 & 0 & 0 & 0 & 0 & 0 & 0 & 0 & 0 \end{pmatrix} \begin{pmatrix} u_1 \\ v_1 \\ u_2 \\ v_2 \\ u_3 \\ v_3 \\ u_4 \\ v_4 \\ u_5 \\ v_5 \\ u_6 \\ v_6 \end{pmatrix} = \begin{pmatrix} 0 \\ 0 \\ 0 \\ 0 \\ 0 \\ 0 \\ 0 \\ 0 \\ 0 \\ 0 \\ 0 \\ 0 \end{pmatrix}$$

The equations for element 2 are as follows:

$h = 0.25$; $E = 10000$; $\nu = 0.2$

Plane stress constitutive matrix, $\boldsymbol{C} = \begin{pmatrix} 10416.7 & 2083.33 & 0 \\ 2083.33 & 10416.7 & 0 \\ 0 & 0 & 4166.67 \end{pmatrix}$

Nodal coordinates:

Element Node	Global Node Number	x	y
1	4	2.	1.5
2	2	0.	2.
3	1	0.	0.

$x_1 = 2.$; $x_2 = 0.$; $x_3 = 0.$

$y_1 = 1.5$; $y_2 = 2.$; $y_3 = 0.$

Using these values, we get

$$b_1 = 2.; \quad b_2 = -1.5; \quad b_3 = -0.5$$
$$c_1 = 0.; \quad c_2 = 2.; \quad c_3 = -2.$$

Element area, $A = 2.$

$$B^T = \begin{pmatrix} 0.5 & 0 & -0.375 & 0 & -0.125 & 0 \\ 0 & 0. & 0 & 0.5 & 0 & -0.5 \\ 0. & 0.5 & 0.5 & -0.375 & -0.5 & -0.125 \end{pmatrix}$$

Thus the element stiffness matrix is

$$k = hABCB^T = \begin{pmatrix} 1302.08 & 0 & -976.563 & 260.417 & -325.521 & -260.417 \\ 0 & 520.833 & 520.833 & -390.625 & -520.833 & -130.208 \\ -976.563 & 520.833 & 1253.26 & -585.938 & -276.693 & 65.1042 \\ 260.417 & -390.625 & -585.938 & 1595.05 & 325.521 & -1204.43 \\ -325.521 & -520.833 & -276.693 & 325.521 & 602.214 & 195.313 \\ -260.417 & -130.208 & 65.1042 & -1204.43 & 195.313 & 1334.64 \end{pmatrix}$$

Load vector due to distributed load on side 1 (nodes {4, 2}):

Specified load components: $q_n = -20$; $q_t = 0$

End nodal coordinates: ({2., 1.5} {0., 2.}), giving side length $L = 2.06155$

Components of unit normal to the side: $n_x = 0.242536$; $n_y = 0.970143$

Using these values, we get

$$r_q^T = \begin{pmatrix} -1.25 & -5. & -1.25 & -5. & 0 & 0 \end{pmatrix}$$

Complete element equations:

$$\begin{pmatrix} 1302.08 & 0 & -976.563 & 260.417 & -325.521 & -260.417 \\ 0 & 520.833 & 520.833 & -390.625 & -520.833 & -130.208 \\ -976.563 & 520.833 & 1253.26 & -585.938 & -276.693 & 65.1042 \\ 260.417 & -390.625 & -585.938 & 1595.05 & 325.521 & -1204.43 \\ -325.521 & -520.833 & -276.693 & 325.521 & 602.214 & 195.313 \\ -260.417 & -130.208 & 65.1042 & -1204.43 & 195.313 & 1334.64 \end{pmatrix} \begin{pmatrix} u_4 \\ v_4 \\ u_2 \\ v_2 \\ u_1 \\ v_1 \end{pmatrix} = \begin{pmatrix} -1.25 \\ -5. \\ -1.25 \\ -5. \\ 0. \\ 0. \end{pmatrix}$$

The element contributes to {7, 8, 3, 4, 1, 2} degrees of freedom.

Locations for element contributions to a global vector: $\begin{pmatrix} 7 \\ 8 \\ 3 \\ 4 \\ 1 \\ 2 \end{pmatrix}$

and to a global matrix: $\begin{pmatrix} [7,7] & [7,8] & [7,3] & [7,4] & [7,1] & [7,2] \\ [8,7] & [8,8] & [8,3] & [8,4] & [8,1] & [8,2] \\ [3,7] & [3,8] & [3,3] & [3,4] & [3,1] & [3,2] \\ [4,7] & [4,8] & [4,3] & [4,4] & [4,1] & [4,2] \\ [1,7] & [1,8] & [1,3] & [1,4] & [1,1] & [1,2] \\ [2,7] & [2,8] & [2,3] & [2,4] & [2,1] & [2,2] \end{pmatrix}$

Adding element equations into appropriate locations, we have

$$\begin{pmatrix} 1578.78 & 195.313 & -276.693 & 325.521 & -976.563 & 260.417 & -325.521 & -781.25 & 0 & 0 & 0 & 0 \\ 195.313 & 1725.26 & 65.1042 & -1204.43 & 520.833 & -390.625 & -781.25 & -130.208 & 0 & 0 & 0 & 0 \\ -276.693 & 65.1042 & 1253.26 & -585.938 & 0 & 0 & -976.563 & 520.833 & 0 & 0 & 0 & 0 \\ 325.521 & -1204.43 & -585.938 & 1595.05 & 0 & 0 & 260.417 & -390.625 & 0 & 0 & 0 & 0 \\ -976.563 & 520.833 & 0 & 0 & 1671.01 & -781.25 & -694.444 & 260.417 & 0 & 0 & 0 & 0 \\ 260.417 & -390.625 & 0 & 0 & -781.25 & 2126.74 & 520.833 & -1736.11 & 0 & 0 & 0 & 0 \\ -325.521 & -781.25 & -976.563 & 260.417 & -694.444 & 520.833 & 1996.53 & 0 & 0 & 0 & 0 & 0 \\ -781.25 & -130.208 & 520.833 & -390.625 & 260.417 & -1736.11 & 0 & 2256.94 & 0 & 0 & 0 & 0 \\ 0 & 0 & 0 & 0 & 0 & 0 & 0 & 0 & 0 & 0 & 0 & 0 \\ 0 & 0 & 0 & 0 & 0 & 0 & 0 & 0 & 0 & 0 & 0 & 0 \\ 0 & 0 & 0 & 0 & 0 & 0 & 0 & 0 & 0 & 0 & 0 & 0 \\ 0 & 0 & 0 & 0 & 0 & 0 & 0 & 0 & 0 & 0 & 0 & 0 \end{pmatrix} \begin{pmatrix} u_1 \\ v_1 \\ u_2 \\ v_2 \\ u_3 \\ v_3 \\ u_4 \\ v_4 \\ u_5 \\ v_5 \\ u_6 \\ v_6 \end{pmatrix} = \begin{pmatrix} 0 \\ 0 \\ -1.25 \\ -5. \\ 0 \\ 0 \\ -1.25 \\ -5. \\ 0 \\ 0 \\ 0 \\ 0 \end{pmatrix}$$

The remaining elements can be processed in exactly the same manner. After assembling all elements, the global equations for the model are as follows:

$$\begin{pmatrix} 1578.78 & 195.313 & -276.693 & 325.521 & -976.563 & 260.417 & -325.521 & -781.25 & 0 & 0 & 0 & 0 \\ 195.313 & 1725.26 & 65.1042 & -1204.43 & 520.833 & -390.625 & -781.25 & -130.208 & 0 & 0 & 0 & 0 \\ -276.693 & 65.1042 & 1253.26 & -585.938 & 0 & 0 & -976.563 & 520.833 & 0 & 0 & 0 & 0 \\ 325.521 & -1204.43 & -585.938 & 1595.05 & 0 & 0 & 260.417 & -390.625 & 0 & 0 & 0 & 0 \\ -976.563 & 520.833 & 0 & 0 & 3125. & -520.833 & -1171.88 & 520.833 & -651.042 & 260.417 & -325.521 & -781.25 \\ 260.417 & -390.625 & 0 & 0 & -520.833 & 4166.67 & 520.833 & -3385.42 & 520.833 & -260.417 & -781.25 & -130.208 \\ -325.521 & -781.25 & -976.563 & 260.417 & -1171.88 & 520.833 & 3125. & -520.833 & 0 & 0 & -651.042 & 520.833 \\ -781.25 & -130.208 & 520.833 & -390.625 & 520.833 & -3385.42 & -520.833 & 4166.67 & 0 & 0 & 260.417 & -260.417 \\ 0 & 0 & 0 & 0 & -651.042 & 520.833 & 0 & 0 & 1692.71 & -781.25 & -1041.67 & 260.417 \\ 0 & 0 & 0 & 0 & 260.417 & -260.417 & 0 & 0 & -781.25 & 2864.58 & 520.833 & -2604.17 \\ 0 & 0 & 0 & 0 & -325.521 & -781.25 & -651.042 & 260.417 & -1041.67 & 520.833 & 2018.23 & 0 \\ 0 & 0 & 0 & 0 & -781.25 & -130.208 & 520.833 & -260.417 & 260.417 & -2604.17 & 0 & 2994.79 \end{pmatrix} \begin{pmatrix} u_1 \\ v_1 \\ u_2 \\ v_2 \\ u_3 \\ v_3 \\ u_4 \\ v_4 \\ u_5 \\ v_5 \\ u_6 \\ v_6 \end{pmatrix} = \begin{pmatrix} 0 \\ 0 \\ -1.25 \\ -5. \\ 0 \\ 0 \\ -2.5 \\ -10. \\ 0 \\ 0 \\ -1.25 \\ -5. \end{pmatrix}$$

◆ ***Mathematica*/MATLAB Implementation 1.7 on the Book Web Site:**
Plane stress example assembly

1.3 BOUNDARY CONDITIONS AND NODAL SOLUTION

A variety of boundary conditions were specified for the problems considered in the previous two sections. For the truss problem, some nodes were located at the supports, which means that the displacements at these nodes must be zero. For the heat flow problem conditions to be satisfied along the boundaries involved specified temperature, convection, insulation, or heat flux. For the stress analysis problem, on some surfaces loading was applied while other surfaces were attached to fixed supports.

In the finite element analysis, some of the boundary conditions are incorporated directly into the element equations. These are known as the *natural boundary conditions*. Examples of natural boundary conditions encountered in the previous sections are convection, insulation, and heat flux boundary conditions for the heat flow problem and applied surface loading for the structural problems. A review of element equations and assembly procedures discussed in the previous sections should make it clear that these types of boundary conditions have been taken into account while formulating and assembling the element equations.

The boundary conditions that are not incorporated directly into the element equations are known as the *essential boundary conditions*. For heat flow problems these involve specified temperatures along some boundaries. For stress analysis problems nodal supports represent essential boundary conditions. Once again, a review of element equations and assembly procedures discussed in the previous sections should clearly indicate that these types of boundary conditions have not been taken into account while formulating and assembling the element equations.

Precise reasoning for splitting the boundary conditions into two categories—natural and essential—will become clear after studying the mathematical foundations of the finite element method in the following chapter. At this stage our main concern is to see how to actually incorporate boundary conditions into the finite element equations. Obviously we must take care of all applicable boundary conditions when solving a given problem. We start with the element equations and their assembly. The boundary conditions that can be incorporated through the element equations are handled first. The remaining boundary conditions are the essential boundary conditions that are taken into account using procedures discussed in this section.

In the assembly process it is assumed that all nodal degrees of freedom (dof) are unknown. However, due to essential boundary conditions, some of these degrees of freedom may have a zero or some other specified value. Therefore, introduction of essential boundary conditions into the global equations involves substituting the known values into the vector of nodal variables and making appropriate changes to the equations.

There are three methods for accomplishing this. The first method involves rearrangement of equations but has an advantage that the final system of equations is smaller than the global system. The second method keeps the original order of variables but does require several modifications to the equations. The third method is an approximate method but requires the fewest changes to the equations. On some computers rearranging a large system of equations is very inefficient and hence the approximate method may be more economical. For hand calculations obviously the first method is preferred. Most examples presented in this text use this first approach. The details of these methods follow.

1.3.1 Essential Boundary Conditions by Rearranging Equations

As an illustration, assume that a finite element model from a heat flow problem results in the following global system of equations:

$$\begin{pmatrix} 1000 & 10 & 20 & 30 & 40 & 50 \\ 10 & 2000 & 21 & 31 & 41 & 51 \\ 20 & 21 & 3000 & 32 & 42 & 52 \\ 30 & 31 & 32 & 4000 & 43 & 53 \\ 40 & 41 & 42 & 43 & 5000 & 54 \\ 50 & 51 & 52 & 53 & 54 & 6000 \end{pmatrix} \begin{pmatrix} T_1 \\ T_2 \\ T_3 \\ T_4 \\ T_5 \\ T_6 \end{pmatrix} = \begin{pmatrix} 100 \\ 110 \\ 120 \\ 130 \\ 140 \\ 150 \end{pmatrix}$$

Consider the situation when some of the nodes are located on sides over which the temperatures are known. As an illustration, say $T_1 = 5$, $T_4 = -7$, and $T_5 = 0$. Thus we only have three nodal unknown temperatures remaining, namely T_2, T_3, and T_6. These three unknowns can be obtained by solving the corresponding equations, 2, 3, and 6. Introducing the known values in the nodal degrees of freedom vector and retaining only the equations

2, 3, and 6, we have

$$\begin{pmatrix} 10 & 2000 & 21 & 31 & 41 & 51 \\ 20 & 21 & 3000 & 32 & 42 & 52 \\ 50 & 51 & 52 & 53 & 54 & 6000 \end{pmatrix} \begin{pmatrix} 5 \\ T_2 \\ T_3 \\ -7 \\ 0 \\ T_6 \end{pmatrix} = \begin{pmatrix} 110 \\ 120 \\ 150 \end{pmatrix}$$

Since there are only three unknowns, these equations can be rearranged by moving the known values to the right-hand side. The first column is multiplied by 5, the fourth column by −7, and the fifth column by 0, and hence we get

$$\begin{pmatrix} 2000 & 21 & 51 \\ 21 & 3000 & 52 \\ 51 & 52 & 6000 \end{pmatrix} \begin{pmatrix} T_2 \\ T_3 \\ T_6 \end{pmatrix} + 5 \begin{pmatrix} 10 \\ 20 \\ 50 \end{pmatrix} - 7 \begin{pmatrix} 31 \\ 32 \\ 53 \end{pmatrix} + 0 \begin{pmatrix} 41 \\ 42 \\ 54 \end{pmatrix} = \begin{pmatrix} 110 \\ 120 \\ 150 \end{pmatrix}$$

Moving all constant vectors to the right-hand side, we have

$$\begin{pmatrix} 2000 & 21 & 51 \\ 21 & 3000 & 52 \\ 51 & 52 & 6000 \end{pmatrix} \begin{pmatrix} T_2 \\ T_3 \\ T_6 \end{pmatrix} = \begin{pmatrix} 110 \\ 120 \\ 150 \end{pmatrix} + \begin{pmatrix} -50 \\ -100 \\ -250 \end{pmatrix} + \begin{pmatrix} 217 \\ 224 \\ 371 \end{pmatrix}$$

Adding all terms on the right-hand-side, the final 3×3 system of equations is as follows:

$$\begin{pmatrix} 2000 & 21 & 51 \\ 21 & 3000 & 52 \\ 51 & 52 & 6000 \end{pmatrix} \begin{pmatrix} T_2 \\ T_3 \\ T_6 \end{pmatrix} = \begin{pmatrix} 277 \\ 244 \\ 271 \end{pmatrix}$$

The solution of this system of equations gives

$$\{T_2 \to 0.136559, T_3 \to 0.0796266, T_6 \to 0.0433158\}$$

Note that in these computations, when the specified value is zero, nothing gets added to the right-hand side. Thus imposing a zero value for a nodal degree of freedom simply requires removing both the corresponding row and the column from the system of equations. In this case it is not necessary to even assemble the corresponding rows and columns and therefore such degrees of freedom can be eliminated even before the assembly process begins. This is an important observation to save computational effort, especially for structural problems where zero nodal displacements are common due to supports.

You may be curious as to why we retained equations 2, 3, and 6 and removed the other three. We have three unknowns left but why can we *not use any three equations* to solve for the three remaining unknowns? For a precise answer to this question one needs to study the mathematical basis of the finite element method presented in the following chapter. However, a simple reasoning based on physical grounds is as follows. When a temperature is specified at a node, there must be a corresponding unknown heat source at that node to maintain that temperature. Similarly, for a structural problem, when a displacement is specified at a node, there actually is an unknown reaction corresponding to this degree of freedom. Thus, when we insert known values into the nodal variables vector, for equations to remain consistent, unknown heat source or reaction terms must be inserted into the right-hand-side vector for every specified nodal degree of freedom. Denoting theses unknowns as R_1, R_4, and R_5 for the illustrative example, a consistent system of equations after introducing given essential boundary conditions is as follows:

$$\begin{pmatrix} 1000 & 10 & 20 & 30 & 40 & 50 \\ 10 & 2000 & 21 & 31 & 41 & 51 \\ 20 & 21 & 3000 & 32 & 42 & 52 \\ 30 & 31 & 32 & 4000 & 43 & 53 \\ 40 & 41 & 42 & 43 & 5000 & 54 \\ 50 & 51 & 52 & 53 & 54 & 6000 \end{pmatrix} \begin{pmatrix} 5 \\ T_2 \\ T_3 \\ -7 \\ 0 \\ T_6 \end{pmatrix} = \begin{pmatrix} R_1 + 100 \\ 110 \\ 120 \\ R_4 + 130 \\ R_5 + 140 \\ 150 \end{pmatrix}$$

This system represents six equations in six unknowns, and if we want, we can solve these equations directly for the six unknowns. However, since we do not need the newly introduced unknowns, it is advantageous to reduce the system of equations to three for computing the three unknown temperatures. However, the only way to solve for the remaining nodal unknowns is to use equations 2, 3, and 6. Any other combination of three equations will not yield the desired result. The first three equations, as an example, will have four unknowns R_1, T_2, T_3, and T_6 and thus are not suitable for the required solution.

◆ *Mathematica*/MATLAB Implementation 1.8 on the Book Web Site: *Imposing essential boundary conditions*

1.3.2 Essential Boundary Conditions by Modifying Equations

The method of rearrangement of equations to handle essential boundary conditions is attractive because the final system of equations is smaller than the global system. However, on some computers rearranging a large system of equations is very inefficient, and thus it is desirable to develop a method of incorporating essential boundary conditions that does not require rearrangement of equations.

A simple alternative is to modify each equation that corresponds to a specified nodal value as follows:

(i) Insert 1 at the diagonal location and zero at all off-diagonal locations in the row corresponding to the specified degree of freedom.
(ii) Insert the known nodal value to the corresponding location in the right-hand side.

Using this approach for the illustrative example used in the previous section, we get the following system of global equations:

$$\begin{pmatrix} 1 & 0 & 0 & 0 & 0 & 0 \\ 10 & 2000 & 21 & 31 & 41 & 51 \\ 20 & 21 & 3000 & 32 & 42 & 52 \\ 0 & 0 & 0 & 1 & 0 & 0 \\ 0 & 0 & 0 & 0 & 1 & 0 \\ 50 & 51 & 52 & 53 & 54 & 6000 \end{pmatrix} \begin{pmatrix} T_1 \\ T_2 \\ T_3 \\ T_4 \\ T_5 \\ T_6 \end{pmatrix} = \begin{pmatrix} 5 \\ 110 \\ 120 \\ -7 \\ 0 \\ 150 \end{pmatrix}$$

Solving this system of equations for all six nodal values gives the following solution that is exactly the same as before:

$$\{T_1 = 5., \; T_2 = 0.136559, \; T_3 = 0.0796266, \; T_4 = -7., \; T_5 = 0, \; T_6 = 0.0433158\}$$

One drawback of the procedure so far is that the resulting coefficient matrix is not symmetric anymore. For large problems this is a serious drawback. It is possible to rectify the situation by making the corresponding column entries 0 and subtracting the products of terms in these columns with the known values from the right-hand side in a manner similar to that done in the previous case:

$$\begin{pmatrix} 1 & 0 & 0 & 0 & 0 & 0 \\ 0 & 2000 & 21 & 0 & 0 & 51 \\ 0 & 21 & 3000 & 0 & 0 & 52 \\ 0 & 0 & 0 & 1 & 0 & 0 \\ 0 & 0 & 0 & 0 & 1 & 0 \\ 0 & 51 & 52 & 0 & 0 & 6000 \end{pmatrix} \begin{pmatrix} T_1 \\ T_2 \\ T_3 \\ T_4 \\ T_5 \\ T_6 \end{pmatrix} = \begin{pmatrix} 5 \\ 110 \\ 120 \\ -7 \\ 0 \\ 150 \end{pmatrix} - 5 \times \begin{pmatrix} 0 \\ 10 \\ 20 \\ 0 \\ 0 \\ 50 \end{pmatrix} - (-7) \times \begin{pmatrix} 0 \\ 31 \\ 32 \\ 0 \\ 0 \\ 53 \end{pmatrix} - 0 \times \begin{pmatrix} 0 \\ 41 \\ 42 \\ 0 \\ 0 \\ 54 \end{pmatrix} = \begin{pmatrix} 5 \\ 277 \\ 244 \\ -7 \\ 0 \\ 271 \end{pmatrix}$$

$$\{T_1 = 5., \; T_2 = 0.136559, \; T_3 = 0.0796266, \; T_4 = -7., \; T_5 = 0, \; T_6 = 0.0433158\}$$

For imposing zero values of the nodal variables the process works very well because no modifications to the right-hand side are necessary. However, for nonzero values the procedure loses some of its simplicity. The following procedure, though approximate, is much simpler to implement.

1.3.3 Approximate Treatment of Essential Boundary Conditions

The previous two methods may be inefficient to implement on some computers because they require several modifications to the system of equations. A simple but approximate alternative is to modify each equation corresponding to a specified nodal value as follows:

(i) Insert a large number at the diagonal corresponding to the known degree of freedom.

(ii) Add a value equal to this large number times the known nodal value to the right-hand side.

Using this approach for the illustrative example used in the previous section and with 10^{10} as an arbitrarily chosen large number, we get the following system of global equations:

$$\begin{pmatrix} 10^{10} & 10 & 20 & 30 & 40 & 50 \\ 10 & 2000 & 21 & 31 & 41 & 51 \\ 20 & 21 & 3000 & 32 & 42 & 52 \\ 30 & 31 & 32 & 10^{10} & 43 & 53 \\ 40 & 41 & 42 & 43 & 10^{10} & 54 \\ 50 & 51 & 52 & 53 & 54 & 6000 \end{pmatrix} \begin{pmatrix} T_1 \\ T_2 \\ T_3 \\ T_4 \\ T_5 \\ T_6 \end{pmatrix} = \begin{pmatrix} 5 \times 10^{10} \\ 110 \\ 120 \\ -7 \times 10^{10} \\ 0 \\ 150 \end{pmatrix}$$

Solving this system of equations for all six nodal values gives the following solution that is essentially the same as before:

$$\{T_1 = 5., \; T_2 = 0.136559, \; T_3 = 0.0796266, \; T_4 = -7., \; T_5 = 8.97177{\times}10^{-9}, \; T_6 = 0.0433158\}$$

To see the reason why this procedure works, consider the first equation of the modified system:

$$10000000000T_1 + 10T_2 + 20T_3 + 30T_4 + 40T_5 + 50T_6 = 50000000000$$

The two large terms dominate the equation and contribution of remaining terms is negligible. Thus the equation is effectively

$$10000000000T_1 \approx 50000000000$$

giving

$$T_1 = 5$$

Similar reasoning shows that the fourth equation gives $T_4 = -7$ and the fifth equation $T_5 = 0$. The other equations give the remaining unknowns. The method gives reasonable answers as long as the large number added to the diagonal is at least an order of magnitude greater than the largest element in the coefficient matrix. Using the number 10^6 instead of 10^{10}, we get the following solution that shows little more error in the specified values but may still be considered reasonable:

$$\{T_1 = 5.0002, \; T_2 = 0.136559, \; T_3 = 0.0796257,$$
$$T_4 = -7.00016, \; T_5 = 0.0000897165, \; T_6 = 0.0433147\}$$

1.3.4 Computation of Reactions to Verify Overall Equilibrium

For structural problems, reactions at the supports provide a quick and simple check on the validity of the analysis. For these problems the finite element equations represent equilibrium equations. Thus for overall equilibrium the sum of applied loads must be equal and opposite to the sum of all reactions at the supports. Common blunders, such as employing wrong or inconsistent units or careless computational errors, can easily be discovered if reactions are available.

The concept of equilibrium is also applicable to heat flow problems where the total heat flowing into the body must be equal to that going out of the body.

After solving for the nodal unknowns, the reactions can be computed by using the equations that were deleted when the essential boundary conditions were being imposed. Thus for the illustrative example the reactions can be computed from the first, fourth, and fifth equations. Retaining these rows and inserting all nodal values, the reactions are obtained from the following computation:

$$\begin{pmatrix} 1000 & 10 & 20 & 30 & 40 & 50 \\ 30 & 31 & 32 & 4000 & 43 & 53 \\ 40 & 41 & 42 & 43 & 5000 & 54 \end{pmatrix} \begin{pmatrix} 5 \\ 0.136559 \\ 0.0796266 \\ -7 \\ 0 \\ 0.0433158 \end{pmatrix} = \begin{pmatrix} R_1 + 100 \\ R_4 + 130 \\ R_5 + 140 \end{pmatrix}$$

Rearranging and performing computations, we get the following reactions:

$$\begin{pmatrix} R_1 \\ R_4 \\ R_5 \end{pmatrix} = \begin{pmatrix} 1000 & 10 & 20 & 30 & 40 & 50 \\ 30 & 31 & 32 & 4000 & 43 & 53 \\ 40 & 41 & 42 & 43 & 5000 & 54 \end{pmatrix} \begin{pmatrix} 5 \\ 0.136559 \\ 0.0796266 \\ -7 \\ 0 \\ 0.0433158 \end{pmatrix} - \begin{pmatrix} 100 \\ 130 \\ 140 \end{pmatrix} = \begin{pmatrix} 4695.12 \\ -27970.9 \\ -229.718 \end{pmatrix}$$

The following examples illustrate the use of reactions in validating solution consistency.

◆ *Mathematica*/MATLAB Implementation 1.9 on the Book Web Site: *Imposing essential boundary conditions*

Example 1.7 Five-Bar Truss The following global equations for five-bar plane truss were developed in Example 1.4. Incorporate essential boundary conditions into the system and obtain the final reduced system of equations in terms of the remaining unknowns. Solve for nodal values. Compute reactions and verify overall equilibrium.

$$\begin{pmatrix} 32600.2 & 76067.2 & -32600.2 & -76067.2 & 0 & 0 & 0 & 0 \\ 76067.2 & 297490. & -76067.2 & -177490. & 0 & -120000 & 0 & 0 \\ -32600.2 & -76067.2 & 243089. & 119136. & -32998.3 & 32998.3 & -177490. & -76067.2 \\ -76067.2 & -177490. & 119136. & 243089. & 32998.3 & -32998.3 & -76067.2 & -32600.2 \\ 0 & 0 & -32998.3 & 32998.3 & 152998. & -32998.3 & -120000 & 0 \\ 0 & -120000 & 32998.3 & -32998.3 & -32998.3 & 152998. & 0 & 0 \\ 0 & 0 & -177490. & -76067.2 & -120000 & 0 & 297490. & 76067.2 \\ 0 & 0 & -76067.2 & -32600.2 & 0 & 0 & 76067.2 & 32600.2 \end{pmatrix} \begin{pmatrix} u_1 \\ v_1 \\ u_2 \\ v_2 \\ u_3 \\ v_3 \\ u_4 \\ v_4 \end{pmatrix} = \begin{pmatrix} 0 \\ 0 \\ 0 \\ -150000. \\ 0 \\ 0 \\ 0 \\ 0 \end{pmatrix}$$

Nodes 1 and 4 of the truss have pin supports. Thus the essential boundary conditions are as follows:

Node	dof	Value
1	u_1	0
	v_1	0
4	u_4	0
	v_4	0

After deleting equations {1, 2, 7, 8}, we have

$$\begin{pmatrix} -32600.2 & -76067.2 & 243089. & 119136. & -32998.3 & 32998.3 & -177490. & -76067.2 \\ -76067.2 & -177490. & 119136. & 243089. & 32998.3 & -32998.3 & -76067.2 & -32600.2 \\ 0 & 0 & -32998.3 & 32998.3 & 152998. & -32998.3 & -120000 & 0 \\ 0 & -120000 & 32998.3 & -32998.3 & -32998.3 & 152998. & 0 & 0 \end{pmatrix} \begin{pmatrix} 0 \\ 0 \\ u_2 \\ v_2 \\ u_3 \\ v_3 \\ 0 \\ 0 \end{pmatrix} = \begin{pmatrix} 0. \\ -150000. \\ 0. \\ 0. \end{pmatrix}$$

Since all entries in columns {1, 2, 7, 8} are multiplied by zero nodal values, they can be removed. The final global system of equations is thus as follows:

$$\begin{pmatrix} 243089. & 119136. & -32998.3 & 32998.3 \\ 119136. & 243089. & 32998.3 & -32998.3 \\ -32998.3 & 32998.3 & 152998. & -32998.3 \\ 32998.3 & -32998.3 & -32998.3 & 152998. \end{pmatrix} \begin{pmatrix} u_2 \\ v_2 \\ u_3 \\ v_3 \end{pmatrix} = \begin{pmatrix} 0 \\ -150000. \\ 0 \\ 0 \end{pmatrix}$$

Solving the final system of global equations, we get the following nodal values:

$$\{u_2 = 0.538954,\ v_2 = -0.953061,\ u_3 = 0.264704,\ v_3 = -0.264704\}$$

Complete Table of Nodal Values

	u	v
1	0	0
2	0.538954	−0.953061
3	0.264704	−0.264704
4	0	0

Computation of Reactions

Equation numbers of dof with specified values: {1, 2, 7, 8}

Extracting equations {1, 2, 7, 8} from the global system, we have

$$\begin{pmatrix} 32600.2 & 76067.2 & -32600.2 & -76067.2 & 0 & 0 & 0 & 0 \\ 76067.2 & 297490. & -76067.2 & -177490. & 0 & -120000 & 0 & 0 \\ 0 & 0 & -177490. & -76067.2 & -120000 & 0 & 297490. & 76067.2 \\ 0 & 0 & -76067.2 & -32600.2 & 0 & 0 & 76067.2 & 32600.2 \end{pmatrix} \begin{pmatrix} u_1 \\ v_1 \\ u_2 \\ v_2 \\ u_3 \\ v_3 \\ u_4 \\ v_4 \end{pmatrix} = \begin{pmatrix} R_1 + 0. \\ R_2 + 0. \\ R_3 + 0. \\ R_4 + 0. \end{pmatrix}$$

Substituting the nodal values and rearranging,

$$\begin{pmatrix} R_1 \\ R_2 \\ R_3 \\ R_4 \end{pmatrix} = \begin{pmatrix} 32600.2 & 76067.2 & -32600.2 & -76067.2 & 0 & 0 & 0 & 0 \\ 76067.2 & 297490. & -76067.2 & -177490. & 0 & -120000 & 0 & 0 \\ 0 & 0 & -177490. & -76067.2 & -120000 & 0 & 297490. & 76067.2 \\ 0 & 0 & -76067.2 & -32600.2 & 0 & 0 & 76067.2 & 32600.2 \end{pmatrix} \begin{pmatrix} 0 \\ 0 \\ 0.538954 \\ -0.953061 \\ 0.264704 \\ -0.264704 \\ 0 \\ 0 \end{pmatrix}$$

Carrying out computations, the reactions are as follows:

Label	dof	Reaction
R_1	u_1	54926.7
R_2	v_1	159927.
R_3	u_4	−54926.7
R_4	v_4	−9926.67

Sum of reactions:

dof: u	0
dof: v	150000.

There is no applied load in the x direction. Since the sum of reactions in the horizontal direction is zero, the equilibrium is satisfied in this direction. There is an applied load of 150000 in the $-y$ direction which balances with the sum of reactions in the y direction, indicating that equilibrium is satisfied in this direction as well. Hence the solution satisfies the overall equilibrium.

Example 1.8 Heat Flow through a Square Duct The following global equations for heat flow through a square duct are developed in Example 1.5. Incorporate essential boundary conditions into the system and obtain the final reduced system of equations in terms of the remaining unknowns. Solve for nodal values. Compute reactions and verify overall energy balance:

$$\begin{pmatrix} 1.4 & 0 & 0 & -0.7 & -0.7 \\ 0 & 4.56667 & 1.58333 & 0 & -2.1 \\ 0 & 1.58333 & 3.51667 & 0.35 & -1.4 \\ -0.7 & 0 & 0.35 & 3.15 & -2.8 \\ -0.7 & -2.1 & -1.4 & -2.8 & 7. \end{pmatrix} \begin{pmatrix} T_1 \\ T_2 \\ T_3 \\ T_4 \\ T_5 \end{pmatrix} = \begin{pmatrix} 0 \\ 81. \\ 81. \\ 0 \\ 0 \end{pmatrix}$$

Nodes 1 and 4 are on the inside face that is maintained at a temperature of 300°C. Thus the essential boundary conditions are as follows:

Node	dof	Value
1	T_1	300
4	T_4	300

Incorporating these boundary conditions means setting $T_1 = T_4 = 300$ and removing equations 1 and 4 from the global system:

$$\begin{pmatrix} 0 & 4.56667 & 1.58333 & 0 & -2.1 \\ 0 & 1.58333 & 3.51667 & 0.35 & -1.4 \\ -0.7 & -2.1 & -1.4 & -2.8 & 7 \end{pmatrix} \begin{pmatrix} 300 \\ T_2 \\ T_3 \\ 300 \\ T_5 \end{pmatrix} = \begin{pmatrix} 81 \\ 81 \\ 0 \end{pmatrix}$$

Multiplying the first column by 300 and the fourth column by 300, the equations can be written as follows:

$$\begin{pmatrix} 4.56667 & 1.58333 & -2.1 \\ 1.58333 & 3.51667 & -1.4 \\ -2.1 & -1.4 & 7 \end{pmatrix} \begin{pmatrix} T_2 \\ T_3 \\ T_5 \end{pmatrix} + 300 \begin{pmatrix} 0 \\ 0 \\ -0.7 \end{pmatrix} + 300 \begin{pmatrix} 0 \\ 0.35 \\ -2.8 \end{pmatrix} = \begin{pmatrix} 81 \\ 81 \\ 0 \end{pmatrix}$$

Moving constant vectors to the right-hand side, we get the final system of equations as follows:

$$\begin{pmatrix} 4.56667 & 1.58333 & -2.1 \\ 1.58333 & 3.51667 & -1.4 \\ -2.1 & -1.4 & 7. \end{pmatrix} \begin{pmatrix} T_2 \\ T_3 \\ T_5 \end{pmatrix} = \begin{pmatrix} 81. \\ -24. \\ 1050. \end{pmatrix}$$

Solving the final system of global equations, we get the following nodal values:

$$\{T_2 = 93.5466, \; T_3 = 23.8437, \; T_5 = 182.833\}$$

Complete Table of Nodal Values

	T
1	300
2	93.5466
3	23.8437
4	300
5	182.833

Computation of Reactions

Equation numbers of dof with specified values: {1, 4}

Extracting these equations from the global system (the one prior to adjustment for essential boundary conditions), we have

$$\begin{pmatrix} 1.4 & 0 & 0 & -0.7 & -0.7 \\ -0.7 & 0 & 0.35 & 3.15 & -2.8 \end{pmatrix} \begin{pmatrix} T_1 \\ T_2 \\ T_3 \\ T_4 \\ T_5 \end{pmatrix} = \begin{pmatrix} R_1 \\ R_2 \end{pmatrix},$$

Substituting the nodal values and rearranging,

$$\begin{pmatrix} R_1 \\ R_2 \end{pmatrix} = \begin{pmatrix} 1.4 & 0 & 0 & -0.7 & -0.7 \\ -0.7 & 0 & 0.35 & 3.15 & -2.8 \end{pmatrix} \begin{pmatrix} 300 \\ 93.5466 \\ 23.8437 \\ 300 \\ 182.833 \end{pmatrix}$$

Table of reactions:

Label	dof	Reaction
R_1	T_1	82.0171
R_2	T_4	231.414

Sum of reactions: 313.431

The sum of reactions represents the total heat energy put into the system. This must be balanced by the heat loss due to convection. The convection heat loss takes place only on side 2–3 of the model. The average temperature of this side is

$$T_{avg} = \tfrac{1}{2}(T_2 + T_3) = 58.695$$

Assuming this temperature to be constant for the entire 0.3-m length of the side,

$$\text{Convection heat loss from side 2–3} = hL(T_{avg} - T_\infty) = 313.43$$

This heat loss is exactly the same as the sum of reactions, indicating that the solution satisfies overall energy balance.

Example 1.9 Stress Analysis of a Bracket The following global equations for stress analysis of a bracket are developed in Example 1.6. Incorporate essential boundary conditions into the system and obtain the final reduced system of equations in terms of the remaining unknowns. Solve for nodal values. Compute reactions and verify overall equilibrium:

$$\begin{pmatrix}
1578.78 & 195.313 & -276.693 & 325.521 & -976.563 & 260.417 & -325.521 & -781.25 & 0 & 0 & 0 & 0 \\
195.313 & 1725.26 & 65.1042 & -1204.43 & 520.833 & -390.625 & -781.25 & -130.208 & 0 & 0 & 0 & 0 \\
-276.693 & 65.1042 & 1253.26 & -585.938 & 0 & 0 & -976.563 & 520.833 & 0 & 0 & 0 & 0 \\
325.521 & -1204.43 & -585.938 & 1595.05 & 0 & 0 & 260.417 & -390.625 & 0 & 0 & 0 & 0 \\
-976.563 & 520.833 & 0 & 0 & 3125. & -520.833 & -1171.88 & 520.833 & -651.042 & 260.417 & -325.521 & -781.25 \\
260.417 & -390.625 & 0 & 0 & -520.833 & 4166.67 & 520.833 & -3385.42 & 520.833 & -260.417 & -781.25 & -130.208 \\
-325.521 & -781.25 & -976.563 & 260.417 & -1171.88 & 520.833 & 3125. & -520.833 & 0 & 0 & -651.042 & 520.833 \\
-781.25 & -130.208 & 520.833 & -390.625 & 520.833 & -3385.42 & -520.833 & 4166.67 & 0 & 0 & 260.417 & -260.417 \\
0 & 0 & 0 & 0 & -651.042 & 520.833 & 0 & 0 & 1692.71 & -781.25 & -1041.67 & 260.417 \\
0 & 0 & 0 & 0 & 260.417 & -260.417 & 0 & 0 & -781.25 & 2864.58 & 520.833 & -2604.17 \\
0 & 0 & 0 & 0 & -325.521 & -781.25 & -651.042 & 260.417 & -1041.67 & 520.833 & 2018.23 & 0 \\
0 & 0 & 0 & 0 & -781.25 & -130.208 & 520.833 & -260.417 & 260.417 & -2604.17 & 0 & 2994.79
\end{pmatrix}
\begin{pmatrix} u_1 \\ v_1 \\ u_2 \\ v_2 \\ u_3 \\ v_3 \\ u_4 \\ v_4 \\ u_5 \\ v_5 \\ u_6 \\ v_6 \end{pmatrix}
=
\begin{pmatrix} 0 \\ 0 \\ -1.25 \\ -5. \\ 0 \\ 0 \\ -2.5 \\ -10. \\ 0 \\ 0 \\ -1.25 \\ -5. \end{pmatrix}$$

Nodes 1 and 2 are on the fixed end of the bracket. Thus the essential boundary conditions are as follows:

Node	dof	Value
1	u_1	0
	v_1	0
2	u_2	0
	v_2	0

Incorporating these boundary conditions means removing equations 1, 2, 3, and 4 from the global system. Furthermore, since the specified values are 0, the corresponding columns can also be removed. Thus, after incorporating the essential boundary conditions, the final system of equations is as follows:

$$\begin{pmatrix} 3125. & -520.833 & -1171.88 & 520.833 & -651.042 & 260.417 & -325.521 & -781.25 \\ -520.833 & 4166.67 & 520.833 & -3385.42 & 520.833 & -260.417 & -781.25 & -130.208 \\ -1171.88 & 520.833 & 3125. & -520.833 & 0 & 0 & -651.042 & 520.833 \\ 520.833 & -3385.42 & -520.833 & 4166.67 & 0 & 0 & 260.417 & -260.417 \\ -651.042 & 520.833 & 0 & 0 & 1692.71 & -781.25 & -1041.67 & 260.417 \\ 260.417 & -260.417 & 0 & 0 & -781.25 & 2864.58 & 520.833 & -2604.17 \\ -325.521 & -781.25 & -651.042 & 260.417 & -1041.67 & 520.833 & 2018.23 & 0 \\ -781.25 & -130.208 & 520.833 & -260.417 & 260.417 & -2604.17 & 0 & 2994.79 \end{pmatrix} \begin{pmatrix} u_3 \\ v_3 \\ u_4 \\ v_4 \\ u_5 \\ v_5 \\ u_6 \\ v_6 \end{pmatrix} = \begin{pmatrix} 0 \\ 0 \\ -2.5 \\ -10. \\ 0 \\ 0 \\ -1.25 \\ -5. \end{pmatrix}$$

Solving the final system of global equations, we get the following nodal values:

$$\{u_3 = -0.0103553,\ v_3 = -0.0255297,\ u_4 = 0.00472765,\ v_4 = -0.0247357,$$
$$u_5 = -0.0131394,\ v_5 = -0.0554931,\ u_6 = 0.0000838902,\ v_6 = -0.0555664\}$$

Complete Table of Nodal Values

	u	v
1	0	0
2	0	0
3	−0.0103553	−0.0255297
4	0.00472765	−0.0247357
5	−0.0131394	−0.0554931
6	0.0000838902	−0.0555664

Computation of Reactions

Equation numbers of dof with specified values: {1, 2, 3, 4}

Extracting equations {1, 2, 3, 4} from the global system, we have

$$\begin{pmatrix} 1578.78 & 195.313 & -276.693 & 325.521 & -976.563 & 260.417 & -325.521 & -781.25 & 0 & 0 & 0 & 0 \\ 195.313 & 1725.26 & 65.1042 & -1204.43 & 520.833 & -390.625 & -781.25 & -130.208 & 0 & 0 & 0 & 0 \\ -276.693 & 65.1042 & 1253.26 & -585.938 & 0 & 0 & -976.563 & 520.833 & 0 & 0 & 0 & 0 \\ 325.521 & -1204.43 & -585.938 & 1595.05 & 0 & 0 & 260.417 & -390.625 & 0 & 0 & 0 & 0 \end{pmatrix} \begin{pmatrix} u_1 \\ v_1 \\ u_2 \\ v_2 \\ u_3 \\ v_3 \\ u_4 \\ v_4 \\ u_5 \\ v_5 \\ u_6 \\ v_6 \end{pmatrix} = \begin{pmatrix} R_1 + 0. \\ R_2 + 0. \\ R_3 - 1.25 \\ R_4 - 5. \end{pmatrix}$$

Substituting the nodal values and rearranging,

$$\begin{pmatrix} R_1 \\ R_2 \\ R_3 \\ R_4 \end{pmatrix} = \begin{pmatrix} 1578.78 & 195.313 & -276.693 & 325.521 & -976.563 & 260.417 & -325.521 & -781.25 & 0 & 0 & 0 & 0 \\ 195.313 & 1725.26 & 65.1042 & -1204.43 & 520.833 & -390.625 & -781.25 & -130.208 & 0 & 0 & 0 & 0 \\ -276.693 & 65.1042 & 1253.26 & -585.938 & 0 & 0 & -976.563 & 520.833 & 0 & 0 & 0 & 0 \\ 325.521 & -1204.43 & -585.938 & 1595.05 & 0 & 0 & 260.417 & -390.625 & 0 & 0 & 0 & 0 \end{pmatrix} \begin{pmatrix} 0 \\ 0 \\ 0 \\ 0 \\ -0.0103553 \\ -0.0255297 \\ 0.00472765 \\ -0.0247357 \\ -0.0131394 \\ -0.0554931 \\ 0.0000838902 \\ -0.0555664 \end{pmatrix} - \begin{pmatrix} 0. \\ 0. \\ -1.25 \\ -5. \end{pmatrix}$$

Table of reactions:

Label	dof	Reaction
R_1	u_1	21.25
R_2	v_1	4.10648
R_3	u_2	−16.25
R_4	v_2	15.8935

Sum of reactions:

dof: u	5.
dof: v	20.

The sum of reactions must be balanced by the applied loading. The loading is applied normal to the top surface. To determine the total load and its components in the horizontal and vertical directions, we need its length and normal vector. The end nodal coordinates of this line are

$$x_1 = 4; \qquad y_1 = 1; \qquad x_2 = 0; \qquad y_2 = 2$$

The length and components of the unit normal to the line are computed as follows:

$$L = \sqrt{(x_2 - x_1)^2 + (y_2 - y_1)^2} = 4.12311$$

$$n_x = (y_2 - y_1)/L = 0.242536; \qquad n_y = -(x_2 - x_1)/L = 0.970143$$

The total applied load and its components in the coordinate directions can now be determined as follows:

$$\text{Total load} = qhL = -20.6155$$

$$\text{Horizontal component} = qhLn_x = -5.$$

$$\text{Vertical component} = qhLn_y = -20.$$

These are equal and opposite to the sum of reactions. Hence the solution satisfies the overall equilibrium.

1.4 ELEMENT SOLUTIONS AND MODEL VALIDITY

Once the solution for all nodal degrees of freedom is available, we can go back to the element equations and develop a complete solution over each element. The process simply requires substituting computed nodal values into the assumed element solution defined during derivation of element equations. Any secondary quantities, such as derivatives or integrals involving the solution, can be obtained by performing necessary computations on the element solutions. The examples presented in the following sections clarify the procedure for different elements introduced earlier in this chapter.

1.4.1 Plane Truss Element

As will be seen in Chapters 2 and 4, the fundamental assumption made in deriving the plane truss element equations is that the axial displacement varies linearly over the element. This means that, in terms of nodal degrees of freedom, the displacement along the element axis is written as follows:

$$u(s) = \begin{pmatrix} 1 - \frac{s}{L} & \frac{s}{L} \end{pmatrix} \begin{pmatrix} d_1 \\ d_2 \end{pmatrix}; \qquad 0 \le s \le L$$

where, as shown in Figure 1.21, s is a local coordinate that runs along the element length with $s = 0$ at the first node of the element to $s = L$ at the second node and $L =$ length of the element. The terms d_1 and d_2 are the axial displacements at the element ends. The relationship between the local axial displacements and the global nodal degrees of freedom is as follows.

Axial displacements: $$\begin{pmatrix} d_1 \\ d_2 \end{pmatrix} = \begin{pmatrix} l_s & m_s & 0 & 0 \\ 0 & 0 & l_s & m_s \end{pmatrix} \begin{pmatrix} u_1 \\ v_1 \\ u_2 \\ v_2 \end{pmatrix} \equiv \boldsymbol{Td}$$

Transformation matrix: $$\boldsymbol{T} = \begin{pmatrix} l_s & m_s & 0 & 0 \\ 0 & 0 & l_s & m_s \end{pmatrix}$$

where l_s and m_s are the direction cosines of the element axis as defined earlier. The axial strain is simply the first derivative of the axial displacement, giving constant strain over the element as follows:

$$\epsilon = \frac{du}{ds} = \begin{pmatrix} -\frac{1}{L} & \frac{1}{L} \end{pmatrix} \begin{pmatrix} d_1 \\ d_2 \end{pmatrix} \equiv \frac{1}{L}(-d_1 + d_2)$$

Using Hooke's law, the axial stress is obtained by multiplying strain by the modulus of elasticity:

$$\sigma = E\epsilon$$

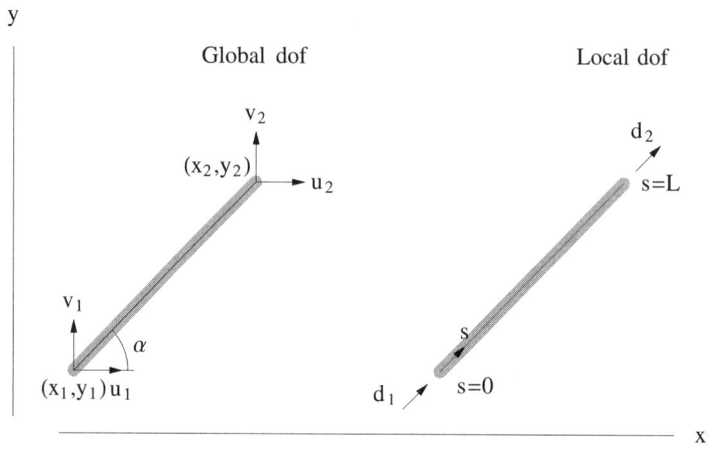

Figure 1.21. Global and local degrees of freedom for a plane truss element

Since axial stress is equal to axial force divided by the area, the axial force in the element is

$$F = \sigma A$$

The sign convention used in these equations assumes that the tension is positive and the compression is negative.

Example 1.10 Five-Bar Truss The following nodal values for a five-bar plane truss were obtained in Example 1.7. Determine axial strains, axial stresses, and axial forces in different elements of the truss.

	u	v
1	0	0
2	0.538954	−0.953061
3	0.264704	−0.264704
4	0	0

The displacements are in inches, loads in pounds, and stresses in pounds per inch squared.

The solution for element 1 is as follows:

Element nodal coordinates: first node (node #1): {0, 0}; Second node (node #2): {1500., 3500.}

$x_1 = 0; y_1 = 0; x_2 = 1500.; y_2 = 3500.$

$L = \sqrt{(x_2 - x_1)^2 + (y_2 - y_1)^2} = 3807.89$

Direction cosines: $l_s = \dfrac{x_2 - x_1}{L} = 0.393919; m_s = \dfrac{y_2 - y_1}{L} = 0.919145$

Global to local transformation matrix,

$$T = \begin{pmatrix} 0.393919 & 0.919145 & 0 & 0 \\ 0 & 0 & 0.393919 & 0.919145 \end{pmatrix}$$

Element nodal displacements in global coordinates, $d = \begin{pmatrix} u_1 \\ v_1 \\ u_2 \\ v_2 \end{pmatrix} = \begin{pmatrix} 0 \\ 0 \\ 0.538954 \\ -0.953061 \end{pmatrix}$

Element nodal displacements in local coordinates, $d_l = Td = \begin{pmatrix} 0. \\ -0.663697 \end{pmatrix}$

Axial displacements at element ends, $d_1 = 0.$; $d_2 = -0.663697$

Axial strain, $\epsilon = (d_2 - d_1)/L = -0.000174295$

$E = 200000$; $A = 4000$

Axial stress, $\sigma = E\epsilon = -34.8591$; Axial force $= \sigma A = -139436.$

For any other element the calculations follow exactly the same pattern.

The solution summary is as follows:

	Stress	Axial force
1	−34.8591	−139436.
2	−6.29994	−25199.8
3	−10.5881	−31764.4
4	−10.5881	−31764.4
5	22.4608	44921.7

◆ *Mathematica*/MATLAB Implementation 1.10 on the Book Web Site: *Plane truss element results*

◆ *Mathematica*/MATLAB Implementation 1.11 on the Book Web Site: *Complete solution of a plane truss*

1.4.2 Triangular Element for Two-Dimensional Heat Flow

The finite element equations for a triangular element for two-dimensional steady-state heat flow are based on the assumption of linear temperature distribution over the element. In terms of nodal temperatures, the temperature distribution over a typical element is written as follows:

$$T(x, y) = N_1 T_1 + N_2 T_2 + N_3 T_3 = (N_1 \quad N_2 \quad N_3) \begin{pmatrix} T_1 \\ T_2 \\ T_3 \end{pmatrix} \equiv N^T d$$

where

$$\mathbf{N} = \begin{pmatrix} N_1 \\ N_2 \\ N_3 \end{pmatrix} = \frac{1}{2A} \begin{pmatrix} xb_1 + yc_1 + f_1 \\ xb_2 + yc_2 + f_2 \\ xb_3 + yc_3 + f_3 \end{pmatrix}$$

$$b_1 = y_2 - y_3; \quad b_2 = y_3 - y_1; \quad b_3 = y_1 - y_2;$$

$$c_1 = x_3 - x_2; \quad c_2 = x_1 - x_3; \quad c_3 = x_2 - x_1;$$

$$f_1 = x_2 y_3 - x_3 y_2; \quad f_2 = x_3 y_1 - x_1 y_3; \quad f_3 = x_1 y_2 - x_2 y_1;$$

The quantities N_i, $i = 1, 2, 3$, are linear triangle interpolation functions. The vector $\boldsymbol{d}$ is the vector of nodal quantities that are all known at this stage. Knowing the coordinates at the three element nodes, the interpolation functions can easily be written for each element. Multiplying these interpolation functions by the nodal temperature computed from the solution of global equations gives the temperature distribution $T(x, y)$ over each element. This expression can be differentiated or integrated in a usual manner to obtain other quantities of interest.

Note that for each element a different expression for the solution $T(x, y)$ is obtained. The finite element assembly process matches the degrees of freedom at the common nodes between the elements. Therefore, T must be continuous across common element boundaries. However, since T is a linear function of x and y, its derivatives are constant. Thus over each element the derivative values could be different, and in general two different derivative values will be computed at a common interface between the two elements. The magnitude of this discontinuity can be used as a guide to determine if the finite element discretization is suitable or needs further refinement.

Example 1.11 Heat Flow through a Square Duct The following nodal temperature values for heat flow through a square duct were obtained in Example 1.8. Determine the temperature distribution over each element and its x and y derivatives. By comparing the magnitudes of these derivatives across common element interfaces, comment on suitability of the finite element mesh.

Node	Temperature
1	300
2	93.5466
3	23.8437
4	300
5	182.833

The temperature is in degrees Celsius and heat flow in watts per meter.

The solution for element 1 is as follows:

Element nodes: First node (node #1): {0., 0.}; Second node (node #2): {0.2, 0.}; Third node (node #5): {0.1, 0.1}

$x_1 = 0.; x_2 = 0.2; x_3 = 0.1$

$y_1 = 0.; y_2 = 0.; y_3 = 0.1$

Using these values, we get

$$
\begin{array}{lll}
b_1 = -0.1; & b_2 = 0.1; & b_3 = 0. \\
c_1 = -0.1; & c_2 = -0.1; & c_3 = 0.2 \\
f_1 = 0.02; & f_2 = 0.; & tf_3 = 0.
\end{array}
$$

Element area, $A = 0.01$

Substituting these into the formulas for triangle interpolation functions, we get

$$\text{Interpolation functions, } \boldsymbol{N}^T = \{-5.x - 5.y + 1., 5.x - 5.y, 10.y\}$$

From a global solution the temperatures at the element nodes are

$$(\text{from nodes } \{1, 2, 5\}), \boldsymbol{d}^T = \{300, 93.5466, 182.833\}$$

Thus the temperature distribution over the element, $T(x, y) = \boldsymbol{N}^T\boldsymbol{d} = -1032.27x - 139.406y + 300.$

Differentiating with respect to x and y, $\partial T/\partial x = -1032.27$ and $\partial T/\partial y = -139.406$

For any other element the calculations follow exactly the same pattern.

The solution summary is as follows:

	$T(x, y)$	$\partial T/\partial x$	$\partial T/\partial y$
1	$-1032.27x - 139.406y + 300.$	-1032.27	-139.406
2	$-1125.2x - 232.343y + 318.587$	-1125.2	-232.343
3	$-1171.67x - 209.109y + 320.911$	-1171.67	-209.109
4	$300. - 1171.67x$	-1171.67	0

As seen from the model shown in Figure 1.18, elements 1 and 2 have a common side between them. However, the derivative values for the two elements show significant difference. Elements 3 and 4 also share a side. The x derivatives for the two elements are the same; however, their y derivatives are vastly different. This large difference in the derivative values shows that the finite element mesh used for the analysis is very coarse. A much finer mesh is needed for any meaningful results. As mentioned earlier, this coarse mesh is used here to illustrate all necessary computations in detail.

◆ *Mathematica*/MATLAB Implementation 1.12 on the Book Web Site: *Heat flow element results*

◆ *Mathematica*/MATLAB Implementation 1.13 on the Book Web Site: *Complete solution of a heat flow problem*

1.4.3 Triangular Element for Two-Dimensional Stress Analysis

The derivation of the plane stress triangular element equations is based on the assumption of linear displacements over the element. In terms of nodal quantities, the displacements over an element can be written as follows.

$$u(x, y) = N_1 u_1 + N_2 u_2 + N_3 u_3$$
$$v(x, y) = N_1 v_1 + N_2 v_2 + N_3 v_3$$

$$\begin{pmatrix} u(x, y) \\ v(x, y) \end{pmatrix} = \begin{pmatrix} N_1 & 0 & N_2 & 0 & N_3 & 0 \\ 0 & N_1 & 0 & N_2 & 0 & N_3 \end{pmatrix} \begin{pmatrix} u_1 \\ v_1 \\ u_2 \\ v_2 \\ u_3 \\ v_3 \end{pmatrix} \equiv \boldsymbol{N}^T \boldsymbol{d}$$

where N_i, $i = 1, 2, 3$, are the triangle interpolation functions (same as those used for the triangular element for heat flow). Knowing the coordinates at the three element nodes, the interpolation functions can easily be calculated for each element. Multiplying these interpolation functions by the nodal displacements computed from the solution of global equations gives the horizontal and vertical displacement distributions $u(x, y)$ and $v(x, y)$ over each element. These expression can be differentiated or integrated in a usual manner to obtain other quantities of interest.

As discussed later in Chapter 7, the three components of in-plane element strains are computed directly by differentiating displacement functions as follows:

$$\epsilon = \begin{pmatrix} \epsilon_x \\ \epsilon_y \\ \gamma_{xy} \end{pmatrix} = \begin{pmatrix} \frac{\partial u}{\partial x} \\ \frac{\partial v}{\partial y} \\ \frac{\partial u}{\partial y} + \frac{\partial v}{\partial x} \end{pmatrix} = \begin{pmatrix} b_1 & 0 & b_2 & 0 & b_3 & 0 \\ 0 & c_1 & 0 & c_2 & 0 & c_3 \\ c_1 & b_1 & c_2 & b_2 & c_3 & b_3 \end{pmatrix} \begin{pmatrix} u_1 \\ v_1 \\ u_2 \\ v_2 \\ \vdots \end{pmatrix} = \boldsymbol{B}^T \boldsymbol{d}$$

The stresses can be computed as follows:

$$\boldsymbol{\sigma} = \boldsymbol{C\epsilon}$$

where $\boldsymbol{C}$ is the appropriate constitutive matrix. For homogeneous, isotropic elastic materials under plane stress conditions

$$\boldsymbol{C} = \frac{E}{1 - \nu^2} \begin{pmatrix} 1 & \nu & 0 \\ \nu & 1 & 0 \\ 0 & 0 & \frac{1-\nu}{2} \end{pmatrix}$$

where E = Young's modulus and ν = Poisson's ratio.

For plane stress problems, the three out-of-plane strain and stress components are as follows:

$$\epsilon_z = -\frac{\nu(\sigma_x + \sigma_y)}{E}; \qquad \gamma_{yz} = 0; \qquad \gamma_{zx} = 0$$

$$\sigma_z = 0; \qquad \tau_{yz} = 0; \qquad \tau_{zx} = 0$$

The complete vectors of strains and stresses are ordered as follows:

$$\epsilon = (\epsilon_x \quad \epsilon_y \quad \epsilon_z \quad \gamma_{xy} \quad \gamma_{yz} \quad \gamma_{zx})^T$$

$$\boldsymbol{\sigma} = (\sigma_x \quad \sigma_y \quad \sigma_z \quad \tau_{xy} \quad \tau_{yz} \quad \tau_{zx})^T$$

From a design point of view, one is often interested in principal stresses or effective (von Mises) stress. The principal stresses can be computed by determining eigenvalues of the following 3×3 matrix of stress components:

$$\boldsymbol{S} = \begin{pmatrix} \sigma_x & \tau_{xy} & \tau_{zx} \\ \tau_{xy} & \sigma_y & \tau_{yz} \\ \tau_{zx} & \tau_{yz} & \sigma_z \end{pmatrix}; \qquad \text{Principal stresses} = \text{eigenvalues}[\boldsymbol{S}]$$

The effective (von Mises) stress can be computed from the component stresses using the following equation:

$$\sigma_e = \frac{1}{\sqrt{2}}\sqrt{(\sigma_x - \sigma_y)^2 + (\sigma_y - \sigma_z)^2 + (\sigma_z - \sigma_x)^2 + 6(\tau_{xy}^2 + \tau_{yz}^2 + \tau_{zx}^2)}$$

Similar to the heat flow problem, for each element different expressions for displacements $u(x, y)$ and $v(x, y)$ are obtained. The finite element assembly process matches the degrees of freedom at the common nodes between the elements. Therefore, the displacements must be continuous across common element boundaries. However, since displacements are linear functions of x and y, their derivatives, and hence strains and stresses, are constant. Thus over each element the stress values could be different, and in general two different stress values will be computed at a common interface between the two elements. The magnitude of this discontinuity can be used as a guide to determine if the finite element discretization is suitable or needs further refinement.

Example 1.12 Stress Analysis of a Bracket The following nodal displacement values for the bracket problem were obtained in Example 1.9. Determine displacement distribution over each element. Using these, compute element strains, stresses, principal stresses, and effective stress. By comparing the magnitudes of stresses across common element interfaces, comment on the suitability of the finite element mesh.

	u	v
1	0	0
2	0	0
3	−0.0103553	−0.0255297
4	0.00472765	−0.0247357
5	−0.0131394	−0.0554931
6	0.0000838902	−0.0555664

Solution for Element 1

$h = 0.25$; $E = 10000$; $\nu = 0.2$

Plane stress constitutive matrix, $\boldsymbol{C} = \begin{pmatrix} 10416.7 & 2083.33 & 0 \\ 2083.33 & 10416.7 & 0 \\ 0 & 0 & 4166.67 \end{pmatrix}$

Element nodes: first node (node 1): {0., 0.}; second node (node 3): {2., 0.}; third node (node 4): {2., 1.5}

$$x_1 = 0.; \quad x_2 = 2.; \quad x_3 = 2.$$
$$y_1 = 0.; \quad y_2 = 0.; \quad y_3 = 1.5$$

Using these values, we get

$$b_1 = -1.5; \quad b_2 = 1.5; \quad b_3 = 0.$$
$$c_1 = 0.; \quad c_2 = -2.; \quad c_3 = 2.$$
$$f_1 = 3.; \quad f_2 = 0.; \quad f_3 = 0.$$

Element area: $A = 1.5$

$$\boldsymbol{B}^T = \begin{pmatrix} -0.5 & 0 & 0.5 & 0 & 0. & 0 \\ 0 & 0. & 0 & -0.666667 & 0 & 0.666667 \\ 0. & -0.5 & -0.666667 & 0.5 & 0.666667 & 0. \end{pmatrix}$$

Substituting these into the formulas for triangle interpolation functions, we get

Interpolation functions, $\{1. - 0.5x, 0.5x - 0.666667y, 0.666667y\}$

$$\boldsymbol{N}^T = \begin{pmatrix} 1. - 0.5x & 0 & 0.5x - 0.666667y & 0 & 0.666667y & 0 \\ 0 & 1. - 0.5x & 0 & 0.5x - 0.666667y & 0 & 0.666667y \end{pmatrix}$$

From a global solution the displacements at the element nodes are

(displacements at nodes {1, 3, 4}):

$$\boldsymbol{d}^T = \{0, 0, -0.0103553, -0.0255297, 0.00472765, -0.0247357\}$$

The displacement distribution over the element is

$$\begin{pmatrix} \mathrm{u(x, y)} \\ \mathrm{v(x, y)} \end{pmatrix} = \boldsymbol{N}^T\boldsymbol{d} = \begin{pmatrix} 0.0100553y - 0.00517764x \\ 0.000529362y - 0.0127648x \end{pmatrix}$$

In-plane strain components, $\boldsymbol{\epsilon} = \boldsymbol{B}^T\boldsymbol{d} = (-0.00517764 \quad 0.000529362 \quad -0.00270956)$

In-plane stress components, $\boldsymbol{\sigma} = \boldsymbol{C}\boldsymbol{\epsilon} = (-52.8309 \quad -5.27256 \quad -11.2898)$

Computing out-of-plane strain and stress components using appropriate formulas, the complete strain and stress vectors are as follows:

$$\epsilon^T = (-0.00517764 \quad 0.000529362 \quad 0.00116207 \quad -0.00270956 \quad 0 \quad 0)$$

$$\sigma^T = (-52.8309 \quad -5.27256 \quad 0 \quad -11.2898 \quad 0 \quad 0)$$

Substituting these stress components into appropriate formulas,

$$\text{Principal stresses} = (0 \quad -2.72856 \quad -55.3749)$$

$$\text{Effective stress (von Mises)} = 54.0623$$

Proceeding in exactly the same manner, we can get solutions for all elements. A summary of the solution at element centers follows:

	Coordinate	Displacement	Stresses	Principal Stresses	Effective Stress
1	1.33333	−0.00187588	−52.8309	0	54.0623
	0.5	−0.0167551	−5.27256	−2.72856	
			0	−55.3749	
			−11.2898		
			0	0	
2	0.666667	0.00157588	24.6232	67.2393	92.0659
	1.16667	−0.00824522	4.92464	0	
			0	−37.6915	
			−51.5326		
			0		
			0		
3	3.33333	−0.0078036	−14.6533	0	18.3167
	0.333333	−0.0455297	−3.66334	0	
			0	−18.3167	
			−7.32667		
			0		
			0		
4	2.66667	−0.00184791	3.10223	26.3357	38.0742
	0.833333	−0.0352772	5.91407	0	
			0	−17.3194	
			−21.7822		
			0		
			0		

As seen from the model shown in Figure 1.26, elements 1 and 4 have a common side between them. However, the effective stress values for the two elements show significant difference. Elements 3 and 4 also share a side. The effective stresses for these two elements are quite different as well. This large difference in the stress values shows that the finite element mesh used for the analysis is very coarse. A much finer mesh is needed for any meaningful results. As mentioned earlier, this coarse mesh is used here to illustrate all necessary computations in detail.

◆ *Mathematica*/MATLAB Implementation 1.14 on the Book Web Site: *Plane stress element results*

◆ *Mathematica*/MATLAB Implementation 1.15 on the Book Web Site: *Complete solution of a plane stress problem*

1.5 SOLUTION OF LINEAR EQUATIONS

As seen from the previous section, the global system of equations after boundary conditions consists of n equations in n unknowns:

$$\boldsymbol{K}\boldsymbol{d} = \boldsymbol{R}$$

Most scientific calculators have functions for solving linear system of equations. These are useful for solving small textbook problems. Computer algebra systems (*Mathematica*, MATLAB, MAPLE, etc.) have sophisticated built-in functions for solving large systems of linear equations. For example, using *Mathematica*, a system of equations can be solved by first defining matrix $\boldsymbol{K}$ and right-hand-side vector $\boldsymbol{R}$ and then issuing the following command:

```
LinearSolve[K,R]
```

For completeness two basic methods for solution of a large-scale linear system of equations are briefly discussed in this section. Detailed treatment of these methods is beyond the scope of this text.

1.5.1 Solution Using Choleski Decomposition

An efficient method to solve the system of equations is to first decompose the matrix $\boldsymbol{K}$ into a product of lower and upper triangular matrices. For symmetric matrices the so-called Choleski decomposition is very efficient. In this decomposition a symmetric matrix $\boldsymbol{K}$ is decomposed into a product of a lower triangular matrix $\boldsymbol{L}$ and its transpose $\boldsymbol{L}^T$ in the following form:

$$\boldsymbol{K} = \boldsymbol{L}\boldsymbol{L}^T$$

$$\boldsymbol{L} = \begin{pmatrix} l_{11} & 0 & 0 & \cdots & 0 \\ l_{21} & l_{22} & 0 & \cdots & 0 \\ l_{31} & l_{32} & l_{33} & \cdots & 0 \\ \vdots & \vdots & \vdots & \ddots & 0 \\ l_{n1} & l_{n2} & l_{n3} & \cdots & l_{nn} \end{pmatrix}; \qquad \boldsymbol{L}^T = \begin{pmatrix} l_{11} & l_{21} & l_{31} & \cdots & l_{n1} \\ 0 & l_{22} & l_{32} & \cdots & l_{n2} \\ 0 & 0 & l_{33} & \cdots & l_{n3} \\ \vdots & \vdots & \vdots & \ddots & \vdots \\ 0 & 0 & 0 & \cdots & l_{nn} \end{pmatrix}$$

Such a decomposition is always possible for nonsingular symmetric matrices. Assuming this decomposition is known, we can solve the system of equations in two steps as follows:

$$\boldsymbol{K d} = \boldsymbol{R} \Longrightarrow \boldsymbol{L L}^T \boldsymbol{d} = \boldsymbol{R}$$

If we define $\boldsymbol{L}^T \boldsymbol{d} = \boldsymbol{y}$, then

$$\boldsymbol{L L}^T \boldsymbol{d} = \boldsymbol{R} \Longrightarrow \boldsymbol{L y} = \boldsymbol{R} \Longrightarrow \begin{pmatrix} l_{11} & 0 & 0 & \cdots & 0 \\ l_{21} & l_{22} & 0 & \cdots & 0 \\ l_{31} & l_{32} & l_{33} & \cdots & 0 \\ \vdots & \vdots & \vdots & \ddots & 0 \\ l_{n1} & l_{n2} & l_{n3} & \cdots & l_{nn} \end{pmatrix} \begin{pmatrix} y_1 \\ y_2 \\ y_3 \\ \vdots \\ y_n \end{pmatrix} = \begin{pmatrix} r_1 \\ r_2 \\ r_3 \\ \vdots \\ r_n \end{pmatrix}$$

Since $\boldsymbol{L}$ is a lower triangular matrix, the first row gives the solution for the first variable as $y_1 = r_1/l_{11}$. Knowing this, we can solve for $y_2 = (r_2 - l_{21}y_1)/l_{22}$ from the second equation. Thus, proceeding forward from the first equation, we can solve for all intermediate variables y_i. This is known as *forward elimination*.

$$y_i = \frac{r_i - \sum_{k=1}^{i-1} l_{ik} y_k}{l_{ii}}; \qquad i = 1, 2, \ldots, n$$

Now we can solve for the actual unknowns d_i as follows:

$$\boldsymbol{L}^T \boldsymbol{d} = \boldsymbol{y} \Longrightarrow \begin{pmatrix} l_{11} & l_{21} & l_{31} & \cdots & l_{n1} \\ 0 & l_{22} & l_{32} & \cdots & l_{n2} \\ 0 & 0 & l_{33} & \cdots & l_{n3} \\ \vdots & \vdots & \vdots & \ddots & \vdots \\ 0 & 0 & 0 & \cdots & l_{nn} \end{pmatrix} \begin{pmatrix} d_1 \\ d_2 \\ d_3 \\ \vdots \\ d_n \end{pmatrix} = \begin{pmatrix} y_1 \\ y_2 \\ y_3 \\ \vdots \\ y_n \end{pmatrix}$$

In this system the last equation has only one unknown, giving $d_n = y_n/l_{nn}$. The second to the last equation is used next to get d_{n-1}, and thus, working backward from the last equation, we can solve for all unknowns. This is known as *backward substitution*.

$$d_i = \frac{y_i - \sum_{k=i+1}^{n} l_{ki} d_k}{l_{ii}}; \qquad i = n, n-1, \ldots, 1$$

The procedure is especially efficient if solutions for several different right-hand sides are needed. Since in this case the major computational task of decomposing the matrix into $\boldsymbol{L L}^T$ form is not necessary for every right-hand side, one simply needs to perform a forward elimination and a backward substitution for each different right-hand side.

Choleski Decomposition The Choleski decomposition algorithm can be developed by considering direct multiplication of terms in the $\boldsymbol{L}$ and $\boldsymbol{L}^T$ matrices and equating them to the corresponding terms in the $\boldsymbol{K}$ matrix:

$$\boldsymbol{K} = \boldsymbol{L}\boldsymbol{L}^T \Longrightarrow \begin{pmatrix} k_{11} & k_{21} & k_{31} & \cdots & k_{n1} \\ k_{21} & k_{22} & k_{32} & \cdots & k_{n2} \\ k_{31} & k_{32} & k_{33} & \cdots & k_{n3} \\ \vdots & \vdots & \vdots & \ddots & \vdots \\ k_{n1} & k_{n2} & k_{n3} & \cdots & k_{nn} \end{pmatrix} = \begin{pmatrix} l_{11} & 0 & 0 & \cdots & 0 \\ l_{21} & l_{22} & 0 & \cdots & 0 \\ l_{31} & l_{32} & l_{33} & \cdots & 0 \\ \vdots & \vdots & \vdots & \ddots & \vdots \\ l_{n1} & l_{n2} & l_{n3} & \cdots & l_{nn} \end{pmatrix} \begin{pmatrix} l_{11} & l_{21} & l_{31} & \cdots & l_{n1} \\ 0 & l_{22} & l_{32} & \cdots & l_{n2} \\ 0 & 0 & l_{33} & \cdots & l_{n2} \\ \vdots & \vdots & \vdots & \ddots & \vdots \\ 0 & 0 & 0 & \cdots & l_{nn} \end{pmatrix}$$

By direct multiplication we can see that the entries in the first row of the $\boldsymbol{L}$ matrix can be established as follows:

$$\text{Row 1:} \qquad k_{11} = l_{11}^2 \Longrightarrow l_{11} = \sqrt{k_{11}}$$

$$k_{21} = l_{11}l_{21} \Longrightarrow l_{21} = \frac{k_{21}}{l_{11}}$$

$$k_{i1} = l_{11}l_{i1} \Longrightarrow l_{i1} = \frac{k_{i1}}{l_{11}}; \qquad i = 2, \ldots, n$$

Once entries in row 1 are known, we can proceed to row 2 and establish the entries there by direct multiplication as follows:

$$\text{Row 2:} \qquad k_{22} = l_{21}^2 + l_{22}^2 \Longrightarrow l_{22} = \sqrt{k_{22} - l_{21}^2}$$

$$k_{32} = l_{21}l_{31} + l_{22}l_{32} \Longrightarrow l_{32} = \frac{k_{32} - l_{21}l_{31}}{l_{22}}$$

$$k_{i2} = l_{21}l_{i1} + l_{22}l_{i2} \Longrightarrow l_{i2} = \frac{k_{i2} - l_{21}l_{i1}}{l_{22}}; \qquad i = 3, \ldots, n$$

Proceeding in a similar manner, we can write the following general formulas for entries in the ith row of the $\boldsymbol{L}$ matrix:

$$\text{Row } i\text{:} \qquad k_{ii} = l_{i1}^2 + l_{i2}^2 + \cdots + l_{ii}^2 \Longrightarrow l_{ii} = \sqrt{k_{ii} - \textstyle\sum_{k=1}^{i-1} l_{ik}^2}$$

$$k_{ij} = l_{i1}l_{j1} + l_{i2}l_{j2} + \cdots + l_{ij}l_{jj} \Longrightarrow l_{ij} = \frac{k_{ij} - \sum_{k=1}^{j-1} l_{ik}l_{jk}}{l_{jj}}$$

Notice that the procedure requires taking the square root of terms coming from the diagonal of the $\boldsymbol{K}$ matrix and then division by this term to establish other terms in a given row. This is called the pivot element. Clearly these operations will fail if there is a negative or zero pivot. However, as long as the matrix $\boldsymbol{K}$ is nonsingular, it is always possible to rearrange the system of equations to avoid a negative or zero pivot.

The computational steps can be written in the form of the following algorithm:

Choleski Decomposition Algorithm

For $i = 1, n$

  For $j = 1, i - 1$

$$l_{ij} = \left(k_{ij} - \sum_{k=1}^{j-1} l_{ik} l_{jk}\right) / l_{jj}$$

end

$$l_{ii} = \sqrt{k_{ii} - \sum_{k=1}^{i-1} l_{ik}^2}$$

end

Example 1.13 Find the solution of the following system of equations using Choleski decomposition:

$$\begin{pmatrix} 9 & 3 & 6 & 3 \\ 3 & 12 & 2 & -10 \\ 6 & 2 & 11 & -5 \\ 3 & -10 & -5 & 21 \end{pmatrix} \begin{pmatrix} d_1 \\ d_2 \\ d_3 \\ d_4 \end{pmatrix} = \begin{pmatrix} 5 \\ 6 \\ 7 \\ 8 \end{pmatrix}$$

We first generate the $\boldsymbol{LL}^T$ decomposition of the coefficient matrix using the above algorithm:

$$\boldsymbol{K} = \boldsymbol{LL}^T = \begin{pmatrix} 3 & 0 & 0 & 0 \\ 1 & \sqrt{11} & 0 & 0 \\ 2 & 0 & \sqrt{7} & 0 \\ 1 & -\sqrt{11} & -\sqrt{7} & \sqrt{2} \end{pmatrix} \begin{pmatrix} 3 & 1 & 2 & 1 \\ 0 & \sqrt{11} & 0 & -\sqrt{11} \\ 0 & 0 & \sqrt{7} & -\sqrt{7} \\ 0 & 0 & 0 & \sqrt{2} \end{pmatrix} = \begin{pmatrix} 9 & 3 & 6 & 3 \\ 3 & 12 & 2 & -10 \\ 6 & 2 & 11 & -5 \\ 3 & -10 & -5 & 21 \end{pmatrix}$$

Using the $\boldsymbol{L}$ matrix, the solution for intermediate variables (forward elimination) is as follows:

$$\begin{pmatrix} 3 & 0 & 0 & 0 \\ 1 & \sqrt{11} & 0 & 0 \\ 2 & 0 & \sqrt{7} & 0 \\ 1 & -\sqrt{11} & -\sqrt{7} & \sqrt{2} \end{pmatrix} \begin{pmatrix} y_1 \\ y_2 \\ y_3 \\ y_4 \end{pmatrix} = \begin{pmatrix} 5 \\ 6 \\ 7 \\ 8 \end{pmatrix}$$

Row 1: $3y_1 = 5 \Longrightarrow y_1 = \frac{5}{3}$

Row 2: $1y_1 + \sqrt{11}y_2 = 6 \Longrightarrow y_2 = \frac{13}{3\sqrt{11}}$

Row 3: $2y_1 + 0y_2 + \sqrt{7}y_3 = 7 \Longrightarrow y_3 = \frac{11}{3\sqrt{7}}$

Row 4: $1y_1 + (-\sqrt{11})y_2 + (-\sqrt{7})y_3 + \sqrt{2}y_4 = 8 \Longrightarrow y_4 = \frac{43}{3\sqrt{2}}$

Using the $\boldsymbol{L}^T$ matrix, the solution for actual variables (backward substitution) is as follows:

$$\begin{pmatrix} 3 & 1 & 2 & 1 \\ 0 & \sqrt{11} & 0 & -\sqrt{11} \\ 0 & 0 & \sqrt{7} & -\sqrt{7} \\ 0 & 0 & 0 & \sqrt{2} \end{pmatrix} \begin{pmatrix} d_1 \\ d_2 \\ d_3 \\ d_4 \end{pmatrix} = \begin{pmatrix} \frac{5}{3} \\ \frac{13}{3\sqrt{11}} \\ \frac{11}{3\sqrt{7}} \\ \frac{43}{3\sqrt{2}} \end{pmatrix}$$

$$\begin{aligned}
&\text{Row 4:} && \sqrt{2}d_4 = \tfrac{43}{3\sqrt{2}} \Longrightarrow d_4 = \tfrac{43}{6}\\
&\text{Row 3:} && \sqrt{7}d_3 + (-\sqrt{7})d_4 = \tfrac{11}{3\sqrt{7}} \Longrightarrow d_3 = \tfrac{323}{42}\\
&\text{Row 2:} && \sqrt{11}d_2 + 0d_3 + (-\sqrt{11})d_4 = \tfrac{13}{3\sqrt{11}} \Longrightarrow d_2 = \tfrac{499}{66}\\
&\text{Row 1:} && 3d_1 + 1d_2 + 2d_3 + 1d_4 = \tfrac{5}{3} \Longrightarrow d_1 = -\tfrac{730}{77}
\end{aligned}$$

1.5.2 Conjugate Gradient Method

The well-known conjugate gradient method for solution of unconstrained optimization problems can be used to develop an iterative method for solution of a linear system of equations. The method has been used effectively for solution of nonlinear finite element problems that require repeated solutions of large systems of equations. Consider a symmetric system of $n \times n$ linear equations expressed in matrix form as follows:

$$\boldsymbol{Kd} = \boldsymbol{R}$$

$$\begin{pmatrix} k_{11} & k_{12} & \cdots & k_{1n} \\ k_{21} & k_{22} & \cdots & k_{2n} \\ \vdots & \vdots & \ddots & \vdots \\ k_{n1} & k_{n2} & \cdots & k_{nn} \end{pmatrix} \begin{pmatrix} d_1 \\ d_2 \\ \vdots \\ d_n \end{pmatrix} = \begin{pmatrix} r_1 \\ r_2 \\ \vdots \\ r_n \end{pmatrix}$$

Finding the solution of this system of equations is equivalent to minimizing the following quadratic function:

$$\min f(\boldsymbol{d}) = \tfrac{1}{2}\boldsymbol{d}^T\boldsymbol{Kd} - \boldsymbol{d}^T\boldsymbol{R}$$

This can easily be seen by using the necessary conditions for the minimum, namely, the gradient of the function must be zero:

$$\text{grad } f(\boldsymbol{d}) \equiv g(\boldsymbol{d}) = \begin{pmatrix} \partial f/\partial d_1 \\ \partial f/\partial d_2 \\ \vdots \\ \partial f/\partial d_n \end{pmatrix} = \begin{pmatrix} \frac{1}{2}\boldsymbol{d}^T\boldsymbol{Ke}_1 + \frac{1}{2}\boldsymbol{e}_1^T\boldsymbol{Kd} - \boldsymbol{e}_1^T\boldsymbol{R} \\ \frac{1}{2}\boldsymbol{d}^T\boldsymbol{Ke}_2 + \frac{1}{2}\boldsymbol{e}_2^T\boldsymbol{Kd} - \boldsymbol{e}_2^T\boldsymbol{R} \\ \vdots \\ \frac{1}{2}\boldsymbol{d}^T\boldsymbol{Ke}_n + \frac{1}{2}\boldsymbol{e}_n^T\boldsymbol{Kd} - \boldsymbol{e}_n^T\boldsymbol{R} \end{pmatrix}$$

where

$$\boldsymbol{e}_1 = \begin{pmatrix} 1 \\ 0 \\ \vdots \\ 0 \end{pmatrix}; \qquad \boldsymbol{e}_2 = \begin{pmatrix} 0 \\ 1 \\ \vdots \\ 0 \end{pmatrix}; \qquad \text{etc.}$$

Since $\boldsymbol{K}$ is a symmetric matrix, the transpose of the first term in each row is same as the second term. Thus the gradient can be expressed as follows:

$$g(d) = \begin{pmatrix} e_1^T Kd - e_1^T R \\ e_2^T Kd - e_2^T R \\ \vdots \\ e_n^T Kd - e_n^T R \end{pmatrix} = \begin{pmatrix} e_1^T \\ e_2^T \\ \vdots \\ e_n^T \end{pmatrix} Kd - \begin{pmatrix} e_1^T \\ e_2^T \\ \vdots \\ e_n^T \end{pmatrix} R$$

Noting that

$$\begin{pmatrix} e_1^T \\ e_2^T \\ \vdots \\ e_n^T \end{pmatrix} = \begin{pmatrix} 1 & 0 & \cdots & 0 \\ 0 & 1 & \cdots & 0 \\ \vdots & \vdots & \ddots & \vdots \\ 0 & 0 & \cdots & 1 \end{pmatrix} = \text{identity matrix}$$

we see that the gradient of the quadratic function is

$$g(d) = Kd - R$$

A condition of zero gradient implies $Kd - R = 0$, which clearly is the given linear system of equations. Thus solution of a linear system of equations is equivalent to finding a minimum of the quadratic function f.

Basic Conjugate Gradient Method In the conjugate gradient method, the minimum of f is located by starting from an assumed solution $d^{(0)}$ (say with all variables equal to zero) and performing iterations as follows:

$$d^{(k+1)} = d^{(k)} + \alpha^{(k)} h^{(k)}; \qquad k = 0, 1, \ldots$$

where $\alpha^{(k)}$ is known as the step length and $h^{(k)}$ is the search direction. At the kth iteration the search direction $h^{(k)}$ is determined as follows:

$$h^{(k)} = -g^{(k)} + \beta^{(k)} h^{(k-1)}$$

The scalar multiplier β determines the portion of the previous direction to be added to determine the new direction. According to the Fletcher-Reeves formula,

$$\beta^{(k)} = \frac{[g^{(k)}]^T g^{(k)}}{[g^{(k-1)}]^T g^{(k-1)}}$$

After establishing the search direction, the minimization problem reduces to finding $\alpha^{(k)}$ in order to

$$\text{Minimize } f(d^{(k+1)}) = f(d^{(k)} + \alpha^{(k)} h^{(k)})$$

For the quadratic function f we have

$$f(d^{(k)} + \alpha^{(k)} h^{(k)}) = \tfrac{1}{2}(d^{(k)} + \alpha^{(k)} h^{(k)})^T K (d^{(k)} + \alpha^{(k)} h^{(k)}) - (d^{(k)} + \alpha^{(k)} h^{(k)})^T R$$

Expanding,

$$f(\boldsymbol{d}^{(k)} + \alpha^{(k)}\boldsymbol{h}^{(k)}) = \frac{1}{2}(\boldsymbol{d}^{(k)})^T \boldsymbol{K}\boldsymbol{d}^{(k)} + \frac{1}{2}\alpha^{(k)}(\boldsymbol{d}^{(k)})^T \boldsymbol{K}\boldsymbol{h}^{(k)} + \frac{1}{2}\alpha^{(k)}(\boldsymbol{h}^{(k)})^T \boldsymbol{K}\boldsymbol{d}^{(k)} + \frac{1}{2}(\alpha^{(k)})^2(\boldsymbol{h}^{(k)})^T \boldsymbol{K}\boldsymbol{h}^{(k)} - (\boldsymbol{d}^{(k)})^T \boldsymbol{R} - \alpha^{(k)}(\boldsymbol{h}^{(k)})^T \boldsymbol{R}$$

For the minimum the derivative of this expression with respect to $\alpha^{(k)}$ must be equal to zero, giving the following equation for the step length:

$$\frac{df(\boldsymbol{d}^{(k)} + \alpha^{(k)}\boldsymbol{h}^{(k)})}{d\alpha^{(k)}} = 0 \Longrightarrow \frac{1}{2}(\boldsymbol{d}^{(k)})^T \boldsymbol{K}\boldsymbol{h}^{(k)} + \frac{1}{2}(\boldsymbol{h}^{(k)})^T \boldsymbol{K}\boldsymbol{d}^{(k)} + \alpha^{(k)}(\boldsymbol{h}^{(k)})^T \boldsymbol{K}\boldsymbol{h}^{(k)} - (\boldsymbol{h}^{(k)})^T \boldsymbol{R} = 0$$

Noting the symmetry of $\boldsymbol{K}$, the first two terms can be combined and moved to the right-hand side to give

$$\alpha^{(k)}(\boldsymbol{h}^{(k)})^T \boldsymbol{K}\boldsymbol{h}^{(k)} = (\boldsymbol{h}^{(k)})^T \boldsymbol{R} - (\boldsymbol{h}^{(k)})^T \boldsymbol{K}\boldsymbol{d}^{(k)} = -(\boldsymbol{h}^{(k)})^T (\boldsymbol{K}\boldsymbol{d}^{(k)} - \boldsymbol{R}) = -(\boldsymbol{h}^{(k)})^T \boldsymbol{g}^{(k)}$$

giving

$$\alpha^{(k)} = -\frac{(\boldsymbol{h}^{(k)})^T \boldsymbol{g}^{(k)}}{(\boldsymbol{h}^{(k)})^T \boldsymbol{K}\boldsymbol{h}^{(k)}}$$

Since β is usually small, $\boldsymbol{h}^{(k)} \approx -\boldsymbol{g}^{(k)}$, and thus for computational efficiency the step length expression is slightly modified as follows:

$$\alpha^{(k)} = \frac{(\boldsymbol{g}^{(k)})^T \boldsymbol{g}^{(k)}}{(\boldsymbol{h}^{(k)})^T \boldsymbol{K}\boldsymbol{h}^{(k)}}$$

Basic Conjugate Gradient Algorithm The computational steps can be organized into the following algorithm.

Choose a starting point $\boldsymbol{d}^{(0)}$. Compute $\boldsymbol{g}^{(0)} = \boldsymbol{K}\boldsymbol{d}^{(0)} - \boldsymbol{R}$. Set $\boldsymbol{h}^{(0)} = -\boldsymbol{g}^{(0)}$. Compute $r^{(0)} = (\boldsymbol{g}^{(0)})^T \boldsymbol{g}^{(0)}$. Choose a convergence tolerance parameter ϵ. Set $k = 0$.

1. If $r^{(k)} \leq \epsilon$, stop; $\boldsymbol{d}^{(k)}$ is the desired solution. Otherwise, continue.
2. Compute the step length:

$$\boldsymbol{z}^{(k)} = \boldsymbol{K}\boldsymbol{h}^{(k)}; \qquad \alpha^{(k)} = \frac{r^{(k)}}{(\boldsymbol{h}^{(k)})^T \boldsymbol{z}^{(k)}}$$

3. Next point:

$$\boldsymbol{d}^{(k+1)} = \boldsymbol{d}^{(k)} + \alpha^{(k)}\boldsymbol{h}^{(k)}$$

4. Update the quantities for the next iteration:

$$g^{(k+1)} = K(d^{(k)} + \alpha^{(k)}h^{(k)}) - R \equiv g^{(k)} + \alpha^{(k)}z^{(k)}$$

$$r^{(k+1)} = (g^{(k+1)})^T g^{(k+1)}$$

$$\beta^{(k)} = r^{(k+1)}/r^{(k)}$$

$$h^{(k+1)} = -g^{(k+1)} + \beta^{(k)}h^{(k)}$$

5. Set $k = k + 1$ and go to step 1.

Example 1.14 Find the solution of the following system of equations using the basic conjugate gradient method:

$$\begin{pmatrix} 9 & 3 & 6 & 3 \\ 3 & 12 & 2 & -10 \\ 6 & 2 & 11 & -5 \\ 3 & -10 & -5 & 21 \end{pmatrix} \begin{pmatrix} d_1 \\ d_2 \\ d_3 \\ d_4 \end{pmatrix} = \begin{pmatrix} 5 \\ 6 \\ 7 \\ 8 \end{pmatrix}$$

Starting point, $d^{(0)} = \{0, 0, 0, 0\}$
$g^{(0)} = Kd^{(0)} - R = \{-5., -6., -7., -8.\}$
$h^{(0)} = -g^{(0)} = \{5., 6., 7., 8.\}$
$r^{(0)} = g^{(0)T}g^{(0)} = 174.;\ \beta^{(0)} = 0$

Iteration 1:

$z^{(0)} = Kh^{(0)} = \{129., 21., 79., 88.\}$
Step length, $\alpha^{(0)} = r^{(0)}/h^{(0)T}z^{(0)} = 0.0857988$
Next point, $d^{(1)} = d^{(0)} + \alpha^{(0)}h^{(0)} = \{0.428994, 0.514793, 0.600592, 0.686391\}$
$g^{(1)} = g^{(0)} + \alpha^{(0)}z^{(0)} = \{6.06805, -4.19822, -0.221893, -0.449704\}$
$r^{(1)} = g^{(1)T}g^{(1)} = 54.6978$
$\beta^{(1)} = r^{(1)}/r^{(0)} = 0.314355$
$h^{(1)} = -g^{(1)} + \beta^{(0)}h^{(0)} = \{-4.49627, 6.08435, 2.42238, 2.96454\}$

Iteration 2:

$z^{(1)} = Kh^{(1)} = \{1.21451, 34.7228, -2.98549, -24.1888\}$
Step length, $\alpha^{(1)} = r^{(1)}/h^{(1)T}z^{(1)} = 0.431153$
Next point, $d^{(2)} = d^{(1)} + \alpha^{(1)}h^{(1)} = \{-1.50959, 3.13808, 1.64501, 1.96456\}$
$g^{(2)} = g^{(1)} + \alpha^{(1)}z^{(1)} = \{6.59169, 10.7726, -1.50909, -10.8788\}$
$r^{(2)} = g^{(2)T}g^{(2)} = 280.125$
$\beta^{(2)} = r^{(2)}/r^{(1)} = 5.12132$
$h^{(2)} = -g^{(2)} + \beta^{(1)}h^{(1)} = \{-29.6185, 20.3873, 13.9149, 26.0612\}$

Iteration 3:

$z^{(2)} = Kh^{(2)} = \{-43.7321, -76.9897, -114.179, 184.981\}$
Step length, $\alpha^{(2)} = r^{(2)}/h^{(2)T}z^{(2)} = 0.0947098$

Next point, $\boldsymbol{d}^{(3)} = \boldsymbol{d}^{(2)} + \alpha^{(2)}\boldsymbol{h}^{(2)} = \{-4.31475, 5.06896, 2.96288, 4.43281\}$

$\boldsymbol{g}^{(3)} = \boldsymbol{g}^{(2)} + \alpha^{(2)}z^{(2)} = \{2.44982, 3.48091, -12.323, 6.64076\}$

$r^{(3)} = \boldsymbol{g}^{(3)T}\boldsymbol{g}^{(3)} = 214.073$

$\beta^{(3)} = r^{(3)}/r^{(2)} = 0.764207$

$\boldsymbol{h}^{(3)} = -\boldsymbol{g}^{(3)} + \beta^{(2)}\boldsymbol{h}^{(2)} = \{-25.0845, 12.0992, 22.9568, 13.2754\}$

Iteration 4:

$z^{(3)} = \boldsymbol{K}\boldsymbol{h}^{(3)} = \{-11.8961, -16.903, 59.8392, -32.2469\}$

Step length, $\alpha^{(3)} = r^{(3)}/\boldsymbol{h}^{(3)T}z^{(3)} = 0.205934$

Next point, $\boldsymbol{d}^{(4)} = \boldsymbol{d}^{(3)} + \alpha^{(3)}\boldsymbol{h}^{(3)} = \{-9.48052, 7.56061, 7.69048, 7.16667\}$

$\boldsymbol{g}^{(4)} = \boldsymbol{g}^{(3)} + \alpha^{(3)}z^{(3)} = \{7.10543 \times 10^{-15}, -9.76996 \times 10^{-15}, 0., 2.57572 \times 10^{-14}\}$

$r^{(4)} = \boldsymbol{g}^{(4)T}\boldsymbol{g}^{(4)} = 8.09371 \times 10^{-28}$

$\beta^{(4)} = r^{(4)}/r^{(3)} = 3.78081 \times 10^{-30}$

$\boldsymbol{h}^{(4)} = -\boldsymbol{g}^{(4)} + \beta^{(3)}\boldsymbol{h}^{(3)}$

$= \{-7.10543 \times 10^{-15}, 9.76996 \times 10^{-15}, 8.67954 \times 10^{-29}, -2.57572 \times 10^{-14}\}$

Solution converged after four iterations:

$\boldsymbol{g} = \{7.10543 \times 10^{-15}, -9.76996 \times 10^{-15}, 0., 2.57572 \times 10^{-14}\}$

$r = \boldsymbol{g}^T\boldsymbol{g} = 8.09371 \times 10^{-28}$

Solution, $\boldsymbol{d} = \{-9.48052, 7.56061, 7.69048, 7.16667\}$

Preconditioned Conjugate Gradient When the coefficient matrix $\boldsymbol{K}$ is ill-conditioned, the convergence of the conjugate gradient method could be very slow. Introducing a suitably chosen preconditioning matrix $\boldsymbol{P}$ improves convergence. To see this, we multiply both sides of the system of equations by the inverse of the preconditioning matrix to get

$$\boldsymbol{P}^{-1}\boldsymbol{K}\boldsymbol{d} = \boldsymbol{P}^{-1}\boldsymbol{R}$$

Clearly, if $\boldsymbol{P} = \boldsymbol{K}$, we have the solution of the system of equations. Thus, intuitively, we can expect that, if $\boldsymbol{P}$ is a matrix that is reasonably close to $\boldsymbol{K}$, then the method will converge faster. The matrix $\boldsymbol{P}$ is known as the preconditioning matrix. Several different techniques have been proposed for establishing this matrix. Two popular choices will be discussed later in the section. First we develop the computational procedure for the solution of equations when the preconditioning matrix is included.

In general, the product $\boldsymbol{P}^{-1}\boldsymbol{K}$ does not produce a symmetric matrix, even if both matrices are individually symmetric. In order to get a symmetric system of equations, we introduce the following decomposition of the $\boldsymbol{P}$ matrix,

$$\boldsymbol{P} = \boldsymbol{E}\boldsymbol{E}^T \Longrightarrow \boldsymbol{P}^{-1} = \boldsymbol{E}^{-T}\boldsymbol{E}^{-1}$$

and a new vector of intermediate variables $\boldsymbol{y}$,

$$\boldsymbol{E}^T\boldsymbol{d} = \boldsymbol{y} \Longrightarrow \boldsymbol{E}^{-T}\boldsymbol{y} = \boldsymbol{d}$$

Introducing these, the original system of equations is modified as follows:

$$\boldsymbol{E}^{-T}\boldsymbol{E}^{-1}\boldsymbol{K}\boldsymbol{E}^{-T}\boldsymbol{y} = \boldsymbol{E}^{-T}\boldsymbol{E}^{-1}\boldsymbol{R} \Longrightarrow \boldsymbol{E}^{-1}\boldsymbol{K}\boldsymbol{E}^{-T}\boldsymbol{y} = \boldsymbol{E}^{-1}\boldsymbol{R}$$

Thus we have a system of equations as follows:

$$\tilde{\boldsymbol{K}}\boldsymbol{y} = \tilde{\boldsymbol{R}}$$

where $\tilde{\boldsymbol{K}} = \boldsymbol{E}^{-1}\boldsymbol{K}\boldsymbol{E}^{-T}$ and $\tilde{\boldsymbol{R}} = \boldsymbol{E}^{-1}\boldsymbol{R}$.

The matrix $\tilde{\boldsymbol{K}}$ is a symmetric matrix. At this point we can use the standard conjugate gradient method to find a solution for intermediate variables $\boldsymbol{y}$:

$$\tilde{\boldsymbol{g}}^{(k)} = \tilde{\boldsymbol{K}}\boldsymbol{y}^{(k)} - \tilde{\boldsymbol{R}}; \qquad \tilde{\boldsymbol{h}}^{(k)} = -\tilde{\boldsymbol{g}}^{(k)} + \beta^{(k)}\tilde{\boldsymbol{h}}^{(k-1)}$$

$$\beta^{(k)} = \frac{[\tilde{\boldsymbol{g}}^{(k)}]^T\tilde{\boldsymbol{g}}^{(k)}}{[\tilde{\boldsymbol{g}}^{(k-1)}]^T\tilde{\boldsymbol{g}}^{(k-1)}}; \qquad \alpha^{(k)} = \frac{(\tilde{\boldsymbol{g}}^{(k)})^T\tilde{\boldsymbol{g}}^{(k)}}{(\tilde{\boldsymbol{h}}^{(k)})^T\tilde{\boldsymbol{A}}\tilde{\boldsymbol{h}}^{(k)}}$$

$$\boldsymbol{y}^{(k+1)} = \boldsymbol{y}^{(k)} + \alpha^{(k)}\tilde{\boldsymbol{h}}^{(k)}$$

Multiplying vector $\boldsymbol{y}^{(k+1)}$ by $\boldsymbol{E}^{-T}$ gives the solution of the original problem:

$$\boldsymbol{d}^{(k+1)} = \boldsymbol{E}^{-T}\boldsymbol{y}^{(k+1)}$$

A serious drawback of the procedure so far is that it requires a decomposition of the preconditioning matrix $\boldsymbol{P}^{-1} = \boldsymbol{E}^{-T}\boldsymbol{E}^{-1}$, which is a very time consuming task for a general matrix. Fortunately, it is possible to rewrite the above expressions so that only $\boldsymbol{P}^{-1}$ appears in the equations and thus there is no need to explicitly carry out the decomposition.

Substituting for $\tilde{\boldsymbol{K}}$ and $\tilde{\boldsymbol{R}}$ into the $\tilde{\boldsymbol{g}}^{(k)}$ expression, we have

$$\tilde{\boldsymbol{g}}^{(k)} = \boldsymbol{E}^{-1}\boldsymbol{K}\boldsymbol{E}^{-T}\boldsymbol{y}^{(k)} - \boldsymbol{E}^{-1}\boldsymbol{R} = \boldsymbol{E}^{-1}(\boldsymbol{K}\boldsymbol{d}^{(k)} - \boldsymbol{R}) = \boldsymbol{E}^{-1}\boldsymbol{g}^{(k)}$$

Using this, we can write $\tilde{\boldsymbol{h}}^{(k)}$ as follows:

$$\tilde{\boldsymbol{h}}^{(k)} = -\tilde{\boldsymbol{g}}^{(k)} + \beta^{(k)}\tilde{\boldsymbol{h}}^{(k-1)} = -\boldsymbol{E}^{-1}\boldsymbol{g}^{(k)} + \beta^{(k)}\tilde{\boldsymbol{h}}^{(k-1)}$$

Multiplying both sides by $\boldsymbol{E}^{-T}$,

$$\boldsymbol{E}^{-T}\tilde{\boldsymbol{h}}^{(k)} \equiv \boldsymbol{h}^{(k)} = -\boldsymbol{E}^{-T}\boldsymbol{E}^{-1}\boldsymbol{g}^{(k)} + \beta^{(k)}\boldsymbol{E}^{-T}\tilde{\boldsymbol{h}}^{(k-1)} \Longrightarrow \boldsymbol{h}^{(k)} = -\boldsymbol{P}^{-1}\boldsymbol{g}^{(k)} + \beta^{(k)}\boldsymbol{h}^{(k-1)}$$

Similarly, other expressions in the algorithm can be rewritten as follows:

$$\beta^{(k)} = \frac{[\tilde{\boldsymbol{g}}^{(k)}]^T\tilde{\boldsymbol{g}}^{(k)}}{[\tilde{\boldsymbol{g}}^{(k-1)}]^T\tilde{\boldsymbol{g}}^{(k-1)}} = \frac{[\boldsymbol{g}^{(k)}]^T\boldsymbol{E}^{-T}\boldsymbol{E}^{-1}\boldsymbol{g}^{(k)}}{[\boldsymbol{g}^{(k-1)}]^T\boldsymbol{E}^{-T}\boldsymbol{E}^{-1}\boldsymbol{g}^{(k-1)}} = \frac{[\boldsymbol{g}^{(k)}]^T\boldsymbol{P}^{-1}\boldsymbol{g}^{(k)}}{[\boldsymbol{g}^{(k-1)}]^T\boldsymbol{P}^{-1}\boldsymbol{g}^{(k-1)}}$$

$$\alpha^{(k)} = \frac{(\tilde{\boldsymbol{g}}^{(k)})^T\tilde{\boldsymbol{g}}^{(k)}}{(\tilde{\boldsymbol{h}}^{(k)})^T\boldsymbol{E}^{-1}\boldsymbol{K}\boldsymbol{E}^{-T}\tilde{\boldsymbol{d}}^{(k)}} = \frac{[\boldsymbol{g}^{(k)}]^T\boldsymbol{P}^{-1}\boldsymbol{g}^{(k)}}{(\boldsymbol{E}^{-T}\tilde{\boldsymbol{h}}^{(k)})^T\boldsymbol{K}\boldsymbol{E}^{-T}\tilde{h}^{(k)}} = \frac{[\boldsymbol{g}^{(k)}]^T\boldsymbol{P}^{-1}\boldsymbol{g}^{(k)}}{(\boldsymbol{h}^{(k)})^T\boldsymbol{K}\boldsymbol{h}^{(k)}}$$

$$\boldsymbol{y}^{(k+1)} = \boldsymbol{y}^{(k)} + \alpha^{(k)}\tilde{\boldsymbol{h}}^{(k)}$$

Multiplying the expression for the intermediate variables by $\boldsymbol{E}^{-T}$, we have

$$\boldsymbol{E}^{-T}\boldsymbol{y}^{(k+1)} \equiv \boldsymbol{d}^{(k+1)} = \boldsymbol{E}^{-T}\boldsymbol{y}^{(k)} + \alpha^{(k)}\boldsymbol{E}^{-T}\tilde{\boldsymbol{h}}^{(k)} = \boldsymbol{d}^{(k)} + \alpha^{(k)}\boldsymbol{h}^{(k)}$$

Preconditioned Conjugate Gradient (PCG) Algorithm The computational steps can be organized into the following algorithm:

Initialization

Choose a starting point $\boldsymbol{d}^{(0)}$. Set $\boldsymbol{g}^{(0)} = \boldsymbol{K}\boldsymbol{d}^{(0)} - \boldsymbol{R}$.
Solve for $\boldsymbol{w}^{(0)}$: $\boldsymbol{P}\boldsymbol{w}^{(0)} = \boldsymbol{g}^{(0)}$.
Set $\boldsymbol{h}^{(0)} = -\boldsymbol{w}^{(0)}$.
Compute $r^{(0)} = (\boldsymbol{g}^{(0)})^T\boldsymbol{w}^{(0)}$.
Choose a convergence tolerance parameter ϵ. Set $k = 0$.

1. If $r^{(k)} \leq \epsilon$, stop; $\boldsymbol{d}^{(k)}$ is the desired solution. Otherwise, continue.
2. Compute the step length:

$$\boldsymbol{z}^{(k)} = \boldsymbol{K}\boldsymbol{h}^{(k)}; \qquad \alpha^{(k)} = \frac{r^{(k)}}{(\boldsymbol{h}^{(k)})^T\boldsymbol{z}^{(k)}}$$

3. Next point:

$$\boldsymbol{d}^{(k+1)} = \boldsymbol{d}^{(k)} + \alpha^{(k)}\boldsymbol{h}^{(k)}$$

4. Update the quantities for the next iteration:

$\boldsymbol{g}^{(k+1)} = \boldsymbol{g}^{(k)} + \alpha^{(k)}\boldsymbol{z}^{(k)}$
Solve for $\boldsymbol{w}^{(k+1)}$: $\boldsymbol{P}\boldsymbol{w}^{(k+1)} = \boldsymbol{g}^{(k+1)}$
$r^{(k+1)} = (\boldsymbol{g}^{(k+1)})^T\boldsymbol{w}^{(k+1)}$
$\beta^{(k)} = r^{(k+1)}/r^{(k)}$
$\boldsymbol{h}^{(k+1)} = -\boldsymbol{w}^{(k+1)} + \beta^{(k)}\boldsymbol{h}^{(k)}$

5. Set $k = k + 1$ and go to step 1.

Jacobi Preconditioning One of the simplest choices of the preconditioning matrices is the so-called Jacobi preconditioning. Here the $\boldsymbol{P}$ matrix is a diagonal matrix with entries equal to the diagonal elements of the $\boldsymbol{K}$ matrix. Thus $\boldsymbol{P}^{-1}$ is established as follows:

$$p_{ii}^{-1} = \frac{1}{k_{ii}}; \qquad p_{ij}^{-1} = 0; \qquad i \neq j; \qquad i, j = 1, \ldots, n$$

Example 1.15 Find the solution of the following system of equations using the PCG method with Jacobi preconditioning:

$$\begin{pmatrix} 9 & 3 & 6 & 3 \\ 3 & 12 & 2 & -10 \\ 6 & 2 & 11 & -5 \\ 3 & -10 & -5 & 21 \end{pmatrix} \begin{pmatrix} x_1 \\ x_2 \\ x_3 \\ x_4 \end{pmatrix} = \begin{pmatrix} 5 \\ 6 \\ 7 \\ 8 \end{pmatrix}$$

$$\text{Preconditioning matrix, } \boldsymbol{P} = \begin{pmatrix} 9. & 0 & 0 & 0 \\ 0 & 12. & 0 & 0 \\ 0 & 0 & 11. & 0 \\ 0 & 0 & 0 & 21. \end{pmatrix}$$

Starting point, $\boldsymbol{d}^{(0)} = \{0, 0, 0, 0\}$

$\boldsymbol{g}^{(0)} = \boldsymbol{Kd}^{(0)} - \boldsymbol{R} = \{-5, -6, -7, -8\}$

Solving $\boldsymbol{Pw}^{(0)} = \boldsymbol{g}^{(0)}$, we have $\boldsymbol{w}^{(0)} = \{-0.555556, -0.5, -0.636364, -0.380952\}$

$\boldsymbol{h}^{(0)} = -\boldsymbol{w}^{(0)} = \{0.555556, 0.5, 0.636364, 0.380952\}$

$r^{(0)} = \boldsymbol{g}^{(0)T}\boldsymbol{w}^{(0)} = 1.10873; \beta^{(0)} = 0$

Iteration 1:

$z^{(0)} = \boldsymbol{Kh}^{(0)} = \{11.461, 5.12987, 9.42857, 1.48485\}$

Step length, $\alpha^{(0)} = r^{(0)}/\boldsymbol{h}^{(0)T}z^{(0)} = 0.85689$

Next point, $\boldsymbol{d}^{(1)} = \boldsymbol{d}^{(0)} + \alpha^{(0)}\boldsymbol{h}^{(0)} = \{0.47605, 0.428445, 0.545294, 0.326434\}$

$\boldsymbol{g}^{(1)} = \boldsymbol{g}^{(0)} + \alpha^{(0)}z^{(0)} = \{4.82085, -1.60426, 1.07925, -6.72765\}$

Solving $\boldsymbol{Pw}^{(1)} = \boldsymbol{g}^{(1)}$, we have $\boldsymbol{w}^{(1)} = \{0.53565, -0.133689, 0.0981136, -0.320364\}$

$r^{(1)} = \boldsymbol{g}^{(1)T}\boldsymbol{w}^{(1)} = 5.05795$

$\beta^{(1)} = r^{(1)}/r^{(0)} = 0.380871$

$\boldsymbol{h}^{(1)} = -\boldsymbol{g}^{(1)} + \beta^{(0)}\boldsymbol{h}^{(0)} = \{-0.324055, 0.324124, 0.144259, 0.465458\}$

Iteration 2:

$z^{(1)} = \boldsymbol{Kh}^{(1)} = \{0.317806, -1.44873, -2.03652, 4.83991\}$

Step length, $\alpha^{(1)} = r^{(1)}/\boldsymbol{h}^{(1)T}z^{(1)} = 3.64817$

Next point, $\boldsymbol{d}^{(2)} = \boldsymbol{d}^{(1)} + \alpha^{(1)}\boldsymbol{h}^{(1)} = \{-0.706158, 1.61091, 1.07158, 2.02451\}$

$\boldsymbol{g}^{(2)} = \boldsymbol{g}^{(1)} + \alpha^{(1)}z^{(1)} = \{5.98026, -6.8895, -6.35033, 10.9292\}$

Solving $\boldsymbol{Pw}^{(2)} = \boldsymbol{g}^{(2)}$, we have $\boldsymbol{w}^{(2)} = \{0.664474, -0.574125, -0.577303, 0.520438\}$

$r^{(2)} = \boldsymbol{g}^{(2)T}\boldsymbol{w}^{(2)} = 17.2832$

$\beta^{(2)} = r^{(2)}/r^{(1)} = 3.41704$

$\boldsymbol{h}^{(2)} = -\boldsymbol{g}^{(2)} + \beta^{(1)}\boldsymbol{h}^{(1)} = \{-1.77178, 1.68167, 1.07024, 1.07005\}$

Iteration 3:

$z^{(2)} = \boldsymbol{Kh}^{(2)} = \{-1.26943, 6.30468, -0.84494, -5.01222\}$

Step length, $\alpha^{(2)} = r^{(2)}/\boldsymbol{h}^{(2)T}z^{(2)} = 2.62505$

Next point, $\boldsymbol{d}^{(3)} = \boldsymbol{d}^{(2)} + \alpha^{(2)}\boldsymbol{h}^{(2)} = \{-5.35718, 6.02538, 3.88102, 4.83344\}$

$g^{(3)} = g^{(2)} + \alpha^{(2)} z^{(2)} = \{2.64795, 9.66064, -8.56835, -2.22814\}$
Solving $Pw^{(3)} = g^{(3)}$, we have $w^{(3)} = \{0.294216, 0.805053, -0.778941, -0.106102\}$
$r^{(3)} = g^{(3)T} w^{(3)} = 15.467$
$\beta^{(3)} = r^{(3)}/r^{(2)} = 0.894918$
$h^{(3)} = -g^{(3)} + \beta^{(2)} h^{(2)} = \{-1.87982, 0.699903, 1.73672, 1.06371\}$

Iteration 4:

$z^{(3)} = Kh^{(3)} = \{-1.20719, -4.40425, 3.90628, 1.0158\}$
Step length, $\alpha^{(3)} = r^{(3)}/h^{(3)T} z^{(3)} = 2.19348$
Next point, $d^{(4)} = d^{(3)} + \alpha^{(3)} h^{(3)} = \{-9.48052, 7.56061, 7.69048, 7.16667\}$

$$g^{(4)} = g^{(3)} + \alpha^{(3)} z^{(3)}$$
$$= \{-7.99361 \times 10^{-15}, 1.42109 \times 10^{-14}, 8.88178 \times 10^{-15}, -3.73035 \times 10^{-14}\}$$

Solving $Pw^{(4)} = g^{(4)}$, we have $w^{(4)} = \{-8.88178 \times 10^{-16}, 1.18424 \times 10^{-15}, 8.07435 \times 10^{-16}, -1.77636 \times 10^{-15}\}$
$r^{(4)} = g^{(4)T} w^{(4)} = 9.73646 \times 10^{-29}$
$\beta^{(4)} = r^{(4)}/r^{(3)} = 6.29497 \times 10^{-30}$

$$h^{(4)} = -g^{(4)} + \beta^{(3)} h^{(3)}$$
$$= \{8.88178 \times 10^{-16}, -1.18424 \times 10^{-15}, -8.07435 \times 10^{-16}, 1.77636 \times 10^{-15}\}$$

The solution converged after four iterations:

$g = \{-7.99361 \times 10^{-15}, 1.42109 \times 10^{-14}, 8.88178 \times 10^{-15}, -3.73035 \times 10^{-14}\}$
$r = g^T g = 9.73646 \times 10^{-29}$
Solution, $d = \{-9.48052, 7.56061, 7.69048, 7.16667\}$

Incomplete Choleski Preconditioning Currently the most popular preconditioning method is that based on partial or incomplete Choleski factorization of the K matrix. In this method a lower triangular matrix L is constructed by considering only elements in a limited band around the main diagonal of matrix K. Denoting this band by m, the incomplete Choleski decomposition algorithm is as follows:

For $i = 1, n$
    For $j = 1, i - 1$
        If $(i - j \leq m)$ then
            $l_{ij} = \left(k_{ij} - \sum_{k=1}^{j-1} l_{ik} l_{jk}\right)/l_{jj}$
        Otherwise $l_{ij} = 0$
    end
    $l_{ii} = \sqrt{k_{ii} - \sum_{k=1}^{i-1} l_{ik}^2}$
end

With this lower triangular L matrix, the vector h in the PCG algorithm is established using forward elimination and backward substitution as follows:

$\boldsymbol{Ph} = \boldsymbol{g} \Longrightarrow \boldsymbol{LL}^T\boldsymbol{h} = \boldsymbol{g}$

Forward elimination for intermediate variables $\boldsymbol{y}$: $\boldsymbol{Ly} = \boldsymbol{g}$

Back substitution for $\boldsymbol{h}$: $\boldsymbol{L}^T\boldsymbol{h} = \boldsymbol{y}$

Example 1.16 Find the solution of the following system of equations using the PCG method with incomplete Choleski decomposition:

$$\begin{pmatrix} 9 & 3 & 6 & 3 \\ 3 & 12 & 2 & -10 \\ 6 & 2 & 11 & -5 \\ 3 & -10 & -5 & 21 \end{pmatrix} \begin{pmatrix} x_1 \\ x_2 \\ x_3 \\ x_4 \end{pmatrix} = \begin{pmatrix} 5 \\ 6 \\ 7 \\ 8 \end{pmatrix}$$

The solution using a band $m = 2$ (two terms in each row around the diagonal) for generating the preconditioning matrix is as follows:

Lower triangular factor of preconditioning matrix,

$$\boldsymbol{L} = \begin{pmatrix} 3. & 0 & 0 & 0 \\ 1. & 3.31662 & 0 & 0 \\ 2. & 0. & 2.64575 & 0 \\ 0 & -3.01511 & -1.88982 & 2.8875 \end{pmatrix}$$

Starting point, $\boldsymbol{d}^{(0)} = \{0, 0, 0, 0\}$

$\boldsymbol{g}^{(0)} = \boldsymbol{Kd}^{(0)} - \boldsymbol{R} = \{-5, -6, -7, -8\}$

Solving $\boldsymbol{LL}^T\boldsymbol{w}^{(0)} = \boldsymbol{g}^{(0)}$, we have $\boldsymbol{w}^{(0)} = \{1.28557, -1.98131, -1.77103, -1.74611\}$

$\boldsymbol{h}^{(0)} = -\boldsymbol{w}^{(0)} = \{-1.28557, 1.98131, 1.77103, 1.74611\}$

$r^{(0)} = \boldsymbol{g}^{(0)T}\boldsymbol{w}^{(0)} = 11.7637; \beta^{(0)} = 0$

Iteration 1:

$\boldsymbol{z}^{(0)} = \boldsymbol{Kh}^{(0)} = \{10.2383, 6., 7., 4.1433\}$

Step length, $\alpha^{(0)} = r^{(0)}/\boldsymbol{h}^{(0)T}\boldsymbol{z}^{(0)} = 1.73367$

Next point, $\boldsymbol{d}^{(1)} = \boldsymbol{d}^{(0)} + \alpha^{(0)}\boldsymbol{h}^{(0)} = \{-2.22874, 3.43493, 3.07037, 3.02717\}$

$\boldsymbol{g}^{(1)} = \boldsymbol{g}^{(0)} + \alpha^{(0)}\boldsymbol{z}^{(0)} = \{12.7498, 4.402, 5.13567, -0.816894\}$

Solving $\boldsymbol{LL}^T\boldsymbol{w}^{(1)} = \boldsymbol{g}^{(1)}$, we have $\boldsymbol{w}^{(1)} = \{2.02045, -0.322185, -0.744609, -0.369609\}$

$r^{(1)} = \boldsymbol{g}^{(1)T}\boldsymbol{w}^{(1)} = 20.82$

$\beta^{(1)} = r^{(1)}/r^{(0)} = 0.654181$

$\boldsymbol{h}^{(1)} = -\boldsymbol{g}^{(1)} + \beta^{(0)}\boldsymbol{h}^{(0)} = \{-2.86144, 1.61832, 1.90318, 1.51188\}$

Iteration 2:

$\boldsymbol{z}^{(1)} = \boldsymbol{Kh}^{(1)} = \{-4.9433, -0.476917, -0.556404, -2.53399\}$

Step length, $\alpha^{(1)} = r^{(1)}/\boldsymbol{h}^{(1)T}\boldsymbol{z}^{(1)} = 2.45428$

Next point, $\boldsymbol{d}^{(2)} = \boldsymbol{d}^{(1)} + \alpha^{(1)}\boldsymbol{h}^{(1)} = \{-9.25152, 7.40674, 7.74131, 6.73774\}$

$\boldsymbol{g}^{(2)} = \boldsymbol{g}^{(1)} + \alpha^{(1)}\boldsymbol{z}^{(1)} = \{0.617597, 3.23151, 3.7701, -7.036\}$

Solving $\boldsymbol{LL}^T\boldsymbol{w}^{(2)} = \boldsymbol{g}^{(2)}$, we have $\boldsymbol{w}^{(2)} = \{-0.166593, 0.0693565, 0.318143, -0.226273\}$

$r^{(2)} = \boldsymbol{g}^{(2)T}\boldsymbol{w}^{(2)} = 2.91273$
$\beta^{(2)} = r^{(2)}/r^{(1)} = 0.1399$
$\boldsymbol{h}^{(2)} = -\boldsymbol{g}^{(2)} + \beta^{(1)}\boldsymbol{h}^{(1)} = \{-0.233724, 0.157047, -0.0518876, 0.437785\}$

Iteration 3:

$z^{(2)} = \boldsymbol{K}\boldsymbol{h}^{(2)} = \{-0.630348, -3.29823, -3.84794, 7.18128\}$
Step length, $\alpha^{(2)} = r^{(2)}/\boldsymbol{h}^{(2)T}z^{(2)} = 0.979771$
Next point, $\boldsymbol{d}^{(3)} = \boldsymbol{d}^{(2)} + \alpha^{(2)}\boldsymbol{h}^{(2)} = \{-9.48052, 7.56061, 7.69048, 7.16667\}$

$$\boldsymbol{g}^{(3)} = \boldsymbol{g}^{(2)} + \alpha^{(2)}z^{(2)}$$
$$= \{1.77636 \times 10^{-15}, -1.77636 \times 10^{-15}, 8.88178 \times 10^{-16}, 4.44089 \times 10^{-15}\}$$

Solving $\boldsymbol{L}\boldsymbol{L}^T\boldsymbol{w}^{(3)} = \boldsymbol{g}^{(3)}$, we have $\boldsymbol{w}^{(3)} = \{1.03298 \times 10^{-16}, 1.10676 \times 10^{-17}, 1.35579 \times 10^{-16}, 2.49022 \times 10^{-16}\}$
$r^{(3)} = \boldsymbol{g}^{(3)T}\boldsymbol{w}^{(3)} = 1.39013 \times 10^{-30}$
$\beta^{(3)} = r^{(3)}/r^{(2)} = 4.77262 \times 10^{-31}$

$$\boldsymbol{h}^{(3)} = -\boldsymbol{g}^{(3)} + \beta^{(2)}\boldsymbol{h}^{(2)}$$
$$= \{-1.03298 \times 10^{-16}, -1.10676 \times 10^{-17}, -1.35579 \times 10^{-16}, -2.49022 \times 10^{-16}\}$$

The solution converged after three iterations:

$\boldsymbol{g} = \{1.77636 \times 10^{-15}, -1.77636 \times 10^{-15}, 8.88178 \times 10^{-16}, 4.44089 \times 10^{-15}\}$
$r = \boldsymbol{g}^T\boldsymbol{g} = 1.39013 \times 10^{-30}$
Solution, $\boldsymbol{d} = \{-9.48052, 7.56061, 7.69048, 7.16667\}$

1.6 MULTIPOINT CONSTRAINTS

In some finite element modeling situations it becomes necessary to introduce constraints between several different degrees of freedoms. Such constraints are known as multipoint constraints and in general are expressed as follows:

$$c_{11}d_1 + c_{12}d_2 + \cdots + c_{1n}d_n = q_1$$
$$c_{21}d_1 + c_{22}d_2 + \cdots + c_{2n}d_n = q_2$$
$$\ldots$$

where c_{ij}, $i, j = 1, 2, \ldots$, and q_i, $i = 1, 2, \ldots$, are specified constants and d_i, $i = 1, 2, \ldots$, are the nodal degrees of freedom. In matrix form the constraints equations can be expressed as follows:

$$\boldsymbol{C}\boldsymbol{d} = \boldsymbol{q}$$

where with m constraints $\boldsymbol{C}$ is an $m \times n$ matrix and $\boldsymbol{q}$ is an $m \times 1$ vector.

A common situation is modeling an inclined roller support. As is clear from Figure 1.22, in this case the displacement component normal to the incline is zero but neither

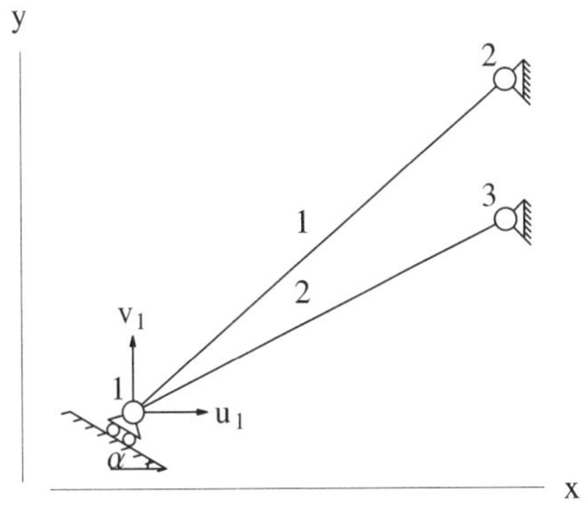

Figure 1.22. Inclined roller support

its horizontal or vertical displacement is zero, and hence this boundary condition cannot be incorporated using techniques discussed earlier. With the support making an angle α with the horizontal, the component of u_1 normal to the support is $u_1 \sin(\alpha)$ and that of v_1 is $v_1 \cos(\alpha)$. Thus we must solve the global system of equations with the following multipoint constraint between the degrees of freedom at this node:

$$u_1 \sin(\alpha) + v_1 \cos(\alpha) = 0$$

Another common use of multipoint constraints is in creating valid meshes when transitioning from a coarse to a fine mesh. As discussed earlier, the mesh consisting of three four-node quadrilateral elements in Figure 1.23 is invalid because node 4, which forms a corner of elements 2 and 3, is not attached to one of the four corners of element 1. A valid mesh can be obtained if the displacements at node 4 are related to the displacements at nodes 3 and 5. For the case when the node 4 is located at the midway point on line 3–5 the mesh compatibility can be achieved by defining two multipoint constraints defining displacements at node 4 as averages of the displacements at nodes 3 and 5. Thus we solve the system of equations with the two following multipoint constraints:

$$u_4 = \frac{u_3 + u_5}{2} \Longrightarrow u_3 - 2u_4 + u_5 = 0$$

$$v_4 = \frac{v_3 + v_5}{2} \Longrightarrow v_3 - 2v_4 + v_5 = 0$$

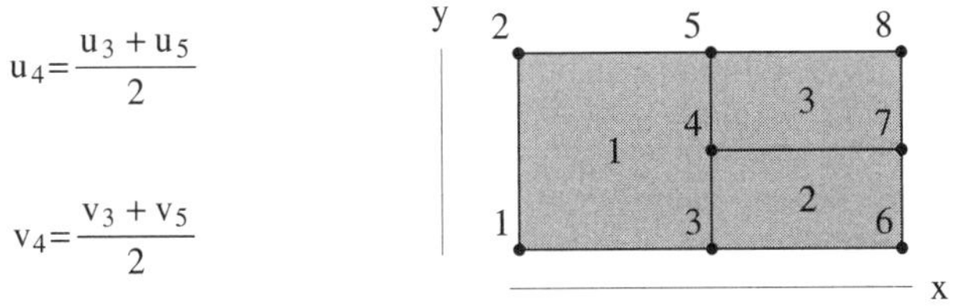

Figure 1.23. Multipoint constraints to create valid mesh in transition region

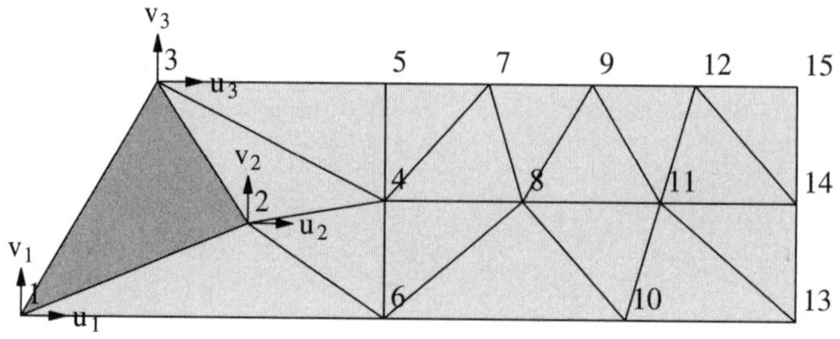

Figure 1.24. Rigid element in a finite element mesh

As a final example that requires use of multipoint constraints, we consider a situation in which some portions of a finite element model are much stiffer as compared to the rest. To avoid numerical difficulties in solving the resulting system of equations, it generally is more efficient to treat the stiff regions as completely rigid. Figure 1.24 shows a simple example in which the dark shaded triangle is assumed to be rigid. The nodes at the corners of this triangle must displace in such a manner that the original shape is maintained. A planar rigid body has only three degrees of freedom (two translations and a planar rotation). Thus the six degrees of freedom of the element must be related by three constraint equations.

For a triangle the three constraint equations can be written by recognizing the fact that the lengths of the sides of the triangle must remain unchanged between the initial configuration and the deformed configuration. In the initial configuration coordinates of the nodes of the rigid triangle are (x_1, y_1), (x_2, y_2), and (x_3, y_3). In the deformed configuration these coordinates will be $(x_1 + u_1, y_1 + v_1)$, $(x_2 + u_2, y_2 + v_2)$, and $(x_3 + u_3, y_3 + v_3)$. Equating lengths of sides obtained from these coordinates gives us the following three equations:

$$(-u_1 + u_2 - x_1 + x_2)^2 + (-v_1 + v_2 - y_1 + y_2)^2 = (x_2 - x_1)^2 + (y_2 - y_1)^2$$
$$(-u_2 + u_3 - x_2 + x_3)^2 + (-v_2 + v_3 - y_2 + y_3)^2 = (x_3 - x_2)^2 + (y_3 - y_2)^2$$
$$(u_1 - u_3 + x_1 - x_3)^2 + (v_1 - v_3 + y_1 - y_3)^2 = (x_1 - x_3)^2 + (y_1 - y_3)^2$$

Expanding these equations and for small displacements neglecting the displacement squared terms $(u_1^2, u_1 u_2, \ldots)$, we get the following equations:

$$2u_1x_1 - 2u_2x_1 - 2u_1x_2 + 2u_2x_2 + 2v_1y_1 - 2v_2y_1 - 2v_1y_2 + 2v_2y_2 = 0$$
$$2u_2x_2 - 2u_3x_2 - 2u_2x_3 + 2u_3x_3 + 2v_2y_2 - 2v_3y_2 - 2v_2y_3 + 2v_3y_3 = 0$$
$$2u_1x_1 - 2u_3x_1 - 2u_1x_3 + 2u_3x_3 + 2v_1y_1 - 2v_3y_1 - 2v_1y_3 + 2v_3y_3 = 0$$

Choosing u_1, v_1, and u_2 as the three independent degrees of freedom (usually called the *master* degrees of freedom), we can solve these equations for the remaining three degrees of freedom (called the *slave* degrees of freedom). The resulting equations are written in the

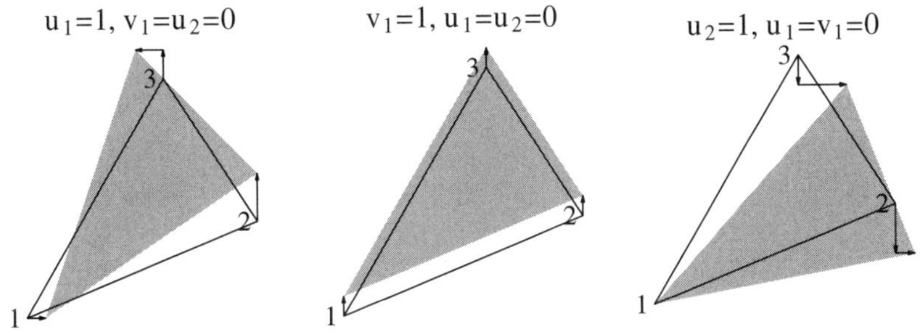

Figure 1.25. Rigid-body modes for a plane triangle

form of the following multipoint constraints:

$$\begin{pmatrix} v_2 \\ u_3 \\ v_3 \end{pmatrix} = \begin{pmatrix} \frac{x_1-x_2}{y_1-y_2} & 1 & \frac{x_2-x_1}{y_1-y_2} \\ \frac{y_3-y_2}{y_1-y_2} & 0 & \frac{y_1-y_3}{y_1-y_2} \\ \frac{x_1-x_3}{y_1-y_2} & 1 & \frac{x_3-x_1}{y_1-y_2} \end{pmatrix} \begin{pmatrix} u_1 \\ v_1 \\ u_2 \end{pmatrix}$$

or

$$\frac{x_1-x_2}{y_1-y_2}u_1 + v_1 + \frac{x_2-x_1}{y_1-y_2}u_2 - v_2 = 0$$
$$\frac{y_3-y_2}{y_1-y_2}u_1 + \frac{y_1-y_3}{y_1-y_2}u_2 - u_3 = 0$$
$$\frac{x_1-x_3}{y_1-y_2}u_1 + v_1 + \frac{x_3-x_1}{y_1-y_2}u_2 - v_3 = 0$$

The first column of this transformation matrix represents a rigid-body mode that sets $u_1 = 1$ and $v_1 = u_2 = 0$. Similarly, the second column represents a mode with $v_1 = 1, u_1 = u_2 = 0$ and the third column $u_2 = 1, u_1 = v_1 = 0$. These three rigid-body modes are shown in Figure 1.25.

An arbitrary rigid zone in a finite element model can be handled by breaking the shape into rigid triangles and then using the constraints for each of the triangles. The idea can also be extended to three-dimensional tetrahedrals each consisting of four triangular faces.

1.6.1 Solution Using Lagrange Multipliers

In the presence of multipoint constraints, the solution of global equations obviously becomes complicated. The minimization form introduced in the previous section allows us to use standard optimization techniques for constrained problems. Thus we set up the following optimization problem for solution of nodal degrees of freedom:

Find $\boldsymbol{d}$ to
Minimize $f = \frac{1}{2}\boldsymbol{d}^T\boldsymbol{K}\boldsymbol{d} - \boldsymbol{d}^T\boldsymbol{R}$
Subject to $\boldsymbol{C}\boldsymbol{d} - \boldsymbol{q} = \boldsymbol{0}$

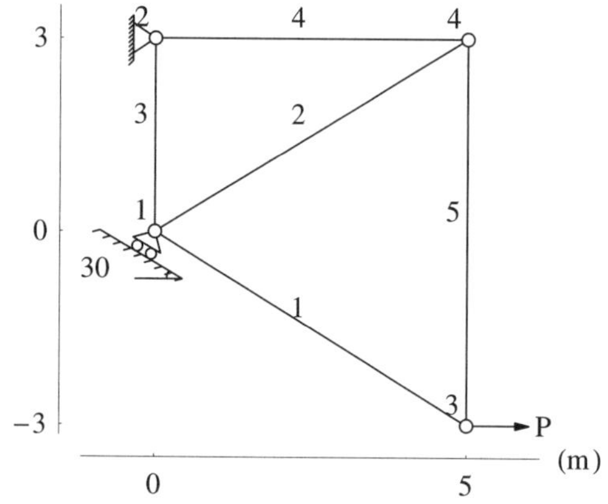

Figure 1.26. Five-bar truss with inclined support

A standard optimization technique for solution of this problem is to introduce m Lagrange multipliers λ_i, one for each equality constraint, and define an equivalent unconstrained problem as follows (for details see Bhatti [95]):

Find $\boldsymbol{d}$ and $\boldsymbol{\lambda}$ to

Minimize $L = \frac{1}{2}\boldsymbol{d}^T \boldsymbol{K}\boldsymbol{d} - \boldsymbol{d}^T \boldsymbol{R} + \boldsymbol{\lambda}^T(\boldsymbol{C}\boldsymbol{d} - \boldsymbol{q})$

where L is known as the Lagrangian. The necessary conditions for a minimum of the Lagrangian are that its derivatives with respect to d_i and λ_i be zero. Carrying out differentiations, the resulting equations are as follows:

$$\frac{\partial L}{\partial \boldsymbol{d}} = 0 \Longrightarrow \boldsymbol{K}\boldsymbol{d} - \boldsymbol{R} + \boldsymbol{C}^T\boldsymbol{\lambda} = \boldsymbol{0}$$

$$\frac{\partial L}{\partial \boldsymbol{\lambda}} = 0 \Longrightarrow \boldsymbol{C}\boldsymbol{d} - \boldsymbol{q} = \boldsymbol{0}$$

Writing the two sets of equations together, the complete system of linear equations can be expressed as follows:

$$\begin{pmatrix} \boldsymbol{K} & \boldsymbol{C}^T \\ \boldsymbol{C} & \boldsymbol{0} \end{pmatrix} \begin{pmatrix} \boldsymbol{d} \\ \boldsymbol{\lambda} \end{pmatrix} = \begin{pmatrix} \boldsymbol{R} \\ \boldsymbol{q} \end{pmatrix}$$

This system of linear equations can be solved using any of the methods discussed previously in this chapter. For structural problems a Lagrange multiplier can be interpreted as the force necessary to apply the specified constraint.

Example 1.17 Truss with an Inclined Support Consider the five-bar pin-jointed structure shown in Figure 1.26. All members have the same cross-sectional area and are of the same material, $E = 70$ GPa, and $A = 10^{-3}$ m^2. The load $P = 20$ kN. The dimensions in meters are shown in the figure.

For numerical calculations, the N · mm units are convenient. The displacements will be in mm and the stresses in MPa. The complete computations are as follows:

Specified Nodal Loads

Node	dof	Value
3	u_3	20000.

Equations for Element 1

$E = 70000; A = 1000$

Element Node	Global Node Number	x	y
1	1	0	0
2	3	5000.	−3000.

$x_1 = 0; \quad y_1 = 0; \quad x_2 = 5000.; \quad y_2 = -3000.$

$L = \sqrt{(x_2 - x_1)^2 + (y_2 - y_1)^2} = 5830.95$

Direction cosines:

$$l_s = \frac{x_2 - x_1}{L} = 0.857493; \qquad m_s = \frac{y_2 - y_1}{L} = -0.514496$$

Substituting into the truss element equations, we get

$$\begin{pmatrix} 8827.13 & -5296.28 & -8827.13 & 5296.28 \\ -5296.28 & 3177.77 & 5296.28 & -3177.77 \\ -8827.13 & 5296.28 & 8827.13 & -5296.28 \\ 5296.28 & -3177.77 & -5296.28 & 3177.77 \end{pmatrix} \begin{pmatrix} u_1 \\ v_1 \\ u_3 \\ v_3 \end{pmatrix} = \begin{pmatrix} 0. \\ 0. \\ 0. \\ 0. \end{pmatrix}$$

The element contributes to {1, 2, 5, 6} global degrees of freedom.

Adding element equations into appropriate locations, we have

$$\begin{pmatrix} 8827.13 & -5296.28 & 0 & 0 & -8827.13 & 5296.28 & 0 & 0 \\ -5296.28 & 3177.77 & 0 & 0 & 5296.28 & -3177.77 & 0 & 0 \\ 0 & 0 & 0 & 0 & 0 & 0 & 0 & 0 \\ 0 & 0 & 0 & 0 & 0 & 0 & 0 & 0 \\ -8827.13 & 5296.28 & 0 & 0 & 8827.13 & -5296.28 & 0 & 0 \\ 5296.28 & -3177.77 & 0 & 0 & -5296.28 & 3177.77 & 0 & 0 \\ 0 & 0 & 0 & 0 & 0 & 0 & 0 & 0 \\ 0 & 0 & 0 & 0 & 0 & 0 & 0 & 0 \end{pmatrix} \begin{pmatrix} u_1 \\ v_1 \\ u_2 \\ v_2 \\ u_3 \\ v_3 \\ u_4 \\ v_4 \end{pmatrix} = \begin{pmatrix} 0 \\ 0 \\ 0 \\ 0 \\ 20000. \\ 0 \\ 0 \\ 0 \end{pmatrix}$$

Processing the other four elements in exactly the same manner, we get the following equations:

$$\begin{pmatrix} 17654.3 & 0 & 0 & 0 & -8827.13 & 5296.28 & -8827.13 & -5296.28 \\ 0 & 29688.9 & 0 & -23333.3 & 5296.28 & -3177.77 & -5296.28 & -3177.77 \\ 0 & 0 & 14000. & 0 & 0 & 0 & -14000. & 0 \\ 0 & -23333.3 & 0 & 23333.3 & 0 & 0 & 0 & 0 \\ -8827.13 & 5296.28 & 0 & 0 & 8827.13 & -5296.28 & 0 & 0 \\ 5296.28 & -3177.77 & 0 & 0 & -5296.28 & 14844.4 & 0 & -11666.7 \\ -8827.13 & -5296.28 & -14000. & 0 & 0 & 0 & 22827.1 & 5296.28 \\ -5296.28 & -3177.77 & 0 & 0 & 0 & -11666.7 & 5296.28 & 14844.4 \end{pmatrix} \begin{pmatrix} u_1 \\ v_1 \\ u_2 \\ v_2 \\ u_3 \\ v_3 \\ u_4 \\ v_4 \end{pmatrix} = \begin{pmatrix} 0 \\ 0 \\ 0 \\ 0 \\ 20000. \\ 0 \\ 0 \\ 0 \end{pmatrix}$$

The essential boundary conditions are

Node	dof	Value
2	u_2	0
	v_2	0

Then we remove {3, 4} rows and columns. After adjusting for essential boundary conditions, we have

$$\begin{pmatrix} 17654.3 & 0 & -8827.13 & 5296.28 & -8827.13 & -5296.28 \\ 0 & 29688.9 & 5296.28 & -3177.77 & -5296.28 & -3177.77 \\ -8827.13 & 5296.28 & 8827.13 & -5296.28 & 0 & 0 \\ 5296.28 & -3177.77 & -5296.28 & 14844.4 & 0 & -11666.7 \\ -8827.13 & -5296.28 & 0 & 0 & 22827.1 & 5296.28 \\ -5296.28 & -3177.77 & 0 & -11666.7 & 5296.28 & 14844.4 \end{pmatrix} \begin{pmatrix} u_1 \\ v_1 \\ u_3 \\ v_3 \\ u_4 \\ v_4 \end{pmatrix} = \begin{pmatrix} 0 \\ 0 \\ 20000. \\ 0 \\ 0 \\ 0 \end{pmatrix}$$

The multipoint constraint due to inclined support at node 1 is $u_1 \sin(\pi/6) + v_1 \cos(\pi/6) = 0$. The augmented global equations with the Lagrange multiplier are as follows:

$$\begin{pmatrix} 17654.3 & 0 & -8827.13 & 5296.28 & -8827.13 & -5296.28 & 1/2 \\ 0 & 29688.9 & 5296.28 & -3177.77 & -5296.28 & -3177.77 & \frac{\sqrt{3}}{2} \\ -8827.13 & 5296.28 & 8827.13 & -5296.28 & 0 & 0 & 0 \\ 5296.28 & -3177.77 & -5296.28 & 14844.4 & 0 & -11666.7 & 0 \\ -8827.13 & -5296.28 & 0 & 0 & 22827.1 & 5296.28 & 0 \\ -5296.28 & -3177.77 & 0 & -11666.7 & 5296.28 & 14844.4 & 0 \\ 1/2 & \frac{\sqrt{3}}{2} & 0 & 0 & 0 & 0 & 0 \end{pmatrix} \begin{pmatrix} u_1 \\ v_1 \\ u_3 \\ v_3 \\ u_4 \\ v_4 \\ \lambda \end{pmatrix} = \begin{pmatrix} 0 \\ 0 \\ 20000. \\ 0 \\ 0 \\ 0 \\ 0 \end{pmatrix}$$

Solving the final system of global equations, we get

$$\{u_1 = 5.14286,\ v_1 = -2.96923,\ u_3 = 16.8629,\ v_3 = 12.788,$$
$$u_4 = -1.42857,\ v_4 = 11.7594,\ \lambda = 80000.\}$$

Solution for Element 1

Nodal coordinates

Element Node	Global Node Number	x	y
1	1	0	0
2	3	5000.	−3000.

$x_1 = 0; \quad y_1 = 0; \quad x_2 = 5000.; \quad y_2 = -3000.$

$L = \sqrt{(x_2 - x_1)^2 + (y_2 - y_1)^2} = 5830.95$

Direction cosines:

$$l_s = \frac{x_2 - x_1}{L} = 0.857493; \qquad m_s = \frac{y_2 - y_1}{L} = -0.514496$$

Global-to-local transformation matrix:

$$\boldsymbol{T} = \begin{pmatrix} 0.857493 & -0.514496 & 0 & 0 \\ 0 & 0 & 0.857493 & -0.514496 \end{pmatrix}$$

Element nodal displacements in global coordinates:

$$\boldsymbol{d} = \begin{pmatrix} u_1 \\ v_1 \\ u_3 \\ v_3 \end{pmatrix} = \begin{pmatrix} 5.14286 \\ -2.96923 \\ 16.8629 \\ 12.788 \end{pmatrix}$$

Element nodal displacements in local coordinates:

$$\boldsymbol{d}_l = \boldsymbol{Td} = \begin{pmatrix} 5.93762 \\ 7.88048 \end{pmatrix}$$

$E = 70000; A = 1000$

Axial strain, $\epsilon = (d_2 - d_1)/L = 0.000333197$

Axial stress, $\sigma = E\epsilon = 23.3238$

Axial force, $\sigma A = 23323.8$

In a similar manner, we can compute the solutions over the remaining elements:

	Stress	Axial Force
1	23.3238	23323.8
2	23.3238	23323.8
3	69.282	69282.
4	−20.	−20000.
5	−12.	−12000.

1.6.2 Solution Using Penalty Function

The Lagrange multiplier method of imposing constraints has two drawbacks. First, it requires adding new rows and columns to the global system of equations that for large systems may be inefficient. Second, the resulting system has zeros on the diagonal corresponding to the constraint equations. Some simple equation solvers that assume nonzero diagonal terms may not work for this system. Another standard technique of imposing constraints, the so-called penalty function approach, does not have these two drawbacks.

In this method a large penalty number μ is chosen and the equivalent unconstrained problem is defined as follows (for details see Bhatti [95]):

Find $\boldsymbol{d}$ to

Minimize $\phi = \frac{1}{2}\boldsymbol{d}^T\boldsymbol{K}\boldsymbol{d} - \boldsymbol{d}^T\boldsymbol{R} + \frac{1}{2}\mu(\boldsymbol{C}\boldsymbol{d} - \boldsymbol{q})^T(\boldsymbol{C}\boldsymbol{d} - \boldsymbol{q})$

With μ being large, the minimization process forces the constraints to be satisfied. The necessary conditions for the minimum result in the following system of equations:

$$\frac{\partial \phi}{\partial \boldsymbol{d}} = 0 \Longrightarrow \boldsymbol{K}\boldsymbol{d} - \boldsymbol{R} + \mu(\boldsymbol{C}^T\boldsymbol{C}\boldsymbol{d} - \boldsymbol{C}^T\boldsymbol{q}) = \boldsymbol{0}$$

Rearranging terms, the system of linear equations can be expressed as follows:

$$(\boldsymbol{K} + \mu\boldsymbol{C}^T\boldsymbol{C})\boldsymbol{d} = \boldsymbol{R} + \mu\boldsymbol{C}^T\boldsymbol{q}$$

This system of linear equations can be solved using any of the methods discussed previously in this section.

The performance of the method depends on the value chosen for the penalty parameter μ. Large values, say of the order of $\mu = 10^{10}$, give accurate solutions; however, the resulting system of equations may be ill-conditioned. If μ values are small as compared to other terms in the global equations, the solution will not satisfy the constraints very accurately. A general rule of thumb is to set μ equal to 10^5 times the largest number in the global K matrix.

Example 1.18 Truss Supporting a Rigid Plate A plane truss is designed to support a rigid triangular plate as shown in Figure 1.27. All members have the same cross-sectional area $A = 1$ in^2 and are of the same material, $E = 29{,}000$ ksi. The load $P = 20$ kips. The dimensions in ft are shown in the figure. Note there is no connection between the diagonal members where they cross each other.

The model consists of five nodes and thus the global system of equations before boundary conditions will be 10×10. The equations for the six truss elements are written as in the previous examples and assembled in the usual manner to give the following system of equations:

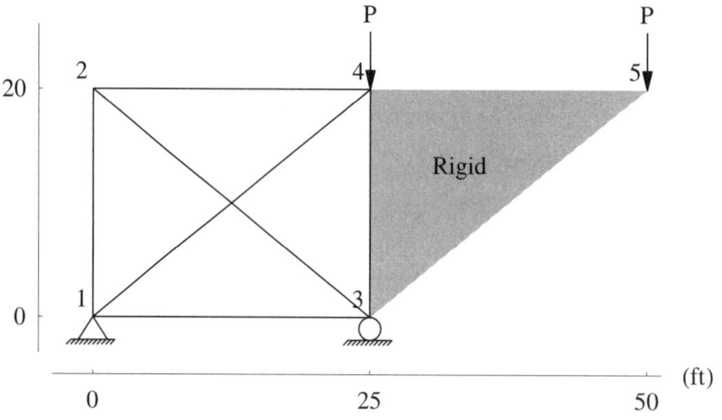

Figure 1.27. Truss supporting a rigid plate

$$
\begin{pmatrix}
142.693 & 36.8215 & 0 & 0 & -96.6667 & 0 & -46.0268 & -36.8215 & 0 & 0 \\
36.8215 & 150.29 & 0 & -120.833 & 0 & 0 & -36.8215 & -29.4572 & 0 & 0 \\
0 & 0 & 142.693 & -36.8215 & -46.0268 & 36.8215 & -96.6667 & 0. & 0 & 0 \\
0 & -120.833 & -36.8215 & 150.29 & 36.8215 & -29.4572 & 0. & 0. & 0 & 0 \\
-96.6667 & 0 & -46.0268 & 36.8215 & 142.693 & -36.8215 & 0. & 0. & 0 & 0 \\
0 & 0 & 36.8215 & -29.4572 & -36.8215 & 150.29 & 0. & -120.833 & 0 & 0 \\
-46.0268 & -36.8215 & -96.6667 & 0. & 0. & 0. & 142.693 & 36.8215 & 0 & 0 \\
-36.8215 & -29.4572 & 0. & 0. & 0. & -120.833 & 36.8215 & 150.29 & 0 & 0 \\
0 & 0 & 0 & 0 & 0 & 0 & 0 & 0 & 0 & 0 \\
0 & 0 & 0 & 0 & 0 & 0 & 0 & 0 & 0 & 0
\end{pmatrix}
\begin{pmatrix} u_1 \\ v_1 \\ u_2 \\ v_2 \\ u_3 \\ v_3 \\ u_4 \\ v_4 \\ u_5 \\ v_5 \end{pmatrix}
=
\begin{pmatrix} 0 \\ 0 \\ 0 \\ 0 \\ 0 \\ 0 \\ 0 \\ -20 \\ 0 \\ -20 \end{pmatrix}
$$

The essential boundary conditions at node 1 ($u_1 = v_1 = 0$) are incorporated by removing the corresponding rows and columns in the usual way. Node 3 also has zero vertical displacement. However, since this node is connected to the rigid plate as well, the boundary condition $v_3 = 0$ will be imposed later as part of the multipoint constraints. Removing the first two rows and columns, the global system of equations is as follows:

$\boldsymbol{Kd} = \boldsymbol{R} \Longrightarrow$

$$
\begin{pmatrix}
142.693 & -36.8215 & -46.0268 & 36.8215 & -96.6667 & 0. & 0 & 0 \\
-36.8215 & 150.29 & 36.8215 & -29.4572 & 0. & 0. & 0 & 0 \\
-46.0268 & 36.8215 & 142.693 & -36.8215 & 0. & 0. & 0 & 0 \\
36.8215 & -29.4572 & -36.8215 & 150.29 & 0. & -120.833 & 0 & 0 \\
-96.6667 & 0. & 0. & 0. & 142.693 & 36.8215 & 0 & 0 \\
0. & 0. & 0. & -120.833 & 36.8215 & 150.29 & 0 & 0 \\
0 & 0 & 0 & 0 & 0 & 0 & 0 & 0 \\
0 & 0 & 0 & 0 & 0 & 0 & 0 & 0
\end{pmatrix}
\begin{pmatrix} u_2 \\ v_2 \\ u_3 \\ v_3 \\ u_4 \\ v_4 \\ u_5 \\ v_5 \end{pmatrix}
=
\begin{pmatrix} 0. \\ 0. \\ 0. \\ 0. \\ 0. \\ -20. \\ 0. \\ -20. \end{pmatrix}
$$

The rigid plate is connected between nodes 3, 5, and 4. Treating u_3, v_3, and u_5 as independent degrees of freedom, the multipoint constraints are as follows:

$$
\begin{pmatrix} v_5 \\ u_4 \\ v_4 \end{pmatrix}
=
\begin{pmatrix}
\frac{x_3-x_5}{y_3-y_5} & 1 & \frac{x_5-x_3}{y_3-y_5} \\
\frac{y_4-y_5}{y_3-y_5} & 0 & \frac{y_3-y_4}{y_3-y_5} \\
\frac{x_3-x_4}{y_3-y_5} & 1 & \frac{x_4-x_3}{y_3-y_5}
\end{pmatrix}
\begin{pmatrix} u_3 \\ v_3 \\ u_5 \end{pmatrix}
=
\begin{pmatrix}
\frac{5}{4} & 1 & -\frac{5}{4} \\
0 & 0 & 1 \\
0 & 1 & 0
\end{pmatrix}
\begin{pmatrix} u_3 \\ v_3 \\ u_5 \end{pmatrix}
$$

Expanding and rearranging, we have

$$
-\frac{5u_3}{4} + \frac{5u_5}{4} - v_3 + v_5 = 0
$$

$$
u_4 - u_5 = 0; \qquad v_4 - v_3 = 0
$$

To this list we must also add the roller support constraint that $v_3 = 0$. Thus the complete set of constraint equations, expanded to include all degrees of freedom present in the global equations, are

$$Cd = q \Longrightarrow \begin{pmatrix} 0 & 0 & -\frac{5}{4} & -1 & 0 & 0 & \frac{5}{4} & 1 \\ 0 & 0 & 0 & 0 & 1 & 0 & -1 & 0 \\ 0 & 0 & 0 & -1 & 0 & 1 & 0 & 0 \\ 0 & 0 & 0 & 1 & 0 & 0 & 0 & 0 \end{pmatrix} \begin{pmatrix} u_2 \\ v_2 \\ u_3 \\ v_3 \\ u_4 \\ v_4 \\ u_5 \\ v_5 \end{pmatrix} = \begin{pmatrix} 0 \\ 0 \\ 0 \\ 0 \end{pmatrix}.$$

To use the penalty function approach, we choose the penalty parameter μ equal to 10^5 times the largest number in the global $\boldsymbol{K}$ matrix:

$$\mu = 150.29 \times 10^5 = 1.5029 \times 10^7$$

Incorporating the constraints into the global equations with this value of μ, the final system of equations is as follows:

$$(\boldsymbol{K} + \mu \boldsymbol{C}^T \boldsymbol{C})\boldsymbol{d} = \boldsymbol{R} + \mu \boldsymbol{C}^T \boldsymbol{q} \Longrightarrow$$

$$10^5 \begin{pmatrix} 0.00142693 & -0.000368215 & -0.000460268 & 0.000368215 & -0.000966667 & 0. & 0 & 0 \\ -0.000368215 & 0.0015029 & 0.000368215 & -0.000294572 & 0. & 0. & 0 & 0 \\ -0.000460268 & 0.000368215 & -234.827 & -187.863 & 0. & 0. & 234.829 & 187.863 \\ 0.000368215 & -0.000294572 & -187.863 & -450.87 & 0. & 150.289 & 187.863 & 150.29 \\ -0.000966667 & 0. & 0. & 0. & -150.289 & 0.000368215 & 150.29 & 0 \\ 0. & 0. & 0. & 150.289 & 0.000368215 & -150.289 & 0 & 0 \\ 0 & 0 & 234.829 & 187.863 & 150.29 & 0 & -385.119 & -187.863 \\ 0 & 0 & 187.863 & 150.29 & 0 & 0 & -187.863 & -150.29 \end{pmatrix} \begin{pmatrix} u_2 \\ v_2 \\ u_3 \\ v_3 \\ u_4 \\ v_4 \\ u_5 \\ v_5 \end{pmatrix} = \begin{pmatrix} 0. \\ 0. \\ 0. \\ 0. \\ 0. \\ -20. \\ 0. \\ -20. \end{pmatrix}$$

Solving this system of linear equations, we get

$$\{u_2 = 0.172845,\ v_2 = 0.076446,\ u_3 = -0.139174,\ v_3 = 3.99227 \times 10^{-6},$$
$$u_4 = 0.292292,\ v_4 = 6.03917 \times 10^{-6},\ u_5 = 0.29229,\ v_5 = -0.539324\}$$

Substituting these values into the constraint equations, we can see that the constraints are reasonably satisfied:

$$Cd = \begin{pmatrix} 1.33076 \times 10^{-6} \\ 1.66345 \times 10^{-6} \\ 2.0469 \times 10^{-6} \\ 3.99227 \times 10^{-6} \end{pmatrix} \approx \begin{pmatrix} 0 \\ 0 \\ 0 \\ 0 \end{pmatrix}$$

Knowing the nodal values, the element solutions can be computed in the usual manner:

	Strain	Stress	Axial Force
1	0.000318525	9.23722	9.23722
2	−0.000463913	−13.4535	−13.4535
3	0.000594098	17.2289	17.2289
4	0.000398156	11.5465	11.5465
5	−0.000509888	−14.7868	−14.7868
6	8.52877×10^{-9}	0.000247334	0.000247334

Using the Lagrange multiplier method, the solution is obtained as follows:

Augmented system of equations:

$$\begin{pmatrix} 142.693 & -36.8215 & -46.0268 & 36.8215 & -96.6667 & 0. & 0 & 0 & 0 & 0 & 0 & 0 \\ -36.8215 & 150.29 & 36.8215 & -29.4572 & 0. & 0. & 0 & 0 & 0 & 0 & 0 & 0 \\ -46.0268 & 36.8215 & 142.693 & -36.8215 & 0. & 0. & 0 & 0 & -\frac{5}{4} & 0 & 0 & 0 \\ 36.8215 & -29.4572 & -36.8215 & 150.29 & 0. & -120.833 & 0 & 0 & -1 & 0 & -1 & 1 \\ -96.6667 & 0. & 0. & 0. & 142.693 & 36.8215 & 0 & 0 & 0 & 1 & 0 & 0 \\ 0. & 0. & 0. & -120.833 & 36.8215 & 150.29 & 0 & 0 & 0 & 0 & 1 & 0 \\ 0 & 0 & 0 & 0 & 0 & 0 & 0 & 0 & \frac{5}{4} & -1 & 0 & 0 \\ 0 & 0 & 0 & 0 & 0 & 0 & 0 & 0 & 1 & 0 & 0 & 0 \\ 0 & 0 & -\frac{5}{4} & -1 & 0 & 0 & \frac{5}{4} & 1 & 0 & 0 & 0 & 0 \\ 0 & 0 & 0 & 0 & 1 & 0 & -1 & 0 & 0 & 0 & 0 & 0 \\ 0 & 0 & 0 & -1 & 0 & 1 & 0 & 0 & 0 & 0 & 0 & 0 \\ 0 & 0 & 0 & 1 & 0 & 0 & 0 & 0 & 0 & 0 & 0 & 0 \end{pmatrix} \begin{pmatrix} d_1 \\ d_2 \\ d_3 \\ d_4 \\ d_5 \\ d_6 \\ d_7 \\ d_8 \\ \lambda_1 \\ \lambda_2 \\ \lambda_3 \\ \lambda_4 \end{pmatrix} = \begin{pmatrix} 0. \\ 0. \\ 0. \\ 0. \\ 0. \\ -20. \\ 0. \\ -20. \\ 0 \\ 0 \\ 0 \\ 0 \end{pmatrix}$$

Solution:

$$\{d_1 = 0.172849,\ d_2 = 0.0764461,\ d_3 = -0.139174,\ d_4 = -4.68418 \times 10^{-18},$$
$$d_5 = 0.292296,\ d_6 = -1.62088 \times 10^{-17},\ d_7 = 0.292296,\ d_8 = -0.539337,$$
$$\lambda_1 = -20.,\ \lambda_2 = -25.,\ \lambda_3 = -30.7628,\ \lambda_4 = -60.\}$$

1.7 UNITS

It is important to use a consistent system of units during finite element calculations. When using the U.S. customary system of units, it is convenient to use the pound-inch (PI) units. Thus, the nodal coordinates will be entered in inches, the modulus of elasticity in pounds per square inch (psi), mass density in pounds per inch cubed (lb_m/in^3), thermal conductivity in British thermal units per hour-inch-degrees Fahrenheit [Btu/(hr · in · °F)], and coefficient of thermal expansion in inch per inch per degrees Fahrenheit (in/in/°F). When using the International System of Units, it is recommended to use the newton-millimeter (N · mm) units. Thus the nodal coordinates will be entered in millimeters, the modulus of elasticity in megapascals (MPa), mass density in metric tonnes per millimeter cubed (mt/mm^3), thermal conductivity in watts per millimeter-degrees Celsius [W/(mm · °C)], and coefficient of thermal expansion in millimeters per millimeter per degrees Celsius (mm/mm/°C). Typical numerical values for these quantities for concrete, steel, and aluminum are as follows:

Concrete (medium strength)

Quantity	PI System	N · mm System
Modulus of elasticity	4.64×10^6 psi	0.32×10^5 MPa
Mass density	$2.24 \times 10^{-4}\ \text{lb}_m/\text{in}^3$	$2.4 \times 10^{-9}\ \text{mt/mm}^3$
Thermal conductivity	0.048 Btu/(hr · in · °F)	0.001 W/(mm · °C)
Coefficient of thermal expansion	6.1×10^{-6} in/in/°F	11×10^{-6}mm/mm/°C

Steel (carbon and alloys):

Quantity	PI System	N · mm system
Modulus of elasticity	30×10^6 psi	2.1×10^5 MPa
Mass density	7.32×10^{-4} lb_m/in^3	7.83×10^{-9} mt/mm^3
Thermal conductivity	2.07 Btu/(hr · in · °F)	0.043 W/(mm · °C)
Coefficient of thermal expansion	6.5×10^{-6} in/in/°F	12×10^{-6} mm/mm/°C

Aluminum alloys:

Quantity	PI System	N · mm System
Modulus of elasticity	10.4×10^6 psi	0.72×10^5 MPa
Mass density	2.62×10^{-4} lb_m/in^3	2.8×10^{-9} mt/mm^3
Thermal conductivity	10.11 Btu/(hr · in · °F)	0.21 W/(mm · °C)
Coefficient of thermal expansion	12.8×10^{-6} in/in/°F	23×10^{-6} mm/mm/°C

Commonly used conversion factors:

	Given in	Multiply by	To get
Length	in	25.4	mm
Force or weight	lb_f	4.4484	N
Weight to mass	lb_f	1/386.4	lb_m
Weight to mass	N	1.02×10^{-4}	mt
Mass	lb_m	0.17524	mt
Mass density	lb_m/in^3	1.0694×10^{-5}	mt/mm^3
Power	Btu/hr	0.2931	W
Temperature	°F	(°F − 32)/1.8	°C
Coefficient of thermal expansion	in/in/°F	1.8	mm/mm/°C
Thermal conductivity	Btu/(hr · in · °F)	0.02077	W/(mm · °C)
Convection coefficient	Btu/(hr · in^2 · °F)	6.411×10^{-3}	W/(mm^2 · °C)
Modulus of elasticity or stress	psi	6.895×10^{-3}	MPa

PROBLEMS

Element Equations

1.1 A triangular finite element for a 2D heat flow problem is shown in Figure 1.28. Write the finite element equations for this element. Assume that the heat conduction coefficient is 1.4 W/m · °C. The convection heat loss takes place along side 2–3 of the element. The average convection heat transfer coefficient is 20 W/m · °C and the surrounding air temperature is 25°C. There is no heat generation inside the element.

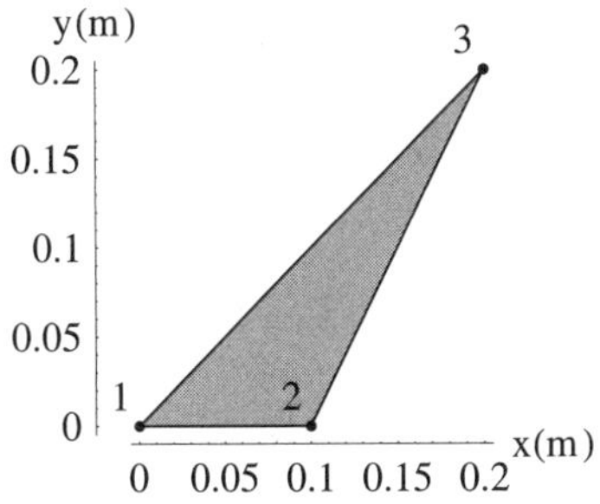

Figure 1.28. Triangular element

1.2 A triangular finite element for a plane stress problem is shown in Figure 1.29. Write the finite element equations for this element. Assume $E = 20 \times 10^6$ N/cm², $\nu = 0.3$, thickness = 2 cm, and a uniform pressure of 300 N/cm² normal to side 1–3 of the element.

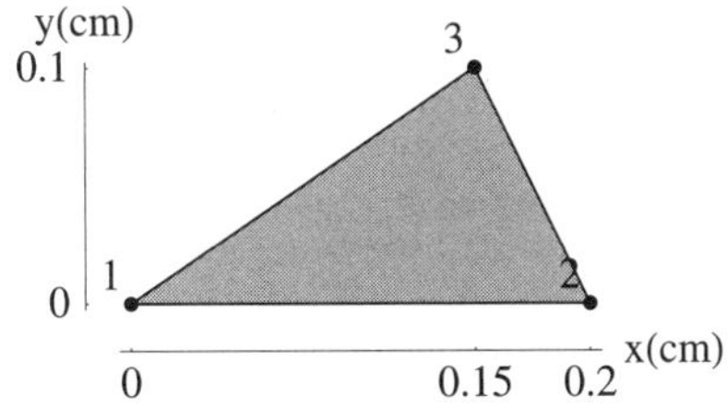

Figure 1.29. Triangular element

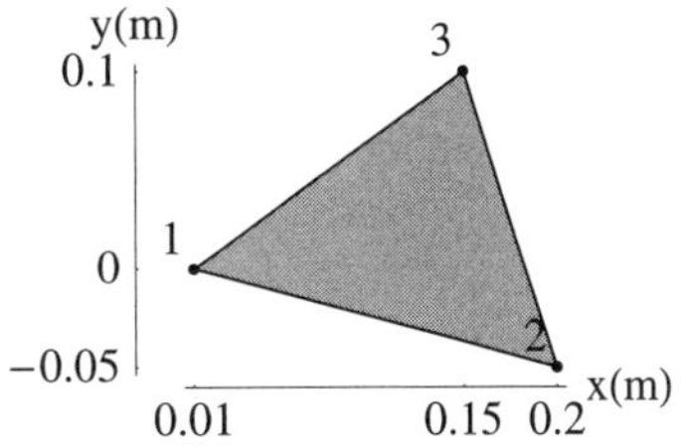

Figure 1.30. Triangular element

1.3 A triangular finite element for a 2D heat flow problem is shown in Figure 1.30. Write the finite element equations for this element. Assume that the heat conduction coefficient is 1.17 W/m · °C. The convection heat loss takes place along side 3–1 of the element. The average convection heat transfer coefficient is 17 W/m · °C and the surrounding air temperature is 32°C. There is no heat generation inside the element.

1.4 Consider two-dimensional steady-state heat flow in a V-grooved solid body with a cross section as shown in Figure 1.31. A hot liquid flowing through the groove maintains the surface at a temperature of 170°F. The thermal conductivity of the material is $k = 0.04$ Btu/hr · in · °F. The bottom surface is insulated. The sides have a convection coefficient of $h = 0.03$ Btu/hr · in² · °F with the outside temperature of 70°F. Determine the temperature distribution in the solid. Taking advantage of symmetry and using only two triangular elements, the model is as shown in the figure. Develop finite element equations for element 1.

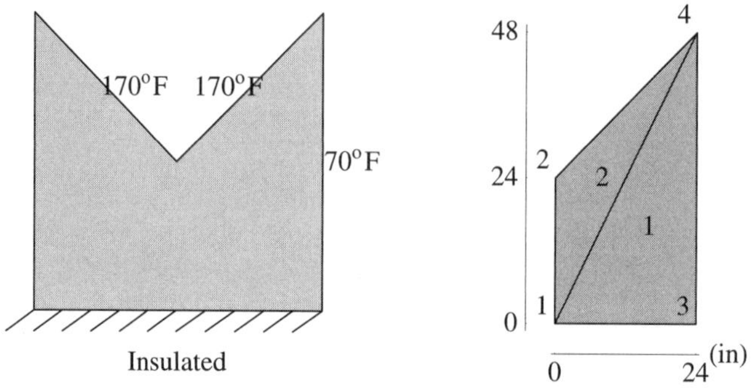

Figure 1.31. V-grooved solid

1.5 Develop finite element equations for element 2 of the model shown in Figure 1.31.

1.6 The cross section of a long concrete dam (thermal conductivity = 0.6 W/m · °C) with base = 3 m and height = 4 m is shown in Figure 1.32. The face of the dam is exposed to heat flux q_0 = 800 W/m^2 from sun. The vertical face is subjected to convection by water at 15°C with convection heat transfer coefficient 150 W/m^2·°C. It is proposed to determine the temperature distribution in the dam using only one triangular element. Set up the system of finite element equations.

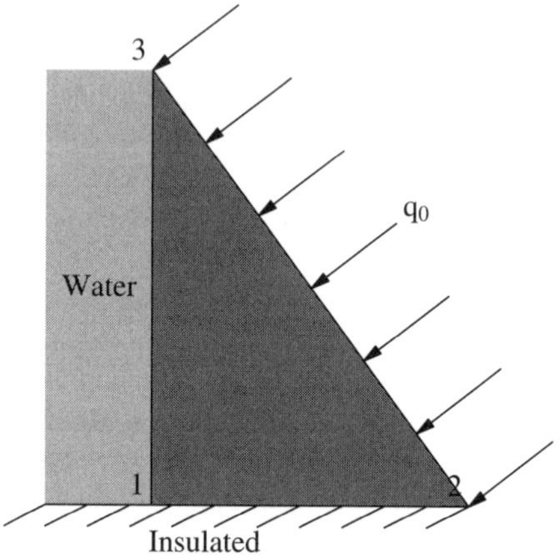

Figure 1.32. Concrete dam

Assembly of Element Equations

1.7 A three-bar truss is shown in Figure 1.33. Write the element equations and carry out the assembly of the element matrices to form the global system of equations. The area of cross section of each element is 500 mm^2 and E = 210 GPa. Use N · mm units in your calculations.

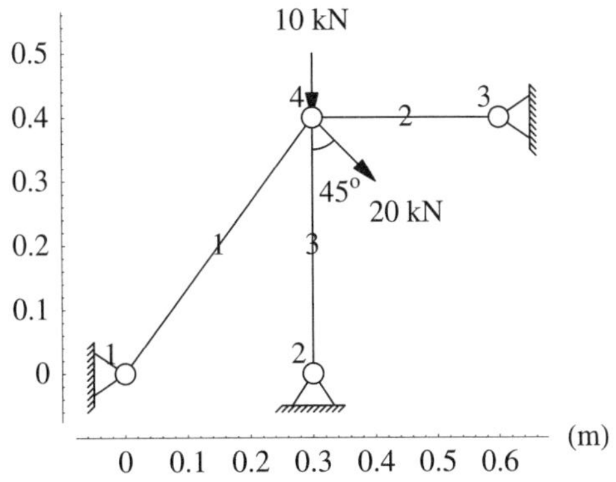

Figure 1.33. Plane truss

1.8 A three-bar truss is shown in Figure 1.34. Write the element equations and carry out the assembly of the element matrices to form the global $\boldsymbol{K}$ matrix and the $\boldsymbol{R}$ vector. The area of cross section of each element is $500\,\text{mm}^2$ and $E = 210\,\text{GPa}$. Use $\text{N} \cdot \text{mm}$ units in your calculations.

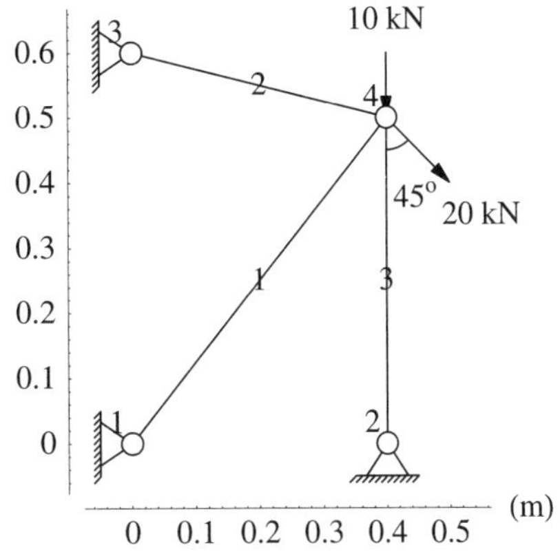

Figure 1.34. Plane truss

1.9 A finite element model employs triangular plane stress elements and consists of only six nodes. Each node has two degrees of freedom. The first element is connected to the model through nodes 1, 4, and 6 and has the following coefficient matrix and the right-hand-side vector:

$$\boldsymbol{k} = \begin{pmatrix} 246 & 198 & 221 & 179 & 188 & 194 \\ 198 & 273 & 182 & 242 & 221 & 186 \\ 221 & 182 & 215 & 167 & 198 & 185 \\ 179 & 242 & 167 & 243 & 225 & 183 \\ 188 & 221 & 198 & 225 & 266 & 180 \\ 194 & 186 & 185 & 183 & 180 & 201 \end{pmatrix}; \qquad \boldsymbol{r} = \begin{pmatrix} 10 \\ 6 \\ 9 \\ 10 \\ 8 \\ 7 \end{pmatrix}$$

Carry out the assembly of the element matrices to form the global $\boldsymbol{K}$ matrix and the $\boldsymbol{R}$ vector. Note that this is the first element that is being assembled, and hence the global matrices are all zeros prior to assembly of this element.

1.10 Form the global system of equations for the two-element finite element model shown in Figure 1.35. The equations for the individual elements are as follows:

$$\text{Element 1:} \quad \begin{pmatrix} 0.04 & -0.04 & 0 \\ -0.04 & 0.53 & 0.23 \\ 0 & 0.23 & 0.49 \end{pmatrix} \begin{pmatrix} T_1 \\ T_3 \\ T_4 \end{pmatrix} = \begin{pmatrix} 0 \\ 50.4 \\ 50.4 \end{pmatrix}$$

$$\text{Element 2:} \quad \begin{pmatrix} 0.04 & 0.02 & -0.06 \\ 0.02 & 0.02 & -0.04 \\ -0.06 & -0.04 & 0.1 \end{pmatrix} \begin{pmatrix} T_1 \\ T_4 \\ T_2 \end{pmatrix} = \begin{pmatrix} 0 \\ 0 \\ 0 \end{pmatrix}$$

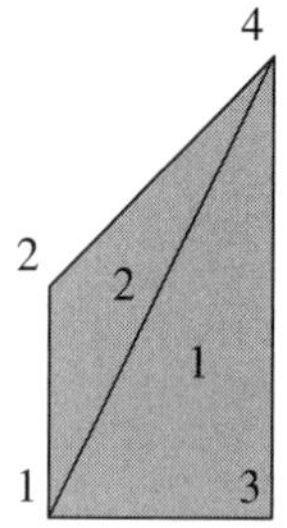

Figure 1.35. Two-element model

Handling Essential Boundary Conditions and Solution for Nodal Unknowns

1.11 In Problem 1.7 you were asked to obtain a global system of equations for a plane truss. Continue further with the analysis and obtain the reduced system of equations after adjusting for specified displacement boundary conditions. Solve the resulting system for the unknown nodal displacements. Compute reactions at the supports and verify that your solution satisfies overall equilibrium.

1.12 The global $\boldsymbol{K}$ matrix and the $\boldsymbol{R}$ vector for a finite element system are as follows:

$$\boldsymbol{K} = \begin{pmatrix} 63993.2 & 79991.4 & 0 & 0 & 0 & 0 & -63993.2 & -79991.4 \\ 79991.4 & 99989.3 & 0 & 0 & 0 & 0 & -79991.4 & -99989.3 \\ 0 & 0 & 0. & 0. & 0 & 0 & 0. & 0. \\ 0 & 0 & 0. & 210000. & 0 & 0 & 0. & -210000. \\ 0 & 0 & 0 & 0 & 239682. & -59920.6 & -239682. & 59920.6 \\ 0 & 0 & 0 & 0 & -59920.6 & 14980.1 & 59920.6 & -14980.1 \\ -63993.2 & -79991.4 & 0. & 0. & -239682. & 59920.6 & 303675. & 20070.9 \\ -79991.4 & -99989.3 & 0. & -210000. & 59920.6 & -14980.1 & 20070.9 & 324969. \end{pmatrix}$$

$$\boldsymbol{R}^T = (0. \quad 0. \quad 0. \quad 0. \quad 0. \quad 0. \quad 14142.1 \quad -24142.1)$$

Obtain the reduced system of equations after imposing the following essential boundary conditions:

$$d_1 = -0.1; \qquad d_2 = 0.2; \qquad d_3 = d_4 = d_5 = d_6 = 0$$

Solve the final system of equations for the remaining two nodal unknowns.

1.13 After assembly a finite element model yields the following global system of equations:

$$\begin{pmatrix} 358815 & -447 & -2060 & -60 & -1053 & 365 \\ -447 & 553547 & -1916 & -85 & -313 & 931 \\ -2060 & -1916 & 182342 & -1 & 1479 & 239 \\ -60 & -85 & -1 & 435603 & -1321 & 1501 \\ -1053 & -313 & 1479 & -1321 & 277735 & -1056 \\ 365 & 931 & 239 & 1501 & -1056 & 33134 \end{pmatrix} \begin{pmatrix} d_1 \\ d_2 \\ d_3 \\ d_4 \\ d_5 \\ d_6 \end{pmatrix} = \begin{pmatrix} 2 \\ 3 \\ -8 \\ 5 \\ -7 \\ -4 \end{pmatrix}$$

Get the reduced system of equations after imposing essential boundary conditions $d_1 = -1$, $d_2 = 0$, and $d_4 = 4$. Solve the reduced system for remaining nodal unknowns.

1.14 A typical truss to support a highway is as shown in Figure 1.36. Because of a construction error, the support at node 6 is made 5 mm too tall. The contractor is considering using jack hammers to force the truss to fit it into place. You are asked to conduct a finite element analysis to make sure that the stresses in the truss elements will remain within allowable limits under the given loading and the forced fit. A colleague of yours has started the analysis already and has obtained the following global stiffness matrix after assembly of the first eight elements (N · mm units are used):

$$\begin{pmatrix} 7560. & -1920. & -5000. & 0 & 0 & 0 & 0 & 0 & -2560. & 1920. & 0 & 0 \\ -1920. & 1440. & 0 & 0 & 0 & 0 & 0 & 0 & 1920. & -1440. & 0 & 0 \\ -5000. & 0 & 12560. & -1920. & -5000. & 0 & 0 & 0 & 0 & 0 & -2560. & 1920. \\ 0 & 0 & -1920. & 8106.67 & 0 & 0 & 0 & 0 & 0 & -6666.67 & 1920. & -1440. \\ 0 & 0 & -5000. & 0 & 10000. & 0 & -5000. & 0 & 0 & 0 & 0 & 0 \\ 0 & 0 & 0 & 0 & 0 & 6666.67 & 0 & 0 & 0 & 0 & 0 & -6666.67 \\ 0 & 0 & 0 & 0 & -5000. & 0 & 5000. & 0 & 0 & 0 & 0 & 0 \\ 0 & 0 & 0 & 0 & 0 & 0 & 0 & 0 & 0 & 0 & 0 & 0 \\ -2560. & 1920. & 0 & 0 & 0 & 0 & 0 & 0 & 7560. & -1920. & -5000. & 0 \\ 1920. & -1440. & 0 & -6666.67 & 0 & 0 & 0 & 0 & -1920. & 8106.67 & 0 & 0 \\ 0 & 0 & -2560. & 1920. & 0 & 0 & 0 & 0 & -5000. & 0 & 7560. & -1920. \\ 0 & 0 & 1920. & -1440. & 0 & -6666.67 & 0 & 0 & 0 & 0 & -1920. & 8106.67 \end{pmatrix} \begin{pmatrix} u_1 \\ v_1 \\ u_2 \\ v_2 \\ u_3 \\ v_3 \\ u_4 \\ v_4 \\ u_5 \\ v_5 \\ u_6 \\ v_6 \end{pmatrix}$$

(a) Form the equations for element 9 and assemble them to form the global system of equations. Use the following numerical values:

$$E = 200000\ \text{N/mm}^2; \qquad A = 100\ \text{mm}^2; \qquad P = 20000\ \text{N}$$

(b) Taking boundary conditions into consideration, determine the final system of equations to solve for nodal displacements.

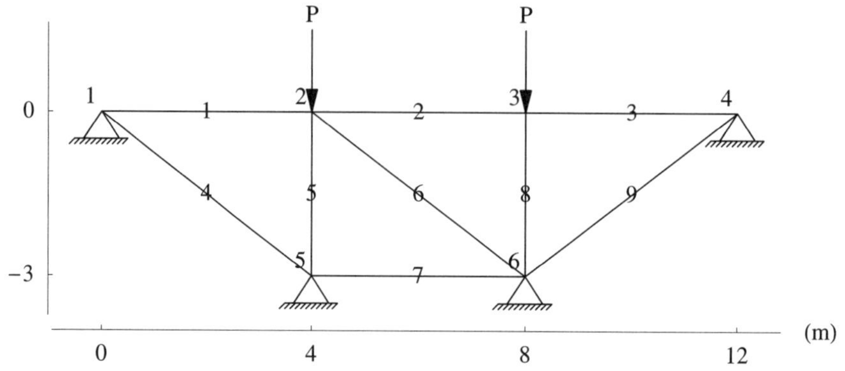

Figure 1.36. Bridge truss

Complete Solution over Each Element

1.15 The following nodal temperatures are obtained for the heat flow problem in the V-grooved solid shown in Figure 1.31. Compute the temperature distribution and its x and y derivatives over each element. By comparing the temperature derivatives over the two elements, comment on the quality of the solution.

Node	Temperature
1	99.4118
2	170
3	28.8235
4	170

1.16 The correct nodal displacements, in mm, for the three-bar in Problem 1.7 are as follows:

	u	v
1	0	0
2	0	0
3	0	0
4	0.0506837	−0.0736988

Compute axial strains, axial stresses, and axial forces in each element. Using the computed axial forces, draw a free-body diagram of all forces acting at node 4. By summing forces in the horizontal and vertical directions, show that the joint equilibrium is satisfied.

1.17 Determine joint deflections, stresses in each member, and support reactions for the plane truss shown in Figure 1.37. The members have a cross-sectional area of 8 cm^2 and are made of steel ($E = 200$ GPa). Show all calculations.

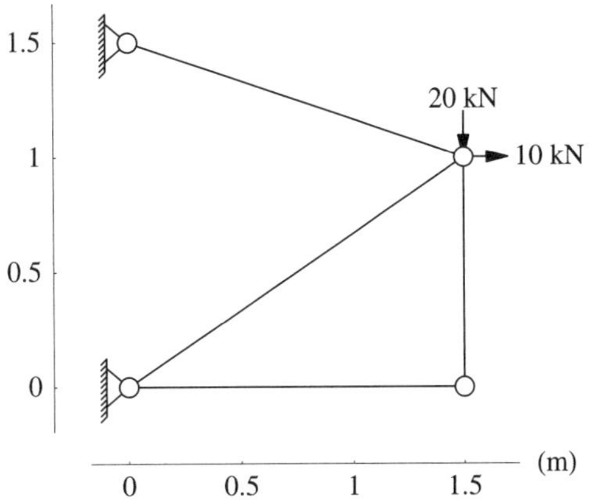

Figure 1.37. Plane truss

1.18 Determine the temperature at the centroid of the concrete dam cross section shown in Figure 1.32, assuming the temperatures at the nodes are 10, 20, and 30°C, respectively.

Solution of Linear Equations

1.19 Solve the following system of equations using Choleski decomposition:

$$\begin{pmatrix} 2 & -2 & 0 & 0 & 0 \\ -2 & 6 & 4 & 0 & 0 \\ 0 & 4 & 9 & -1.5 & 0 \\ 0 & 0 & -1.5 & 1.25 & 0.5 \\ 0 & 0 & 0 & 0.5 & 3.25 \end{pmatrix} \begin{pmatrix} x_1 \\ x_2 \\ x_3 \\ x_4 \\ x_5 \end{pmatrix} = \begin{pmatrix} 2 \\ -2 \\ 8 \\ 0.5 \\ -3.25 \end{pmatrix}$$

1.20 Solve the following system of equations using Choleski decomposition:

$$\begin{pmatrix} 2 & 1 & 0 & 0 \\ 1 & 2 & 1 & 0 \\ 0 & 1 & 2 & 1 \\ 0 & 0 & 1 & 2 \end{pmatrix} \begin{pmatrix} x_1 \\ x_2 \\ x_3 \\ x_4 \end{pmatrix} = \begin{pmatrix} -1 \\ 0 \\ 0 \\ 0 \end{pmatrix}$$

1.21 Factor the following matrix using Choleski decomposition:

$$\begin{pmatrix} 2 & -1 & 0 & 0 & 0 & 0 \\ -1 & 2 & -1 & 0 & 0 & 0 \\ 0 & -1 & 3 & -1 & 0 & 0 \\ 0 & 0 & -1 & 3 & -1 & 0 \\ 0 & 0 & 0 & -1 & 2 & -1 \\ 0 & 0 & 0 & 0 & -1 & 2 \end{pmatrix}$$

Using the factored matrix, obtain solutions for the following right-hand sides:

$$b^{(1)} = \begin{pmatrix} -1 \\ 0 \\ 0 \\ 0 \\ 0 \\ 0 \end{pmatrix}; \quad b^{(2)} = \begin{pmatrix} -1 \\ 2 \\ 0 \\ 0 \\ 0 \\ 0 \end{pmatrix}; \quad b^{(3)} = \begin{pmatrix} -1 \\ 2 \\ 0 \\ 0 \\ 0 \\ -2 \end{pmatrix}$$

1.22 Factor the following matrix using Choleski decomposition:

$$\begin{pmatrix} 1 & 1 & 1 & 1 & 0 & 0 & 1 & 0 & -1 & 0 \\ 1 & 12 & 3 & 1 & 2 & 0 & 1 & 2 & 1 & 2 \\ 1 & 3 & 3 & 1 & 1 & 1 & 2 & 2 & 1 & 2 \\ 1 & 1 & 1 & 5 & 2 & 2 & 1 & 2 & 3 & 2 \\ 0 & 2 & 1 & 2 & 6 & 3 & 2 & 0 & 5 & 4 \\ 0 & 0 & 1 & 2 & 3 & 4 & 2 & 1 & 5 & 3 \\ 1 & 1 & 2 & 1 & 2 & 2 & 7 & 2 & -1 & 4 \\ 0 & 2 & 2 & 2 & 0 & 1 & 2 & 9 & 2 & 5 \\ -1 & 1 & 1 & 3 & 5 & 5 & -1 & 2 & 24 & 4 \\ 0 & 2 & 2 & 2 & 4 & 3 & 4 & 5 & 4 & 7 \end{pmatrix}$$

Using the factored matrix, obtain solutions for the following right-hand sides:

$$b^{(1)} = \begin{pmatrix} 0 \\ -8 \\ -3 \\ 14 \\ 39 \\ 32 \\ 1 \\ -45 \\ 56 \\ 1 \end{pmatrix}; \quad b^{(2)} = \begin{pmatrix} 15 \\ 43 \\ 43 \\ 27 \\ 12 \\ 14 \\ 59 \\ 79 \\ -2 \\ 63 \end{pmatrix}; \quad b^{(3)} = \begin{pmatrix} 6 \\ 10 \\ 8 \\ 12 \\ 7 \\ 0 \\ 29 \\ 19 \\ -24 \\ 21 \end{pmatrix}$$

1.23 Solve the system of equations in Problem 1.19 using the basic conjugate gradient method.

1.24 Solve the system of equations in Problem 1.20 using the basic conjugate gradient method.

1.25 Solve the system of equations with the given coefficient matrix and the first right-hand side in Problem 1.21 using the basic conjugate gradient method.

1.26 Solve the system of equations with the given coefficient matrix and the first right-hand side in Problem 1.22 using the basic conjugate gradient method.

1.27 Solve the system of equations in Problem 1.19 using the conjugate gradient method with Jacobi preconditioning.

1.28 Solve the system of equations in Problem 1.20 using the conjugate gradient method with Jacobi preconditioning.

1.29 Solve the system of equations with the given coefficient matrix and the first right-hand side in Problem 1.21 using the conjugate gradient method with Jacobi preconditioning.

1.30 Solve the system of equations with the given coefficient matrix and the first right-hand side in Problem 1.22 using the conjugate gradient method with Jacobi preconditioning.

1.31 Solve the system of equations with the given coefficient matrix and the first right-hand side in Problem 1.21 using the conjugate gradient method with incomplete Choleski preconditioning with $m = 1$. The solution should converge in one iteration. Why?

1.32 Solve the system of equations with the given coefficient matrix and the first right-hand side in Problem 1.22 using the conjugate gradient method with incomplete Choleski preconditioning with $m = 1$.

Multipoint Constraints

1.33 A finite element system results in the following system of global equations:

$$\begin{pmatrix} 2 & -2 & 0 & 0 & 0 \\ -2 & 6 & 4 & 0 & 0 \\ 0 & 4 & 9 & -1.5 & 0 \\ 0 & 0 & -1.5 & 1.25 & 0.5 \\ 0 & 0 & 0 & 0.5 & 3.25 \end{pmatrix} \begin{pmatrix} d_1 \\ d_2 \\ d_3 \\ d_4 \\ d_5 \end{pmatrix} = \begin{pmatrix} 2 \\ -2 \\ 8 \\ 0.5 \\ -3.25 \end{pmatrix}$$

Using the Lagrange multiplier approach, solve for the nodal unknowns if the model requires the following multipoint constraint:

$$d_1 + d_2 = 1$$

1.34 A finite element system results in the following system of global equations:

$$\begin{pmatrix} 2 & -1 & 0 & 0 & 0 & 0 \\ -1 & 2 & -1 & 0 & 0 & 0 \\ 0 & -1 & 3 & -1 & 0 & 0 \\ 0 & 0 & -1 & 3 & -1 & 0 \\ 0 & 0 & 0 & -1 & 2 & -1 \\ 0 & 0 & 0 & 0 & -1 & 2 \end{pmatrix} \begin{pmatrix} d_1 \\ d_2 \\ d_3 \\ d_4 \\ d_5 \\ d_6 \end{pmatrix} = \begin{pmatrix} -1 \\ 2 \\ 0 \\ 0 \\ 0 \\ 0 \end{pmatrix}$$

Using the Lagrange multiplier approach, solve for the nodal unknowns if the model requires the following multipoint constraints:

$$2d_1 - d_6 + 1 = 0; \qquad d_2 + d_6 - 1 = 0$$

1.35 Solve Problem 1.33 using the penalty function approach.

1.36 Solve Problem 1.34 using the penalty function approach.

Computational Projects

1.37 Using a finer mesh and the available finite element software, determine the temperature distribution through a solid with a V-groove as shown in Figure 1.31. Take advantage of symmetry and model only half of the domain.

1.38 Determine the temperature distribution through a solid with a circular groove as shown in Figure 1.38. A hot fluid flowing through the groove keeps the temperature of the circular surface at 150°C . The bottom surface of the solid is in contact with a coolant that maintains it at 0°C. The top and side surfaces are well insulated. The solid is made of two different materials, with thermal conductivities k_1 = 10W/m·°C and k_2 = 20 W/m · °C. Take advantage of symmetry and model only half of the domain.

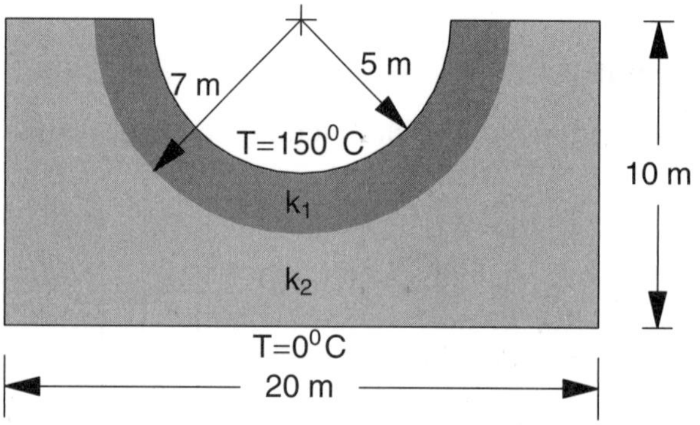

Figure 1.38. Solid with a circular groove

1.39 Heating wires are embedded in a concrete slab, as shown in cross section in Figure 1.39, to keep snow from accumulating on the slab. In a proposed design wires are placed at 2 cm from the top and at 4-cm intervals. The heat generated by each wire is 10 W/m length. The bottom of the slab is insulated. The top is exposed to a convection heat loss with a convection coefficient of h = 30 W/m^2 · °C. The thermal conductivity of concrete is 1.2 W/m · °C. Determine the slab surface temperature when the air temperature is −5°C. Note that, because of symmetry, we need to model

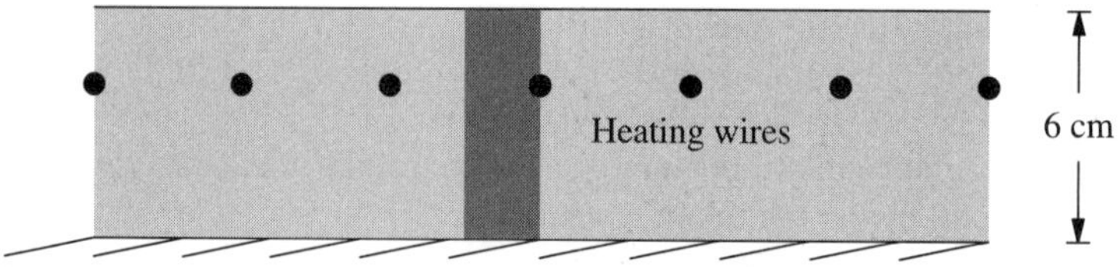

Figure 1.39. Heating wires embedded in concrete

only the 2 cm × 6 cm dark shaded area. There will be no heat flow across the sides of this area. The heating wire represents a concentrated point heat source.

1.40 Steps of a staircase are supported by two identical trusses made of 2-in-nominal-diameter pipes (outside diameter = 2.375 in, wall thickness = 0.154 in). One side truss is shown in Figure 1.40. There are 12 steps each with 7.5 in rise and 11.5 in run. The load from the steps is assumed to be applied as concentrated downward loads of 100 lb at each nodal point. Determine nodal deflections, stresses in each member, and support reactions.

Verify computer results by performing the following calculations by hand:

(a) Use computed reaction forces and applied external forces to check over all equilibrium of forces.

(b) Isolate a typical node in the truss and draw a free-body diagram of the forces acting at that node. Verify that the equilibrium is satisfied at the node.

(c) Pass a vertical section through the truss. Draw the free body diagram of the truss on the right-hand section. Verify that the section is in equilibrium.

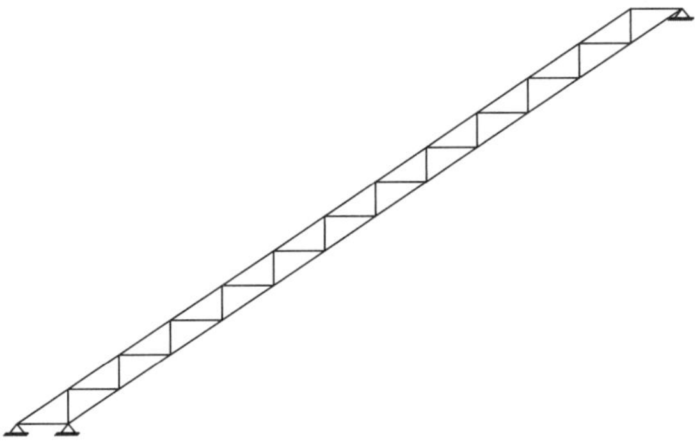

Figure 1.40. Staircase truss

1.41 The lower portion of an aluminum step bracket, shown in Figure 1.41, is subjected to a uniform pressure of 20 N/mm^2. The left end is fixed and the right end is free. The dimensions of the bracket are thickness = 3 mm, L = 150 mm, b = 10 mm,

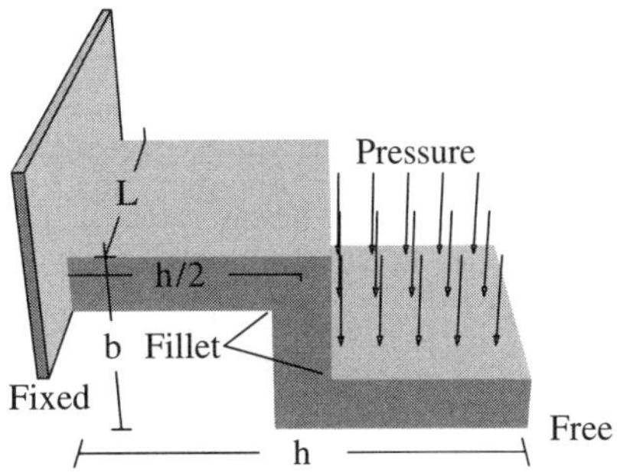

Figure 1.41. Step bracket

and h = 24 mm. Fillets of 3 mm radius are provided at the two reentrant corners. The material properties are E = 70,000 N/mm^2 and ν = 0.3. Determine stresses in the bracket using a planar finite element model (using say Plane42 element of ANSYS). Compare solutions based on plane stress and plane strain assumptions. (The situation can also be modeled using beam/frame elements, three-dimensional solid elements, and plate elements.)

1.42 The top surface of an S-shaped aluminum block, shown in Figure 1.42, is subjected to a uniform pressure of 20 N/mm^2. The bottom is fixed. The dimensions of the block are t = 3 mm, L = 15 mm, b = 10 mm, and h = 24 mm. The material properties are E = 70,000N/mm^2 and ν = 0.3. Determine stresses in the block using a planar finite element model (using say Plane42 element of ANSYS). Compare solutions based on plane stress and plane strain assumptions. (The situation can also be modeled using beam/frame elements, three-dimensional solid elements, and plate elements.)

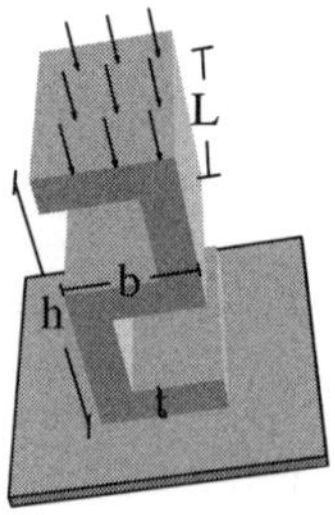

Figure 1.42. S-shaped aluminum block

1.43 *Design optimization.* Aluminum fins are to be used to cool an integrated circuit (IC) chip as shown in Figure 1.43. One side of the chip is insulated. The chip dissipates electrical energy at a rate of 150 W/m. The thermal conductivity of aluminum is 170 W/m · K. The convection coefficient of the fin surface is 55 W/m^2 · K, and the ambient air temperature is 30°C.

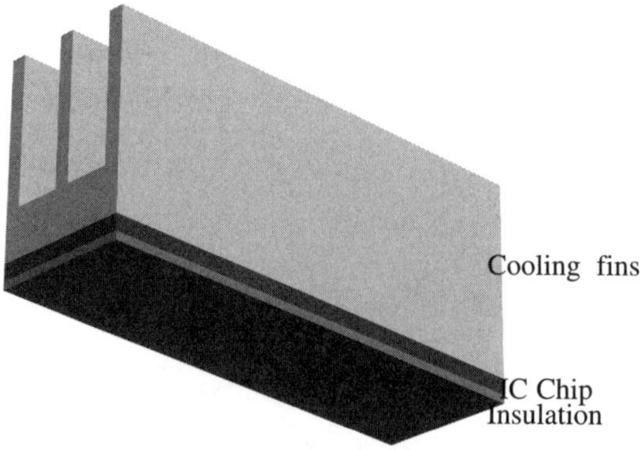

Figure 1.43. Aluminum fins for cooling IC chip

Since the length of the fins is long, we can take a cross section and model as a plane problem. Furthermore, taking advantage of symmetry, we need to model half of the section, as shown in Figure 1.44. The bottom of the fin receives heat flux from the IC chip. Because of symmetry, no heat can flow across the left side. Convection heat loss takes place along the remaining exposed sides.

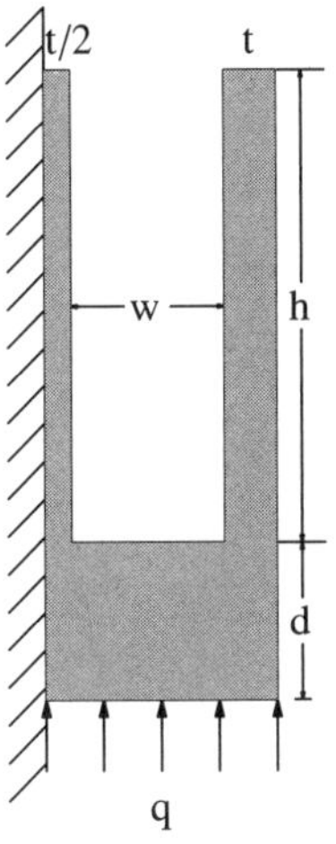

Figure 1.44. Planar model for aluminum fins for cooling IC chip

The dimensions d, w, and t are fixed as follows:

$$d = 3 \text{ mm}; \qquad w = 3 \text{ mm}; \qquad t = 1 \text{ mm}$$

The design goal is to determine the optimum height h of the fins so that the temperature of the IC chip does not exceed 70°C. Assume a starting value of $h = 5$ mm.

CHAPTER TWO

MATHEMATICAL FOUNDATION OF THE FINITE ELEMENT METHOD

From a mathematical point of view the finite element method is a special form of the well-known Galerkin and Rayleigh-Ritz methods for finding approximate solution of differential equations. In both methods the governing differential equation first is converted into an equivalent integral form. The Rayleigh-Ritz method employs calculus of variations to define an equivalent variational or energy functional. A function that minimizes this energy functional represents a solution of the governing differential equation. The Galerkin method uses a more direct approach. An approximate solution, with one or more unknown parameters, is chosen. In general, this assumed solution will not satisfy the differential equation. The integral form represents the *residual* obtained by integrating the error over the solution domain. Employing a criteria adopted to minimize the residual gives equations for finding the unknown parameters.

For most practical problems solutions of differential equations are required to satisfy not only the differential equation but also the specified boundary conditions at one or more points along the boundary of the solution domain. In both methods some of the boundary conditions must be satisfied explicitly by the assumed solutions while others are satisfied implicitly through the minimization process. The boundary conditions are thus divided into two categories, *essential* and *natural*. The essential boundary conditions are those that must explicitly be satisfied while the natural boundary conditions are incorporated into the integral formulation. In general, therefore, the approximate solutions will not satisfy the natural boundary conditions exactly.

The basic concepts will be explained with reference to the problem of axial deformation of bars. The derivation of the governing differential equation is considered in the next section. Approximate solutions using the classical form of Galerkin and Rayleigh-Ritz methods are presented. Finally the methods are cast into the form that is suitable for developing finite element equations.

2.1 AXIAL DEFORMATION OF BARS

2.1.1 Differential Equation for Axial Deformations

Consider a bar of any arbitrary cross section subjected to loads in the axial direction only, as shown in Figure 2.1. The area of cross section is denoted by A and it could vary over the length of the bar. The modulus of elasticity is denoted by E. The bar may be subjected to a distributed axial load $q(x, t)$ along its length. The load can vary over the bar length and may also be a function of time. The axial displacement is denoted by $u(x, t)$.

The governing differential equation can be written by considering equilibrium of a differential element as shown in Figure 2.2. Note the sign convention adopted in drawing the free-body diagram assumes that the tension in the bar is positive. The axial force at x is denoted by F. The axial force at $x + dx$ is $F + (\partial F/\partial x)\, dx$ based on the Taylor series expansion. The acceleration is indicated by $\ddot{u} \equiv \partial^2 u/\partial t^2$. From Newton's second law of motion a force, called the inertia force, is produced that is proportional to the mass of the element and acts in the direction opposite to the direction of motion. Denoting the mass density by ρ, the mass per unit length is $m = A\rho\, dx$ and thus the inertia force is $m\ddot{u} = A\rho\, dx\, \ddot{u}$. The only other force acting on the element is the applied axial load $q(x, t)$. Considering summation of forces in the x direction, we have

$$A(x)\rho\, dx\, \ddot{u}(x, t) + F = q\, dx + F + \frac{\partial F}{\partial x} dx \Longrightarrow A\rho\ddot{u} = q + \frac{\partial F}{\partial x}$$

The axial force is related to the axial stress σ_x as follows:

$$F = A\sigma_x$$

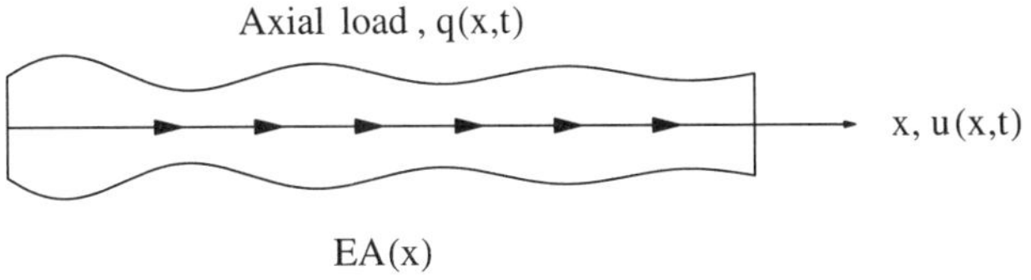

Figure 2.1. Axially loaded bar

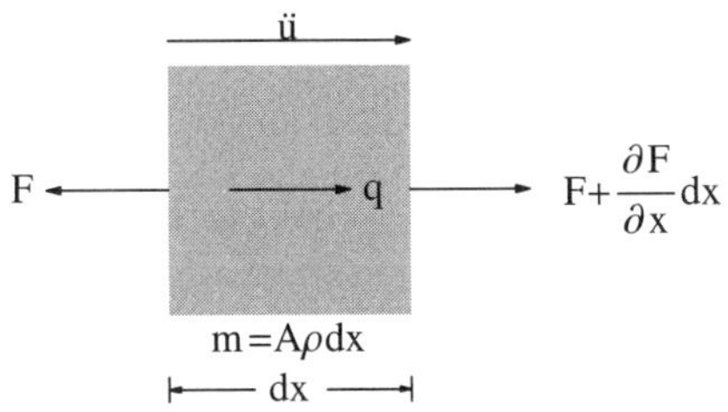

Figure 2.2. Forces acting on a differential element in an axially loaded bar

Assuming linear elastic material, axial stress is related to the axial strain ϵ_x through the modulus of elasticity. Thus

$$F = AE\epsilon_x$$

Assuming small displacements, the axial strain is related to the first derivative of the axial displacement. Thus the axial force is related to displacement as follows:

$$F = AE\frac{\partial u}{\partial x}$$

Substituting this into the equilibrium equation, the governing differential equation for axial deformation of bars is as follows:

$$\frac{\partial}{\partial x}\left(AE\frac{\partial u}{\partial x}\right) + q = A\rho\ddot{u}$$

This is a second-order partial differential equation. Since the equation involves second-order derivatives with respect to both x and t, we need two boundary conditions and two initial conditions for a proper solution. The initial conditions are specified displacement and velocity along the bar at time $t = 0$:

$$u(x, 0) = u_0; \qquad \dot{u}(x, 0) = v_0$$

where u_0 and v_0 are the specified values. The boundary conditions involve specification of displacement or its first derivative at the ends. Since the first derivative of displacement is related to the axial force, the derivative boundary condition is expressed in terms of the applied force. The possible boundary conditions at the right end of the bar $x = x_l$ are as follows:

$$u(x_l, t) = u_{x_l} \qquad \text{or} \qquad A(x_l)E(x_l)\frac{\partial u(x_l, t)}{\partial x} = P_{x_l}(t); \qquad x_l \rightarrow \text{right end of the bar}$$

where u_{x_l} is a specified displacement and P_{x_l} is a specified force. Both these quantities could be functions of time. Care must be exercised in assigning proper signs to force boundary condition terms. Applied forces are considered positive when they act in the positive coordinate direction. From the free-body diagram in Figure 2.2, it should be clear that, if a force is specified at the left end of a bar, then the appropriate force boundary condition must include a negative sign as follows:

$$-A(x_0)E(x_0)\frac{\partial u(x_0, t)}{\partial x} = P_{x_0}(t); \qquad x_0 \rightarrow \text{left end of the bar}$$

If the forces and displacements do not vary with time, we have a static analysis situation and the equilibrium equation is an ordinary second-order differential equation as follows:

$$\frac{d}{dx}\left(AE\frac{du}{dx}\right) + q(x) = 0; \qquad x_0 < x < x_l$$

with the boundary conditions of the form

$$u(x_0) = u_{x_0} \qquad \text{or} \qquad -A(x_0)E(x_0)\frac{du(x_0)}{dx} = P_{x_0}$$

$$u(x_l) = u_{x_l} \qquad \text{or} \qquad A(x_l)E(x_l)\frac{\partial u(x_l)}{\partial x} = P_{x_l}$$

An integration of the second-order differential equation, if possible, would yield two arbitrary constants, and we would need two conditions to evaluate these constants. Thus at least two boundary conditions must be specified for a unique solution. On physical grounds, it is easy to see that we must specify displacement at least at one point along the bar to prevent the whole bar from moving as a rigid body. The second boundary condition could be either of the displacement or the force type. Note that at the same point both a displacement and a force cannot be specified independently. The reason for this is that, if a displacement is specified, then the corresponding force represents a reaction at the point and in general is unknown.

2.1.2 Exact Solutions of Some Axial Deformation Problems

It is possible to directly integrate the differential equation to obtain exact solutions for the axial deformation problems in which the loading and the area of cross section are simple functions of x. Solutions for a uniform bar and a tapered bar subjected to linearly varying load are presented in this section. In later sections these exact solutions will be used to check the quality of the approximate solutions that are obtained using the Galerkin, the Rayleigh-Ritz, and the finite element methods.

Axial Deformation of a Uniform Bar Consider a uniform bar fixed at one end and subjected to a static point load at the other end, as shown in Figure 2.3. The bar is also subjected to a linearly varying axial load $q(x) = cx$, where c is a given constant. The problem is described in terms of the following boundary value problem:

$$EA\frac{d^2u}{dx^2} + cx = 0; \qquad 0 < x < L$$

$$u(0) = 0; \qquad EA\frac{du(L)}{dx} = P$$

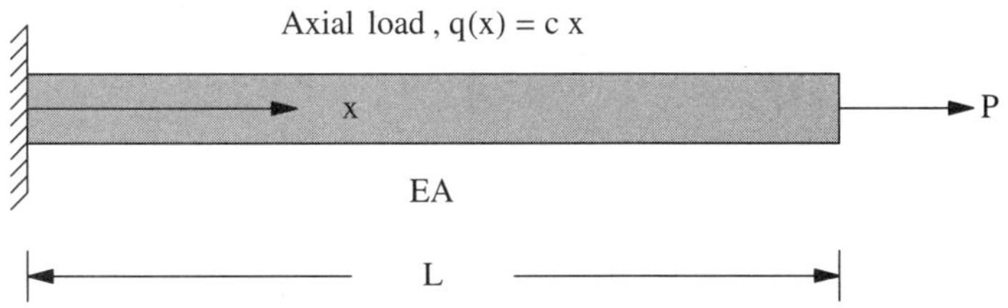

Figure 2.3. Uniform axially loaded bar

An exact solution of the problem can easily be obtained by integrating the differential equation twice and then using the boundary conditions to evaluate the resulting integration constants. Integrating both sides of the differential equation once, we get

$$EA\frac{du}{dx} + \frac{cx^2}{2} = C_1$$

where C_1 is an integration constant. Integrating once again, we get

$$EAu(x) + \frac{cx^3}{6} = C_1x + C_2$$

where C_2 is another integration constant. Rearranging terms,

$$u(x) = \frac{1}{EA}\left(C_1x + C_2 - \frac{cx^3}{6}\right)$$

Using the boundary conditions

$$u(0) = 0 \Longrightarrow C_2 = 0$$

and

$$EA\frac{du(L)}{dx} = P \Longrightarrow \tfrac{1}{6}(-3cL^2 + 6C_1) = P \Longrightarrow C_1 = \tfrac{1}{2}(2P + cL^2)$$

Thus the exact solution of the problem is

$$u(x) = \frac{x(6P + 3cL^2 - cx^2)}{6EA}$$

Axial Deformation of a Tapered Bar Consider a tapered bar fixed at one end and subjected to a static point load at the other end, as shown in Figure 2.4. The bar is also subjected to a linearly varying axial load $q(x) = cx$, where c is a given constant. The problem is described in terms of the following boundary value problem:

$$\frac{d}{dx}\left(EA(x)\frac{du}{dx}\right) + cx = 0; \qquad 0 < x < L$$

$$A(x) = A_0 - \frac{A_0 - A_L}{L}x$$

$$u(0) = 0; \qquad EA_L\frac{du(L)}{dx} = P$$

where A_0 is the area of cross section at $x = 0$ and A_L is that at $x = L$.

Integrating both sides of the differential equation once, we get

$$E\left(A_0 - \frac{A_0 - A_L}{L}x\right)\frac{du}{dx} + \frac{cx^2}{2} = C_1$$

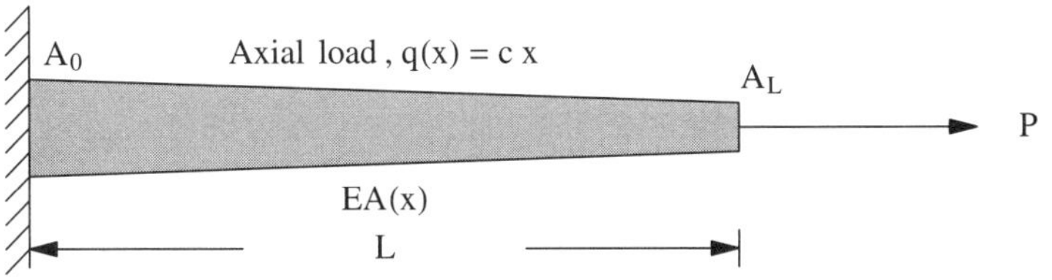

Figure 2.4. Tapered axially loaded bar

where C_1 is an integration constant. Straightforward integration once again is not possible because of the presence of the $x\,du/dx$ term. However, using more advanced techniques for solution of ordinary differential equations, it is possible to obtain the following general solution in terms of two integration constants C_1 and C_2:

$$\begin{aligned} u(x) = {} & \frac{cL^2 x A_0}{2E(A_0 - A_L)^2} + \frac{cLx^2}{4E(A_0 - A_L)} \\ & + \frac{\log[-LA_0 + xA_0 - xA_L](cL^3 A_0^2 + 2EA_0^2 C_1 - 4EA_0 A_L C_1 + 2EA_L^2 C_1)}{2E(A_0 - A_L)^3} + C_2 \end{aligned}$$

Using the boundary conditions, the integration constants are evaluated as follows:

$$C_1 = -\frac{L(2P + cL^2)}{2E}$$

$$C_2 = -\frac{cL^3 \log[-LA_0] A_0^2}{2E(A_0 - A_L)^3} + \frac{L(2P + cL^2)\log[-LA_0](2EA_0^2 - 4EA_0A_L + 2EA_L^2)}{4E^2(A_0 - A_L)^3}$$

The exact solution of the problem can now be written as follows:

$$\begin{aligned} u(x) = {} & -\frac{1}{4(-1 + r)^3 EA_0}\Big(L\big(c(-1 + r)x(-2L + (-1 + r)x) \\ & + 2(2P(-1 + r)^2 + cL^2(-2 + r)r)\log[L] \\ & - 2(2P(-1 + r)^2 + cL^2(-2 + r)r)\log[L + (-1 + r)x]\big)\Big) \end{aligned}$$

where $r = A_L/A_0$, the ratio of areas at two ends of the bar.

The only difference between this example and the previous one is that here the area of cross section of the bar varies linearly. The solution is much more difficult to obtain and the final solution expression is quite complicated as well. This example demonstrates how quickly analytical solutions become difficult to obtain and shows the importance of numerical methods for practical problems.

As should be expected, by taking the limit as $A_0 \to A_L \equiv A$, i.e., $r \to 1$, it can be shown that this solution reduces to the one for a uniform bar presented in the previous section. Also note that, when $c = 0$ (no distributed axial load), the axial force at any point in the bar is equal to P, which obviously should be the case from statics.

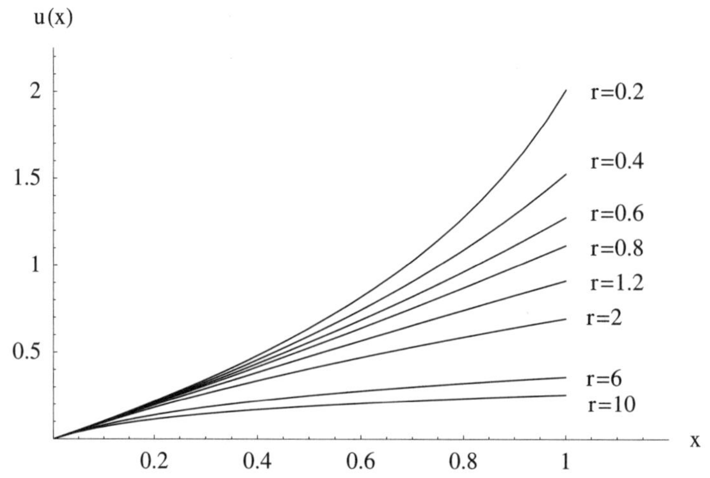

Figure 2.5. Solutions of tapered axially loaded bar subjected to end load

Solutions for several values of r are plotted in Figure 2.5 with the following numerical values:

$$c = 0; \qquad P = 1; \qquad A_0 = 1; \qquad L = 1; \qquad E = 1$$

2.2 AXIAL DEFORMATION OF BARS USING GALERKIN METHOD

The governing differential equation for axial deformation of bars is the following second-order differential equation:

$$\frac{d}{dx}\left(AE\frac{du}{dx}\right) + q = 0; \qquad x_0 < x < x_l$$

where x_0 and x_l are the coordinates of the ends of the bar. The primary unknown is the axial displacement $u(x)$. Once the displacement is known, axial strain, stress, and force can be computed from the following relationships:

$$\epsilon_x = \frac{du}{dx}; \qquad \sigma_x = E\epsilon_x; \qquad F = A\sigma_x$$

At the ends either a displacement or an axial force can be specified. Thus the boundary conditions for the problem are of the following form:

$$u(x_0) = u_{x_0} \qquad \text{or} \qquad -A(x_0)E(x_0)\frac{du(x_0)}{dx} = P_{x_0}$$

$$u(x_l) = u_{x_l} \qquad \text{or} \qquad A(x_l)E(x_l)\frac{\partial u(x_l)}{\partial x} = P_{x_l}$$

where $u_{x0}, P_{x0}, \ldots$ are appropriate specified values. Mathematically speaking, for a second-order differential equation either u is specified or its first derivative du/dx is specified. Both u and du/dx cannot have specified values at the same point.

2.2.1 Weak Form for Axial Deformations

In the Galerkin method we assume a general form of the solution. This assumed solution must contain some unknown parameters whose values are determined so that the error between the assumed solution and the exact solution is as small as possible. The assumed solution can be of any form. As an example, we will consider a solution in the form of a polynomial:

$$\tilde{u}(x) = a_0 + a_1 x + a_2 x^2 + \cdots + a_n x^n$$

where $a_0, a_1, \ldots$ are the unknown parameters. In general, this assumed solution will not satisfy the differential equation for all values of x. When this assumed solution is substituted into the differential equation, the error in satisfying the governing differential equation is

$$e(x) = \frac{d}{dx}\left(AE\frac{d\tilde{u}}{dx}\right) + q(x) \neq 0$$

The tilde (~) over u is used to emphasize that it is an assumed solution and not necessarily the same as the exact solution of the differential equation. In the following development we will have to perform several mathematical operations, such as differentiation and integration, on the assumed solution. Carrying the tilde symbol through all these manipulations becomes tedious. Since we are dealing primarily with the approximate solutions, for notational convenience, we will drop the tilde and use only u instead to indicate an approximate solution. An occasional reference to the exact solution will be indicated by using the word *exact* explicitly.

The total error, called the residual, for the entire solution domain can be obtained by integrating $e(x)$ over the domain. However, in a straight integration of e, the negative and positive errors at different points may cancel each other. To avoid error cancellation, $e(x)$ is multiplied by a suitable weighting function and then integrated over the solution domain. Since there are n unknown parameters, we need n weighting functions to establish weighted residual equations as follows:

$$\int_{x_0}^{x_l} e(x) w_i(x)\, dx = 0; \qquad i = 0, 1, \ldots, n$$

where $w_i(x)$ are suitable weighting functions. This form is known as the *weak form*. The differential equation is the *strong form* because it requires error term $e(x)$ to vanish at every x. The integral form makes the total error go to zero but does not necessarily satisfy the governing differential equation for all values of x.

In a method known as the *least-squared weighted residual method* the error term is squared to define the total squared error as follows:

$$e_T = \int_{x_0}^{x_l} e^2\, dx$$

The necessary conditions for the minimum of the total squared error give n equations that can be solved for the unknown parameters:

$$\frac{\partial e_T}{\partial a_i} = 2\int_{x_0}^{x_l} e\frac{\partial e}{\partial a_i} dx = 0; \qquad i = 0, 1, \ldots$$

Thus in the least-squares method the weighting functions are the partial derivatives of the error term $e(x)$.

> Least-squares weighting functions:
>
> $$w_i(x) = \frac{\partial e}{\partial a_i}; \qquad i = 0, 1, \ldots, n$$

A more popular method in the finite element applications is the *Galerkin method.* In this method, instead of taking partial derivatives of the error function, the weighting functions are defined as the partial derivatives of the assumed solution.

> Galerkin weighting functions:
>
> $$w_i(x) = \frac{\partial u}{\partial a_i}; \qquad i = 0, 1, \ldots, n$$

Thus the Galerkin weighted residual method defines the following n equations to solve for the unknown parameters:

$$\int_{x_0}^{x_l} e\frac{\partial u}{\partial a_i} dx = 0; \qquad i = 0, 1, \ldots$$

It turns out that for a large number of engineering applications the Galerkin method gives the same solution as another popular method, the Rayleigh-Ritz method, presented in a later section. Furthermore, since the least-squares method has no particular advantage over the Galerkin method for the kinds of problems discussed in this book, only the Galerkin method is presented in detail.

So far in developing the residual we have considered error in satisfying the differential equation alone. A solution must also satisfy boundary conditions. In order to be able to introduce the boundary conditions into the weighted residual, we use mathematical manipulations involving integration by parts.

The *integration-by-parts* formula is used to rewrite an integral of a product of a derivative of a function, say $f(x)$, and another function, say $g(x)$, as follows:

$$\int_{x_0}^{x_l}\left[\frac{d}{dx}(f(x))\right]g(x)\,dx = f(x_l)g(x_l) - f(x_0)g(x_0) - \int_{x_0}^{x_l}\left[\frac{d}{dx}(g(x))\right]f(x)\,dx$$

Note that the integrand must involve the product of a function and the derivative of another function. The application of the formula produces two terms that are evaluated at the ends of integration domain and another integral in which the derivative shifts from one function to the other.

For a second-order differential equation, we know that the boundary conditions specify either u or du/dx at the ends. Thus we are looking for a way of introducing these terms into the weighted residual. Since the integration-by-parts formula gives rise to boundary terms, it is precisely the tool that we need. Note that the highest derivative term in the residual $e(x)$ involves a second derivative on u. Using integration by parts on this term will result in two terms evaluated at the integration limits that are used to incorporate the boundary conditions into the residual.

For the axial deformation problem, the weighted residual is

$$\int_{x_0}^{x_l} e(x)w_i(x)\,dx = \int_{x_0}^{x_l}\left(\frac{d}{dx}\left(AE\frac{du}{dx}\right) + q(x)\right)w_i(x)\,dx = 0; \qquad i = 0, 1, \ldots$$

Note that, in general, A and E could be functions of x, and therefore, care must be taken when carrying out differentiation and integration. Also $q(x)$ and $w_i(x)$ are arbitrary functions of x at this stage and cannot be taken out of the integral. Writing the two terms in the weighted residual as separate integrals, we have

$$\int_{x_0}^{x_l}\frac{d}{dx}\left(AE\frac{du}{dx}\right)w_i(x)\,dx + \int_{x_0}^{x_l} q(x)w_i(x)\,dx = 0; \qquad i = 0, 1, \ldots$$

The first integral contains the second-order derivative on u and is written exactly in the form to which integration by parts is applicable with $f(x) = AE(du/dx)$ and $g(x) = w_i(x)$. Thus the application of integration by parts to the first integral gives

$$A(x_l)E(x_l)\frac{du(x_l)}{dx}w_i(x_l) - A(x_0)E(x_0)\frac{du(x_0)}{dx}w_i(x_0) - \int_{x_0}^{x_l}AE\frac{du}{dx}\frac{dw_i}{dx}dx + \int_{x_0}^{x_l}q(x)w_i(x)\,dx = 0$$

The first two terms in the weak form give us a way to incorporate the specified force or derivative boundary conditions into the weak form. If a force P_{x0} is applied at end x_0, then

$$-A(x_0)E(x_0)\frac{du(x_0)}{dx} = P_{x_0} \Longrightarrow \text{Second term in the weak form: } P_{x_0}w_i(x_0)$$

If a force P_{xl} is applied at end x_l, then

$$A(x_l)E(x_l)\frac{du(x_l)}{dx} = P_{x_l} \Longrightarrow \text{First term in the weak form: } P_{x_l}w_i(x_l)$$

Thus, when a force is specified, the boundary condition can be *naturally* incorporated into the weak form. Hence this type of boundary condition is called a *natural boundary condition* (NBC).

The situation is very different if the u is specified at one or both ends. The derivative du/dx at the corresponding point is unknown. (Physically $AE\, du/dx$ at the point represents the unknown reaction.) The specified displacement boundary condition therefore cannot be incorporated into the weak form directly. The assumed solutions must satisfy this type of boundary condition explicitly, and thus such a boundary condition is called the *essential boundary condition* (EBC). An assumed solution that satisfies the EBC is known as an *admissible solution*. Since the weighting functions are partial derivatives of the assumed solution, with an admissible assumed solution, all weighting functions corresponding to the location of an EBC are zero. Therefore, the boundary term in the weak form vanishes at the point where an EBC is specified. Thus we must deal with the boundary terms as follows:

Boundary Condition Term	Specified Value	Boundary Term in the Weak Form	Requirement on the Assumed Solution	Type
$-A(x_0)E(x_0)\dfrac{du(x_0)}{dx}$	P_{x_0}	$P_{x_0} w_i(x_0)$	None	NBC
$A(x_l)E(x_l)\dfrac{du(x_l)}{dx}$	P_{x_l}	$P_{x_l} w_i(x_l)$	None	NBC
$u(x_0)$	u_{x_0}	None	Must satisfy	EBC
$u(x_l)$	u_{x_l}	None	Must satisfy	EBC

For structural problems, the weak form can be interpreted as the well-known *principle of virtual displacements*. To see this, assume we have specified force boundary conditions at the ends. Then the weak form can be written as follows:

$$P_{x_l} w_i(x_l) + P_{x_0} w_i(x_0) + \int_{x_0}^{x_l} \left(-AE\frac{du}{dx}\frac{dw_i}{dx} + q(x)w_i(x) \right) dx = 0$$

Rearranging terms, we have

$$\int_{x_0}^{x_l} E\frac{du}{dx}\frac{dw_i}{dx}A\, dx = \int_{x_0}^{x_l} q(x)w_i(x)\, dx + P_{x_l} w_i(x_l) + P_{x_0} w_i(x_0)$$

If we interpret $w_i(x)$ as a virtual displacement, then the right-hand side is the virtual work done by the applied forces. The left-hand side is the total internal virtual work, since $E\, du/dx \equiv \sigma_x$ is the axial stress and dw_i/dx is axial virtual strain. Thus the weak form implies that when a bar is given a virtual displacement, then the external virtual work is equal to the total internal virtual work, which is a statement of the principle of virtual displacements. Since this principle is widely used in structural mechanics, it is one of the main reasons why the Galerkin weighted residual method is more popular in developing finite element equations. Also recall that the virtual displacements are required to satisfy the displacement (i.e., essential) boundary conditions.

From the final weak form we note that another advantage of the integration by parts is that the order of the derivative on the terms remaining inside the integral sign is reduced by 1. Thus for a second-order problem the weak form involves only first-order derivatives. It may not appear to be a big deal here, but it has important consequences in developing simple finite elements for practical problems. It will be seen in a later example that, when dealing with a fourth-order problem, integration by parts must be carried out twice to reduce the highest order derivative to 2.

2.2.2 Uniform Bar Subjected to Linearly Varying Axial Load

We now use the Galerkin method to find approximate solutions for a uniform bar (EA constant) fixed at one end and subjected to a static point load at the other end, as shown in Figure 2.3. The bar is also subjected to a linearly varying axial load $q(x) = cx$, where c is a given constant. The problem is described in terms of the following boundary value problem:

$$EA\frac{d^2u}{dx^2} + cx = 0; \qquad 0 < x < L$$

$$u(0) = 0; \qquad EA\frac{du(L)}{dx} = P$$

The following exact solution of the problem was obtained in Section 2.1:

$$u(x) = \frac{x(6P + 3cL^2 - cx^2)}{6EA}$$

With the solution domain from $(0, L)$, the EA constant, $q(x) = cx$, the essential boundary condition $u(0) = 0 \Rightarrow w(0) = 0$, and the natural boundary condition $EAu'(L) = P$, the weak form specific to this problem is as follows:

$$Pw_i(L) + \int_0^L \left(-AE\frac{du}{dx}\frac{dw_i}{dx} + cxw_i(x)\right) dx = 0$$

$$\text{EBC:} \qquad u(0) = 0$$

This weak form can now be used to find a variety of approximate solutions to the problem.

Linear Solution The simplest possible solution that we can assume is a linear polynomial. An approximate solution of the problem is obtained using the following starting solution:

$$u(x) = a_0 + xa_1$$

To satisfy the essential boundary condition, we must have

$$u(0) = 0 \Longrightarrow a_0 + 0a_1 = 0 \Longrightarrow a_0 = 0$$

Thus the admissible assumed solution is

$$u(x) = xa_1 \qquad \text{giving} \qquad \frac{du}{dx} = a_1$$

Substituting into the weak form, we have

$$Pw_i(L) + \int_0^L \left(-AE(a_1)\frac{dw_i}{dx} + cxw_i\right) dx = 0$$

There is only one unknown parameter left in the solution and therefore we need only one equation to find it. The Galerkin weighting function is

$$w_1 = \frac{\partial u}{\partial a_1} = x; \qquad \frac{\partial w_1}{\partial x} = 1; \qquad w_1(0) = 0; \qquad w_1(L) = L$$

Substituting this into the weak form, we have

$$PL + \int_0^L (-AEa_1 + cx^2)\, dx = 0$$

Carrying out integration and simplifying, we get

$$\frac{cL^3}{3} + PL - EAa_1L = 0$$

Solving this equation for a_1, we get

$$a_1 = -\frac{-cL^2 - 3P}{3EA}$$

Thus a linear approximate solution for the problem is as follows:

$$u(x) = a_1x = \frac{(cL^2 + 3P)x}{3EA}$$

Quadratic Solution A better solution can be obtained if we start with a quadratic polynomial:

$$u(x) = a_2x^2 + a_1x + a_0$$

To satisfy the essential boundary condition, we must have

$$u(0) = 0 \Longrightarrow a_2 0 + a_1 0 + a_0 = 0 \Longrightarrow a_0 = 0$$

Thus the admissible assumed solution is

$$u(x) = a_2x^2 + a_1x; \qquad \frac{du}{dx} = 2a_2x + a_1$$

Substituting into the weak form, we have

$$Pw_i(L) + \int_0^L \left(-AE(2a_2x + a_1)\frac{dw_i}{dx} + cxw_i\right) dx = 0$$

There are two unknown parameters left in the solution and therefore we need two equations to find them. We get these equations by using the two Galerkin weighting functions

$$w_1 = \frac{\partial u}{\partial a_1} = x; \qquad \frac{\partial w_1}{\partial x} = 1; \qquad w_1(0) = 0; \qquad w_1(L) = L$$

$$w_2 = \frac{\partial u}{\partial a_2} = x^2; \qquad \frac{\partial w_2}{\partial x} = 2x; \qquad w_2(0) = 0; \qquad w_2(L) = L^2$$

Substituting w_1 into the weak form, we have

$$PL + \int_0^L (-AE(2a_2x + a_1) + cx^2)\, dx = 0 \Longrightarrow \frac{cL^3}{3} - EAa_2L^2 + PL - EAa_1L = 0$$

Substituting w_2 into the weak form, we have

$$PL^2 + \int_0^L (-AE(2a_2x + a_1)(2x) + cx^3)\, dx = 0 \Longrightarrow \frac{cL^4}{4} - \frac{4}{3}EAa_2L^3 + PL^2 - EAa_1L^2 = 0$$

Solving the two equations, we have

$$a_1 = -\frac{-7cL^2 - 12P}{12EA}; \qquad a_2 = -\frac{cL}{4EA}$$

Thus a quadratic approximate solution is as follows:

$$u(x) = a_2x^2 + a_1x = -\frac{cL}{4EA}x^2 - \frac{-7cL^2 - 12P}{12EA}x = \frac{(12P + cL(7L - 3x))x}{12EA}$$

Cubic Solution The exact solution of the problem is a cubic polynomial. As demonstrated below, the Galerkin method finds an exact solution if we start with a cubic polynomial:

$$u(x) = a_3x^3 + a_2x^2 + a_1x + a_0$$

To satisfy the essential boundary condition, we must have

$$u(0) = 0 \Longrightarrow a_3 0 + a_2 0 + a_1 0 + a_0 = 0 \Longrightarrow a_0 = 0$$

Thus the admissible assumed solution is

$$u(x) = a_3x^3 + a_2x^2 + a_1x; \qquad \frac{du}{dx} = 3a_3x^2 + 2a_2x + a_1$$

Substituting into the weak form, we have

$$Pw_i(L) + \int_0^L \left(-AE(3a_3x^2 + 2a_2x + a_1)\frac{dw_i}{dx} + cxw_i\right) dx = 0$$

There are three unknown parameters left, which we find by using the three Galerkin weighting functions into the weak form, giving the following equations:

$$w_1 = \frac{\partial u}{\partial a_1} = x; \quad \frac{\partial w_1}{\partial x} = 1; \quad w_1(0) = 0; \quad w_1(L) = L$$

$$w_2 = \frac{\partial u}{\partial a_2} = x^2; \quad \frac{\partial w_2}{\partial x} = 2x; \quad w_2(0) = 0; \quad w_2(L) = L^2$$

$$w_3 = \frac{\partial u}{\partial a_3} = x^3; \quad \frac{\partial w_3}{\partial x} = 3x^2; \quad w_3(0) = 0; \quad w_3(L) = L^3$$

Substitute admissible solution and weights into the weak form and perform integrations to get the following:

	Weight	Equation
1	x	$\frac{1}{3}(c - 3EAa_3)L^3 - EAa_2L^2 + PL - EAa_1L = 0$
2	x^2	$\frac{1}{4}(c - 6EAa_3)L^4 - \frac{4}{3}EAa_2L^3 + PL^2 - EAa_1L^2 = 0$
3	x^3	$\frac{1}{5}(c - 9EAa_3)L^5 - \frac{3}{2}EAa_2L^4 + PL^3 - EAa_1L^3 = 0$

Solving these equations,

$$a_1 = -\frac{-cL^2 - 2P}{2EA}, \qquad a_2 = 0, \qquad a_3 = -\frac{c}{6EA}$$

Thus a cubic solution is as follows:

$$u(x) = a_3x^3 + a_2x^2 + a_1x = -\frac{c}{6EA}x^3 - \frac{-cL^2 - 2P}{2EA}x = \frac{x(3cL^2 - cx^2 + 6P)}{6EA}$$

Since this is the exact solution, trying any higher order polynomial will not make any difference.

Carefully note the distinction between the two types of boundary conditions. To make the assumed solution admissible, we required it to satisfy only the displacement boundary condition $[u(0) = 0]$. We never explicitly require the assumed solution to satisfy the force boundary condition. We can easily see that for this example the linear and quadratic solutions in fact do not satisfy this boundary condition. Only the cubic solution satisfies this condition exactly.

	$u(x)$	$\partial u/\partial x$	$\partial u(L)/\partial x$	$EA\, \partial u(L)/\partial x$
Linear	$\frac{(cL^2 + 3P)x}{3EA}$	$\frac{cL^2 + 3P}{3EA}$	$\frac{cL^2 + 3P}{3EA}$	$\frac{cL^2}{3} + P$
Quadratic	$\frac{(12P + cL(7L - 3x))x}{12EA}$	$\frac{12P + cL(7L - 3x)}{12EA} - \frac{cLx}{4EA}$	$\frac{4cL^2 + 12P}{12EA} - \frac{cL^2}{4EA}$	$\frac{cL^2}{12} + P$
Cubic	$\frac{x(3cL^2 - cx^2 + 6P)}{6EA}$	$\frac{3cL^2 - cx^2 + 6P}{6EA} - \frac{cx^2}{3EA}$	$\frac{2cL^2 + 6P}{6EA} - \frac{cL^2}{3EA}$	P

The logic in using the terminology *essential* and *natural* boundary conditions should now be clear. Essential boundary conditions are those that must explicitly be satisfied by the assumed solution while the natural boundary conditions are only implicitly satisfied through the weak form. A solution that satisfies the differential equation and all boundary conditions is obviously the exact solution.

2.2.3 Tapered Bar Subjected to Linearly Varying Axial Load

Consider now the tapered bar fixed at one end and subjected to a static point load at the other end, as shown in Figure 2.4. The bar is also subjected to a linearly varying axial load $q(x) = cx$, where c is a given constant. The problem is described in terms of the following boundary value problem:

$$\frac{d}{dx}\left(EA\frac{du}{dx}\right) + cx = 0; \quad 0 < x < L$$

$$A(x) = A_0 - \frac{A_0 - A_L}{L}x = \frac{(L + (-1 + r)x)A_0}{L}$$

$$u(0) = 0; \qquad ErA_0\frac{du(L)}{dx} = P$$

where A_0 is the area of cross section at $x = 0$, A_L is that at $x = L$, and $r = A_L/A_0$. The weak form specific to this problem is the same as that for the uniform bar in the previous section, except that A is now a function of x:

$$Pw_i(L) + \int_0^L \left(-AE\frac{du}{dx}\frac{dw_i}{dx} + cxw_i(x)\right)dx = 0$$

$$\text{EBC:} \qquad u(0) = 0$$

Linear Solution Starting assumed solution: $u(x) = a_0 + xa_1$

The admissible solution must satisfy EBC:

EBC	Equation
$u(0) = 0$	$a_0 = 0$

Thus the admissible assumed solution is $u(x) = xa_1$.

Weighting function $\rightarrow \{x\}$

Substitute into the weak form and perform integrations to get:

Weight	Equation
x	$LP + \dfrac{cL^4/3 - (1/2)rEa_1A_0L^2 - (1/2)Ea_1A_0L^2}{L} = 0$

Solving this equation,

$$a_1 = \frac{-(cL^3/3) - PL}{-(1/2)LEA_0 - (1/2)LrEA_0}$$

Substituting into the admissible solution, we get the following solution of the problem:

$$u(x) = \frac{2cxL^2 + 6Px}{3rEA_0 + 3EA_0}$$

Quadratic Solution Starting assumed solution: $u(x) = a_2x^2 + a_1x + a_0$

The admissible solution must satisfy the EBC:

EBC	Equation
$u(0) = 0$	$a_0 = 0$

Thus the admissible assumed solution is $u(x) = a_2x^2 + a_1x$.
Weighting functions $\to \{x, x^2\}$
Substitute into the weak form and perform integrations to get:

Weight	Equation
x	$LP + \frac{(cL^4/3) - (2/3)rEa_2A_0L^3 - (1/3)Ea_2A_0L^3 - (1/2)rEa_1A_0L^2 - (1/2)Ea_1A_0L^2}{L} = 0$
x^2	$PL^2 + \frac{(cL^5/4) - rEa_2A_0L^4 - (1/3)Ea_2A_0L^4 - (2/3)rEa_1A_0L^3 - (1/3)Ea_1A_0L^3}{L} = 0$

Solving these equations,

$$a_1 = -\frac{-cL^2 - 6crL^2 - 12Pr}{2(r^2 + 4r + 1)EA_0}; \qquad a_2 = -\frac{-cL^2 + 7crL^2 - 12P + 12Pr}{4L(r^2 + 4r + 1)EA_0}$$

Substituting into the admissible solution, we get the following solution of the problem:

$$u(x) = \frac{x(c(2L(6r + 1) - 7rx + x)L^2 + 12P(2Lr - xr + x))}{4L(r^2 + 4r + 1)EA_0}$$

The exact solution of the problem was derived earlier. The linear and quadratic solutions are compared with the exact solution in Figure 2.6. The following numerical values are used:

$$c = 0; \quad P = 1; \quad A_0 = 1; \quad L = 1; \quad E = 1; \quad r = \tfrac{1}{2}$$

The quadratic solution is reasonably close to the exact solution. Of course the solution can be improved further by starting with a cubic or a higher-order polynomial.

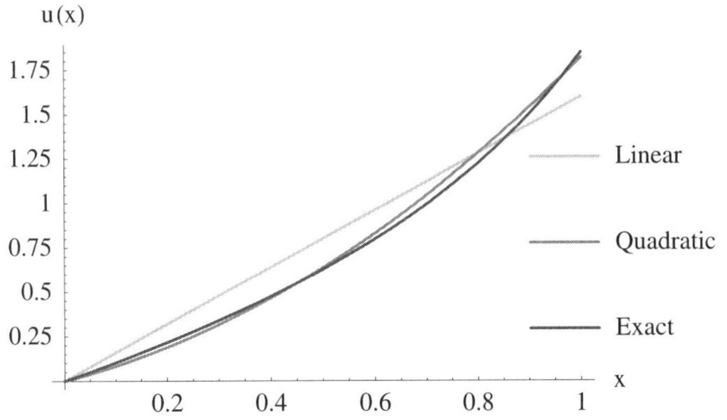

Figure 2.6. Solutions of tapered axially loaded bar subjected to end load

2.3 ONE-DIMENSIONAL BVP USING GALERKIN METHOD

So far only the axial deformation problem has been used to illustrate the Galerkin method. The method in fact is applicable to any differential equation. It is particularly well suited to boundary value problems (BVPs) in which a solution must satisfy differential equations and several boundary conditions over the domain. This section first summarizes the overall solution procedure and then presents several examples of finding approximate solutions of one-dimensional boundary value problems.

2.3.1 Overall Solution Procedure Using Galerkin Method

Given a boundary value problem, the overall procedure for obtaining an approximate solution using the Galerkin method is summarized here. Several examples in this section further clarify the details.

(i) Construct the weak form.

1. Move all terms in the differential equation to the left-hand side. With an assumed solution the left-hand side now represents the residual $e(x)$.
2. Define the total weighted residual. It consists of integral of the left-hand side of the differential equation multiplied by a weighting function w_i.
3. Use integration by parts to incorporate boundary conditions. For a second-order problem, all terms involving second-order derivatives are integrated by parts once. The highest order of remaining terms in the residual should therefore be 1. For a fourth-order problem integration by parts is used twice on all terms involving fourth-order derivatives and once on the terms involving third-order derivatives. The highest order of remaining terms in the residual should therefore be 2.
4. Identify essential and natural boundary conditions. For second-order problems boundary conditions involving first-order derivatives are natural because they

can be incorporated into the weak form. For fourth-order problems boundary conditions involving second- and third-order derivatives are natural. All other boundary conditions are essential. Incorporate natural boundary conditions into the weak form.

(ii) Construct an admissible assumed solution.

1. Start with an assumed solution with some unknown parameters and enough terms in it so that it is possible to evaluate the residual. Any suitable function can be used for the assumed solution. However, since polynomials are easy to differentiate and integrate, it is convenient to start with a polynomial of desired order.
2. Set some of the parameter values such that the essential boundary conditions are satisfied regardless of the values of the remaining unknown parameters. If there are no essential boundary conditions, then the solution already is an admissible solution and we can go to the next step. Otherwise set up equations to satisfy essential boundary conditions. Solve these equations for some of the parameters. Substitute these parameter values back into the starting solution to get an admissible assumed solution.

(iii) Set up equations and solve for the unknown parameters.

1. Determine a set of weighting functions by differentiating the admissible solution with respect to the unknown parameters.
2. Set up a system of equations by substituting each weighting function in turn into the weak form.
3. Solve the system of equations for the unknown parameters.

(iv) Approximate solution.

1. Substitute computed values of the parameters into the admissible solution to get the approximate solution.
2. Check the quality of the solution by substituting it into the differential equation and the natural boundary conditions to see how well it satisfies them.

For convenience the *prime* and superscript notations for derivatives will be used frequently in the following examples. Thus the first and second derivatives of a function $u(x)$ will be denoted as $u'(x) \equiv du(x)/dx$ and $u''(x) \equiv d^2u/dx^2$. When third- or higher order derivatives are written, the prime notation becomes cumbersome and therefore a superscript notation will be adopted. For example, the third and fourth derivatives of a function $u(x)$ will be denoted as $u^{(3)}(x) \equiv d^3u/dx^3$ and $u^{(4)}(x) \equiv d^4u/dx^4$.

Example 2.1 Second-Order Equation Find an approximate solution of the following boundary value problem:

$$x^2u'' + 2xu' + x = 1; \qquad 1 < x < 2$$

$$u(1) = 2; \qquad u'(2) + 2u(2) = 5$$

The weak form is derived first followed by detailed calculations of quadratic and cubic approximate solutions.

With $u(x)$ as an assumed solution, the residual is

$$e(x) = u''(x)x^2 + 2u'(x)x + x - 1$$

Multiplying by $w_i(x)$ and writing integral over the given limits, the Galerkin weighted residual is

$$\int_1^2 (w_i u'' x^2 + w_i x + 2w_i u' x - w_i)\, dx = 0$$

Using integration by parts, the order of derivative in $x^2 w_i u''$ can be reduced to 1 as follows:

$$\int_1^2 (x^2 w_i u'')\, dx = 4w_i(2)u'(2) - w_i(1)u'(1) + \int_1^2 (-xu'(2w_i + xw_i'))\, dx$$

Combining all terms, the weighted residual now is as follows:

$$4w_i(2)u'(2) - w_i(1)u'(1) + \int_1^2 ((x-1)w_i - x^2 u' w_i')\, dx = 0$$

Consider the boundary terms

$$4w_i(2)u'(2) - w_i(1)u'(1)$$

Given the NBC for the problem:

$2u(2) + u'(2) - 5 = 0$

Rearranging: $(u'(2) \to 5 - 2u(2))$

Given the EBC for the problem:

$u(1) - 2 = 0$

Therefore with admissible solutions (those satisfying the EBC): $(w_i(1) \to 0)$

Thus the boundary terms in the weak form reduce to

$$4(5 - 2u(2))w_i(2)$$

and assuming admissible solutions, the final weak form is as follows:

$$4(5 - 2u(2))w_i(2) + \int_1^2 ((x-1)w_i - x^2 u' w_i')dx = 0$$

EBC: $u(1) - 2 = 0$

Quadratic Solution Starting assumed solution: $u(x) = a_2x^2 + a_1x + a_0$

The admissible solution must satisfy the EBC:

EBC	Equation
$u(1) - 2 = 0$	$a_0 + a_1 + a_2 - 2 = 0$

Solving this equation, $a_0 = -a_1 - a_2 + 2$
Thus the admissible assumed solution is $u(x) = a_2x^2 + a_1x - a_1 - a_2 + 2$.
Weighting functions → $\{x - 1, x^2 - 1\}$
Substitute into the weak form and perform integrations to get:

Weight	Equation
$x - 1$	$-\frac{7a_1}{3} - \frac{15a_2}{2} + 4(5 - 2(a_1 + 3a_2 + 2)) + \frac{1}{3} = 0$
$x^2 - 1$	$-\frac{15a_1}{2} - \frac{124a_2}{5} + 12(5 - 2(a_1 + 3a_2 + 2)) + \frac{11}{12} = 0$

Solving these equations, $a_1 = \frac{1511}{962}$; $a_2 = -\frac{545}{1443}$
Substituting into the admissible solution, we get the following solution of the problem:

$$u(x) = \frac{-1090x^2 + 4533x + 2329}{2886}$$

Cubic Solution Starting assumed solution: $u(x) = a_3x^3 + a_2x^2 + a_1x + a_0$

The admissible solution must satisfy the EBC:

EBC	Equation
$u(1) - 2 = 0$	$a_0 + a_1 + a_2 + a_3 - 2 = 0$

Solving this equation, $a_0 = -a_1 - a_2 - a_3 + 2$
Thus the admissible assumed solution is $u(x) = a_3x^3 + a_2x^2 + a_1x - a_1 - a_2 - a_3 + 2$.
Weighting functions → $\{x - 1, x^2 - 1, x^3 - 1\}$
Substitute into the weak form and perform integrations to get:

Weight	Equation
$x - 1$	$-\frac{7a_1}{3} - \frac{15a_2}{2} - \frac{93a_3}{5} + 4(5 - 2(a_1 + 3a_2 + 7a_3 + 2)) + \frac{1}{3} = 0$
$x^2 - 1$	$-\frac{15a_1}{2} - \frac{124a_2}{5} - 63a_3 + 12(5 - 2(a_1 + 3a_2 + 7a_3 + 2)) + \frac{11}{12} = 0$
$x^3 - 1$	$-\frac{93a_1}{5} - 63a_2 - \frac{1143a_3}{7} + 28(5 - 2(a_1 + 3a_2 + 7a_3 + 2)) + \frac{39}{20} = 0$

Solving these equations, $a_1 = \frac{26897}{8951}$; $a_2 = -\frac{72035}{53706}$; $a_3 = \frac{7455}{35804}$

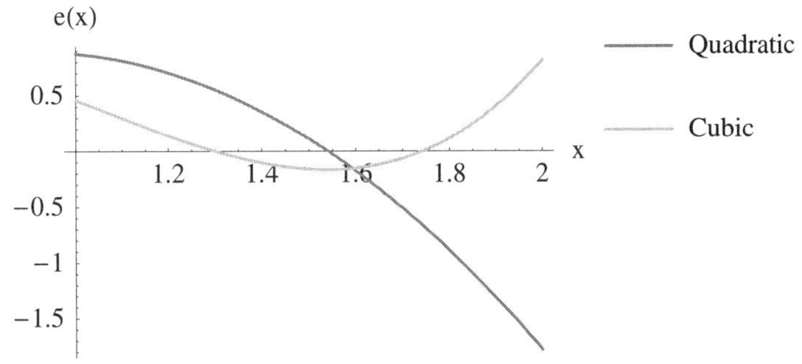

Figure 2.7. Error in satisfying the differential equation

Substituting into the admissible solution, we get the following solution of the problem:

$$u(x) = \frac{22365x^3 - 144070x^2 + 322764x + 13765}{107412}$$

Substituting these approximate solutions into the differential equation, we get the error in satisfying the differential equation. This error is plotted in Figure 2.7.

	$u(x)$	Error, $e(x)$
Quadratic	$-\frac{545x^2}{1443} + \frac{1511x}{962} + \frac{2329}{2886}$	$-\frac{1090x^2}{481} + \frac{1992x}{481} - 1$
Cubic	$\frac{7455x^3}{35804} - \frac{72035x^2}{53706} + \frac{26897x}{8951} + \frac{13765}{107412}$	$\frac{22365x^3}{8951} - \frac{72035x^2}{8951} + \frac{62745x}{8951} - 1$

To demonstrate convergence, solutions up to fifth order are computed. All solutions and their derivatives are compared in Figure 2.8. The higher order solutions are almost on top of each other, indicating the solution convergence. The first derivatives of solutions show greater discrepancy, but they also converge as more terms are included in the assumed solution.

◆ *Mathematica*/MATLAB Implementation 2.1 on the Book Web Site: *Second-order BVP using the Galerkin method*

2.3.2 Higher Order Boundary Value Problems

The example problems considered so far in this chapter have all been second-order differential equations. They are called second order because the highest derivative present in the differential equation is 2. For the second-order problems, a boundary condition in which only u is specified is essential, and the one involving first derivative of u is called natural. The assumed solutions are not required to satisfy the natural boundary conditions. They are actually incorporated into the weak form and are satisfied approximately with the approximation getting better as the number of parameters in the assumed solution is increased.

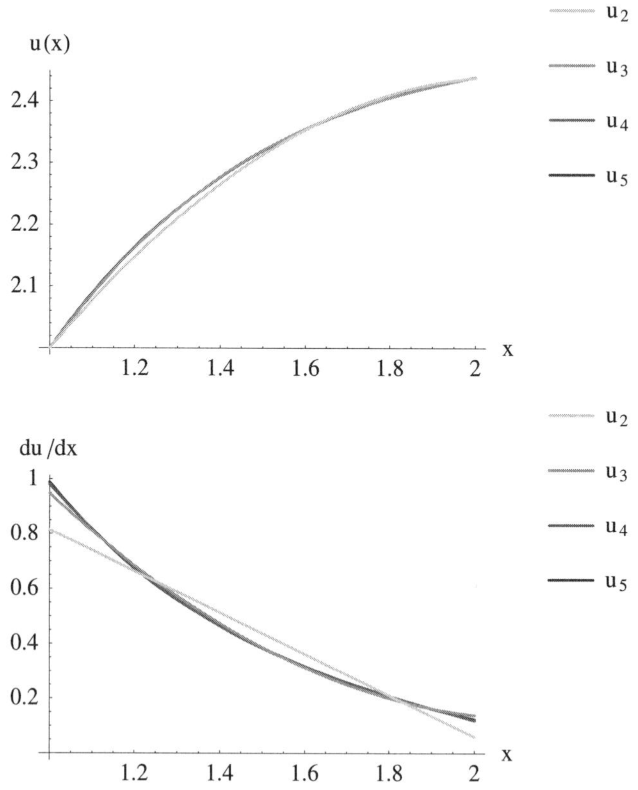

Figure 2.8. Comparison of solutions and their first derivatives

The Galerkin method applies to higher order differential equations as well. For example, consider a fourth-order differential equation of the following form:

$$\frac{d^4u}{dx^4} + 1 = 0; \qquad x_0 < x < x_l$$

or

$$u^{(4)}(x) + 1 = 0$$

The superscript (4) over $u(x)$ indicates fourth derivative of $u(x)$ with respect to x.

With $u(x)$ an assumed solution, the residual is

$$e(x) = u^{(4)}(x) + 1$$

Multiplying by $w_i(x)$ and writing the integral over the given limits, the Galerkin weighted residual is

$$\int_{x_0}^{x_l} (u^{(4)} w_i + w_i)\, dx = 0$$

Using integration by parts, the order of derivative in $w_i u^{(4)}$ can be reduced to 2 as follows:

$$\int_{x_0}^{x_l} (w_i u^{(4)})\, dx = w_i(x_l)u^{(3)}(x_l) - w_i(x_0)u^{(3)}(x_0) + \int_{x_0}^{x_l} (-w_i' u^{(3)})\, dx$$

$$\int_{x_0}^{x_l} (-w_i' u^{(3)})\, dx = w_i'(x_0)u''(x_0) - w_i'(x_l)u''(x_l) + \int_{x_0}^{x_l} (u'' w_i'')\, dx$$

Combining all terms, the weighted residual now is as follows:

$$w_i'(x_0)u''(x_0) - w_i'(x_l)u''(x_l) - w_i(x_0)u^{(3)}(x_0) + w_i(x_l)u^{(3)}(x_l) + \int_{x_0}^{x_l} (w_i + u'' w_i'')\, dx = 0$$

Consider the boundary terms

$$w_i'(x_0)u''(x_0) - w_i'(x_l)u''(x_l) - w_i(x_0)u^{(3)}(x_0) + w_i(x_l)u^{(3)}(x_l)$$

Each one of these terms gives rise to two possibilities:

$-w_i(x_0)u^{(3)}(x_0)$	Either $-u^{(3)}(x_0)$ is known or $w_i(x_0) = 0$
$w_i'(x_0)u''(x_0)$	Either $u''(x_0)$ is known or $w_i'(x_0) = 0$
$w_i(x_l)u^{(3)}(x_l)$	Either $u^{(3)}(x_l)$ is known or $w_i(x_l) = 0$
$-w_i'(x_l)u''(x_l)$	Either $-u''(x_l)$ is known or $w_i'(x_l) = 0$

From these requirements the possible boundary conditions are as follows:

	NBC		EBC
1	$-u^{(3)}(x_0)$ is given	or	$w_i(x_0) = 0 \Longrightarrow$ Must satisfy $u(x_0)$ boundary condition
2	$u''(x_0)$ is given	or	$w_i'(x_0) = 0 \Longrightarrow$ Must satisfy $u'(x_0)$ boundary condition
3	$u^{(3)}(x_l)$ is given	or	$w_i(x_l) = 0 \Longrightarrow$ Must satisfy $u(x_l)$ boundary condition
4	$-u''(x_l)$ is given	or	$w_i'(x_l) = 0 \Longrightarrow$ Must satisfy $u'(x_l)$ boundary condition

Thus for a fourth-order problem boundary conditions involving u and u' both are essential while those involving u'' and $u^{(3)}$ are natural. A practical problem governed by a fourth-order differential equation is that of beam bending. As will be seen in Chapter 4, for beams, the bending moment is proportional to u'' while the shear force is proportional to $u^{(3)}$. Thus specified bending moments and shear forces represent natural boundary conditions and displacements and slopes represent essential boundary conditions.

The generalization to further higher order differential equations should now be obvious. For a general boundary value problem in which the highest derivative present is of the order $2p$ (an even order), we should be able to carry out integration by parts p times. Each integration by parts introduces boundary terms that give rise to essential and natural boundary conditions. For the general case the classification of boundary conditions is as follows:

- Those with the order from 0 to $p - 1$ are the essential boundary conditions.
- Those with the order from p to $2p - 1$ are the natural boundary conditions.

The assumed solutions must satisfy all essential boundary conditions for any value of the unknown parameters. Solutions that satisfy the essential boundary conditions and have the necessary continuity for required derivatives are called admissible solutions.

Example 2.2 Fourth-Order Equation Find an approximate solution of the following fourth-order boundary value problem:

$$u^{(4)}(x) + 8u''(x) + 4u(x) = 10; \qquad 0 < x < 5$$

with the boundary conditions $u(0) = 0$; $u'(0) = 1$; $u(5) = 2$; $u'(5) = 0$. The superscript (4) on u indicates its fourth derivative. The weak form is derived first followed by detailed calculations of fourth- and fifth-order solutions.

With $u(x)$ as an assumed solution, the residual is

$$e(x) = 4u(x) + 8u''(x) + u^{(4)}(x) - 10$$

Multiplying by $w_i(x)$ and writing the integral over the given limits, the Galerkin weighted residual is

$$\int_0^5 (4uw_i + 8u''w_i + u^{(4)}w_i - 10w_i)\, dx = 0$$

Using integration by parts, the order of derivative in $w_i u^{(4)}$ can be reduced to 2 as follows:

$$\int_0^5 (w_i u^{(4)})\, dx = w_i(5)u^{(3)}(5) - w_i(0)u^{(3)}(0) + \int_0^5 (-w_i' u^{(3)})\, dx$$

$$\int_0^5 (-w_i' u^{(3)})\, dx = w_i'(0)u''(0) - w_i'(5)u''(5) + \int_0^5 (u'' w_i'')\, dx$$

Combining all terms, the weighted residual now is as follows:

$$w_i'(0)u''(0) - w_i'(5)u''(5) - w_i(0)u^{(3)}(0) + w_i(5)u^{(3)}(5) + \int_0^5 (2w_i(2u + 4u'' - 5) + u'' w_i'')\, dx = 0$$

Consider the boundary terms

$$w_i'(0)u''(0) - w_i'(5)u''(5) - w_i(0)u^{(3)}(0) + w_i(5)u^{(3)}(5)$$

Given the EBC for the problem:

$$u(0) = 0; \qquad u'(0) - 1 = 0; \qquad u(5) - 2 = 0; \qquad u'(5) = 0$$

Therefore, with admissible solutions (those satisfying EBC)

$$w_i(0) \to 0; \qquad w_i'(0) \to 0; \qquad w_i(5) \to 0; \qquad w_i'(5) \to 0$$

Assuming admissible solutions, the final weak form is as follows:

$$\int_0^5 (2w_i(2u + 4u'' - 5) + u''w_i'')\,dx = 0$$

Fourth-Order Solution Starting assumed solution: $u(x) = a_4x^4 + a_3x^3 + a_2x^2 + a_1x + a_0$

The admissible solution must satisfy the EBC:

EBC	Equation
$u(0) = 0$	$a_0 = 0$
$u'(0) - 1 = 0$	$a_1 - 1 = 0$
$u(5) - 2 = 0$	$a_0 + 5a_1 + 25a_2 + 125a_3 + 625a_4 - 2 = 0$
$u'(5) = 0$	$a_1 + 10a_2 + 75a_3 + 500a_4 = 0$

Solving these equations, $a_0 = 0$; $a_1 = 1$; $a_2 = 25a_4 - \frac{4}{25}$; $a_3 = \frac{1}{125} - 10a_4$

Thus the admissible assumed solution is

$$u(x) = a_4x^4 - 10a_4x^3 + \frac{x^3}{125} + 25a_4x^2 - \frac{4x^2}{25} + x$$

Weighting function $\to \{x^4 - 10x^3 + 25x^2\}$

Substitute into the weak form and perform integrations to get:

Weight	Equation
$x^4 - 10x^3 + 25x^2$	$\frac{188750a_4}{63} - \frac{23875}{42} = 0$

Solving this equation, $a_4 = \frac{573}{3020}$

Substituting into the admissible solution, we get the following solution of the problem:

$$u(x) = \frac{x(14325x^3 - 142646x^2 + 346045x + 75500)}{75500}$$

Fifth-Order Solution Starting assumed solution: $u(x) = a_5x^5 + a_4x^4 + a_3x^3 + a_2x^2 + a_1x + a_0$

The admissible solution must satisfy the EBC:

EBC	Equation
$u(0) = 0$	$a_0 = 0$
$u'(0) - 1 = 0$	$a_1 - 1 = 0$
$u(5) - 2 = 0$	$a_0 + 5a_1 + 25a_2 + 125a_3 + 625a_4 + 3125a_5 - 2 = 0$
$u'(5) = 0$	$a_1 + 10a_2 + 75a_3 + 500a_4 + 3125a_5 = 0$

Solving these equations,

$$\left\{a_0 \to 0, a_1 \to 1, a_2 \to 25a_4 + 250a_5 - \tfrac{4}{25}, a_3 \to -10a_4 - 75a_5 + \tfrac{1}{125}\right\}$$

Thus the admissible assumed solution is

$$u(x) = a_5x^5 + a_4x^4 - 10a_4x^3 - 75a_5x^3 + \frac{x^3}{125} + 25a_4x^2 + 250a_5x^2 - \frac{4x^2}{25} + x$$

Weighting functions $\to \{x^4 - 10x^3 + 25x^2, x^5 - 75x^3 + 250x^2\}$
Substitute into the weak form and perform integrations to get:

Weight	Equation
$x^4 - 10x^3 + 25x^2$	$\frac{188750a_4}{63} + \frac{2359375a_5}{63} - \frac{23875}{42} = 0$
$x^5 - 75x^3 + 250x^2$	$\frac{2359375a_4}{63} + \frac{319843750a_5}{693} - \frac{873625}{126} = 0$

Solving these equations, $a_4 = -\frac{120581}{883350}$; $a_5 = \frac{3817}{146250}$
Substituting into the admissible solution, we get the following solution of the problem:

$$u(x) = \frac{x(576367x^4 - 3014525x^3 - 12905605x^2 + 65195225x + 22083750)}{22083750}$$

Substituting these approximate solutions into the differential equation, we get the error in satisfying the differential equation. A plot of the error shown in Figure 2.9 indicates that both solutions have significant error. Since the differential equation is quite complicated, we need a fairly high-order polynomial to get reasonable solution. To demonstrate convergence, solutions up to 12th order are computed. These solutions are compared in Figure

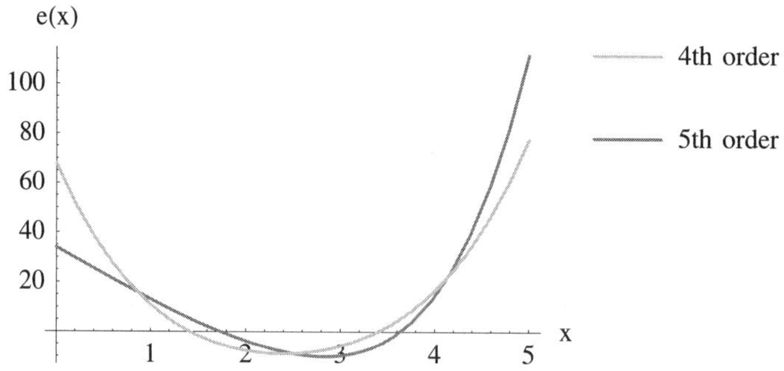

Figure 2.9. Error in satisfying differential equation

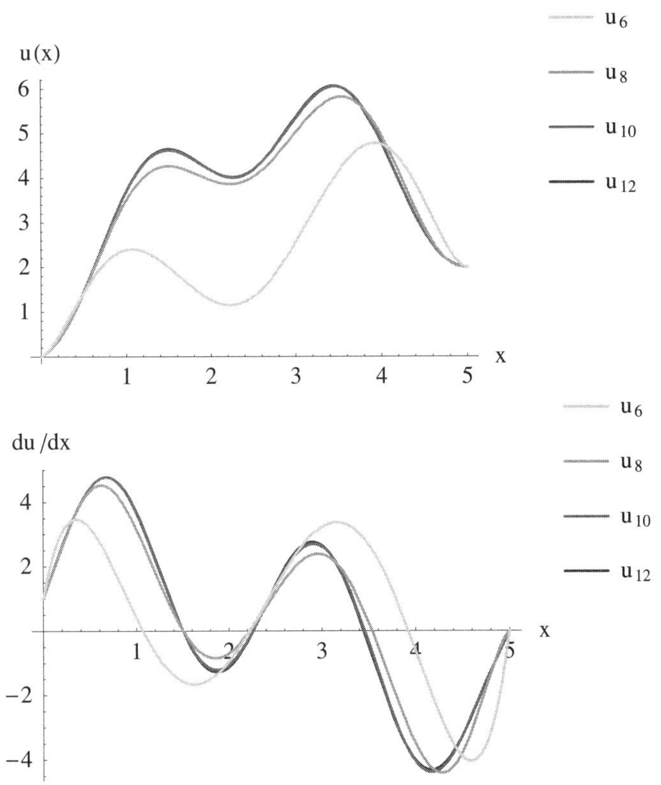

Figure 2.10. Comparison of solutions and their first derivatives

2.10. The 10th- and 12th-order solutions are fairly close to each other, indicating that they represent good solutions to the problem.

	Error, $e(x)$
Fourth-order	$\frac{14325x^4-142646x^3+689845x^2-1636252x+1281380}{18875}$
Fifth-order	$\frac{2(576367x^5-3014525x^4+10149075x^3-7153375x^2-115492500x+187484375)}{11041875}$

◆ *Mathematica*/MATLAB Implementation 2.2 on the Book Web Site: *Fourth-order BVP using the Galerkin method*

Example 2.3 Third-Order Equation The even-order equations, such as the second and the fourth order, are common in engineering applications. However, the Galerkin method can be applied to odd-order equations as well. To demonstrate this, we consider the following third-order boundary value problem:

$$u^{(3)}(x) + 2u(x) = x^2; \qquad 0 < x < 1$$

with the boundary conditions $u(0) = 1$; $u(1) = 2$; $u'(1) = 3$. The superscript (3) on u indicates its third derivative.

The weak form is derived first followed by detailed calculations of a quadratic solution in *Mathematica*.

With $u(x)$ as an assumed solution the residual is

$$e(x) = -x^2 + 2u(x) + u^{(3)}(x)$$

Multiplying by $w_i(x)$ and writing the integral over the given limits, the Galerkin weighted residual is

$$\int_0^1 (-w_i x^2 + 2uw_i + w_i u^{(3)})\, dx = 0$$

Using integration by parts twice on the $w_i u^{(3)}$ term, we have

$$\int_0^1 (w_i u^{(3)})\, dx = w_i(1)u''(1) - w_i(0)u''(0) + \int_0^1 (-w_i' u'')\, dx$$

$$\int_0^1 (-w_i' u'')\, dx = u'(0)w_i'(0) - u'(1)w_i'(1) + \int_0^1 (u' w_i'')\, dx$$

Combining all terms, the weighted residual now is as follows:

$$u'(0)w_i'(0) - u'(1)w_i'(1) - w_i(0)u''(0) + w_i(1)u''(1) + \int_0^1 (u' w_i'' - (x^2 - 2u)w_i)\, dx = 0$$

The boundary condition $u'(1) = 3$ can be directly incorporated into the weak form, giving

$$u'(0)w_i'(0) - 3w_i'(1) - w_i(0)u''(0) + w_i(1)u''(1) + \int_0^1 (u' w_i'' - (x^2 - 2u)w_i)\, dx = 0$$

The other two boundary conditions are considered essential:

```
WeakForm[w_, u_] :=
    Simplify[((D[u, x] D[w, x])/.x→0) - ((3 D[w, x])/.x→1)
    - ((D[u, {x, 2}] w)/.x→0) +((D[u, {x, 2}] w)/.x→1)
    + Integrate[D[u, x] D[w, {x, 2}] - (x^2 - 2u) w, {x, 0, 1}]]
```

We start with a quadratic polynomial ux with three parameters a_0, a_1, and a_2 as follows:

```
ux = a0 + a1 x + a2 x^2;
```

The first task is to make it an admissible solution. We must satisfy the two essential boundary conditions:

```
ebc1 = (ux/.x→0) == 1
```

a0 == 1

```
ebc2 = (ux/.x→1) == 2
```

a0 + a1 + a2 == 2

Solving these equations, we get

```
sol = Solve[{ebc1, ebc2}, {a0, a1}]
```

$\{\{a0 \to 1, a1 \to 1 - a2\}\}$

Substituting the solution into *ux* gives an admissible solution as follows:

```
u = ux/.sol[[1]]
```

$a2\, x^2 + (1 - a2)x + 1$

Now we can set up the equation needed to find the remaining coefficient:

```
w = D[u, a2]; eq1 = WeakForm[w, u] == 0
```

$$\frac{1}{60}(64\, a2 - 147) == 0$$

Solving this equation,

```
sol = Solve[{eq1}, {a2}]
```

$$\left\{\left\{a2 \to \frac{147}{64}\right\}\right\}$$

Substituting the solution into the assumed admissible solution, we have a quadratic approximate solution of the problem as follows:

```
ux = Simplify[u/.sol[[1]]]
```

$$\frac{1}{64}(147x^2 - 83x + 64)$$

It is possible to obtain an exact analytical solution of the problem as follows:

```
Clear[u]; sol = DSolve[uj'[x] + 2u[x] == x^2, u[x], x][[1, 1, 2]];
exactSol = Simplify[sol/.Solve[{(sol/.x → 0) == 1,
  (sol/.x → 1) == 2, D[sol, x]/.x → 1) == 3}][[1]]]
```

$$-\left(e^{-\frac{2x+1}{2^{2/3}}}\left(\sqrt[3]{2}\sqrt{3}\,e^{\sqrt[3]{2}(x+2)}x^2 - 3\sqrt[3]{2}\,e^{\frac{2x+1}{2^{2/3}}}\sin\left(\frac{\sqrt{3}}{2^{2/3}}\right)x^2 - \sqrt[3]{2}\sqrt{3}\,e^{\frac{2x+1}{2^{2/3}}}\cos\left(\frac{\sqrt{3}}{2^{2/3}}\right)x^2\right.\right.$$
$$+ \sqrt[3]{2}\sqrt{3}\left(3e^{\frac{3(x+1)}{2^{2/3}}} - 2e^{\frac{3x+1}{2^{2/3}}}\right)\cos\left(\frac{\sqrt{3}(x-1)}{2^{2/3}}\right) + e^{\frac{3x+1}{2^{2/3}}}\left(6\sqrt[3]{2} + \left(8 - 3\sqrt[3]{2}\right)e^{\sqrt[3]{2}}\right)\sin\left(\frac{\sqrt{3}(x-1)}{2^{2/3}}\right)$$
$$- 6\sqrt[3]{2}\,e^{\frac{3x}{2^{2/3}}}\sin\left(\frac{\sqrt{3}x}{2^{2/3}}\right) - 8e^{\frac{3x}{2^{2/3}}}\sin\left(\frac{\sqrt{3}x}{2^{2/3}}\right) - 3\sqrt[3]{2}\,e^{\frac{3}{2^{2/3}}}\sin\left(\frac{\sqrt{3}}{2^{2/3}}\right)$$
$$\left.\left. + 8e^{\frac{3}{2^{2/3}}}\sin\left(\frac{\sqrt{3}}{2^{2/3}}\right) - 3\sqrt[3]{2}\sqrt{3}\,e^{\frac{3}{2^{2/3}}}\cos\left(\frac{\sqrt{3}}{2^{2/3}}\right) + 2\sqrt[3]{2}\sqrt{3}\,e^{2\sqrt[3]{2}}\right)\right)\Big/$$
$$\left(2\sqrt[3]{2}\left(-\sqrt{3}\,e^{\frac{3}{2^{2/3}}} + \sqrt{3}\cos\left(\frac{\sqrt{3}}{2^{2/3}}\right) + 3\sin\left(\frac{\sqrt{3}}{2^{2/3}}\right)\right)\right)$$

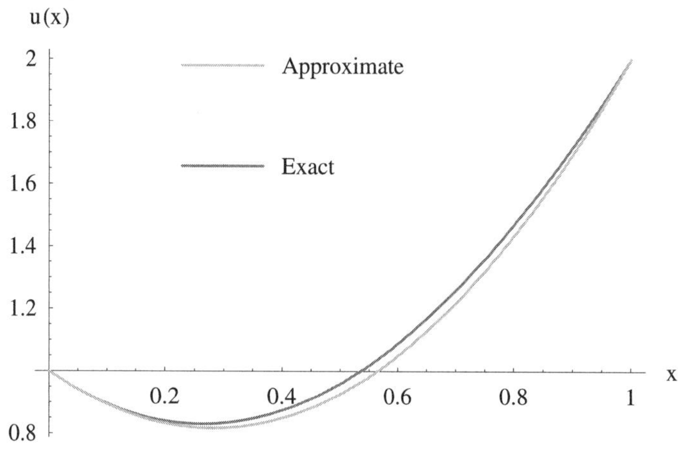

Figure 2.11. Comparison of the approximate and the exact solutions

```
{D[exactSol, {x, 3}] + 2 exactSol, exactSol/.x → 0,
  exactSol/.x → 1, D[exactSol, x]/.x → 1}//Simplify
```

$\{x^2, 1, 2, 3\}$

As can be seen from Figure 2.11, the quadratic approximate solution compares very well with the exact solution. A cubic solution will almost be indistinguishable from the exact solution:

```
Needs["Graphics`Legend`"];
Plot[{ux, exactSol}, {x, 0, 1}, PlotStyle → {Hue[0.6], Hue[0.9]},
  AxesLabel → {"x", "u(x)"}, PlotLegend → {"Approx", "Exact"},
  LegendPosition → {-1/2, 0}, LegendShadow → None,
  PlotRange → All];
```

2.4 RAYLEIGH-RITZ METHOD

As is the case with the Galerkin method, the Rayleigh-Ritz method also requires an equivalent integral formulation. However, the derivation of this equivalent integral form is based on fundamentally different concepts. For many practical problems, the integral form is based on energy considerations. For structural problems, the potential energy is an example of such an integral form that has been used successfully in developing a large number of finite elements. Similar energy-based integral forms are available for other applications. If the energy function is not known from physical considerations, it may still be possible to derive an equivalent integral formulation using mathematical manipulations. However, such manipulations involve a branch of calculus known as the *calculus of variations*. A review of basic calculus of variations and mathematical manipulations to derive integral formulations is presented in Appendix B. In the main body of this text the use of the Rayleigh-Ritz method will be restricted to structural problems for which the potential energy function is well known.

Once the equivalent integral form is known, the steps in the Rayleigh-Ritz method are similar to those in the Galerkin method. A solution is assumed with some unknown parameters. The assumed solution is made admissible by requiring that it explicitly satisfy the essential boundary conditions. The energy form reduces to a function of unknown parameters once the assumed solution is substituted into the energy function. The necessary conditions for the minimum of this function then give the required equations for finding suitable values for the unknown parameters.

2.4.1 Potential Energy for Axial Deformation of Bars

Consider a linear elastic bar of any arbitrary cross section subjected to loads in the axial direction only, as shown in Figure 2.1. The area of cross section is denoted by $A(x)$ and could vary over the length of the bar. The modulus of elasticity is denoted by E. The bar may be subjected to distributed axial load $q(x)$ along its length. In addition, there may be concentrated axial loads P_i applied at some locations x_i on the bar. The axial displacement is denoted by $u(x)$. Denoting the axial stress by σ_x and corresponding strain by ϵ_x, the axial strain energy is defined as follows:

$$U = \frac{1}{2}\int_V \sigma_x \epsilon_x \, dV$$

where the integration is over V, the volume of the bar. For axial deformations stress and strain are assumed to be constant over a cross section, and thus the volume integral can be expressed as a one-dimensional integral

$$U = \frac{1}{2}\int_{x_0}^{x_l} \sigma_x \epsilon_x A \, dx$$

where x_0 and x_l are end coordinates of the bar. Using Hooke's law and the strain-displacement relationship $\sigma_x = E\epsilon_x = E\, du/dx$ yields

$$U = \frac{1}{2}\int_{x_0}^{x_l} EA\left(\frac{du}{dx}\right)^2 dx$$

The potential of the applied loads is equal to the negative of the work done by all external applied forces:

$$W = \int_{x_0}^{x_l} qu \, dx + \sum P_i u(x_i)$$

where $u(x_i)$ is the axial displacement at the location of the concentrated applied load P_i. Using these expressions, the potential energy for axial deformation of bars is defined as follows:

$$\Pi = U - W = \frac{1}{2}\int_{x_0}^{x_l} EA(x)\left(\frac{du}{dx}\right)^2 dx - \int_{x_0}^{x_l} q(x)u(x)\, dx - \sum P_i u(x_i)$$

Using calculus of variations, it is possible to show that a function $u(x)$ that minimizes the potential energy is a solution of the differential equation governing the axial deformation of bars.

2.4.2 Overall Solution Procedure Using the Rayleigh-Ritz Method

Given a boundary value problem, the overall procedure for obtaining an approximate solution using the Rayleigh-Ritz method is as follows:

(i) Obtain the equivalent energy form. For structural problems use the potential energy. For general boundary value problems use calculus of variations to construct an equivalent energy form (see Appendix B). Clearly identify the essential and natural boundary conditions.

(ii) Construct an admissible assumed solution.

1. Start with an assumed solution with some unknown parameters and enough terms in it so that it is possible to evaluate the terms in the energy form. Any suitable function can be used for the assumed solution. However, since polynomials are easy to differentiate and integrate, it is convenient to start with a polynomial of desired order.
2. Set some of the parameter values such that the essential boundary conditions are satisfied regardless of the values of the remaining unknown parameters. If there are no essential boundary conditions, then the solution already is admissible and we can go to the next step. Otherwise set up equations to satisfy essential boundary conditions. Solve these equations for some of the parameters. Substitute these parameter values back into the starting solution to get an admissible assumed solution.

(iii) Set up equations and solve for the unknown parameters.

1. Substitute the admissible assumed solution into the energy form and carry out the integration. The resulting expression should be a function only of the unknown parameters.
2. Set up a system of equations by using necessary conditions for the minimum of the energy form. Since the minimum of the energy function corresponds to the solution of the boundary value problem, we get a system of equations by setting to zero the partial derivatives of the energy function with respect to unknown parameters.
3. Solve the system of equations for the unknown parameters.

(iv) Approximate solution.

1. Substitute computed values of parameters into the admissible solution to get the approximate solution.
2. Check the quality of the solution by substituting it into the differential equation and the natural boundary conditions to see how well it satisfies them.

2.4.3 Uniform Bar Subjected to Linearly Varying Axial Load

We now use the Rayleigh-Ritz method to find approximate solutions of a uniform bar (EA constant) fixed at one end and subjected to a static point load at the other end, as shown in Figure 2.3. The bar is also subjected to a linearly varying axial load $q(x) = cx$, where c is a given constant. The potential energy for the problem is as follows:

$$\Pi = \frac{1}{2}EA \int_0^L \left(\frac{du}{dx}\right)^2 dx - c \int_0^L xu\, dx - Pu(L)$$

$$\text{EBC:} \qquad u(0) = 0.$$

The simplest possible solution that we can assume is a linear polynomial. Using this as the starting solution, an approximate solution of the problem is obtained as follows:

Linear Solution Starting assumed solution: $u(x) = a_0 + xa_1$

The admissible solution must satisfy the EBC:

EBC	Equation
$u(0) = 0$	$a_0 = 0$

Thus the admissible assumed solution is $u(x) = xa_1$.
Substituting the admissible solution into Π and carrying out integration,

$$\Pi = -\tfrac{1}{3}ca_1L^3 + \tfrac{1}{2}EAa_1^2L - Pa_1L$$

For the minimum of Π, set its derivatives with respect to parameters to 0:

$$\frac{\partial \Pi}{\partial a_1} = 0: \qquad -\frac{cL^3}{3} - PL + EAa_1L = 0$$

Solving this equation,

$$a_1 = \frac{(cL^3/3) + PL}{EA L}$$

Substituting into the admissible solution, we get the following solution of the problem:

$$u = \frac{(cL^2 + 3P)x}{3EA}$$

A better solution can be obtained if we start with a quadratic polynomial:

Quadratic Solution Starting assumed solution: $u(x) = a_2x^2 + a_1x + a_0$

The admissible solution must satisfy the EBC:

EBC	Equation
$u(0) = 0$	$a_0 = 0$

Thus the admissible assumed solution is $u(x) = a_2x^2 + a_1x$.
Substituting the admissible solution into Π and carrying out integration,

$$\Pi = -\tfrac{1}{4}ca_2L^4 + \tfrac{1}{3}(2EAa_2^2 - ca_1)L^3 + EAa_1a_2L^2 + \tfrac{1}{2}EAa_1^2L - P(a_2L^2 + a_1L)$$

For the minimum of Π, set its derivatives with respect to parameters to 0:

$$\frac{\partial\Pi}{\partial a_1} = 0: \qquad -\frac{cL^3}{3} + EAa_2L^2 - PL + EAa_1L = 0$$

$$\frac{\partial\Pi}{\partial a_2} = 0: \qquad -\frac{cL^4}{4} + \tfrac{4}{3}EAa_2L^3 - PL^2 + EAa_1L^2 = 0$$

Solving these equations,

$$a_1 = -\frac{-7cL^2 - 12P}{12EA}; \qquad a_2 = -\frac{cL}{4EA}$$

Substituting into the admissible solution, we get the following solution of the problem:

$$u = \frac{(12P + cL(7L - 3x))x}{12EA}$$

The exact solution is a cubic polynomial. As demonstrated below, if we start with a cubic polynomial, the Galerkin method finds this exact solution:

Cubic Solution Starting assumed solution: $u(x) = a_3x^3 + a_2x^2 + a_1x + a_0$

The admissible solution must satisfy the EBC:

EBC	Equation
$u(0) = 0$	$a_0 = 0$

Thus the admissible assumed solution is $u(x) = a_3x^3 + a_2x^2 + a_1x$.
Substituting the admissible solution into Π and carrying out integration,

$$\Pi = \tfrac{1}{10}a_3(9EAa_3 - 2c)L^5 + \tfrac{1}{4}a_2(6EAa_3 - c)L^4 + \tfrac{1}{3}(2EAa_2^2 - ca_1 + 3EAa_1a_3)L^3 + EAa_1a_2L^2 + \tfrac{1}{2}EAa_1^2L - P(a_3L^3 + a_2L^2 + a_1L)$$

For the minimum of Π, set its derivatives with respect to parameters to 0:

$$\frac{\partial \Pi}{\partial a_1} = 0: \qquad \tfrac{1}{3}(3EAa_3 - c)L^3 + EAa_2L^2 - PL + EAa_1L = 0$$

$$\frac{\partial \Pi}{\partial a_2} = 0: \qquad \tfrac{1}{4}(6EAa_3 - c)L^4 + \tfrac{4}{3}EAa_2L^3 - PL^2 + EAa_1L^2 = 0$$

$$\frac{\partial \Pi}{\partial a_3} = 0: \qquad \tfrac{9}{10}EAa_3L^5 + \tfrac{1}{10}(9EAa_3 - 2c)L^5 + \tfrac{3}{2}EAa_2L^4 - PL^3 + EAa_1L^3 = 0$$

Solving these equations,

$$a_1 = -\frac{-cL^2 - 2P}{2EA}; \qquad a_2 = 0; \qquad a_3 = -\frac{c}{6EA}$$

Substituting into the admissible solution, we get the following solution of the problem:

$$u = \frac{x(3cL^2 - cx^2 + 6P)}{6EA}$$

Since this is the exact solution, trying any higher order polynomial will not make any difference. Also note that exactly the same solutions were obtained by using the Galerkin method earlier.

2.4.4 Tapered Bar Subjected to Linearly Varying Axial Load

Consider a tapered bar fixed at one end and subjected to a static point load at the other end, as shown in Figure 2.4. The bar is also subjected to a linearly varying axial load $q(x) = cx$, where c is a given constant. The potential energy for the problem is as follows:

$$\Pi = \frac{1}{2}E\int_0^L A(x)\left(\frac{du}{dx}\right)^2 dx - c\int_0^L xu\,dx - Pu(L)$$

EBC: $u(0) = 0$

where

$$A(x) = A_0 - \frac{A_0 - A_L}{L}x = \frac{(L + (-1 + r)x)A_0}{L}$$

where A_0 is the area of cross section at $x = 0$, A_L is that at $x = L$, and $r = A_L/A_0$.

Linear Solution Starting assumed solution: $u(x) = a_0 + xa_1$

The admissible solution must satisfy the EBC:

EBC	Equation
$u(0) = 0$	$a_0 = 0$

Thus the admissible assumed solution is $u(x) = xa_1$.
Substituting the admissible solution into Π and carrying out integration,

$$\Pi = -\tfrac{1}{3}ca_1L^3 - Pa_1L + \tfrac{1}{4}rEa_1^2A_0L + \tfrac{1}{4}Ea_1^2A_0L$$

For the minimum of Π, set its derivatives with respect to parameters to 0:

$$\frac{\partial\Pi}{\partial a_1} = 0: \qquad -\frac{cL^3}{3} - PL + \frac{1}{2}rEa_1A_0L + \frac{1}{2}Ea_1A_0L = 0$$

Solving this equation,

$$\left\{a_1 \to \frac{(cL^3/3) + PL}{(1/2)LEA_0 + (1/2)LrEA_0}\right\}$$

Substituting into the admissible solution, we get the following solution of the problem:

$$u = \frac{2(cL^2 + 3P)x}{3(rEA_0 + EA_0)}$$

Quadratic Solution Starting assumed solution: $u(x) = a_2x^2 + a_1x + a_0$

The admissible solution must satisfy the EBC:

EBC	Equation
$u(0) = 0$	$a_0 = 0$

Thus the admissible assumed solution is $u(x) = a_2x^2 + a_1x$.
Substituting the admissible solution into Π and carrying out integration,

$$\Pi = -\tfrac{1}{4}ca_2L^4 - \tfrac{1}{3}ca_1L^3 + \tfrac{1}{2}rEa_2^2A_0L^3 + \tfrac{1}{6}Ea_2^2A_0L^3 + \tfrac{2}{3}rEa_1a_2A_0L^2$$
$$+ \tfrac{1}{3}Ea_1a_2A_0L^2 + \tfrac{1}{4}rEa_1^2A_0L + \tfrac{1}{4}Ea_1^2A_0L - P(a_2L^2 + a_1L)$$

For the minimum of Π, set its derivatives with respect to parameters to 0:

$$\frac{\partial\Pi}{\partial a_1} = 0: \qquad -\frac{cL^3}{3} + \frac{2}{3}rEa_2A_0L^2 + \frac{1}{3}Ea_2A_0L^2 - PL + \frac{1}{2}rEa_1A_0L + \frac{1}{2}Ea_1A_0L = 0$$

$$\frac{\partial\Pi}{\partial a_2} = 0: \qquad -\frac{cL^4}{4} + rEa_2A_0L^3 + \frac{1}{3}Ea_2A_0L^3 - PL^2 + \frac{2}{3}rEa_1A_0L^2 + \frac{1}{3}Ea_1A_0L^2 = 0$$

Solving these equations,

$$a_1 = -\frac{-cL^2 - 6crL^2 - 12Pr}{2(r^2 + 4r + 1)EA_0}; \qquad a_2 = -\frac{-cL^2 + 7crL^2 - 12P + 12Pr}{4L(r^2 + 4r + 1)EA_0}$$

Substituting into the admissible solution, we get the following solution of the problem:

$$u = \frac{x(c(2L(6r + 1) - 7rx + x)L^2 + 12P(2Lr - xr + x))}{4L(r^2 + 4r + 1)EA_0}$$

Exactly the same solutions were obtained by using the Galerkin method and are compared with the exact solution in Figure 2.6.

◆ *Mathematica*/MATLAB Implementation 2.3 on the Book Web Site: *Tapered bar using the Rayleigh-Ritz method*

2.5 COMMENTS ON GALERKIN AND RAYLEIGH-RITZ METHODS

2.5.1 Admissible Assumed Solution

It has been mentioned several times already that the assumed solutions must satisfy the essential boundary conditions. This requirement is not difficult to fulfill for one-dimensional problems. As was done in several examples, we simply start with a polynomial of a chosen degree and use the essential boundary conditions to evaluate as many coefficients as possible. Substituting these coefficients into the starting polynomial gives us an admissible assumed solution for the problem.

Unfortunately this straightforward procedure does not work for two- and three-dimensional problems. For example, consider a simple situation of a rectangular plate fixed at left the end and subjected to loads at the right end, as shown in Figure 2.12. The problem is two dimensional and therefore the displacements are functions of x and y. Similar to the case for one-dimensional problems, we start with an assumed solution for horizontal displacement $u(x, y)$ as a two-dimensional polynomial as follows:

$$u(x, y) = a_0 + a_1 x + a_2 y + a_3 x^2 + a_4 xy + a_5 y^2 + \cdots$$

The essential boundary condition is that displacements along the vertical line at $x = 0$ must be zero. To satisfy this requirement, we get the following equation:

$$u(0, y) = a_0 + a_2 y + a_5 y^2 + \cdots = 0$$

This is the only equation we have, and thus there is no unique way to find coefficients that will satisfy the equation. One possibility for this example is to set $a_0 = a_2 = a_5 = \cdots = 0$. This will give us an admissible solution but it may not result in a good approximate solution, because it eliminates a large number of y terms from the assumed solution.

The problem of finding suitable admissible assumed solutions is one of the biggest drawback of the classical methods. This example is still very simple. Imagine if the support was along an inclined line. How would you find suitable values of coefficients to satisfy the essential boundary condition? As will be seen in the following section, the finite element form of the assumed solution overcomes this difficulty by expressing the solution in terms of nodal unknowns.

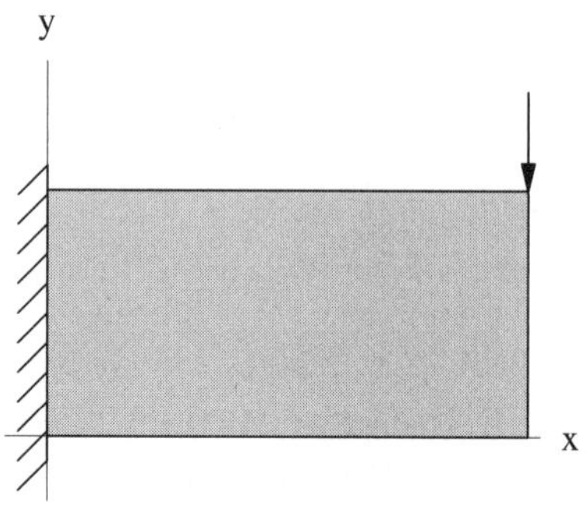

Figure 2.12. Rectangular plate

2.5.2 Solution Convergence—the Completeness Requirement

Several examples presented in this chapter have demonstrated that the solutions obtained by using the Galerkin and the Rayleigh-Ritz methods converge to the exact solution as more terms are added to the assumed solution. One of the requirements for convergence is that the assumed solutions must be capable of representing all terms in an exact solution. The solution will not converge if a term that is present in the exact solution is not included in the assumed solution. For example, the exact solution for a uniform bar subjected to linearly varying axial load and concentrated tip loading is a cubic function of x. In Section 2.2, the solution of this problem converged to the exact solution when we used a cubic assumed solution. Any complete polynomial of degree larger than 3 would also give exactly the same solution. However, if we start with an assumed solution that skips over, say, the cubic term, as more terms are added, the solution may get better but will never converge to the exact solution. Thus one of the fundamental convergence requirements is that polynomials used in the assumed solutions must be complete.

It should also be pointed out that, even though we have considered only polynomial assumed solutions in the examples, there is no reason to restrict ourselves to polynomials. We can use any complete set of functions that satisfies essential boundary conditions. Given enough terms and with a complete set of functions, the approximate solutions should converge to the exact solution.

Example 2.4 As an illustration of lack of convergence, when an incomplete polynomial is assumed, consider the following example:

$$u''(x) + x = 0; \qquad 0 < x < 1$$

$$u(0) = u(1) = 0$$

It can easily be verified that the following is an appropriate weak form for the problem:

$$\int_0^1 (xw_i - u'w_i')\,dx = 0$$

We first obtain a solution using a complete cubic polynomial:

Starting assumed solution: $u(x) = a_3x^3 + a_2x^2 + a_1x + a_0$
The admissible solution must satisfy the EBC:

EBC	Equation
$u(0) = 0$	$a_0 = 0$
$u(1) = 0$	$a_0 + a_1 + a_2 + a_3 = 0$

Solving these equations, $\{a_0 \to 0, a_1 \to -a_2 - a_3\}$
Thus the admissible assumed solution is $u(x) = a_3x^3 + a_2x^2 - a_2x - a_3x$.
Weighting functions $\to \{x^2 - x, x^3 - x\}$
Substitute into the weak form and perform integrations to get:

Weight	Equation
$x^2 - x$	$\frac{1}{12}(-4a_2 - 6a_3 - 1) = 0$
$x^3 - x$	$\frac{1}{30}(-15a_2 - 24a_3 - 4) = 0$

Solving these equations, we get $a_2 = 0$; $a_3 = -\frac{1}{6}$
Substituting into the admissible solution, we get the following solution of the problem:

$$u(x) = \frac{1}{6}(x - x^3)$$

By substituting it into the differential equation and the boundary conditions, it can easily be verified that this is the exact solution of the problem. Using complete polynomials, any higher order polynomial will give the same solution. However, if we use an incomplete polynomial, the solution will not be exact. For example, consider a fourth-order polynomial with the linear term missing:

Starting assumed solution: $u(x) = a_4x^4 + a_3x^3 + a_2x^2 + a_0$
The admissible solution must satisfy the EBC:

EBC	Equation
$u(0) = 0$	$a_0 = 0$
$u(1) = 0$	$a_0 + a_2 + a_3 + a_4 = 0$

Solving these equations, $\{a_0 \to 0, a_2 \to -a_3 - a_4\}$
Thus, the admissible assumed solution is $u(x) = a_4x^4 + a_3x^3 - a_3x^2 - a_4x^2$.
Weighting functions $\to \{x^3 - x^2, x^4 - x^2\}$
Substitute into the weak form and perform integrations to get:

Weight	Equation
$x^3 - x^2$	$\frac{1}{60}(-8a_3 - 14a_4 - 3) = 0$
$x^4 - x^2$	$\frac{1}{420}(-98a_3 - 176a_4 - 35) = 0$

Solving these equations, we get $a_3 = -\frac{19}{18}$; $a_4 = \frac{7}{18}$

Substituting into the admissible solution, we get the following solution of the problem:

$$u(x) = \frac{1}{18}x^2(7x^2 - 19x + 12)$$

Clearly this is not the exact solution. For this example, since we know the exact solution, it is obvious that we can obtain an exact solution by starting with a polynomial having linear and cubic terms only. However, in general, we do not know what terms are present in the exact solution. Therefore, we must use complete polynomials for solutions to converge to the exact solution as more terms are included in the polynomial.

2.5.3 Galerkin versus Rayleigh-Ritz

The Rayleigh-Ritz method requires that given differential equations must first be converted to their equivalent energy form. However, it may not always be possible to obtain this equivalent variational form. For example, the third-order problem in Example 2.3 does not have an equivalent energy form. This apparent disadvantage is not of great concern in actual applications. As pointed out earlier, for a large class of engineering problems the functional form is well known. In fact, in the energy methods of structural mechanics, the functional (usually the potential energy) is used as a basis for developing the governing differential equations.

The Galerkin method is clearly the more general of the two methods. It can be applied to any boundary value problem. Furthermore, for those problems for which an equivalent variational form does exist, the Galerkin method gives the same solution as the Rayleigh-Ritz method. Thus the Galerkin method is the obvious choice for general applications. The variational methods remain popular in practice because of the large body of existing literature on the energy methods. For stress analysis applications the governing differential equations are fairly complex. However, the equivalent functional is very straightforward to write. In these situations the Rayleigh-Ritz method is preferred because it can be applied directly. The Galerkin method would require one to first develop the weak form starting from fairly complex governing differential equations.

2.6 FINITE ELEMENT FORM OF ASSUMED SOLUTIONS

Both the Galerkin and the Rayleigh-Ritz methods are powerful tools for finding approximate solutions to boundary value problems. The quality of the approximation depends on the choice of the assumed solution. However, there are no guidelines for an appropriate choice of assumed solutions, especially for two- and three-dimensional problems. Furthermore, to be admissible, these assumed solutions must satisfy essential boundary conditions. As mentioned before, for complicated problems it may be difficult, if not impossible, to come up with appropriate admissible solutions. Furthermore, since the assumed solutions are defined over the entire solution domain, often a large number of terms must be included in order to represent the solution accurately.

The finite element method overcomes these difficulties by introducing two fundamental concepts:

(i) *Solution Domain Is Discretized into Elements.* The solution domain is divided into several simpler subdomains called *elements*. Each element has a simple geometry so that appropriate assumed solutions can easily be written for an element. The only restriction on the element shape is that it should be possible to carry out the required differentiations and integrations of the assumed solutions over each element. Since each element covers only a small portion of the solution domain, a low-degree polynomial can usually be used to describe the solution over an element. The integral in the weak or the functional form can be evaluated separately over each element and added (assembled) together as follows:

$$\int_0^1 (\cdots)\,dx = \int_{x_1=0}^{x_2} (\cdots)\,dx + \int_{x_2}^{x_3} (\cdots)\,dx + \cdots$$

(ii) *Coefficients in Assumed Solution over an Element Represent Solution and Its Appropriate Derivatives at the Nodes.* In the classical methods the unknown coefficients in the assumed solution do not have any physical meaning. They are just mathematical quantities which when substituted into the assumed solution give an approximate solution. In the finite element method the polynomial coefficients are defined in terms of unknown solutions at some selected points over the element. The locations chosen to define the assumed solution are called *nodes*. Usually element ends and corners are chosen as nodal locations. The solutions at the nodes are called *nodal degrees of freedom*. The choice of these nodal degrees of freedom is dictated by the order of derivatives in the essential boundary conditions. Since for a second-order problem the essential boundary conditions do not involve any derivatives, the nodal unknowns for these problems are simply the solution variables. For a fourth-order differential equation, at each node we must choose the solution and its first derivative as degrees of freedom because these are the terms in the essential boundary conditions for this problem. With this choice of nodal degrees of freedom it becomes almost trivial to make the assumed solutions admissible. All that one has to do is to set the corresponding nodal value equal to the value specified by the essential boundary condition.

Nodal degrees of freedom: quantities needed for EBC
- Second-order problem: u
- Fourth-order problem: u and u'
- Sixth-order problem: u, u', and u''

In the following sections, these ideas are illustrated by writing assumed element solutions for second- and fourth-order ordinary differential equations. The solution domain for these one-dimensional problems is a line; therefore the elements for these problems are simply line elements.

2.6.1 Linear Interpolation Functions for Second-Order Problems

A two-node line element for a second-order problem is shown in Figure 2.13. The element extends from x_1 to x_2 and has a length $l = x_2 - x_1$. The circles represent nodes. For a second-order problem the nodal degrees of freedom are the unknown solutions at the nodes and are indicated by u_1 and u_2.

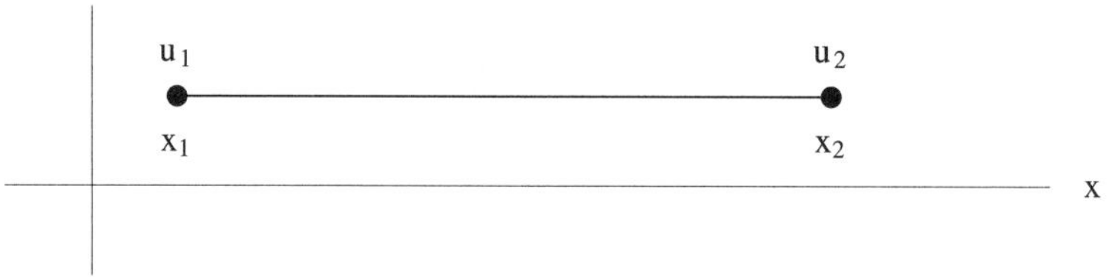

Figure 2.13. A simple two-node element for second-order problems

Since there are two nodes on the element, a linear solution can be written over the element by starting from a linear polynomial with two coefficients and then evaluating these coefficients in terms of the nodal unknowns as follows:

$$u(x) = a_0 + a_1 x$$

$$\text{at } x = x_1; \qquad u(x_1) = u_1 \Longrightarrow u_1 = a_0 + a_1 x_1$$

$$\text{at } x = x_2; \qquad u(x_2) = u_2 \Longrightarrow u_2 = a_0 + a_1 x_2$$

Solving the two equations for a_0 and a_1, we get

$$a_0 = -\frac{u_1 x_2 - u_2 x_1}{x_1 - x_2}; \qquad a_1 = -\frac{u_2 - u_1}{x_1 - x_2}$$

Thus linear solution over the element in terms of nodal degrees of freedom is as follows:

$$u(x) = -\frac{u_1 x_2 - u_2 x_1}{x_1 - x_2} - \frac{u_2 - u_1}{x_1 - x_2} x$$

Collecting together terms involving u_1 and u_2, we can write this solution as follows:

$$u(x) = \frac{x - x_2}{x_1 - x_2} u_1 + \frac{x - x_1}{x_2 - x_1} u_2$$

This is the finite element form of a linear solution. It clearly is equivalent to the linear polynomial. However, unlike the polynomial coefficients a_0 and a_1, which do not have any physical meaning, the coefficients u_1 and u_2 are the solutions at the two nodes of the element. This is the key feature of the finite element form of the assumed solution and has the following three major advantages:

(i) In order to make the solution admissible in the polynomial form, we had to use the essential boundary conditions to set up equations and then solve them to find values for one or more parameters. However, in the finite element form, making the solution admissible is trivial. For example, if the essential boundary condition is $u(x_2) = 5$, then all we have to do is to set $u_2 = 5$ and we are done.

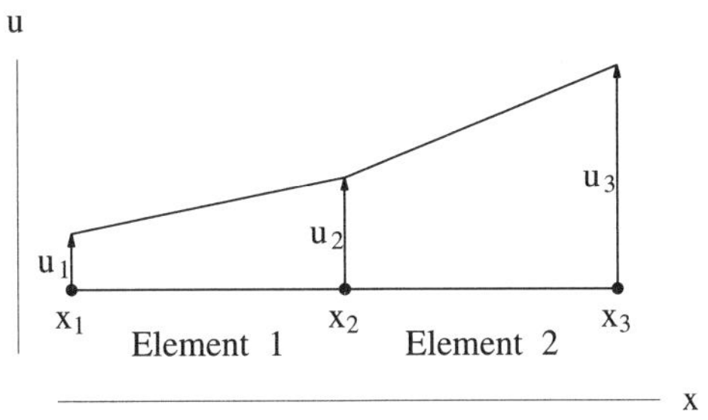

Figure 2.14. A simple two-element model

(ii) The finite element form is suitable for direct assembly of element equations. Assume we are using two elements to model a solution domain, as shown in Figure 2.14. The node 2 is common between the two elements. We can get a piecewise linear solution for the entire domain just by making sure that at the common node the two elements use the same nodal value. This simple observation makes it possible to perform computations independently for each element and then simply add the contributions from each element. During this so-called assembly process the nodes common between different elements are assigned the same nodal degree of freedom.

(iii) In order to add more terms to the assumed solution, we have two options now. We can use higher order polynomials to derive higher order finite element solutions by following exactly the same procedure used for deriving the linear solution. Alternatively, we can simply use a large number of linear elements. By using a sufficiently large number of simple elements, we can get reasonably accurate solutions to even complicated differential equations.

The finite element assumed solutions are usually written using a matrix notation. A matrix notation is not necessary for this simple element. However, for more complicated elements it is almost impossible to write element solutions concisely without using the matrix notation. The linear finite element solution can be written as follows:

$$u(x) = \frac{x - x_2}{x_1 - x_2} u_1 + \frac{x - x_1}{x_2 - x_1} u_2 = \sum_{i=1}^{2} N_i(x) u_i = \begin{pmatrix} N_1 & N_2 \end{pmatrix} \begin{pmatrix} u_1 \\ u_2 \end{pmatrix} \equiv \boldsymbol{N}^T \boldsymbol{d}$$

where

$$N_1 = \frac{x - x_2}{x_1 - x_2}; \qquad N_2 = \frac{x - x_1}{x_2 - x_1}$$

The $N_i(x)$ are known as interpolation or shape functions and u_i are the nodal unknowns and $\boldsymbol{N}$ is a 2×1 column vector of interpolation functions. The superscript T denotes the transpose of vector $\boldsymbol{N}$ to make it into a row vector for matrix multiplication with the 2×1 vector $\boldsymbol{d}$ of nodal degrees of freedom.

Note that the interpolation function N_1 is 1 at node 1 and 0 at node 2 while N_2 is 1 at node 2 and 0 at node 1. The function N_1 therefore defines the influence of u_1 on the assumed solution and N_2 that of u_2. Because of this, the interpolation functions are also known as influence functions.

2.6.2 Lagrange Interpolation

An interpolation function is a smooth function that passes through all data points in a given data set. Thus, treating the nodal locations and unknown solutions at these locations as data points, the problem of writing a finite element solution for second-order differential equations boils down to choosing an appropriate interpolation technique.

There are a variety of interpolation methods available in the literature. One simple and well known technique is the *Lagrange interpolation*. If we have n data points $\{x_i, u_i\}$, $i = 1, 2, \ldots, n$ [where $u(x_i) \equiv u_i$], then the Lagrange interpolation formula for passing a polynomial of degree $n - 1$ through these points can be written as follows:

$$u(x) = \sum_{i=1}^{n} L_i(x) u_i \equiv \begin{pmatrix} L_1 & L_2 & L_3 & \cdots \end{pmatrix} \begin{pmatrix} u_1 \\ u_2 \\ u_3 \\ \vdots \end{pmatrix} \equiv \boldsymbol{N}^T \boldsymbol{d}$$

where $L_i(x)$ are the Lagrange interpolation functions given by the following formula:

$$L_i(x) \equiv N_i = \prod_{\substack{j=1 \\ j \neq i}}^{n} \frac{x - x_j}{x_i - x_j} \equiv \frac{x - x_1}{x_i - x_1} \times \frac{x - x_2}{x_i - x_2} \times \cdots \times \frac{x - x_{i-1}}{x_i - x_{i-1}} \times \frac{x - x_{i+1}}{x_i - x_{i+1}} \times \cdots \times \frac{x - x_n}{x_i - x_n}$$

The symbol Π indicates a product of terms. For writing the ith interpolation function, the product includes all terms except the ith. Thus there are $n - 1$ terms in the product and each interpolation function is of degree $n - 1$, where n is the number of data points.

Using this formula, a linear interpolation ($n = 2$) can be written as follows:

$$u(x) = N_1(x) u_1 + N_2(x) u_2$$
$$N_1(x) = \frac{x - x_2}{x_1 - x_2}; \qquad N_2(x) = \frac{x - x_1}{x_2 - x_1}$$

These are the same shape functions derived in the previous section. A quadratic polynomial ($n = 3$) can be written as follows:

$$u(x) = N_1(x) u_1 + N_2(x) u_2 + N_3(x) u_3$$
$$N_1(x) = \frac{x - x_2}{x_1 - x_2} \times \frac{x - x_3}{x_1 - x_3}; \qquad N_2(x) = \frac{x - x_1}{x_2 - x_1} \times \frac{x - x_3}{x_2 - x_3}; \qquad N_3(x) = \frac{x - x_1}{x_3 - x_1} \times \frac{x - x_2}{x_3 - x_2}$$

Example 2.5 Write Lagrange interpolation functions to pass a polynomial through four points. Plot the resulting function if the nodal values and coordinates are as follows:

Point	Data Value
0	$\frac{1}{2}$
1	1
2	3
$\frac{5}{2}$	$-\frac{1}{2}$

We have four data points; therefore $n = 4$. Each Lagrange interpolation function will be a cubic function. Using the Lagrange interpolation formula, the four interpolation function are

$$N_1 = \frac{(x - x_2)(x - x_3)(x - x_4)}{(x_1 - x_2)(x_1 - x_3)(x_1 - x_4)}; \qquad N_2 = \frac{(x - x_1)(x - x_3)(x - x_4)}{(x_2 - x_1)(x_2 - x_3)(x_2 - x_4)}$$

$$N_3 = \frac{(x - x_1)(x - x_2)(x - x_4)}{(x_3 - x_1)(x_3 - x_2)(x_3 - x_4)}; \qquad N_4 = \frac{(x - x_1)(x - x_2)(x - x_3)}{(x_4 - x_1)(x_4 - x_2)(x_4 - x_3)}$$

Substituting the numerical values of the coordinates, $x_1 = 0$, $x_2 = 1$, $x_3 = 2$, and $x_4 = \frac{5}{2}$, we get

$$N_1 = -\tfrac{1}{5}\left(x - \tfrac{5}{2}\right)(x - 2)(x - 1); \qquad N_2 = \tfrac{2}{3}\left(x - \tfrac{5}{2}\right)(x - 2)x$$

$$N_3 = -\left(x - \tfrac{5}{2}\right)(x - 1)x; \qquad N_4 = \tfrac{8}{15}(x - 2)(x - 1)x$$

Multiplying these interpolation functions with the given nodal values, $u_1 = \frac{1}{2}$, $u_2 = 1$, $u_3 = 3$, and $u_4 = -\frac{1}{2}$, we get the following cubic function that passes through the given data point:

$$u(x) = u_1N_1 + u_2N_2 + u_3N_3 + u_4N_4 = \frac{N_1}{2} + N_2 + 3N_3 - \frac{N_4}{2}$$

$$u(x) = -\frac{27x^3}{10} + \frac{177x^2}{20} - \frac{113x}{20} + \frac{1}{2}$$

The interpolated function is plotted in Figure 2.15. The given data points are shown on the graph with large dots. The function clearly passes through all four data points.

2.6.3 Galerkin Weighting Functions in Finite Element Form

As seen in the previous sections, the finite element assumed solution is expressed in the following form:

$$u(x) = \begin{pmatrix} N_1 & N_2 & \cdots \end{pmatrix} \begin{pmatrix} u_1 \\ u_2 \\ \vdots \end{pmatrix} \equiv \boldsymbol{N}^T \boldsymbol{d}$$

The weighting functions in the Galerkin method are the derivatives of the assumed solution with respect to the unknown coefficients. The unknown coefficients in the finite element

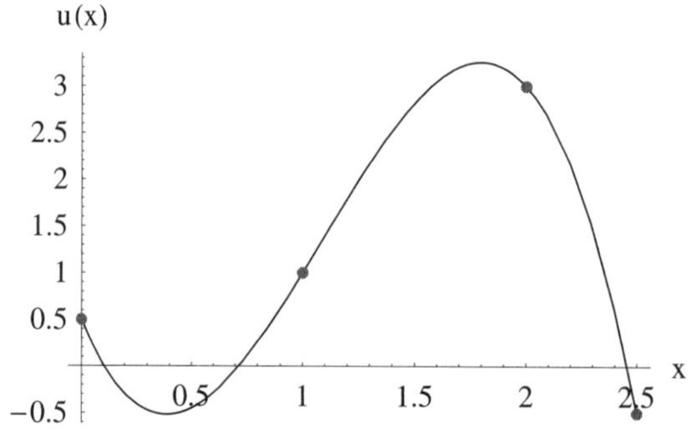

Figure 2.15. Interpolated function passing through given data points

form are the nodal degrees of freedom. Therefore

$$w_i = \frac{du}{du_i} \equiv N_i$$

Thus the interpolation functions also play the role of weighting functions when using the Galerkin method.

From this definition of weighting functions, it is easy to see that, if a particular nodal value is specified because of an essential boundary condition, then that degree of freedom is not a variable and hence there is no weighting function corresponding to that degree of freedom. As discussed in Chapter 1, the essential boundary conditions are typically ignored at the element level, and thus all interpolation functions for an element are used as weighting functions to obtain the element equations. It is only after assembly that we turn our attention to imposing essential boundary conditions. The equation corresponding to a specified nodal value clearly does not make sense because of the way the weighting functions are defined. Hence the equations corresponding to specified nodal values must be removed from the global equations before solving for the actual nodal unknowns.

2.6.4 Hermite Interpolation for Fourth-Order Problems

Essential boundary conditions for a fourth-order differential equation consist of the solution and its first derivative. To satisfy these essential boundary conditions at the ends of an element, we must choose the solution and its first derivative as nodal degrees of freedom. Thus the simplest two-node line element for a fourth-order problem has a total of four degrees of freedom, as shown in Figure 2.16. The element extends from x_1 to x_2 and has a length $l = x_2 - x_1$. For simplicity the left end of the element is assumed at $x = 0$ and the right end is at $x = l$.

The Lagrange interpolation formula cannot be used to interpolate when we have the solution and its first derivative specified at each data point. We can generate two independent interpolations between $\{(x_1, u_1), (x_2, u_2)\}$ and $\{(x_1, u_1'), (x_2, u_2')\}$ if we like. However, then there will be no connection between the solution and its first derivative. The appropriate method to interpolate between data values and their first derivatives at each point is known

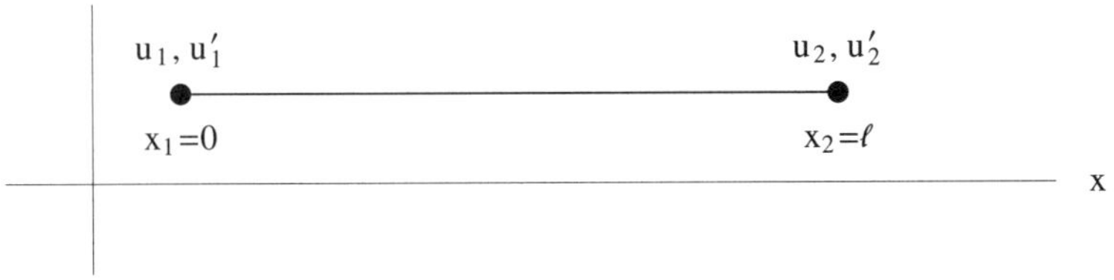

Figure 2.16. A two-node element for fourth-order problems

as *Hermite interpolation*. For this two-node element this interpolation is derived by using the basic principles. A general Hermite interpolation formula is given later.

Derivation of Hermite Interpolation for a Two-Node Element Since we have four nodal unknowns, we can derive the finite element form of the solution (i.e., the shape functions) by starting from a cubic polynomial with four coefficients and then evaluating these coefficients in terms of nodal unknowns as follows:

$$u(x) = a_0 + a_1 x + a_2 x^2 + a_3 x^3$$

$$u'(x) = a_1 + 2xa_2 + 3x^2 a_3$$

$$\text{at } x = 0: \quad \Longrightarrow u_1 = a_0; \qquad u_1' = a_1$$

$$\text{at } x = l: \quad \Longrightarrow u_2 = a_0 + la_1 + l^2 a_2 + l^3 a_3; \qquad u_2' = a_1 + 2la_2 + 3l^2 a_3$$

The first two equations give two of the coefficients. Solving the other two equations for a_2 and a_3, we get

$$a_2 = -\frac{3u_1 - 3u_2 + 2lu_1' + lu_2'}{l^2}; \qquad a_3 = -\frac{-2u_1 + 2u_2 - lu_1' - lu_2'}{l^3}$$

Thus a cubic solution over the element in terms of nodal degrees of freedom is as follows:

$$u(x) = u_1 + xu_1' - \frac{x^3(-2u_1 + 2u_2 - lu_1' - lu_2')}{l^3} - \frac{x^2(3u_1 - 3u_2 + 2lu_1' + lu_2')}{l^2}$$

Collecting terms involving the degrees of freedom together, we can write this solution as follows:

$$u(x) = u_1\left(1 + \frac{2x^3}{l^3} - \frac{3x^2}{l^2}\right) + u_1'\left(x + \frac{x^3}{l^2} - \frac{2x^2}{l}\right) + u_2\left(-\frac{2x^3}{l^3} + \frac{3x^2}{l^2}\right) + u_2'\left(\frac{x^3}{l^2} - \frac{x^2}{l}\right)$$

$$u(x) = \begin{pmatrix} N_1 & N_2 & N_3 & N_4 \end{pmatrix} \begin{pmatrix} u_1 \\ u_1' \\ u_2 \\ u_2' \end{pmatrix} \equiv \boldsymbol{N}^T \boldsymbol{d}$$

The four interpolation functions are

$$N_1 = 1 + \frac{2x^3}{l^3} - \frac{3x^2}{l^2}; \qquad N_2 = x + \frac{x^3}{l^2} - \frac{2x^2}{l}; \qquad N_3 = -\frac{2x^3}{l^3} + \frac{3x^2}{l^2}; \qquad N_4 = \frac{x^3}{l^2} - \frac{x^2}{l}$$

General Formula for Hermite Interpolation Given function values u_i and their first derivatives u_i' at n data points $x_1, x_2, \ldots, x_n$, a polynomial of degree $2n - 1$ that fits all data can be written as follows:

$$u(x) = \sum_{i=1}^{n} P_i(x)u_i + \sum_{i=1}^{n} Q_i(x)u_i' \equiv (P_1 \quad Q_1 \quad \cdots \quad P_n \quad Q_n)\begin{pmatrix} u_1 \\ u_1' \\ \vdots \\ u_n \\ u_n' \end{pmatrix} = \boldsymbol{N}^T\boldsymbol{d}$$

where $P_i(x)$ and $Q_i(x)$ are polynomials of degree $2n - 1$. These polynomials are expressed in terms of Lagrange interpolation functions $L_i(x)$ as follows:

$$P_i(x) = (1 - 2L_i'(x_i)(x - x_i))L_i^2(x)$$
$$Q_i(x) = (x - x_i)L_i^2(x)$$

Recall that the Lagrange interpolation functions are as follows:

$$L_i(x) = \prod_{\substack{j=1 \\ j\neq i}}^{n} \frac{x - x_j}{x_i - x_j} \equiv \frac{x - x_1}{x_i - x_1} \times \frac{x - x_2}{x_i - x_2} \times \cdots \times \frac{x - x_{i-1}}{x_i - x_{i-1}} \times \frac{x - x_{i+1}}{x_i - x_{i+1}} \times \cdots \times \frac{x - x_n}{x_i - x_n}$$

The complete vector of interpolation functions is denoted by $\boldsymbol{N}$ and the corresponding vector of nodal values by $\boldsymbol{d}$. The functions P_i interpolate between the data values and Q_i between the derivative values. Each interpolation function is of degree $2n - 1$, where n is the number of data points. Note that functions $P_i(x)$ are 1 at the ith data point and zero at all other data points. The first derivatives of P_i, $i = 1, \ldots,$ are zero at all data points. On the other hand, the functions $Q_i(x)$ are zero at all data points while their first derivatives are 1 at ith point and zero everywhere else. That is,

$$P_i(x_j) = \delta_{ij}; \qquad P_i'(x_j) = 0$$
$$Q_i(x_j) = 0; \qquad Q_i'(x_j) = \delta_{ij}$$

where δ_{ij} is the so-called dirac delta function, which is equal to 1 when $i = j$ and equal to 0 otherwise.

Example 2.6 Using Hermite interpolation, obtain a function that passes through the following two points with the slopes at these points as given in the following table:

Point, x_i	Data Value, u_i	Derivative Value, u'_i
0	1	0
1	2	−1

Data point locations: {0, 1}

Lagrange interpolation functions for these points:

$$\{L_1 = 1 - x, L_2 = x\}$$

Derivatives of Lagrange interpolation functions:

$$\{L'_1 = -1, L'_2 = 1\}$$

Derivatives at the given points:

$$\{L'_1(0) = -1, L'_2(1) = 1\}$$

Using these, the Hermite interpolations for data values are

$$\{P_1 = 2x^3 - 3x^2 + 1, P_2 = 3x^2 - 2x^3\}$$

Those for derivative values are

$$\{Q_1 = x^3 - 2x^2 + x, Q_2 = x^3 - x^2\}$$

Vector of Hermite interpolation functions

$$\boldsymbol{N}^T = \{2x^3 - 3x^2 + 1, x^3 - 2x^2 + x, 3x^2 - 2x^3, x^3 - x^2\}$$

These are the same interpolation functions derived earlier with $l = 1$. The vector of given nodal data values and their derivatives is

$$\boldsymbol{d}^T = (1 \quad 0 \quad 2 \quad -1)$$

Thus the function that interpolates between these values is as follows:

$$u(x) = (2x^3 - 3x^2 + 1 \quad x^3 - 2x^2 + x \quad 3x^2 - 2x^3 \quad x^3 - x^2)\begin{pmatrix} 1 \\ 0 \\ 2 \\ -1 \end{pmatrix} = -3x^3 + 4x^2 + 1$$

The interpolated function is plotted in Figure 2.17. The given data points are shown on the graph with large dots. The function clearly passes through both data points and has the specified slope of 0 at $x = 0$ and -1 at $x = 1$.

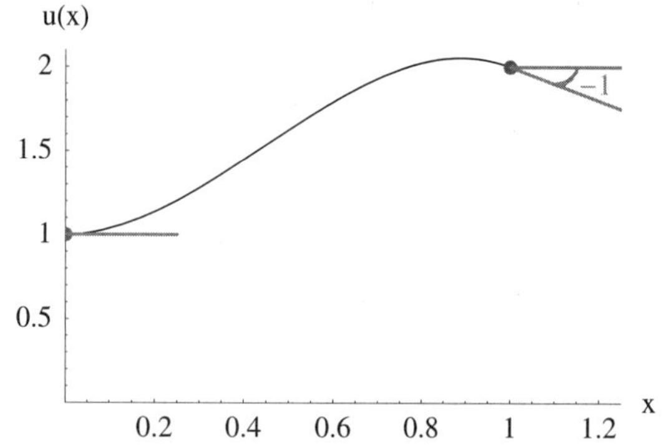

Figure 2.17. Interpolated function

Example 2.7 Using Hermite interpolation, obtain a function for the following set of data:

Point, x_i	Data Value, u_i	Derivative Value, u_i'
0	1	0
1	2	0
3	0	1

The Hermite interpolation functions to pass a polynomial through three data points and their first derivatives located at $x = 0, 1, 3$ are as follows:

Data point locations: {0, 1, 3}

Lagrange interpolation functions for these points:

$$\left\{L_1 = \tfrac{1}{3}(x-3)(x-1), L_2 = -\tfrac{1}{2}(x-3)x, L_3 = \tfrac{1}{6}(x-1)x\right\}$$

Derivatives of Lagrange interpolation functions:

$$\left\{L_1' = \frac{x-3}{3} + \frac{x-1}{3}, L_2' = \frac{3-x}{2} - \frac{x}{2}, L_3' = \frac{x-1}{6} + \frac{x}{6}\right\}$$

Derivatives at the given points:

$$\left\{L_1'(0) = -\tfrac{4}{3}, L_2'(1) = \tfrac{1}{2}, L_3'(3) = \tfrac{5}{6}\right\}$$

Using these, the Hermite interpolations for data values are

$$\left\{P_1 = \frac{8x^5}{27} - \frac{61x^4}{27} + \frac{152x^3}{27} - \frac{14x^2}{3} + 1, P_2 = -\frac{x^5}{4} + 2x^4 - \frac{21x^3}{4} + \frac{9x^2}{2},\right.$$

$$\left. P_3 = -\frac{5x^5}{108} + \frac{7x^4}{27} - \frac{41x^3}{108} + \frac{x^2}{6}\right\}$$

Those for derivative values are

$$\left\{Q_1 = \frac{x^5}{9} - \frac{8x^4}{9} + \frac{22x^3}{9} - \frac{8x^2}{3} + x, Q_2 = \frac{x^5}{4} - \frac{7x^4}{4} + \frac{15x^3}{4} - \frac{9x^2}{4}, \right.$$

$$\left. Q_3 = \frac{x^5}{36} - \frac{5x^4}{36} + \frac{7x^3}{36} - \frac{x^2}{12}\right\}$$

Vector of Hermite interpolation functions:

$$\boldsymbol{N}^T = \left\{\frac{8x^5}{27} - \frac{61x^4}{27} + \frac{152x^3}{27} - \frac{14x^2}{3} + 1, \frac{x^5}{9} - \frac{8x^4}{9} + \frac{22x^3}{9} - \frac{8x^2}{3} + x, \right.$$

$$-\frac{x^5}{4} + 2x^4 - \frac{21x^3}{4} + \frac{9x^2}{2}, \frac{x^5}{4} - \frac{7x^4}{4} + \frac{15x^3}{4} - \frac{9x^2}{4},$$

$$\left. -\frac{5x^5}{108} + \frac{7x^4}{27} - \frac{41x^3}{108} + \frac{x^2}{6}, \frac{x^5}{36} - \frac{5x^4}{36} + \frac{7x^3}{36} - \frac{x^2}{12}\right\}$$

The vector of function values and their derivatives is $(1 \quad 0 \quad 2 \quad 0 \quad 0 \quad 1)$. Thus we have

$$u(x) = N_1 + 2N_3 + N_6 = -\frac{19x^5}{108} + \frac{173x^4}{108} - \frac{505x^3}{108} + \frac{17x^2}{4} + 1$$

The interpolated function is plotted in Figure 2.18. The given data points are shown on the graph with large dots. The function clearly passes through all three data points and has the specified slope at these points.

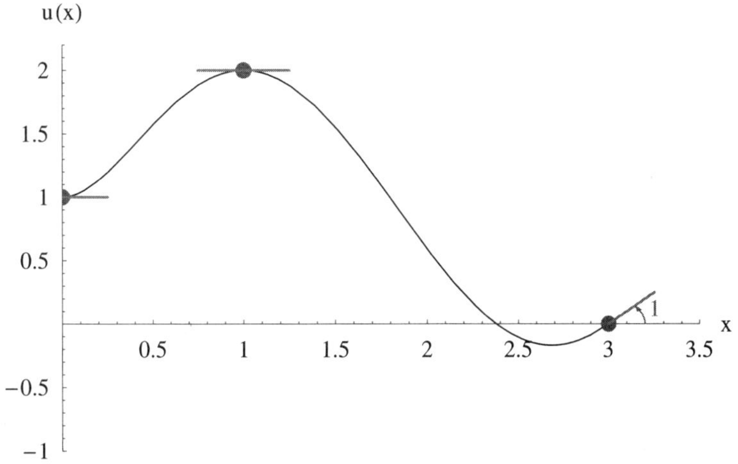

Figure 2.18. Interpolated function

2.7 FINITE ELEMENT SOLUTION OF AXIAL DEFORMATION PROBLEMS

As mentioned earlier, using the assumed solution in the form of interpolation functions, it is possible to perform computations independently for each element and then simply add the contributions from different elements. Thus we can employ the standard six-step finite element procedure to obtain approximate solutions:

1. Development of element equations
2. Discretization of solution domain into a finite element mesh
3. Assembly of element equations
4. Introduction of boundary conditions
5. Solution for nodal unknowns
6. Computation of solution and related quantities over each element

Except for the derivation of element equations, all steps have been discussed in detail in Chapter 1. In this section the procedure for deriving element equations is illustrated considering a two-node element for axial deformation problem. Several numerical examples are then presented that use these equations to solve a variety of axial deformation problems.

2.7.1 Two-Node Uniform Bar Element for Axial Deformations

The simplest element for axial deformations of bars is a two-node line element, as shown in Figure 2.19. The element extends from x_1 to x_2 and has a length $L = x_2 - x_1$. The circles represent nodes. The unknown solutions at the nodes are indicated by u_1 and u_2. In addition to the distributed load $q(x)$, the element may be subjected to concentrated axial loads P_i applied at the ends of the element.

With these assumptions the governing differential equation over an element is

$$\frac{d}{dx}\left(AE\frac{du}{dx}\right) + q = 0; \qquad x_1 < x < x_2$$

Any possible concentrated loads at element ends form natural boundary conditions. With the sign convention explained earlier we have

$$-AE\frac{du(x_1)}{dx} = P_1; \qquad AE\frac{du(x_2)}{dx} = P_2$$

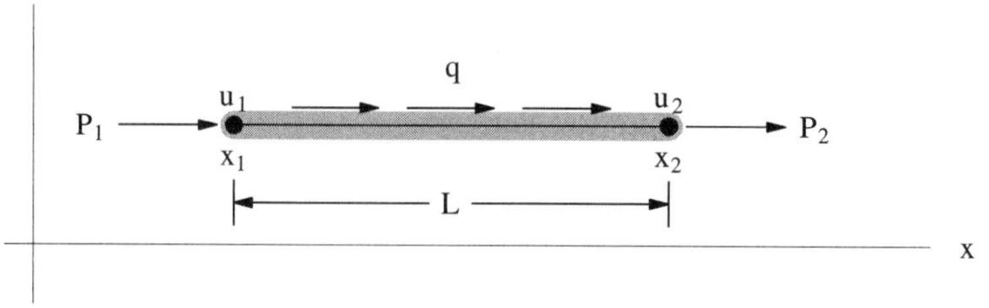

Figure 2.19. A simple two-node element for axial deformation

The primary unknown is the axial displacement u. Once the displacement is known, axial strain, stress, and force can be computed from the following relationships:

$$\epsilon_x = \frac{du}{dx}; \qquad \sigma_x = E\epsilon_x; \qquad F = A\sigma_x$$

Linear Assumed Solution The assumed solution is a linear interpolation between the nodal unknowns. Thus

$$u(x) = \begin{pmatrix} \frac{x - x_2}{x_1 - x_2} & \frac{x - x_1}{x_2 - x_1} \end{pmatrix} \begin{pmatrix} u_1 \\ u_2 \end{pmatrix} = (N_1 \quad N_2) \begin{pmatrix} u_1 \\ u_2 \end{pmatrix} \equiv \boldsymbol{N}^T \boldsymbol{d}$$

$$N_1 = \frac{x - x_2}{x_1 - x_2} = -\frac{x - x_2}{L}; \qquad N_2 = \frac{x - x_1}{x_2 - x_1} = \frac{x - x_1}{L}$$

We will need $u'(x)$ in the later derivation. Differentiating with respect to x, we get

$$u'(x) = \frac{du(x)}{dx} = \begin{pmatrix} \frac{1}{x_1 - x_2} & \frac{1}{x_2 - x_1} \end{pmatrix} \begin{pmatrix} u_1 \\ u_2 \end{pmatrix} = (N_1' \quad N_2') \begin{pmatrix} u_1 \\ u_2 \end{pmatrix} \equiv \boldsymbol{B}^T \boldsymbol{d}$$

$$N_1' = \frac{1}{x_1 - x_2} = -\frac{1}{L}; \qquad N_2' = \frac{1}{x_2 - x_1} = \frac{1}{L}$$

Element Equations Using Galerkin Method The weak form can be written using standard steps of writing the weighted residual, integration by parts, and incorporating the natural boundary conditions. The weighting functions are the same as the interpolation functions N_i.

With $u(x)$ an assumed solution the residual is

$$e(x) = q + \frac{d}{dx}(AEu')$$

Multiplying by $N_i(x)$ and writing the integral over the given limits, the Galerkin weighted residual is

$$\int_{x_1}^{x_2} \left(qN_i + \frac{d}{dx}(AEu')N_i \right) dx = 0$$

Using integration by parts,

$$AEN_i(x_2)u'(x_2) - AEN_i(x_1)u'(x_1) + \int_{x_1}^{x_2} (qN_i - AEu'N_i')\, dx = 0$$

Given the NBC for the problem,

$$-P_1 - AEu'(x_1) = 0; \qquad AEu'(x_2) - P_2 = 0$$

Rearranging,

$$u'(x_1) \to -\frac{P_1}{AE}; \qquad u'(x_2) \to \frac{P_2}{AE}$$

Thus the boundary terms in the weak form reduce to

$$P_1 N_i(x_1) + P_2 N_i(x_2)$$

Assuming admissible solutions, the final weak form is as follows:

$$P_1 N_i(x_1) + P_2 N_i(x_2) + \int_{x_1}^{x_2} (qN_i - AEu'N_i')\,dx = 0$$

With the two interpolation functions, the two equations for the element are as follows:

$$\int_{x_1}^{x_2} (qN_1(x) - AEu'(x)N_1'(x))\,dx + P_1 N_1(x_1) + P_2 N_1(x_2) = 0$$

$$\int_{x_1}^{x_2} (qN_2(x) - AEu'(x)N_2'(x))\,dx + P_1 N_2(x_1) + P_2 N_2(x_2) = 0$$

Noting the property of the Lagrange interpolation functions that $N_1(x_1) = 1$, $N_1(x_2) = 0$, $N_2(x_1) = 0$, and $N_2(x_2) = 1$ and writing the two equations together using matrix notation, we have

$$\int_{x_1}^{x_2} \left(\begin{pmatrix} N_1 \\ N_2 \end{pmatrix} q - \begin{pmatrix} N_1' \\ N_2' \end{pmatrix} AEu'(x) \right) dx + \begin{pmatrix} P_1 \\ P_2 \end{pmatrix} = \begin{pmatrix} 0 \\ 0 \end{pmatrix}$$

Substituting for $u'(x)$, we have

$$\int_{x_1}^{x_2} \left(\begin{pmatrix} N_1 \\ N_2 \end{pmatrix} q - \begin{pmatrix} N_1' \\ N_2' \end{pmatrix} AE \begin{pmatrix} N_1' & N_2' \end{pmatrix} \begin{pmatrix} u_1 \\ u_2 \end{pmatrix} \right) dx + \begin{pmatrix} P_1 \\ P_2 \end{pmatrix} = \begin{pmatrix} 0 \\ 0 \end{pmatrix}$$

or

$$\int_{x_1}^{x_2} \boldsymbol{N} q\,dx - \int_{x_1}^{x_2} \boldsymbol{B} AE \boldsymbol{B}^T \boldsymbol{d}\,dx + \boldsymbol{r}_p = \boldsymbol{0}$$

where $\boldsymbol{r}_p$ is the 2×1 vector of element nodal loads $(P_1, P_2)^T$ and $\boldsymbol{0}$ is a 2×1 vector of zeros. Rearranging terms and taking nodal degrees of freedom out of the integral sign because they are not functions of x, we have

$$\int_{x_1}^{x_2} \boldsymbol{B} AE \boldsymbol{B}^T\,dx\,\boldsymbol{d} = \int_{x_1}^{x_2} \boldsymbol{N} q\,dx + \boldsymbol{r}_p \Longrightarrow \boldsymbol{kd} = \boldsymbol{r}_q + \boldsymbol{r}_p$$

where

$$\boldsymbol{k} = \int_{x_1}^{x_2} \boldsymbol{B} AE \boldsymbol{B}^T\,dx; \qquad \boldsymbol{r}_q = \int_{x_1}^{x_2} \boldsymbol{N} q\,dx; \qquad \boldsymbol{r}_p = \begin{pmatrix} P_1 \\ P_2 \end{pmatrix}$$

Note the careful arrangement of terms when the two equations were written together. Since $AEu'(x)$ is a scalar, the second term inside the integral of the weak form can be written in two different ways as follows:

$$\int_{x_1}^{x_2} \begin{pmatrix} N_1' \\ N_2' \end{pmatrix} AEu'(x)\,dx \qquad \text{or} \qquad \int_{x_1}^{x_2} AEu'(x) \begin{pmatrix} N_1' \\ N_2' \end{pmatrix} dx$$

The first form was used above to get the equations in the convenient form presented. With the second form we have the following situation:

$$\int_{x_1}^{x_2} AEu'(x) \begin{pmatrix} N_1' \\ N_2' \end{pmatrix} dx = \int_{x_1}^{x_2} AE\boldsymbol{B}^T\boldsymbol{dB}\,dx$$

Notice that now $\boldsymbol{d}$ cannot be taken out of the integral sign (recall that with matrices $\boldsymbol{dB} \neq \boldsymbol{Bd}$) and thus the form is not suitable for writing element equations in a compact form as a system of linear equations. We must use the first form and take unknown nodal values out so that the remaining terms can be integrated to give element equations.

To write explicit equations, we must substitute derivatives of the interpolation functions and carry out integrations. For simplicity, it is assumed that EA and q are constant over an element. Thus we have

$$\boldsymbol{k} = \int_{x_1}^{x_2} \boldsymbol{B}AE\boldsymbol{B}^T\,dx = \begin{pmatrix} \int_{x_1}^{x_2} AE\frac{1}{L^2}\,dx & -\int_{x_1}^{x_2} AE\frac{1}{L^2}\,dx \\ -\int_{x_1}^{x_2} AE\frac{1}{L^2}\,dx & \int_{x_1}^{x_2} AE\frac{1}{L^2}\,dx \end{pmatrix} = \frac{AE}{L}\begin{pmatrix} 1 & -1 \\ -1 & 1 \end{pmatrix}$$

$$\boldsymbol{r}_q = \int_{x_1}^{x_2} \boldsymbol{N}q\,dx = \begin{pmatrix} \int_{x_1}^{x_2} -\frac{x-x_2}{L}q\,dx \\ \int_{x_1}^{x_2} \frac{x-x_1}{L}q\,dx \end{pmatrix} = \frac{qL}{2}\begin{pmatrix} 1 \\ 1 \end{pmatrix}$$

We now have the two-node element equations for the axial deformation problem:

$$\frac{AE}{L}\begin{pmatrix} 1 & -1 \\ -1 & 1 \end{pmatrix}\begin{pmatrix} u_1 \\ u_2 \end{pmatrix} = \frac{qL}{2}\begin{pmatrix} 1 \\ 1 \end{pmatrix} + \begin{pmatrix} P_1 \\ P_2 \end{pmatrix}$$

$$\boldsymbol{kd} = \boldsymbol{r}_q + \boldsymbol{r}_p$$

$$\boldsymbol{k} = \frac{AE}{L}\begin{pmatrix} 1 & -1 \\ -1 & 1 \end{pmatrix}; \qquad \boldsymbol{r}_q = \frac{qL}{2}\begin{pmatrix} 1 \\ 1 \end{pmatrix}; \qquad \boldsymbol{r}_p = \begin{pmatrix} P_1 \\ P_2 \end{pmatrix}$$

Element Equations Using Rayleigh-Ritz Method The Galerkin method was used to derive element equations in the previous section. The same equations can also be derived using the Rayleigh-Ritz method. The potential energy function for the element is as follows:

$$\Pi = U - W = \frac{1}{2}\int_{x_1}^{x_2} AE\,(u'(x))^2\,dx - \int_{x_1}^{x_2} qu\,dx - P_1u_1 - P_2u_2$$

The square of the first derivative of the assumed solution must be written carefully so that the nodal unknowns can be taken out of the integral sign. In order to achieve this goal we proceed as follows:

$$u'(x) = \boldsymbol{B}^T \boldsymbol{d}$$
$$(u'(x))^2 = (u'(x))^T u'(x) = (\boldsymbol{B}^T \boldsymbol{d})^T \boldsymbol{B}^T \boldsymbol{d} = \boldsymbol{d}^T \boldsymbol{B}\boldsymbol{B}^T \boldsymbol{d}$$

where we have used the rule for the transpose of a product of two matrices $(\boldsymbol{AB})^T = \boldsymbol{B}^T \boldsymbol{A}^T$. The strain energy term can now be evaluated as follows:

$$U = \frac{1}{2}\int_{x_1}^{x_2} AE\boldsymbol{d}^T \boldsymbol{B}\boldsymbol{B}^T \boldsymbol{d}\, dx = \frac{1}{2}\boldsymbol{d}^T \int_{x_1}^{x_2} AE\boldsymbol{B}\boldsymbol{B}^T\, dx\, \boldsymbol{d} \equiv \frac{1}{2}\boldsymbol{d}^T \boldsymbol{k}\boldsymbol{d}$$

where

$$\boldsymbol{k} = \int_{x_1}^{x_2} AE\boldsymbol{B}\boldsymbol{B}^T\, dx = \begin{pmatrix} \int_{x_1}^{x_2} AE\frac{1}{L^2}\, dx & -\int_{x_1}^{x_2} AE\frac{1}{L^2}\, dx \\ -\int_{x_1}^{x_2} AE\frac{1}{L^2}\, dx & \int_{x_1}^{x_2} AE\frac{1}{L^2}\, dx \end{pmatrix} = \frac{AE}{L}\begin{pmatrix} 1 & -1 \\ -1 & 1 \end{pmatrix}$$

The work done by the distributed load can be evaluated as follows:

$$W_q = \int_{x_1}^{x_2} qu\, dx = \int_{x_1}^{x_2} q\boldsymbol{N}^T \boldsymbol{d}\, dx = \boldsymbol{r}_q^T \boldsymbol{d} \equiv \boldsymbol{d}^T \boldsymbol{r}_q$$

where

$$\boldsymbol{r}_q^T = \int_{x_1}^{x_2} q\boldsymbol{N}^T\, dx \Longrightarrow \boldsymbol{r}_q = \int_{x_1}^{x_2} \boldsymbol{N}q\, dx = \begin{pmatrix} \int_{x_1}^{x_2} -\frac{x-x_2}{L} q\, dx \\ \int_{x_1}^{x_2} \frac{x-x_1}{L} q\, dx \end{pmatrix} = \frac{qL}{2}\begin{pmatrix} 1 \\ 1 \end{pmatrix}$$

The work done by the concentrated loads applied at element ends is

$$W_p = P_1 u_1 - P_2 u_2 = (P_1 \quad P_2)\begin{pmatrix} u_1 \\ u_2 \end{pmatrix} \equiv \boldsymbol{r}_p^T \boldsymbol{d} \equiv \boldsymbol{d}^T \boldsymbol{r}_p$$

The potential energy can now be written as follows:

$$\Pi = U - W = \frac{1}{2}\boldsymbol{d}^T \boldsymbol{k}\boldsymbol{d} - \boldsymbol{r}_q^T \boldsymbol{d} - \boldsymbol{r}_p^T \boldsymbol{d} = \boldsymbol{d}^T\left(\frac{1}{2}\boldsymbol{k}\boldsymbol{d} - \boldsymbol{r}_q - \boldsymbol{r}_p\right)$$

The necessary conditions for the minimum of the potential energy give (see details below)

$$\frac{\partial \Pi}{\partial \boldsymbol{d}} = \left(\frac{1}{2}\boldsymbol{k}\boldsymbol{d} - \boldsymbol{r}_q - \boldsymbol{r}_p\right) + \frac{1}{2}\boldsymbol{k}\boldsymbol{d} = 0$$

Rearranging terms, we get the same element equations as those obtained by using the Galerkin method:

$$\boldsymbol{k}\boldsymbol{d} = \boldsymbol{r}_q + \boldsymbol{r}_p \Longrightarrow \frac{AE}{L}\begin{pmatrix} 1 & -1 \\ -1 & 1 \end{pmatrix}\begin{pmatrix} u_1 \\ u_2 \end{pmatrix} = \frac{qL}{2}\begin{pmatrix} 1 \\ 1 \end{pmatrix} + \begin{pmatrix} P_1 \\ P_2 \end{pmatrix}$$

The detailed justification for the way the differentiation of the potential energy with respect to the nodal variables is carried out is as follows:

$$\Pi = \boldsymbol{d}^T\left(\tfrac{1}{2}\boldsymbol{kd} - \boldsymbol{r}_q - \boldsymbol{r}_p\right) = (u_1 \quad u_2)\left(\frac{1}{2}\boldsymbol{k}\begin{pmatrix}u_1\\u_2\end{pmatrix} - \boldsymbol{r}_q - \boldsymbol{r}_p\right)$$

Using the product rule of differentiation,

$$\frac{\partial\Pi}{\partial u_1} = (1 \quad 0)\left(\frac{1}{2}\boldsymbol{k}\begin{pmatrix}u_1\\u_2\end{pmatrix} - \boldsymbol{r}_q - \boldsymbol{r}_p\right) + (u_1 \quad u_2)\left(\frac{1}{2}\boldsymbol{k}\begin{pmatrix}1\\0\end{pmatrix}\right) = 0$$

$$\frac{\partial\Pi}{\partial u_2} = (0 \quad 1)\left(\frac{1}{2}\boldsymbol{k}\begin{pmatrix}u_1\\u_2\end{pmatrix} - \boldsymbol{r}_q - \boldsymbol{r}_p\right) + (u_1 \quad u_2)\left(\frac{1}{2}\boldsymbol{k}\begin{pmatrix}0\\1\end{pmatrix}\right) = 0$$

Combining the two equations together, we have

$$\begin{pmatrix}\frac{\partial\Pi}{\partial u_1}\\ \frac{\partial\Pi}{\partial u_2}\end{pmatrix} = \begin{pmatrix}1 & 0\\0 & 1\end{pmatrix}\left(\frac{1}{2}\boldsymbol{k}\begin{pmatrix}u_1\\u_2\end{pmatrix} - \boldsymbol{r}_q - \boldsymbol{r}_p\right) + (u_1 \quad u_2)\left(\frac{1}{2}\boldsymbol{k}\begin{pmatrix}1 & 0\\0 & 1\end{pmatrix}\right) = \begin{pmatrix}0\\0\end{pmatrix}$$

or

$$\frac{\partial\Pi}{\partial\boldsymbol{d}} = \left(\frac{1}{2}\boldsymbol{k}\begin{pmatrix}u_1\\u_2\end{pmatrix} - \boldsymbol{r}_q - \boldsymbol{r}_p\right) + (u_1 \quad u_2)\left(\frac{1}{2}\boldsymbol{k}\right) = \begin{pmatrix}0\\0\end{pmatrix}$$

$$\frac{\partial\Pi}{\partial\boldsymbol{d}} = \left(\frac{1}{2}\boldsymbol{kd} - \boldsymbol{r}_q - \boldsymbol{r}_p\right) + \boldsymbol{d}^T\left(\frac{1}{2}\boldsymbol{k}\right) = \boldsymbol{0}$$

Since $\boldsymbol{k}$ is a symmetric matrix, $\boldsymbol{d}^T\left(\frac{1}{2}\boldsymbol{k}\right) = \frac{1}{2}\boldsymbol{kd}$, and we have

$$\frac{\partial\Pi}{\partial\boldsymbol{d}} = \left(\frac{1}{2}\boldsymbol{kd} - \boldsymbol{r}_q - \boldsymbol{r}_p\right) + \frac{1}{2}\boldsymbol{kd} = \boldsymbol{0} \Longrightarrow \boldsymbol{kd} = \boldsymbol{r}_q + \boldsymbol{r}_p$$

Thus, even though $\boldsymbol{d}$ is a vector, in the compact matrix form, the expression for the derivative of Π with respect to $\boldsymbol{d}$ appears as if $\boldsymbol{d}$ is a scalar variable. Based on this observation, in the remainder of the book, after writing the energy function, its derivatives with respect to the nodal variables will be written directly without going through a detailed justification every time.

2.7.2 Numerical Examples

Example 2.8 Column in a Multistory Building An interior column in a multistory building is subjected to axial loads from different floors, as shown in Figure 2.20. Determine axial displacements at the story levels. Using these displacements, compute the axial strain and stress distribution in the column. Use the following data:

Story height, $h = 15$ ft

Story loads: $P_1 = 50{,}000$ lb, $P_2 = P_3 = 40{,}000$ lb, $P_4 = 35{,}000$ lb

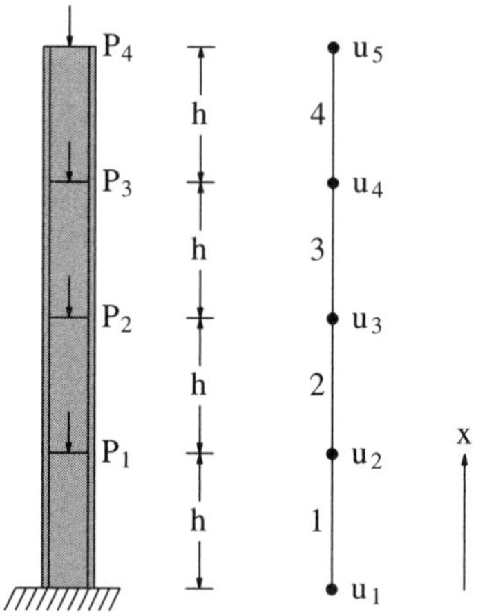

Figure 2.20. Column in a multistory building and a four-element model

Modulus of elasticity, $E = 29 \times 10^6$ lb/in^2

Area of cross section, $A = 21.8$ in^2

Since during the derivation of element equations it was assumed that the concentrated loads can only be applied at the element ends, we must divide the column into elements at the story levels. Thus the simplest possible finite element model is the four-element model shown in Figure 2.20.

The concentrated nodal loads will be added directly to the global equations after assembly. All elements have the same length and other properties. Thus the element equations are as follows:

$$\frac{EA}{h}\begin{pmatrix} 1 & -1 \\ -1 & 1 \end{pmatrix}\begin{pmatrix} u_1 \\ u_2 \end{pmatrix} = \begin{pmatrix} 0 & 0 \end{pmatrix}$$

Use pound-inches. The computed nodal displacements will be in inches and the stresses in pounds per square inch.

Nodal locations: {0, 180, 360, 540, 720}

Element 1:

$$\begin{pmatrix} 3.51222 \times 10^6 & -3.51222 \times 10^6 \\ -3.51222 \times 10^6 & 3.51222 \times 10^6 \end{pmatrix}\begin{pmatrix} u_1 \\ u_2 \end{pmatrix} = \begin{pmatrix} 0 \\ 0 \end{pmatrix}$$

Element 2:

$$\begin{pmatrix} 3.51222 \times 10^6 & -3.51222 \times 10^6 \\ -3.51222 \times 10^6 & 3.51222 \times 10^6 \end{pmatrix}\begin{pmatrix} u_2 \\ u_3 \end{pmatrix} = \begin{pmatrix} 0 \\ 0 \end{pmatrix}$$

Element 3:

$$\begin{pmatrix} 3.51222 \times 10^6 & -3.51222 \times 10^6 \\ -3.51222 \times 10^6 & 3.51222 \times 10^6 \end{pmatrix}\begin{pmatrix} u_3 \\ u_4 \end{pmatrix} = \begin{pmatrix} 0 \\ 0 \end{pmatrix}$$

Element 4:

$$\begin{pmatrix} 3.51222 \times 10^6 & -3.51222 \times 10^6 \\ -3.51222 \times 10^6 & 3.51222 \times 10^6 \end{pmatrix} \begin{pmatrix} u_4 \\ u_5 \end{pmatrix} = \begin{pmatrix} 0 \\ 0 \end{pmatrix}$$

Global equations after assembly and incorporating the NBC (placing concentrated applied loads):

$$\begin{pmatrix} 3.51222 \times 10^6 & -3.51222 \times 10^6 & 0 & 0 & 0 \\ -3.51222 \times 10^6 & 7.02444 \times 10^6 & -3.51222 \times 10^6 & 0 & 0 \\ 0 & -3.51222 \times 10^6 & 7.02444 \times 10^6 & -3.51222 \times 10^6 & 0 \\ 0 & 0 & -3.51222 \times 10^6 & 7.02444 \times 10^6 & -3.51222 \times 10^6 \\ 0 & 0 & 0 & -3.51222 \times 10^6 & 3.51222 \times 10^6 \end{pmatrix} \begin{pmatrix} u_1 \\ u_2 \\ u_3 \\ u_4 \\ u_5 \end{pmatrix} = \begin{pmatrix} 0 \\ -50000. \\ -40000. \\ -40000. \\ -35000. \end{pmatrix}$$

Essential boundary conditions:

dof	Value
u_1	0

Since the EBC is 0, we simply remove the first row and column to get the final system of equations as follows:

$$\begin{pmatrix} 7.02444 \times 10^6 & -3.51222 \times 10^6 & 0 & 0 \\ -3.51222 \times 10^6 & 7.02444 \times 10^6 & -3.51222 \times 10^6 & 0 \\ 0 & -3.51222 \times 10^6 & 7.02444 \times 10^6 & -3.51222 \times 10^6 \\ 0 & 0 & -3.51222 \times 10^6 & 3.51222 \times 10^6 \end{pmatrix} \begin{pmatrix} u_2 \\ u_3 \\ u_4 \\ u_5 \end{pmatrix} = \begin{pmatrix} -50000. \\ -40000. \\ -40000. \\ -35000. \end{pmatrix}$$

Solution for nodal unknowns:

dof	x	Solution
u_1	0	0
u_2	180	−0.0469788
u_3	360	−0.0797216
u_4	540	−0.101076
u_5	720	−0.111041

Once the nodal displacements are known, the displacement over any element is obtained from the linear interpolation functions as follows:

$$u(x) = \begin{pmatrix} \frac{x - x_2}{x_1 - x_2} & \frac{x - x_1}{x_2 - x_1} \end{pmatrix} \begin{pmatrix} u_1 \\ u_2 \end{pmatrix}$$

Element 1:

Nodes: $\{x_1 \to 0, x_2 \to 180\}$

Interpolation functions: $N^T = \{1 - \frac{x}{180}, \frac{x}{180}\}$

Nodal values: $\boldsymbol{d}^T = \{0, -0.0469788\}$

Solution: $\boldsymbol{N}^T\boldsymbol{d} = -0.000260993x$

Element 2:

Nodes: $\{x_1 \to 180, x_2 \to 360\}$

Interpolation functions: $\boldsymbol{N}^T = \left\{2 - \frac{x}{180}, \frac{x}{180} - 1\right\}$

Nodal values: $\boldsymbol{d}^T = \{-0.0469788, -0.0797216\}$

Solution: $\boldsymbol{N}^T\boldsymbol{d} = -0.000181904x - 0.014236$

Element 3:

Nodes: $\{x_1 \to 360, x_2 \to 540\}$

Interpolation functions: $\boldsymbol{N}^T = \left\{3 - \frac{x}{180}, \frac{x}{180} - 2\right\}$

Nodal values: $\boldsymbol{d}^T = \{-0.0797216, -0.101076\}$

Solution: $\boldsymbol{N}^T\boldsymbol{d} = -0.000118633\,x - 0.0370136$

Element 4:

Nodes: $\{x_1 \to 540, x_2 \to 720\}$

Interpolation functions: $\boldsymbol{N}^T = \left\{4 - \frac{x}{180}, \frac{x}{180} - 3\right\}$

Nodal values: $\boldsymbol{d}^T = \{-0.101076, -0.111041\}$

Solution: $\boldsymbol{N}^T\boldsymbol{d} = -0.0000553622x - 0.07118$

Solution summary:

	Range	$u(x)$
1	$0 \le x \le 180$	$-0.000260993x$
2	$180 \le x \le 360$	$-0.000181904x - 0.014236$
3	$360 \le x \le 540$	$-0.000118633x - 0.0370136$
4	$540 \le x \le 720$	$-0.0000553622x - 0.07118$

From these displacements we get the following axial strains, stresses, and axial forces:

$$\epsilon_x = \frac{du}{dx}; \qquad \sigma_x = E\epsilon_x; \qquad F = A\sigma_x$$

	Range	ϵ	σ_x	F
1	$0 \le x \le 180$	−0.000260993	−7568.81	−165000.
2	$180 \le x \le 360$	−0.000181904	−5275.23	−115000.
3	$360 \le x \le 540$	−0.000118633	−3440.37	−75000.
4	$540 \le x \le 720$	−0.0000553622	−1605.5	−35000.

The negative sign indicates compression. The problem is statically determinate, and thus simple application of the static equilibrium condition indicates that the computed axial forces are exact.

Example 2.9 Tapered Bar Consider solution of the tapered axially loaded bar shown in Figure 2.21. Use a two-node uniform axial deformation element to model the bar and determine the axial stress and force distribution in the bar. Compute the support reactions

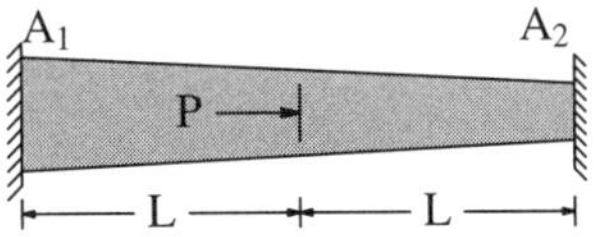

Figure 2.21. Tapered bar

from the axial force and see whether the overall equilibrium is satisfied. Comment on the quality of the finite element solution. Use the following numerical data:

$E = 70\,\text{GPa}$; $F = 20\,\text{kN}$; $L = 300\,\text{mm}$; $A_1 = 2400\,\text{mm}^2$; $A_2 = 600\,\text{mm}^2$

where A_1 and A_2 are the areas of cross section at the two ends of the bar. The area of cross section can be expressed as a linear function of x using the Lagrange interpolation formula as follows:

$$A(x) = \frac{x - 2L}{-2L} A_1 + \frac{x}{2L} A_2$$

Since the displacements are generally small, numerically it is convenient to use newton-millimeters. Then the computed nodal displacements are in millimeters and the stresses in megapascals.

A four-element model is as shown in Figure 2.22. Since the element equations derived earlier are based on the assumption of a uniform cross section, we must assign an average area to each element in the finite element model. Denoting the area of cross section at the left end of an element by A_l and that at the right end by A_r, the average area for each element is $(A_l + A_r)/2$. The concentrated nodal loads will be added directly to the global equations after assembly. Thus, with l the length of an element, the element equations are as follows:

$$\frac{(A_l + A_r)E}{2l} \begin{pmatrix} 1 & -1 \\ -1 & 1 \end{pmatrix} \begin{pmatrix} u_l \\ u_r \end{pmatrix} = \begin{pmatrix} 0 \\ 0 \end{pmatrix}$$

Nodal locations: {0, 150., 300., 450., 600.}

Areas at the nodes: {2400., 1950., 1500., 1050., 600.}

Average area for each element: {2175., 1725., 1275., 825.}

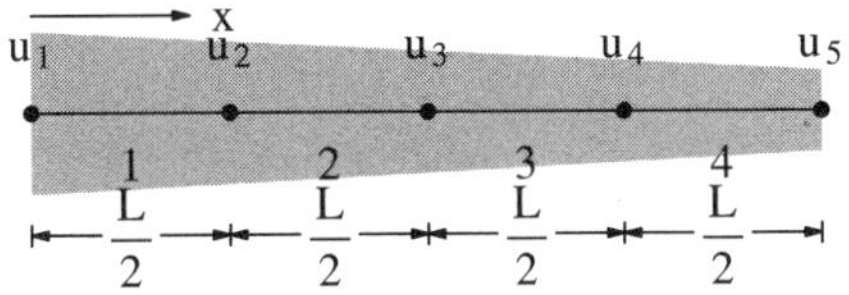

Figure 2.22. Four-element model for the tapered bar

Element 1:

$$\begin{pmatrix} 1.015\times 10^6 & -1.015\times 10^6 \\ -1.015\times 10^6 & 1.015\times 10^6 \end{pmatrix}\begin{pmatrix} u_1 \\ u_2 \end{pmatrix} = \begin{pmatrix} 0 \\ 0 \end{pmatrix}$$

Element 2:

$$\begin{pmatrix} 805000. & -805000. \\ -805000. & 805000. \end{pmatrix}\begin{pmatrix} u_2 \\ u_3 \end{pmatrix} = \begin{pmatrix} 0 \\ 0 \end{pmatrix}$$

Element 3:

$$\begin{pmatrix} 595000. & -595000. \\ -595000. & 595000. \end{pmatrix}\begin{pmatrix} u_3 \\ u_4 \end{pmatrix} = \begin{pmatrix} 0 \\ 0 \end{pmatrix}$$

Element 4:

$$\begin{pmatrix} 385000. & -385000. \\ -385000. & 385000. \end{pmatrix}\begin{pmatrix} u_4 \\ u_5 \end{pmatrix} = \begin{pmatrix} 0 \\ 0 \end{pmatrix}$$

Global equations after incorporating the NBC:

$$\begin{pmatrix} 1.015\times 10^6 & -1.015\times 10^6 & 0 & 0 & 0 \\ -1.015\times 10^6 & 1.82\times 10^6 & -805000. & 0 & 0 \\ 0 & -805000. & 1.4\times 10^6 & -595000. & 0 \\ 0 & 0 & -595000. & 980000. & -385000. \\ 0 & 0 & 0 & -385000. & 385000. \end{pmatrix}\begin{pmatrix} u_1 \\ u_2 \\ u_3 \\ u_4 \\ u_5 \end{pmatrix} = \begin{pmatrix} 0 \\ 0 \\ 20000 \\ 0 \\ 0 \end{pmatrix}$$

Essential boundary conditions:

dof	Value
u_1	0
u_5	0

Incorporating the EBC, the final system of equations is

$$\begin{pmatrix} 1.82\times 10^6 & -805000. & 0 \\ -805000. & 1.4\times 10^6 & -595000. \\ 0 & -595000. & 980000. \end{pmatrix}\begin{pmatrix} u_2 \\ u_3 \\ u_4 \end{pmatrix} = \begin{pmatrix} 0. \\ 20000. \\ 0. \end{pmatrix}$$

Solution for nodal unknowns:

dof	x	Solution
u_1	0	0
u_2	150.	0.0129577
u_3	300.	0.0292958
u_4	450.	0.0177867
u_5	600.	0

Solution over element 1:

Nodes: $\{x_1 \to 0, x_2 \to 150.\}$

Interpolation functions: $\boldsymbol{N}^T = \{1. - 0.00666667x, 0.00666667x\}$

Nodal values: $\boldsymbol{d}^T = \{0, 0.0129577\}$

Solution: $\boldsymbol{N}^T\boldsymbol{d} = 0.000086385x$

Solution over element 2:

Nodes: $\{x_1 \to 150., x_2 \to 300.\}$

Interpolation functions: $\boldsymbol{N}^T = \{2. - 0.00666667x, 0.00666667x - 1.\}$

Nodal values: $\boldsymbol{d}^T = \{0.0129577, 0.0292958\}$

Solution: $\boldsymbol{N}^T\boldsymbol{d} = 0.00010892x - 0.00338028$

Solution over element 3:

Nodes: $\{x_1 \to 300., x_2 \to 450.\}$

Interpolation functions: $\boldsymbol{N}^T = \{3. - 0.00666667x, 0.00666667x - 2.\}$

Nodal values: $\boldsymbol{d}^T = \{0.0292958, 0.0177867\}$

Solution: $\boldsymbol{N}^T\boldsymbol{d} = 0.0523139 - 0.000076727x$

Solution over element 4:

Nodes: $\{x_1 \to 450., x_2 \to 600.\}$

Interpolation functions: $\boldsymbol{N}^T = \{4. - 0.00666667x, 0.00666667x - 3.\}$

Nodal values: $\boldsymbol{d}^T = \{0.0177867, 0\}$

Solution: $\boldsymbol{N}^T\boldsymbol{d} = 0.0711469 - 0.000118578x$

Solution summary:

	Range	Solution
1	$0 \le x \le 150.$	$0.000086385x$
2	$150. \le x \le 300.$	$0.00010892x - 0.00338028$
3	$300. \le x \le 450.$	$0.0523139 - 0.000076727x$
4	$450. \le x \le 600.$	$0.0711469 - 0.000118578x$

From these displacements we get the following axial strains, stresses, and axial forces:

$$\epsilon_x = \frac{du}{dx}; \qquad \sigma_x = E\epsilon_x; \qquad F = A\sigma_x$$

	Range	ϵ	σ	F
1	$0 \le x \le 150.$	0.000086385	6.04695	13152.1
2	$150. \le x \le 300.$	0.00010892	7.62441	13152.1
3	$300. \le x \le 450.$	−0.000076727	−5.37089	−6847.89
4	$450. \le x \le 600.$	−0.000118578	−8.30047	−6847.89

The axial forces at the ends must balance the support reactions. With the sign convention for axial forces discussed earlier, the reaction at the left support is the negative of the axial

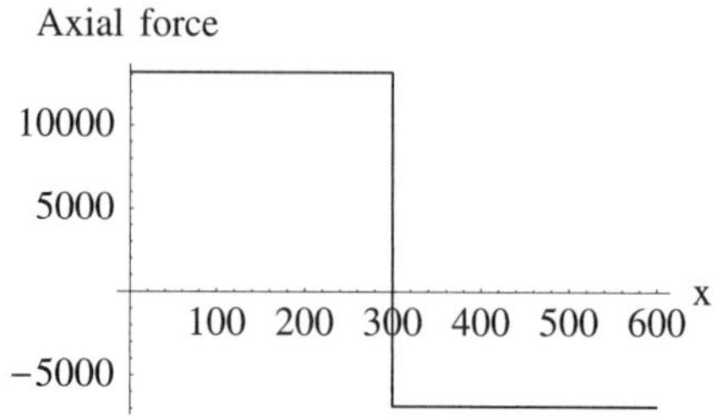

Figure 2.23. Four-element solution for the tapered bar—axial force distribution

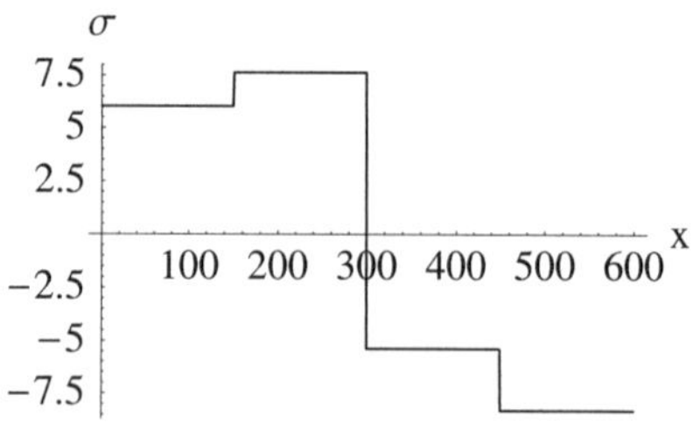

Figure 2.24. Four-element solution for the tapered bar—stress distribution

force at this point and that at the right support is equal to the axial force. Thus from the axial forces we get the support reactions as

$$\text{Reactions} = \{-13152.1, -6847.89\}$$

$$\text{Sum of reactions} = -20000.$$

The sum of reactions is equal and opposite to the applied load, and therefore the overall equilibrium is satisfied. Plots of the axial stress and axial force are shown in Figures 2.23 and 2.24. The axial force plot looks reasonable. In the stress plot we expect a discontinuity at the middle because of the concentrated applied force. However, the stress at nodes 2 and 4 should be continuous. A large discontinuity in the stress at these locations indicates that the solution is not very accurate.

The following solution is obtained using eight equal-length elements:

	Range	ϵ	σ	F
1	$0 \leq x \leq 75.$	0.0000824431	5.77101	13201.2
2	$75. \leq x \leq 150.$	0.0000914369	6.40058	13201.2
3	$150. \leq x \leq 225.$	0.000102633	7.18432	13201.2
4	$225. \leq x \leq 300.$	0.000116954	8.18679	13201.2
5	$300. \leq x \leq 375.$	−0.0000700005	−4.90004	−6798.8
6	$375. \leq x \leq 450.$	−0.000083549	−5.84843	−6798.8
7	$450. \leq x \leq 525.$	−0.000103601	−7.25206	−6798.8
8	$525. \leq x \leq 600.$	−0.000136317	−9.54218	−6798.8

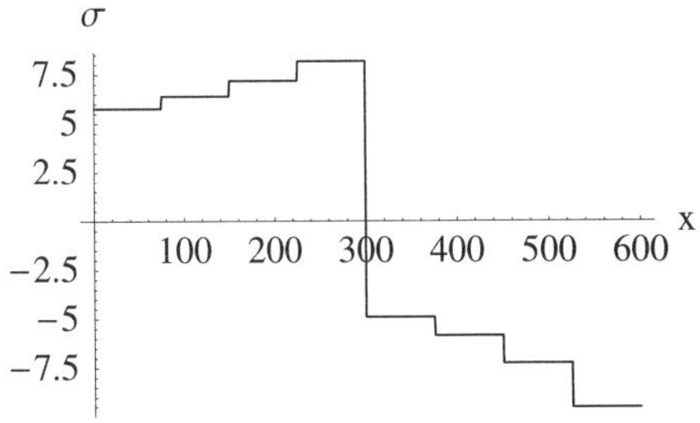

Figure 2.25. Eight-element solution for the tapered bar—stress distribution

The stresses are plotted in Figure 2.25. Because of the constant-area assumption over each element, we still have discontinuities in the stress. However, if we take the average of the stresses from the two elements at common nodes, the solution is very close to the exact solution:

```
avg = Map[Apply[Plus, #]/2&,
    {σ[[{1, 2}]], σ[[{2, 3}]], σ[[{3, 4}]], σ[[{5, 6}]],
    σ[[{6, 7}]], σ[[{7, 8}]]}]
```

{6.0858, 6.79245, 7.68556, −5.37424, −6.55024, −8.39712}

◆ *Mathematica*/MATLAB Implementation 2.4 on the Book Web Site: *Solution of axial deformation problems*

PROBLEMS

Exact Solution of Differential Equations

2.1 A uniform bar is subjected to uniform axial load along its length. The bar is fixed at the left end but there is a gap g between the support and the right end before the load is applied, as shown in the Figure 2.26. Assuming that the applied load is large enough to close the gap, write the governing differential equation with appropriate boundary conditions. Use direct integration to determine the exact solution of the

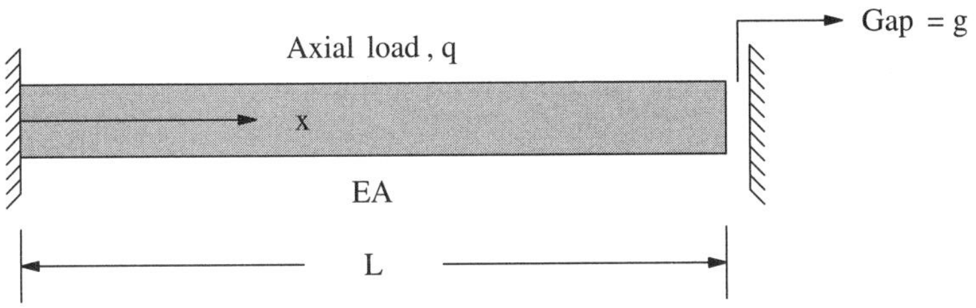

Figure 2.26. Axially loaded bar

problem. Plot the axial force along the bar. Use the following numerical values:

$$L = 800\text{ mm}; \quad g = 0.25\text{ mm}; \quad E = 200\text{ GPa}; \quad A = 200\text{ mm}^2; \quad q = 100\text{ N/mm}$$

2.2 Consider a large plane wall, shown in Figure 2.27, with thickness = 0.4 m, surface area = 20 m^2, and thermal conductivity $k = 2.3$ W/m · °C. The inside of the wall is maintained at a constant temperature of $T_0 = 80$°C while the outside loses heat by convection to the surrounding air at $T_\infty = 15$°C with a heat transfer coefficient of $h = 24$ W/m^2 · °C. The governing differential equation for the problem is

$$k\frac{d^2T}{dx^2} = 0; \qquad 0 < x < L = 0.4$$

$$T(0) = T_0; \qquad -k\frac{dT(L)}{dx} = h(T(L) - T_\infty)$$

Use direct integration to determine an expression for temperature distribution through the wall. What is the rate of heat transfer through the wall?

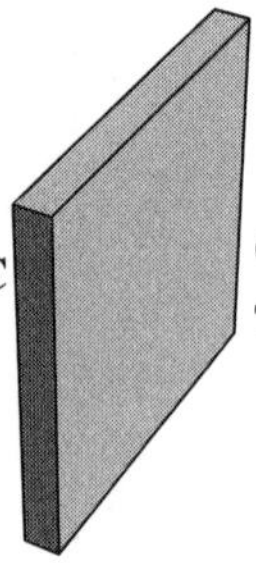

Figure 2.27. Heat conduction through a wall

2.3 Steady-state heat flow through long hollow circular cylinders can be described by the following ordinary differential equation:

$$\frac{d}{dr}\left(kA\frac{dT(r)}{dr}\right) + AQ = 0; \qquad r_i < r < r_o$$

$$T(r_i) = T_i; \qquad T(r_o) = T_0$$

where r is the radial coordinate, $T(r)$ is the temperature, k is the thermal conductivity, Q is the heat generation per unit area, $A = 2\pi rL$ is the surface area, L is the length of the cylinder, r_i is the inner radius, and r_o is the outer radius. The boundary conditions specify the temperature on the inside and outside of the cylinder, respectively.

(a) Show that the following represents an exact solution for the problem for the case when $Q = 0$:

$$T(r) = T_i - (T_i - T_o)\frac{\ln(r/r_i)}{\ln(r_o/r_i)}$$

(b) Show that the following is an appropriate weak form for obtaining an approximate solution using the Galerkin weighted residual method:

$$\int_{r_i}^{r_o} \left(-kA \frac{dT}{dr} \frac{dw_j}{dr} + AQw_j \right) dr = 0$$

(c) Using the weak form given in (b), find an approximate solution of the problem with a trial solution of the form $T(r) = a_0 + a_1 r + a_2 r^2$. Assume the following numerical values:

$$r_i = 1 \text{ in}; \quad r_o = 4 \text{ in}; \quad T_i = 400^\circ\text{F}; \quad T_o = 100^\circ\text{F}$$

$$L = 100 \text{ in}; \quad Q = 0; \quad k = 0.04 \text{ Btu/(h} \cdot \text{ft}^\circ\text{F)}$$

Approximate Solutions Using Galerkin Method

2.4 An engineering analysis problem is formulated in terms of the following ordinary differential equation:

$$\frac{d^2u}{dx^2} + x\frac{du}{dx} = u - 1; \qquad 0 < x < 1$$

$$u(0) = 2; \qquad u(1) = \frac{du(1)}{dx}$$

(a) What is the order of the differential equation?

(b) Is the boundary condition at $x = 0$ a natural or an essential boundary condition?

(c) Is the boundary condition at $x = 1$ a natural or an essential boundary condition?

(d) Obtain a suitable weak form for the problem.

(e) Starting with a linear polynomial, obtain an approximate solution of the problem.

2.5 An engineering analysis problem is formulated in terms of the following ordinary differential equation:

$$-u'' + \pi^2 u - 2\pi^2 \sin(\pi x) = 0; \qquad 0 < x < 1$$

$$u(0) = u(1) = 0$$

$$\text{Exact solution:} \qquad u(x) = \sin(\pi x)$$

Starting with a quadratic polynomial, obtain an approximate solution of the problem. Compare your solution with the exact solution.

2.6 An engineering analysis problem is formulated in terms of the following second-order boundary value problem:

$$2xu'' + 3u' = 4; \qquad 1 < x < 2$$
$$u'(1) = 1 \quad \text{and} \quad u(2) - 2 = 0$$

(a) Derive a suitable weak form for use with the Galerkin method. Clearly indicate how the boundary conditions will be handled.

(b) Starting with a solution of the form $u(x) = a_0 + a_1 x$, obtain an approximate solution of the problem.

2.7 Consider the following boundary value problem:

$$u'' \sin(x) + u' \cos(x) + u \sin(x) = 0; \qquad \pi/4 < x < \pi/2$$
$$u(\pi/4) = 1 \quad \text{and} \quad u'(\pi/2) = 2$$

(a) Verify that the differential equation can be written as follows:

$$\frac{d}{dx}[u' \sin(x)] + u \sin(x) = 0$$

(b) Using the Galerkin weighted residual method, obtain an approximate solution of the boundary value problem using a trial solution of the form $u(x) = a_0 + a_1 x$.

2.8 Consider the following boundary value problem:

$$u'' + xu' - 2u = 1; \qquad 0 < x < 1$$
$$u'(0) - u(0) = 1; \qquad u'(1) = 2$$

Construct a suitable weak form. Using this weak form, obtain a quadratic approximate solution of the problem.

2.9 Consider the following boundary value problem:

$$u'' + u + x = 0$$

with the boundary conditions $u(0) = 0$, $u(1) = 0$. Construct a suitable weak form. Using this weak form, obtain a quadratic approximate solution of the problem.

2.10 Consider the following boundary value problem:

$$u'' + xu' - u = 1; \qquad 0 < x < 1$$
$$u'(0) - u(0) = 1; \qquad u'(1) = 1$$

Construct a suitable weak form. Using this weak form, obtain a quadratic approximate solution of the problem.

2.11 An engineering analysis problem is formulated in terms of the following second-order boundary value problem:

$$u(x) - u'(x) + u''(x) = 1; \qquad 0 < x < 2$$

$$u'(0) = u(0) \qquad \text{and} \qquad u(2) = 1$$

Derive a suitable weak form for use with the Galerkin method. Clearly indicate how the boundary conditions will be handled. Starting with a solution of the form $u(x) = a_0 + a_1 x$, obtain an approximate solution of the problem.

2.12 Use the Galerkin method to find a quadratic solution of the axially loaded bar of Problem 2.1. Verify that this is the exact solution of the problem.

2.13 Use the Galerkin method to find an approximate solution of a uniform bar fixed at both ends and subjected to an axial load $q(x) = cx^2$ kN/m, as shown in Figure 2.28. Assume a quadratic polynomial solution. Determine the axial force distribution in the bar. Plot this force distribution. Determine the support reactions from the force distribution and see whether the overall equilibrium is satisfied. Comment on the quality of the solution. Use the following numerical data:

$$E = 70 \text{ GPa}; \qquad c = 200; \qquad L = 600 \text{ mm}; \qquad A = 600 \text{ mm}^2$$

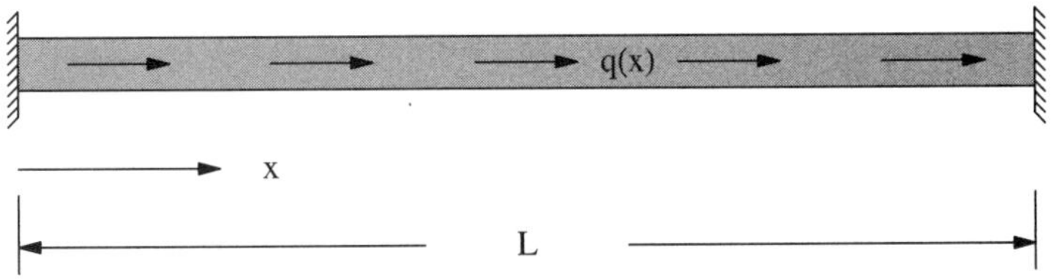

Figure 2.28. Axially loaded bar

2.14 Consider the problem of finding the axial displacement of a truncated solid cone of length L hanging under its own weight and subjected to a downward load F at the tip, as shown in Figure 2.29. The diameter at the top is d_0 and changes linearly to d_L at the bottom. The problem is described in terms of the following boundary value problem:

$$\frac{d}{dx}\left(EA(x)\frac{du}{dx}\right) + \rho A(x) = 0; \qquad 0 < x < L$$

$$A(x) = \frac{\pi}{4}d(x)^2; \qquad d(x) = d_0 - \frac{d_0 - d_L}{L}x$$

$$u(0) = 0; \qquad EA\frac{du(L)}{dx} = F$$

where ρ is the weight density and E is the modulus of elasticity. Obtain a quadratic approximate solution using the Galerkin method. Compare this solution with the exact solution (which can be obtained using *Mathematica*'s NDSolve function). Use the following numerical data:

$$d_0 = 1 \text{ in}; \quad d_L = 1/4 \text{ in}; \quad L = 100 \text{ in}; \quad F = 1000 \text{ lb};$$
$$E = 10^6 \text{ lb/in}^2; \quad \rho = 1 \text{ lb/in}^3$$

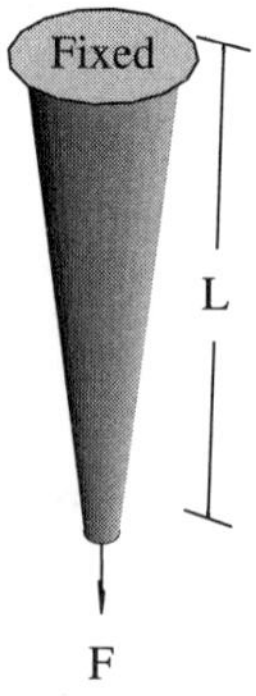

Figure 2.29. Hanging bar

2.15 Derive a weak form for the following sixth-order differential equation. Clearly identify the essential and natural boundary conditions.

$$\frac{d^6u(x)}{dx^6} + \frac{d^4u(x)}{dx^4} + x = 0; \qquad x_0 < x < x_l$$

2.16 The governing differential equation for the torsion of a thin-walled section with warping restraint can be written as follows:

$$EJ_w\phi'''' - GJ_0\phi'' - t(x) = 0; \qquad 0 \leq x \leq L$$

where E is the modulus of elasticity, G is the shear modulus, J_w is the warping constant, J_0 is the torsional constant, t is the thickness of the section, and ϕ is the angle through which the cross section rotates. Derive an equivalent weak form and indicate appropriate natural and essential boundary conditions.

2.17 A slider bearing with linear profile is shown in Figure 2.30. The two surfaces shown are moving relative to each other at a constant velocity U and are separated by a viscous fluid with viscosity μ. The governing differential equation for determining pressure in the fluid under the bearing can be written as follows:

$$\frac{d}{dx}\left(h^3\frac{dp}{dx}\right) + 6\mu U\frac{dh}{dx} = 0; \qquad 0 < x < L$$
$$p(0) = p(L) = p_0$$

where $p(x)$ is pressure in the fluid and $h(x)$ is the thickness of the fluid film. Derive a suitable weak form. Starting with a cubic polynomial, obtain an approximate solution using the Galerkin method. Use the following numerical data:

$$\mu = 10^{-4}\ \text{lb}\cdot\text{s/ft}^2; \qquad L = 8\ \text{in}; \quad h_0 = 0.0001\ \text{in}; \quad U = 12\ \text{in/s}; \quad p_0 = 14.7\ \text{psi}$$

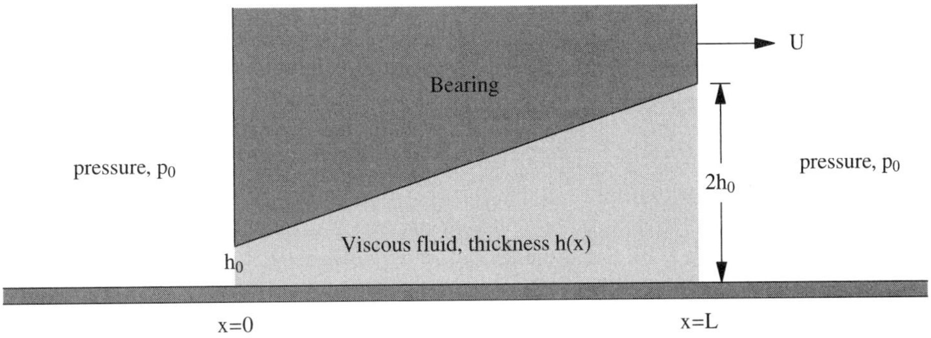

Figure 2.30. Slider bearing with linear profile

Approximate Solutions Using Rayleigh-Ritz Method

2.18 Repeat Problem 2.12 using the Rayleigh-Ritz method.

2.19 Repeat Problem 2.13 using the Rayleigh-Ritz method.

2.20 Repeat Problem 2.14 using the Rayleigh-Ritz method. Note that the potential energy for the problem is as follows:

$$\Pi_p = U - W$$

$$U = \frac{1}{2}\int_0^L EA\left(\frac{du}{dx}\right)^2 dx; \qquad W = \int_0^L \rho A u\, dx + Fu(L)$$

Lagrange and Hermite Interpolations

2.21 For a viscous fluid the velocities at various depths are measured as follows:

Velocity, m/s	3.75	4	3.75	3	1.75
Depth below the surface, m	0	1	1.4	1.9	2.3

Use Lagrange interpolation to obtain an expression for fluid velocity as a function of depth. Plot this relationship and verify that the function passes through all data points.

2.22 The thickness of a concrete dam at various heights, starting from the bottom, is measured as follows:

Thickness, m	2.81	1.79	1.39	1.07	0.8
Height, m	0	7.9	13.7	18.3	22

Use Lagrange interpolation to obtain a relationship for dam thickness as a function of height. Plot this relationship and verify that the function passes through all data points.

2.23 Use Hermite interpolation to obtain a function for the following set of data:

Point, x_i	Function Value, u_i	Derivative Value, u_i'
0	0.1	0
2	0	−1.3

Plot the resulting function and show that it passes through the given points and has the desired slopes at these points.

2.24 Use Hermite interpolation to obtain a function for the following set of data:

Point, x_i	Function Value, u_i	Derivative Value, u_i'
0	1	0
2.5	0	1.5

Plot the resulting function and show that it passes through the given points and has the desired slopes at these points.

2.25 A flat aluminum bar has constant thickness of 10 mm and a variable profile as shown in Figure 2.31. Use Lagrange interpolation to determine an expression for its area of cross section.

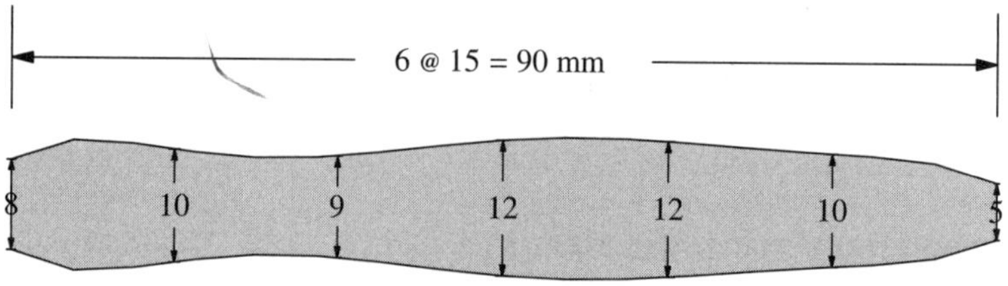

Figure 2.31.

Two-Node Uniform Bar Element for Axial Deformations

2.26 Find the axial force distribution using only two linear axial deformation elements for the axially loaded bar of Problem 2.1.

2.27 A bar of constant cross section A and modulus of elasticity E is attached at both ends to rigid supports and is loaded axially by force P as shown in Figure 2.32. Find axial displacement and force distribution using only two linear axial deformation elements. Use the following numerical values:

$$a = 300 \text{ mm}; \quad b = 600 \text{ mm}; \quad E = 200 \text{ GPa}; \quad A = 200 \text{ mm}^2; \quad P = 10 \text{ kN}$$

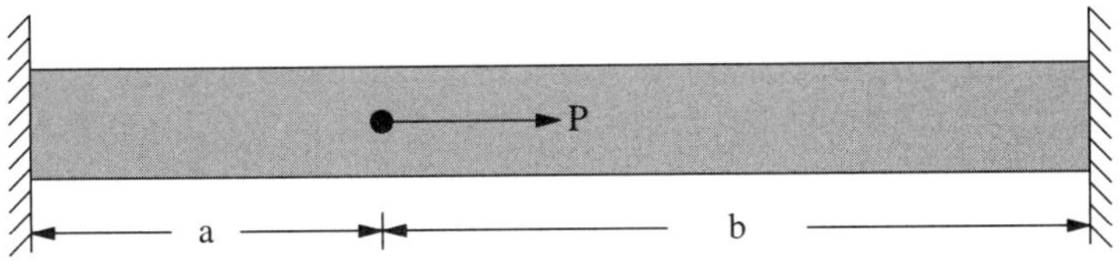

Figure 2.32. Axially loaded bar

2.28 Find the axial force distribution using five linear axial deformation elements for the axially loaded bar of Problem 2.13. Assume the load for each element is constant and is equal to the average of the loads at the two ends of the element.

2.29 Find the axial force distribution using three linear axial deformation elements for the axially loaded bar of Problem 2.14. Assume the area for each element is constant and is equal to the average of the areas at the two ends of the element.

2.30 Using the Galerkin method, derive finite element equations for a two-node axial deformation element assuming that the axial load q varies linearly over the element as shown in Figure 2.33. Assume that EA is constant over the element and that there may be concentrated axial loads P_i applied at the ends of the element.

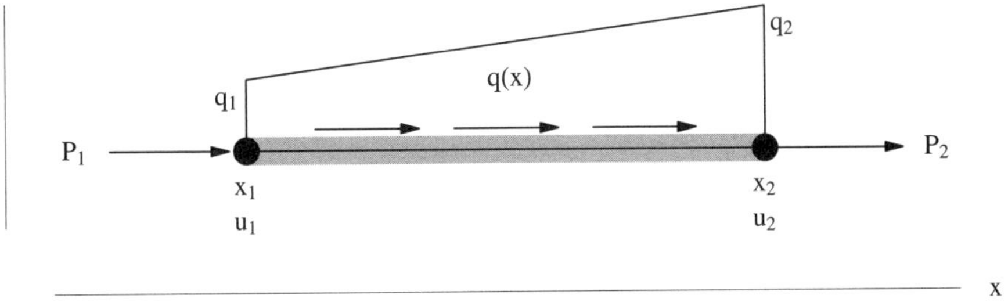

Figure 2.33. Bar element with linearly varying axial load

2.31 A typical finite element for an axial deformation problem involving tapered bars is shown in Figure 2.34. The area of cross section $A(x)$ varies linearly from A_l to A_r over the length of the element. Similarly, the applied axial load $q(x)$ varies linearly from q_l to q_r over the length of the element. Derive equations for a linear (two-node) finite element using the Rayleigh-Ritz method.

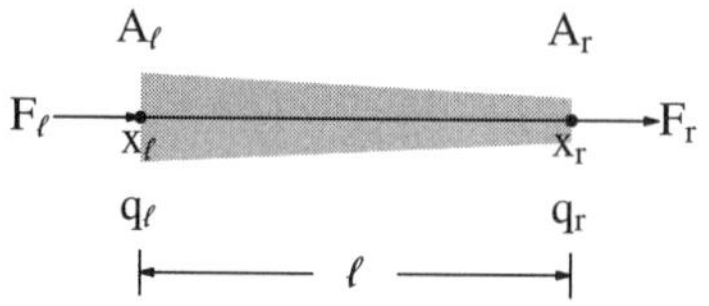

Figure 2.34. Bar element with linearly varying cross section and load

2.32 A flat aluminum bar has constant thickness of 10 mm and a variable profile as shown in Figure 2.35. The left end of the bar is fixed. A tensile load of 6 kN is applied at the right end. Determine the axial stress distribution in the bar. Assume the area of cross section changes linearly over each element. Perform a solution convergence study as the number of elements is increased.

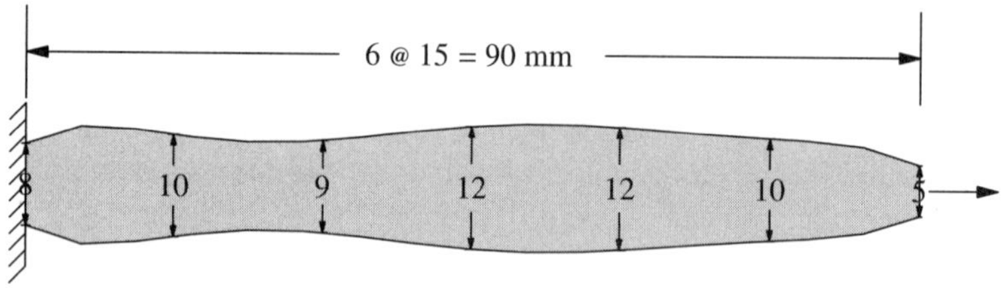

Figure 2.35.

CHAPTER THREE

ONE-DIMENSIONAL BOUNDARY VALUE PROBLEM

A large number of practical problems are governed by the one-dimensional boundary value problem (1D BVP) of the following form:

$$\frac{d}{dx}\left(k(x)\frac{du(x)}{dx}\right) + p(x)u(x) + q(x) = 0; \qquad x_0 < x < x_L$$

where $k(x)$, $p(x)$, and $q(x)$ are given functions of x and $u(x)$ is the solution variable. Since the differential equation is of second order, for a unique solution, there must be at least two specified boundary conditions. The boundary conditions of the following form may be specified at one or more points:

Essential boundary conditions (EBC):

$$u = \text{specified}$$

Natural boundary conditions (NBC):

$$-\frac{du}{dx} = \alpha u + \beta \qquad \text{if specified at the left end } (x = x_0)$$

$$\frac{du}{dx} = \alpha u + \beta \qquad \text{if specified at the right end } (x = x_L)$$

where α and β are some specified constants and u is the unknown solution at the point where the NBC is specified. Observe that in the natural boundary condition at the left end the negative of the derivative is specified. The same thing was done in the axial deformation problem in Chapter 2 where the choice was based on the sign convention for the applied nodal forces. In the present case the decision to put a negative sign for the derivative is arbitrary. However, with this decision, the finite element equations will come out in the standard form with all terms having consistent signs. Furthermore, when applying this element to physical problems, the sign will make sense with the usual sign convention adopted for those problems.

A few selected applications that are governed by the differential equation of this form are presented in the first section. The second section presents a general finite element formulation for the problem. The remaining sections in this chapter use this finite element formulation to obtain approximate solutions for a variety of problems.

3.1 SELECTED APPLICATIONS OF 1D BVP

3.1.1 Steady-State Heat Conduction

A simple one-dimensional heat conduction situation is encountered when considering heat flow through large panels or walls. Considering a unit slice of the panel, the heat flow problem through the thickness results in a situation shown in Figure 3.1. The governing differential equation is as follows:

$$\frac{d}{dx}\left(k_x A \frac{dT}{dx}\right) + QA = 0; \qquad x_0 < x < x_L$$

where k_x = thermal conductivity, A = area of cross section perpendicular to the direction of heat flow, and Q = internal heat source per unit volume. This equation is a special case of the general form with

$$k(x) = k_x A; \qquad p(x) = 0; \qquad q(x) = QA$$

The boundary conditions of the following form may be specified at one or more points:

$$T = \text{specified temperature}$$

$$-kA\frac{dT}{dx} = q \text{ specified heat flow}$$

The specified heat flow is considered positive when the heat is flowing into the body and negative if it is flowing out of the body. Thus with a heat flow in the positive x direction we

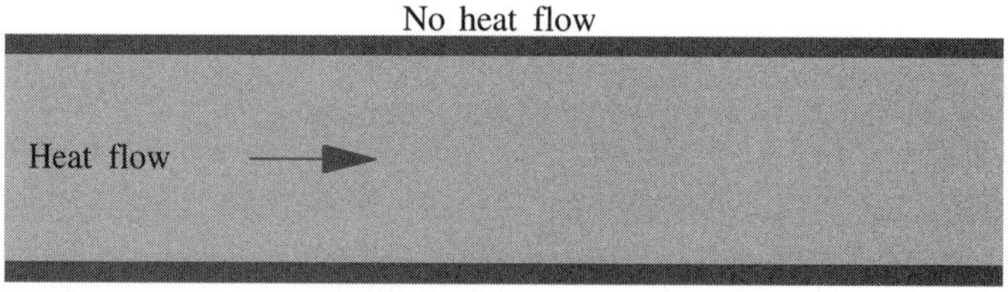

Figure 3.1. One-dimensional heat conduction

have the following two situations:

At the left end: $$-kA\frac{dT}{dx} = q \Longrightarrow -\frac{dT}{dx} = \frac{q}{kA}$$

At the right end: $$-kA\frac{dT}{dx} = -q \Longrightarrow \frac{dT}{dx} = \frac{q}{kA}$$

These boundary conditions are equivalent to the general form with $\alpha = 0$ and $\beta = q/kA$.

3.1.2 Heat Flow through Thin Fins

Thin fins are employed to dissipate heat from a hot body into the surrounding fluid (usually air). A typical situation with uniform fins is illustrated in Figure 3.2. The main mechanism of heat transfer is convection from the large surface area provided by the fins.

A single fin is isolated, and assuming temperature is uniform over the width w of the fin, the problem can be described in terms of the following differential equation:

$$\frac{d}{dx}\left(k_x A\frac{dT}{dx}\right) - hPT + QA + hPT_\infty = 0; \qquad x_0 < x < x_L$$

where k_x = thermal conductivity, h = convection coefficient, P = surface area per unit length over which convection takes place, A = area of cross section, T_∞ = temperature of surrounding fluid, and Q = internal heat source per unit volume. For the situation shown in Figure 3.2, $A = wt$, and since the convection takes place at the top, bottom, and sides of the fin, $P = 2W + 2t$, where t is fin thickness. Generally, Q is zero for heat flow through fins. Compared to the general form,

$$k(x) = k_x A; \qquad p(x) = -hP; \qquad q(x) = QA + hPT_\infty$$

The boundary conditions of the following form may be specified at one or more points:

$T = T_b$ (specified temperature at the base)

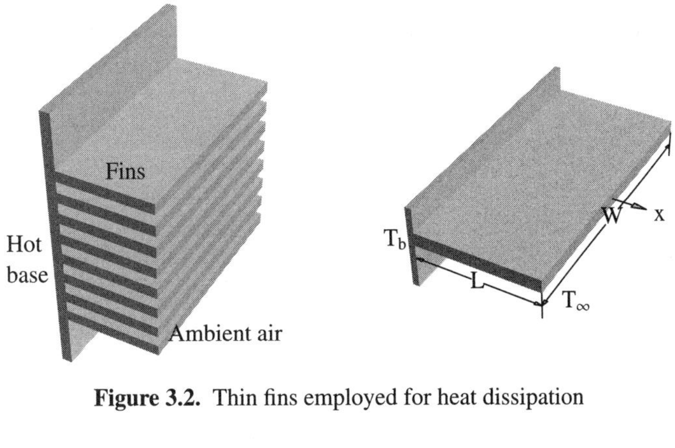

Figure 3.2. Thin fins employed for heat dissipation

Convection heat loss at the free end:

$$-kA\frac{dT}{dx} = -hA(T - T_\infty) \Longrightarrow \frac{dT}{dx} = \frac{hA}{kA}T - \frac{hAT_\infty}{kA}$$

Compared to the general form,

$$\alpha = \frac{h}{k} \quad \text{and} \quad \beta = -\frac{hT_\infty}{k}$$

When designing fins, a key quantity of interest is the total heat dissipated by a fin. Once the temperature distribution is known, the total heat loss from a fin can be computed by integrating the heat lost due to convection from the surface of the fin. Thus we have

$$\text{Heat loss} = \int_{x_0}^{x_L} hP(T(x) - T_\infty)\, dx$$

3.1.3 Viscous Fluid Flow between Parallel Plates—Lubrication Problem

Viscous fluids are used to reduce friction between moving parts. Actual lubrication problems involve quite complex geometries. However, for a simple one-dimensional analysis the problem can be viewed as a viscous fluid flow between two surfaces that are moving relative to each other. As shown in Figure 3.3, we can assume that the bottom surface is fixed and the top surface is moving at a constant velocity U which is equal to the relative velocity between the two surfaces. The two surfaces are usually at different temperatures, and since the viscosity varies with temperature, the fluid viscosity is typically a function of y. The analysis objective is to determine the fluid velocity profile $u(y)$.

The governing differential equation for determining fluid velocity profile $u(y)$ can be expressed as follows:

$$\frac{d}{dy}\left(\mu(y)\frac{du}{dy}\right) - \frac{dP}{dx} = 0$$

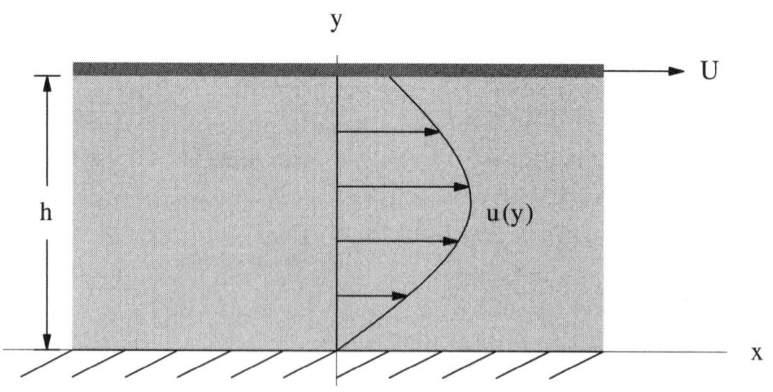

Figure 3.3. Velocity profile of a viscous fluid flow between two surfaces

where $\mu(y)$ is the fluid viscosity and dP/dx is the pressure drop in the direction of flow. The pressure drop is measured and must be known in order to determine the velocity profile. Compared to the general form,

$$k = \mu; \qquad p = 0; \qquad q = -\frac{dP}{dx}$$

The boundary conditions come from the assumption of no slip at the contact between plates and fluid. Thus we have the following boundary conditions:

$$u(0) = 0 \qquad \text{and} \qquad u(h) = U$$

Note that both boundary conditions are of the EBC type.

3.1.4 Slider Bearing

A slider bearing with linear profile is shown in Figure 3.4. The two surfaces shown are moving relative to each other at a constant velocity U and are separated by a viscous fluid with viscosity μ. The governing differential equation for determining pressure in the fluid under the bearing can be written as follows:

$$\frac{d}{dx}\left(h^3 \frac{dp}{dx}\right) + 6\mu U \frac{dh}{dx} = 0; \qquad 0 < x < L$$

where $p(x)$ is pressure in the fluid and $h(x)$ is the thickness of the fluid film. Compared to the general form,

$$k(x) = h^3; \qquad p(x) = 0; \qquad q(x) = 6\mu U \frac{dh}{dx}$$

Outside of the bearing the pressure is equal to the atmospheric pressure, p_0. Thus the boundary conditions are as follows (both are of the EBC type):

$$p(0) = p(L) = p_0$$

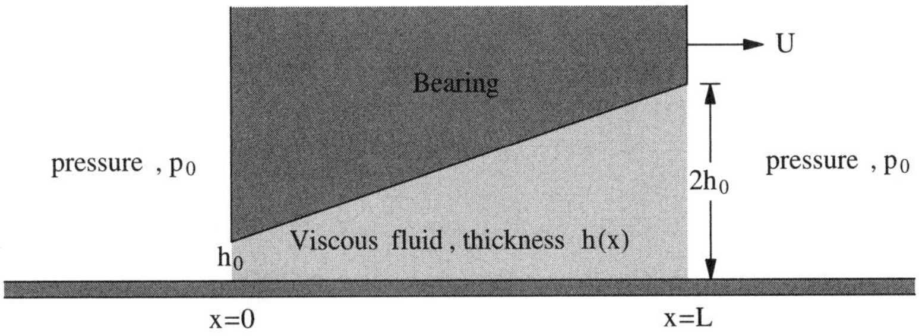

Figure 3.4. Linear slider bearing

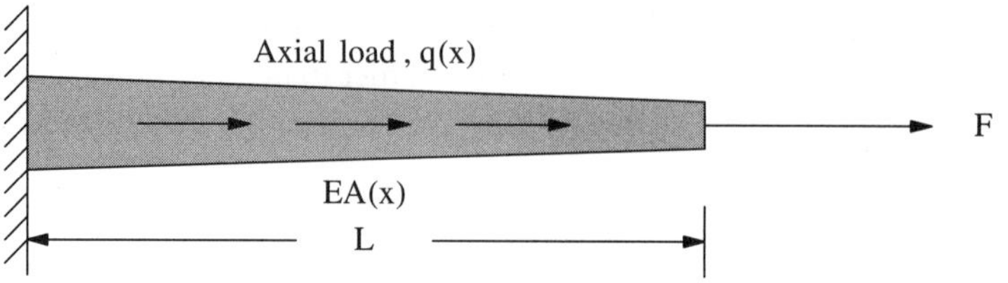

Figure 3.5. Nonuniform axially loaded bar

3.1.5 Axial Deformation of Bars

As seen from Chapter 2, axial deformation of bars is also governed by a differential equation that has the same general form as the 1D BVP presented in the first section. For example, the governing differential equation for a nonuniform bar shown in Figure 3.5 is as follows:

$$\frac{d}{dx}\left(EA(x)\frac{du}{dx}\right) + q(x) = 0; \qquad 0 < x < L$$

where $A(x)$ is the area of cross section, E is Young's modulus, and $q(x)$ is applied distributed load in the axial direction. Compared to the general form,

$$k(x) = EA; \qquad p(x) = 0; \qquad q(x) = q$$

The boundary conditions are zero displacement at $x = 0$ and applied axial load F at $x = L$:

$$u(0) = 0; \qquad EA(L)\frac{du(L)}{dx} = F$$

Compared to the general form, the natural boundary condition at $x = L$ implies $\alpha = 0$ and $\beta = F/EA$.

3.1.6 Elastic Buckling of Long Slender Bars

Consider a long slender bar pinned at one end and simply supported at the other and subjected to an axial load P as shown in Figure 3.6. Under the usual small-displacement assumption, an axial load produces axial displacements only. However, because of small imperfections, long slender bars tend to buckle in a transverse deflection mode. Taking this phenomenon into consideration, the governing differential equation for transverse deflection $v(x)$ of slender bars subjected to end axial load P is as follows:

$$\frac{d^2}{dx^2}\left(EI\frac{d^2v}{dx^2}\right) + P\left(\frac{d^2v}{dx^2}\right) = 0$$

where E is Young's modulus and I is the moment of inertia. This is a fourth-order differential equation in v. It can, however, be converted into a second-order differential equation

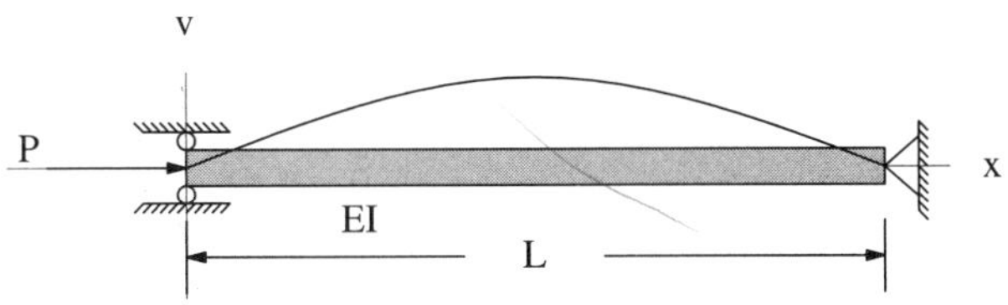

Figure 3.6. Buckling of a long slender bar

by defining a new variable $y = d^2v/dx^2$. In terms of y we have the following second-order differential equation:

$$\frac{d^2}{dx^2}(EIy) + Py = 0$$

If EI is a function of x, then this equation is *not* in the form of the general boundary value problem being considered in this chapter. This can be seen more clearly by expanding the first term by using the chain rule of differentiation, which gives

$$EI\left(\frac{d^2y}{dx^2}\right) + 2\frac{d(EI)}{dx}\frac{dy}{dx} + \left(P + \frac{d^2(EI)}{dx^2}\right)y = 0$$

The factor 2 in the second term is not present in the general form since, expanding the first term of the general form of boundary value problem, we have

$$\frac{d}{dx}\left(k\frac{du}{dx}\right) = k\left(\frac{d^2u}{dx^2}\right) + \frac{dk}{dx}\frac{du}{dx}$$

If EI is constant, we have

$$EI\left(\frac{d^2y}{dx^2}\right) + Py = 0$$

This is in the form of the general boundary value problem. Thus for a uniform bar compared to the general form

$$k = EI; \qquad p = P; \qquad q = 0$$

It is shown in Chapter 4 that the bending moment is related to the second derivative of the transverse displacement as follows:

$$M = EI\frac{d^2v}{dx^2} \equiv EIy$$

Since the moment is zero at a simply supported end, we get the following boundary conditions for the second-order problem:

$$y(0) = 0; \qquad y(L) = 0$$

If the axial load P is given, then the problem can be solved in a usual manner to get $y(x)$. However, a more interesting application is to find load P that will cause the bar to buckle. To find the buckling load P, the finite element equations are written and assembled in the usual manner. However, the global equations result in an eigenvalue problem. The eigenvalues are the buckling load values. The corresponding eigenvectors are the buckling modes. The procedure is illustrated through a numerical example in a later section.

3.2 FINITE ELEMENT FORMULATION FOR SECOND-ORDER 1D BVP

Element equations for a general n-node element for the second-order 1D BVP are derived in this section. The procedure is exactly the same as that used in Chapter 2 for deriving equations for a two-node element for axial deformations. Each element can have up to n nodes, where $n \geq 2$. The element extends from x_1 to x_n and has a length $L = x_n - x_1$. The circles represent nodes. The unknown solutions at the nodes are indicated by $u_1, u_2, \ldots, u_n$, as shown in Figure 3.7. Thus over an element we must satisfy

$$\frac{d}{dx}\left(k(x)\frac{du(x)}{dx}\right) + p(x)u(x) + q(x) = 0; \qquad x_1 < x < x_n$$

As discussed in Chapters 1 and 2, any essential boundary conditions will be imposed after assembling all element equations by setting the corresponding nodal degrees of freedom to the specified values. During the derivation of element equations we therefore do not have to worry about the essential boundary conditions. Any specified natural boundary conditions are included in the weak form, and thus during the derivation of element equations we must allow for the possibility of an NBC at the ends of an element. Therefore, during derivation of element equations we consider the following conditions at the ends of an element:

$$-\frac{du(x_1)}{dx} = \alpha_1 u(x_1) + \beta_1; \qquad \frac{du(x_n)}{dx} = \alpha_n u(x_n) + \beta_n$$

where we have used subscripts on the α and β terms to indicate the node number at which an NBC is specified. It should be recognized that in actual numerical solutions of problems any specified NBC will affect the first node of the first element and the last node of the last element. Most other elements in the model will have no contributions from these NBC terms.

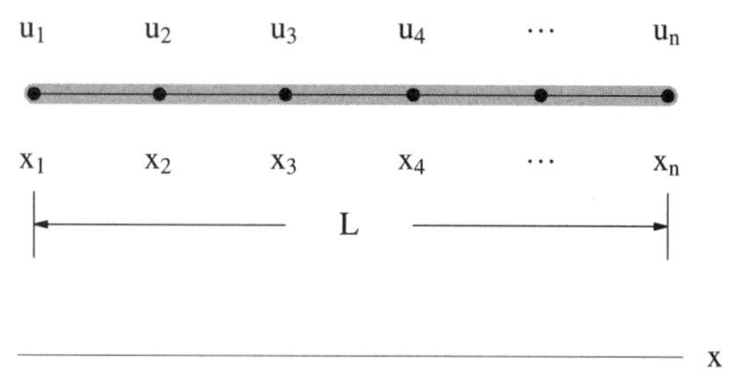

Figure 3.7. An n-node element for second-order BVP

Assumed Solution The assumed solution can easily be written using the Lagrange interpolation:

$$u(x) = \sum_{i=1}^{n} N_i(x)u(x_i) = \begin{pmatrix} N_1 & N_2 & \cdots & N_n \end{pmatrix}\begin{pmatrix} u_1 \\ u_2 \\ \vdots \\ u_n \end{pmatrix} \equiv \boldsymbol{N}^T\boldsymbol{d}$$

where $N_i(x)$ are the Lagrange interpolation functions

$$N_i(x) = \prod_{\substack{j=1 \\ j\neq i}}^{n} \frac{x-x_j}{x_i-x_j} \equiv \frac{x-x_1}{x_i-x_1} \times \frac{x-x_2}{x_i-x_2} \times \cdots \times \frac{x-x_{i-1}}{x_i-x_{i-1}} \times \frac{x-x_{i+1}}{x_i-x_{i+1}} \times \cdots \times \frac{x-x_n}{x_i-x_n}$$

We will need $u'(x)$ in the later derivation. Differentiating with respect to x, we get

$$u'(x) = \begin{pmatrix} N_1' & N_2' & \cdots & N_n' \end{pmatrix}\begin{pmatrix} u_1 \\ u_2 \\ \vdots \\ u_n \end{pmatrix} \equiv \boldsymbol{B}^T\boldsymbol{d}$$

Element Equations Using Galerkin Method The weak form can be written using standard steps of writing the weighted residual, integration by parts, and incorporating the natural boundary conditions. The weighting functions are the same as the interpolation functions N_i.

With $u(x)$ an assumed solution, the residual is

$$e(x) = q(x) + p(x)u(x) + k'(x)u'(x) + k(x)u''(x)$$

Multiplying by $N_i(x)$ and writing the integral over the given limits, the Galerkin weighted residual is

$$\int_{x_1}^{x_n} (qN_i + puN_i + k'u'N_i + ku''N_i)\,dx = 0$$

Using integration by parts, the order of derivative in kN_iu'' can be reduced to 1 as follows:

$$\int_{x_1}^{x_n} (kN_iu'')\,dx = k(x_n)N_i(x_n)u'(x_n) - k(x_1)N_i(x_1)u'(x_1) + \int_{x_1}^{x_n} (-u'(N_ik' + k\,N_i'))\,dx$$

Combining all terms, the weighted residual now is as follows:

$$k(x_n)N_i(x_n)u'(x_n) - k(x_1)N_i(x_1)u'(x_1) + \int_{x_1}^{x_n} (qN_i + puN_i - ku'N_i')\,dx = 0$$

Specified values of u' at the end nodes of an element can be directly substituted into the boundary terms to give

$$k(x_n)(\beta_n + \alpha_n u(x_n))N_i(x_n) - k(x_1)(-\beta_1 - \alpha_1 u(x_1))N_i(x_1) + \int_{x_1}^{x_n} (qN_i + puN_i - ku'N_i')\,dx = 0$$

With the n interpolation functions $N_1, N_2, \ldots, N_n$, the system of n equations for the element, written in the matrix form, is as follows:

$$k(x_n)\beta_n \begin{pmatrix} N_1(x_n) \\ N_2(x_n) \\ \vdots \\ N_n(x_n) \end{pmatrix} + \alpha_n k(x_n) u(x_n) \begin{pmatrix} N_1(x_n) \\ N_2(x_n) \\ \vdots \\ N_n(x_n) \end{pmatrix} + k(x_1)\beta_1 \begin{pmatrix} N_1(x_1) \\ N_2(x_1) \\ \vdots \\ N_n(x_1) \end{pmatrix} + \alpha_1 k(x_1) u(x_1) \begin{pmatrix} N_1(x_1) \\ N_2(x_1) \\ \vdots \\ N_n(x_1) \end{pmatrix}$$

$$+ \int_{x_1}^{x_n} q(x) \begin{pmatrix} N_1(x) \\ N_2(x) \\ \vdots \\ N_n(x) \end{pmatrix} + p(x) \begin{pmatrix} N_1(x) \\ N_2(x) \\ \vdots \\ N_n(x) \end{pmatrix} u(x) - k(x) \begin{pmatrix} N_1'(x) \\ N_2'(x) \\ \vdots \\ N_n'(x) \end{pmatrix} u'(x)\, dx = \begin{pmatrix} 0 \\ 0 \\ \vdots \\ 0 \end{pmatrix}$$

As noted earlier, the interpolation functions are such that $N_i = 1$ at x_i and 0 at all other nodes. Thus in the boundary terms $N_1(x_1) = N_n(x_n) = 1$ and all other interpolation functions are zero, giving

$$\begin{pmatrix} k(x_1)\beta_1 \\ 0 \\ \vdots \\ k(x_n)\beta_n \end{pmatrix} + \begin{pmatrix} \alpha_1 k(x_1) u(x_1) \\ 0 \\ \vdots \\ \alpha_n k(x_n) u(x_n) \end{pmatrix} + \int_{x_1}^{x_n} q(x) \begin{pmatrix} N_1(x) \\ N_2(x) \\ \vdots \\ N_n(x) \end{pmatrix} + p(x) \begin{pmatrix} N_1(x) \\ N_2(x) \\ \vdots \\ N_n(x) \end{pmatrix} u(x) - k(x) \begin{pmatrix} N_1'(x) \\ N_2'(x) \\ \vdots \\ N_n'(x) \end{pmatrix} u'(x)\, dx = \begin{pmatrix} 0 \\ 0 \\ \vdots \\ 0 \end{pmatrix}$$

Rearranging terms and moving all quantities not involving unknown u to the right-hand side, we have

$$\int_{x_1}^{x_n} k(x) \begin{pmatrix} N_1'(x) \\ N_2'(x) \\ \vdots \\ N_n'(x) \end{pmatrix} u'(x)\, dx - \int_{x_1}^{x_n} p(x) \begin{pmatrix} N_1(x) \\ N_2(x) \\ \vdots \\ N_n(x) \end{pmatrix} u(x)\, dx - \begin{pmatrix} \alpha_1 k(x_1) u(x_1) \\ 0 \\ \vdots \\ \alpha_n k(x_n) u(x_n) \end{pmatrix} = \int_{x_1}^{x_n} q(x) \begin{pmatrix} N_1(x) \\ N_2(x) \\ \vdots \\ N_n(x) \end{pmatrix} dx + \begin{pmatrix} k(x_1)\beta_1 \\ 0 \\ \vdots \\ k(x_n)\beta_n \end{pmatrix}$$

Substituting the assumed solution $u(x)$ and its first derivative $u'(x)$, we have

$$\int_{x_1}^{x_n} k(x) \begin{pmatrix} N_1' \\ N_2' \\ \vdots \\ N_n' \end{pmatrix} (N_1' \quad N_2' \quad \cdots \quad N_n')\, dx \begin{pmatrix} u_1 \\ u_2 \\ \vdots \\ u_n \end{pmatrix}$$

$$- \int_{x_1}^{x_n} p(x) \begin{pmatrix} N_1 \\ N_2 \\ \vdots \\ N_n \end{pmatrix} (N_1 \quad N_2 \quad \cdots \quad N_n)\, dx \begin{pmatrix} u_1 \\ u_2 \\ \vdots \\ u_n \end{pmatrix} - \begin{pmatrix} \alpha_1 k(x_1) u_1 \\ 0 \\ \vdots \\ \alpha_n k(x_n) u_n \end{pmatrix}$$

$$= \int_{x_1}^{x_n} q(x) \begin{pmatrix} N_1 \\ N_2 \\ \vdots \\ N_n \end{pmatrix} dx + \begin{pmatrix} k(x_1)\beta_1 \\ 0 \\ \vdots \\ k(x_n)\beta_n \end{pmatrix}$$

Each term in this equation is simplified separately as follows:

First term:

$$\int_{x_1}^{x_n} k(x) \begin{pmatrix} N_1' \\ N_2' \\ \vdots \\ N_n' \end{pmatrix} (N_1' \quad N_2' \quad \cdots \quad N_n')\, dx \begin{pmatrix} u_1 \\ u_2 \\ \vdots \\ u_n \end{pmatrix} = \int_{x_1}^{x_n} k\boldsymbol{B}\boldsymbol{B}^T\, dx\, \boldsymbol{d} = \boldsymbol{k}_k \boldsymbol{d}$$

where

$$\boldsymbol{k}_k = \begin{pmatrix} \int_{x_1}^{x_n} kN_1'N_1'\, dx & \int_{x_1}^{x_n} kN_1'N_2'\, dx & \cdots \\ \int_{x_1}^{x_n} kN_2'N_1'\, dx & \int_{x_1}^{x_n} kN_2'N_2'\, dx & \cdots \\ \vdots & \vdots & \vdots \end{pmatrix}$$

Second term:

$$-\int_{x_1}^{x_n} p(x) \begin{pmatrix} N_1 \\ N_2 \\ \vdots \\ N_n \end{pmatrix} (N_1 \quad N_2 \quad \cdots \quad N_n)\, dx \begin{pmatrix} u_1 \\ u_2 \\ \vdots \\ u_n \end{pmatrix} = \int_{x_1}^{x_n} (-p)\boldsymbol{N}\boldsymbol{N}^T\, dx\, \boldsymbol{d} = \boldsymbol{k}_p \boldsymbol{d}$$

where

$$\boldsymbol{k}_p = \begin{pmatrix} -\int_{x_1}^{x_n} pN_1N_1\, dx & -\int_{x_1}^{x_n} pN_1N_2\, dx & \cdots \\ -\int_{x_1}^{x_n} pN_2N_1\, dx & -\int_{x_1}^{x_n} pN_2N_2\, dx & \cdots \\ \vdots & \vdots & \vdots \end{pmatrix}$$

Third term:

$$-\begin{pmatrix} \alpha_1 k(x_1) u_1 \\ 0 \\ \vdots \\ \alpha_n k(x_n) u_n \end{pmatrix} = \begin{pmatrix} -\alpha_1 k(x_1) & 0 & \cdots & 0 \\ 0 & 0 & \cdots & 0 \\ \vdots & \vdots & \vdots & \vdots \\ 0 & 0 & \cdots & -\alpha_n k(x_n) \end{pmatrix} \begin{pmatrix} u_1 \\ u_2 \\ \vdots \\ u_n \end{pmatrix} = \boldsymbol{k}_\alpha \boldsymbol{d}$$

where

$$\boldsymbol{k}_\alpha = \begin{pmatrix} -\alpha_1 k(x_1) & 0 & \cdots & 0 \\ 0 & 0 & \cdots & 0 \\ \vdots & \vdots & \vdots & \vdots \\ 0 & 0 & \cdots & -\alpha_n k(x_n) \end{pmatrix}$$

First term on the right side:

$$\int_{x_1}^{x_n} q(x) \begin{pmatrix} N_1 \\ N_2 \\ \vdots \\ N_n \end{pmatrix} dx = \begin{pmatrix} \int_{x_1}^{x_n} qN_1\, dx \\ \int_{x_1}^{x_n} qN_2\, dx \\ \vdots \\ \int_{x_1}^{x_n} qN_n\, dx \end{pmatrix} \equiv \boldsymbol{r}_q$$

Second term on the right side:

$$\begin{pmatrix} k(x_1)\beta_1 \\ 0 \\ \vdots \\ k(x_n)\beta_n \end{pmatrix} \equiv \boldsymbol{r}_\beta$$

Thus the element equations are

$$\boldsymbol{k}_k\boldsymbol{d} + \boldsymbol{k}_p\boldsymbol{d} + \boldsymbol{k}_\alpha\boldsymbol{d} = \boldsymbol{r}_q + \boldsymbol{r}_\beta$$

As mentioned at the beginning of this section, the $\boldsymbol{k}_\alpha$ and $\boldsymbol{r}_\beta$ terms will be present in the element equations only if a natural boundary condition is specified for an element. Also for most elements only either the first term or the last term will be nonzero because typically the NBC will be specified at only one end of an element. The equations indicate that, if there is a natural boundary condition specified at node 1 of the element, then a term $-\alpha_1 k(x_1)$ must be added to the first diagonal term in the element $\boldsymbol{k}$ matrix and $k(x_1)\,\beta_1$ is added to the first entry in the element $\boldsymbol{r}$ vector. The same treatment applies for a natural boundary condition specified at the last node of the element except that last diagonal term is modified. Thus, instead of dealing with the natural boundary condition terms at the element level, it is easier to simply assign natural boundary conditions to the global nodes. The element equations are therefore simply

$$(\boldsymbol{k}_k + \boldsymbol{k}_p)\boldsymbol{d} = \boldsymbol{r}_q \Longrightarrow \boldsymbol{kd} = \boldsymbol{r}$$

During assembly, the natural boundary condition terms are incorporated directly into the global equations as follows:

NBC at dof i:

Add to global $\boldsymbol{k}$:	$k(i, i) = k(i, i) - \alpha_i k(x_i)$
Add to global $\boldsymbol{r}$:	$r(i) = r(i) + k(x_i)\beta_i$

This is a good place to reflect on the decision to arbitrarily assign a negative sign to the derivative term specified at the left end. If it was not done, then it can easily be seen from the above derivation that there will be a positive sign associated with the $\alpha_1 k(x_1)$ term and a negative sign with the $\alpha_n k(x_n)$ term. This would mean that while writing element equations we would have to keep track of whether we are considering an NBC at the left end or the right end of an element. By choosing the signs of derivatives the way we did, at the element level, there is no need to switch signs in the NBC terms. Furthermore, as seen from the applications presented in the previous section, the sign convention adopted

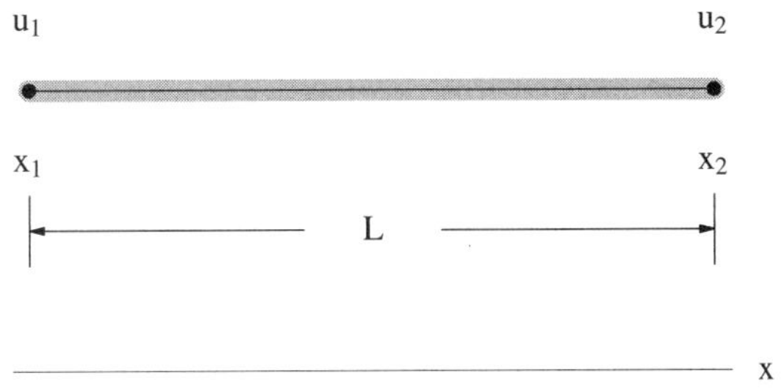

Figure 3.8. A two-node linear element for second-order BVP

here conforms to the usually accepted physical notions (such as heat flowing into a body is positive and that out of the body is negative).

Explicit integration of the terms in the element equations is possible after one has chosen the number of nodes for an element and the functions $k(x)$, $p(x)$, and $q(x)$ are known.

Explicit Equations for a Two-Node Linear Element For a two-node element, shown in Figure 3.8, the interpolation functions are simple linear functions of x. Furthermore, if we assume that k, p, and q are constant over an element, then it is easy to carry out integrations and write explicit formulas for element equations:

Element nodes: $\{x_1, x_1 + L\}$

Interpolation functions, $\boldsymbol{N}^T = \begin{pmatrix} \frac{x_1+L-x}{L} & -\frac{x_1 - x}{L} \end{pmatrix}$

$\boldsymbol{B}^T = \frac{d\boldsymbol{N}^T}{dx} = \begin{pmatrix} -\frac{1}{L} & \frac{1}{L} \end{pmatrix}$

$$\boldsymbol{k}_k = \int_{x_1}^{x_1+L} (k\boldsymbol{B}\boldsymbol{B}^T)\, dx = \begin{pmatrix} \frac{k}{L} & -\frac{k}{L} \\ -\frac{k}{L} & \frac{k}{L} \end{pmatrix}$$

$$\boldsymbol{k}_p = \int_{x_1}^{x_1+L} (-p\boldsymbol{N}\boldsymbol{N}^T)\, dx = \begin{pmatrix} -\frac{Lp}{3} & -\frac{Lp}{6} \\ -\frac{Lp}{6} & -\frac{Lp}{3} \end{pmatrix}$$

$$\boldsymbol{r}_q^T = \int_{x_1}^{x_1+L} (qN)\, dx = \left\{ \frac{Lq}{2}, \frac{Lq}{2} \right\}$$

Thus explicit equations for a linear element are as follows:

$$\begin{pmatrix} \frac{k}{L} - \frac{Lp}{3} & -\frac{k}{L} - \frac{Lp}{6} \\ -\frac{k}{L} - \frac{Lp}{6} & \frac{k}{L} - \frac{Lp}{3} \end{pmatrix} \begin{pmatrix} u_1 \\ u_2 \end{pmatrix} = \begin{pmatrix} \frac{Lq}{2} \\ \frac{Lq}{2} \end{pmatrix}$$

Explicit Equations for a Three-Node Quadratic Element With reasonable effort, it is possible to obtain explicit expressions for a three-node element, shown in Figure 3.9, with the middle node placed at the center of the element. As was the case for the linear element, it is also assumed that k, p, and q are constant over an element:

Element nodes: $\left\{x_1, \frac{1}{2}(2x_1 + L), x_1 + L\right\}$

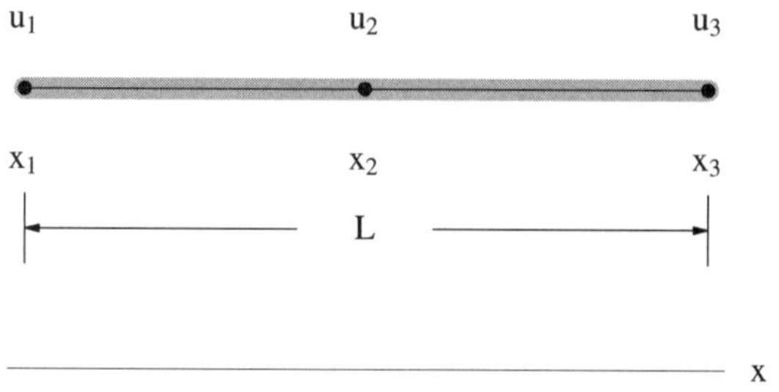

Figure 3.9. A three-node quadratic element for second-order BVP

Interpolation functions, $\boldsymbol{N}^T = \left(\frac{(2x_1+L-2x)(x_1+L-x)}{L^2} \quad \frac{4(x_1-x)(-x_1-L+x)}{L^2} \quad \frac{(2x_1+L-2x)(x_1-x)}{L^2} \right)$

$$\boldsymbol{B}^T = \frac{d\boldsymbol{N}^T}{dx} = \left(-\frac{x_1+3(x_1+L)-4x}{L^2} \quad \frac{4(2x_1+L-2x)}{L^2} \quad -\frac{4x_1+L-4x}{L^2} \right)$$

$$\boldsymbol{k}_k = \int_{x_1}^{x_1+L} (k\boldsymbol{B}\boldsymbol{B}^T)\, dx = \begin{pmatrix} \frac{7k}{3L} & -\frac{8k}{3L} & \frac{k}{3L} \\ -\frac{8k}{3L} & \frac{16k}{3L} & -\frac{8k}{3L} \\ \frac{k}{3L} & -\frac{8k}{3L} & \frac{7k}{3L} \end{pmatrix}$$

$$\boldsymbol{k}_p = \int_{x_1}^{x_1+L} (-p\boldsymbol{N}\boldsymbol{N}^T)\, dx = \begin{pmatrix} -\frac{2Lp}{15} & -\frac{Lp}{15} & \frac{Lp}{30} \\ -\frac{Lp}{15} & -\frac{8Lp}{15} & -\frac{Lp}{15} \\ \frac{Lp}{30} & -\frac{Lp}{15} & -\frac{2Lp}{15} \end{pmatrix}$$

$$\boldsymbol{r}_q^T = \int_{x_1}^{x_1+L} (q\boldsymbol{N})\, dx = \left\{ \frac{Lq}{6}, \frac{2Lq}{3}, \frac{Lq}{6} \right\}$$

Thus explicit equations quadratic element are as follows:

$$\begin{pmatrix} \frac{7k}{3L} - \frac{2Lp}{15} & -\frac{8k}{3L} - \frac{Lp}{15} & \frac{k}{3L} + \frac{Lp}{30} \\ -\frac{8k}{3L} - \frac{Lp}{15} & \frac{16k}{3L} - \frac{8Lp}{15} & -\frac{8k}{3L} - \frac{Lp}{15} \\ \frac{k}{3L} + \frac{Lp}{30} & -\frac{8k}{3L} - \frac{Lp}{15} & \frac{7k}{3L} - \frac{2Lp}{15} \end{pmatrix} \begin{pmatrix} u_1 \\ u_2 \\ u_3 \end{pmatrix} = \begin{pmatrix} \frac{Lq}{6} \\ \frac{2Lq}{3} \\ \frac{Lq}{6} \end{pmatrix}$$

3.2.1 Complete Solution Procedure

The approximate solution of any problem governed by a differential equation of the type considered in this chapter can be obtained by using the element equations derived in this section by following the usual finite element steps:

1. Development of element equations
2. Discretization of solution domain into a finite element mesh
3. Assembly of element equations
4. Introduction of boundary conditions
5. Solution for nodal unknowns
6. Computation of solution and related quantities over each element

The simplest element is obviously the two-node linear element. Since the first derivative of the assumed solution is constant over an element, one must use a fairly large number of linear elements to get accurate solutions. With three-node quadratic elements one typically needs fewer elements. If desired, one can develop a four-node cubic element as well that will require even fewer elements for accurate solution. However, each higher order element obviously requires more effort, and thus, in practice, problems are solved almost exclusively using linear or quadratic elements. Because of well-known difficulties associated with the higher order Lagrange interpolation functions (see Chapter 9), it is not recommended to use elements higher than a cubic based on the formulation presented here.

Solutions to many practical problems can be obtained by using the explicit element equations which are based on the assumption of constant coefficients k, p, and q over an element. When these coefficients are not constant, there are two choices. The simplest option is to assign average coefficient values to each element and use the explicit equations given. As the number of elements is increased, the discrepancy between the averaged values and the actual coefficient values should diminish. Alternatively, one can include the actual coefficients as functions of x and carry out integrations to establish appropriate matrices in the element equations:

$$\boldsymbol{k}_k = \int_{x_1}^{x_n} k(x)\boldsymbol{B}\boldsymbol{B}^T\,dx; \qquad \boldsymbol{k}_p = \int_{x_1}^{x_n} (-p(x))\boldsymbol{N}\boldsymbol{N}^T\,dx; \qquad \boldsymbol{r}_q = \int_{x_1}^{x_n} q(x)\boldsymbol{N}\,dx$$

Several examples in the following sections illustrate the procedure.

Mathematica Functions for Solution of One-Dimensional Boundary Value Problem For *Mathematica* solution of one-dimensional boundary value problems using a linear or quadratic element, the following functions can be defined and used in the manner shown in Chapter 2 for the axial deformation problem:

```
BVP1DLinElement[k_, p_, q_, {x1_, x2_}] := Module[{L = x2 − x1, ke, re},
    ke = {{k/L − (L ∗ p)/3, −(k/L) − (L ∗ p)/6}, {−(k/L) − (L ∗ p)/6, k/L − (L ∗ p)/3}};
    re = {(L ∗ q)/2, (L ∗ q)/2}; {ke, re}
  ]

BVP1DLinSolution[{x1_, x2_}, {d1_, d2_}] := Module[{L = x2 − x1, u},
    n = {(x1 + L − x)/L, −(x1 − x)/L}; u = Expand[n.{d1, d2}];
    {x1 ≤ x ≤ x2, u}]

BVP1DQuadElement[k_, p_, q_, {x1_, x2_, x3_}] := Module[{L = x3 − x1, ke, re},
    ke = {{(7 ∗ k)/(3 ∗ L) − (2 ∗ L ∗ p)/15, (−8 ∗ k)/(3 ∗ L) − (L ∗ p)/15,
        k/(3 ∗ L) + (L ∗ p)/30}, {(−8 ∗ k)/(3 ∗ L) − (L ∗ p)/15,
        (16 ∗ k)/(3 ∗ L) − (8 ∗ L ∗ p)/15, (−8 ∗ k)/(3 ∗ L) − (L ∗ p)/15},
        {k/(3 ∗ L) + (L ∗ p)/30, (−8 ∗ k)/(3 ∗ L) − (L ∗ p)/15,
        (7 ∗ k)/(3 ∗ L) − (2 ∗ L ∗ p)/15}};
    re = {(L ∗ q)/6, (2 ∗ L ∗ q)/3, (L ∗ q)/6};
    {ke, re}
  ]
```

```
BVP1DQuadSolution[{x1_, x2_, x3_}, {d1_, d2_, d3_}] := Module[{L = x3 − x1, u},
    n = {(2 ∗ x1 + L − 2 ∗ x) ∗ (x1 + L − x)/L^2, 4 ∗ (x1 − x) ∗ (−x1 − L + x)/L^2,
        (2 ∗ x1 + L − 2 ∗ x) ∗ (x1 − x)/L^2};
    u = Expand[n.{d1, d2, d3}];
    {x1 ≤ x ≤ x3, u}]
```

MATLAB Functions for Solution of One-Dimensional Boundary Value Problem For MATLAB solution of one-dimensional boundary value problems using linear or quadratic elements, the following functions can be defined and used in the manner shown in Chapter 2 for the axial deformation problem:

MatlabFiles\Chap3\BVP1DLinElement.m

```
function[ke, re] = BVP1DLinElement(k, p, q, coord)
% [ke, re] = BVP1DLinElement(k, p, q, coord)
% Generates equations for a linear element for 1D BVP
% k,p,q = parameters defining the BVP
% coord = coordinates at the element ends

L = coord(2) − coord(1);
ke = [k/L − (L ∗ p)/3, −(k/L) − (L ∗ p)/6; −(k/L) − (L ∗ p)/6, k/L − (L ∗ p)/3];
re = [(L ∗ q)/2; (L ∗ q)/2];
```

MatlabFiles\Chap3\BVP1DQuadElement.m

```
function[ke, re] = BVP1DQuadElement(k, p, q, coord)
% [ke, re] = BVP1DQuadElement(k, p, q, coord)
% Generates equations for a quadratic element for 1D BVP
% k,p,q = parameters defining the BVP
% coord = coordinates at the element ends

L = coord(3) − coord(1);
ke = [(7 ∗ k)/(3 ∗ L) − (2 ∗ L ∗ p)/15, (−8 ∗ k)/(3 ∗ L) − (L ∗ p)/15,...
      k/(3 ∗ L) + (L ∗ p)/30;
    (−8 ∗ k)/(3 ∗ L) − (L ∗ p)/15, (16 ∗ k)/(3 ∗ L) − (8 ∗ L ∗ p)/15,...
      (−8 ∗ k)/(3 ∗ L) − (L ∗ p)/15;
    k/(3 ∗ L) + (L ∗ p)/30, (−8 ∗ k)/(3 ∗ L) − (L ∗ p)/15,...
      (7 ∗ k)/(3 ∗ L) − (2 ∗ L ∗ p)/15];
re = [(L ∗ q)/6; (2 ∗ L ∗ q)/3; (L ∗ q)/6];
```

3.3 STEADY-STATE HEAT CONDUCTION

Example 3.1 Consider heat conduction through a 5-mm-thick base plate of a 900-W household iron. The base area is 150 cm^2 and the thermal conductivity $k = 20$ W/m · °C. The inner surface of the base plate is subjected to uniform heat flux generated by resistance heaters inside, as shown in Figure 3.10. During steady state the outer surface temperature

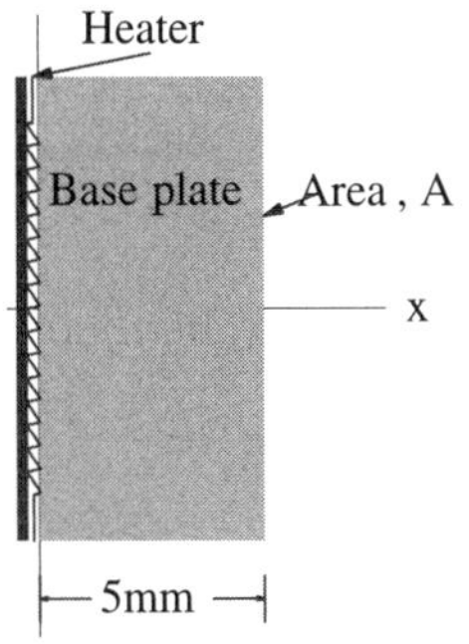

Figure 3.10. Base plate of a household iron

is measured to be 84°C. Determine the temperature through the base plate. What is the temperature at the inside surface?

Heat generation in the base plate is zero. The outside surface has an essential boundary condition. The inside surface has a natural boundary condition. The problem is described in terms of the following differential equation and boundary conditions:

$$kA\left(\frac{d^2T}{dx^2}\right) = 0; \qquad 0 < x < L$$

$$k = 20\ \text{W/m}\cdot{}^\circ\text{C}; \qquad A = \frac{150}{100^2}\ \text{m}^2; \qquad L = \frac{5}{1000}\ \text{m}$$

$$T(L) = 84 \qquad \text{and} \qquad -kA\frac{dT(0)}{dx} = 900$$

Compared to the general form, we have

$$k(x) = kA = \frac{3}{10}; \qquad p(x) = 0; \qquad q(x) = 0$$

$$\text{NBC at } x = 0\text{:} \qquad \alpha = 0; \qquad \beta = \frac{900}{kA} = 3000$$

The problem is very simple and an exact solution can readily be obtained by direct integration. However, for illustration purposes, a finite element solution is obtained using only one two-node linear element. The finite element model is as shown in Figure 3.11. The element equations are written by substituting numerical values into the explicit linear element equations presented earlier. The complete one-element solution is as follows:

Figure 3.11. One two-node finite element model

Element nodes: $\{x_1 \to 0, x_2 \to 0.005\}$

$$\begin{pmatrix} 60. & -60. \\ -60. & 60. \end{pmatrix} \begin{pmatrix} T_1 \\ T_2 \end{pmatrix} = \begin{pmatrix} 0 \\ 0 \end{pmatrix}$$

Global equations before boundary conditions are the same as the element equations.
Natural boundary conditions:

dof	α	β
T_1	0	3000

dof	$k(x)$	$-k(x)\alpha$	$k(x)\beta$
T_1	$\frac{3}{10}$	0	900

Global equations after incorporating the NBC:

$$\begin{pmatrix} 60. & -60. \\ -60. & 60. \end{pmatrix} \begin{pmatrix} T_1 \\ T_2 \end{pmatrix} = \begin{pmatrix} 900 \\ 0 \end{pmatrix}$$

Essential boundary conditions:

dof	Value
T_2	84

Incorporating the EBC, the final system of equations is

$$(60.)(T_1) = (5940.)$$

Solution for nodal unknowns:

dof	x	Solution
T_1	0	99.
T_2	0.005	84

Solution over element 1:

Nodes: $\{x_1 \to 0, x_2 \to 0.005\}$

Interpolation functions: $\boldsymbol{N}^T = \{1. - 200.\, x, 200.\, x\}$

Nodal values: $\boldsymbol{d}^T = \{99., 84\}$

Solution: $T(x) = \boldsymbol{N}^T \boldsymbol{d} = 99. - 3000.\, x$

A plot of the solution is shown in Figure 3.12. It can easily be verified that the solution satisfies the differential equation and both the boundary conditions. Thus it is an exact solution and there is no need to increase the number of elements.

3.4 STEADY-STATE HEAT CONDUCTION AND CONVECTION

Example 3.2 A 2-m-high and 3-m-wide plate is at a temperature of 100°C. It is to be cooled by using 20-cm-long and 0.3-cm-thick aluminum fins ($k = 237$ W/m · °C). The fins extend over the entire width of the plate, as shown in Figure 3.13. Along the plate

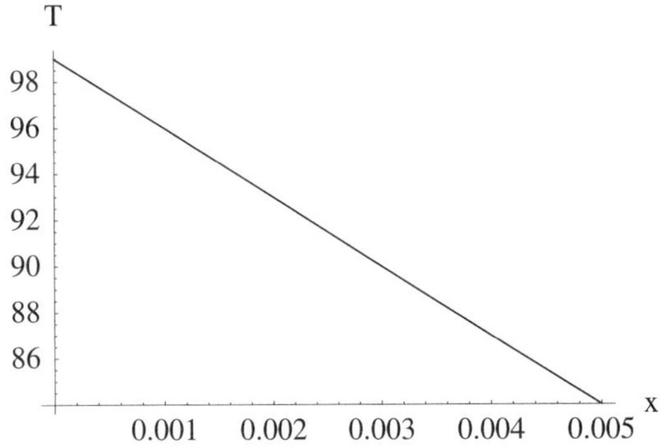

Figure 3.12. Temperature distribution in the base plate

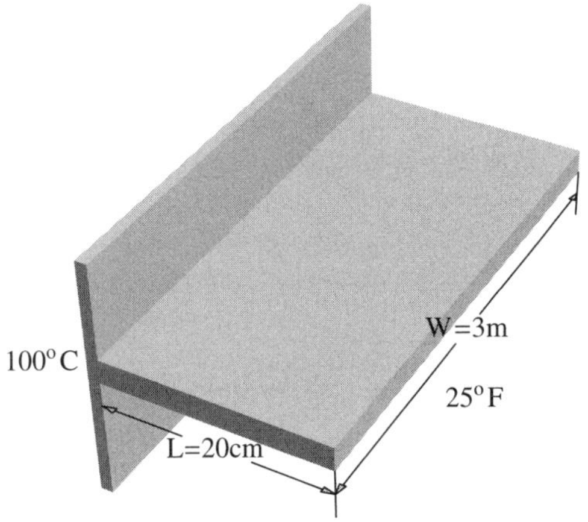

Figure 3.13. One-dimensional heat transfer through a fin

height the clear spacing between the fins is 0.4 cm. The ambient air temperature is 25°C and average convection coefficient is 30 W/m$^2 \cdot$ °C. Determine the temperature distribution through the fins. What is the rate of heat transfer from the entire finned surface of the plate?

Assuming that over the width of a fin the temperature is uniform and that it varies only along its length, the problem can be treated as one dimensional. Heat generation in the fin is zero. Since the convection takes place at the top, bottom, and sides of the fin, $P = 2W + 2t$, where t is fin thickness. The base has an essential boundary condition. The outside end of the fin surface has a convection boundary condition. The governing differential equation is as follows:

$$\frac{d}{dx}\left(kA\frac{dT}{dx}\right) - hPT + hPT_{\infty} = 0; \qquad 0 < x < L$$

$$P = 2(W + t); \qquad A = Wt;$$

$$T(0) = 100 \qquad \text{and} \qquad kA\frac{dT(L)}{dx} = -hA(T - T_{\infty})$$

Compared to the general form, we have

$$k(x) = kA; \qquad p(x) = -hP; \qquad q(x) = hPT_\infty$$

$$\text{NBC at } x = L: \qquad \alpha = -\frac{hA}{kA}; \qquad \beta = \frac{hAT_\infty}{kA}$$

Using the given numerical data,

$$W = 3 \text{ m}; \quad t = 0.3/100 \text{ m}; \quad L = 20/100 \text{ m}; \quad k = 237; \quad h = 30; \quad T_\infty = 30$$

$$P = 6.006; \qquad A = 0.009$$

$$kA = 2.133; \qquad hP = 180.18; \qquad hPT_\infty = 4504.5$$

$$\alpha = -0.126582; \qquad \beta = 3.16456$$

(i) Solution Using Four Linear Elements

A finite element solution using four two-node linear elements is presented. The finite element model is as shown in Figure 3.14. For each element the equations are written by substituting numerical values into the explicit linear element equations presented earlier. The complete four-element solution is as follows:

Nodal locations: {0, 0.05, 0.1, 0.15, 0.2}

Element 1:

$$\begin{pmatrix} 45.663 & -41.1585 \\ -41.1585 & 45.663 \end{pmatrix} \begin{pmatrix} T_1 \\ T_2 \end{pmatrix} = \begin{pmatrix} 112.613 \\ 112.613 \end{pmatrix}$$

Element 2:

$$\begin{pmatrix} 45.663 & -41.1585 \\ -41.1585 & 45.663 \end{pmatrix} \begin{pmatrix} T_2 \\ T_3 \end{pmatrix} = \begin{pmatrix} 112.613 \\ 112.613 \end{pmatrix}$$

Element 3:

$$\begin{pmatrix} 45.663 & -41.1585 \\ -41.1585 & 45.663 \end{pmatrix} \begin{pmatrix} T_3 \\ T_4 \end{pmatrix} = \begin{pmatrix} 112.613 \\ 112.613 \end{pmatrix}$$

Element 4:

$$\begin{pmatrix} 45.663 & -41.1585 \\ -41.1585 & 45.663 \end{pmatrix} \begin{pmatrix} T_4 \\ T_5 \end{pmatrix} = \begin{pmatrix} 112.613 \\ 112.613 \end{pmatrix}$$

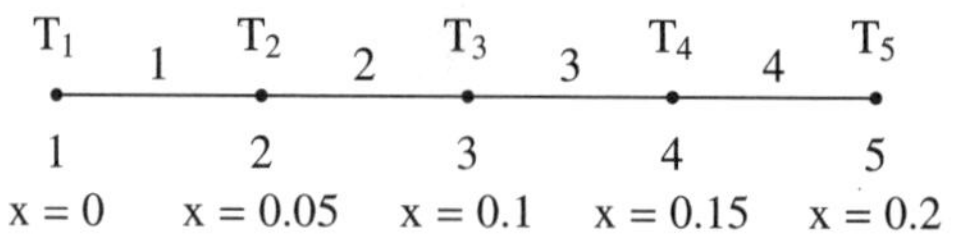

Figure 3.14. Four-element model for the fin

Global equations before boundary conditions:

$$\begin{pmatrix} 45.663 & -41.1585 & 0 & 0 & 0 \\ -41.1585 & 91.326 & -41.1585 & 0 & 0 \\ 0 & -41.1585 & 91.326 & -41.1585 & 0 \\ 0 & 0 & -41.1585 & 91.326 & -41.1585 \\ 0 & 0 & 0 & -41.1585 & 45.663 \end{pmatrix} \begin{pmatrix} T_1 \\ T_2 \\ T_3 \\ T_4 \\ T_5 \end{pmatrix} = \begin{pmatrix} 112.613 \\ 225.225 \\ 225.225 \\ 225.225 \\ 112.613 \end{pmatrix}$$

Natural boundary conditions:

dof	α	β	$k(x)$	$-k(x)\,\alpha$	$k(x)\beta$
T_5	−0.126582	3.16456	2.133	0.27	6.75

Global equations after incorporating the NBC:

$$\begin{pmatrix} 45.663 & -41.1585 & 0 & 0 & 0 \\ -41.1585 & 91.326 & -41.1585 & 0 & 0 \\ 0 & -41.1585 & 91.326 & -41.1585 & 0 \\ 0 & 0 & -41.1585 & 91.326 & -41.1585 \\ 0 & 0 & 0 & -41.1585 & 45.933 \end{pmatrix} \begin{pmatrix} T_1 \\ T_2 \\ T_3 \\ T_4 \\ T_5 \end{pmatrix} = \begin{pmatrix} 112.613 \\ 225.225 \\ 225.225 \\ 225.225 \\ 119.363 \end{pmatrix}$$

Essential boundary conditions:

dof	Value
T_1	100

Incorporating the EBC, the final system of equations is

$$\begin{pmatrix} 91.326 & -41.1585 & 0 & 0 \\ -41.1585 & 91.326 & -41.1585 & 0 \\ 0 & -41.1585 & 91.326 & -41.1585 \\ 0 & 0 & -41.1585 & 45.933 \end{pmatrix} \begin{pmatrix} T_2 \\ T_3 \\ T_4 \\ T_5 \end{pmatrix} = \begin{pmatrix} 4341.08 \\ 225.225 \\ 225.225 \\ 119.363 \end{pmatrix}$$

Solution for nodal unknowns:

dof	x	Solution
T_1	0	100
T_2	0.05	73.8496
T_3	0.1	58.3916
T_4	0.15	50.2425
T_5	0.2	47.6187

Solution over element 1:

Nodes: $\{x_1 \to 0, x_2 \to 0.05\}$

Interpolation functions: $\boldsymbol{N}^T = \{1. - 20.\,x, 20.\,x\}$

Nodal values: $d^T = \{100, 73.8496\}$

Solution: $T(x) = N^T d = 100. - 523.009\,x$

Solution over element 2:

Nodes: $\{x_1 \to 0.05, x_2 \to 0.1\}$

Interpolation functions: $N^T = \{2. - 20.\,x, 20.\,x - 1.\}$

Nodal values: $d^T = \{73.8496, 58.3916\}$

Solution: $T(x) = N^T d = 89.3075 - 309.16\,x$

Solution over element 3:

Nodes: $\{x_1 \to 0.1, x_2 \to 0.15\}$

Interpolation functions: $N^T = \{3. - 20.\,x, 20.\,x - 2.\}$

Nodal values: $d^T = \{58.3916, 50.2425\}$

Solution: $T(x) = N^T d = 74.6897 - 162.981\,x$

Solution over element 4:

Nodes: $\{x_1 \to 0.15, x_2 \to 0.2\}$

Interpolation functions: $N^T = \{4. - 20.\,x, 20.\,x - 3.\}$

Nodal values: $d^T = \{50.2425, 47.6187\}$

Solution: $T(x) = N^T d = 58.114 - 52.4766\,x$

A plot of the solution is shown in Figure 3.15. The heat loss over a fin is determined by computing heat loss over each element and then summing element contributions:

$$\text{Heat loss} = \sum \int_{x_i}^{x_{i+1}} hP(T(x) - T_\infty)\,dx$$

	Range	Temperature	Heat Loss
1	$0 \leq x \leq 0.05$	$100. - 523.009\,x$	557.88
2	$0.05 \leq x \leq 0.1$	$89.3075 - 309.16\,x$	370.455
3	$0.1 \leq x \leq 0.15$	$74.6897 - 162.981\,x$	264.117
4	$0.15 \leq x \leq 0.2$	$58.114 - 52.4766\,x$	215.591

Total heat loss → 1408.04

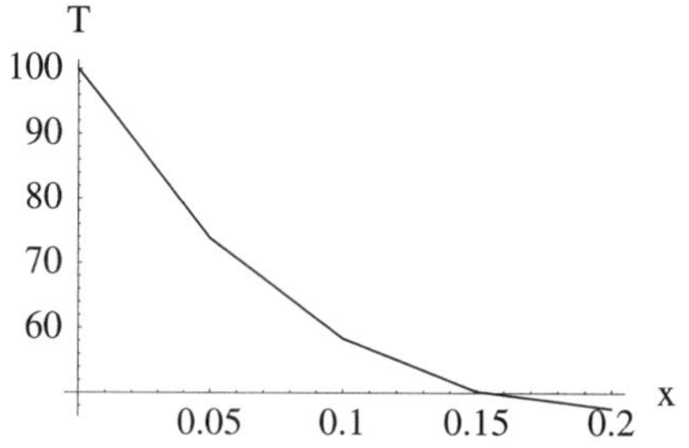

Figure 3.15. Temperature distribution in the fin

To compute the heat loss from the entire plate surface, we need to determine the total number of fins and multiply the heat loss per fin by this number:

$$\text{Number of fins} = \frac{\text{plate height}}{\text{fin thickness} + \text{fin spacing}} = \frac{2}{0.3/100 + 0.4/100}$$
$$= 285.714, \text{ say } 286 \text{ fins}$$

$$\text{Heat loss through all fins} = 286 \times 1408.04 = 402700 \text{ W}$$

(ii) Solution Using Two Quadratic Elements

To see the effect of higher order elements, now consider a finite element solution using only two quadratic elements. The finite element model still looks the same as shown in Figure 3.14. However, the first three nodes define the first element and the next three the second element. For each element the equations are written by substituting numerical values into the explicit quadratic element equations presented earlier. The complete two-element solution is as follows:

Nodal locations: {0, 0.05, 0.1, 0.15, 0.2}

Element 1:

$$\begin{pmatrix} 52.1724 & -55.6788 & 6.5094 \\ -55.6788 & 123.37 & -55.6788 \\ 6.5094 & -55.6788 & 52.1724 \end{pmatrix} \begin{pmatrix} T_1 \\ T_2 \\ T_3 \end{pmatrix} = \begin{pmatrix} 75.075 \\ 300.3 \\ 75.075 \end{pmatrix}$$

Element 2:

$$\begin{pmatrix} 52.1724 & -55.6788 & 6.5094 \\ -55.6788 & 123.37 & -55.6788 \\ 6.5094 & -55.6788 & 52.1724 \end{pmatrix} \begin{pmatrix} T_3 \\ T_4 \\ T_5 \end{pmatrix} = \begin{pmatrix} 75.075 \\ 300.3 \\ 75.075 \end{pmatrix}$$

Global equations before boundary conditions:

$$\begin{pmatrix} 52.1724 & -55.6788 & 6.5094 & 0 & 0 \\ -55.6788 & 123.37 & -55.6788 & 0 & 0 \\ 6.5094 & -55.6788 & 104.345 & -55.6788 & 6.5094 \\ 0 & 0 & -55.6788 & 123.37 & -55.6788 \\ 0 & 0 & 6.5094 & -55.6788 & 52.1724 \end{pmatrix} \begin{pmatrix} T_1 \\ T_2 \\ T_3 \\ T_4 \\ T_5 \end{pmatrix} = \begin{pmatrix} 75.075 \\ 300.3 \\ 150.15 \\ 300.3 \\ 75.075 \end{pmatrix}$$

Natural boundary conditions:

dof	α	β	$k(x)$	$-k(x)\alpha$	$k(x)\beta$
T_5	−0.126582	3.16456	2.133	0.27	6.75

Global equations after incorporating the NBC:

$$\begin{pmatrix} 52.1724 & -55.6788 & 6.5094 & 0 & 0 \\ -55.6788 & 123.37 & -55.6788 & 0 & 0 \\ 6.5094 & -55.6788 & 104.345 & -55.6788 & 6.5094 \\ 0 & 0 & -55.6788 & 123.37 & -55.6788 \\ 0 & 0 & 6.5094 & -55.6788 & 52.4424 \end{pmatrix} \begin{pmatrix} T_1 \\ T_2 \\ T_3 \\ T_4 \\ T_5 \end{pmatrix} = \begin{pmatrix} 75.075 \\ 300.3 \\ 150.15 \\ 300.3 \\ 81.825 \end{pmatrix}$$

Essential boundary conditions:

dof	Value
T_1	100

Incorporating the EBC, the final system of equations is

$$\begin{pmatrix} 123.37 & -55.6788 & 0 & 0 \\ -55.6788 & 104.345 & -55.6788 & 6.5094 \\ 0 & -55.6788 & 123.37 & -55.6788 \\ 0 & 6.5094 & -55.6788 & 52.4424 \end{pmatrix} \begin{pmatrix} T_2 \\ T_3 \\ T_4 \\ T_5 \end{pmatrix} = \begin{pmatrix} 5868.18 \\ -500.79 \\ 300.3 \\ 81.825 \end{pmatrix}$$

Solution for nodal unknowns:

dof	x	Solution
T_1	0	100
T_2	0.05	74.074
T_3	0.1	58.7352
T_4	0.15	50.6042
T_5	0.2	47.9969

Solution over element 1:

Nodes: $\{x_1 \to 0, x_2 \to 0.05, x_3 \to 0.1\}$

Interpolation functions: $\boldsymbol{N}^T = \{200.\,x^2 - 30.\,x + 1.,\ 40.\,x - 400.\,x^2,\ 200.\,x^2 - 10.\,x\}$

Nodal values: $\boldsymbol{d}^T = \{100, 74.074, 58.7352\}$

Solution: $T(x) = \boldsymbol{N}^T\boldsymbol{d} = 2117.42x^2 - 624.39x + 100.$

Solution over element 2:

Nodes: $\{x_1 \to 0.1, x_2 \to 0.15, x_3 \to 0.2\}$

Interpolation functions:

$$\boldsymbol{N}^T = \{200.\,x^2 - 70.\,x + 6.,\ -400.\,x^2 + 120.\,x - 8.,\ 200.\,x^2 - 50.\,x + 3.\}$$

Nodal values: $\boldsymbol{d}^T = \{58.7352, 50.6042, 47.9969\}$

Solution: $T(x) = \boldsymbol{N}^T\boldsymbol{d} = 1104.76x^2 - 438.809x + 91.5686$

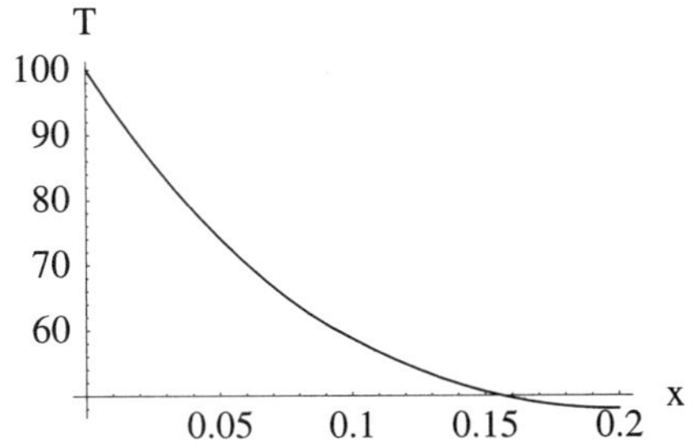

Figure 3.16. Temperature distribution in the fin

A plot of the solution is shown in Figure 3.16. The heat loss over a fin is determined by computing heat loss over each element and then summing element contributions:

$$\text{Heat loss} = \sum \int_{x_1}^{x_2} hP(T(x) - T_\infty)\, dx$$

	Range	Temperature	Heat Loss
1	$0 \le x \le 0.1$	$2117.42x^2 - 624.39x + 100.$	916.009
2	$0.1 \le x \le 0.2$	$1104.76x^2 - 438.809x + 91.5686$	477.924

Total heat loss → 1393.93

(iii) Solution Convergence

The solution convergence can be demonstrated by evaluating solutions using increasing numbers of elements. Figure 3.17 shows solutions obtained using one through five linear (two-node) elements. Figure 3.18 shows solutions obtained using quadratic (three-node) elements. The convergence is much more rapid in the case of quadratic elements, where, except for the one-element solution, all other solutions are practically indistinguishable from each other.

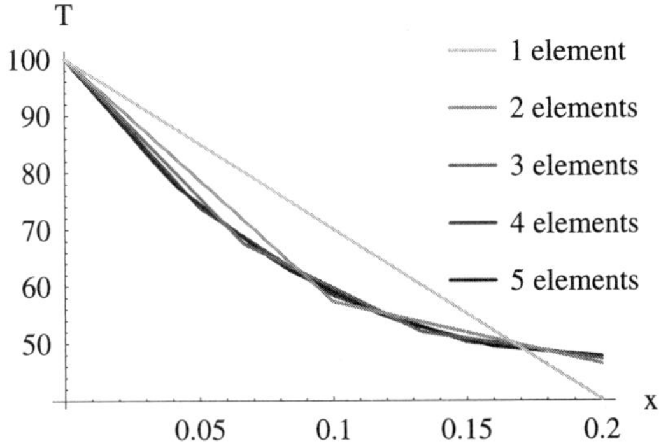

Figure 3.17. Comparison of solutions using linear elements

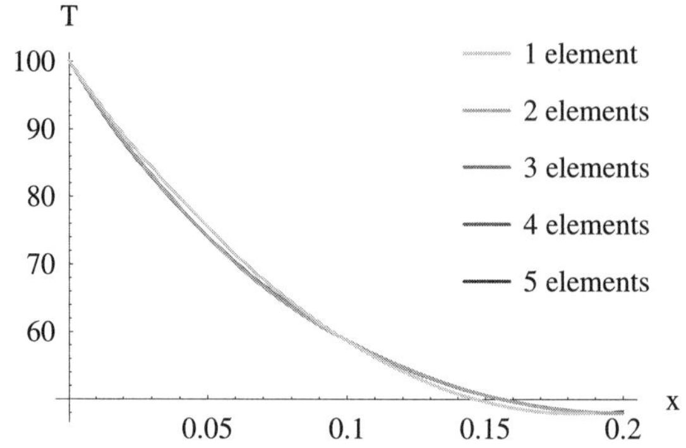

Figure 3.18. Comparison of solutions using quadratic elements

◆ *Mathematica*/MATLAB Implementation 3.1 on the Book Web Site: *Four-quadratic-element solution*

3.5 VISCOUS FLUID FLOW BETWEEN PARALLEL PLATES

Example 3.3 Determine the velocity profile for flow between two fixed plates shown in Figure 3.19. The fluid temperature varies linearly from 80°F at the bottom to 200°F at the top. The fluid viscosity at different temperatures is as follows:

Temperature, °F	$\mu \times 10^{-6}$
80	16
120	14
160	11
200	7

The governing differential equation for determining a fluid velocity profile $u(y)$ is as follows:

$$\frac{d}{dy}\left(\mu(y)\frac{du}{dy}\right) - \frac{dP}{dx} = 0; \qquad 0 < y < h$$

$$u(0) = 0 \qquad \text{and} \qquad u(h) = 0$$

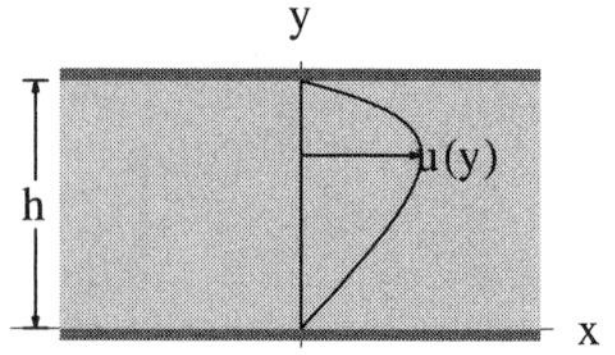

Figure 3.19. Velocity profile of a viscous fluid flow between two plates

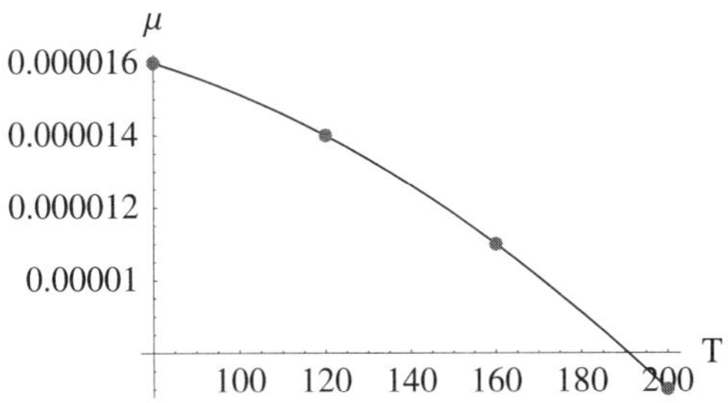

Figure 3.20. Viscosity as a function of temperature

Use the following numerical data:

$$h = 0.3 \text{ ft}; \qquad \frac{dP}{dx} = -2 \times 10^{-5} \text{ lb/ft}^3$$

In order to proceed with the solution, we need an expression for $\mu(y)$. However, the given data show viscosity as a function of temperature. Thus we first use curve fitting to get the following equation for viscosity as a function of temperature:

```
tmData = {{80, 16 * 10^-6}, {120, 14 * 10^-6}, {160, 11 * 10^-6}, {200, 7 * 10^-6}};
μ[T] = Fit[tmData, {1, T, T^2}, T]
```

$-3.125 \times 10^{-10} T^2 + 1.25 \times 10^{-8} T + 0.000017$

The plot in Figure 3.20 shows very good agreement between this function and the given data.

Assuming the temperature varies linearly from 80°F at the bottom to 200°F at the top, the temperature as a function of y is as follows:

$$T(y) = 400y + 80$$

Substituting this into the viscosity-temperature relationship, we get the following expression for viscosity as a function of height:

$$\mu(y) = -0.00005y^2 - 0.000015y + 0.000016$$

A finite element solution using four two-node linear elements is presented. The finite element model is as shown in Figure 3.21. For convenience the model is shown horizontally.

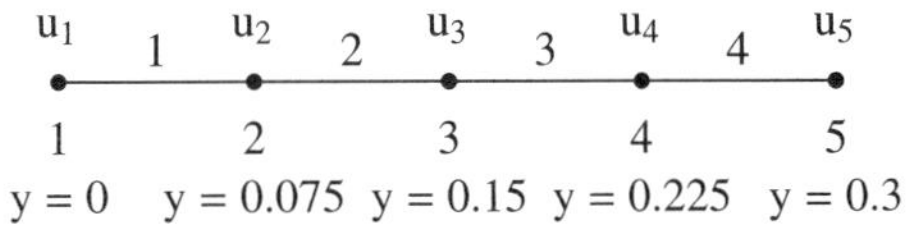

Figure 3.21. Four-element model

Since μ is not constant, we cannot use the explicit linear element equations given earlier. We must either assign average values to the elements or derive a new $\boldsymbol{k}$ matrix by substituting the $\mu(y)$ expression and carrying out integrations. Here we use the latter approach and derive the specific element equations first.

Derivation of Element Equations

Element nodes: $\{y_1, y_2\}$

Interpolation functions, $\boldsymbol{N}^T = \begin{pmatrix} \frac{y_2-y}{y_2-y_1} & \frac{y_1-y}{y_1-y_2} \end{pmatrix}$

$\boldsymbol{B}^T = d\boldsymbol{N}^T/dx = \begin{pmatrix} \frac{1}{y_1-y_2} & \frac{1}{y_2-y_1} \end{pmatrix}$

$k(y) = \mu(y); \qquad p(y) = 0; \qquad q(y) = -dP/dx$

$$\boldsymbol{k}_k = \int_{y_1}^{y_2} (\mu(y)\boldsymbol{B}\boldsymbol{B}^T)dy = \begin{pmatrix} \frac{\int_{y_1}^{y_2} \mu(y)\,dy}{(y_1-y_2)^2} & \frac{\int_{y_1}^{y_2} \mu(y)\,dy}{(y_1-y_2)(y_2-y_1)} \\ \frac{\int_{y_1}^{y_2} \mu(y)\,dy}{(y_1-y_2)(y_2-y_1)} & \frac{\int_{y_1}^{y_2} \mu(y)\,dy}{(y_1-y_2)^2} \end{pmatrix}$$

$$\boldsymbol{r}_q^T = \int_{y_1}^{y_2} (-\tfrac{dP}{dx}\boldsymbol{N})\,dy = \left\{ \frac{1}{2}\frac{dP}{dx}(y_1 - y_2), \frac{1}{2}\frac{dP}{dx}(y_1 - y_2) \right\}$$

The complete element equations are as follows:

$$\begin{pmatrix} \frac{\int_{y_1}^{y_2} \mu(y)\,dy}{(y_1-y_2)^2} & \frac{\int_{y_1}^{y_2} \mu(y)\,dy}{(y_1-y_2)(y_2-y_1)} \\ \frac{\int_{y_1}^{y_2} \mu(y)\,dy}{(y_1-y_2)(y_2-y_1)} & \frac{\int_{y_1}^{y_2} \mu(y)\,dy}{(y_2-y_1)^2} \end{pmatrix} \begin{pmatrix} u_1 \\ u_2 \end{pmatrix} = \begin{pmatrix} \frac{1}{2}\frac{dP}{dx}(y_1 - y_2) \\ \frac{1}{2}\frac{dP}{dx}(y_1 - y_2) \end{pmatrix}$$

The numerator in each term of the coefficient matrix is the same and can be written as follows:

$$\int_{y_1}^{y_2} \mu(y)\,dy = 0.0000166667y_1^3 + 7.5 \times 10^{-6}y_1^2 - 0.000016y_1 \\ - 0.0000166667y_2^3 - 7.5 \times 10^{-6}y_2^2 + 0.000016y_2$$

Using these element equations, the complete four-element solution is as follows:

Nodal locations: $\{0, 0.075, 0.15, 0.225, 0.3\}$

Element 1:

$$\begin{pmatrix} 0.000204583 & -0.000204583 \\ -0.000204583 & 0.000204583 \end{pmatrix} \begin{pmatrix} u_1 \\ u_2 \end{pmatrix} = \begin{pmatrix} 7.5 \times 10^{-7} \\ 7.5 \times 10^{-7} \end{pmatrix}$$

Element 2:

$$\begin{pmatrix} 0.000182083 & -0.000182083 \\ -0.000182083 & 0.000182083 \end{pmatrix} \begin{pmatrix} u_2 \\ u_3 \end{pmatrix} = \begin{pmatrix} 7.5 \times 10^{-7} \\ 7.5 \times 10^{-7} \end{pmatrix}$$

Element 3:

$$\begin{pmatrix} 0.000152083 & -0.000152083 \\ -0.000152083 & 0.000152083 \end{pmatrix} \begin{pmatrix} u_3 \\ u_4 \end{pmatrix} = \begin{pmatrix} 7.5 \times 10^{-7} \\ 7.5 \times 10^{-7} \end{pmatrix}$$

Element 4:

$$\begin{pmatrix} 0.000114583 & -0.000114583 \\ -0.000114583 & 0.000114583 \end{pmatrix} \begin{pmatrix} u_4 \\ u_5 \end{pmatrix} = \begin{pmatrix} 7.5 \times 10^{-7} \\ 7.5 \times 10^{-7} \end{pmatrix}$$

Essential boundary conditions:

dof	Value
u_1	0
u_5	0

Incorporating the EBC, the final system of equations is

$$\begin{pmatrix} 0.000386667 & -0.000182083 & 0 \\ -0.000182083 & 0.000334167 & -0.000152083 \\ 0 & -0.000152083 & 0.000266667 \end{pmatrix} \begin{pmatrix} u_2 \\ u_3 \\ u_4 \end{pmatrix} = \begin{pmatrix} 1.5 \times 10^{-6} \\ 1.5 \times 10^{-6} \\ 1.5 \times 10^{-6} \end{pmatrix}$$

Solution for nodal unknowns:

dof	y	Solution
u_1	0	0
u_2	0.075	0.0127967
u_3	0.15	0.0189367
u_4	0.225	0.0164248
u_5	0.3	0

Solution over element 1:

Nodes: $\{y_1 \to 0, y_2 \to 0.075\}$

Interpolation functions: $\boldsymbol{N}^T = \{1. - 13.3333\,y, 13.3333\,y\}$

Nodal values: $\boldsymbol{d}^T = \{0, 0.0127967\}$

Solution: $u(y) = \boldsymbol{N}^T\boldsymbol{d} = 0.170623\,y$

Solution over element 2:

Nodes: $\{y_1 \to 0.075, y_2 \to 0.15\}$

Interpolation functions: $\boldsymbol{N}^T = \{2. - 13.3333\,y, 13.3333\,y - 1.\}$

Nodal values: $\boldsymbol{d}^T = \{0.0127967, 0.0189367\}$

Solution: $u(y) = \boldsymbol{N}^T\boldsymbol{d} = 0.0818665\,y + 0.0066567$

Solution over element 3:

Nodes: $\{y_1 \to 0.15, y_2 \to 0.225\}$

Interpolation functions: $\boldsymbol{N}^T = \{3. - 13.3333\,y, 13.3333\,y - 2.\}$

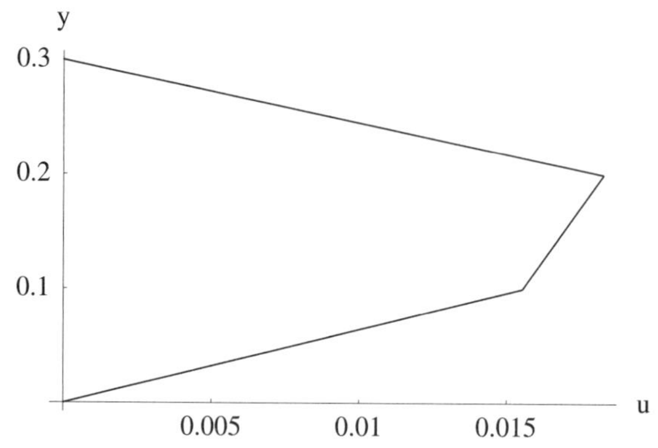

Figure 3.22. Computed velocity profile using four linear elements

Nodal values: $\boldsymbol{d}^T = \{0.0189367, 0.0164248\}$

Solution: $u(y) = \boldsymbol{N}^T\boldsymbol{d} = 0.0239604 - 0.0334914\,y$

Solution over element 4:

Nodes: $\{y_1 \to 0.225, y_2 \to 0.3\}$

Interpolation functions: $\boldsymbol{N}^T = \{4. - 13.3333\,y, 13.3333\,y - 3.\}$

Nodal values: $\boldsymbol{d}^T = \{0.0164248, 0\}$

Solution: $u(y) = \boldsymbol{N}^T\boldsymbol{d} = 0.0656993 - 0.218998\,y$

Solution summary:

	Range	Solution
1	$0 \le y \le 0.075$	$0.170623\,y$
2	$0.075 \le y \le 0.15$	$0.0818665\,y + 0.0066567$
3	$0.15 \le y \le 0.225$	$0.0239604 - 0.0334914\,y$
4	$0.225 \le y \le 0.3$	$0.0656993 - 0.218998\,y$

The computed velocity profile is plotted in Figure 3.22. To demonstrate convergence, the velocity profiles are computed using two to five linear and quadratic elements. The resulting profiles are plotted in Figures 3.23 and 3.24. As expected, the quadratic element solution converges much more rapidly.

3.6 ELASTIC BUCKLING OF BARS

Example 3.4 Compute buckling load for a column simply supported at both ends as shown in Figure 3.25. The governing differential equation is as follows:

$$EI\left(\frac{d^2y}{dx^2}\right) + Py = 0; \qquad y(0) = y(L) = 0$$

The length $L = 10$ ft and $EI = 10^6$ lb · in^2.

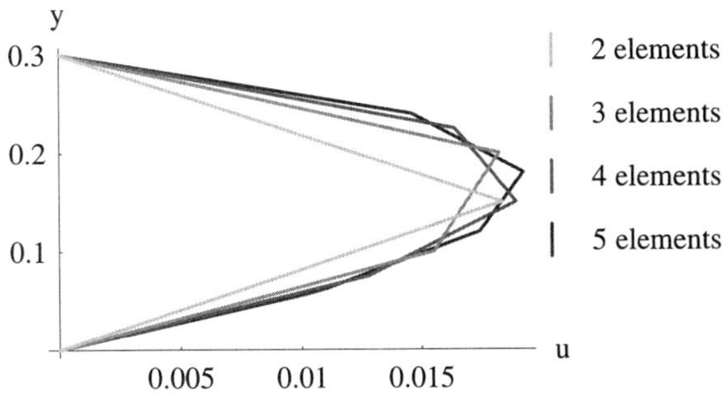

Figure 3.23. Computed velocity profile using linear elements

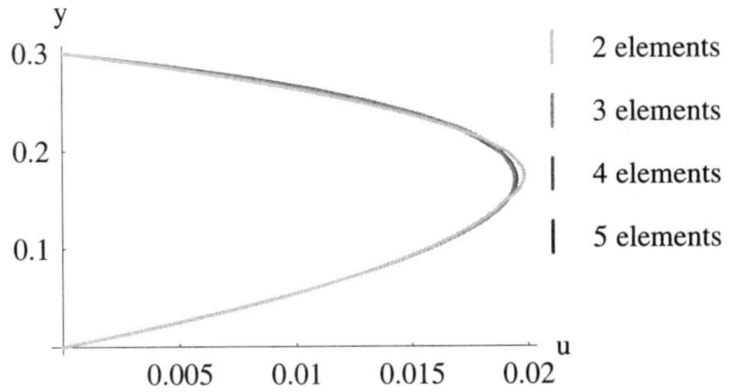

Figure 3.24. Computed velocity profile using quadratic elements

Compared to the general form,

$$k = EI; \qquad p = P; \qquad q = 0$$

Since the buckling load P is unknown, the element matrices $\boldsymbol{k}_k$ and $\boldsymbol{k}_p$ are computed separately. The matrix $\boldsymbol{k}_p$ is multiplied by the unknown load P. Each of the two matrices are assembled in the usual manner to form global matrices. The final global equations are written in the following form:

$$\boldsymbol{k}_k \boldsymbol{d} + P\boldsymbol{k}_p \boldsymbol{d} = \boldsymbol{0} \Longrightarrow (\boldsymbol{k}_k + P\boldsymbol{k}_p)\boldsymbol{d} = \boldsymbol{0}$$

This is a homogeneous system of equations and can be recognized as a generalized eigenvalue problem. A nonzero solution for the nodal degrees of freedom $\boldsymbol{d}$ is possible only

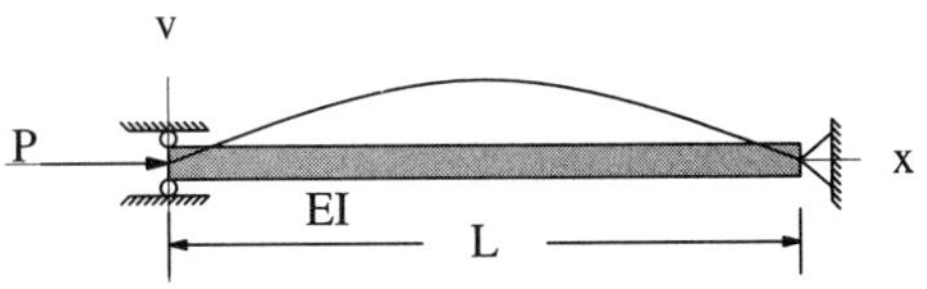

Figure 3.25. Buckling of a long slender bar

if the coefficient matrix cannot be inverted. Thus a necessary condition for a non-trivial solution of the equations is that the determinant of the coefficient matrix be zero:

$$\det(\boldsymbol{k}_k + P\boldsymbol{k}_p) = 0$$

For an n-degree-of-freedom system, evaluating this determinant gives an equation called the *characteristic equation* that is a polynomial of degree n in terms of P. The roots of the characteristic equation represent different buckling loads (eigenvalues). Substituting each buckling load in the global equations, one can compute the corresponding nodal displacements. These represent the buckling modes (eigenvectors). Generally one is interested in the lowest buckling load and the corresponding mode shape.

(i) Solution Using Four Linear Elements

For a two-node linear element the finite element equations are derived in the previous section. Taking the unknown P outside as a common factor, the element equations are as follows:

$$\begin{pmatrix} \frac{k}{L} & -\frac{k}{L} \\ -\frac{k}{L} & \frac{k}{L} \end{pmatrix} \begin{pmatrix} v_1 \\ v_2 \end{pmatrix} + P \begin{pmatrix} -\frac{L}{3} & -\frac{L}{6} \\ -\frac{L}{6} & -\frac{L}{3} \end{pmatrix} \begin{pmatrix} v_1 \\ v_2 \end{pmatrix} = \begin{pmatrix} 0 \\ 0 \end{pmatrix}$$

A finite element solution using four two-node linear elements is presented. The finite element model is as shown in Figure 3.26. The element equations are written by substituting numerical values into these element equations. The complete four-element solution is as follows:

Nodal locations: {0, 30., 60., 90., 120.}

Element 1:

$$\begin{pmatrix} 33333.3 & -33333.3 \\ -33333.3 & 33333.3 \end{pmatrix} \begin{pmatrix} v_1 \\ v_2 \end{pmatrix} - P \begin{pmatrix} 10. & 5. \\ 5. & 10. \end{pmatrix} \begin{pmatrix} v_1 \\ v_2 \end{pmatrix} = \begin{pmatrix} 0 \\ 0 \end{pmatrix}$$

Element 2:

$$\begin{pmatrix} 33333.3 & -33333.3 \\ -33333.3 & 33333.3 \end{pmatrix} \begin{pmatrix} v_2 \\ v_3 \end{pmatrix} - P \begin{pmatrix} 10. & 5. \\ 5. & 10. \end{pmatrix} \begin{pmatrix} v_2 \\ v_3 \end{pmatrix} = \begin{pmatrix} 0 \\ 0 \end{pmatrix}$$

Element 3:

$$\begin{pmatrix} 33333.3 & -33333.3 \\ -33333.3 & 33333.3 \end{pmatrix} \begin{pmatrix} v_3 \\ v_4 \end{pmatrix} - P \begin{pmatrix} 10. & 5. \\ 5. & 10. \end{pmatrix} \begin{pmatrix} v_3 \\ v_4 \end{pmatrix} = \begin{pmatrix} 0 \\ 0 \end{pmatrix}$$

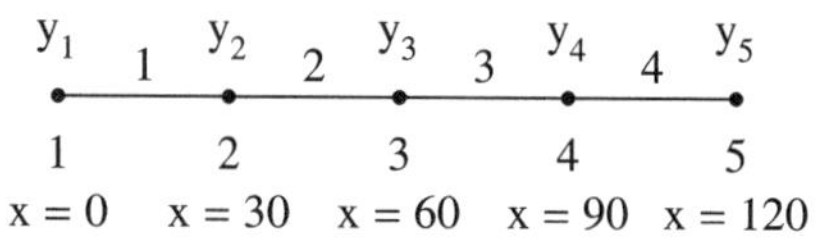

Figure 3.26. Four-element model

Element 4:

$$\begin{pmatrix} 33333.3 & -33333.3 \\ -33333.3 & 33333.3 \end{pmatrix}\begin{pmatrix} v_4 \\ v_5 \end{pmatrix} - P\begin{pmatrix} 10. & 5. \\ 5. & 10. \end{pmatrix}\begin{pmatrix} v_4 \\ v_5 \end{pmatrix} = \begin{pmatrix} 0 \\ 0 \end{pmatrix}$$

Essential boundary conditions:

dof	Value
v_1	0
v_5	0

Assembling and incorporating the EBC, the final system of equations is

$$\boldsymbol{k}_k = \begin{pmatrix} 66666.7 & -33333.3 & 0 \\ -33333.3 & 66666.7 & -33333.3 \\ 0 & -33333.3 & 66666.7 \end{pmatrix}; \qquad \boldsymbol{k}_p = \begin{pmatrix} 20. & 5. & 0 \\ 5. & 20. & 5. \\ 0 & 5. & 20. \end{pmatrix}$$

Solution of the eigenvalue problem:

Characteristic equation: $\det[\boldsymbol{k}_k - P\boldsymbol{k}_p] = 0$ gives

$$-7000.P^3 + 9. \times 10^7 P^2 - 2.66667 \times 10^{11} P + 1.48148 \times 10^{14} = 0$$

Computing roots of the characteristic equation, we get the eigenvalues

$$\{P_1 = 721.295, P_2 = 3333.33, P_3 = 8802.51\}$$

Substituting these eigenvalues into the global equations, the corresponding eigenvectors are as follows:

	Eigenvalue	Eigenvector
1	721.295	(0 −0.5 −0.707107 −0.5 0)
2	3333.33	(0 −0.707107 0 0.707107 0)
3	8802.51	(0 −0.5 0.707107 −0.5 0)

Using the first eigenvalue,

$$\text{Lowest buckling load} = 721.295$$

The example corresponds to the classical Euler buckling situation. The buckling load using Euler's equation is

$$\frac{\pi^2 EI}{L^2} = 685.389$$

The four-linear-element finite element solution is clearly not very accurate. To get a better solution, we must either use more linear elements or employ higher order elements.

(ii) Solution Using Four Quadratic Elements

For a three-node quadratic element the finite element equations are derived in the previous section. Taking the unknown P outside as a common factor, the element equations are as follows:

$$\begin{pmatrix} \frac{7k}{3L} & -\frac{8k}{3L} & \frac{k}{3L} \\ -\frac{8k}{3L} & \frac{16k}{3L} & -\frac{8k}{3L} \\ \frac{k}{3L} & -\frac{8k}{3L} & \frac{7k}{3L} \end{pmatrix} \begin{pmatrix} v_1 \\ v_2 \\ v_3 \end{pmatrix} + P \begin{pmatrix} -\frac{2L}{15} & -\frac{L}{15} & \frac{L}{30} \\ -\frac{L}{15} & -\frac{8L}{15} & -\frac{L}{15} \\ \frac{L}{30} & -\frac{L}{15} & -\frac{2L}{15} \end{pmatrix} \begin{pmatrix} v_1 \\ v_2 \\ v_3 \end{pmatrix} = \begin{pmatrix} 0 \\ 0 \\ 0 \end{pmatrix}$$

Using four quadratic elements we get a solution that is practically the same as that given by Euler's formula:

Nodal locations: {0, 15., 30., 45., 60., 75., 90., 105., 120.}

Element 1:

$$\begin{pmatrix} 77777.8 & -88888.9 & 11111.1 \\ -88888.9 & 177778. & -88888.9 \\ 11111.1 & -88888.9 & 77777.8 \end{pmatrix} \begin{pmatrix} v_1 \\ v_2 \\ v_3 \end{pmatrix} - P \begin{pmatrix} 4. & 2. & -1. \\ 2. & 16. & 2. \\ -1. & 2. & 4. \end{pmatrix} \begin{pmatrix} v_1 \\ v_2 \\ v_3 \end{pmatrix} = \begin{pmatrix} 0 \\ 0 \\ 0 \end{pmatrix}$$

Element 2:

$$\begin{pmatrix} 77777.8 & -88888.9 & 11111.1 \\ -88888.9 & 177778. & -88888.9 \\ 11111.1 & -88888.9 & 77777.8 \end{pmatrix} \begin{pmatrix} v_3 \\ v_4 \\ v_5 \end{pmatrix} - P \begin{pmatrix} 4. & 2. & -1. \\ 2. & 16. & 2. \\ -1. & 2. & 4. \end{pmatrix} \begin{pmatrix} v_3 \\ v_4 \\ v_5 \end{pmatrix} = \begin{pmatrix} 0 \\ 0 \\ 0 \end{pmatrix}$$

Element 3:

$$\begin{pmatrix} 77777.8 & -88888.9 & 11111.1 \\ -88888.9 & 177778. & -88888.9 \\ 11111.1 & -88888.9 & 77777.8 \end{pmatrix} \begin{pmatrix} v_5 \\ v_6 \\ v_7 \end{pmatrix} - P \begin{pmatrix} 4. & 2. & -1. \\ 2. & 16. & 2. \\ -1. & 2. & 4. \end{pmatrix} \begin{pmatrix} v_5 \\ v_6 \\ v_7 \end{pmatrix} = \begin{pmatrix} 0 \\ 0 \\ 0 \end{pmatrix}$$

Element 4:

$$\begin{pmatrix} 77777.8 & -88888.9 & 11111.1 \\ -88888.9 & 177778. & -88888.9 \\ 11111.1 & -88888.9 & 77777.8 \end{pmatrix} \begin{pmatrix} v_7 \\ v_8 \\ v_9 \end{pmatrix} - P \begin{pmatrix} 4. & 2. & -1. \\ 2. & 16. & 2. \\ -1. & 2. & 4. \end{pmatrix} \begin{pmatrix} v_7 \\ v_8 \\ v_9 \end{pmatrix} = \begin{pmatrix} 0 \\ 0 \\ 0 \end{pmatrix}$$

Essential boundary conditions:

dof	Value
v_1	0
v_9	0

Assembling and incorporating the EBC, the final system of equations is

$$k_k = \begin{pmatrix} 177778. & -88888.9 & 0 & 0 & 0 & 0 & 0 \\ -88888.9 & 155556. & -88888.9 & 11111.1 & 0 & 0 & 0 \\ 0 & -88888.9 & 177778. & -88888.9 & 0 & 0 & 0 \\ 0 & 11111.1 & -88888.9 & 155556. & -88888.9 & 11111.1 & 0 \\ 0 & 0 & 0 & -88888.9 & 177778. & -88888.9 & 0 \\ 0 & 0 & 0 & 11111.1 & -88888.9 & 155556. & -88888.9 \\ 0 & 0 & 0 & 0 & 0 & -88888.9 & 177778. \end{pmatrix}$$

$$k_p = \begin{pmatrix} 16. & 2. & 0 & 0 & 0 & 0 & 0 \\ 2. & 8. & 2. & -1. & 0 & 0 & 0 \\ 0 & 2. & 16. & 2. & 0 & 0 & 0 \\ 0 & -1. & 2. & 8. & 2. & -1. & 0 \\ 0 & 0 & 0 & 2. & 16. & 2. & 0 \\ 0 & 0 & 0 & -1. & 2. & 8. & 2. \\ 0 & 0 & 0 & 0 & 0 & 2. & 16. \end{pmatrix}$$

Solution of the eigenvalue problem:

Characteristic equation: $\det[k_k - Pk_p] = 0$ gives

$$-2.6112 \times 10^7 P^7 + 3.48046 \times 10^{12} P^6 - 1.67258 \times 10^{17} P^5 + 3.64352 \times 10^{21} P^4$$
$$- 3.75246 \times 10^{25} P^3 + 1.75502 \times 10^{29} P^2 - 3.19639 \times 10^{32} P + 1.47981 \times 10^{35} = 0$$

Computing roots of the characteristic equation we get the eigenvalues

$$\{P_1 = 685.74, P_2 = 2762.18, P_3 = 6373.94, P_4 = 11111.1,$$
$$P_5 = 21406.4, P_6 = 35756.3, P_7 = 55194.\}$$

Substituting these eigenvalues into the global equations, the corresponding eigenvectors are as follows:

	Eigenvalue	Eigenvector
1	685.74	(0 −0.191327 −0.353581 −0.461903 −0.50004 −0.461903 −0.353581 −0.191327 0)
2	2762.18	(0 −0.353471 −0.500117 −0.353471 0 0.353471 0.500117 0.353471 0)
3	6373.94	(0 −0.467899 −0.348932 0.19381 0.493465 0.19381 −0.348932 −0.467899 0)
4	11111.1	(0 −0.5 0 0.5 0 −0.5 0 0.5 0)
5	21406.4	(0 0.340986 −0.426486 −0.141241 0.603142 −0.141241 −0.426486 0.340986 0)
6	35756.3	(0 −0.249323 0.612924 −0.249323 0 0.249323 −0.612924 0.249323 0)
7	55194.	(0 −0.125228 0.443236 −0.302327 0.62683 −0.302327 0.443236 −0.125228 0)

Buckling load using four quadratic elements = 685.74

This value compares very well with the buckling load obtained from Euler's formula.

◆ *Mathematica*/MATLAB Implementation 3.2 on the Book Web Site:

Solution of buckling problem

3.7 SOLUTION OF SECOND-ORDER 1D BVP

As a final example, we consider solution of a second-order ordinary differential equation that does not necessarily represent a physical system. If the coefficients in the differential equation are constant, then we can use the explicit linear and quadratic element equations given earlier. For variable coefficients, we can either use approximate values of the coefficients by using average values over an element or derive specific element equations by using the actual coefficients in the element equations and then carrying out exact integrations. The latter procedure is used in the following example.

Example 3.5 Find the approximate solution of the following boundary value problem:

$$x^2u'' + 2xu' + u + 1 = \frac{d}{dx}(x^2u') + u + 1 = 0; \qquad 1 < x < 2$$

with the boundary conditions $u(1) = 1$, $u'(2) = 1$.

Compared to the general form, we have

$$k(x) = x^2; \qquad p(x) = 1; \qquad q(x) = 1$$

$$\text{NBC at } x = 2: \qquad \alpha = 0; \qquad \beta = 1$$

For this example $k(x)$ is not constant; therefore, we cannot use the explicit expressions of the finite element equations derived in Section 3.2. We must carry out integrations with $k = x^2$ to establish the $\boldsymbol{k}_k$ matrix in the element equations.

(i) Solution Using Two Linear Elements

A finite element solution using two linear elements is presented in detail. The finite element model is as shown in Figure 3.27. The complete solution details are as follows:

Derivation of element equations:

Element nodes: $\{x_1, x_2\}$

Interpolation functions, $\boldsymbol{N}^T = \begin{pmatrix} \frac{x_2-x}{x_2-x_1} & \frac{x_1-x}{x_1-x_2} \end{pmatrix}$

$\boldsymbol{B}^T = \frac{d\boldsymbol{N}^T}{dx} = \begin{pmatrix} \frac{1}{x_1-x_2} & \frac{1}{x_2-x_1} \end{pmatrix}$

$k(x) = x^2; \qquad p(x) = 1; \qquad q(x) = 1$

$$\boldsymbol{k}_k = \int_{x_1}^{x_2} (x^2\boldsymbol{B}\boldsymbol{B}^T)\, dx = \begin{pmatrix} \frac{x_1^2+x_2x_1+x_2^2}{3x_2-3x_1} & \frac{x_1^2+x_2x_1+x_2^2}{3x_1-3x_2} \\ \frac{x_1^2+x_2x_1+x_2^2}{3x_1-3x_2} & \frac{x_1^2+x_2x_1+x_2^2}{3x_2-3x_1} \end{pmatrix}$$

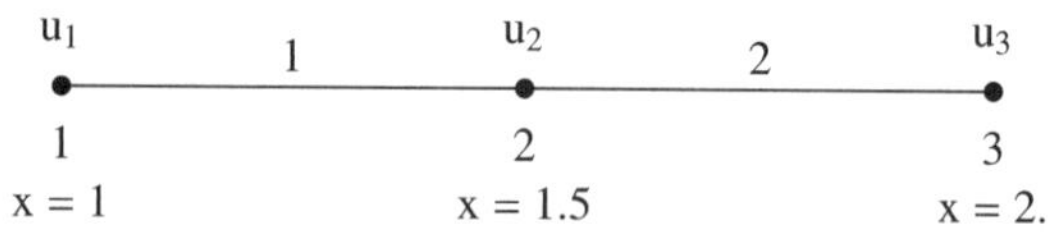

Figure 3.27. Two-element model

$$\boldsymbol{k}_p = \int_{x_1}^{x_2} (-1\boldsymbol{N}\boldsymbol{N}^T)\,dx = \begin{pmatrix} \frac{x_1-x_2}{3} & \frac{x_1-x_2}{6} \\ \frac{x_1-x_2}{6} & \frac{x_1-x_2}{3} \end{pmatrix}$$

$$\boldsymbol{r}_q^T = \int_{x_1}^{x_2} (1\boldsymbol{N})\,dx = \left\{\frac{x_2-x_1}{2}, \frac{x_2-x_1}{2}\right\}$$

The complete element equations are as follows:

$$\begin{pmatrix} \frac{x_1 x_2}{x_2-x_1} & \frac{x_1{}^2+x_2{}^2}{2x_1-2x_2} \\ \frac{x_1{}^2+x_2{}^2}{2x_1-2x_2} & \frac{x_1 x_2}{x_2-x_1} \end{pmatrix} \begin{pmatrix} u_1 \\ u_2 \end{pmatrix} = \begin{pmatrix} \frac{x_2-x_1}{2} \\ \frac{x_2-x_1}{2} \end{pmatrix}$$

Two-element solution:

Nodal locations: {1, 1.5, 2}

Element 1:

Element nodes: $\{x_1 \to 1, x_2 \to 1.5\}$

$$\begin{pmatrix} 3. & -3.25 \\ -3.25 & 3. \end{pmatrix} \begin{pmatrix} u_1 \\ u_2 \end{pmatrix} = \begin{pmatrix} 0.25 \\ 0.25 \end{pmatrix}$$

Element 2:

Element nodes: $\{x_2 \to 1.5, x_3 \to 2\}$

$$\begin{pmatrix} 6. & -6.25 \\ -6.25 & 6. \end{pmatrix} \begin{pmatrix} u_2 \\ u_3 \end{pmatrix} = \begin{pmatrix} 0.25 \\ 0.25 \end{pmatrix}$$

Global equations before boundary conditions:

$$\begin{pmatrix} 3. & -3.25 & 0 \\ -3.25 & 9. & -6.25 \\ 0 & -6.25 & 6. \end{pmatrix} \begin{pmatrix} u_1 \\ u_2 \\ u_3 \end{pmatrix} = \begin{pmatrix} 0.25 \\ 0.5 \\ 0.25 \end{pmatrix}$$

Natural boundary conditions:

dof	α	β
u_3	0	1

dof	$k(x)$	$-k(x)\,\alpha$	$k(x)\beta$
u_3	4	0	4

Global equations after incorporating the NBC:

$$\begin{pmatrix} 3. & -3.25 & 0 \\ -3.25 & 9. & -6.25 \\ 0 & -6.25 & 6. \end{pmatrix} \begin{pmatrix} u_1 \\ u_2 \\ u_3 \end{pmatrix} = \begin{pmatrix} 0.25 \\ 0.5 \\ 4.25 \end{pmatrix}$$

Essential boundary conditions:

dof	Value
u_1	1

Incorporating the EBC, the final system of equations is

$$\begin{pmatrix} 9. & -6.25 \\ -6.25 & 6. \end{pmatrix} \begin{pmatrix} u_2 \\ u_3 \end{pmatrix} = \begin{pmatrix} 3.75 \\ 4.25 \end{pmatrix}$$

Solution for nodal unknowns:

dof	x	Solution
u_1	1	1
u_2	1.5	3.28452
u_3	2	4.12971

Solution over element 1:

Nodes: $\{x_1 \to 1, x_2 \to 1.5\}$

Interpolation functions: $\boldsymbol{N}^T = \{3. - 2.\, x,\ 2.\, x - 2.\}$

Nodal values: $\boldsymbol{d}^T = \{1, 3.28452\}$

Solution: $u(x) = \boldsymbol{N}^T\boldsymbol{d} = 4.56904x - 3.56904$

Solution over element 2:

Nodes: $\{x_1 \to 1.5, x_2 \to 2\}$

Interpolation functions: $\boldsymbol{N}^T = \{4. - 2.\, x,\ 2.\, x - 3.\}$

Nodal values: $\boldsymbol{d}^T = \{3.28452, 4.12971\}$

Solution: $u(x) = \boldsymbol{N}^T\boldsymbol{d} = 1.69038x + 0.748954$

Solution summary:

	Range	Solution
1	$1 \le x \le 1.5$	$4.56904x - 3.56904$
2	$1.5 \le x \le 2$	$1.69038x + 0.748954$

A plot of the solution is shown in Figure 3.28.

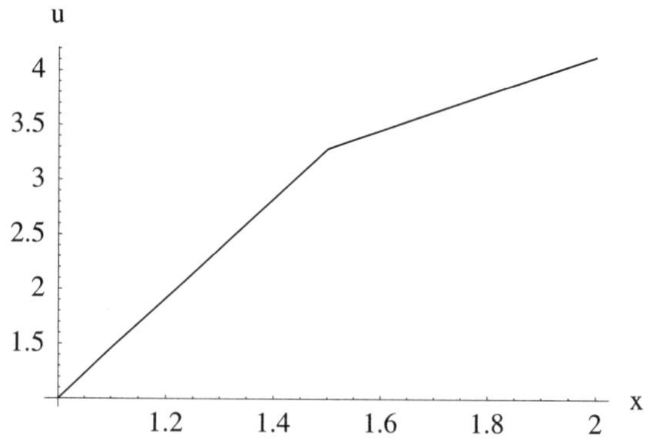

Figure 3.28. Two-linear-element solution

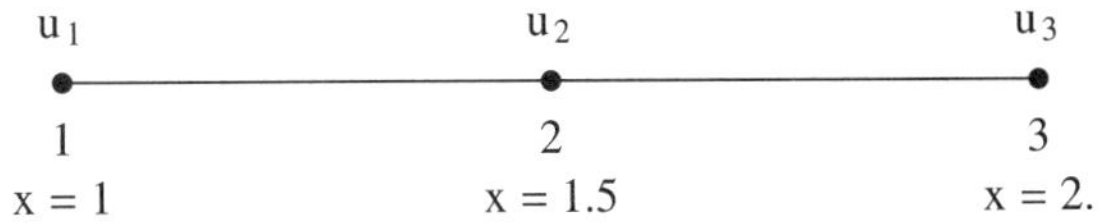

Figure 3.29. One-quadratic-element model

(ii) Solution Using One Quadratic Element

A finite element solution using only one quadratic element is presented in detail. The finite element model is as shown in Figure 3.29. Note that the model consists of only one element and all three nodes belong to the same quadratic element.

Derivation of element equations:

Element nodes: $\left\{x_1, \frac{x_1+x_3}{2}, x_3\right\}$

Interpolation functions, $\boldsymbol{N}^T = \begin{pmatrix} \frac{(x_1+x_3-2x)(x_3-x)}{(x_1-x_3)^2} & \frac{4(x_1-x)(x-x_3)}{(x_1-x_3)^2} & \frac{(x_1+x_3-2x)(x_1-x)}{(x_1-x_3)^2} \end{pmatrix}$

$\boldsymbol{B}^T = \frac{d\boldsymbol{N}^T}{dx} = \begin{pmatrix} -\frac{x_1+3x_3-4x}{(x_1-x_3)^2} & \frac{4(x_1+x_3-2x)}{(x_1-x_3)^2} & -\frac{3x_1+x_3-4x}{(x_1-x_3)^2} \end{pmatrix}$

$k(x) = x^2;\ p(x) = 1;\ q(x) = 1$

$$\boldsymbol{k}_k = \int_{x_1}^{x_3} (x^2 \boldsymbol{B}\boldsymbol{B}^T)\, dx = \begin{pmatrix} \frac{23x_1{}^2+9x_3x_1+3x_3{}^2}{15x_3-15x_1} & \frac{26x_1{}^2+8x_3x_1+6x_3{}^2}{15x_1-15x_3} & \frac{3x_1{}^2-x_3x_1+3x_3{}^2}{15x_3-15x_1} \\ \frac{26x_1{}^2+8x_3x_1+6x_3{}^2}{15x_1-15x_3} & -\frac{16(2x_1{}^2+x_3x_1+2x_3{}^2)}{15(x_1-x_3)} & \frac{6x_1{}^2+8x_3x_1+26x_3{}^2}{15x_1-15x_3} \\ \frac{3x_1{}^2-x_3x_1+3x_3{}^2}{15x_3-15x_1} & \frac{6x_1{}^2+8x_3x_1+26x_3{}^2}{15x_1-15x_3} & \frac{3x_1{}^2+9x_3x_1+23x_3{}^2}{15x_3-15x_1} \end{pmatrix}$$

$$\boldsymbol{k}_p = \int_{x_1}^{x_3} (-1\boldsymbol{N}\boldsymbol{N}^T)\, dx = \begin{pmatrix} \frac{2(x_1-x_3)}{15} & \frac{x_1-x_3}{15} & \frac{x_3-x_1}{30} \\ \frac{x_1-x_3}{15} & \frac{8(x_1-x_3)}{15} & \frac{x_1-x_3}{15} \\ \frac{x_3-x_1}{30} & \frac{x_1-x_3}{15} & \frac{2(x_1-x_3)}{15} \end{pmatrix}$$

$$\boldsymbol{r}_q^T = \int_{x_1}^{x_3} (1\boldsymbol{N})\, dx = \left\{\frac{x_3-x_1}{6}, -\frac{2}{3}(x_1-x_3), \frac{x_3-x_1}{6}\right\}$$

The complete element equations are as follows:

$$\begin{pmatrix} \frac{21x_1{}^2+13x_3x_1+x_3{}^2}{15x_3-15x_1} & \frac{27x_1{}^2+6x_3x_1+7x_3{}^2}{15x_1-15x_3} & \frac{7x_1{}^2-4x_3x_1+7x_3{}^2}{30x_3-30x_1} \\ \frac{27x_1{}^2+6x_3x_1+7x_3{}^2}{15x_1-15x_3} & -\frac{8(3x_1{}^2+4x_3x_1+3x_3{}^2)}{15(x_1-x_3)} & \frac{7x_1{}^2+6x_3x_1+27x_3{}^2}{15x_1-15x_3} \\ \frac{7x_1{}^2-4x_3x_1+7x_3{}^2}{30x_3-30x_1} & \frac{7x_1{}^2+6x_3x_1+27x_3{}^2}{15x_1-15x_3} & \frac{x_1{}^2+13x_3x_1+21x_3{}^2}{15x_3-15x_1} \end{pmatrix} \begin{pmatrix} u_1 \\ u_2 \\ u_3 \end{pmatrix} = \begin{pmatrix} \frac{x_3-x_1}{6} \\ -\frac{2}{3}(x_1-x_3) \\ \frac{x_3-x_1}{6} \end{pmatrix}$$

One-element solution:

Nodal locations: $\{1, 1.5, 2\}$

Element nodes: $\{x_1 \to 1, x_2 \to 1.5, x_3 \to 2\}$

$$\begin{pmatrix} \frac{17}{5} & -\frac{67}{15} & \frac{9}{10} \\ -\frac{67}{15} & \frac{184}{15} & -\frac{127}{15} \\ \frac{9}{10} & -\frac{127}{15} & \frac{37}{5} \end{pmatrix} \begin{pmatrix} u_1 \\ u_2 \\ u_3 \end{pmatrix} = \begin{pmatrix} \frac{1}{6} \\ \frac{2}{3} \\ \frac{1}{6} \end{pmatrix}$$

The global equations are the same as the element equations.

Natural boundary conditions:

dof	α	β
u_3	0	1

dof	$k(x)$	$-k(x)\alpha$	$k(x)\beta$
u_3	4	0	4

Global equations after incorporating the NBC:

$$\begin{pmatrix} \frac{17}{5} & -\frac{67}{15} & \frac{9}{10} \\ -\frac{67}{15} & \frac{184}{15} & -\frac{127}{15} \\ \frac{9}{10} & -\frac{127}{15} & \frac{37}{5} \end{pmatrix} \begin{pmatrix} u_1 \\ u_2 \\ u_3 \end{pmatrix} = \begin{pmatrix} \frac{1}{6} \\ \frac{2}{3} \\ \frac{25}{6} \end{pmatrix}$$

Essential boundary conditions:

dof	Value
u_1	1

Incorporating the EBC, the final system of equations is

$$\begin{pmatrix} \frac{184}{15} & -\frac{127}{15} \\ -\frac{127}{15} & \frac{37}{5} \end{pmatrix} \begin{pmatrix} u_2 \\ u_3 \end{pmatrix} = \begin{pmatrix} \frac{77}{15} \\ \frac{49}{15} \end{pmatrix}$$

Solution for nodal unknowns:

dof	x	Solution
u_1	1	1
u_2	1.5	$\frac{2954}{859}$
u_3	2	$\frac{3759}{859}$

Solution over elements:

Nodes: $\{x_1 \to 1, x_2 \to 1.5, x_3 \to 2\}$

Interpolation functions: $N^T = \{2x^2 - 7x + 6, -4x^2 + 12x - 8, 2x^2 - 5x + 3\}$

Nodal values: $d^T = \left\{1, \frac{2954}{859}, \frac{3759}{859}\right\}$

Solution: $u(x) = N^T d = -\frac{2580x^2}{859} + \frac{10640x}{859} - \frac{7201}{859}$

Figure 3.30 shows a plot of this solution.

(iii) Solution Convergence

To demonstrate convergence, the solutions are computed using one to four linear and quadratic elements. These solutions are plotted in Figures 3.31 and 3.32. As expected, the quadratic element solution converges much more rapidly.

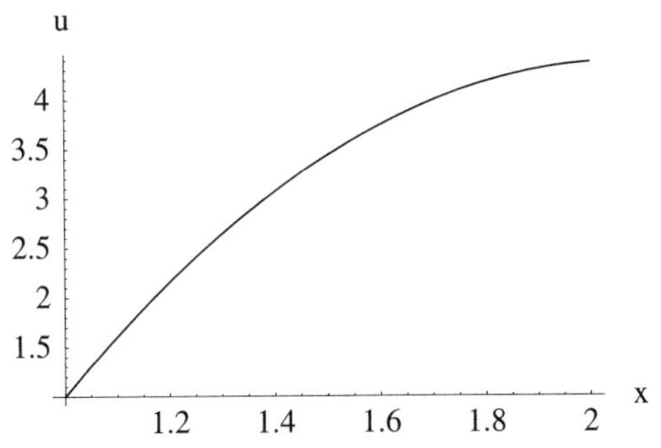

Figure 3.30. One-quadratic-element solution

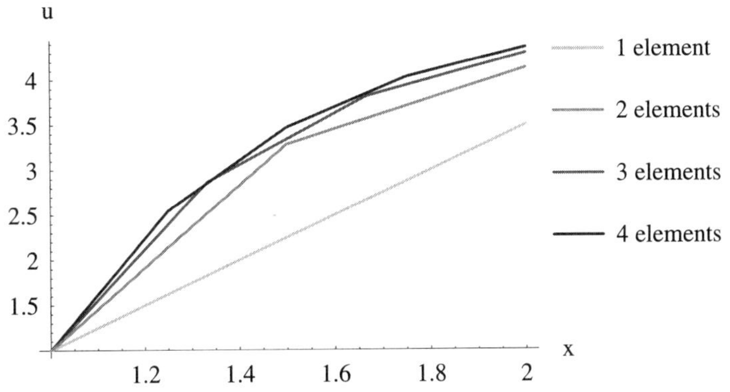

Figure 3.31. Comparison of solutions with one to four linear elements

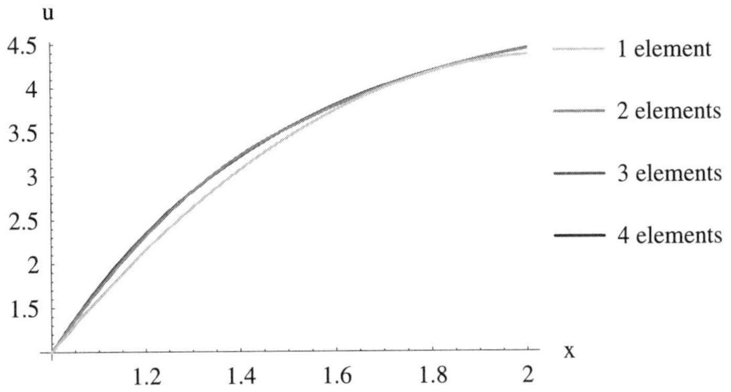

Figure 3.32. Comparison of solutions with one to four quadratic elements

3.8 A CLOSER LOOK AT THE INTERELEMENT DERIVATIVE TERMS

A sharp reader probably has noticed that in the finite element solutions presented in this and the previous chapter, from a mathematical point of view, the boundary terms arising from the integration by parts are not handled precisely. This was done deliberately to avoid a potentially confusing discussion that actually does not impact the finite element solution procedure. For completeness this discussion is taken up in this section.

Recall from Section 3.2 that, after integration by parts, the weighted residual over a general n-node element is expressed as follows:

$$k(x_n)N_i(x_n)u'(x_n) - k(x_1)N_i(x_1)u'(x_1) + \int_{x_1}^{x_n} (qN_i + puN_i - ku'N_i')\, dx = 0$$

If there are specified values of u' at the end nodes of an element, then we have

$$-u'(x_1) = \beta_1 + \alpha_1 u(x_1); \qquad u'(x_n) = \beta_n + \alpha_n u(x_n)$$

Substituting these into the boundary terms, we obtained the following weak form:

$$k(x_n)(\beta_n + \alpha_n u(x_n))N_i(x_n) - k(x_1)(-\beta_1 - \alpha_1 u(x_1))N_i(x_1) + \int_{x_1}^{x_n} (qN_i + puN_i - ku'N_i')\, dx = 0$$

Precisely speaking, this form is valid only if *each element* has natural boundary conditions specified at both ends. If we look back at all the examples presented in the previous sections, it is not true even for one example. Some examples do not have any NBC terms at all while others have an NBC specified at only one end of one element. Thus the question is how do we justify this treatment of interelement boundary terms? The answer is that, during assembly, when element equations are added together, these interelement terms from adjoining elements cancel each other and we are left with terms only at the ends of the solution domain. If an NBC is specified at an end, then we simply use that in the equations and we are in good shape. If an EBC is specified at the end, then the derivative term remains unknown. However, since we do not use the equation that corresponds to the degree of freedom with specified EBC, the term that is left out does not cause any problems. For structural problems this unknown term is in fact the reaction corresponding to a node where displacement is prescribed. This point was discussed using physical arguments in Chapter 1 where we introduced the procedure for handling essential boundary conditions. Here you see the mathematical reasoning for this treatment.

To see this more clearly, consider a two-node linear element with nodes at x_1 and x_2 and degrees of freedom u_1 and u_2. Without introducing the NBC terms in the weighted residual, the two equations for the element are as follows:

$$k(x_2)u'(x_2)\begin{pmatrix} N_1(x_2) \\ N_2(x_2) \end{pmatrix} - k(x_1)u'(x_1)\begin{pmatrix} N_1(x_1) \\ N_2(x_1) \end{pmatrix}$$
$$+ \int_{x_1}^{x_2} q\begin{pmatrix} N_1 \\ N_2 \end{pmatrix} + p\begin{pmatrix} N_1 \\ N_2 \end{pmatrix} u(x) - k\begin{pmatrix} N_1' \\ N_2' \end{pmatrix} u'(x)\, dx = \begin{pmatrix} 0 \\ 0 \end{pmatrix}$$

Since $N_1(x_1) = N_2(x_2) = 1$ and $N_1(x_2) = N_2(x_1) = 0$, we have

$$\begin{pmatrix} -k(x_1)u'(x_1) \\ k(x_2)u'(x_2) \end{pmatrix} + \int_{x_1}^{x_2} q\begin{pmatrix} N_1 \\ N_2 \end{pmatrix} + p\begin{pmatrix} N_1 \\ N_2 \end{pmatrix} u(x) - k\begin{pmatrix} N_1' \\ N_2' \end{pmatrix} u'(x)\, dx = \begin{pmatrix} 0 \\ 0 \end{pmatrix}$$

The integral terms have been evaluated earlier in the explicit form, and thus we have the following element equations:

$$\begin{pmatrix} \frac{k}{L} - \frac{Lp}{3} & -\frac{k}{L} - \frac{Lp}{6} \\ -\frac{k}{L} - \frac{Lp}{6} & \frac{k}{L} - \frac{Lp}{3} \end{pmatrix} \begin{pmatrix} u_1 \\ u_2 \end{pmatrix} = \begin{pmatrix} \frac{Lq}{2} \\ \frac{Lq}{2} \end{pmatrix} + \begin{pmatrix} -k(x_1)u'(x_1) \\ k(x_2)u'(x_2) \end{pmatrix}$$

where L is element length. To see cancellation of the boundary terms at intermediate nodes, we use only two such elements to obtain a solution of the following boundary value problem:

$$\begin{aligned} u''(x) + 1 = 0; &\qquad 0 < x < 1 \\ u'(0) - 2u(0) = 1; &\qquad u(1) = 1 \end{aligned}$$

We clearly have a natural boundary condition specified at $x = 0$ and an essential boundary condition at $x = 1$. Compared to the general form, we have

$$k = 1; \qquad p = 0; \qquad q = 1$$

$$\text{NBC at } x = 0: \qquad -u'(0) = -2u(0) - 1 \Longrightarrow \alpha = -2, \qquad \beta = -1$$

The two-element finite element model is as shown in Figure 3.33. Substituting the numerical values, the finite element equations for the two elements are as follows:

$$\text{Element 1:} \qquad \begin{pmatrix} \frac{1}{1/2} & -\frac{1}{1/2} \\ -\frac{1}{1/2} & \frac{1}{1/2} \end{pmatrix} \begin{pmatrix} u_1 \\ u_2 \end{pmatrix} = \begin{pmatrix} \frac{1/2}{2} \\ \frac{1/2}{2} \end{pmatrix} + \begin{pmatrix} -u'(0) \\ u'(\frac{1}{2}) \end{pmatrix}$$

$$\text{Element 2:} \qquad \begin{pmatrix} \frac{1}{1/2} & -\frac{1}{1/2} \\ -\frac{1}{1/2} & \frac{1}{1/2} \end{pmatrix} \begin{pmatrix} u_2 \\ u_3 \end{pmatrix} = \begin{pmatrix} \frac{1/2}{2} \\ \frac{1/2}{2} \end{pmatrix} + \begin{pmatrix} -u'(\frac{1}{2}) \\ u'(1) \end{pmatrix}$$

Assembling the element equations, we get the following global system of equations:

$$\begin{pmatrix} 2 & -2 & 0 \\ -2 & 4 & -2 \\ 0 & -2 & 2 \end{pmatrix} \begin{pmatrix} u_1 \\ u_2 \\ u_3 \end{pmatrix} = \begin{pmatrix} \frac{1}{4} \\ \frac{1}{2} \\ \frac{1}{4} \end{pmatrix} + \begin{pmatrix} -u'(0) \\ u'(\frac{1}{2}) - u'(\frac{1}{2}) \\ u'(1) \end{pmatrix}$$

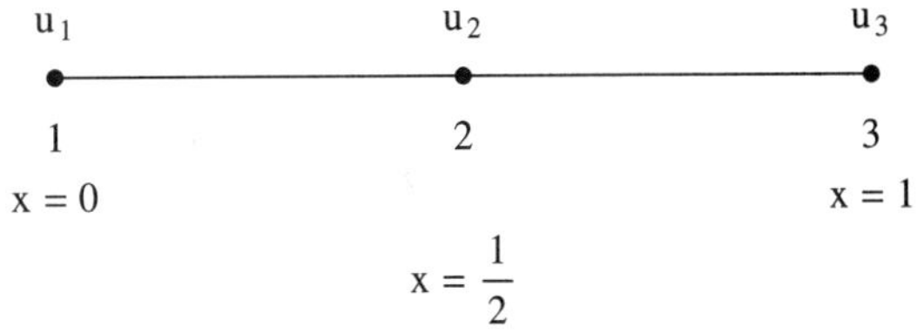

Figure 3.33. Two-element model

Clearly at node 2 the derivative terms from the two elements cancel each other. From the EBC we know $u(1) \equiv u_3 = 1$. However, $u'(1)$ is not known. From the NBC at $x = 0$ we know that $-u'(0) = -2u(0) - 1 \equiv -2u_1 - 1$. Thus we have

$$\begin{pmatrix} 2 & -2 & 0 \\ -2 & 4 & -2 \\ 0 & -2 & 2 \end{pmatrix} \begin{pmatrix} u_1 \\ u_2 \\ 1 \end{pmatrix} = \begin{pmatrix} \frac{1}{4} \\ \frac{1}{2} \\ \frac{1}{4} \end{pmatrix} + \begin{pmatrix} -2u_1 - 1 \\ 0 \\ u'(1) \end{pmatrix}$$

These equations involve only two unknown nodal degrees of freedom which can be solved for by considering only the first two equations:

$$\begin{pmatrix} 2 & -2 & 0 \\ -2 & 4 & -2 \end{pmatrix} \begin{pmatrix} u_1 \\ u_2 \\ 1 \end{pmatrix} = \begin{pmatrix} \frac{1}{4} \\ \frac{1}{2} \end{pmatrix} + \begin{pmatrix} -2u_1 - 1 \\ 0 \end{pmatrix}$$

This clearly shows that in order to solve for the nodal unknowns we do not need to know the derivative term at the location that corresponds to an essential boundary condition. For actually solving for $u_1$and u_2, we rearrange the two equations as follows:

$$\begin{pmatrix} 2 & -2 \\ -2 & 4 \end{pmatrix} \begin{pmatrix} u_1 \\ u_2 \end{pmatrix} + \begin{pmatrix} 0 \\ -2 \end{pmatrix} = \begin{pmatrix} \frac{1}{4} \\ \frac{1}{2} \end{pmatrix} - \begin{pmatrix} 2u_1 \\ 0 \end{pmatrix} - \begin{pmatrix} 1 \\ 0 \end{pmatrix}$$

Moving unknowns to the left side,

$$\begin{pmatrix} 2 & -2 \\ -2 & 4 \end{pmatrix} \begin{pmatrix} u_1 \\ u_2 \end{pmatrix} + \begin{pmatrix} 2u_1 \\ 0 \end{pmatrix} = \begin{pmatrix} \frac{1}{4} \\ \frac{1}{2} \end{pmatrix} - \begin{pmatrix} 1 \\ 0 \end{pmatrix} + \begin{pmatrix} 0 \\ 2 \end{pmatrix}$$

Thus the final system of equations, after incorporating all boundary conditions, is as follows:

$$\begin{pmatrix} 4 & -2 \\ -2 & 4 \end{pmatrix} \begin{pmatrix} u_1 \\ u_2 \end{pmatrix} = \begin{pmatrix} -\frac{3}{4} \\ \frac{5}{2} \end{pmatrix}$$

This example clearly demonstrates the validity of the procedures used in the finite element solution and should dispel any lingering doubts that a reader might have about the nature of finite element computations. Exactly the same arguments can be made by considering two- and three-dimensional problems. However, in those situations we are obviously dealing with line and surface integrals to evaluate boundary terms, and thus it is not easy to demonstrate these details through simple calculations.

PROBLEMS

Second-Order 1D BVP

3.1 An engineering problem is formulated in terms of the following boundary value problem:

$$\tfrac{1}{2}x^2u'' + xu' - xu = 4; \qquad 1 < x < 3$$

with the boundary conditions $u'(1) = 1$ and $u(3) = 1$. Express the problem in the form of the general boundary value problem presented in this chapter. (*Hint:* By expanding the derivative, you can easily see that $(d/dx)(\frac{1}{2}x^2u') = \frac{1}{2}x^2u'' + xu'$.) Clearly identify $k(x)$, $p(x)$, and $q(x)$ terms. Classify the boundary conditions into the essential and the natural types. For the natural boundary conditions identify the α and β terms.

Selected Applications of 1D BVP

3.2 The door of an industrial gas furnace is to be designed to reduce heat loss to no more than 1200 W/m^2. A preliminary design calls for a 200-mm-thick insulation sandwiched between a 6.25-mm-thick stainless steel sheet on the interior surface and a 9.5-mm sheet on the outside. The thermal conductivity of steel is 25 W/m · K and that of insulation is 0.27 W/m · K. The convection coefficient of the inside surface is 20 W/m^2 · K and that of the outside surface is 5 W/m^2 · K. The inside temperature of the furnace is at 1200°C while the surroundings are at 20°C (Figure 3.34). Use three linear elements, one through each layer, to determine the temperature through the door. Does the design meet the stated heat loss goal? Use the simplest finite element model possible based on linear elements.

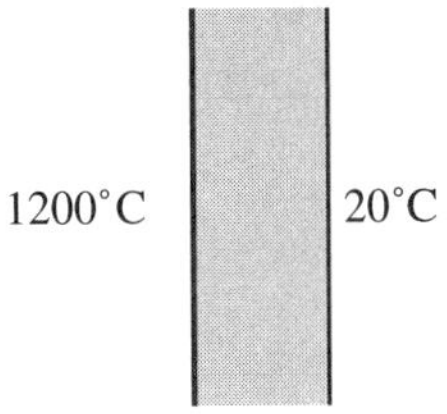

Figure 3.34. Furnace door

3.3 Circular pin fins are used to dissipate heat from an electronic device. A typical fin is as shown in Figure 3.35. The length of the fin is 2.5 cm and its diameter is 0.25 cm. The thermal conductivity of fin material is 396 W/m · K and the convection coefficient around the circumference and the end is 10 W/m^2 · K. The base of the fin is at a temperature of 95°C and the surrounding air temperature is 25°C. Determine temperature distribution in the fin. Calculate the heat loss through the fin.

(a) Use four linear elements.

(b) Use two quadratic elements.

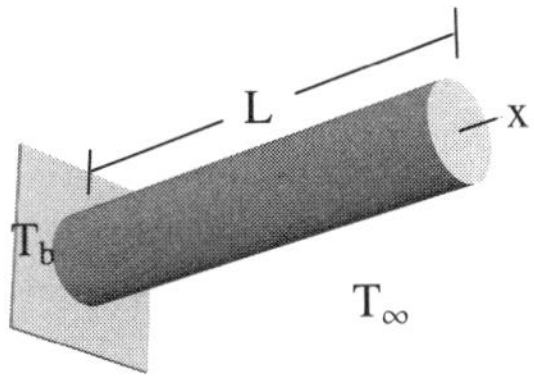

Figure 3.35. Pin fin

3.4 A 5-cm-long turbine blade is exposed to combustion gases at 900°C (Figure 3.36). The area of cross section of the blade is 4.5 cm^2 and its perimeter is 12 cm. The blade is made of high-alloy steel with thermal conductivity $k = 25$ W/m · K. The convection coefficient for the blade surface and the end is 500 $\text{W/m}^2 \cdot \text{K}$. The temperature of the blade at the attachment point is 500°C. Determine the temperature distribution in the blade.

(a) Use three linear elements.

(b) Use one quadratic element.

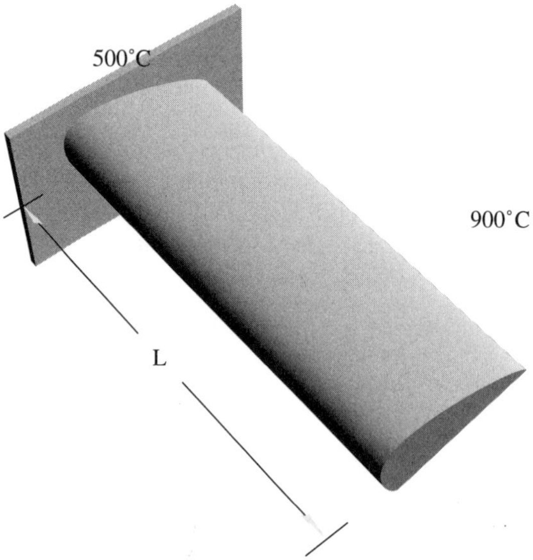

Figure 3.36. Turbine blade

3.5 Consider solution of the tapered axially loaded bar shown in Figure 3.37. Divide the bar into two finite elements and determine the axial force distribution in the bar. Plot this force distribution. Compute the support reactions from the force distribution and

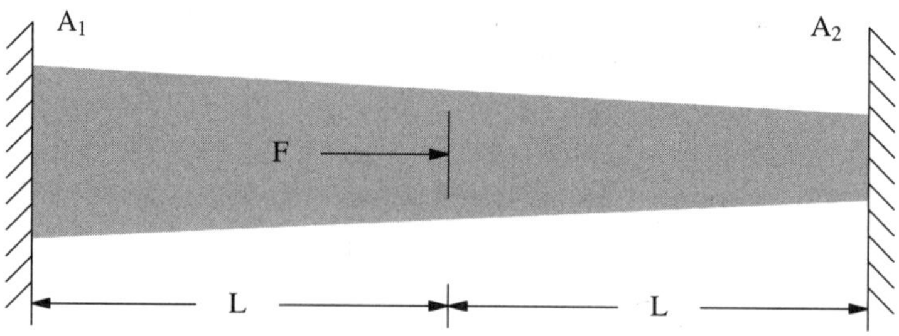

Figure 3.37. Axially loaded bar

see whether the overall equilibrium is satisfied. Comment on the quality of the finite element solution. Use the following numerical data:

$$E = 70\ \text{GPa};\quad F = 20\ \text{kN};\quad L = 300\ \text{mm};\quad A_1 = 2400\ \text{mm}^2;\quad A_2 = 600\ \text{mm}^2$$

(a) Use two linear elements.

(b) Use two quadratic elements.

3.6 Consider solution of the axially loaded bar shown in Figure 3.38. The two end segments have linearly varying areas of cross section and the middle segment is of uniform cross section. Divide the bar into four equal-length finite elements and determine the axial stress and force distribution in the bar. Compute the support reactions from the axial forces and check to see if the overall equilibrium is satisfied. Plot the stress distribution and comment on the accuracy of the finite element solution. Use the following numerical data:

$$E = 70\ \text{GPa};\quad P = 20\ \text{kN};\quad L = 300\ \text{mm};\quad A_1 = 2400\ \text{mm}^2;\quad A_2 = 600\ \text{mm}^2$$

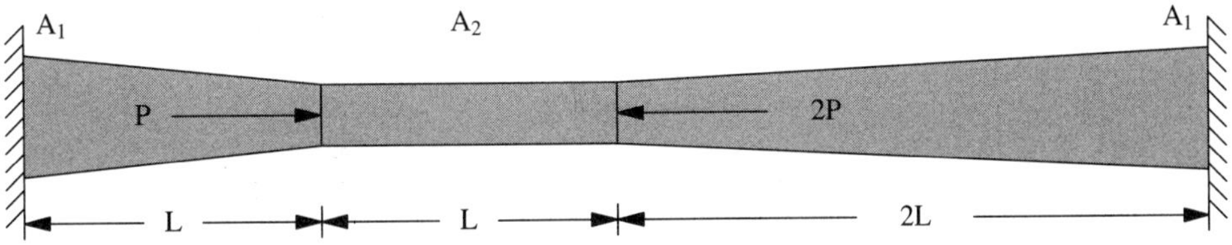

Figure 3.38. Axially loaded bar

3.7 Determine the axial stress distribution in a bar that is rotating at 500 rpm as shown in Figure 3.39. The problem can be treated as one dimensional with the governing differential equation as follows:

$$\frac{d}{dx}\left(EA\frac{du}{dx}\right) + \rho A x \omega^2 = 0; \qquad 0 < x < L$$

$$u(0) = 0; \qquad EA\frac{du(L)}{dx} = 0$$

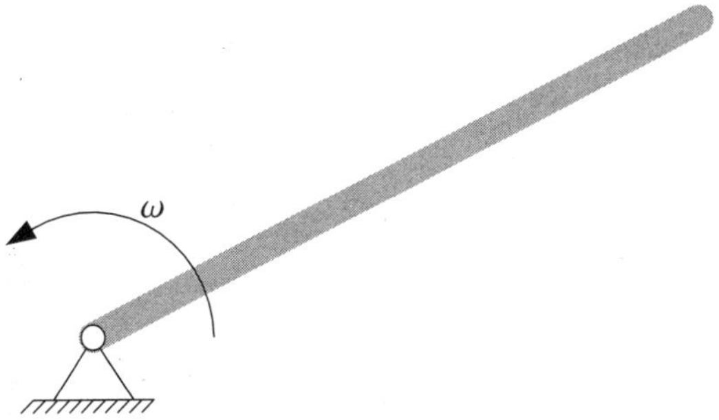

Figure 3.39. Rotating bar

where x is the coordinate along the axis of the bar, $u(x)$ is the axial displacement, L = length of the bar, E = Young's modulus, A = area of cross section, ρ = mass density, and ω = angular velocity in rad/s. The axial stress is $\sigma_x = E\, du/dx$. An exact analytical solution of the problem is

$$\sigma_{x,\text{Exact}} = \frac{\rho\omega^2}{2}(L^2 - x^2)$$

Compare solutions with one, two, and three elements. Use the following numerical data:

$$L = 80\ \text{cm}; \qquad E = 200\ \text{GPa}; \qquad A = 250\ \text{mm}^2; \qquad \rho = 7850\ \text{kg/m}^3$$

3.8 A slider bearing with stepped profile is shown in Figure 3.40. The two surfaces shown are moving relative to each other at a constant velocity U and are separated by a viscous fluid with viscosity μ. The governing differential equation for determining pressure in the fluid under the bearing can be written as follows:

$$\frac{d}{dx}\left(h^3 \frac{dp}{dx}\right) + 6\mu U \frac{dh}{dx} = 0; \qquad 0 < x < L$$

$$p(0) = p(L) = p_0$$

where $p(x)$ is the pressure in the fluid and $h(x)$ is the thickness of the fluid film. Obtain an approximate solution using two linear elements. Use the following numerical values:

$$\mu = 10^{-4}\ \text{lb}\cdot\text{s/ft}^2; \quad L = 8\ \text{in}; \quad h_0 = 0.0001\ \text{in}; \quad U = 12\ \text{in/s}; \quad p_0 = 14.7\ \text{psi}$$

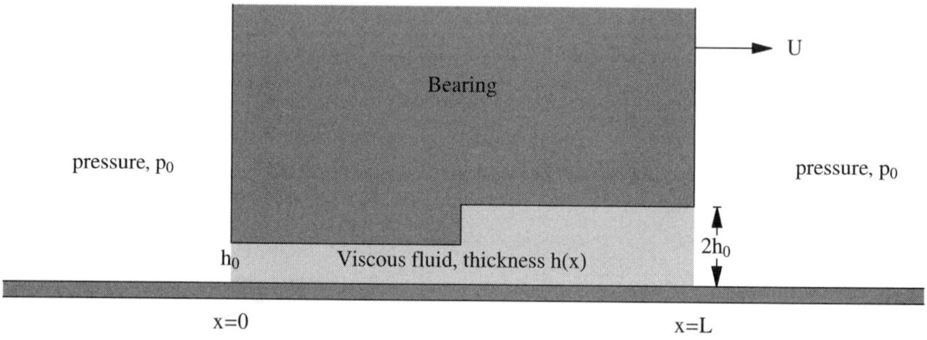

Figure 3.40. Slider bearing with stepped profile

Solution of Second-Order 1D BVP

3.9 Use three linear finite elements to find an approximate solution of the following boundary value problem:

$$\tfrac{1}{2}u'' - u - 4 = 0; \qquad 1 < x < 4$$

with the boundary conditions $u'(1) = 1$, $u(4) = 1$.

3.10 Find an approximate solution of the following boundary value problem:

$$x^2u'' + 2xu' - xu + 4 = 0 \quad \text{or} \quad \frac{d(x^2u')}{dx} - xu + 4 = 0; \qquad 1 < x < 3$$

with the boundary conditions $u(1) = 1$, $u'(3) - 2u(3) = 2$.

3.11 Find an approximate solution of the following boundary value problem using linear and quadratic finite elements:

$$-u'' + \pi^2 u - 2\pi^2 \sin(\pi x) = 0; \qquad 0 < x < 1$$
$$u(0) = u(1) = 0$$
Exact solution: $u(x) = \sin(\pi x)$

3.12 In spherical coordinates the electromagnetic potential V is expressed as a function of θ by the following boundary value problem:

$$\frac{d}{d\theta}\left(\sin\theta \frac{dV}{d\theta}\right) = 0; \qquad \alpha < \theta < \pi/2$$
$$V(\alpha) = V_0; \qquad V(\pi/2) = 0$$

where α is a given angle less than π and V_0 is the specified potential at this angle. Assuming $\alpha = \pi/4$ and $V_0 = 1$ and using *two linear finite elements*, determine the potential V at $\theta = 3\pi/8$.

CHAPTER FOUR

TRUSSES, BEAMS, AND FRAMES

Many structural systems used in practice consist of long slender members of various shapes. Common examples are roof trusses, bridge supports, crane booms, and antenna towers. Structural systems that are arranged so that each member primarily resists axial forces only are usually known as *trusses*. Long slender members that are subjected to loading normal to their longitudinal axis must resist bending and shear forces and are called *beams*. A structural *frame* consists of members that must resist both bending and axial forces.

Finite element equations for analysis of trusses are obtained by a simple coordinate transformation of the two-node axial deformation element developed in Chapter 2. Elements for both plane trusses and three-dimensional space trusses are generated using this technique in the following two sections. The third section considers analysis of trusses subjected to initial strain due to temperature changes or fabrication errors. For modeling supports and various other constraints, it is often useful to introduce spring elements. The axial and torsional spring elements are presented in the fourth section. Beams are governed by a fourth-order differential equation. Section 4.5 contains a detailed derivation of this equation, discussion of appropriate boundary conditions, and exact solutions for few simple cases. A two-node finite element for beam bending is developed in Section 4.6 using the Hermitian interpolation functions. For beams subjected to distributed loading, Section 4.7 describes a superposition procedure so that exact solutions for bending moment and shear force can be obtained using only one element per span. In Sections 4.8 and 4.9, the axial deformation and beam bending elements are combined together to develop elements suitable for analysis of general structural frames.

4.1 PLANE TRUSSES

A truss is a structure in which members are arranged in such a way that they are subjected to axial loads only. The joints in trusses are considered pinned. Plane trusses, where all members are assumed to be in the *x*-*y* plane, are considered in this section. The general case of space trusses is considered in the following section.

A plane truss element is an axial deformation element oriented arbitrarily in a two-dimensional space. As shown in Figure 4.1, in a local coordinate s that runs along its axis ($0 \leq s \leq L$), the element is exactly the same as the two-node axial deformation element developed in Chapter 2. Thus in terms of s the assumed displacement over the element is

$$u(s) = \begin{pmatrix} \frac{L-s}{L} & \frac{s}{L} \end{pmatrix} \begin{pmatrix} d_1 \\ d_2 \end{pmatrix}; \qquad 0 \leq s \leq L$$

and the element equations in the local coordinate system are as follows:

$$\frac{EA}{L} \begin{pmatrix} 1 & -1 \\ -1 & 1 \end{pmatrix} \begin{pmatrix} d_1 \\ d_2 \end{pmatrix} = \begin{pmatrix} P_1 \\ P_2 \end{pmatrix} \Longrightarrow \boldsymbol{k}_l \boldsymbol{d}_l = \boldsymbol{r}_l$$

Local degrees of freedom: $\boldsymbol{d}_l = \begin{pmatrix} d_1 \\ d_2 \end{pmatrix}$

Local applied nodal forces: $\boldsymbol{r}_l = \begin{pmatrix} P_1 \\ P_2 \end{pmatrix}$

where E = elastic modulus of the material, A = area of cross section of the element, L = length of the element, d_1 and d_2 are the displacements along the axis of the element, and P_1 and P_2 are possible axial loads applied at the bar ends. It is assumed that there is no distributed axial load along the length of the element.

Using these expressions equations for any truss element can be written. However, when there are several truss elements oriented arbitrarily, each element will have its own local axis and the direct assembly procedure discussed in Chapter 1 will not work. To be able to

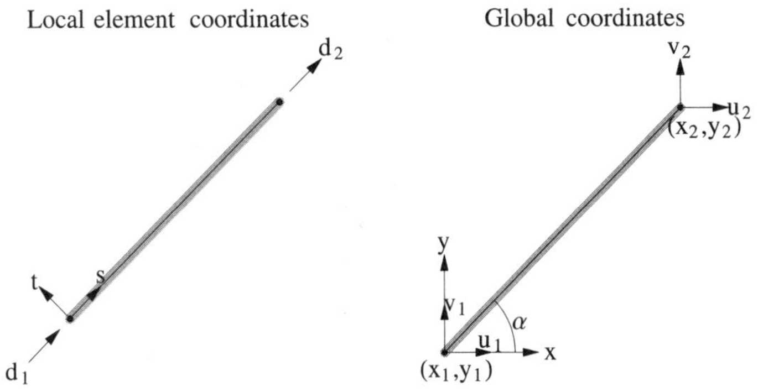

Figure 4.1. Local and global coordinates for an axially loaded bar

assemble element equations, we must refer all elements to one common reference coordinate system. Thus we define a global x-y coordinate system and locate all elements with respect to this system. The components of the axial displacements in the global coordinate system are the x displacements denoted by u and y displacements denoted by v. Each node thus has two degrees of freedom in the global coordinate system. The possible applied loads at the element ends are also decomposed into their x and y components. Thus in the global coordinates we have

$$\text{Nodal degrees of freedom:} \qquad \boldsymbol{d} = \begin{pmatrix} u_1 \\ v_1 \\ u_2 \\ v_2 \end{pmatrix}$$

$$\text{Applied nodal forces:} \qquad \boldsymbol{r} = \begin{pmatrix} F_{x1} \\ F_{y1} \\ F_{x2} \\ F_{y2} \end{pmatrix}$$

From Figure 4.1, it can easily be seen the vector d_1 is related to vectors u_1 and v_1 as follows:

$$u_1 = d_1 \cos\alpha \equiv d_1 l_s$$
$$v_1 = d_1 \cos(90 - \alpha) \equiv d_1 \sin\alpha \equiv d_1 m_s$$
$$\Longrightarrow \begin{pmatrix} u_1 \\ v_1 \end{pmatrix} = \begin{pmatrix} l_s \\ m_s \end{pmatrix} d_1$$

where l_s is the cosine of the angle between the element s axis and the global x axis and m_s is the cosine of the angle between the element axis and the global y axis and are called the direction cosines. The angle is positive when measured from the positive x axis in the counterclockwise direction. Denoting the coordinates of the ends of the element by (x_1, y_1) and (x_2, y_2), we can determine the element length and the direction cosines as follows:

$$L = \sqrt{(x_2 - x_1)^2 + (y_2 - y_1)^2}; \qquad l_s = \cos\alpha = \frac{x_2 - x_1}{L};$$
$$m_s = \cos(90 - \alpha) = \sin\alpha = \frac{y_2 - y_1}{L}$$

Multiplying u_1 by l_s and v_1 by m_s and adding the two equations together give transformation from global to local degrees of freedom as follows:

$$u_1 l_s + v_1 m_s = d_1 \cos^2\alpha + d_1 \sin^2\alpha \equiv d_1$$

The same relationship holds for the degrees of freedom at node 2. Thus the transformation between the global and the local degrees of freedom can be written as follows:

From global to local:
$$\begin{pmatrix} d_1 \\ d_2 \end{pmatrix} = \begin{pmatrix} l_s & m_s & 0 & 0 \\ 0 & 0 & l_s & m_s \end{pmatrix} \begin{pmatrix} u_1 \\ v_1 \\ u_2 \\ v_2 \end{pmatrix} \Longrightarrow \boldsymbol{d}_l = \boldsymbol{Td}$$

From local to global:
$$\begin{pmatrix} u_1 \\ v_1 \\ u_2 \\ v_2 \end{pmatrix} = \begin{pmatrix} l_s & 0 \\ m_s & 0 \\ 0 & l_s \\ 0 & m_s \end{pmatrix} \begin{pmatrix} d_1 \\ d_2 \end{pmatrix} \Longrightarrow \boldsymbol{d} = \boldsymbol{T}^T \boldsymbol{d}_l$$

Using this transformation, the element equations in the local coordinates can be related to the global coordinate system as follows:

$$\boldsymbol{k}_l \boldsymbol{d}_l = \boldsymbol{r}_l \Longrightarrow \boldsymbol{k}_l \boldsymbol{Td} = \boldsymbol{r}_l$$

Multiplying both sides by $\boldsymbol{T}^T$, we get

$$\boldsymbol{T}^T \boldsymbol{k}_l \boldsymbol{Td} = \boldsymbol{T}^T \boldsymbol{r}_l$$

Noting that $\boldsymbol{T}^T \boldsymbol{r}_l$ is the transformation of applied loads from the local to the global coordinates, we have the following element equations in terms of global degrees of freedom and applied nodal loads in the global directions:

$$\boldsymbol{kd} = \boldsymbol{r}$$

where

$$\boldsymbol{k} = \boldsymbol{T}^T \boldsymbol{k}_l \boldsymbol{T} \qquad \text{and} \qquad \boldsymbol{r} = \boldsymbol{T}^T \boldsymbol{r}_l$$

Carrying out matrix multiplication, we get the following element equations:

$$\frac{EA}{L} \begin{pmatrix} l_s^2 & l_s m_s & -l_s^2 & -l_s m_s \\ l_s m_s & m_s^2 & -l_s m_s & -m_s^2 \\ -l_s^2 & -l_s m_s & l_s^2 & l_s m_s \\ -l_s m_s & -m_s^2 & l_s m_s & m_s^2 \end{pmatrix} \begin{pmatrix} u_1 \\ v_1 \\ u_2 \\ v_2 \end{pmatrix} = \begin{pmatrix} F_{1x} \\ F_{1y} \\ F_{2x} \\ F_{2y} \end{pmatrix}$$

As discussed in Chapter 1, it is convenient to add the concentrated loads directly to the global equations at the start of the assembly. Thus in the element equations the right-hand-side load vector is taken as all zeros:

$$\frac{EA}{L} \begin{pmatrix} l_s^2 & l_s m_s & -l_s^2 & -l_s m_s \\ l_s m_s & m_s^2 & -l_s m_s & -m_s^2 \\ -l_s^2 & -l_s m_s & l_s^2 & l_s m_s \\ -l_s m_s & -m_s^2 & l_s m_s & m_s^2 \end{pmatrix} \begin{pmatrix} u_1 \\ v_1 \\ u_2 \\ v_2 \end{pmatrix} = \begin{pmatrix} 0 \\ 0 \\ 0 \\ 0 \end{pmatrix}$$

These equations can be used to analyze any plane assemblage of bars in which individual elements can be assumed to resist axial loads. The assembly and solution procedures have been discussed in detail in Chapter 1. After computing nodal displacements for each element, the element solution is computed by first transforming the nodal displacements back to the local coordinates as follows:

$$\text{Axial displacements:} \qquad \begin{pmatrix} d_1 \\ d_2 \end{pmatrix} = \begin{pmatrix} l_s & m_s & 0 & 0 \\ 0 & 0 & l_s & m_s \end{pmatrix} \begin{pmatrix} u_1 \\ v_1 \\ u_2 \\ v_2 \end{pmatrix}$$

The axial displacement at any point along the element can be computed as follows:

$$u(s) = \begin{pmatrix} \frac{L-s}{L} & \frac{s}{L} \end{pmatrix} \begin{pmatrix} d_1 \\ d_2 \end{pmatrix}; \qquad 0 \le s \le L$$

The axial strain is simply the first derivative of the axial displacement, giving constant strain over the element as follows:

$$\epsilon = \frac{du}{ds} = \begin{pmatrix} -\frac{1}{L} & \frac{1}{L} \end{pmatrix} \begin{pmatrix} d_1 \\ d_2 \end{pmatrix} = \frac{1}{L}(-d_1 + d_2)$$

The axial stress is $\sigma = E\epsilon$ and axial force in the element is $F = \sigma A$. The sign convention used in these equations assumes that the tension is positive and the compression is negative.

Example 4.1 Six-Bar Truss Consider the simple six-bar pin-jointed structure shown in Figure 4.2. All members have the same cross-sectional area and are of the same material, $E = 200$ GPa and $A = 0.001$ m^2. The load $P = 20$ kN and acts at an angle $\alpha = 30°$. The dimensions in meters are shown in the figure.

Each node in the model has two displacement degrees of freedom and thus there are a total of ten degrees of freedom as shown in Figure 4.3. The displacements are identified by the letters u and v with a subscript indicating the corresponding node number.

The calculations are similar to the truss examples presented in Chapter 1. Complete calculations for this example can be found on the book web site.

The solution summary is as follows:

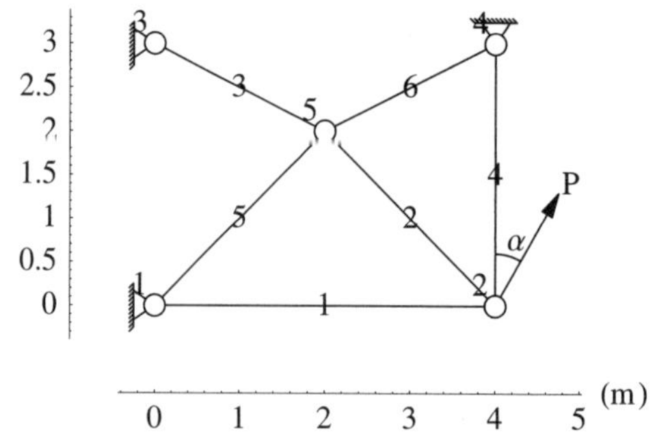

Figure 4.2. Six-bar truss

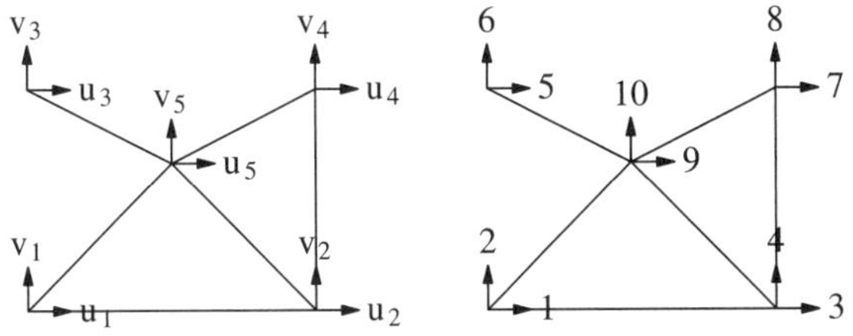

Figure 4.3. Six-bar truss nodal degrees of freedom

Nodal Solution

	x Coordinate	*y* Coordinate	*u*	*v* (mm)
1	0	0	0	0
2	4000.	0	0.213105	0.249979
3	0	3000.	0	0
4	4000.	3000.	0	0
5	2000.	2000.	−0.00609705	0.0122424

Element Solution

	Stress (Mpa)	Axial Force (N)
1	10.6553	10655.3
2	−0.926689	−926.689
3	−0.97746	−977.46
4	−16.6652	−16665.2
5	0.307267	307.267
6	−0.00193181	−1.93181

◆ *Mathematica*/MATLAB Implementation 4.1 on the Book Web Site:
Analysis of a plane truss

4.2 SPACE TRUSSES

By considering an axially loaded element to be arbitrarily oriented in a three-dimensional space, we can easily develop an element that can be used to find joint displacements and axial forces in any space pin-jointed framework. With the local s axis along the axis of the element, the local element equations are still the same as the two-node axial deformation element:

$$\frac{EA}{L}\begin{pmatrix} 1 & -1 \\ -1 & 1 \end{pmatrix}\begin{pmatrix} d_1 \\ d_2 \end{pmatrix} = \begin{pmatrix} P_1 \\ P_2 \end{pmatrix} \Longrightarrow \boldsymbol{k}_l \boldsymbol{d}_l = \boldsymbol{r}_l$$

where E = elastic modulus of the material, A = area of cross section of the element, L = length of the element, d_1 and d_2 are the displacements along the axis of the element,

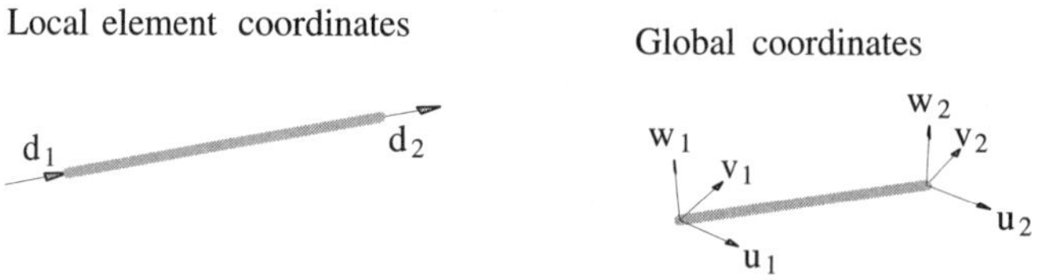

Figure 4.4. Local and global coordinates for an axially loaded bar in three dimensions

and P_1 and P_2 are possible applied axial loads at the ends. In the global coordinates the nodal displacements and applied forces are as shown in Figure 4.4 and are as follows:

$$\text{Nodal degrees of freedom:} \quad \boldsymbol{d} = \begin{pmatrix} u_1 \\ v_1 \\ w_1 \\ u_2 \\ v_2 \\ w_2 \end{pmatrix} \qquad \text{Applied nodal forces:} \quad \boldsymbol{r} = \begin{pmatrix} F_{x1} \\ F_{y1} \\ F_{z1} \\ F_{x2} \\ F_{y2} \\ F_{z2} \end{pmatrix}$$

The transformation between the global to local degrees of freedom and vice versa is as follows:

$$\begin{pmatrix} d_1 \\ d_2 \end{pmatrix} = \begin{pmatrix} l_s & m_s & n_s & 0 & 0 & 0 \\ 0 & 0 & 0 & l_s & m_s & n_s \end{pmatrix} \begin{pmatrix} u_1 \\ v_1 \\ w_1 \\ u_2 \\ v_2 \\ w_2 \end{pmatrix} \Longrightarrow \boldsymbol{d}_l = \boldsymbol{T}\boldsymbol{d}$$

$$\begin{pmatrix} u_1 \\ v_1 \\ w_1 \\ u_2 \\ v_2 \\ w_2 \end{pmatrix} = \begin{pmatrix} l_s & 0 \\ m_s & 0 \\ n_s & 0 \\ 0 & l_s \\ 0 & m_s \\ 0 & n_s \end{pmatrix} \begin{pmatrix} d_1 \\ d_2 \end{pmatrix} \Longrightarrow \boldsymbol{d} = \boldsymbol{T}^T \boldsymbol{d}_l$$

where the direction cosines and the length can be computed from the nodal coordinates as follows:

$$l_s = \frac{x_2 - x_1}{L}; \qquad m_s = \frac{y_2 - y_1}{L}; \qquad n_s = \frac{z_2 - z_1}{L}$$

$$L = \sqrt{(x_2 - x_1)^2 + (y_2 - y_1)^2 + (z_2 - z_1)^2}$$

Using this transformation, the element equations in the local coordinates can be related to the global coordinate system as follows:

$$\boldsymbol{k}_l \boldsymbol{d}_l = \boldsymbol{r}_l \Longrightarrow \boldsymbol{k}_l \boldsymbol{T} \boldsymbol{d} = \boldsymbol{r}_l$$

Multiplying both sides by $\boldsymbol{T}^T$, we get

$$\boldsymbol{T}^T \boldsymbol{k}_l \boldsymbol{T} \boldsymbol{d} = \boldsymbol{T}^T \boldsymbol{r}_l$$

Noting that $\boldsymbol{T}^T\boldsymbol{r}_l$ is the transformation of applied loads from the local to the global coordinates, we have the following element equations in terms of global degrees of freedom and applied nodal loads in the global directions:

$$\boldsymbol{kd} = \boldsymbol{r}$$

where

$$\boldsymbol{k} = \boldsymbol{T}^T\boldsymbol{k}_l\boldsymbol{T} \qquad \text{and} \qquad \boldsymbol{r} = \boldsymbol{T}^T\boldsymbol{r}_l$$

Carrying out matrix multiplication, we get the following element equations for a space truss element:

$$\frac{EA}{L}\begin{pmatrix} l_s^2 & m_sl_s & n_sl_s & -l_s^2 & -m_sl_s & -n_sl_s \\ m_sl_s & m_s^2 & m_sn_s & -m_sl_s & -m_s^2 & -m_sn_s \\ n_sl_s & m_sn_s & n_s^2 & -n_sl_s & -m_sn_s & -n_s^2 \\ -l_s^2 & -m_sl_s & -n_sl_s & l_s^2 & m_sl_s & n_sl_s \\ -m_sl_s & -m_s^2 & -m_sn_s & m_sl_s & m_s^2 & m_sn_s \\ -n_sl_s & -m_sn_s & -n_s^2 & n_sl_s & m_sn_s & n_s^2 \end{pmatrix}\begin{pmatrix} u_1 \\ v_1 \\ w_1 \\ u_2 \\ v_2 \\ w_2 \end{pmatrix} = \begin{pmatrix} F_{1x} \\ F_{1y} \\ F_{1z} \\ F_{2x} \\ F_{2y} \\ F_{2z} \end{pmatrix}$$

Assuming that the concentrated loads are added directly to the global equations at the start of the assembly, the element equations are

$$\frac{EA}{L}\begin{pmatrix} l_s^2 & m_sl_s & n_sl_s & -l_s^2 & -m_sl_s & -n_sl_s \\ m_sl_s & m_s^2 & m_sn_s & -m_sl_s & -m_s^2 & -m_sn_s \\ n_sl_s & m_sn_s & n_s^2 & -n_sl_s & -m_sn_s & -n_s^2 \\ -l_s^2 & -m_sl_s & -n_sl_s & l_s^2 & m_sl_s & n_sl_s \\ -m_sl_s & -m_s^2 & -m_sn_s & m_sl_s & m_s^2 & m_sn_s \\ -n_sl_s & -m_sn_s & -n_s^2 & n_sl_s & m_sn_s & n_s^2 \end{pmatrix}\begin{pmatrix} u_1 \\ v_1 \\ w_1 \\ u_2 \\ v_2 \\ w_2 \end{pmatrix} = \begin{pmatrix} 0 \\ 0 \\ 0 \\ 0 \\ 0 \\ 0 \end{pmatrix}$$

The assembly and solution procedures are standard and are discussed in detail in Chapter 1. After computing the nodal displacements, the element quantities can be computed in the usual manner. The axial displacement at any point along the element is computed by first transforming the global displacements into the local coordinates as follows:

$$\text{Axial displacements:} \qquad \begin{pmatrix} d_1 \\ d_2 \end{pmatrix} = \begin{pmatrix} l_s & m_s & n_s & 0 & 0 & 0 \\ 0 & 0 & 0 & l_s & m_s & n_s \end{pmatrix}\begin{pmatrix} u_1 \\ v_1 \\ w_1 \\ u_2 \\ v_2 \\ w_2 \end{pmatrix}$$

The axial displacement over the element is

$$u(s) = \begin{pmatrix} 1 - \frac{s}{L} & \frac{s}{L} \end{pmatrix}\begin{pmatrix} d_1 \\ d_2 \end{pmatrix}; \qquad 0 \le s \le L$$

The axial strain is simply the first derivative of the axial displacement, giving constant strain over the element as follows:

$$\epsilon = \frac{du}{dt} = \begin{pmatrix} -\frac{1}{L} & \frac{1}{L} \end{pmatrix} \begin{pmatrix} d_1 \\ d_2 \end{pmatrix} = \frac{1}{L}(-d_1 + d_2)$$

The axial stress is $\sigma = E\epsilon$ and axial force in the element is $F = \sigma A$. The sign convention used in these equations is that tension is positive and compression is negative.

Example 4.2 Three-Bar Truss Consider the simple three-bar pin-jointed structure shown in Figure 4.5. The cross-sectional area of elements 1 and 2 is 200 mm^2 and that of element 3 is 600 mm^2. All elements are made of the same material with $E = 200$ GPa. The applied load is $P = 20$ kN. The nodal coordinates are as follows:

Node	$x(m)$	$y(m)$	$z(m)$
1	0.96	1.92	0
2	−1.44	1.44	0
3	0	0	0
4	0	0	2

Each node in the model has three displacement degrees of freedom, and thus there are a total of 12 degrees of freedom, as shown in Figure 4.5. The displacements are identified by the letters u, v, and w with a subscript indicating the corresponding node number.

The calculation are similar to the plane truss examples presented in Chapter 1. Complete calculations for this example can be found on the web site.

The solution summary is as follows:

Nodal Solution

	x Coordinate	y Coordinate	z Coordinate	u	v	w (mm)
1	960.	1920.	0	0	0	0
2	−1440.	1440.	0	0	0	0
3	0	0	0	0	0	0
4	0	0	2000.	−0.178143	−2.46857	−0.367431

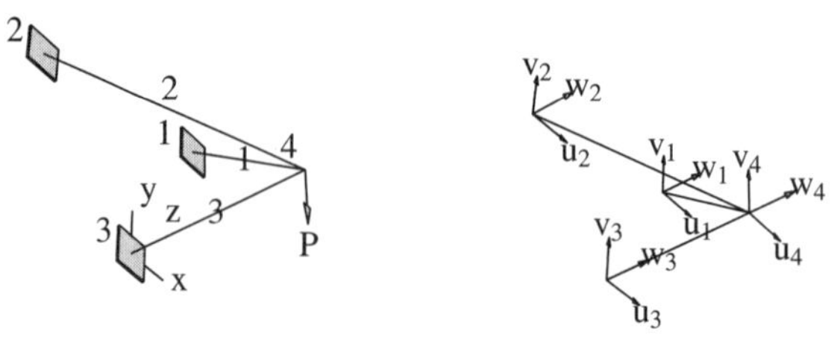

Figure 4.5. Three-bar space truss

Element Solution

	Stress (MPa)	Axial force (N)
1	101.873	20374.6
2	66.0725	13214.5
3	−38.5802	−23148.1

◆ *Mathematica*/MATLAB Implementation 4.2 on the Book Web Site:
Analysis of a space truss

4.3 TEMPERATURE CHANGES AND INITIAL STRAINS IN TRUSSES

Stresses may be induced in indeterminate trusses due to temperature changes or because of forced fit during construction if a truss member is fabricated too short or too long. If an element of length L is fabricated too short by an amount ΔL and is then stretched to make it fit during construction, the element is clearly subjected to the following initial strain:

$$\epsilon_0 = \frac{\Delta L}{L}$$

Similarly, a temperature increase of ΔT in a long slender bar causes a uniform expansion, in which case the element is subjected to the following initial axial strain:

$$\epsilon_0 = \alpha \Delta T$$

where α is the coefficient of thermal expansion of the material. Thus in both situations an element is subjected to an initial strain before any external loads are applied.

In order to account for these effects in trusses, we consider an axial deformation element with an initial strain ϵ_0. The axial strain obtained from differentiating the axial displacement $u(x)$ is clearly the total strain. The stresses in the element will be caused by the difference in the total strains and the initial strains. That is,

$$\sigma_x = E\left(\frac{du}{dx} - \epsilon_0\right)$$

The strain energy function for the element is now as follows:

$$U = \frac{1}{2}\int_{x_1}^{x_2} EA\left(\frac{du}{dx} - \epsilon_0\right)^2 dx = \frac{1}{2}\int_{x_1}^{x_2} EA\left(\frac{du}{dx}\right)^2 dx - \int_{x_1}^{x_2} EA\frac{du}{dx}\epsilon_0\, dx + \frac{1}{2}\int_{x_1}^{x_2} EA(\epsilon_0)^2\, dx$$

Following the derivation in Section 2.7 of Chapter 2, the assumed solution and its derivative are

$$u(x) = (N_1 \quad N_2)\begin{pmatrix} u_1 \\ u_2 \end{pmatrix} \equiv \boldsymbol{N}^T\boldsymbol{d}; \qquad N_1 = -\frac{x - x_2}{L}; \qquad N_2 = \frac{x - x_1}{L}$$

$$\frac{du(x)}{dx} = (N_1' \quad N_2')\begin{pmatrix} u_1 \\ u_2 \end{pmatrix} \equiv \boldsymbol{B}^T\boldsymbol{d}; \qquad N_1' = \frac{1}{x_1 - x_2} = -\frac{1}{L}; \qquad N_2' = \frac{1}{x_2 - x_1} = \frac{1}{L}$$

The strain energy term can now be evaluated as follows:

$$U = \frac{1}{2}\int_{x_1}^{x_2} AE\boldsymbol{d}^T\boldsymbol{B}\boldsymbol{B}^T\boldsymbol{d}\,dx - \boldsymbol{d}^T\int_{x_1}^{x_2} EA\boldsymbol{B}\epsilon_0\,dx + \frac{1}{2}\int_{x_1}^{x_2} EA(\epsilon_0)^2\,dx$$

or

$$U = \frac{1}{2}\boldsymbol{d}^T\boldsymbol{k}\boldsymbol{d} - \boldsymbol{d}^T\boldsymbol{r}_\epsilon + \frac{1}{2}\int_{x_1}^{x_2} \mathrm{EA}(\epsilon_0)^2\,dx$$

where $\boldsymbol{k}$ is the usual stiffness matrix of an axial deformation element

$$\boldsymbol{k} = \int_{x_1}^{x_2} AE\boldsymbol{B}\boldsymbol{B}^T\,dx = \frac{AE}{L}\begin{pmatrix} 1 & -1 \\ -1 & 1 \end{pmatrix}$$

and $\boldsymbol{r}_\epsilon$ is the equivalent nodal load vector due to initial strains ϵ_0:

$$\boldsymbol{r}_\epsilon = \int_{x_1}^{x_2} EA\boldsymbol{B}\epsilon_0\,dx = \begin{pmatrix} \int_{x_1}^{x_2} EA\epsilon_0\left(-\frac{1}{L}\right)dx \\ \int_{x_1}^{x_2} EA\epsilon_0\left(\frac{1}{L}\right)dx \end{pmatrix} \equiv EA\epsilon_0\begin{pmatrix} -1 \\ 1 \end{pmatrix}$$

The last term in the strain energy involves the known initial strain and does not depend on the assumed solution. This term will drop out when writing the necessary conditions for the minimum of the potential energy. Thus the last term can be ignored, and therefore the presence of initial strains results in simply an equivalent nodal load vector $\boldsymbol{r}_\epsilon$. For analysis of plane and space trusses, this equivalent load vector is first transformed to the global coordinates using the appropriate transformation matrix:

$$\boldsymbol{r} = \boldsymbol{T}^T\boldsymbol{r}_\epsilon$$

For a plane truss,

$$\boldsymbol{r} = EA\epsilon_0\begin{pmatrix} -l_s \\ -m_s \\ l_s \\ m_s \end{pmatrix}$$

For a space truss,

$$\boldsymbol{r} = EA\epsilon_0\begin{pmatrix} -l_s \\ -m_s \\ -n_s \\ l_s \\ m_s \\ n_s \end{pmatrix}$$

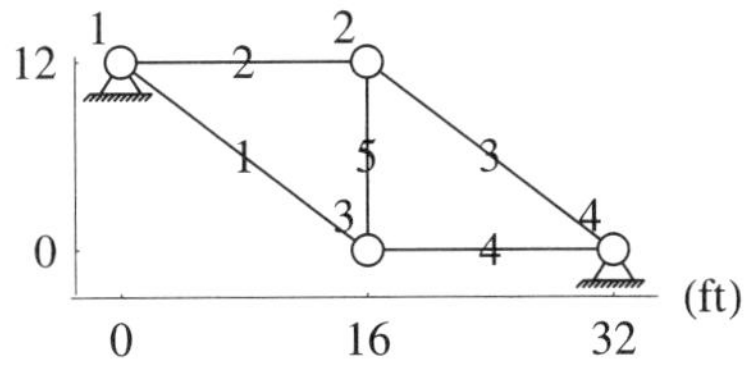

Figure 4.6. Plane truss

The equivalent nodal load vectors from all elements that are subjected to initial strain are assembled in the usual manner and global equations solved for nodal unknowns. The element stresses are obtained from the net strains:

$$\sigma_x = E\left(\frac{du}{dx} - \epsilon_0\right) = E(\boldsymbol{B}^T \boldsymbol{d} - \epsilon_0) = E\left(\frac{-d_1 + d_2}{L} - \epsilon_0\right)$$

Example 4.3 Plane Truss with Temperature Change Consider the five-bar pin-jointed structure shown in Figure 4.6. All members have the same cross-sectional area $A = \frac{1}{2}\,\text{in}^2$ and are of the same material with $E = 29{,}000\,\text{ksi}$ and $\alpha = 6.5 \times 10^{-6}/^\circ\text{F}$. The first element undergoes a temperature rise of 100°F. The dimensions are shown in the figure.

Complete calculations for this example can be found on the web site.

The solution summary is as follows:

Nodal Solution

	u	v (in)
1	0	0
2	−0.0308148	−0.121333
3	0.0308148	−0.138667
4	0	0

Element Solution

	Stress (ksi)	Axial force (kip)
1	−5.8179	−2.90895
2	−4.65432	−2.32716
3	−5.8179	−2.90895
4	−4.65432	−2.32716
5	3.49074	1.74537

4.4 SPRING ELEMENTS

An axial deformation element can also be thought of as a linear spring that is designed to resist axial tension or compression. The spring constant, usually denoted by k, is defined as the load that causes unit extension or compression in the spring. In the axial deformation element, if we assume $u_2 = 0$ and $u_1 = 1$, then we have a unit extension of the bar, and

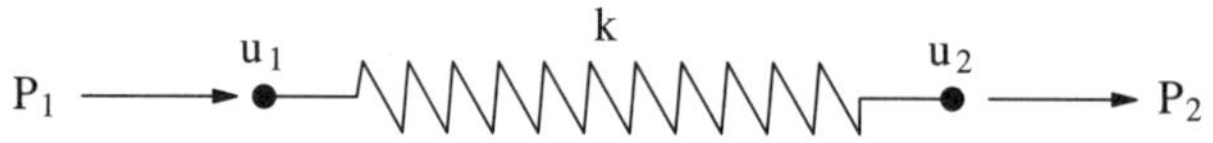

Figure 4.7. An axial spring element

the finite element equations show that the axial load is $P_1 = AE/L$. Thus the term AE/L is equivalent to the spring constant k for a linear spring. Therefore, we can incorporate linear spring elements, such as the one shown in Figure 4.7, into a finite element model through the following element equations:

$$k\begin{pmatrix} 1 & -1 \\ -1 & 1 \end{pmatrix}\begin{pmatrix} u_1 \\ u_2 \end{pmatrix} = \begin{pmatrix} P_1 \\ P_2 \end{pmatrix}$$

For an arbitrarily oriented element these equations can be transformed to global coordinates using the same transformation matrices as those for the truss elements.

Rotational or torsional spring elements can be defined in an analogous manner. Denoting the torsional spring constant by k_T, the end rotations by θ_1 and θ_2, and any applied twisting moments at the ends by M_{T_1}, M_{T_2}, the equations for a torsional spring element are as follows:

$$k_T\begin{pmatrix} 1 & -1 \\ -1 & 1 \end{pmatrix}\begin{pmatrix} \theta_1 \\ \theta_2 \end{pmatrix} = \begin{pmatrix} M_{T_1} \\ M_{T_2} \end{pmatrix}$$

Example 4.4 Find axial forces in the spring-bar assembly shown in shown in Figure 4.8. Use the following numerical values:

$L = 30$ in; $F = 15{,}000$ lb; Spring constant, $k = 100{,}000$ lb/in

For the steel bar: $E = 30 \times 10^6$ lb/in^2; $A = 0.5$ in^2

For the aluminum bar: $E = 10^7$ lb/in^2; $A = 1.2$ in^2

The assembly is modeled using one spring and two axial deformation elements as shown in Figure 4.9. There are three nodes, each with one displacement degree of freedom.

Using pound-inches and following the usual steps, the solution is as follows:

Specified nodal loads:

Node	dof	Value
2	u_2	15000.

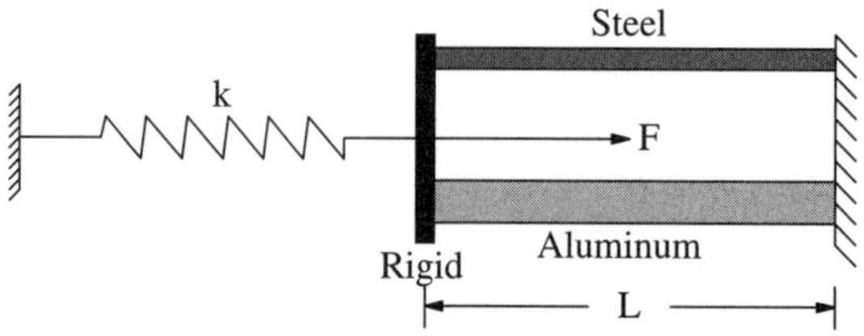

Figure 4.8. Spring-bar assembly

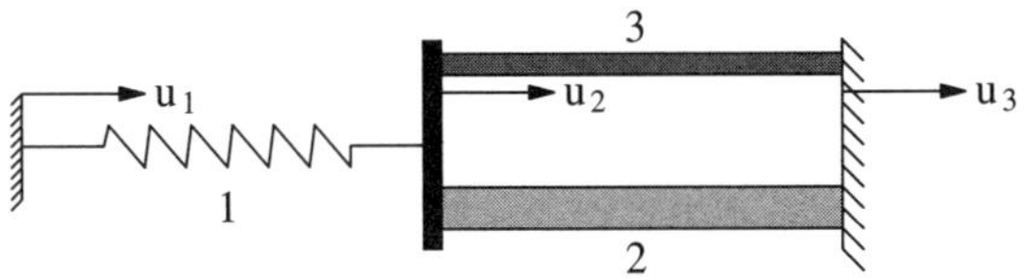

Figure 4.9. Three-element model for the spring-bar assembly

Element 1:

$$\begin{pmatrix} 100000 & -100000 \\ -100000 & 100000 \end{pmatrix} \begin{pmatrix} u_1 \\ u_2 \end{pmatrix} = \begin{pmatrix} 0 \\ 0 \end{pmatrix}$$

Element 2:

$$\begin{pmatrix} 400000. & -400000. \\ -400000. & 400000. \end{pmatrix} \begin{pmatrix} u_2 \\ u_3 \end{pmatrix} = \begin{pmatrix} 0 \\ 0 \end{pmatrix}$$

Element 3:

$$\begin{pmatrix} 500000. & -500000. \\ -500000. & 500000. \end{pmatrix} \begin{pmatrix} u_2 \\ u_3 \end{pmatrix} = \begin{pmatrix} 0 \\ 0 \end{pmatrix}$$

$$\text{Global equations: } \begin{pmatrix} 100000 & -100000 & 0 \\ -100000 & 1. \times 10^6 & -900000. \\ 0 & -900000. & 900000. \end{pmatrix} \begin{pmatrix} u_1 \\ u_2 \\ u_3 \end{pmatrix} = \begin{pmatrix} 0 \\ 15000 \\ 0 \end{pmatrix}$$

Essential BC →

Node	dof	Value
1	1	0
3	3	0

Global equations after EBC: $1. \times 10^6\, u_2 = 15000$

Nodal solution: $\{u_2 \to 0.015\}$

The force in the spring is equal to the spring constant times its change in length:

$$\text{Force in spring} = k(u_2 - u_1) = 1500.\ \text{lb (Tension)}$$

The axial strain in the bars is

$$\text{Axial strain} = (u_3 - u_2)/L = -0.0005$$

The axial forces in each bar is equal to $EA\epsilon$, giving

$$\text{In the steel bar} = -7500.\ \text{lb (Compression)}$$

$$\text{In the aluminum bar} = -6000.\ \text{lb (Compression)}$$

4.5 TRANSVERSE DEFORMATION OF BEAMS

A beam is a structural member that carries a load normal to its axis. The length of a beam is large as compared to its cross-sectional dimensions. In general, the cross section can be of any arbitrary shape and can vary along its length. However, the following discussion is limited to beams whose cross sections are symmetric with respect to the plane of bending. The x axis passes through the centroid of the cross section. Under these conditions, the beam is subjected only to bending and there are no axial or twisting forces.

4.5.1 Differential Equation for Beam Bending

Consider a beam with a symmetric cross section that is subjected to load in the transverse direction, as shown in Figure 4.10. The moment of inertia is denoted by $I(x)$ and the mass of the beam per unit length is denoted by $m(x)$. This mass is due to self-weight and any additional weight distributed along the beam. The beam may be subjected to distributed load $q(x, t)$ along its length.

The governing differential equation can be written by considering the equilibrium of a differential element, as shown in Figure 4.11. The shear force and moment at x are denoted by V and M, respectively. Using the Taylor series expansion, these quantities at $x + dx$ are $V + (\partial V/\partial x)\, dx$ and $M + (\partial M/\partial x)\, dx$. The transverse displacement is denoted by $v(x, t)$. The acceleration is indicated by $\ddot{v} \equiv \partial^2 v/\partial t^2$. From Newton's second law of motion, a force, called the inertia force, is produced that is proportional to the mass of the element and acts in the direction opposite to the direction of motion. With the mass per unit length being $m(x)$, the inertia force is $m\, dx\, \ddot{v}$. The only other force acting on the element is the applied transverse load $q(x, t)$.

Note that the sign convention adopted in drawing the free-body diagram in Figure 4.11 assumes that the internal moments causing compression on the top of the beam are positive. This means that along the left side of an element a clockwise moment is positive while on the right side a counterclockwise moment is positive. On the left side the shear force along the y axis is positive while on the right side the positive shear acts in the $-y$ direction. As far as the external loading is concerned, an external applied moment in the counterclockwise direction is assumed positive while applied force is positive if it is acting along the $+y$ axis.

Considering summation of forces in the y direction, we have

$$m\, dx\, \ddot{v}(x, t) = q(x, t)\, dx + V - \left(V + \frac{\partial V}{\partial x}\, dx\right) \Longrightarrow m\ddot{v}(x, t) + \frac{\partial V}{\partial x} = q(x, t)$$

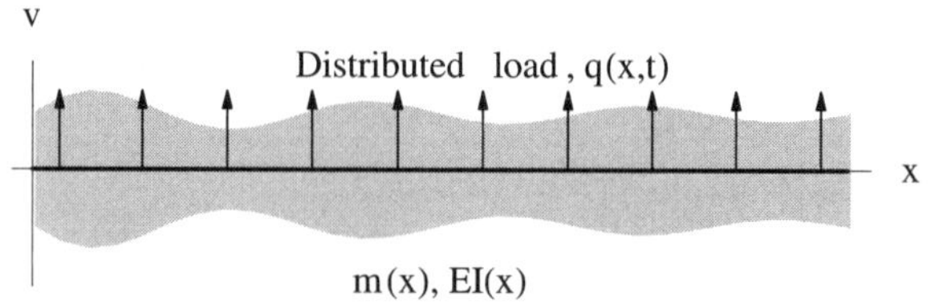

Figure 4.10. Beam bending

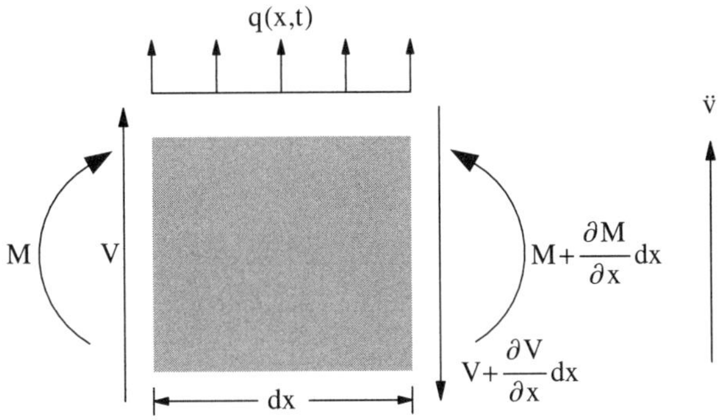

Figure 4.11. Forces acting on a differential beam element

Summation of moments about the right-hand side gives

$$-q(x,t)\,dx\,dx/2 - M - V\,dx + M + \frac{\partial M}{\partial x}\,dx = 0$$

Neglecting the higher order dx^2 term, we get

$$V = \frac{\partial M}{\partial x}$$

In order to relate forces to deformations, a fundamental assumption in beam bending is that a plane section before bending remains plane and normal to the neutral axis after bending. This assumption implies that the axial deformation at a point y above the neutral axis is given by

$$u(x) = -y\frac{dv}{dx}$$

The axial strain ϵ_x is

$$\epsilon_x = \frac{du}{dx} = -y\frac{d^2v}{dx^2}$$

For a linear elastic material the axial stress σ_x and strain are related by Young's modulus E, giving

$$\sigma_x = E\epsilon_x = -Ey\frac{d^2v}{dx^2}$$

Thus the axial stress due to bending varies linearly over the cross section. The maximum compression is at the top and the maximum tension is at the bottom. Taking the moment of forces acting on a cross section, we get

$$M(x) = -\int_A \sigma_x y\,dA = \int_A Ey\frac{d^2v}{dx^2}y\,dA = E\frac{d^2v}{dx^2}\int_A y^2\,dA = EI\frac{d^2v}{dx^2}$$

where A is the area of the cross section and $I = \int_A y^2\, dA$ is the moment of inertia or second moment of the area. Using this equation, we can relate moment and axial stress to get

$$\sigma_x = -Ey\left(\frac{d^2v}{dx^2}\right) \equiv -\frac{My}{I}$$

Differentiating the moment, we can relate the shear force to displacement as follows:

$$V = \frac{\partial M}{\partial x} = \frac{\partial}{\partial x}\left(EI\frac{\partial^2 v}{\partial x^2}\right)$$

Substituting the derivative of V into the first equilibrium equation, the governing differential equation can be written as follows:

$$m\ddot{v}(x, t) + \frac{\partial^2}{\partial x^2}\left(EI\frac{\partial^2 v}{\partial x^2}\right) = q(x, t)$$

This is a fourth-order partial differential equation. The equation involves a fourth-order derivative with respect to x and a second-order derivative with respect to t. We need four boundary conditions and two initial conditions for a proper solution. The beam boundary conditions are discussed in the following section. The initial conditions are the specified displacement and velocity along the beam at time $t = 0$.

$$v(x, 0) = v_0; \qquad \dot{v}(x, 0) = \dot{v}_0$$

where v_0 and $\dot{v}_0$ are specified displacement and velocity values.

For a static analysis situation, the inertia force is zero, and we have the following governing differential equation:

$$\frac{\partial^2}{\partial x^2}\left(EI\frac{\partial^2 v}{\partial x^2}\right) = q(x)$$

4.5.2 Boundary Conditions for Beams

The boundary conditions for beam bending involve specification of displacement v or any of its first three derivatives with respect to x. Some common beam boundary conditions are shown in Figure 4.12. Since the second derivative of displacement is related to the bending

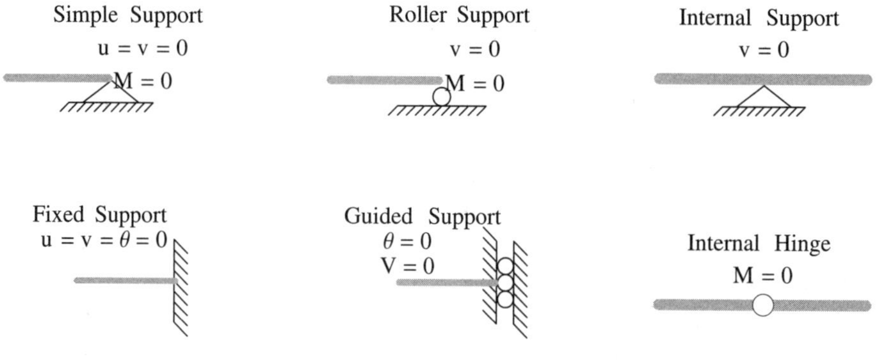

Figure 4.12. Typical beam boundary conditions

moment and the third derivative to shear force, the boundary conditions at a point x_0 along a beam are as shown in the following table. Carefully note the convention used in showing these boundary conditions in the figures. Also note that the signs for internal moments and shears depend on whether we are considering a left side or a right side of a section.

Specified Quantity	Left Side of Section at x_0	Right Side of Section at x_0
Displacement	$v(x_0) = v_{x0}$	$v(x_0) = v_{x0}$
Rotation	$\theta(x_0) \equiv \frac{\partial v(x_0)}{\partial x} = \theta_{x0}$	$\theta(x_0) \equiv \frac{\partial v(x_0)}{\partial x} = \theta_{x0}$
Moment	$-EI\frac{\partial^2 v(x_0)}{\partial x^2} = M_{x0}$	$EI\frac{\partial^2 v(x_0)}{\partial x^2} = M_{x0}$
Shear	$EI\frac{\partial^3 v(x_0)}{\partial x^3} = F_{x0}$	$-EI\frac{\partial^3 v(x_0)}{\partial x^3} = F_{x0}$

Boundary conditions at a free end:

Bending moment: $M = EI\dfrac{\partial^2 v}{\partial x^2} = 0;$ Shear force: $V = \dfrac{\partial}{\partial x}\left[EI\dfrac{\partial^2 v}{\partial x^2}\right] = 0$

Boundary conditions at a fixed end:

Deflection: $v = 0;$ Rotation: $\theta = \dfrac{\partial v}{\partial x} = 0$

Boundary conditions at a simple (pin) support:

Deflection: $v = 0;$ Bending moment: $M = EI\dfrac{\partial^2 v}{\partial x^2} = 0$

Boundary conditions at a guided (sliding) support:

Rotation: $\theta = \dfrac{\partial v}{\partial x} = 0;$ Shear force: $V = \dfrac{\partial}{\partial x}\left[EI\dfrac{\partial^2 v}{\partial x^2}\right] = 0$

More complicated boundary conditions are possible when beams are supported by other elastic members. Such supports are typically represented by extensional or rotational springs. The force in an extensional spring located at x_0, with the spring constant k, is

$$F_e(x_0) = kv(x_0)$$

This force must be balanced by the shear at the support. Using the free-body diagram shown in Figure 4.13 and the sign convention for shear at the right end of a segment, the appropriate boundary condition is

Right-end extensional spring support: $\dfrac{\partial}{\partial x}\left[EI\dfrac{\partial^2 v}{\partial x^2}\right] - kv = 0$

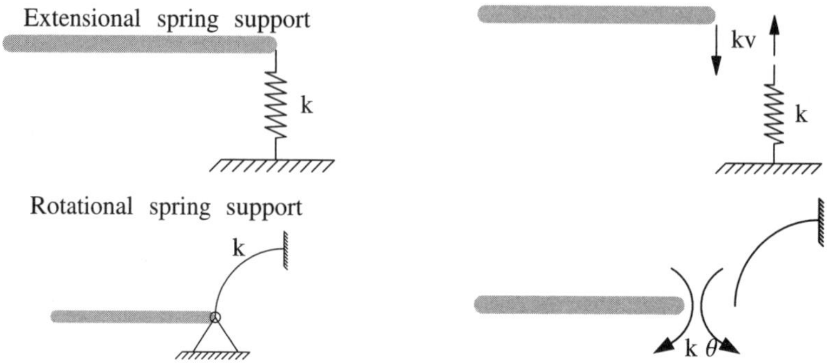

Figure 4.13. Spring-supported boundary conditions

A rotational spring generates a moment equal to $k\theta$, and therefore the boundary condition at the right end of a beam supported by a rotational spring is

$$\text{Right-end rotational spring support:} \qquad EI\frac{\partial^2 v}{\partial x^2} + k\theta = 0$$

From the free-body diagrams it is easy to see that, if the elastic supports are at the left end, the appropriate boundary conditions are as follows:

$$\text{Left-end extensional spring support:} \qquad \frac{\partial}{\partial x}\left[EI\frac{\partial^2 v}{\partial x^2}\right] + kv = 0$$

$$\text{Left-end rotational spring support:} \qquad EI\frac{\partial^2 v}{\partial x^2} - k\theta = 0$$

4.5.3 Shear Stresses in Beams

The basic beam theory does not take into account the deformations due to shear forces. Thus the shear stress does not enter into the governing differential equation. However, knowing the shear forces, we can compute the corresponding shear stresses using the following equation from the mechanics of deformable bodies:

$$\tau = \frac{VQ}{It}$$

where V is the shear force, I is the moment of inertia, and Q and t are related to the point in the cross section where the shear stress in being computed, t being the width of the cross section at this point and Q the moment of the area above this point about the neutral axis. See Example 4.5 for an illustration.

4.5.4 Potential Energy for Beam Bending

For a linear elastic beam the axial strain and stress due to bending are

$$\epsilon_x = \frac{du}{dx} = -y\left(\frac{d^2 v}{dx^2}\right); \qquad \sigma_x = E\epsilon_x = -Ey\left(\frac{d^2 v}{dx^2}\right)$$

Thus the strain energy for a beam element with end coordinates at x_0 and x_l can be written as follows:

$$U = \frac{1}{2}\int_V \sigma_x \epsilon_x \, dV = \frac{1}{2}\int_{x_0}^{x_l}\int_A Ey\left(\frac{d^2v}{dx^2}\right)y\left(\frac{d^2v}{dx^2}\right)dA\,dx = \frac{1}{2}\int_{x_0}^{x_l} EI\left(\frac{d^2v}{dx^2}\right)^2 dx$$

The potential of the applied loads is equal to the negative of the work done by the external applied forces:

$$W = \int_{x_0}^{x_l} qv\,dx + \sum P_i v(x_i)$$

where $v(x_i)$ is the transverse displacement at the location of the concentrated applied load P_i. Thus the potential energy for beam bending can be written as follows:

$$\Pi = U - W = \frac{1}{2}\int_{x_0}^{x_l} EI\left(\frac{d^2v}{dx^2}\right)^2 dx - \int_{x_0}^{x_l} qv\,dx - \sum P_i v(x_i)$$

Using calculus of variations, it is possible to show that a function $v(x)$ that minimizes the potential energy is a solution of the differential equation governing the bending of beams.

4.5.5 Transverse Deformation of a Uniform Beam

Consider a uniform beam simply supported at the left end and spring supported at the right end, as shown in Figure 4.14. The beam is subjected to a linearly increasing load $q(x) = ax/L$, where a is a given constant. The problem is described in terms of the following boundary value problem:

$$EI\frac{d^4v}{dx^4} = \frac{ax}{L}; \qquad 0 < x < L$$

$$v(0) = 0; \qquad EI\frac{d^2v(0)}{dx^2} = 0$$

$$EI\frac{d^3v(L)}{dx^3} - kv(L) = 0; \qquad EI\frac{d^2v(L)}{dx^2} = 0$$

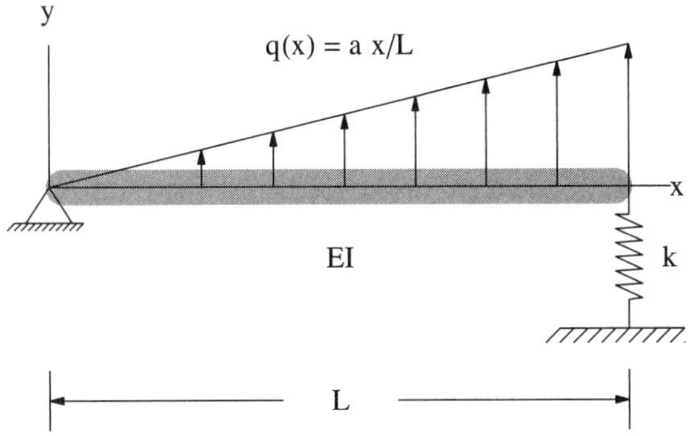

Figure 4.14. Uniform beam subjected to a linearly increasing load

An exact solution of the problem can be obtained by integrating the differential equation four times and then using the boundary conditions to evaluate the resulting integration constants. Integrating both sides of the differential equation, we get

$$EI\frac{d^3v}{dx^3} - \frac{ax^2}{2L} = c_1$$

$$EI\frac{d^2v}{dx^2} - \frac{ax^3}{6L} = c_1x + c_2$$

$$EI\frac{dv}{dx} - \frac{ax^4}{24L} = c_1\frac{x^2}{2} + c_2x + c_3$$

$$EIv - \frac{ax^5}{120L} = c_1\frac{x^3}{6} + c_2\frac{x^2}{2} + c_3x + c_4$$

where $c_1, \ldots, c_4$ are integration constants. Using the boundary conditions,

$$v(0) = 0 \Longrightarrow c_4 = 0$$

$$EI\frac{d^2v(0)}{dx^2} = 0 \Longrightarrow c_2 = 0$$

$$EI\frac{d^2v(L)}{dx^2} = 0 \Longrightarrow -\frac{-20aL^3 - 120L^2c_1}{120L} = 0 \Longrightarrow c_1 = -\frac{aL}{6}$$

$$EI\frac{d^3v(L)}{dx^3} - kv(L) = 0 \Longrightarrow -\frac{-60aL^2 - 120Lc_1}{120L} + \frac{k(-aL^5 - 20L^4c_1 - 120L^2c_3)}{120EIL} = 0$$

$$\Longrightarrow c_3 = -\frac{-120aEI - 7akL^3}{360k}$$

Thus the exact solution of the problem is

$$v(x) = \frac{ax(120EIL + k(7L^4 - 10L^2x^2 + 3x^4))}{360EIkL}$$

As $k \to \infty$, the solution approaches that of a simply supported beam:

$$v(x) = \lim_{k\to\infty}\left(\frac{ax(120EIL + k(7L^4 - 10L^2x^2 + 3x^4))}{360EIkL}\right) = \frac{ax(7L^4 - 10x^2L^2 + 3x^4)}{360EIL}$$

Assuming $EI = 1$, $L = 1$, and $a = 1$, the solution for various k values is as shown in Figure 4.15. With increasing k the spring-supported end displacement is getting small, as expected.

4.5.6 Transverse Deformation of a Tapered Beam Fixed at Both Ends

Consider a nonuniform beam in which the moment of inertia varies linearly along the span, $I(x) = I_0(1 + bx/L)$, where I_0 is a reference moment of inertia and b is a constant. The beam

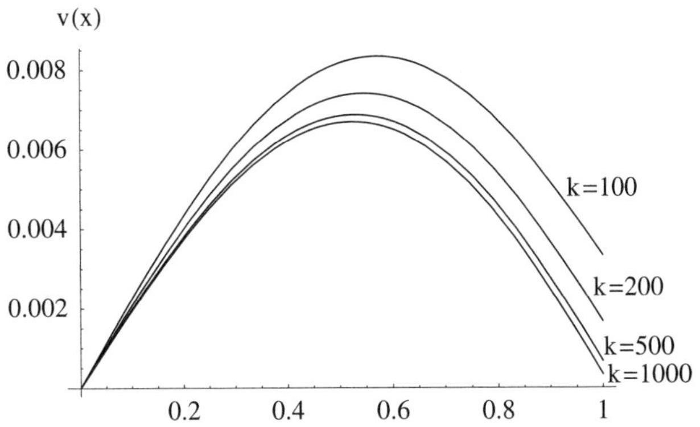

Figure 4.15. Exact solutions for spring-supported beam with various spring constants

is fixed at both ends, as shown in Figure 4.16. The beam is subjected to a linearly increasing load $q(x) = ax/L$, where a is a given constant. The problem is described in terms of the following boundary value problem:

$$\frac{d^2}{dx^2}\left(EI(x)\frac{d^2v}{dx^2}\right) = \frac{ax}{L}; \qquad 0 < x < L$$

$$EI(x) = EI_0\left(1 + \frac{bx}{L}\right)$$

$$v(0) = 0; \qquad \frac{dv(0)}{dx} = 0$$

$$v(L) = 0; \qquad \frac{dv(L)}{dx} = 0$$

The actual derivation of the following solution is quite tedious. However, by direct substitution, it can be verified that the solution satisfies the governing differential equation:

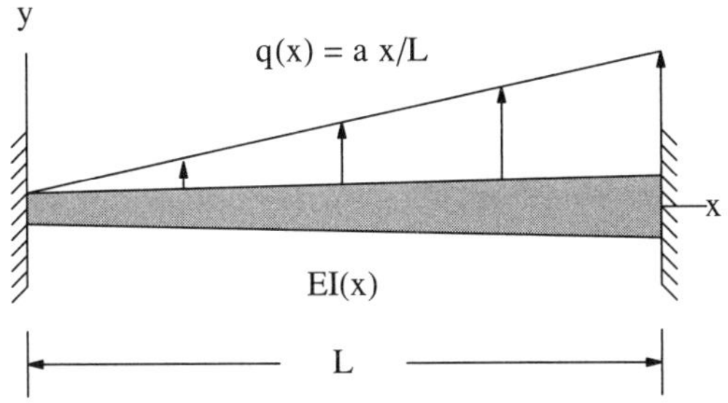

Figure 4.16. Nonuniform beam subjected to a linearly increasing load

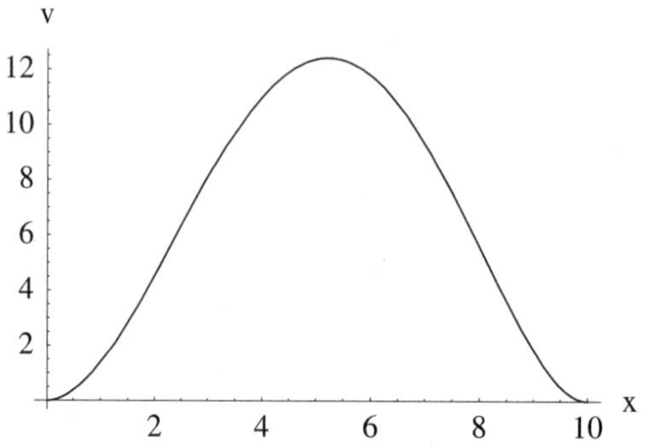

Figure 4.17. Solution of nonuniform beam

$$v(x) = c_2 + xc_3 + x^2c_4 - \frac{aLx^3}{36b^2EI_0} + \frac{ax^4}{72bEI_0} + \frac{x(aL^3 + 6b^2c_1EI_0)}{6b^4EI_0}$$
$$- \frac{x\log[L + bx](aL^3 + 6b^2c_1EI_0)}{6b^4EI_0} + \frac{\log[L + bx](-aL^4 - 6b^2Lc_1EI_0)}{6b^5EI_0}$$

where $c_1, \ldots, c_4$ are integration constants. Using the boundary conditions, the constants can be evaluated and the solution expressed as follows:

$$\begin{aligned}v(x) = {} & (a((L - x)^2(2(-1 + b)L^2 + 2(-2 + b + b^2)Lx + b(2 + b)x^2)\log[L] \\ & + x^2(-6L^2 + 3b^2L^2 + 4Lx + 2bLx - 2bx^2 - b^2x^2)\log[(1 + b)L] \\ & - 2(bL^3x - b^2L^3x - 3bL^2x^2 + 2b^2L^2x^2 + 2bLx^3 - b^2x^4 - L^4\log[L + bx] \\ & + bL^4\log[L + bx] - bL^3x\log[L + bx] + b^2L^3x\log[L + bx])))/ \\ & (72b^2(2b + (2 + b)\log[L] - (2 + b)\log[(1 + b)L])EI_0)\end{aligned}$$

Figure 4.17 shows a plot of this solution with $\{b = 0.1, a = 1, L = 10, EI_0 = 1\}$.

4.6 TWO-NODE BEAM ELEMENT

Recall from the discussion in Chapter 2 that for a fourth-order problem the essential boundary conditions are the solution and its first derivative. For beam bending, therefore, both the transverse displacement v and rotation $\theta \equiv dv/dx$ are essential boundary conditions. In order to be able to satisfy these boundary conditions during the assembly and solution phase, we must choose both v and θ as degrees of freedom for each node. Thus the simplest two-node beam element has four degrees of freedom, as shown in Figure 4.18. For simplicity in integrations the moment of inertia and the distributed load are assumed constant over each element. Furthermore, concentrated loads F_1, F_2 and moments M_1, M_2 are allowed only at the ends of an element.

As demonstrated through examples later in the section, even this simple element is enough to model most practical beam problems. For uniform beams only one element per

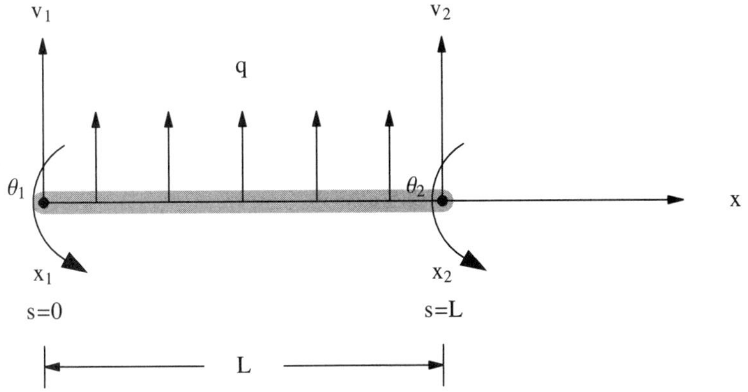

Figure 4.18. A two-node element for beam bending

span is needed to get exact solutions. For concentrated loads along the span or changes in section dimensions, one simply needs to start a new element at the locations where such situations occur. Even beams made up of different materials can be modeled, simply by starting and ending elements at the material interfaces. Only for variable-cross-section beams is it necessary to use a large number of constant moment-of-inertia elements to get a good solution.

4.6.1 Cubic Assumed Solution

Since both displacement and rotation are used as degrees of freedom for each node, the appropriate assumed solution is written using the Hermite interpolation. With the general formula given in Chapter 2 it is possible to write shape functions for an element with arbitrary end coordinates x_1 and x_2. However, the resulting expressions are very messy. More convenient expressions are obtained by introducing the following change in coordinates:

$$s = x - x_1; \qquad x_1 \le x \le x_2; \qquad 0 \le s \le L$$
$$ds = dx \Longrightarrow \frac{dv}{dx} = \frac{dv}{ds}$$

The four interpolation functions in terms of s can easily be written as follows:

Data point locations: $\{0, L\}$

Lagrange interpolation functions for these points:

$$\left\{L_1 = -\frac{s-L}{L}, L_2 = \frac{s}{L}\right\}$$

Derivatives of Lagrange interpolation functions:

$$\left\{L_1' = -\frac{1}{L}, L_2' = \frac{1}{L}\right\}$$

Derivatives at the given points:

$$\left\{L_1'(0) = -\frac{1}{L}, L_2'(L) = \frac{1}{L}\right\}$$

Using these, the Hermite interpolations for data values are

$$\left\{P_1 = \frac{2s^3}{L^3} - \frac{3s^2}{L^2} + 1, P_2 = \frac{3s^2}{L^2} - \frac{2s^3}{L^3}\right\}$$

and those for derivative values are

$$\left\{Q_1 = \frac{s^3}{L^2} - \frac{2s^2}{L} + s, Q_2 = \frac{s^3}{L^2} - \frac{s^2}{L}\right\}$$

Vector of Hermite interpolation functions:

$$N^T = \left\{\frac{2s^3}{L^3} - \frac{3s^2}{L^2} + 1, \frac{s^3}{L^2} - \frac{2s^2}{L} + s, \frac{3s^2}{L^2} - \frac{2s^3}{L^3}, \frac{s^3}{L^2} - \frac{s^2}{L}\right\}$$

Thus the assumed solution is

$$v(s) = \begin{pmatrix} \frac{2s^3}{L^3} - \frac{3s^2}{L^2} + 1 & \frac{s^3}{L^2} - \frac{2s^2}{L} + s & \frac{3s^2}{L^2} - \frac{2s^3}{L^3} & \frac{s^3}{L^2} - \frac{s^2}{L} \end{pmatrix} \begin{pmatrix} v_1 \\ \theta_1 \\ v_2 \\ \theta_2 \end{pmatrix} \equiv \boldsymbol{N}^T \boldsymbol{d}$$

$$N_1 = 1 - \frac{3s^2}{L^2} + \frac{2s^3}{L^3}; \qquad N_2 = s - \frac{2s^2}{L} + \frac{s^3}{L^2}; \qquad N_3 = \frac{3s^2}{L^2} - \frac{2s^3}{L^3}; \qquad N_4 = -\frac{s^2}{L} + \frac{s^3}{L^2}$$

4.6.2 Element Equations Using Rayleigh-Ritz Method

The potential energy for the beam element is as follows:

$$\Pi = U - W = \frac{1}{2}\int_{x_1}^{x_2} EI\left(\frac{d^2v}{dx^2}\right)^2 dx - \int_{x_1}^{x_2} qv\,dx - F_1v_1 - M_1\theta_1 - F_2v_2 - M_2\theta_2$$

or

$$\Pi = \frac{1}{2}\int_0^L EI\left(\frac{d^2v}{ds^2}\right)^2 ds - \int_0^L qv\,ds - F_1v_1 - M_1\theta_1 - F_2v_2 - M_2\theta_2$$

Differentiating the assumed solution twice with respect to s, we have

$$\frac{d^2v}{ds^2} = \begin{pmatrix} -\frac{6}{L^2} + \frac{12s}{L^3} & -\frac{4}{L} + \frac{6s}{L^2} & \frac{6}{L^2} - \frac{12s}{L^3} & -\frac{2}{L} + \frac{6s}{L^2} \end{pmatrix} \begin{pmatrix} v_1 \\ \theta_1 \\ v_2 \\ \theta_2 \end{pmatrix} = \boldsymbol{B}^T \boldsymbol{d}$$

Squaring this (making sure that vectors $\boldsymbol{d}$ are on the outside),

$$\left(\frac{d^2v}{ds^2}\right)^2 = (\boldsymbol{B}^T\boldsymbol{d})^T\boldsymbol{B}^T\boldsymbol{d} = \boldsymbol{d}^T\boldsymbol{B}\boldsymbol{B}^T\boldsymbol{d}$$

where

$$\boldsymbol{B}\boldsymbol{B}^T = \begin{pmatrix} \frac{36(L-2s)^2}{L^6} & \frac{12(2L-3s)(L-2s)}{L^5} & -\frac{36(L-2s)^2}{L^6} & \frac{12(L-3s)(L-2s)}{L^5} \\ \frac{12(2L-3s)(L-2s)}{L^5} & \frac{4(2L-3s)^2}{L^4} & -\frac{12(2L-3s)(L-2s)}{L^5} & \frac{4(L-3s)(2L-3s)}{L^4} \\ -\frac{36(L-2s)^2}{L^6} & -\frac{12(2L-3s)(L-2s)}{L^5} & \frac{36(L-2s)^2}{L^6} & -\frac{12(L-3s)(L-2s)}{L^5} \\ \frac{12(L-3s)(L-2s)}{L^5} & \frac{4(L-3s)(2L-3s)}{L^4} & -\frac{12(L-3s)(L-2s)}{L^5} & \frac{4(L-3s)^2}{L^4} \end{pmatrix}$$

Using this, the strain energy term can be written as follows:

$$U = \frac{1}{2}\int_0^L EI\boldsymbol{d}^T\boldsymbol{B}\boldsymbol{B}^T\boldsymbol{d}\,ds = \frac{1}{2}\boldsymbol{d}^T\int_0^L EI\boldsymbol{B}\boldsymbol{B}^T\,ds\,\boldsymbol{d} \equiv \frac{1}{2}\boldsymbol{d}^T\boldsymbol{k}\boldsymbol{d}$$

where

$$\boldsymbol{k} = \int_0^L EI\boldsymbol{B}\boldsymbol{B}^T\,ds = EI\begin{pmatrix} \int_0^L\left(\frac{36(L-2s)^2}{L^6}\right)ds & \int_0^L\left(\frac{12(2L-3s)(L-2s)}{L^5}\right)ds & \cdots \\ \int_0^L\left(\frac{12(2L-3s)(L-2s)}{L^5}\right)ds & \int_0^L\left(\frac{4(2L-3s)^2}{L^4}\right)ds & \cdots \\ \vdots & \vdots & \vdots \end{pmatrix}$$

Carrying out integrations, we get

$$\boldsymbol{k} = \frac{EI}{L^3}\begin{pmatrix} 12 & 6L & -12 & 6L \\ 6L & 4L^2 & -6L & 2L^2 \\ -12 & -6L & 12 & -6L \\ 6L & 2L^2 & -6L & 4L^2 \end{pmatrix}$$

The work done by the distributed load can be evaluated as follows:

$$W_q = \int_0^L qv\,ds = \int_0^L q\boldsymbol{N}^T\boldsymbol{d}\,ds = \boldsymbol{r}_q^T\boldsymbol{d}\boldsymbol{d}^T\boldsymbol{r}_q$$

where

$$\boldsymbol{r}_q^T = \int_0^L q\boldsymbol{N}^T\,ds \Longrightarrow \boldsymbol{r}_q = \int_0^L \boldsymbol{N}q\,ds = q\begin{pmatrix} \int_0^L\left(1-\frac{3s^2}{L^2}+\frac{2s^3}{L^3}\right)ds \\ \int_0^L\left(s-\frac{2s^2}{L}+\frac{s^3}{L^2}\right)ds \\ \int_0^L\left(\frac{3s^2}{L^2}-\frac{2s^3}{L^3}\right)ds \\ \int_0^L\left(-\frac{s^2}{L}+\frac{s^3}{L^2}\right)ds \end{pmatrix} = \frac{qL}{2}\begin{pmatrix} 1 \\ \frac{L}{6} \\ 1 \\ -\frac{L}{6} \end{pmatrix}$$

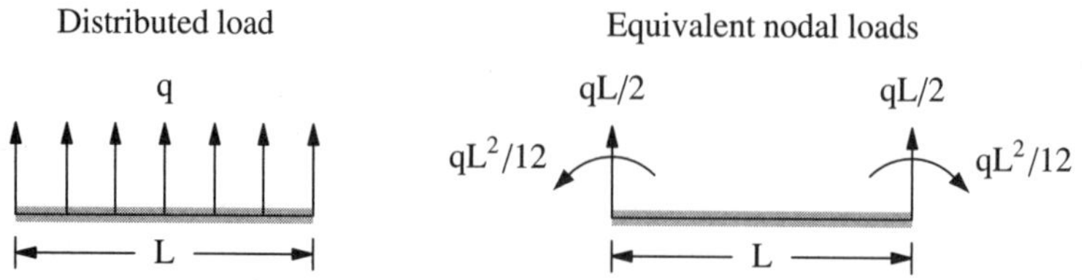

Figure 4.19. Equivalent nodal loads due to a uniformly distributed load on an element

Interpreting terms in the vector r_q, we see that a uniformly distributed load on an element is actually applied as nodal loads $qL/2$ and moments $qL^2/12$. With the sign convention being used, the equivalent nodal values are as shown in Figure 4.19.

The work done by the concentrated loads is

$$W_p = F_1 v_1 + M_1\theta_1 + F_2 v_2 + M_2\theta_2 = \begin{pmatrix} F_1 & M_1 & F_2 & M_2 \end{pmatrix} \begin{pmatrix} v_1 \\ \theta_1 \\ v_2 \\ \theta_2 \end{pmatrix} \equiv \boldsymbol{r}_p^T \boldsymbol{d} \equiv \boldsymbol{d}^T \boldsymbol{r}_p$$

The potential energy can now be written as follows:

$$\Pi = U - W = \tfrac{1}{2}\boldsymbol{d}^T\boldsymbol{k}\boldsymbol{d} - \boldsymbol{r}_q^T\boldsymbol{d} - \boldsymbol{r}_p^T\boldsymbol{d} = \boldsymbol{d}^T\left(\tfrac{1}{2}\boldsymbol{k}\boldsymbol{d} - \boldsymbol{r}_q - \boldsymbol{r}_p\right)$$

The necessary conditions for the minimum of the potential energy give

$$\frac{\partial \Pi}{\partial d} = \left(\tfrac{1}{2}\boldsymbol{k}\boldsymbol{d} - \boldsymbol{r}_q - \boldsymbol{r}_p\right) + \tfrac{1}{2}\boldsymbol{k}\boldsymbol{d} = 0$$

Thus the beam element equations are as follows:

$$\boldsymbol{k}\boldsymbol{d} = \boldsymbol{r}_q + \boldsymbol{r}_p \Longrightarrow \frac{EI}{L^3}\begin{pmatrix} 12 & 6L & -12 & 6L \\ 6L & 4L^2 & -6L & 2L^2 \\ -12 & -6L & 12 & -6L \\ 6L & 2L^2 & -6L & 4L^2 \end{pmatrix}\begin{pmatrix} v_1 \\ \theta_1 \\ v_2 \\ \theta_2 \end{pmatrix} = \frac{qL}{2}\begin{pmatrix} 1 \\ \frac{L}{6} \\ 1 \\ -\frac{L}{6} \end{pmatrix} + \begin{pmatrix} F_1 \\ M_1 \\ F_2 \\ M_2 \end{pmatrix}$$

Assuming that the concentrated loads are added directly to the global equations at the start of the assembly, the element equations are

$$\frac{EI}{L^3}\begin{pmatrix} 12 & 6L & -12 & 6L \\ 6L & 4L^2 & -6L & 2L^2 \\ -12 & -6L & 12 & -6L \\ 6L & 2L^2 & -6L & 4L^2 \end{pmatrix}\begin{pmatrix} v_1 \\ \theta_1 \\ v_2 \\ \theta_2 \end{pmatrix} = \frac{qL}{2}\begin{pmatrix} 1 \\ \frac{L}{6} \\ 1 \\ -\frac{L}{6} \end{pmatrix}$$

For a given beam problem, the element equations can be assembled and nodal displacements can be obtained in the usual manner. Once the nodal displacements are known, the complete finite element solution can be obtained from the interpolation functions:

$$v(s) = \left(\frac{2s^3}{L^3} - \frac{3s^2}{L^2} + 1 \quad \frac{s^3}{L^2} - \frac{2s^2}{L} + s \quad \frac{3s^2}{L^2} - \frac{2s^3}{L^3} \quad \frac{s^3}{L^2} - \frac{s^2}{L} \right) \begin{pmatrix} v_1 \\ \theta_1 \\ v_2 \\ \theta_2 \end{pmatrix} \equiv \boldsymbol{N}^T \boldsymbol{d}$$

The bending moments and shear forces can be obtained by differentiating the displacement and multiplying by EI:

$$M(s) = EI\frac{d^2v}{ds^2}; \qquad V(s) = EI\frac{d^3v}{ds^3}$$

If desired, the axial and shear stresses can be computed from the bending moment and shear forces by using the equations derived earlier. For a symmetric section with height h the maximum stresses due to a given moment and shear are as follows:

$$\text{Maximum bending stress:} \qquad \sigma_{x,\max} = \frac{M(h/2)}{I}$$

$$\text{Maximum shear stress:} \qquad \tau_{\max} = \frac{VQ_{\max}}{It}$$

where $Q_{\max}$ is the moment of half of the area about the neutral axis and t is the width of the section at the neutral axis.

Example 4.5 Find the deflection, bending moment, and shear force distribution in the three-span continuous beam shown in Figure 4.20. The point load is acting at the center of the middle span. Use the following numerical data:

$$L = 20 \text{ ft}; \qquad F = 20{,}000 \text{ lb}; \qquad E = 29 \times 10^6 \text{ lb/in}^2; \qquad \text{Section: W18} \times 40$$

The W18 × 40 section is a standard steel I-shape section with the moment of inertia $I = 612 \text{ in}^4$ and the dimensions as shown in the figure.

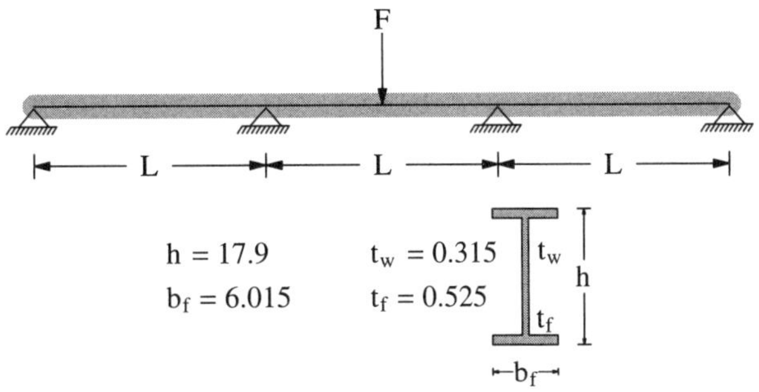

Figure 4.20. Three-span beam with I-shape cross section

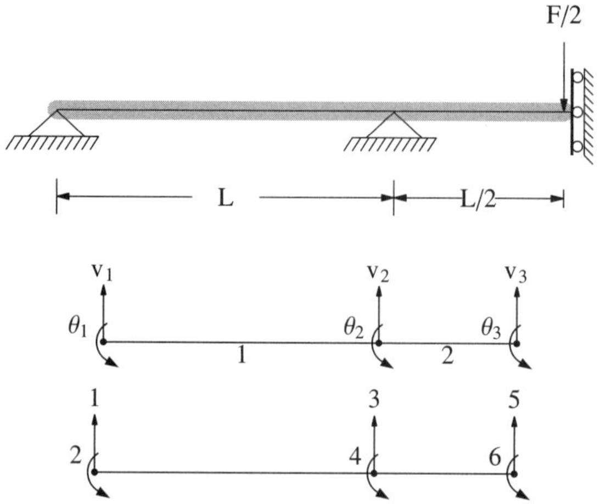

Figure 4.21. Two-element model using symmetry

For maximum shear stress, Q is the moment of the area of half of the I-section about the center. Using the given dimensions, we get

$$Q = \frac{1}{2}\left(\frac{h}{2} - t_f\right)t_w\left(\frac{h}{2} - t_f\right) + \left(\frac{h}{2} - \frac{t_f}{2}\right)b_f t_f = 38.6135\ \text{in}^3$$

Taking advantage of symmetry, we need to model only half of the beam as shown in Figure 4.21. Along the line of symmetry there is unknown vertical displacement but the rotation must be zero. Thus this boundary condition can be represented as a guided support. Also, only half of the load is carried by the symmetric half. The simplest model is a two-beam element model as shown in the figure.

Use kips (1 kip = 1000 lb) and inches. The displacements will be in inches, moments in kip = inches, and stresses in ksi.

Specified nodal loads:

Node	dof	Value
3	v_3	−10

Equations for element 1:

$$\begin{pmatrix} 15.4063 & 1848.75 & -15.4063 & 1848.75 \\ 1848.75 & 295800. & -1848.75 & 147900. \\ -15.4063 & -1848.75 & 15.4063 & -1848.75 \\ 1848.75 & 147900. & -1848.75 & 295800. \end{pmatrix} \begin{pmatrix} v_1 \\ \theta_1 \\ v_2 \\ \theta_2 \end{pmatrix} = \begin{pmatrix} 0 \\ 0 \\ 0 \\ 0 \end{pmatrix}$$

Equations for element 2:

$$\begin{pmatrix} 123.25 & 7395. & -123.25 & 7395. \\ 7395. & 591600. & -7395. & 295800. \\ -123.25 & -7395. & 123.25 & -7395. \\ 7395. & 295800. & -7395. & 591600. \end{pmatrix} \begin{pmatrix} v_2 \\ \theta_2 \\ v_3 \\ \theta_3 \end{pmatrix} = \begin{pmatrix} 0 \\ 0 \\ 0 \\ 0 \end{pmatrix}$$

Adding element equations into appropriate locations, we have

$$\begin{pmatrix} 15.4063 & 1848.75 & -15.4063 & 1848.75 & 0 & 0 \\ 1848.75 & 295800. & -1848.75 & 147900. & 0 & 0 \\ -15.4063 & -1848.75 & 138.656 & 5546.25 & -123.25 & 7395. \\ 1848.75 & 147900. & 5546.25 & 887400. & -7395. & 295800. \\ 0 & 0 & -123.25 & -7395. & 123.25 & -7395. \\ 0 & 0 & 7395. & 295800. & -7395. & 591600. \end{pmatrix} \begin{pmatrix} v_1 \\ \theta_1 \\ v_2 \\ \theta_2 \\ v_3 \\ \theta_3 \end{pmatrix} = \begin{pmatrix} 0 \\ 0 \\ 0 \\ 0 \\ -10 \\ 0 \end{pmatrix}$$

Essential boundary conditions:

Node	dof	Value
1	v_1	0
2	v_2	0
3	θ_3	0

Remove {1, 3, 6} rows and columns:

$$\begin{pmatrix} 295800. & 147900. & 0 \\ 147900. & 887400. & -7395. \\ 0 & -7395. & 123.25 \end{pmatrix} \begin{pmatrix} \theta_1 \\ \theta_2 \\ v_3 \end{pmatrix} = \begin{pmatrix} 0 \\ 0 \\ -10. \end{pmatrix}$$

Solving the final system of global equations, we get

$$\{\theta_1 = 0.000811359, \theta_2 = -0.00162272, v_3 = -0.178499\}$$

Solution for element 1:

Left node: $x_1 = 0$; Right node: $x_2 = 240$.

Local coordinate: $s = x - x_1 = x$; Length, $L = 240$.

Using these s and L values, the interpolation functions in terms of x are

$$\begin{aligned} \boldsymbol{N}^T = \{&1.44676 \times 10^{-7}x^3 - 0.0000520833x^2 + 1, \\ &0.0000173611x^3 - 0.00833333x^2 + x, \\ &0.0000520833x^2 - 1.44676 \times 10^{-7}x^3, \\ &0.0000173611x^3 - 0.00416667x^2\} \end{aligned}$$

Nodal values: $\boldsymbol{d}^T = \{0, 0.000811359, 0, -0.00162272\}$

$v(x) = \boldsymbol{N}^T\boldsymbol{d} = 0.000811359x - 1.40861 \times 10^{-8}x^3$

$E = 29000$; $I = 612$; $h = 17.9$; $Q = 38.6135$; $t = 0.315$

$M(x) = EI\, d^2v/dx^2 = -1.5x$

$V(x) = dM/dx = -1.5$

$\sigma = M(h/2)/I = -0.0219363x$

$\tau = VQ/(It) = -0.300447$

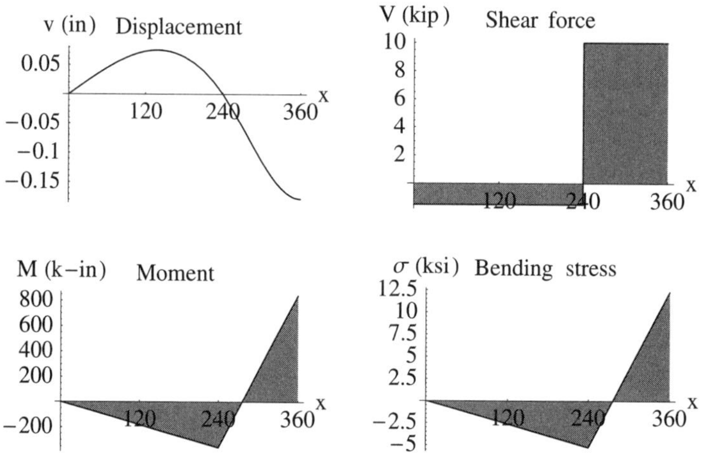

Figure 4.22. Solution of three-span beam

Solution for element 2:

Left node: $x_1 = 240.$; Right node: $x_2 = 360.$

Local coordinate: $s = x - x_1 = x - 240.$; Length, $L = 120.$

Using these s and L values, the interpolation functions in terms of x are

$$N^T = \{1.15741 \times 10^{-6}(x - 240.)^3 - 0.000208333(x - 240.)^2 + 1,$$
$$0.0000694444(x - 240.)^3 - 0.0166667(x - 240.)^2 + x - 240.,$$
$$0.000208333(x - 240.)^2 - 1.15741 \times 10^{-6}(x - 240.)^3,$$
$$0.0000694444(x - 240.)^3 - 0.00833333(x - 240.)^2\}$$

Nodal values: $\boldsymbol{d}^T = \{0, -0.00162272, -0.178499, 0\}$

$v(x) = \boldsymbol{N}^T\boldsymbol{d} = 9.39073 \times 10^{-8}x^3 - 0.0000777552x^2 + 0.0194726x - 1.4929$

$E = 29000$; $I = 612$; $h = 17.9$; $Q = 38.6135$; $t = 0.315$

$M(x) = EI\, d^2v/dx^2 = 10.x - 2760.$

$V(x) = dM/dx = 10.$

$\sigma = M(h/2)/I = 0.146242x - 40.3627$

$\tau = VQ/(It) = 2.00298$

The deflection, shear force, bending moment, and bending stress diagrams are shown in Figure 4.22. From direct integration of the governing differential equation, it can easily be verified that this is the exact solution. Thus there is no need to use a refined model.

Example 4.6 An overhanging beam is connected to a cantilever beam through a simple pin connection as shown in Figure 4.23. Analyze the beam if the right support settles downward by 10 mm. Use $L = 3$ m and $EI = 180$ MN $\cdot$ m^2.

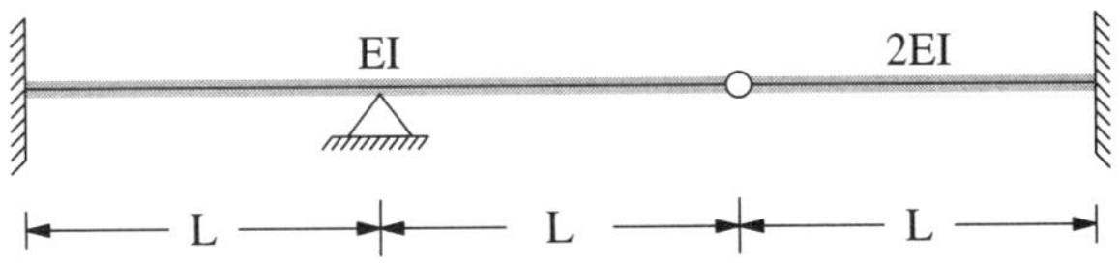

Figure 4.23. Beams connected through a pin connection

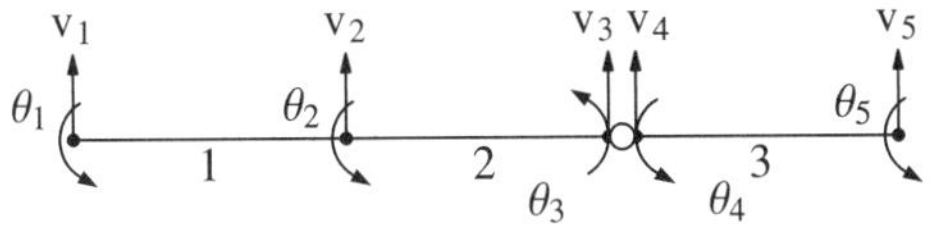

Figure 4.24. Finite element model of beams connected through a pin connection

A three-element model is adequate for this beam. Modeling the internal pin connection requires special consideration because the rotations at either side of the pin are independent. Placing only a single node at this location and using the usual assembly process will imply continuity of rotations across the pin, which obviously is not the behavior of a pin connection. A way to model the pin is to place two separate nodes at the pin location and use a multipoint constraint to link the displacements at these two nodes while keeping the rotations independent. Thus the finite element model consists of five nodes and three elements as shown in Figure 4.24. Nodes 2 and 3 are physically at the same location. Element 2 is connected to nodes 2 and 3 while element 3 is connected to nodes 4 and 5.

We use meganewton-millimeters units for numerical calculations. Substituting given numerical values into the beam element equations, we have the following element equations:

Element 1:
$$\begin{pmatrix} \frac{2}{25} & 120 & -\frac{2}{25} & 120 \\ 120 & 240000 & -120 & 120000 \\ -\frac{2}{25} & -120 & \frac{2}{25} & -120 \\ 120 & 120000 & -120 & 240000 \end{pmatrix} \begin{pmatrix} v_1 \\ \theta_1 \\ v_2 \\ \theta_2 \end{pmatrix} = \begin{pmatrix} 0 \\ 0 \\ 0 \\ 0 \end{pmatrix}$$

Element 2:
$$\begin{pmatrix} \frac{2}{25} & 120 & -\frac{2}{25} & 120 \\ 120 & 240000 & -120 & 120000 \\ -\frac{2}{25} & -120 & \frac{2}{25} & -120 \\ 120 & 120000 & -120 & 240000 \end{pmatrix} \begin{pmatrix} v_2 \\ \theta_2 \\ v_3 \\ \theta_3 \end{pmatrix} = \begin{pmatrix} 0 \\ 0 \\ 0 \\ 0 \end{pmatrix}$$

Element 3:
$$\begin{pmatrix} \frac{4}{25} & 240 & -\frac{4}{25} & 240 \\ 240 & 480000 & -240 & 240000 \\ -\frac{4}{25} & -240 & \frac{4}{25} & -240 \\ 240 & 240000 & -240 & 480000 \end{pmatrix} \begin{pmatrix} v_4 \\ \theta_4 \\ v_5 \\ \theta_5 \end{pmatrix} = \begin{pmatrix} 0 \\ 0 \\ 0 \\ 0 \end{pmatrix}$$

Assembling the elements equations in the usual way, the global equations are as follows:

$$\begin{pmatrix} \frac{2}{25} & 120 & -\frac{2}{25} & 120 & 0 & 0 & 0 & 0 & 0 & 0 \\ 120 & 240000 & -120 & 120000 & 0 & 0 & 0 & 0 & 0 & 0 \\ -\frac{2}{25} & -120 & \frac{4}{25} & 0 & -\frac{2}{25} & 120 & 0 & 0 & 0 & 0 \\ 120 & 120000 & 0 & 480000 & -120 & 120000 & 0 & 0 & 0 & 0 \\ 0 & 0 & -\frac{2}{25} & -120 & \frac{2}{25} & -120 & 0 & 0 & 0 & 0 \\ 0 & 0 & 120 & 120000 & -120 & 240000 & 0 & 0 & 0 & 0 \\ 0 & 0 & 0 & 0 & 0 & 0 & \frac{4}{25} & 240 & -\frac{4}{25} & 240 \\ 0 & 0 & 0 & 0 & 0 & 0 & 240 & 480000 & -240 & 240000 \\ 0 & 0 & 0 & 0 & 0 & 0 & -\frac{4}{25} & -240 & \frac{4}{25} & -240 \\ 0 & 0 & 0 & 0 & 0 & 0 & 240 & 240000 & -240 & 480000 \end{pmatrix} \begin{pmatrix} v_1 \\ \theta_1 \\ v_2 \\ \theta_2 \\ v_3 \\ \theta_3 \\ v_4 \\ \theta_4 \\ v_5 \\ \theta_5 \end{pmatrix} = \begin{pmatrix} 0 \\ 0 \\ 0 \\ 0 \\ 0 \\ 0 \\ 0 \\ 0 \\ 0 \\ 0 \end{pmatrix}$$

The known boundary conditions are $v_1 = \theta_1 = v_2 = 0$, $v_5 = -10$ mm, and $\theta_5 = 0$. The zero boundary conditions are incorporated by simply removing corresponding rows and columns to get

$$\begin{pmatrix} 480000 & -120 & 120000 & 0 & 0 & 0 \\ -120 & \frac{2}{25} & -120 & 0 & 0 & 0 \\ 120000 & -120 & 240000 & 0 & 0 & 0 \\ 0 & 0 & 0 & \frac{4}{25} & 240 & -\frac{4}{25} \\ 0 & 0 & 0 & 240 & 480000 & -240 \\ 0 & 0 & 0 & -\frac{4}{25} & -240 & \frac{4}{25} \end{pmatrix} \begin{pmatrix} \theta_2 \\ v_3 \\ \theta_3 \\ v_4 \\ \theta_4 \\ v_5 \end{pmatrix} = \begin{pmatrix} 0 \\ 0 \\ 0 \\ 0 \\ 0 \\ 0 \end{pmatrix}$$

Setting $v_5 = -10$ means we remove the last row and move -10 times the last column to the right-hand side. Thus the final system of equations is as follows:

$$\begin{pmatrix} 480000 & -120 & 120000 & 0 & 0 \\ -120 & \frac{2}{25} & -120 & 0 & 0 \\ 120000 & -120 & 240000 & 0 & 0 \\ 0 & 0 & 0 & \frac{4}{25} & 240 \\ 0 & 0 & 0 & 240 & 480000 \end{pmatrix} \begin{pmatrix} \theta_2 \\ v_3 \\ \theta_3 \\ v_4 \\ \theta_4 \end{pmatrix} = \begin{pmatrix} 0 \\ 0 \\ 0 \\ -\frac{8}{5} \\ -2400 \end{pmatrix}$$

The solution for nodal unknowns can now be obtained by using the Lagrange multiplier method to impose the multipoint constraint as follows:

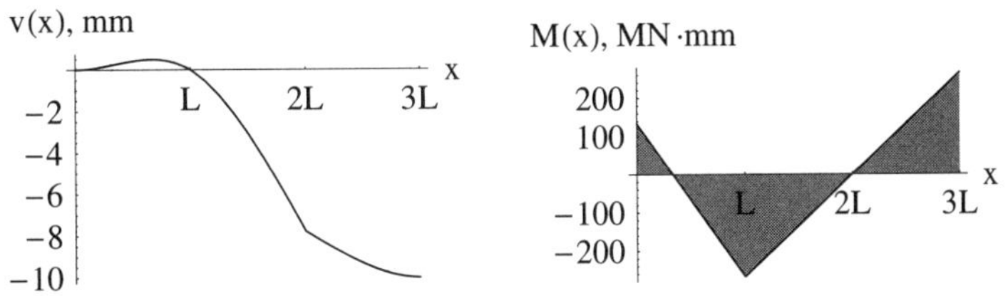

Figure 4.25. Displacement and bending moment plots

Augmented system of equations:

$$\begin{pmatrix} 480000 & -120 & 120000 & 0 & 0 & 0 \\ -120 & \frac{2}{25} & -120 & 0 & 0 & 1 \\ 120000 & -120 & 240000 & 0 & 0 & 0 \\ 0 & 0 & 0 & \frac{4}{25} & 240 & -1 \\ 0 & 0 & 0 & 240 & 480000 & 0 \\ 0 & 1 & 0 & -1 & 0 & 0 \end{pmatrix} \begin{pmatrix} \theta_2 \\ v_3 \\ \theta_3 \\ v_4 \\ \theta_4 \\ \lambda \end{pmatrix} = \begin{pmatrix} 0 \\ 0 \\ 0 \\ -\frac{8}{5} \\ -2400 \\ 0 \end{pmatrix}$$

Solution:

$$\left\{\theta_2 = -\frac{1}{900}, v_3 = -\frac{70}{9}, \theta_3 = -\frac{1}{300}, v_4 = -\frac{70}{9}, \theta_4 = -\frac{1}{900}, \lambda = \frac{4}{45}\right\}$$

The following complete solution over elements can be computed in the usual way. Note that the magnitude of the Lagrange multiplier is equal to the shear force at the pin. This is no coincidence. In structural problems a Lagrange multiplier is interpreted as the force necessary to impose the constraint. Here the constraint is on the transverse displacement across the pin. Thus the Lagrange multiplier will have an interpretation of force to maintain the transverse displacement continuity. This force is exactly the shear force at this location, and hence the value of the Lagrange multiplier is equal to the shear force. If a constraint involved rotations, the corresponding Lagrange multiplier will be equal to the moment at that location. Plots of displacement and bending moment are shown in Figure 4.25. The displacement plot clearly shows that the slope is not continuous across the pin and hence the internal hinge is modeled correctly:

	Displacement, $v(x)$	Moment, $M(x)$	Shear, V
1	$\frac{x^2}{2700000} - \frac{x^3}{8100000000}$	$\frac{400}{3} - \frac{2x}{15}$	$-\frac{2}{15}$
2	$\frac{x^3}{12150000000} - \frac{x^2}{675000} + \frac{x}{180} - \frac{50}{9}$	$\frac{4x}{45} - \frac{1600}{3}$	$\frac{4}{45}$
3	$\frac{x^3}{24300000000} - \frac{x^2}{1350000} + \frac{x}{300} - 10$	$\frac{4x}{45} - \frac{1600}{3}$	$\frac{4}{45}$

Example 4.7 Find the deflection, bending moment, and shear force distribution in the nonuniform beam shown in Figure 4.26. Use the following numerical data:

$L = 2$ m; $F = 18$ kN; $q = 10$ kN/m; $E = 210$ GPa; $I = 4 \times 10^{-4}$ m^4

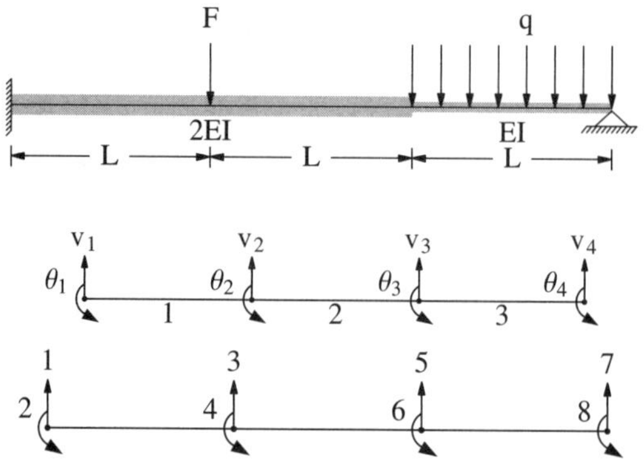

Figure 4.26. Nonuniform beam and three-element model

Placing a node at the concentrated load location and at the location of change in cross section, we have the three-element model shown in the figure.

Using kilonewtons and meters, the solution is as follows:

Specified nodal loads:

Node	dof	Value
2	v_2	−18

Equations for element 1:

$$\begin{pmatrix} 252000. & 252000. & -252000. & 252000. \\ 252000. & 336000. & -252000. & 168000. \\ -252000. & -252000. & 252000. & -252000. \\ 252000. & 168000. & -252000. & 336000. \end{pmatrix} \begin{pmatrix} v_1 \\ \theta_1 \\ v_2 \\ \theta_2 \end{pmatrix} = \begin{pmatrix} 0 \\ 0 \\ 0 \\ 0 \end{pmatrix}$$

Equations for element 2:

$$\begin{pmatrix} 252000. & 252000. & -252000. & 252000. \\ 252000. & 336000. & -252000. & 168000. \\ -252000. & -252000. & 252000. & -252000. \\ 252000. & 168000. & -252000. & 336000. \end{pmatrix} \begin{pmatrix} v_2 \\ \theta_2 \\ v_3 \\ \theta_3 \end{pmatrix} = \begin{pmatrix} 0 \\ 0 \\ 0 \\ 0 \end{pmatrix}$$

Equations for element 3:

$$E = 210000000; \qquad I = \frac{1}{2500}; \qquad q = -10$$

$$\begin{pmatrix} 126000. & 126000. & -126000. & 126000. \\ 126000. & 168000. & -126000. & 84000. \\ -126000. & -126000. & 126000. & -126000. \\ 126000. & 84000. & -126000. & 168000. \end{pmatrix} \begin{pmatrix} v_3 \\ \theta_3 \\ v_4 \\ \theta_4 \end{pmatrix} = \begin{pmatrix} -10. \\ -3.33333 \\ -10. \\ 3.33333 \end{pmatrix}$$

Adding element equations into appropriate locations, we have

$$\begin{pmatrix} 252000. & 252000. & -252000. & 252000. & 0 & 0 & 0 & 0 \\ 252000. & 336000. & -252000. & 168000. & 0 & 0 & 0 & 0 \\ -252000. & -252000. & 504000. & 0 & -252000. & 252000. & 0 & 0 \\ 252000. & 168000. & 0 & 672000. & -252000. & 168000. & 0 & 0 \\ 0 & 0 & -252000. & -252000. & 378000. & -126000. & -126000. & 126000. \\ 0 & 0 & 252000. & 168000. & -126000. & 504000. & -126000. & 84000. \\ 0 & 0 & 0 & 0 & -126000. & -126000. & 126000. & -126000. \\ 0 & 0 & 0 & 0 & 126000. & 84000. & -126000. & 168000. \end{pmatrix} \begin{pmatrix} v_1 \\ \theta_1 \\ v_2 \\ \theta_2 \\ v_3 \\ \theta_3 \\ v_4 \\ \theta_4 \end{pmatrix} = \begin{pmatrix} 0 \\ 0 \\ -18 \\ 0 \\ -10. \\ -3.33333 \\ -10. \\ 3.33333 \end{pmatrix}$$

Essential boundary conditions:

Node	dof	Value
1	v_1	0
	θ_1	0
4	v_4	0

Remove {1, 2, 7} rows and columns:

$$\begin{pmatrix} 504000. & 0 & -252000. & 252000. & 0 \\ 0 & 672000. & -252000. & 168000. & 0 \\ -252000. & -252000. & 378000. & -126000. & 126000. \\ 252000. & 168000. & -126000. & 504000. & 84000. \\ 0 & 0 & 126000. & 84000. & 168000. \end{pmatrix} \begin{pmatrix} v_2 \\ \theta_2 \\ v_3 \\ \theta_3 \\ \theta_4 \end{pmatrix} = \begin{pmatrix} -18. \\ 0 \\ -10. \\ -3.33333 \\ 3.33333 \end{pmatrix}$$

Solving the final system of global equations, we get

$$\{v_2 = -0.000213152, \theta_2 = -0.000131378, v_3 = -0.00034127,$$
$$\theta_3 = 0.0000136054, \theta_4 = 0.000268991\}$$

Solution for element 1:

Left node: $x_1 = 0$; Right node: $x_2 = 2$.

Local coordinate: $s = x - x_1 = x$; Length, $L = 2$.

Using these s and L values, the interpolation functions in terms of x are

$$N^T = \{0.25x^3 - 0.75x^2 + 1, 0.25x^3 - 1.x^2 + x, 0.75x^2 - 0.25x^3, 0.25x^3 - 0.5x^2\}$$

Nodal values: $\boldsymbol{d}^T = \{0, 0, -0.000213152, -0.000131378\}$

$v(x) = \boldsymbol{N}^T\boldsymbol{d} = 0.0000204436x^3 - 0.0000941752x^2$

$M(x) = EI\, d^2v/dx^2 = 20.6071x - 31.6429$

$V(x) = dM/dx = 20.6071$

Solution for element 2:

$v(x) = \boldsymbol{N}^T\boldsymbol{d} = 2.58645 \times 10^{-6}x^3 + 0.0000129677x^2 - 0.000214286x + 0.000142857$

$M(x) = EI\, d^2v/dx^2 = 2.60714x + 4.35714$

$V(x) = dM/dx = 2.60714$

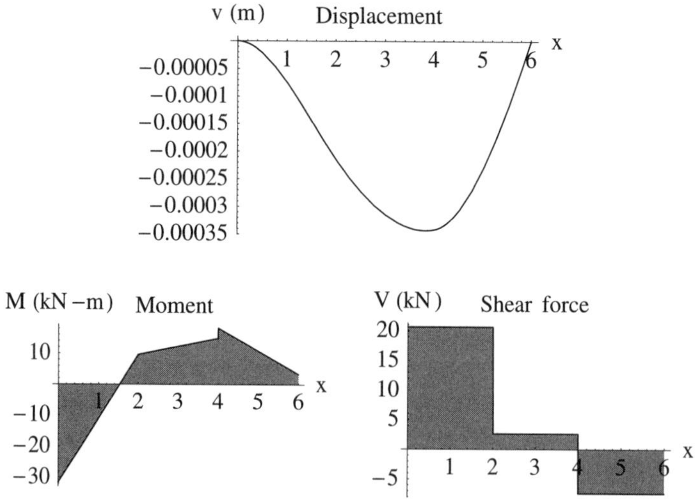

Figure 4.27. Three-element solution of nonuniform beam

Solution for element 3:

$$v(x) = N^T d = -0.0000146684x^3 + 0.000283872x^2 - 0.00155329x + 0.00226871$$

$$M(x) = EI\, d^2v/dx^2 = 47.6905 - 7.39286x$$

$$V(x) = dM/dx = -7.39286$$

The deflection, bending moment, and shear force diagrams are shown in Figure 4.27. The exact solution in the distributed load segment should show a linear shear diagram and a quadratic bending moment diagram. However, since the element is based on a cubic interpolation, it gives a constant shear and a linear moment diagram, and therefore the solution for moments and shears is not very good.

To get accurate results, we could use more elements in the segment with the distributed load. Using five elements in the distributed load segment, we get the moment and shear force diagrams shown in Figure 4.28 that are clearly a lot closer to the expected solution. However, for uniform beams it is not necessary to use more than one element per span. A simple superposition procedure is presented in the following section that gives exact solution without having to use many elements in the loaded span. That procedure obviously is preferred for practical applications.

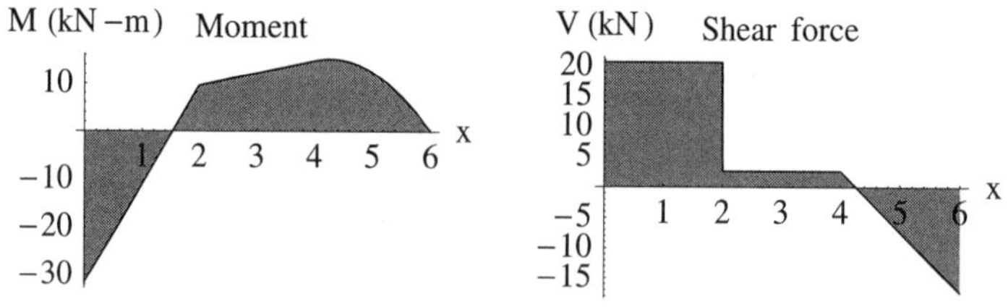

Figure 4.28. Solution with five elements in the distributed load segment

4.7 UNIFORM BEAMS SUBJECTED TO DISTRIBUTED LOADS

As seen in Example 4.7, the simple two-node beam element does not give correct moments and shears for elements with distributed loads. The standard finite element approach for improving a solution is to use more elements. As demonstrated in the previous example, the approach works, and as we increase the number of elements in the loaded span, the results do get closer to the exact solution. However, for uniform beams there is a much simpler alternative based on the principle of superposition.

The idea exploits the fact that the two-node beam element, based on the cubic interpolation functions, gives exact solution for uniform beams subjected to concentrated loads. This was demonstrated in Example 4.5. The finite element formulation converts the distributed load to equivalent loads and moments at the nodes. The finite element solution is therefore exact for these equivalent concentrated loads and moments. The approximations are thus limited to the span that is loaded with the distributed load. This observation suggests a two-step solution procedure. In the first step we follow the usual finite element procedure, with only one element even for spans with distributed loads. In the second step the span with the distributed load is isolated and analyzed separately. Both ends of this isolated span are assigned fixed-end boundary conditions so that the solution will not affect the displacements outside of this span. This is known as the *fixed-end beam solution*. The final solution is the superposition of the finite element solution and the fixed-end beam solutions for spans with distributed loads.

In actual calculations, we do not even have to think of the process as a two-step process. We simply create a finite element model based on the usual considerations of supports, changes in the cross section, and concentrated load locations. After solving for the nodal displacements in the usual way, we compute the displacement solution over an element from the following equation:

$$v(s) = \begin{pmatrix} \frac{2s^3}{L^3} - \frac{3s^2}{L^2} + 1 & \frac{s^3}{L^2} - \frac{2s^2}{L} + s & \frac{3s^2}{L^2} - \frac{2s^3}{L^3} & \frac{s^3}{L^2} - \frac{s^2}{L} \end{pmatrix} \begin{pmatrix} v_1 \\ \theta_1 \\ v_2 \\ \theta_2 \end{pmatrix} + v_f$$

where v_f is a fixed-end beam solution due to a distributed load on an element. In the following sections we derive fixed-end beam solutions for uniform and trapezoidal loading and illustrate the process of computing element solutions.

Fixed-End Beam Solution with Uniform Loading For a fixed-end beam with uniformly distributed load q, shown in Figure 4.29, the exact solution can easily be obtained by direct integration of the governing differential equation as follows:

Governing differential equation: $$EI\frac{d^4v}{ds^4} - q = 0; \qquad 0 < s < L$$

Boundary conditions: $$v(0) = 0; \quad \frac{dv(0)}{dx} = 0; \quad v(L) = 0; \quad \frac{dv(L)}{dx} = 0$$

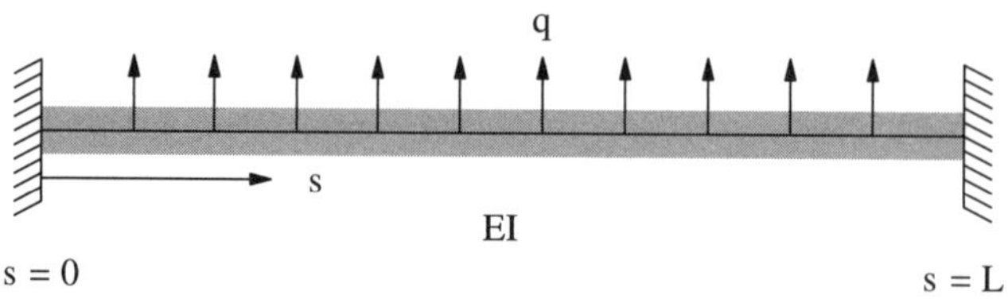

Figure 4.29. A fixed-end beam subjected to uniformly distributed load

Integrating four times and introducing four integration constants $c_1, \ldots, c_4$, we have

$$v(s) = \frac{qs^4}{24EI} + c_4 s^3 + c_3 s^2 + c_2 s + c_1$$

Using the four boundary conditions, we have

$$c_1 = 0; \qquad c_2 = 0; \qquad \frac{qL^4}{24EI} + c_4 L^3 + c_3 L^2 + c_2 L + c_1 = 0; \qquad \frac{qL^3}{6EI} + 3c_4 L^2 + 2c_3 L + c_2 = 0$$

Solving these equations, the integration constants are

$$c_1 = 0; \qquad c_2 = 0; \qquad c_3 = \frac{L^2 q}{24\text{EI}}; \qquad c_4 = -\frac{Lq}{12EI}$$

Thus the exact solution for a fixed-end beam with uniformly distributed loading is as follows:

$$v_f(s) = \frac{q(L-s)^2 s^2}{24EI}$$

The subscript f is used to indicate that this is the fixed-end beam solution.

Example 4.8 For the beam in Example 4.7 use superposition with only three elements to find the exact solution for the deflection, moment, and shear in the beam.

The first step is to perform a standard finite element analysis. The equivalent nodal loads on the finite element model are shown in Figure 4.30. As shown through detailed calculations in Example 4.7, the nodal solution is as follows:

Complete Table of Nodal Values

	v	θ
1	0	0
2	−0.000213152	−0.000131378
3	−0.00034127	0.0000136054
4	0	0.000268991

Using the two-node beam interpolation functions, the following displacements were also computed in Example 4.7 for the three elements:

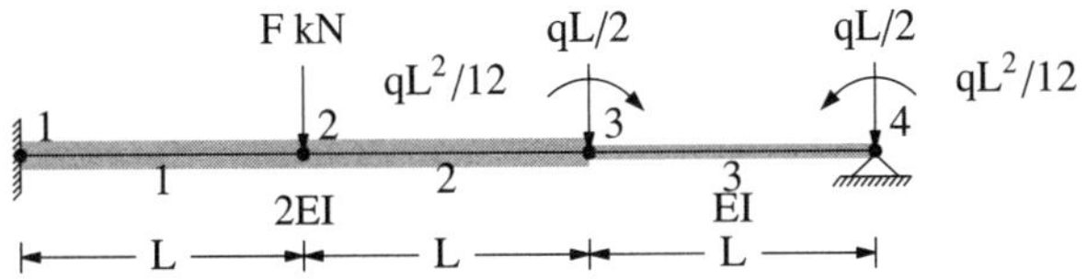

Figure 4.30. Equivalent concentrated loading on the finite element model

Solution for element 1:

$$v(x) = \boldsymbol{N}^T\boldsymbol{d} = 0.0000204436x^3 - 0.0000941752x^2$$

Solution for element 2:

$$v(x) = \boldsymbol{N}^T\boldsymbol{d} = 2.58645 \times 10^{-6}x^3 + 0.0000129677x^2 - 0.000214286x + 0.000142857$$

Solution for element 3:

$$v(x) = \boldsymbol{N}^T\boldsymbol{d} = -0.0000146684x^3 + 0.000283872x^2 - 0.00155329x + 0.00226871$$

For the first two elements these are the exact solutions. For the third element, in order to obtain the exact solution, we must add the fixed-end beam solution. Substituting the numerical data, the fixed-end solution for this span is obtained as follows:

$$v_f(s) = -\frac{(2-s)^2 s^2}{201600}$$

For the third element:

Left node: $x_1 = 4.$; Right node: $x_2 = 6.$

Local coordinate: $s = x - x_1 = x - 4.$; Length, $L = 2.$

Thus

$$v_f(x) = -\frac{(6-x)^2(x-4)^2}{201600}$$

Therefore the exact solution for this element is

$$v(x) = N^T d + v_f(x) = -0.0000146684x^3 + 0.000283872x^2 - 0.00155329x$$
$$+ 0.00226871 + \left(-\frac{(6-x)^2(x-4)^2}{201600}\right)$$

giving

$$v(x) = -4.96032 \times 10^{-6}x^4 + 0.000084538x^3 - 0.000450255x^2$$
$$+ 0.000827664x - 0.000588435$$

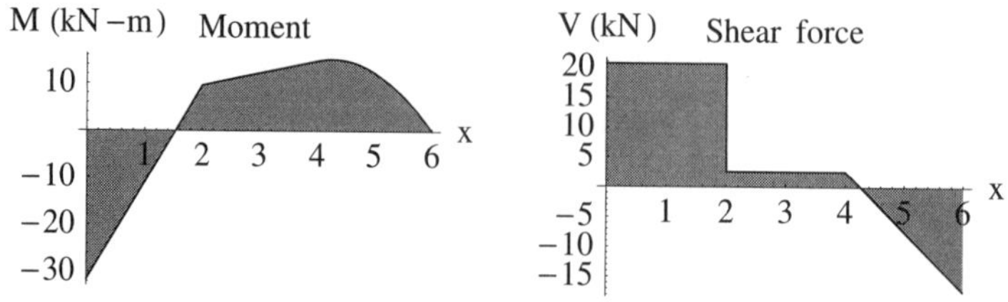

Figure 4.31. Exact three-element solution for beam with distributed load

This is a fourth-order solution and hence will result in a quadratic bending moment diagram and linear shear force:

$$M(x) = EI\, d^2v/dx^2 = -5.x^2 + 42.6071x - 75.6429$$

$$V(x) = EI\, d^3v/dx^3 = 42.6071 - 10.x$$

The bending moment and shear force diagrams are shown in Figure 4.31. They clearly show that the solution is exact.

◆ *Mathematica*/MATLAB Implementation 4.3 on the Book Web Site: *Analysis of beams*

Fixed-End Beam Solution with Trapezoidal Loading In examples so far only the uniformly distributed beam loading has been considered. However, any other loading can easily be handled in exactly the same manner. We must derive an equivalent load vector to use in the finite element analysis. Also we must obtain an exact fixed-end beam solution for use in the superposition to get exact solutions for loaded spans.

Consider a beam element with load varying linearly from a value q_1 at the first node of the element to a value q_2 at the second node, as shown in Figure 4.32. Using linear Lagrange interpolation functions, this loading can be expressed as follows:

$$q(s) = \frac{sq_2}{L} - \frac{(s-L)q_1}{L}$$

Using the beam interpolation functions and the above $q(s)$, the equivalent load vector can be obtained as follows:

$$\boldsymbol{r}_q = \int_0^L \boldsymbol{N} q\, ds = \begin{pmatrix} \int_0^L q(s)\left(1 - \frac{3s^2}{L^2} + \frac{2s^3}{L^3}\right) ds \\ \int_0^L q(s)\left(s - \frac{2s^2}{L} + \frac{s^3}{L^2}\right) ds \\ \int_0^L q(s)\left(\frac{3s^2}{L^2} - \frac{2s^3}{L^3}\right) ds \\ \int_0^L q(s)\left(-\frac{s^2}{L} + \frac{s^3}{L^2}\right) ds \end{pmatrix} = \begin{pmatrix} \frac{1}{20}L(7q_1 + 3q_3) \\ \frac{1}{60}L^2(3q_1 + 2q_2) \\ \frac{1}{20}L(3q_1 + 7q_2) \\ -\frac{1}{60}L^2(2q_1 + 3q_2) \end{pmatrix}$$

Interpreting terms in the vector $\boldsymbol{r}_q$, we see that a trapezoidal loading on an element is actually applied as nodal loads and moments, as shown in Figure 4.32.

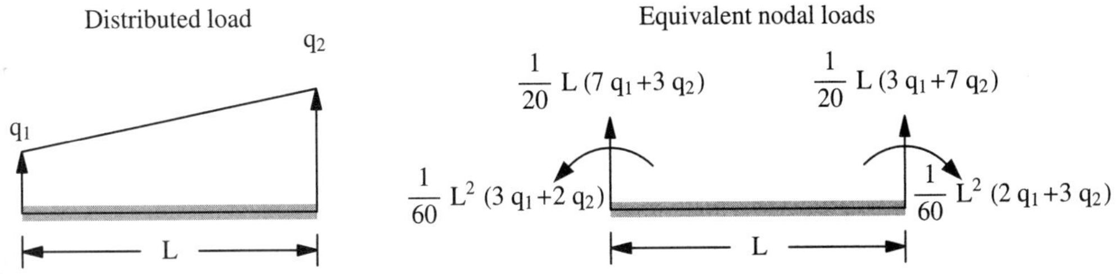

Figure 4.32. Equivalent nodal loads due to a trapezoidal distributed load on an element

For use in superposition, the fixed-end solution with trapezoidal distributed load $q(s)$ can be obtained by solving the governing differential equation as follows:

Governing differential equation: $\quad EI\dfrac{d^4v}{ds^4} - q(s) = 0; \qquad 0 < s < L$

Boundary conditions: $\quad v(0) = 0; \quad \dfrac{dv(0)}{dx} = 0; \quad v(L) = 0; \quad \dfrac{dv(L)}{dx} = 0$

Integrating four times and introducing four integration constants $c_1, \ldots, c_4$, we have

$$v(s) = c_4s^3 + c_3s^2 + c_2s + c_1 - \frac{(1/120)s^5(q_1 - q_2) - (1/24)Ls^4q_1}{EIL}$$

Using the four boundary conditions, we have

$$c_1 = 0; \qquad c_2 = 0$$

$$c_4L^3 + c_3L^2 + c_2L + c_1 - \frac{(1/120)L^5(q_1 - q_2) - (L^5q_1/24)}{EIL} = 0$$

$$3c_4L^2 + 2c_3L + c_2 - \frac{(1/24)L^4(q_1 - q_2) - (L^4q_1/6)}{EIL} = 0$$

Solving these equations, the integration constants are

$$c_1 = 0; \qquad c_2 = 0; \qquad c_3 = -\frac{-3q_1L^2 - 2q_2L^2}{120EI}; \qquad c_4 = -\frac{L(7q_1 + 3q_2)}{120EI}$$

Thus the fixed-end beam solution for trapezoidal distributed loading is as follows:

$$v_f(s) = \frac{(L - s)^2s^2((3L - s)q_1 + (2L + s)q_2)}{120EIL}$$

The subscript f is used to indicate that this is the fixed-end beam solution.

Example 4.9 Consider a uniform beam subjected to triangular loading. The beam is simply supported at left end and is spring supported at the right end, as shown in Figure 4.33.

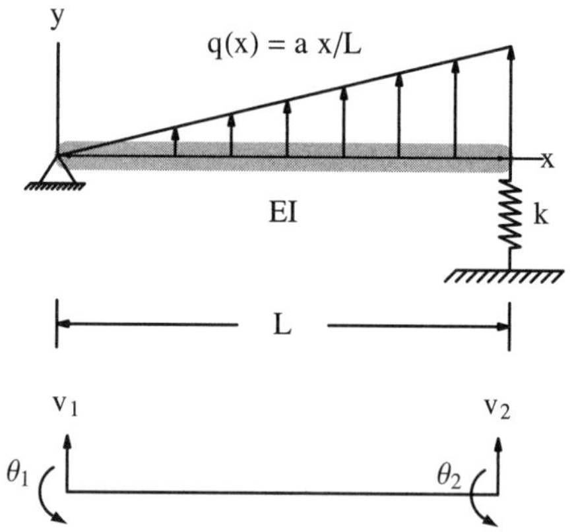

Figure 4.33. Uniform beam subjected to a linearly increasing load

Use superposition to get the exact solution using only one element. Assume the following numerical data:

$$a = 5 \text{ kN/m}; \qquad L = 5 \text{ m}; \qquad E = 200 \text{ GPa}; \qquad I = 10^{-6} \text{ m}^4; \qquad k = 275 \text{ N/mm}$$

Use kilonewton-meters:

$$E = 200 \times 10^6 \text{ kN/m}^2; \qquad I = 10^{-6} \text{ m}^4; \qquad k = 275 \text{ kN/m}$$

For the given triangular load $q_1 = 0$ and $q_2 = 5$. Thus the equivalent nodal load vector is as follows:

$$\frac{(7q_1 + 3q_2)L}{20} = \frac{(7 \times 0 + 3 \times 5)5}{20} = 3.75 \text{ kN}; \qquad \frac{(3q_1 + 7q_2)L}{20} = 8.75 \text{ kN}$$

$$\frac{(3q_1 + 2q_2)L^2}{60} = 4.16667 \text{ kN} \cdot \text{m} \qquad -\frac{(2q_1 + 3q_2)L^2}{60} = -6.25 \text{ kN} \cdot \text{m}$$

The equations for element 1 are as follows:

Element ends = (0, 5.); Length, $L = 5.$

$$\begin{pmatrix} 19.2 & 48. & -19.2 & 48. \\ 48. & 160. & -48. & 80. \\ -19.2 & -48. & 19.2 & -48. \\ 48. & 80. & -48. & 160. \end{pmatrix} \begin{pmatrix} v_1 \\ \theta_1 \\ v_2 \\ \theta_2 \end{pmatrix} = \begin{pmatrix} 3.75 \\ 4.16667 \\ 8.75 \\ -6.25 \end{pmatrix}$$

The specified nodal spring stiffnesses are as follows:

Node	dof	Value
2	v_2	275

The spring contributes to the v_2 degree of freedom. Adding the spring constant to the diagonal term of the third row, we have the global equations as follows:

$$\begin{pmatrix} 19.2 & 48. & -19.2 & 48. \\ 48. & 160. & -48. & 80. \\ -19.2 & -48. & 294.2 & -48. \\ 48. & 80. & -48. & 160. \end{pmatrix} \begin{pmatrix} v_1 \\ \theta_1 \\ v_2 \\ \theta_2 \end{pmatrix} = \begin{pmatrix} 3.75 \\ 4.16667 \\ 8.75 \\ -6.25 \end{pmatrix}$$

After adjusting for essential boundary conditions, we have

$$\begin{pmatrix} 160. & -48. & 80. \\ -48. & 294.2 & -48. \\ 80. & -48. & 160. \end{pmatrix} \begin{pmatrix} \theta_1 \\ v_2 \\ \theta_2 \end{pmatrix} = \begin{pmatrix} 4.16667 \\ 8.75 \\ -6.25 \end{pmatrix}$$

Solving the final system of global equations, we get

$$\{\theta_1 = 0.0668245, v_2 = 0.030303, \theta_2 = -0.0633838\}$$

The solution for element 1 is as follows:

Left node: $x_1 = 0$; Right node: $x_2 = 5$.
Local coordinate: $s = x - x_1 = x$; Length, $L = 5$.
Using these s and L values, the interpolation functions in terms of x are

$$\boldsymbol{N}^T = \{0.016x^3 - 0.12x^2 + 1, 0.04x^3 - 0.4x^2 + x, 0.12x^2 - 0.016x^3, 0.04x^3 - 0.2x^2\}$$

Nodal values: $\boldsymbol{d}^T = \{0, 0.0668245, 0.030303, -0.0633838\}$
$v(x) = \boldsymbol{N}^T\boldsymbol{d} = -0.000347222x^3 - 0.0104167x^2 + 0.0668245x$
Fixed-end beam displacement, $v_f(x) = 0.0000416667(5. - x)^2x^2(x + 10.)$
Adding this value, the complete element solution is

$$v(x) = 0.0000416667x^5 - 0.00347222x^3 + 0.0668245x$$

$$M(x) = EI\, d^2v/dx^2 = 0.166667x^3 - 4.16667x$$

$$V(x) = dM/dx = 0.5x^2 - 4.16667$$

The bending moment and shear force diagrams are shown in Figure 4.34. An exact analytical solution for the problem was presented in an earlier example. Substituting the given numerical data into the analytical solution, we get

$$v(x) = \frac{ax(120EIL + k(7L^4 - 10L^2x^2 + 3x^4))}{360EIkL}$$
$$= 0.0000416667x^5 - 0.00347222x^3 + 0.0668245x$$

which is exactly the same as that obtained using the finite element analysis procedure with superposition.

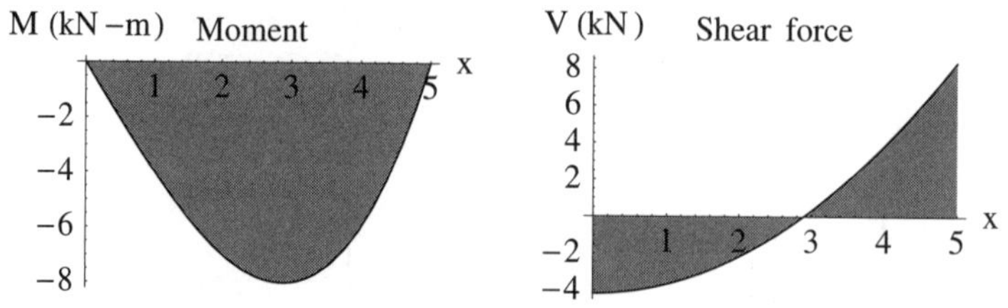

Figure 4.34. One-element solution

4.8 PLANE FRAMES

Members in a plane frame are designed to resist axial and bending deformations. The two-node beam element and the axial deformation element are combined together to form an element that can be used to analyze rigid-jointed planar frameworks. It is assumed that the axial and bending effects are uncoupled from each other, which is a reasonable assumption within the framework of small-deformation theory.

As shown in Figure 4.35, a local coordinate system is established for each element. In this local coordinate system the s axis is along the element axis, with positive direction being from node 1 to node 2. The local t axis is 90° counterclockwise from the s axis. Each node has three degrees of freedom—two translations and one rotation. The moment of inertia and the distributed transverse load are assumed constant over each element. Furthermore, concentrated loads are allowed only at the ends of an element. The distributed load q_s is in the axial direction and q_t is in the transverse direction. These applied loads on an element are positive if they act in the positive directions of the local axes.

In the local coordinate system, the element equations are simply a combination of the axial deformation element and the beam element. With the order of degrees of freedom as shown in Figure 4.35 the local element equations are therefore as follows:

$$\begin{pmatrix} \frac{EA}{L} & 0 & 0 & -\frac{EA}{L} & 0 & 0 \\ 0 & \frac{12EI}{L^3} & \frac{6EI}{L^2} & 0 & -\frac{12EI}{L^3} & \frac{6EI}{L^2} \\ 0 & \frac{6EI}{L^2} & \frac{4EI}{L} & 0 & -\frac{6EI}{L^2} & \frac{2EI}{L} \\ -\frac{EA}{L} & 0 & 0 & \frac{EA}{L} & 0 & 0 \\ 0 & -\frac{12EI}{L^3} & -\frac{6EI}{L^2} & 0 & \frac{12EI}{L^3} & -\frac{6EI}{L^2} \\ 0 & \frac{6EI}{L^2} & \frac{2EI}{L} & 0 & -\frac{6EI}{L^2} & \frac{4EI}{L} \end{pmatrix} \begin{pmatrix} d_1 \\ d_2 \\ d_3 \\ d_4 \\ d_5 \\ d_6 \end{pmatrix} = \begin{pmatrix} \frac{1}{2}q_s L \\ \frac{1}{2}q_t L \\ \frac{1}{12}q_t L^2 \\ \frac{1}{2}q_s L \\ \frac{1}{2}q_t L \\ -\frac{1}{12}q_t L^2 \end{pmatrix} \Longrightarrow \boldsymbol{k}_l \boldsymbol{d}_l = \boldsymbol{r}_l$$

Assuming a common global x-y reference coordinate system for all elements in the frame, the degrees of freedom in the global directions are denoted by u_1, v_1, θ_1, u_2, v_2, and θ_2. To develop the transformation between the global and the local degrees of freedom, we can see from Figure 4.35 that the vector d_1 is related to vectors $u_1$and v_1 as follows:

$$u_1 = d_1 \cos\alpha \equiv d_1 l_s$$

$$v_1 = d_1 \cos(90 - \alpha) \equiv d_1 \sin\alpha \equiv d_1 m_s$$

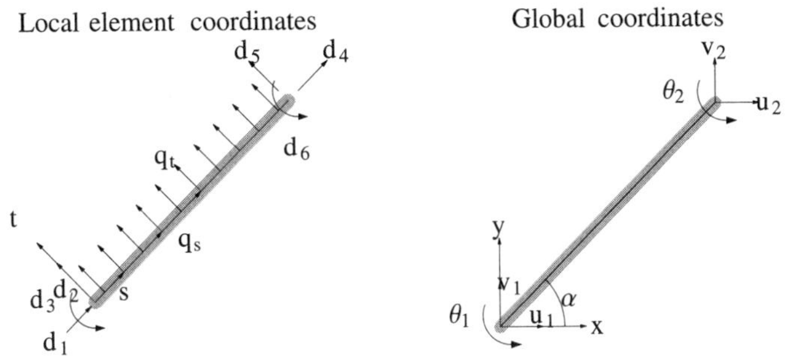

Figure 4.35. Plane frame element

where l_s is the cosine of the angle between the element s axis and the global x axis and m_s is the cosine of the angle between the element axis and the global y axis and are called the direction cosines. The angle is positive when measured from the positive x axis in the counterclockwise direction. Denoting the coordinates of the ends of the element by (x_1, y_1) and (x_2, y_2), we can determine the element length and the direction cosines as follows:

$$L = \sqrt{(x_2 - x_1)^2 + (y_2 - y_1)^2}; \qquad l_s = \cos\alpha = \frac{x_2 - x_1}{L};$$

$$m_s = \cos(90 - \alpha) = \sin\alpha = \frac{y_2 - y_1}{L}$$

Multiplying u_1 by l_s and v_1 by m_s and adding the two equations together give transformation from global to local degrees of freedom as follows:

$$u_1 l_s + v_1 m_s = d_1 \cos^2\alpha + d_1 \sin^2\alpha \equiv d_1$$

Similarly,

$$\begin{aligned} u_1 &= -d_2 \cos(90 - \alpha) \equiv -d_2 m_s \\ v_1 &= d_2 \cos\alpha \equiv d_2 l_s \\ &\Longrightarrow -u_1 m_s + v_1 l_s = d_2 \sin^2\alpha + d_2 \cos^2\alpha \equiv d_2 \end{aligned}$$

The rotation θ is not affected by this transformation and thus

$$\theta_1 = d_3$$

Therefore the three degrees of freedom at node 1 are transformed as follows:

$$\begin{pmatrix} d_1 \\ d_2 \\ d_3 \end{pmatrix} = \begin{pmatrix} l_s & m_s & 0 \\ -m_s & l_s & 0 \\ 0 & 0 & 1 \end{pmatrix} \begin{pmatrix} u_1 \\ v_1 \\ \theta_1 \end{pmatrix}$$

The same relationship holds for the degrees of freedom at node 2. Thus the transformation between the global and the local degrees of freedom can be written as follows:

$$\begin{pmatrix} d_1 \\ d_2 \\ d_3 \\ d_4 \\ d_5 \\ d_6 \end{pmatrix} = \begin{pmatrix} l_s & m_s & 0 & 0 & 0 & 0 \\ -m_s & l_s & 0 & 0 & 0 & 0 \\ 0 & 0 & 1 & 0 & 0 & 0 \\ 0 & 0 & 0 & l_s & m_s & 0 \\ 0 & 0 & 0 & -m_s & l_s & 0 \\ 0 & 0 & 0 & 0 & 0 & 1 \end{pmatrix} \begin{pmatrix} u_1 \\ v_1 \\ \theta_1 \\ u_2 \\ v_2 \\ \theta_2 \end{pmatrix} \Longrightarrow \boldsymbol{d}_l = \boldsymbol{T}\boldsymbol{d}$$

$$L = \sqrt{(x_2 - x_1)^2 + (y_2 - y_1)^2}; \qquad l_s = \cos\alpha = \frac{x_2 - x_1}{L}; \qquad m_s = \sin\alpha = \frac{y_2 - y_1}{L}$$

where α is the angle between the local s axis and the global x axis measured counterclockwise, (x_1, y_1) and (x_2, y_2) are the coordinates of the two nodes at the element ends, and L is the length of the element. It is easy to verify that the inverse transformation (i.e., local to global transformation) is given simply by the transpose of the $\boldsymbol{T}$ matrix,

$$\boldsymbol{d} = \boldsymbol{T}^T \boldsymbol{d}_l$$

Using this transformation, the element equations in the local coordinate system can be related to those in the global coordinate system as follows:

$$\boldsymbol{k}_l \boldsymbol{d}_l = \boldsymbol{r}_l \Longrightarrow \boldsymbol{k}_l \boldsymbol{T}\boldsymbol{d} = \boldsymbol{r}_l$$

Multiplying both sides by $\boldsymbol{T}^T$, we get

$$\boldsymbol{T}^T \boldsymbol{k}_l \boldsymbol{T}\boldsymbol{d} = \boldsymbol{T}^T \boldsymbol{r}_l$$

Noting that $\boldsymbol{T}^T \boldsymbol{r}_l$ is the transformation of applied loads from the local to the global coordinates, we have the following element equations in terms of global degrees of freedom and applied nodal loads in the global directions:

$$\boldsymbol{k}\boldsymbol{d} = \boldsymbol{r}$$

where

$$\boldsymbol{k} = \boldsymbol{T}^T \boldsymbol{k}_l \boldsymbol{T} \qquad \text{and} \qquad \boldsymbol{r} = \boldsymbol{T}^T \boldsymbol{r}_l$$

It is possible to carry out the matrix multiplications and write the element equations in global coordinates explicitly. However, the resulting matrix is quite large and is not written here. The *Mathematica* implementation given at the end of this section shows the explicit expressions. For manual computations it is much easier to write the element equations in local coordinates first and then carry out the matrix multiplications using the numerical data.

The assembly and solution for global nodal unknowns follow the standard procedure. For computing the element solution, the global degrees of freedom for each element are first transformed into the local degrees of freedom by multiplying them by the $\boldsymbol{T}$ matrix for the element. These local degrees of freedom are then used with the axial deformation and beam interpolation functions to get complete displacement solutions. In terms of notation used in this section, the appropriate expressions are as follows:

Global to local transformation:

$$\boldsymbol{d}_l = \boldsymbol{Td}$$

Axial displacement:

$$u(s) = \begin{pmatrix} -\frac{s-L}{L} & \frac{s}{L} \end{pmatrix} \begin{pmatrix} d_1 \\ d_4 \end{pmatrix}; \qquad 0 \le s \le L$$

$$\boldsymbol{N}_u = \begin{pmatrix} -\frac{s-L}{L} & \frac{s}{L} \end{pmatrix}$$

Transverse displacement:

$$v(s) = \begin{pmatrix} \frac{2s^3}{L^3} - \frac{3s^2}{L^2} + 1 & \frac{s^3}{L^2} - \frac{2s^2}{L} & \frac{3s^2}{L^2} - \frac{2s^3}{L^3} & \frac{s^3}{L^2} - \frac{s^2}{L} \end{pmatrix} \begin{pmatrix} d_2 \\ d_3 \\ d_5 \\ d_6 \end{pmatrix}; \qquad 0 \le s \le L$$

$$N_v = \begin{pmatrix} \frac{2s^3}{L^3} - \frac{3s^2}{L^2} + 1 & \frac{s^3}{L^2} - \frac{2s^2}{L} & \frac{3s^2}{L^2} - \frac{2s^3}{L^3} & \frac{s^3}{L^2} - \frac{s^2}{L} \end{pmatrix}$$

Fixed-end solution for uniformly distributed load:

$$v_f(s) = \frac{q_t(L-s)^2 s^2}{24EI}; \qquad 0 \le s \le L$$

To get the exact solution for the distributed load case, the fixed-end solution is added to those elements that are subjected to distributed load q_t. Thus a finite element model of a frame needs only one element between any two joints or between locations of concentrated loads. Finally, the axial forces, bending moments, and shear forces are computed by appropriate differentiation:

Axial force: $$F(s) = EA\frac{du(s)}{ds}; \qquad 0 \le s \le L$$

Bending moment: $$M(s) = EI\frac{d^2v(s)}{ds^2}; \qquad 0 \le s \le L$$

Shear force: $$V(s) = \frac{dM(s)}{ds}; \qquad 0 \le s \le L$$

If desired, the stresses in the elements can be computed using expressions given with axial deformation and beam elements.

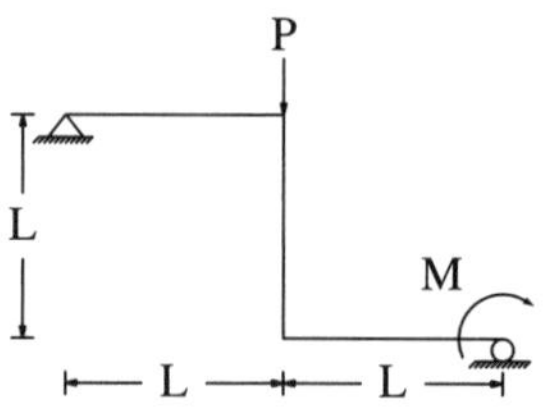

Figure 4.36. Plane frame

Example 4.10 Determine displacements, bending moments, and shear forces in the plane frame shown in Figure 4.36. Draw free-body diagrams for each element clearly showing all element end forces and moments. Use the following numerical data:

$$M = 20\,\text{kN}\cdot\text{m}; \qquad P = 10\,\text{kN}; \qquad L = 1\,\text{m};$$

$$E = 210\,\text{GPa}; \qquad A = 4\times 10^{-2}\,\text{m}^2; \qquad I = 4\times 10^{-4}\,\text{m}^4$$

The simplest possible model for the frame is to use three elements as shown in Figure 4.37. Each node has three degrees of freedom as indicated in the figure.

Use kilonewton-meters for numerical computations. The computed displacements will be in meters and stresses in kilonewtons per square meter.

Specified nodal loads:

Node	dof	Value
2	v_2	−10
4	θ_4	−20

Equations for element 1:

$E = 2.1\times 10^8$; $I = 0.0004$; $A = 0.04$; $q_s = 0.$; $q_t = 0.$

Nodal coordinates:

Element Node	Global Node Number	x	y
1	1	0	0
2	2	1	0

Length = 1; Direction cosines: $l_s = 1$, $m_s = 0$

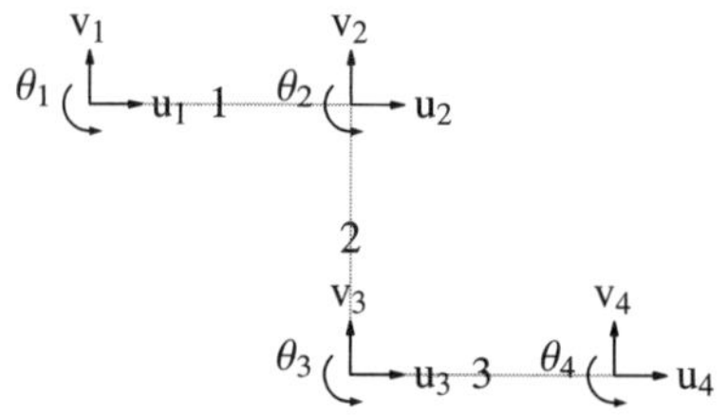

Figure 4.37. Three-element model for the plane frame

Element equations in local coordinates:

$$10^{6.}\begin{pmatrix} 8.4 & 0 & 0 & -8.4 & 0 & 0 \\ 0 & 1.008 & 0.504 & 0 & -1.008 & 0.504 \\ 0 & 0.504 & 0.336 & 0 & -0.504 & 0.168 \\ -8.4 & 0 & 0 & 8.4 & 0 & 0 \\ 0 & -1.008 & -0.504 & 0 & 1.008 & -0.504 \\ 0 & 0.504 & 0.168 & 0 & -0.504 & 0.336 \end{pmatrix}\begin{pmatrix} d_1 \\ d_2 \\ d_3 \\ d_4 \\ d_5 \\ d_6 \end{pmatrix} = \begin{pmatrix} 0. \\ 0. \\ 0. \\ 0. \\ 0. \\ 0. \end{pmatrix}$$

Global to local transformation, $T = \begin{pmatrix} 1 & 0 & 0 & 0 & 0 & 0 \\ 0 & 1 & 0 & 0 & 0 & 0 \\ 0 & 0 & 1 & 0 & 0 & 0 \\ 0 & 0 & 0 & 1 & 0 & 0 \\ 0 & 0 & 0 & 0 & 1 & 0 \\ 0 & 0 & 0 & 0 & 0 & 1 \end{pmatrix}$

Element equations in global coordinates:

$$10^{6.}\begin{pmatrix} 8.4 & 0 & 0 & -8.4 & 0 & 0 \\ 0 & 1.008 & 0.504 & 0 & -1.008 & 0.504 \\ 0 & 0.504 & 0.336 & 0 & -0.504 & 0.168 \\ -8.4 & 0 & 0 & 8.4 & 0 & 0 \\ 0 & -1.008 & -0.504 & 0 & 1.008 & -0.504 \\ 0 & 0.504 & 0.168 & 0 & -0.504 & 0.336 \end{pmatrix}\begin{pmatrix} u_1 \\ v_1 \\ \theta_1 \\ u_2 \\ v_2 \\ \theta_2 \end{pmatrix} = \begin{pmatrix} 0. \\ 0. \\ 0. \\ 0. \\ 0. \\ 0. \end{pmatrix}$$

Equations for element 2:

$$10^{6.}\begin{pmatrix} 1.008 & 0 & 0.504 & -1.008 & 0 & 0.504 \\ 0 & 8.4 & 0 & 0 & -8.4 & 0 \\ 0.504 & 0 & 0.336 & -0.504 & 0 & 0.168 \\ -1.008 & 0 & -0.504 & 1.008 & 0 & -0.504 \\ 0 & -8.4 & 0 & 0 & 8.4 & 0 \\ 0.504 & 0 & 0.168 & -0.504 & 0 & 0.336 \end{pmatrix}\begin{pmatrix} u_2 \\ v_2 \\ \theta_2 \\ u_3 \\ v_3 \\ \theta_3 \end{pmatrix} = \begin{pmatrix} 0. \\ 0. \\ 0. \\ 0. \\ 0. \\ 0. \end{pmatrix}$$

Equations for element 3:

$$10^{6.}\begin{pmatrix} 8.4 & 0 & 0 & -8.4 & 0 & 0 \\ 0 & 1.008 & 0.504 & 0 & -1.008 & 0.504 \\ 0 & 0.504 & 0.336 & 0 & -0.504 & 0.168 \\ -8.4 & 0 & 0 & 8.4 & 0 & 0 \\ 0 & -1.008 & -0.504 & 0 & 1.008 & -0.504 \\ 0 & 0.504 & 0.168 & 0 & -0.504 & 0.336 \end{pmatrix}\begin{pmatrix} u_3 \\ v_3 \\ \theta_3 \\ u_4 \\ v_4 \\ \theta_4 \end{pmatrix} = \begin{pmatrix} 0. \\ 0. \\ 0. \\ 0. \\ 0. \\ 0. \end{pmatrix}$$

Adding element equations into appropriate locations and adjusting for essential boundary conditions, we have

$$10^6 \begin{pmatrix} 0.336 & 0 & -0.504 & 0.168 & 0 & 0 & 0 & 0 & 0 \\ 0 & 9.408 & 0 & 0.504 & -1.008 & 0 & 0.504 & 0 & 0 \\ -0.504 & 0 & 9.408 & -0.504 & 0 & -8.4 & 0 & 0 & 0 \\ 0.168 & 0.504 & -0.504 & 0.672 & -0.504 & 0 & 0.168 & 0 & 0 \\ 0 & -1.008 & 0 & -0.504 & 9.408 & 0 & -0.504 & -8.4 & 0 \\ 0 & 0 & -8.4 & 0 & 0 & 9.408 & 0.504 & 0 & 0.504 \\ 0 & 0.504 & 0 & 0.168 & -0.504 & 0.504 & 0.672 & 0 & 0.168 \\ 0 & 0 & 0 & 0 & -8.4 & 0 & 0 & 8.4 & 0 \\ 0 & 0 & 0 & 0 & 0 & 0.504 & 0.168 & 0 & 0.336 \end{pmatrix} \begin{pmatrix} \theta_1 \\ u_2 \\ v_2 \\ \theta_2 \\ u_3 \\ v_3 \\ \theta_3 \\ u_4 \\ \theta_4 \end{pmatrix} = \begin{pmatrix} 0 \\ 0 \\ -10. \\ 0 \\ 0 \\ 0 \\ 0 \\ 0 \\ -20. \end{pmatrix}$$

Solving the final system of global equations, we get

$$\{\theta_1 = 0.0000784722, u_2 = 0, v_2 = 0.0000685516, \theta_2 = 0.0000487103,$$
$$u_3 = 0.0000189484, v_3 = 0.0000703373, \theta_3 = -0.0000108135,$$
$$u_4 = 0.0000189484, \theta_4 = -0.000159623\}$$

Solution for element 1:

Nodal values in global coordinates,

$$\boldsymbol{d}^T = (0 \quad 0 \quad 0.0000784722 \quad 0 \quad 0.0000685516 \quad 0.0000487103)$$

Nodal values in local coordinates,

$$\boldsymbol{d}_l = \boldsymbol{Td} = (0 \quad 0 \quad 0.0000784722 \quad 0 \quad 0.0000685516 \quad 0.0000487103)$$

Axial displacement interpolation functions, $\boldsymbol{N}_u^T = \{1 - s, s\}$

Axial displacement, $u(s) = \boldsymbol{N}_u^T \begin{pmatrix} d_1 \\ d_4 \end{pmatrix} = 0$

Axial force, $EA\, du(s)/ds = 0$

Beam bending interpolation functions,

$$\boldsymbol{N}_v^T = \{2s^3 - 3s^2 + 1, s^3 - 2s^2 + s, 3s^2 - 2s^3, s^3 - s^2\}$$

Transverse displacement, $v(s) = \boldsymbol{N}_v^T \begin{pmatrix} d_2 \\ d_3 \\ d_5 \\ d_6 \end{pmatrix} = 0.0000784722s - 9.92063 \times 10^{-6} s^3$

Bending moment, $M = EI\, d^2v(s)/ds^2 = -5.s$

Shear force, $V(s) = dM/ds = -5.$

Solution for element 2:

Axial displacement, $u(s) = \boldsymbol{N}_u^T \begin{pmatrix} d_1 \\ d_4 \end{pmatrix} = -1.78571 \times 10^{-6} s - 0.0000685516$

Axial force, $EA\, du(s)/ds = -15.$

Transverse displacement, $v(s) = N_v^T \begin{pmatrix} d_2 \\ d_3 \\ d_5 \\ d_6 \end{pmatrix} = 0.0000487103s - 0.0000297619s^2$

Bending moment, $M = EI\, d^2v(s)/ds^2 = -5.$

Shear force, $V(s) = dM/ds = 0$

Solution for element 3:

Axial displacement, $u(s) = N_u^T \begin{pmatrix} d_1 \\ d_4 \end{pmatrix} = 0.0000189484$

Axial force, $EA\, du(s)/ds = 0$

Total transverse displacement,

$$v(s) = -0.0000297619s^3 - 0.0000297619s^2 - 0.0000108135s + 0.0000703373$$

Bending moment, $M = EI\, d^2v(s)/ds^2 = -15.s - 5.$

Shear force, $V(s) = dM/ds = -15.$

Forces at element ends:

	x	y	Axial Force	Bending Moment	Shear Force
1	0	0	0	0	−5.
	1	0	0	−5.	−5.
2	1	0	−15.	−5.	0
	1	−1	−15.	−5.	0
3	1	−1	0	−5.	−15.
	2	−1	0	−20.	−15.

The element end forces are shown in Figure 4.38. It can easily be seen that each element is in equilibrium. The finite element solution is exact.

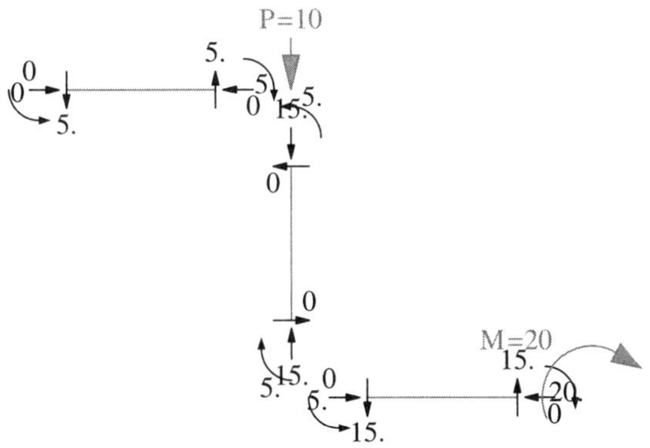

Figure 4.38. Free-body diagram for the plane frame

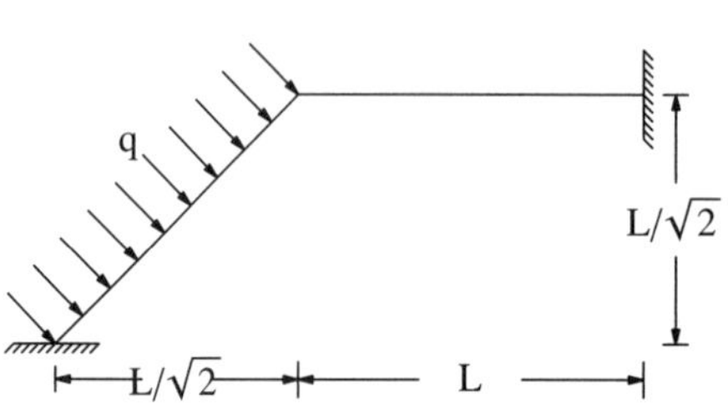

Figure 4.39. Plane frame with distributed load

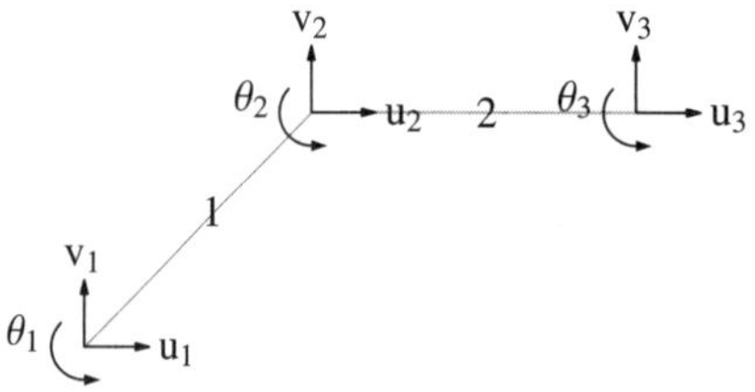

Figure 4.40. Two element model for the plane frame

Example 4.11 Determine displacements, bending moments, and shear forces in the plane frame shown in Figure 4.39. Draw free-body diagrams for each element clearly showing all element end forces and moments. Use the following numerical data: (1 kip = 1000 lb).

$$q = 1 \text{ kip/ft}; \quad L = 15 \text{ ft}; \quad E = 30 \times 10^3 \text{ kip/in}^2; \quad A = 100 \text{ in}^2; \quad I = 1000 \text{ in}^4$$

To model the frame, we need only two elements: one for the inclined member and one for the horizontal as shown in Figure 4.40. Each node has three degrees of freedom as indicated in the figure.

Using kip-inches, the solution is as follows:

Equations for element 1:

$E = 30000; I = 1000; A = 100; q_s = 0.; q_t = -0.0833333$

Nodal coordinates:

Element Node	Global Node Number	x	y
1	1	0.	0.
2	2	127.279	127.279

Length = 180.; Direction cosines: $l_s = 0.707107$, $m_s = 0.707107$

Element equations in local coordinates:

$$\begin{pmatrix} 16666.7 & 0 & 0 & -16666.7 & 0 & 0 \\ 0 & 61.7284 & 5555.56 & 0 & -61.7284 & 5555.56 \\ 0 & 5555.56 & 666667. & 0 & -5555.56 & 333333. \\ -16666.7 & 0 & 0 & 16666.7 & 0 & 0 \\ 0 & -61.7284 & -5555.56 & 0 & 61.7284 & -5555.56 \\ 0 & 5555.56 & 333333. & 0 & -5555.56 & 666667. \end{pmatrix} \begin{pmatrix} d_1 \\ d_2 \\ d_3 \\ d_4 \\ d_5 \\ d_6 \end{pmatrix} = \begin{pmatrix} 0. \\ -7.5 \\ -225. \\ 0. \\ -7.5 \\ 225. \end{pmatrix}$$

Global to local transformation,

$$T = \begin{pmatrix} 0.707107 & 0.707107 & 0 & 0 & 0 & 0 \\ -0.707107 & 0.707107 & 0 & 0 & 0 & 0 \\ 0 & 0 & 1 & 0 & 0 & 0 \\ 0 & 0 & 0 & 0.707107 & 0.707107 & 0 \\ 0 & 0 & 0 & -0.707107 & 0.707107 & 0 \\ 0 & 0 & 0 & 0 & 0 & 1 \end{pmatrix}$$

Element equations in global coordinates:

$$\begin{pmatrix} 8364.2 & 8302.47 & -3928.37 & -8364.2 & -8302.47 & -3928.37 \\ 8302.47 & 8364.2 & 3928.37 & -8302.47 & -8364.2 & 3928.37 \\ -3928.37 & 3928.37 & 666667. & 3928.37 & -3928.37 & 333333. \\ -8364.2 & -8302.47 & 3928.37 & 8364.2 & 8302.47 & 3928.37 \\ -8302.47 & -8364.2 & -3928.37 & 8302.47 & 8364.2 & -3928.37 \\ -3928.37 & 3928.37 & 333333. & 3928.37 & -3928.37 & 666667. \end{pmatrix} \begin{pmatrix} u_1 \\ v_1 \\ \theta_1 \\ u_2 \\ v_2 \\ \theta_2 \end{pmatrix} = \begin{pmatrix} 5.3033 \\ -5.3033 \\ -225. \\ 5.3033 \\ -5.3033 \\ 225. \end{pmatrix}$$

Equations for element 2—element equations in global coordinates:

$$\begin{pmatrix} 16666.7 & 0 & 0 & -16666.7 & 0 & 0 \\ 0 & 61.7284 & 5555.56 & 0 & -61.7284 & 5555.56 \\ 0 & 5555.56 & 666667. & 0 & -5555.56 & 333333. \\ -16666.7 & 0 & 0 & 16666.7 & 0 & 0 \\ 0 & -61.7284 & -5555.56 & 0 & 61.7284 & -5555.56 \\ 0 & 5555.56 & 333333. & 0 & -5555.56 & 666667. \end{pmatrix} \begin{pmatrix} u_2 \\ v_2 \\ \theta_2 \\ u_3 \\ v_3 \\ \theta_3 \end{pmatrix} = \begin{pmatrix} 0 \\ 0 \\ 0 \\ 0 \\ 0 \\ 0 \end{pmatrix}$$

Adding element equations into appropriate locations and adjusting for essential boundary conditions, we have

$$\begin{pmatrix} 25030.9 & 8302.47 & 3928.37 \\ 8302.47 & 8425.93 & 1627.18 \\ 3928.37 & 1627.18 & 1.33333 \times 10^6 \end{pmatrix} \begin{pmatrix} u_2 \\ v_2 \\ \theta_2 \end{pmatrix} = \begin{pmatrix} 5.3033 \\ -5.3033 \\ 225. \end{pmatrix}$$

Solving the final system of global equations, we get

$$\{u_2 = 0.000601607, v_2 = -0.00125474, \theta_2 = 0.000168509\}$$

Solution for element 1:

$E = 30000; I = 1000; A = 100; q_s = 0.; q_t = -0.0833333$

Length = 180.; Direction cosines: $l_s = 0.707107$, $m_s = 0.707107$

Nodal values in global coordinates,

$$d^T = (0 \quad 0 \quad 0 \quad 0.000601607 \quad -0.00125474 \quad 0.000168509)$$

Nodal values in local coordinates,

$$d_l = Td = (0 \quad 0 \quad 0 \quad -0.000461833 \quad -0.00131263 \quad 0.000168509)$$

Axial displacement interpolation functions, $N_u^T = \{1. - 0.00555556s, 0.00555556s\}$

Axial displacement, $u(s) = N_u^T \begin{pmatrix} d_1 \\ d_4 \end{pmatrix} = -2.56574 \times 10^{-6} s$

Axial force, $EA\, du(s)/ds = -7.69721$

Beam bending interpolation functions,

$$N_v^T = \{3.42936 \times 10^{-7} s^3 - 0.0000925926 s^2 + 1, 0.0000308642 s^3 - 0.0111111 s^2 + s,$$
$$0.0000925926 s^2 - 3.42936 \times 10^{-7} s^3, 0.0000308642 s^3 - 0.00555556 s^2\}$$

Transverse displacement, $v(s) = N_v^T \begin{pmatrix} d_2 \\ d_3 \\ d_5 \\ d_6 \end{pmatrix} = 5.65104 \times 10^{-9} s^3 - 1.0577 \times 10^{-6} s^2$

Fixed-end displacement solution $= -1.15741 \times 10^{-10} (180. - s)^2 s^2$

Total transverse displacement, $v(s) = -1.15741 \times 10^{-10} s^4 + 4.73177 \times 10^{-8} s^3 - 4.8077 \times 10^{-6} s^2$

Bending moment, $M = EI\, d^2 v(s)/ds^2 = -0.0416667 s^2 + 8.51719 s - 288.462$

Shear force, $V(s) = dM/ds = 8.51719 - 0.0833333 s$

Solution for element 2:

Axial displacement, $u(s) = N_u^T \begin{pmatrix} d_1 \\ d_4 \end{pmatrix} = 0.000601607 - 3.34226 \times 10^{-6} s$

Axial force, $EA\, du(s)/ds = -10.0268$

Transverse displacement,

$$v(s) = N_v^T \begin{pmatrix} d_2 \\ d_3 \\ d_5 \\ d_6 \end{pmatrix}, = 4.77059 \times 10^{-9} s^3 - 1.75614 \times 10^{-6} s^2 + 0.000168509 s - 0.00125474$$

Bending moment, $M = EI\, d^2 v(s)/ds^2 = 0.858707 s - 105.368$

Shear force, $V(s) = dM/ds = 0.858707$

Forces at element ends:

	x	y	Axial Force	Bending Moment	Shear Force
1	0	0	−7.69721	−288.462	8.51719
	127.279	127.279	−7.69721	−105.368	−6.48281
2	127.279	127.279	−10.0268	−105.368	0.858707
	307.279	127.279	−10.0268	49.1988	0.858707

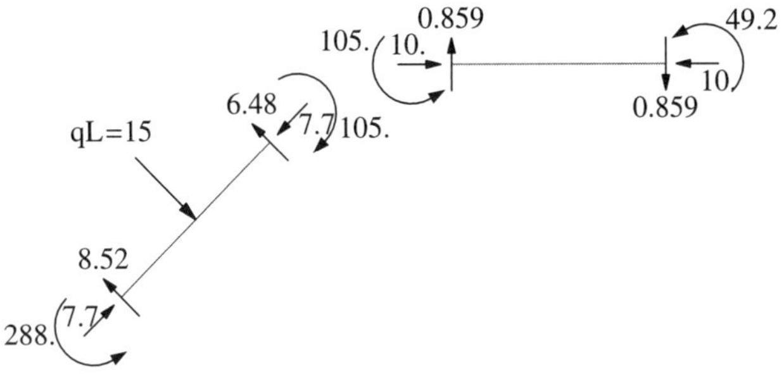

Figure 4.41. Free-body diagram of frame

The element end forces are shown in Figure 4.41. It can easily be seen that the equilibrium is satisfied in each element and hence this is the exact solution.

◆ *Mathematica*/MATLAB Implementation 4.4 on the Book Web Site: *Analysis of plane frames*

Example 4.12 Determine displacements, bending moments, and shear forces in the box frame reinforced with a spring as shown in Figure 4.42. Use the following numerical data (1 kip = 1000 lb):

$$L = 15 \text{ ft}; \quad E = 30{,}000 \text{ kips/in}^2; \quad A = 0.5 \text{ in}^2;$$
$$I = 1000 \text{ in}^4; \quad k = 200 \text{ kips/in}; \quad P = 100 \text{ kip}$$

We can easily create a finite element model of the entire frame using four plane frame elements and one spring element. However, since there are no specified essential boundary conditions, the resulting finite element system of equations will be singular. We can get some boundary conditions by taking advantage of symmetry and modeling the upper half, as shown in Figure 4.43. Note the spring constant and the applied loads are both halved in the model. The ends must have zero rotations and zero vertical displacements due to

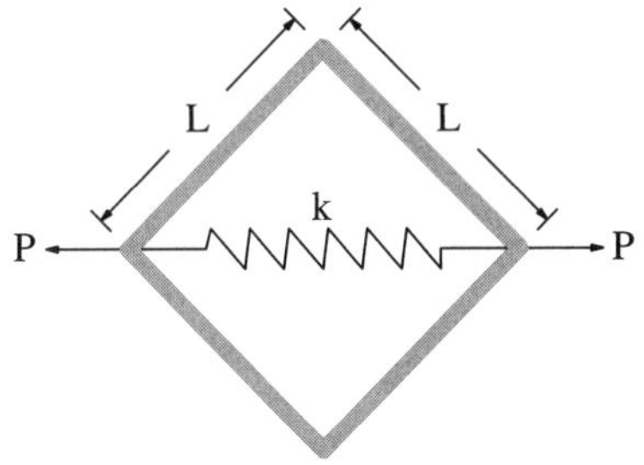

Figure 4.42. Box frame reinforced with a spring

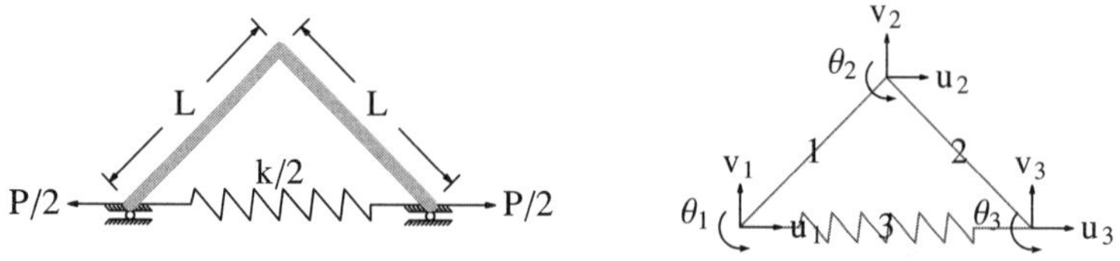

Figure 4.43. Finite element model for the half of the box frame

symmetry. The ends are free to move in the horizontal direction. The system can still move as a rigid body in the horizontal direction, and therefore the finite element equations will still be singular. To get a solvable system of equations, we must assign zero horizontal displacement to at least one of the nodes. For this frame, once again because of symmetry, the top of the frame cannot move in the horizontal direction. Thus we impose the zero horizontal boundary condition at that point.

Using kip-inches, the solution is as follows:

Specified nodal loads:

Node	dof	Value
1	u_1	−50
3	u_3	50

Equations for element 1:

$$\begin{pmatrix} 72.5309 & 10.8025 & -3928.37 & -72.5309 & -10.8025 & -3928.37 \\ 10.8025 & 72.5309 & 3928.37 & -10.8025 & -72.5309 & 3928.37 \\ -3928.37 & 3928.37 & 666667. & 3928.37 & -3928.37 & 333333. \\ -72.5309 & -10.8025 & 3928.37 & 72.5309 & 10.8025 & 3928.37 \\ -10.8025 & -72.5309 & -3928.37 & 10.8025 & 72.5309 & -3928.37 \\ -3928.37 & 3928.37 & 333333. & 3928.37 & -3928.37 & 666667. \end{pmatrix} \begin{pmatrix} u_1 \\ v_1 \\ \theta_1 \\ u_2 \\ v_2 \\ \theta_2 \end{pmatrix} = \begin{pmatrix} 0 \\ 0 \\ 0 \\ 0 \\ 0 \\ 0 \end{pmatrix}$$

Equations for element 2:

$$\begin{pmatrix} 72.5309 & -10.8025 & 3928.37 & -72.5309 & 10.8025 & 3928.37 \\ -10.8025 & 72.5309 & 3928.37 & 10.8025 & -72.5309 & 3928.37 \\ 3928.37 & 3928.37 & 666667. & -3928.37 & -3928.37 & 333333. \\ -72.5309 & 10.8025 & -3928.37 & 72.5309 & -10.8025 & -3928.37 \\ 10.8025 & -72.5309 & -3928.37 & -10.8025 & 72.5309 & -3928.37 \\ 3928.37 & 3928.37 & 333333. & -3928.37 & -3928.37 & 666667. \end{pmatrix} \begin{pmatrix} u_2 \\ v_2 \\ \theta_2 \\ u_3 \\ v_3 \\ \theta_3 \end{pmatrix} = \begin{pmatrix} 0 \\ 0 \\ 0 \\ 0 \\ 0 \\ 0 \end{pmatrix}$$

Equations for element 3—spring element equations in global coordinates:

$$\begin{pmatrix} 100 & -100 \\ -100 & 100 \end{pmatrix} \begin{pmatrix} u_1 \\ u_3 \end{pmatrix} = \begin{pmatrix} 0 \\ 0 \end{pmatrix}$$

Essential boundary conditions:

Node	dof	Value
1	v_1	0
1	θ_1	0
2	u_2	0
3	v_3	0
3	θ_3	0

Adding element equations into appropriate locations and adjusting for essential boundary conditions, we have

$$\begin{pmatrix} 172.531 & -10.8025 & -3928.37 & -100. \\ -10.8025 & 145.062 & 0 & 10.8025 \\ -3928.37 & 0 & 1.33333\times 10^6 & -3928.37 \\ -100. & 10.8025 & -3928.37 & 172.531 \end{pmatrix} \begin{pmatrix} u_1 \\ v_2 \\ \theta_2 \\ u_3 \end{pmatrix} = \begin{pmatrix} -50. \\ 0 \\ 0 \\ 50. \end{pmatrix}$$

Solving the final system of global equations, we get

$$\{u_1 = -0.184555, v_2 = -0.0274869, \theta_2 = 0, u_3 = 0.184555\}$$

Spring force $= k(u_3 - u_1) = 200(0.184555 - (-0.184555)) = 73.822$ (Tension)

4.9 SPACE FRAMES

A space frame element is an extension of the plane frame element to three dimensions. The element loading and cross-sectional properties are described in terms of a local r-s-t coordinate system, as shown in Figure 4.44. The local t axis runs along the centroidal axis of the element. The local r and s axes are the principal moment-of-inertia axes for the cross section; the local r axis is along the axis of the *maximum moment of inertia* and the local s axis is along the axis of the *minimum moment of inertia*. The element includes axial force effects and bending effects due to applied loads in the r-t and s-t planes. In addition, the

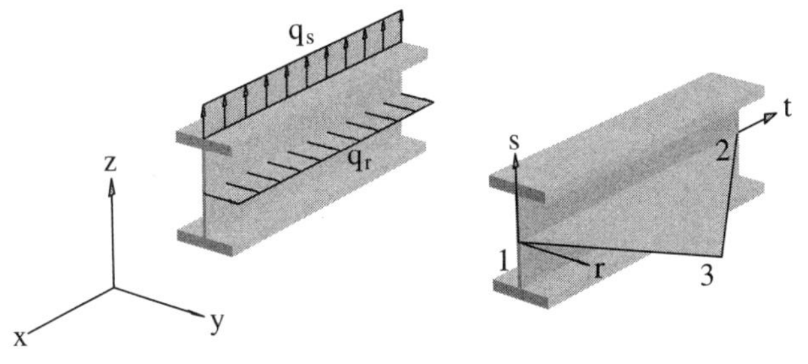

Figure 4.44. Space frame element

element includes torsional effects due to the twisting moment (moments about the t axis). Within the small-displacement theory, all these effects can be assumed to be uncoupled. The finite element equations are thus just a combination of the equations treating these effects individually.

For a space frame element, just knowing the element end coordinates is not enough to establish all three local element coordinates. We need additional information about one of the local principal axes. The simplest method is the so-called *three-node method.* Nodes 1 and 2 are at the element end points. These two nodes define element length and degrees of freedom. For each element a third point must also be identified to define the local r-t plane. Any node in the model can be used as the third point as long as the three points define the local r-t plane for the element. Using the three nodes, the local element axes are established as follows:

(i) The local t axis is first established by creating a vector from node 1 to 2.

(ii) Since the cross product of two vectors is a vector that is normal to the plane defined by the vectors, the local s axis is established by taking the cross product of a vector from node 1 to 2 with a vector from node 1 to 3.

(iii) Finally the local r axis is established by taking the cross product of unit vectors in the s and t directions.

The decision to use the third node to define the r-t plane is obviously arbitrary. We could have used it to define the s-t plane as well. There is no standard convention. Different authors and computer programs use the third node to define either the major bending axis or the minor axis. Thus, when using a computer program, it is very important to find out exactly what convention is being used; otherwise the results obviously will be erroneous.

Detailed expressions for these vectors are developed later in this section.

The following notation is used to describe material and cross-sectional properties:

E	Young's modulus
G	Shear modulus
A	Area of cross section
J	Torsional constant
I_p	Polar moment of inertia
$I_s = I_{\min}$	Moment of inertia of cross section about s axis (minimum principal inertia)
$I_r = I_{\max}$	Moment of inertia of cross section about r axis (maximum principal inertia)
L	Length of element

The torsional constant is sometimes also denoted by k_T or I_t. For circular cross sections $J = I_{\max} + I_{\min} = I_p$, the polar moment of inertia. For other shapes J must be computed using methods of elasticity theory. Formulas for few common shapes are given in Figure 4.45. More formulas for a large number of different cross-sectional shapes can be found in the handbook by W. C. Young and R. G. Budynas, *Roark's Formulas for Stress and Strain*, seventh edition, McGraw-Hill, New York, 2002.

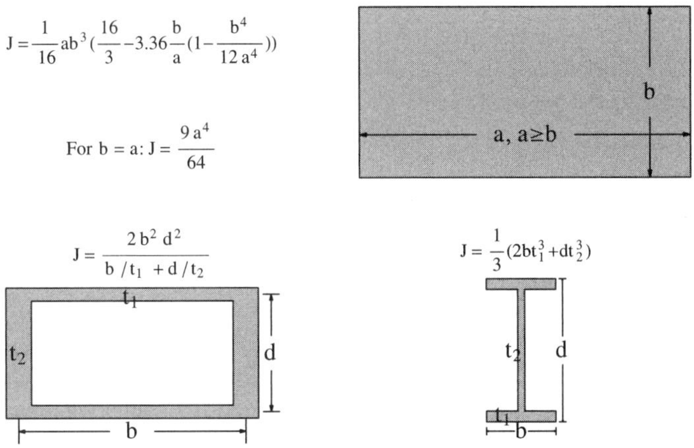

Figure 4.45. Formulas for torsion constant J

4.9.1 Element Equations in Local Coordinate System

Considering axial forces, bending in both planes, and twisting effects, each node of a space frame element has six degrees of freedom; three translations, and three rotations, as shown in Figure 4.46. The nodal displacements and applied forces are positive when these quantities act along the positive coordinate directions. For applied moments and rotations the positive directions are based on the right hand rule. When the thumb of your right hand is pointing toward the positive coordinate direction, the curl of your fingers defines the positive directions for applied moments and rotations in the right-hand rule. The nodal degrees of freedom in the local coordinate system are defined as follows:

d_1, d_2, d_3	Displacements at node 1
d_4, d_5, d_6	Rotations at node 1
d_7, d_8, d_9	Displacements at node 2
d_{10}, d_{11}, d_{12}	Rotations at node 2

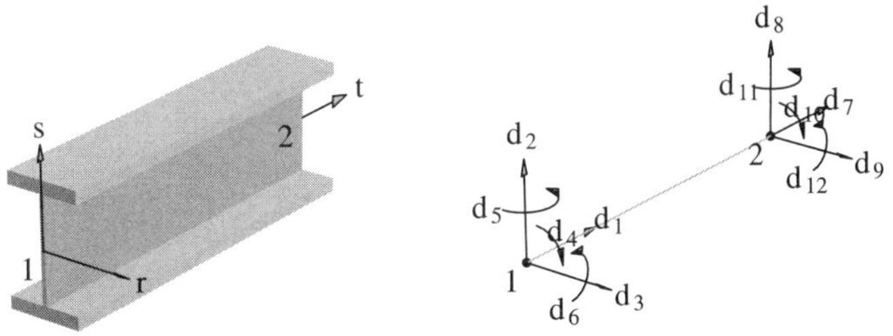

Figure 4.46. Local degrees of freedom for the space frame element

The axial displacements are related to the two axial degrees of freedom d_1 and d_7 as follows:

$$u(t) = \begin{pmatrix} 1 - \frac{t}{L} & \frac{t}{L} \end{pmatrix} \begin{pmatrix} d_1 \\ d_7 \end{pmatrix}; \qquad 0 \le t \le L$$

where L is the length of the element.

The applied loading in the local s-t plane will cause the displacement $v(t)$ in the local s direction and rotations $\theta \equiv dv/dt$ about the r axis. From beam bending the displacement $v(t)$ is related to nodal degrees of freedom d_2, d_6, d_8, and d_{12} as follows:

$$v(t) = \begin{pmatrix} 1 - \frac{3t^2}{L^2} + \frac{2t^3}{L^3} & t - \frac{2t^2}{L} + \frac{t^3}{L^2} & \frac{3t^2}{L^2} - \frac{2t^3}{L^3} & -\frac{t^2}{L} + \frac{t^3}{L^2} \end{pmatrix} \begin{pmatrix} d_2 \\ d_6 \\ d_8 \\ d_{12} \end{pmatrix}; \qquad 0 \le t \le L$$

This loading causes bending moment about the r axis and shear force normal to the t direction:

$$M_r(t) = EI_r \left(\frac{d^2 v}{dt^2} \right); \qquad V_t(t) = \frac{dM_r}{dt} = E_r \left(\frac{d^3 v}{dt^3} \right)$$

For the loading applied in the local r-t plane, the displacement $w(t)$ is in the local r direction and rotations φ are about the s axis. The situation is essentially the same as the beam bending except that a positive displacement in the r direction produces a clockwise rotation about s axis. Thus with the sign convention adopted for beam bending the rotation $\varphi(t)$ is related to the transverse displacement $w(t)$ with a negative sign, i.e.,

$$\varphi(t) \equiv -\frac{\partial w}{\partial t}$$

The finite element shape functions for this situation are the same as those for usual beam bending except for the change in sign for the rotation terms. Thus the displacement $w(t)$ is related to nodal degrees of freedom d_3, d_5, d_9, and d_{11} as follows:

$$w(t) = \begin{pmatrix} 1 - \frac{3t^2}{L^2} + \frac{2t^3}{L^3} & -\left(t - \frac{2t^2}{L} + \frac{t^3}{L^2}\right) & \frac{3t^2}{L^2} - \frac{2t^3}{L^3} & -\left(- \frac{t^2}{L} + \frac{t^3}{L^2}\right) \end{pmatrix} \begin{pmatrix} d_3 \\ d_5 \\ d_9 \\ d_{11} \end{pmatrix}; \qquad 0 \le t \le L$$

This loading causes the bending moment about the s axis and shear force normal to the r direction:

$$M_s(t) = -EI_s \left(\frac{d^2 w}{dt^2} \right); \qquad V_r(t) = -\frac{dM_s}{dt} = EI_s \left(\frac{d^3 w}{dt^3} \right)$$

The elements in a space frame generally are also subjected to twisting moments. The governing differential equation for a bar subjected to twist is a second-order differential equation similar to the one for the axial deformation problem. Assuming twisting moments are applied only at the element ends (no applied distributed twisting moment along the span), the problem can be described in terms of the following differential equation:

$$GJ\frac{d^2\psi}{dt^2} = 0$$

where $\psi(t)$ is the rotation of the section about its t axis, G is the shear modulus, and J is the torsional constant. From the similarity of this equation to the one for the axial deformation problem, it is easy to see that a linear solution in terms of nodal degrees of freedom d_4 and d_{10} is as follows:

$$\psi(t) = \begin{pmatrix} 1 - \frac{t}{L} & \frac{t}{L} \end{pmatrix} \begin{pmatrix} d_4 \\ d_{10} \end{pmatrix}$$

The twisting moment per unit length is related to the angle of twist by the following equation:

$$M_t(t) = GJ\frac{d\psi}{dt}$$

By combining axial deformations, bending in r-t and s-t planes, and effects due to twisting actions, the total strain energy in a frame element can be written as follows:

$$U = \frac{1}{2}\int_0^L EA\left(\frac{du}{dt}\right)^2 dt + \frac{1}{2}\int_0^L EI_r\left(\frac{d^2v}{dt^2}\right)^2 dt + \frac{1}{2}\int_0^L EI_s\left(\frac{d^2w}{dt^2}\right)^2 dt + \frac{1}{2}\int_0^L GJ\left(\frac{d\psi}{dt}\right)^2 dt$$

It is convenient to express this in matrix form as follows:

$$U = \frac{1}{2}\int_0^L \begin{pmatrix} \frac{du}{dt} & \frac{d^2v}{dt^2} & \frac{d^2w}{dt^2} & \frac{d\psi}{dt} \end{pmatrix} \begin{pmatrix} EA & 0 & 0 & 0 \\ 0 & EI_r & 0 & 0 \\ 0 & 0 & EI_s & 0 \\ 0 & 0 & 0 & GJ \end{pmatrix} \begin{pmatrix} \frac{du}{dt} \\ \frac{d^2v}{dt^2} \\ \frac{d^2w}{dt^2} \\ \frac{d\psi}{dt} \end{pmatrix} dt$$

Using the different displacement interpolations given above, the required derivatives can be written as follows:

$$\begin{pmatrix} \frac{du}{dt} \\ \frac{d^2v}{dt^2} \\ \frac{d^2w}{dt^2} \\ \frac{d\psi}{dt} \end{pmatrix} = \begin{pmatrix} -\frac{1}{L} & 0 & 0 & 0 & \cdots \\ 0 & -\frac{6}{L^2} + \frac{12t}{L^3} & 0 & 0 & \cdots \\ 0 & 0 & -\frac{6}{L^2} + \frac{12t}{L^3} & 0 & \cdots \\ 0 & 0 & 0 & -\frac{1}{L} & \cdots \end{pmatrix} \begin{pmatrix} d_1 \\ d_2 \\ \vdots \\ d_{12} \end{pmatrix} \equiv \boldsymbol{B}^T\boldsymbol{d}$$

$$\boldsymbol{B}^T = \begin{pmatrix} -\frac{1}{L} & 0 & 0 & 0 & 0 & 0 & \frac{1}{L} & 0 & 0 & 0 & 0 & 0 \\ 0 & \frac{12t}{L^3} - \frac{6}{L^2} & 0 & 0 & 0 & \frac{6t}{L^2} - \frac{4}{L} & 0 & \frac{6}{L^2} - \frac{12t}{L^3} & 0 & 0 & 0 & \frac{6t}{L^2} - \frac{2}{L} \\ 0 & 0 & \frac{12t}{L^3} - \frac{6}{L^2} & 0 & \frac{4}{L} - \frac{6t}{L^2} & 0 & 0 & 0 & \frac{6}{L^2} - \frac{12t}{L^3} & 0 & \frac{2}{L} - \frac{6t}{L^2} & 0 \\ 0 & 0 & 0 & -\frac{1}{L} & 0 & 0 & 0 & 0 & 0 & \frac{1}{L} & 0 & 0 \end{pmatrix}$$

Thus the strain energy can be written as follows:

$$U = \frac{1}{2}\boldsymbol{d}^T \int_0^L \boldsymbol{BCB}^T \, dt \, \boldsymbol{d} \equiv \frac{1}{2}\boldsymbol{d}^T \boldsymbol{kd}$$

where

$$\boldsymbol{C} = \begin{pmatrix} EA & 0 & 0 & 0 \\ 0 & EI_r & 0 & 0 \\ 0 & 0 & EI_s & 0 \\ 0 & 0 & 0 & GJ \end{pmatrix}$$

$$\boldsymbol{k} = \int_0^L \boldsymbol{BCB}^T \, dt$$

Carrying out matrix multiplications and integration, we get the following space frame stiffness matrix in its local coordinates:

$$\boldsymbol{k} = \begin{pmatrix}
\frac{EA}{L} & 0 & 0 & 0 & 0 & 0 & -\frac{EA}{L} & 0 & 0 & 0 & 0 & 0 \\
0 & \frac{12EI_r}{L^3} & 0 & 0 & 0 & \frac{6EI_r}{L^2} & 0 & -\frac{12EI_r}{L^3} & 0 & 0 & 0 & \frac{6EI_r}{L^2} \\
0 & 0 & \frac{12EI_s}{L^3} & 0 & -\frac{6EI_s}{L^2} & 0 & 0 & 0 & -\frac{12EI_s}{L^3} & 0 & -\frac{6EI_s}{L^2} & 0 \\
0 & 0 & 0 & \frac{GJ}{L} & 0 & 0 & 0 & 0 & 0 & -\frac{GJ}{L} & 0 & 0 \\
0 & 0 & -\frac{6EI_s}{L^2} & 0 & \frac{4EI_s}{L} & 0 & 0 & 0 & \frac{6EI_s}{L^2} & 0 & \frac{2EI_s}{L} & 0 \\
0 & \frac{6EI_r}{L^2} & 0 & 0 & 0 & \frac{4EI_r}{L} & 0 & -\frac{6EI_r}{L^2} & 0 & 0 & 0 & \frac{2EI_r}{L} \\
-\frac{EA}{L} & 0 & 0 & 0 & 0 & 0 & \frac{EA}{L} & 0 & 0 & 0 & 0 & 0 \\
0 & -\frac{12EI_r}{L^3} & 0 & 0 & 0 & -\frac{6EI_r}{L^2} & 0 & \frac{12EI_r}{L^3} & 0 & 0 & 0 & -\frac{6EI_r}{L^2} \\
0 & 0 & -\frac{12EI_s}{L^3} & 0 & \frac{6EI_s}{L^2} & 0 & 0 & 0 & \frac{12EI_s}{L^3} & 0 & \frac{6EI_s}{L^2} & 0 \\
0 & 0 & 0 & -\frac{GJ}{L} & 0 & 0 & 0 & 0 & 0 & \frac{GJ}{L} & 0 & 0 \\
0 & 0 & -\frac{6EI_s}{L^2} & 0 & \frac{2EI_s}{L} & 0 & 0 & 0 & \frac{6EI_s}{L^2} & 0 & \frac{4EI_s}{L} & 0 \\
0 & \frac{6EI_r}{L^2} & 0 & 0 & 0 & \frac{2EI_r}{L} & 0 & -\frac{6EI_r}{L^2} & 0 & 0 & 0 & \frac{4EI_r}{L}
\end{pmatrix}$$

In a similar manner, the equivalent nodal load vector can be established by considering the work done by the distributed applied forces along the local coordinate directions q_r and q_s as follows:

$$W_q = \int_0^L q_s v(t) \, dt + \int_0^L q_r w(t) \, dt \equiv \int_0^L (q_s \quad q_r) \begin{pmatrix} v(t) \\ w(t) \end{pmatrix} dt$$

Writing the displacement expressions in a matrix form, we have

$$\begin{pmatrix} v(t) \\ w(t) \end{pmatrix} = \begin{pmatrix} 0 & \frac{2t^3}{L^3} - \frac{3t^2}{L^2} + 1 & 0 & 0 & 0 & \frac{t^3}{L^2} - \frac{2t^2}{L} + t & \cdots \\ 0 & 0 & \frac{2t^3}{L^3} - \frac{3t^2}{L^2} + 1 & 0 & -\frac{t^3}{L^2} + \frac{2t^2}{L} - t & 0 & \cdots \end{pmatrix} \begin{pmatrix} d_1 \\ d_2 \\ \vdots \\ d_{12} \end{pmatrix}$$

$$\equiv \boldsymbol{N}^T \boldsymbol{d}$$

Using these, the work done can be written as follows:

$$W_q = \int_0^L (q_s \quad q_r) N^T \, dt \, \boldsymbol{d} \equiv \boldsymbol{r}_q^T \boldsymbol{d}$$

where $\boldsymbol{r}_q$ is the equivalent load vector. If all distributed forces are assumed constant over an element, then the integration can easily be carried out to give the following equivalent load vector:

$$\boldsymbol{r}_q^T = \left\{0, \frac{q_s L}{2}, \frac{q_r L}{2}, 0, -\frac{q_r L^2}{12}, \frac{q_s L^2}{12}, 0, \frac{q_s L}{2}, \frac{q_r L}{2}, 0, \frac{q_r L^2}{12}, -\frac{q_s L^2}{12}\right\}$$

4.9.2 Local-to-Global Transformation

The stiffness matrix and the equivalent load vector derived so far have been in terms of a local coordinate system. Since different elements in a space frame will generally have different local axes, before assembly these matrices must be transformed to a global coordinate system that is common to all elements in the frame. In the global x, y, and z coordinate system the nodal displacements are identified as follows:

u_1, v_1, w_1	x, y, and z displacements at node 1
$\theta_{x1}, \theta_{y1}, \theta_{z1}$	Rotations about x, y, z axes at node 1
u_2, v_2, w_2	x, y, and z displacements at node 2
$\theta_{x2}, \theta_{y2}, \theta_{z2}$	Rotations about x, y, z axes at node 2

The corresponding nodal forces and moments in the global x, y, and z coordinate system are as follows:

F_{x1}, F_{y1}, F_{z1}	Applied forces in the global x, y, and z at node 1
M_{x1}, M_{y1}, M_{z1}	Applied moments about the global x, y, and z at node 1
F_{x2}, F_{y2}, F_{z2}	Applied forces in the global x, y, and z at node 2
M_{x2}, M_{y2}, M_{z2}	Applied moments about the global x, y, and z at node 2

The local-to-global transformation matrix is developed by considering three components of displacements and rotations at each node as vector quantities. Thus the complete transformation matrix is a 12×12 matrix consisting of four identical 3×3 rotation matrices as follows:

$$\begin{pmatrix} d_1 \\ d_2 \\ \vdots \\ d_{12} \end{pmatrix} = \begin{pmatrix} \boldsymbol{H} & 0 & 0 & 0 \\ 0 & \boldsymbol{H} & 0 & 0 \\ 0 & 0 & \boldsymbol{H} & 0 \\ 0 & 0 & 0 & \boldsymbol{H} \end{pmatrix} \begin{pmatrix} u_1 \\ v_1 \\ \vdots \\ \theta_{z2} \end{pmatrix} \Longrightarrow \boldsymbol{d}_l = \boldsymbol{T}\boldsymbol{d} \qquad \text{and} \qquad \boldsymbol{d} = \boldsymbol{T}^T \boldsymbol{d}_l$$

where $\boldsymbol{H}$ is a 3×3 three-dimensional rotation matrix and $\boldsymbol{0}$ is a 3×3 zero matrix. The rotation matrix $\boldsymbol{H}$ transforms a vector quantity from the local to the global coordinate system.

Within the small-displacement assumption, the nodal displacements, rotations, forces, and moments are all vector quantities and can all be transformed using this $\boldsymbol{H}$ matrix. The components of a vector along the local s, t, and r coordinates are simply the sum of projections of its x, y, and z components along the local axes. In matrix form the transformation can be written as follows:

$$\begin{pmatrix} t \\ s \\ r \end{pmatrix} = \begin{pmatrix} l_t & m_t & n_t \\ l_s & m_s & n_s \\ l_r & m_r & n_r \end{pmatrix} \begin{pmatrix} x \\ y \\ z \end{pmatrix}; \qquad \boldsymbol{H} = \begin{pmatrix} l_t & m_t & n_t \\ l_s & m_s & n_s \\ l_r & m_r & n_r \end{pmatrix}$$

where l_t is the cosine of the angle between the t and x axes (direction cosine). The other terms have a similar meaning. Thus nine direction cosines are needed to establish the $\boldsymbol{H}$ matrix for each element in a space frame. Explicit expressions for these direction cosines are given later in the section.

Using this transformation matrix, the element equations in the local coordinate system can be related to those in the global coordinate system as follows:

$$\boldsymbol{k}_l \boldsymbol{d}_l = \boldsymbol{r}_l \Longrightarrow \boldsymbol{k}_l \boldsymbol{T} \boldsymbol{d} = \boldsymbol{r}_l$$

Multiplying both sides by $\boldsymbol{T}^T$, we get

$$\boldsymbol{T}^T \boldsymbol{k}_l \boldsymbol{T} \boldsymbol{d} = \boldsymbol{T}^T \boldsymbol{r}_l$$

Noting that $\boldsymbol{T}^T \boldsymbol{r}_l$ is the transformation of applied loads from the local to the global coordinates, we have the following element equations in terms of global degrees of freedom and applied nodal loads in the global directions:

$$\boldsymbol{k}\boldsymbol{d} = \boldsymbol{r}$$

where

$$\boldsymbol{k} = \boldsymbol{T}^T \boldsymbol{k}_l \boldsymbol{T} \qquad \text{and} \qquad \boldsymbol{r} = \boldsymbol{T}^T \boldsymbol{r}_l$$

Three-Node Method for Calculating Direction Cosines Just knowing the element end coordinates is not enough to establish all nine direction cosines in the rotation matrix $\boldsymbol{H}$. We need additional information about one of the local principal axes. The simplest method is the so-called three-node method. Nodes 1 and 2 are at the element end points, as in the case of a plane frame element. A third node is used to define either the local r-t or the s-t plane. In the following development it is assumed that the third node is used to define the local r-t plane, as shown in Figure 4.44. It is important to note that the third node is used for defining the element orientation alone. It does not determine the element length or any other parameter used in the element equations. In a given frame any suitable node can be used to define the third node as long as this node, together with element nodes 1 and 2, determines the r-t local plane for the element. The third node cannot be collinear with element nodes 1 and 2 because in this case the three nodes do not form a plane.

The coordinates of the three nodes for an element are denoted by (x_1, y_1, z_1), (x_2, y_2, z_2), and (x_3, y_3, z_3). Using these coordinates, the nine direction cosines can be computed as follows.

The local t axis is defined first by a vector $\boldsymbol{V}_{12}$ pointing from node 1 to node 2 of the element:

$$\boldsymbol{V}_{12} = (x_2 - x_1)\boldsymbol{i} + (y_2 - y_1)\boldsymbol{j} + (z_2 - z_1)\boldsymbol{k}$$

where $\boldsymbol{i}, \boldsymbol{j}, \boldsymbol{k}$ are unit vectors along the global coordinate directions. The length of the vector $\boldsymbol{V}_{12}$ is the element length $L = \sqrt{(x_2 - x_1)^2 + (y_2 - y_1)^2 + (z_2 - z_1)^2}$. A unit vector along the local t axis is therefore given by

$$\hat{\boldsymbol{V}}_t = \frac{x_2 - x_1}{L}\boldsymbol{i} + \frac{y_2 - y_1}{L}\boldsymbol{j} + \frac{z_2 - z_1}{L}\boldsymbol{k}$$

The three direction cosines defining the local t axis are thus

$$l_t = \frac{x_2 - x_1}{L}; \qquad m_t = \frac{y_2 - y_1}{L}; \qquad n_t = \frac{z_2 - z_1}{L}$$

The local s axis is normal to the plane defined by nodes 1, 2, and 3. A vector along the local s axis is obtained by taking the cross product of the vector $\boldsymbol{V}_{12}$ with vector $\boldsymbol{V}_{13}$ that goes from node 1 to node 3. Thus we have

$$\boldsymbol{V}_{13} = (x_3 - x_1)\boldsymbol{i} + (y_3 - y_1)\boldsymbol{j} + (z_3 - z_1)\boldsymbol{k}$$

$$\boldsymbol{V}_s = \boldsymbol{V}_{13} \times \boldsymbol{V}_{12} = \det\begin{pmatrix} \boldsymbol{i} & \boldsymbol{j} & \boldsymbol{k} \\ x_3 - x_1 & y_3 - y_1 & z_3 - z_1 \\ x_2 - x_1 & y_2 - y_1 & z_2 - z_1 \end{pmatrix} \equiv X_s\boldsymbol{i} + Y_s\boldsymbol{j} + Z_s\boldsymbol{k}$$

$$X_s = y_3(z_2 - z_1) + y_2(z_1 - z_3) + y_1(z_3 - z_2)$$

$$Y_s = x_3(z_1 - z_2) + x_1(z_2 - z_3) + x_2(z_3 - z_1)$$

$$Z_s = x_3(y_2 - y_1) + x_2(y_1 - y_3) + x_1(y_3 - y_2)$$

The length of this vector $\boldsymbol{V}_s$ is

$$L_s = \sqrt{X_s^2 + Y_s^2 + Z_s^2}$$

Therefore a unit vector along the local s axis is given by

$$\hat{\boldsymbol{V}}_s = \frac{X_s}{L_s}\boldsymbol{i} + \frac{Y_s}{L_s}\boldsymbol{j} + \frac{Z_s}{L_s}\boldsymbol{k}$$

The three direction cosines defining the local s axis are thus

$$l_s = \frac{X_s}{L_s}; \qquad m_s = \frac{Y_s}{L_s}; \qquad n_s = \frac{Z_s}{L_s}$$

Finally, the local r axis is normal to the s-t plane. It can be defined by taking the cross product of the unit vector along the s axis with the one along the t axis

$$\hat{V}_r = \hat{V}_t \times \hat{V}_s = \det\begin{pmatrix} i & j & k \\ l_t & m_t & n_t \\ l_s & m_s & n_s \end{pmatrix} \equiv l_r i + m_r j + n_r k$$

where

$$l_r = m_t n_s - m_s n_t; \qquad m_r = n_t l_s - n_s l_t; \qquad n_r = m_s l_t - m_t l_s$$

Example 4.13 An I-shaped frame element is oriented in such a way that its longitudinal axis is along the global z axis and its minor moment of inertia axis makes an angle of 30° with the global x axis. The element is 100 cm long. Establish the 3×3 rotation matrix $\boldsymbol{H}$ for the element.

Assuming node 1 is at (0, 0, 0), node 2 should be at (0, 0, 100). According to the convention adopted here, the third node should define the major moment of inertia axis. The angle between the global x axis and the major moment of inertia axis will be 60°. Thus, to define the orientation of this element, we choose a third point located at 100(− cos(60), sin(60), 0). This will establish the $t - s - r$ coordinate system for the element, as shown in Figure 4.47. The computations of direction cosines using the three-node method are as follows:

Nodal coordinates:

	x	y	z
1	0	0	0
2	0	0	100.
3	−50	$50\sqrt{3}$	0

Vector from node 1 to 2, $\boldsymbol{V}_{12} = \{0, 0, 100.\}$

Element length, $L = 100.$

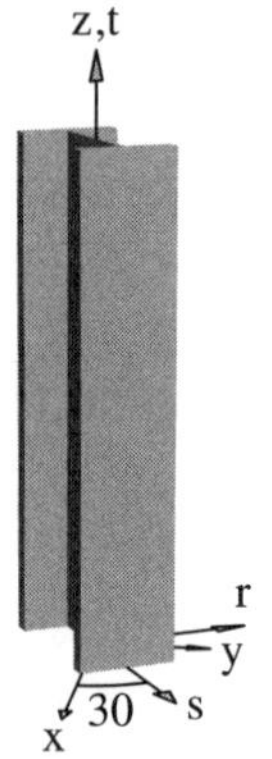

Figure 4.47. Local and global axes for a space frame element

Unit vector, $\hat{V}_t = V_{12}/L = \{0, 0, 1.\}$
Vector from node 1 to 3, $V_{13} = \{-50, 50\sqrt{3}, 0\}$
Vector, $V_s = V_{13} \times V_{12} = \{8660.25, 5000., 0\}$
Length, $L_s = 10000.$
Unit vector, $\hat{V}_s = V_s/L_s = \{0.866025, 0.5, 0\}$
Unit vector, $\hat{V}_r = \hat{V}_t \times \hat{V}_s = \{-0.5, 0.866025, 0.\}$
Thus the rotation matrix is as follows:

$$H = \begin{pmatrix} 0 & 0 & 1. \\ 0.866025 & 0.5 & 0 \\ -0.5 & 0.866025 & 0. \end{pmatrix}$$

4.9.3 Element Solution

For computing element solutions, the global degrees of freedom for each element are first transformed into the local degrees of freedom by multiplying them by the T matrix for the element. These local degrees of freedom are then used with the axial deformation and beam interpolation functions to get complete displacement solutions. Finally the axial forces, bending moments, and shear forces are computed by appropriate differentiation. In terms of notation used in this section the appropriate expressions are as follows:

Global to local transformation:

$$\begin{pmatrix} d_1 \\ d_2 \\ \vdots \\ d_{12} \end{pmatrix} = \begin{pmatrix} H & 0 & 0 & 0 \\ 0 & H & 0 & 0 \\ 0 & 0 & H & 0 \\ 0 & 0 & 0 & H \end{pmatrix} \begin{pmatrix} u_1 \\ v_1 \\ \vdots \\ \theta_{z2} \end{pmatrix} \Longrightarrow d_l = Td$$

Axial displacement:

$$u(t) = \begin{pmatrix} 1 - \frac{t}{L} & \frac{t}{L} \end{pmatrix} \begin{pmatrix} d_1 \\ d_7 \end{pmatrix}; \qquad 0 \le t \le L$$

$$N_u^T = \begin{pmatrix} 1 - \frac{t}{L} & \frac{t}{L} \end{pmatrix}$$

Axial force:

$$F(t) = EA\frac{du(t)}{dt}; \qquad 0 \le s \le L$$

Transverse displacement in the local s direction:

$$v(t) = \begin{pmatrix} 1 - \frac{3t^2}{L^2} + \frac{2t^3}{L^3} & t - \frac{2t^2}{L} + \frac{t^3}{L^2} & \frac{3t^2}{L^2} - \frac{2t^3}{L^3} & -\frac{t^2}{L} + \frac{t^3}{L^2} \end{pmatrix} \begin{pmatrix} d_2 \\ d_6 \\ d_8 \\ d_{12} \end{pmatrix} + \frac{q_s(L-t)^2 t^2}{24EI_r}; \qquad 0 \le t \le L$$

$$N_v^T = \begin{pmatrix} 1 - \frac{3t^2}{L^2} + \frac{2t^3}{L^3} & t - \frac{2t^2}{L} + \frac{t^3}{L^2} & \frac{3t^2}{L^2} - \frac{2t^3}{L^3} & -\frac{t^2}{L} + \frac{t^3}{L^2} \end{pmatrix}$$

Transverse displacement in the local r direction:

$$w(t) = \begin{pmatrix} 1 - \frac{3t^2}{L^2} + \frac{2t^3}{L^3} & -\left(t - \frac{2t^2}{L} + \frac{t^3}{L^2}\right) & \frac{3t^2}{L^2} - \frac{2t^3}{L^3} & -\left(-\frac{t^2}{L} + \frac{t^3}{L^2}\right) \end{pmatrix} \begin{pmatrix} d_3 \\ d_5 \\ d_9 \\ d_{11} \end{pmatrix} + \frac{q_r (L-t)^2 t^2}{24 E I_s}; \qquad 0 \le t \le L$$

$$N_w^T = \begin{pmatrix} 1 - \frac{3t^2}{L^2} + \frac{2t^3}{L^3} & -\left(t - \frac{2t^2}{L} + \frac{t^3}{L^2}\right) & \frac{3t^2}{L^2} - \frac{2t^3}{L^3} & -\left(-\frac{t^2}{L} + \frac{t^3}{L^2}\right) \end{pmatrix}$$

Bending moments:

$$M_r(t) = EI_r \left(\frac{d^2 v}{dt^2} \right); \qquad 0 \le t \le L$$

$$M_s(t) = -EI_s \left(\frac{d^2 w}{dt^2} \right); \qquad 0 \le t \le L$$

Shear forces:

$$V_s(t) = \frac{dM_r}{dt} = EI_r \left(\frac{d^3 v}{dt^3} \right); \qquad 0 \le t \le L$$

$$V_r(t) = -\frac{dM_s}{dt} = EI_s \left(\frac{d^3 w}{dt^3} \right); \qquad 0 \le t \le L$$

Twisting moment:

$$\psi(t) = \begin{pmatrix} 1 - \frac{t}{L} & \frac{t}{L} \end{pmatrix} \begin{pmatrix} d_4 \\ d_{10} \end{pmatrix}; \qquad 0 \le t \le L$$

$$M_t(t) = GJ \frac{d\psi}{dt}$$

Example 4.14 Analyze the one-story three-dimensional frame shown in Figure 4.48. The height of the columns is 12 ft and the length of the beams is 10 ft. Each beam is subjected to a uniformly distributed load of 2 kip/ft in the downward direction. I-shape sections are used for both columns and beams with the arrangement shown in the figure. The columns are connected to the foundation through simple connections that do not resist moments. The material is steel with $E = 29000\ \text{kip/in}^2$ and $G = 11200\ \text{kip/in}^2$. The section properties are as follows:

Beams:	$A = 3.2\ \text{in}^2$;	$J = 43\ \text{in}^4$;	$I_{max} = I_r = 450\ \text{in}^4$;	$I_{min} = I_s = 32\ \text{in}^2$
Columns:	$A = 4\ \text{in}^2$;	$J = 60\ \text{in}^4$;	$I_{max} = I_r = 650\ \text{in}^4$;	$I_{min} = I_s = 54\ \text{in}^2$

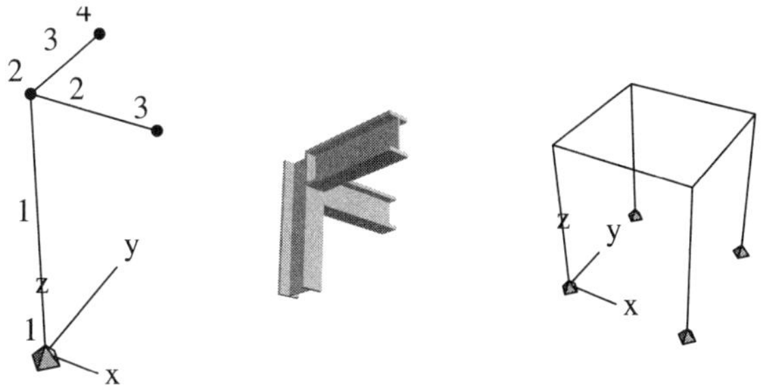

Figure 4.48. One-story space frame

Taking advantage of symmetry, we model a quarter of the frame using three elements. Because of symmetry, the boundary conditions at nodes 3 and 4 are as follows:

Node 3: $u = 0$; $\theta_y = 0$; $\theta_z = 0$

Node 4: $v = 0$; $\theta_x = 0$; $\theta_z = 0$

The distributed load is applied to the elements in their local coordinates. Therefore, to assign proper direction and sign to the distributed loads, we must carefully establish the local coordinates for the elements as follows:

Element 1: Nodes 1, 2, and 4
$\Longrightarrow$ t axis along global z; s axis along global x; r axis along global y

Element 2: Nodes 2, 3, and 4
$\Longrightarrow$ t axis along global x; s axis along global $-z$; r axis along global y
Distributed load: $q_r = 0$; $q_s = \frac{2}{12}$ kip/in

Element 3: Nodes 2, 4, and 3
$\Longrightarrow$ t axis along global y; s axis along global z; r axis along global x
Distributed load: $q_r = 0$; $q_s = -\frac{2}{12}$ kip/in

Matrices for this example are too large for printing. A complete solution can seen on the book website. Solving the final system of global equations, we get

$$\{\theta x_1 = 0.000398634, \theta y_1 = -0.00015917, \theta z_1 = 0, u_2 = 0.000575402, v_2 = 0.000117025,$$
$$w_2 = -0.0248276, \theta x_2 = -0.000799706, \theta y_2 = 0.000330328, \theta z_2 = 0, v_3 = 0.000117025,$$
$$w_3 = -0.041634, \theta x_3 = -0.000799706, u_4 = 0.000575402, w_4 = -0.0557153,$$
$$\theta y_4 = 0.000330328\}$$

The solution for element 1 is as follows:

Nodal values in global coordinates,

$$\boldsymbol{d}^T = \{0, 0, 0, 0.000398634, -0.00015917, 0, 0.000575402, 0.000117025, -0.0248276, -0.000799706, 0.000330328, 0\}$$

Nodal values in local coordinates,

$$\boldsymbol{d}_l = \boldsymbol{Td} = \{0., 0., 0., 0., 0.000398634, -0.00015917, -0.0248276, 0.000575402, 0.000117025, 0., -0.000799706, 0.000330328\}$$

Axial effects:

Interpolation functions, $\boldsymbol{N}_u^T = \{1. - 0.00694444t, 0.00694444t\}$

Axial displacement, $u(t) = \boldsymbol{N}_u^T \begin{pmatrix} d_1 \\ d_7 \end{pmatrix} = -0.000172414t$

Axial force, $EA\, du(t)/dt = -20.$

Torsional effects:

Twist angle, $\psi(t) = \boldsymbol{N}_u^T \begin{pmatrix} d_4 \\ d_{10} \end{pmatrix} = 0$

Twisting moment, $GJ\, d\psi(t)/dt = 0.$

Bending about r axis:

$$N_v^T = \{6.69796 \times 10^{-7}t^3 - 0.000144676t^2 + 1, 0.0000482253t^3 - 0.0138889t^2 + t, 0.000144676t^2 - 6.69796 \times 10^{-7}t^3, 0.0000482253t^3 - 0.00694444t^2\}$$

$$v(t) = \boldsymbol{N}_v^T \begin{pmatrix} d_2 \\ d_6 \\ d_8 \\ d_{12} \end{pmatrix} = 7.86875 \times 10^{-9}t^3 - 0.00015917t$$

Bending moment, $M_r = EI_r\, d^2v(t)/dt^2 = 0.889955t$

Shear force, $V_s = dM_r/dt = 0.889955$

Bending about s axis:

$$\boldsymbol{N}_w^T = \{6.69796 \times 10^{-7}t^3 - 0.000144676t^2 + 1, -0.0000482253t^3 + 0.0138889t^2 - t, 0.000144676t^2 - 6.69796 \times 10^{-7}t^3, 0.00694444t^2 - 0.0000482253t^3\}$$

$$w(t) = \boldsymbol{N}_w^T \begin{pmatrix} d_3 \\ d_5 \\ d_9 \\ d_{11} \end{pmatrix} = -1.94202 \times 10^{-8}t^3 + 3.38615 \times 10^{-8}t^2 + 0.000398634t$$

Bending moment, $M_s = -EI_s\, d^2w(t)/dt^2 = -0.180999t$

Shear force, $V_r = -dM_s/dt = 0.180999$

Solutions over the remaining elements can be obtained in a similar manner:

Forces and Moments at Element Ends

	x	y	z	Axial Force	V_s	V_r	M_r	M_s	M_t
1	0	0	0	−20.	0.889955	0.180999	0	0	0
	0	0	144.	−20.	0.889955	0.180999	128.154	−26.0639	0
2	0	0	144.	−0.889955	−10.	0	128.154	0	0
	60.	0	144.	−0.889955	0	0	−171.846	0	0
3	0	0	144.	−0.180999	10.	0	−26.0639	0	0
	0	60.	144.	−0.180999	0	0	273.936	0	0

◆ ***Mathematica*/MATLAB Implementation 4.5 on the Book Web Site:** *Analysis of space frames*

4.10 FRAMES IN MULTISTORY BUILDINGS

From a structural point of view most modern multistory buildings consist of a grid of structural frames. The floor system, consisting of metal decks and concrete slabs, span across the frames and tie everything together to create a three-dimensional structure, as shown in Figure 4.49. Because of large dimensions, the floor system within its own plane is essentially rigid and is known as a rigid diaphragm. Thus individual frames in a building do not behave as their isolated counterparts. We must analyze the entire three-dimensional system.

With the assumption of an in-plane rigid-floor system, each story has only three independent in-plane degrees of freedom: the two displacements in the x and y directions (u and

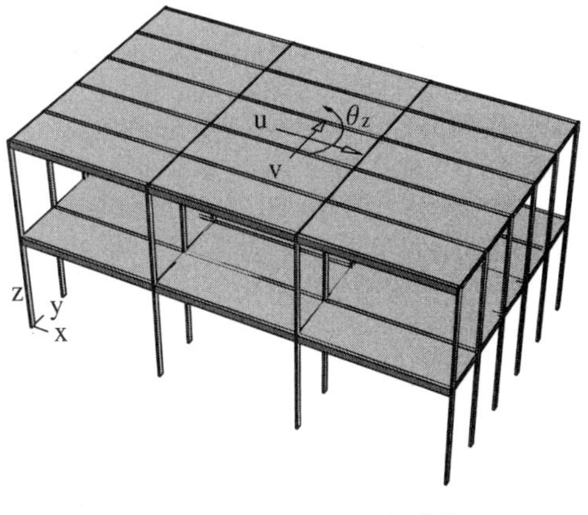

Figure 4.49. Multistory building

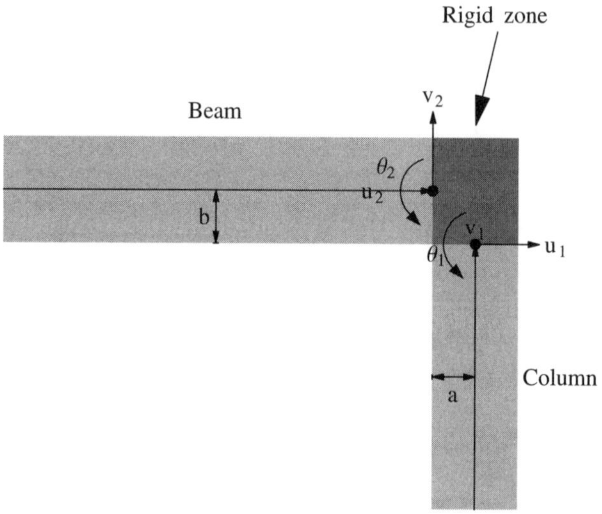

Figure 4.50. Rigid zone at beam-column connections

v) and a rotation about the z axis (θ). Thus, to analyze a building, we use standard frame element equations and assemble contributions from all frames in the building in the usual manner. For each story we define constraints that relate all in-plane degrees of freedom to one node that we call the master node for that story. Identifying the master node degrees of freedom by subscript m, the constraints for degrees of freedom at any other node in the story are expressed as follows:

$$\begin{pmatrix} u_i \\ v_i \\ \theta_i \end{pmatrix} = \begin{pmatrix} 1 & 0 & -(y_i - y_m) \\ 0 & 1 & x_i - x_m \\ 0 & 0 & 1 \end{pmatrix} \begin{pmatrix} u_m \\ v_m \\ \theta_m \end{pmatrix}; \qquad i = 1, 2, \ldots$$

where x_i and y_i are the coordinates of node i.

Joints in a building frame also require special modeling care. The column dimensions in a typical building are of the order 14 to 20 in while the beams may vary in depth from 21 in to 30 in. When creating frame models using centerline dimensions, the joint zone with these large member sizes is fairly large. Very little deformations are expected within this joint zone. Thus rigid joint zones are typically created when analyzing building frames, as shown in Figure 4.50. The nodes are placed outside of the joint region.

The following constraints are defined between the degrees of freedom at these nodes to create the effect of the rigid joint zone:

$$\theta_1 = \theta_2; \qquad u_1 = u_2 + b\theta_2; \qquad v_1 = v_2 + a\theta_2$$

After incorporating these constraints, the rest of the analysis follows the standard procedures used in other examples in this chapter.

PROBLEMS

Plane Trusses

4.1 Determine joint displacements and axial forces in the three-bar pin-jointed structure shown in Figure 4.51. All members have the same cross-sectional area and are of the same material, $A = 1\ \text{in}^2$ and $E = 30 \times 10^6\ \text{lb/in}^2$. The load $P = 15{,}000$ lb. The dimensions in inches are shown in the figure.

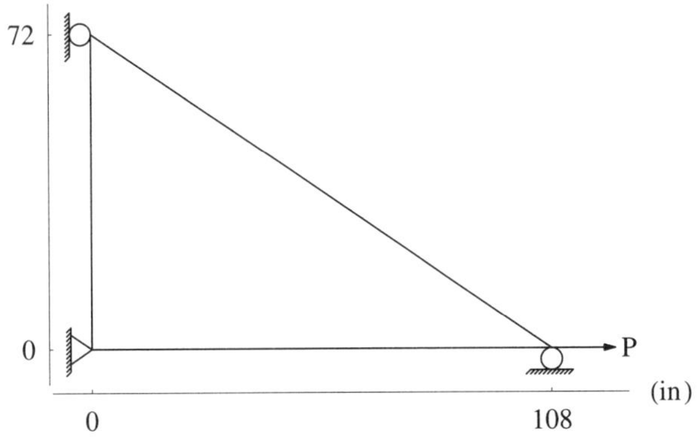

Figure 4.51.

4.2 Determine joint displacements and axial forces in the truss shown in Figure 4.51 if the support where the load is applied settles down by $\frac{1}{2}$ in.

4.3 Taking advantage of symmetry, determine joint displacements and axial forces in the three-bar truss shown in Figure 4.52. All members have the same cross-sectional area and are of the same material, $A = 0.001\ \text{m}^2$ and $E = 200$ GPa. The load $P = 20$ kN. The dimensions in meters are shown in the figure. One of the goals for this problem is for you to learn the correct use of symmetry. Thus no credit will be given if you do not use symmetry even if your solution is correct.

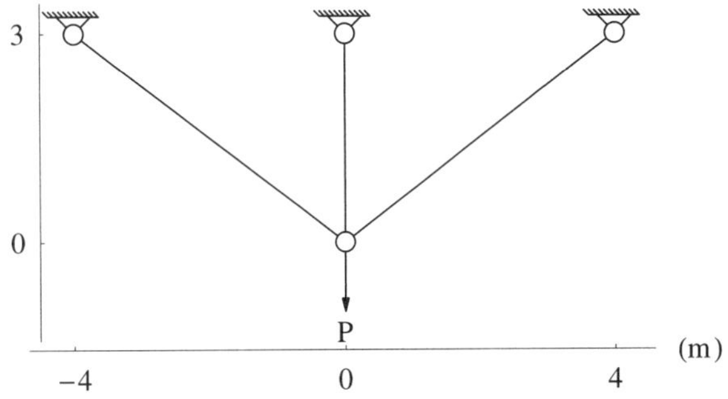

Figure 4.52.

4.4 Determine joint displacements and axial forces in the truss shown in Figure 4.53. All members have the same cross-sectional area and are of the same material, $A = 2\ \text{in}^2$ and $E = 30 \times 10^6\ \text{lb/in}^2$. The load $P = 30{,}000$ lb. The dimensions in feet are shown in the figure.

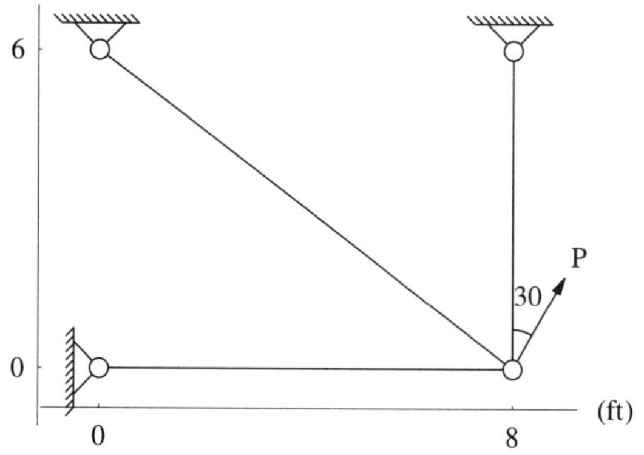

Figure 4.53.

4.5 Determine joint displacements and axial forces in the truss shown in Figure 4.53 if in addition to the applied force the lower left support moves horizontally toward the right by $\frac{1}{4}$ in.

4.6 Determine joint displacements and axial forces in the truss shown in Figure 4.54. Note the diagonals are not connected to each other at their crossing. The cross-sectional area of vertical and horizontal members is $30 \times 10^{-4}\ \text{m}^2$ and that for the diagonals is $10 \times 10^{-4}\ \text{m}^2$. All members are made of steel with $E = 210$ GPa. The load $P = 20$ kN. The dimensions in meters are shown in the figure.

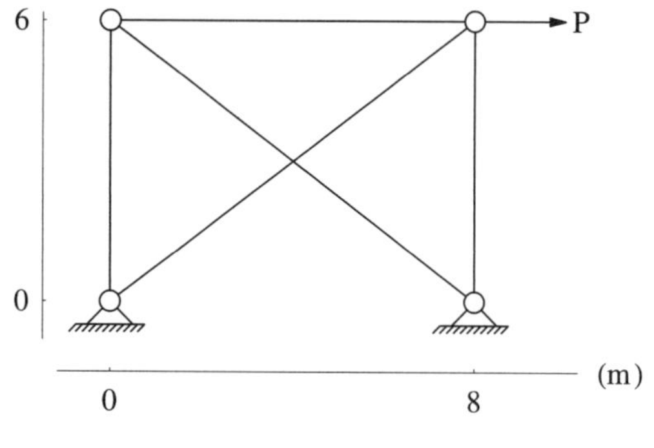

Figure 4.54.

Space Trusses

4.7 Determine joint displacements and axial forces in the space truss shown in Figure 4.55. The cross-sectional area of members is $15 \times 10^{-4}\ \text{m}^2$. All members are made

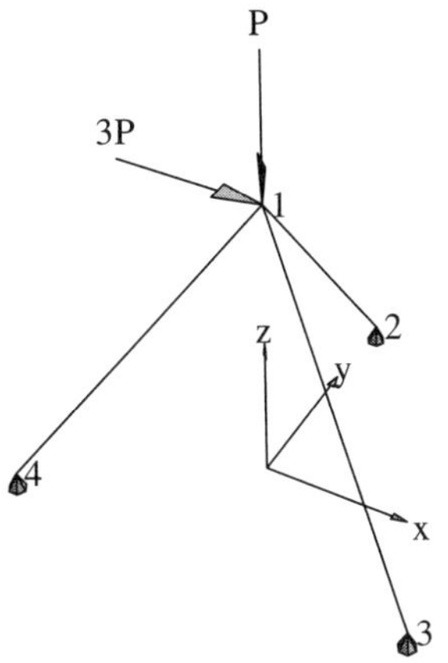

Figure 4.55.

of steel with $E = 210$ GPa. The load $P = 10$ kN is applied in the $-z$ direction and $3P$ in the x direction. The nodal coordinates in meters are as follows:

	x	y	z
1	0.	0.	10.
2	0.	8.	0.
3	6.9282	−4.	0.
4	−6.9282	−4.	0.

4.8 Determine joint displacements and axial forces in the space truss shown in Figure 4.56. The cross-sectional area of members is 15×10^{-4} m^2. All members are made

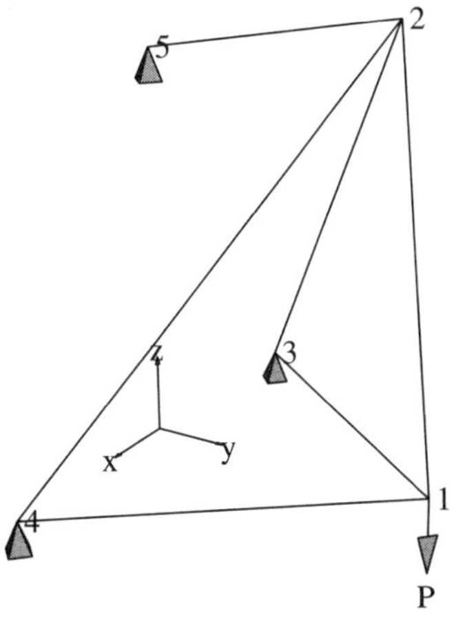

Figure 4.56.

of aluminum with $E = 70$ GPa. The load $P = 10$ kN is applied in the $-z$ direction. The nodal coordinates in meters are as follows:

	x	y	z
1	0.	4.	0.
2	−3.	2.	5.
3	−3.	0.	0.
4	3.	0.	0.
5	0.	0.	5.

Thermal and Initial Strains

4.9 A copper rod 1.4 in in diameter is placed in an aluminum sleeve with inside diameter 1.42 in and wall thickness 0.2 in, as shown in Figure 4.57. The rod is 0.005 in longer than the sleeve. A load $P = 60{,}000$ lb is applied to the assembly through a large bearing plate that can be considered rigid. Determine stresses in the rod and the sleeve. The modulus of elasticity for copper is 17×10^6 psi and that for aluminum is 10×10^6 psi.

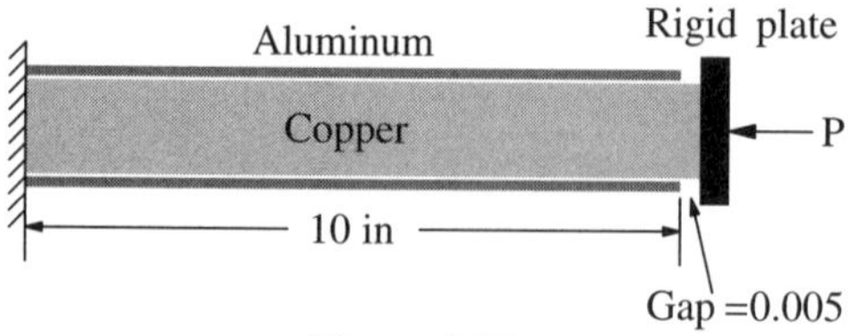

Figure 4.57.

4.10 Determine joint displacements and axial forces in the truss shown in Figure 4.51 if instead of load P the diagonal member experiences a temperature decrease of 70°F. The coefficient of thermal expansion is $\alpha = 6 \times 10^{-6}$/°F.

4.11 Determine joint displacements and axial forces in the truss shown in Figure 4.53. During construction the diagonal member was fabricated $\frac{1}{4}$ in too short and was forced to fit into the assembly through heat treatment.

4.12 Determine joint displacements and axial forces in the space truss shown in Figure 4.56 if instead of load P the member between nodes 1 and 2 experiences a temperature rise of 50°C. The coefficient of thermal expansion is $\alpha = 23 \times 10^{-6}$/°C.

Spring Elements

4.13 Determine support reactions and forces in the springs for the assembly shown in Figure 4.58.

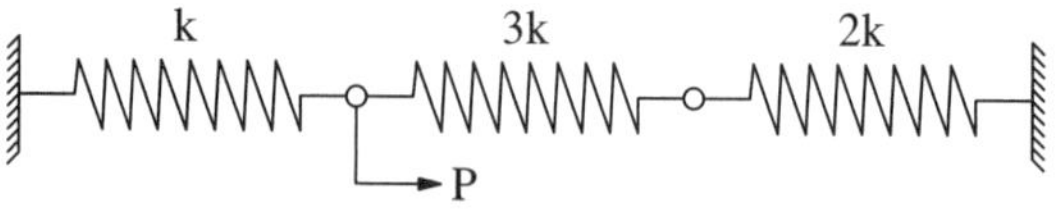

Figure 4.58.

4.14 Determine support reactions and forces in the springs when the assembly is forced to close the gap g, as shown in the Figure 4.59.

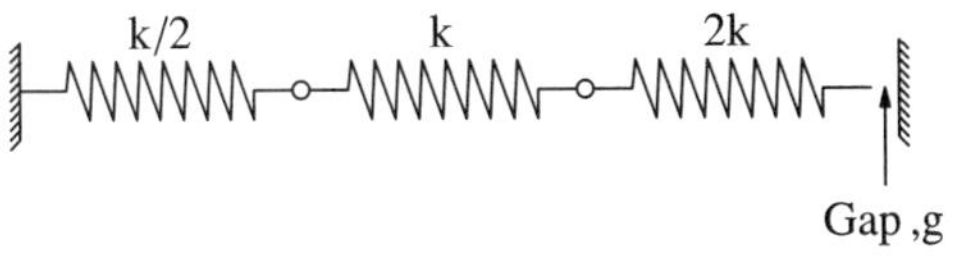

Figure 4.59.

Beam Bending

4.15 A fixed-end beam is subjected to a concentrated load at the midspan $P = 200$ lb, as shown in Figure 4.60. The beam has a rectangular cross section with width = 12 in and height = 1 in. The length of the beam is $L = 200$ in and its modulus of elasticity is $E = 10^7$ lb/in^2. Taking advantage of symmetry, use only one beam element to determine the maximum bending and shear stresses in the beam.

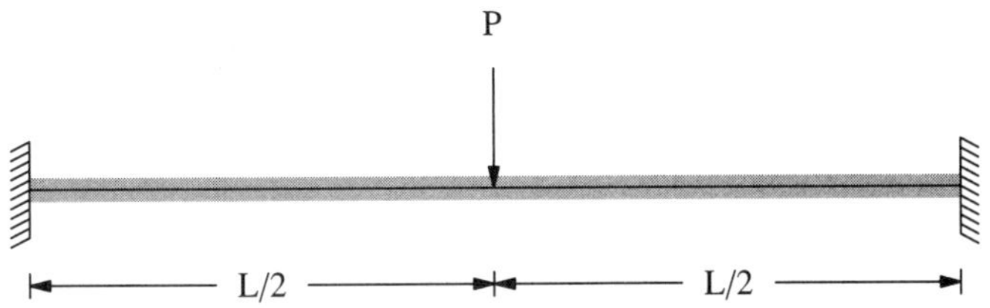

Figure 4.60. Fixed-end beam

4.16 Two cantilever beams are joined together through a simple pin, as shown in Figure 4.61. The concentrated load at the midspan is $P = 200$ lb. The beam has a rectangular cross section with width = 12 in and height = 1 in. The length of the beam is $L = 200$ in and its modulus of elasticity is $E = 10^7$ lb/in^2. Taking advantage of symmetry, use only one beam element to determine maximum bending and shear stresses in the beam.

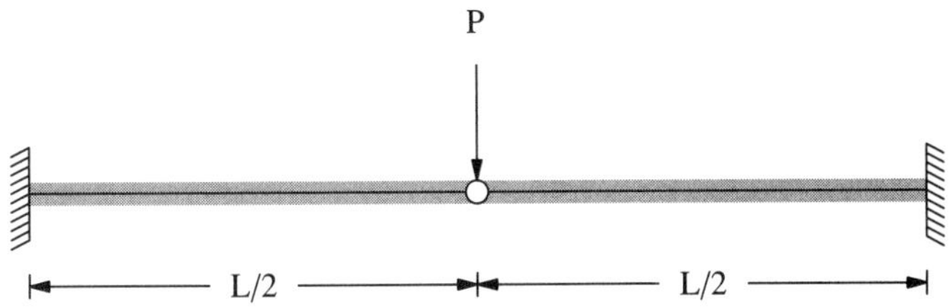

Figure 4.61. Beams connected through a pin

4.17 Two cantilever beams are joined together through a simple pin, as shown in Figure 4.62. The $EI = 10^7$ for the right beam and is twice that for the left beam. The concentrated load at the hinge location is $P = 200$ lb. Determine the displacement and bending moment distribution over the beams.

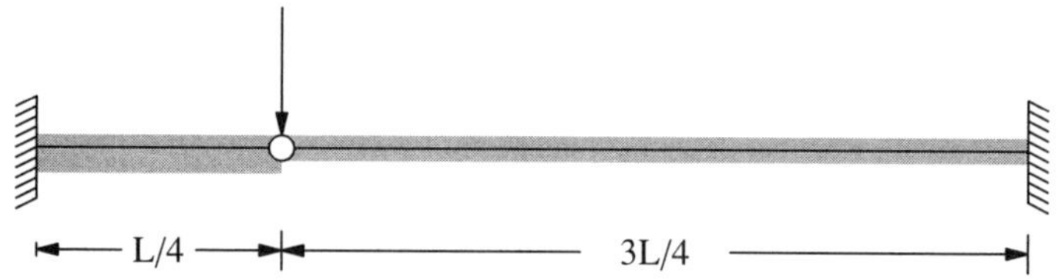

Figure 4.62. Beams connected through a pin

4.18 A two-span beam is subjected to a moment, as shown in the Figure 4.63. Find the resulting displacement, shear force, and bending moments. Compute shear and bending stresses at the middle support in the beam. Use the following numerical data ($1\ k = 1000$ lb).

$$L = 180 \text{ in}; \qquad E = 29{,}000\ k/\text{in}^2; \qquad M = 1000\ k \cdot \text{in}$$

Beam cross section: Standard I-shape (W16 × 26) with $I = 301$ in^4 and dimensions as shown in the figure.

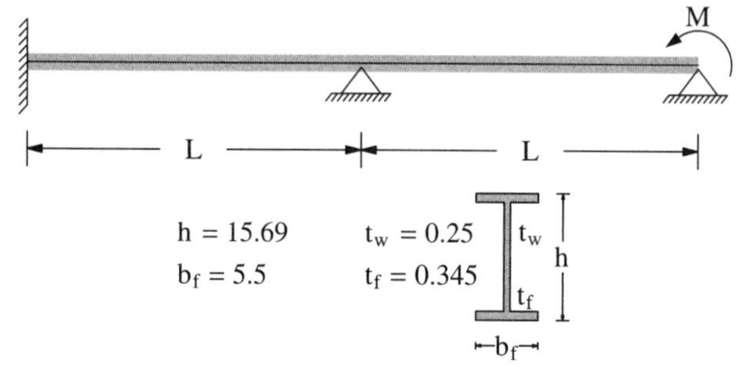

Figure 4.63.

4.19 Immediately after construction the right support of the beam shown in Figure 4.64 undergoes a settlement equal to Δ. Find the resulting displacement, shear force, and bending moments. Compute shear and bending stresses at the middle support in the beam. Use the following numerical data:

$$L = 8 \text{ m}; \qquad E = 200 \text{ GPa}; \qquad \Delta = 10 \text{ mm}$$

Beam cross section: Standard I-shape with $I = 125.3 \times 10^{-6}$ m^4 and dimensions (in mm) as shown in the figure.

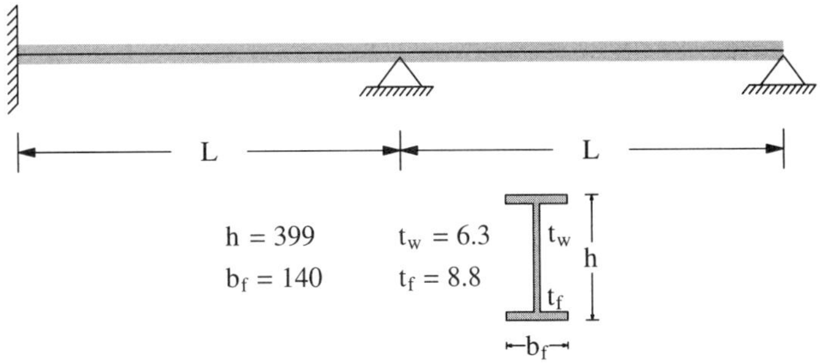

Figure 4.64.

4.20 Find displacements and draw shear force and bending moment diagrams for the beam shown in Figure 4.65. Assume $E = 210$ GPa, $I = 4 \times 10^{-4}$ m^4, $L = 2$ m, $P = 10$ kN, $M = 20$ kN · m.

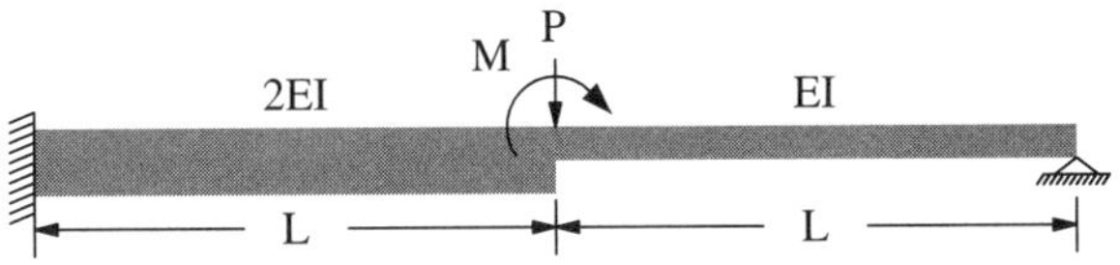

Figure 4.65. Beam with variable section

4.21 Two uniform beams are connected together through a spring as shown in Figure 4.66. Determine deflections, bending moment, and shear forces in the beams. What is the force in the spring? The numerical values are

$E = 10^4$ ksi; $I = 1000$ in^4; $L = 100$ in; $P = 100$ kips; $k = 2000$ kip/in

Use kip · in units in calculations.

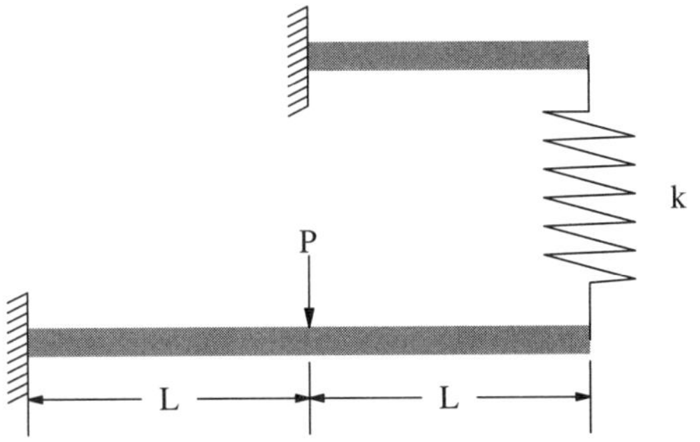

Figure 4.66. Two beams connected through a spring

4.22 Consider a uniform beam simply supported at the ends and spring supported in the middle, as shown in Figure 4.67. Taking advantage of symmetry, analyze the beam using only one element. Assume the following numerical data:

$L = 4$ m; $E = 70$ GPa; $I = 2 \times 10^4$ mm^4; $k = 400$ N/mm; $P = 20$ kN

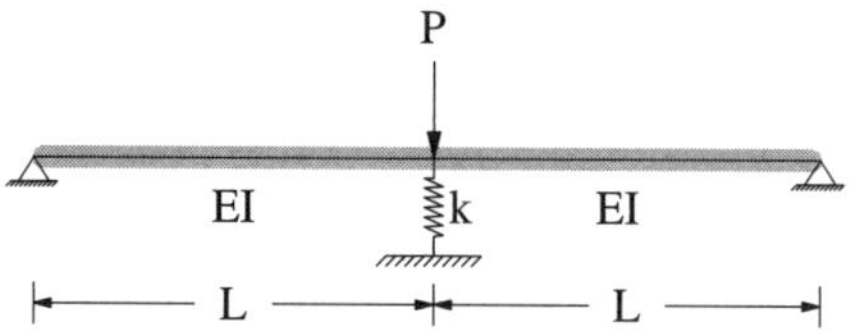

Figure 4.67. Beam supported by a spring

4.23 The left half of a beam is subjected to a uniformly distributed load $q = 1$ lb/in, as shown in Figure 4.68. The beam has a rectangular cross section with width = 12 in and height = 1 in. The length of the beam is $L = 200$ in and its modulus of elasticity is $E = 10^7$ lb/in^2. Create the simplest possible finite element model for this beam. Using this model, determine the bending and shear stresses in the beam at $L/4$ from the left end.

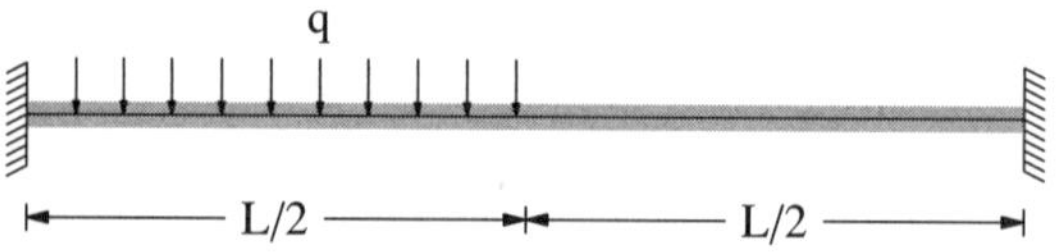

Figure 4.68. Fixed-end beam

4.24 Using only one element, find the *midspan* displacement, bending moment, and shear force for the beam shown in Figure 4.69. Note that the loading is large enough to cause the gap to close. Use the following data: $E = 10^4$ ksi, $I = 1000$ in^4, $L = 100$ in, $q = 1$ k/in, gap = 0.25 in.

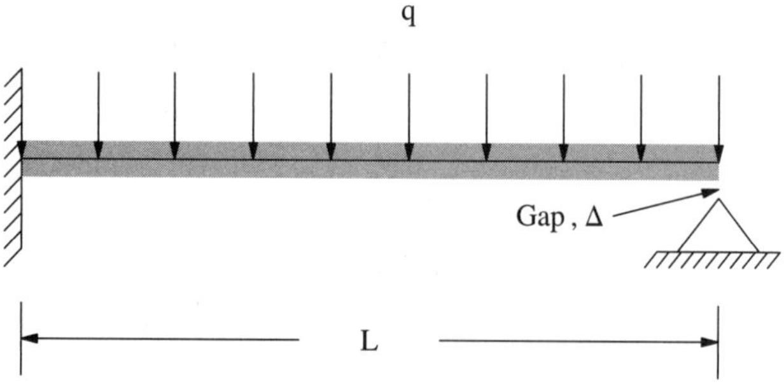

Figure 4.69. Single span beam with a gap

4.25 Compute stresses in the two-span beam shown in Figure 4.70. There is a small gap between the middle support and the beam:

$$L = 6\text{ m};\qquad q = 10\text{ kN/m};\qquad E = 200\text{ GPa};\qquad \text{Gap }\Delta = 5\text{ mm}$$

Beam cross section: rectangular with width = 120 mm and height = 300 mm.

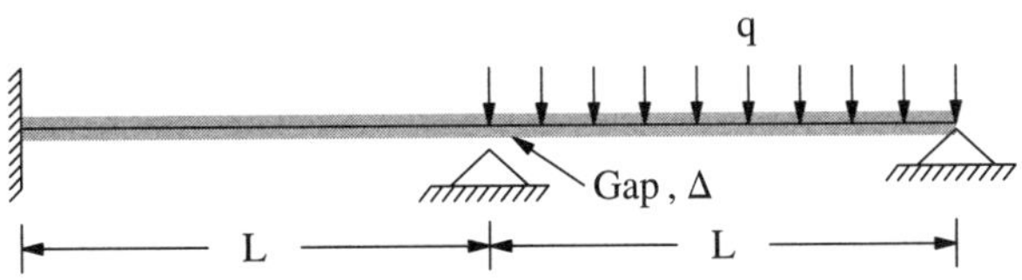

Figure 4.70. Two-span beam

4.26 A beam with an overhang is subjected to a uniformly distributed load $q = 27$ lb/in over the right span and a moment $M = \frac{1}{2}$ lb · in at the free end, as shown in Figure 4.71. The beam has a rectangular cross section with width = 12 in and height = 1 in, giving $A = 12$ in^2 and $I = 1$ in^4. The length of the beam is $L = 1$ in and its modulus of elasticity is $E = 1$ lb/in^2.

(a) Create the simplest possible finite element model for this beam. Write element equations, assemble to form global equations, and obtain the reduced system of equations, taking boundary conditions into consideration.

(b) After a analysis of this beam it is found that the rotation at the free end is $\frac{1}{12}$ rad and that at the left support is $-\frac{1}{12}$ rad. Determine the vertical displacement and bending moment at point a in the beam.

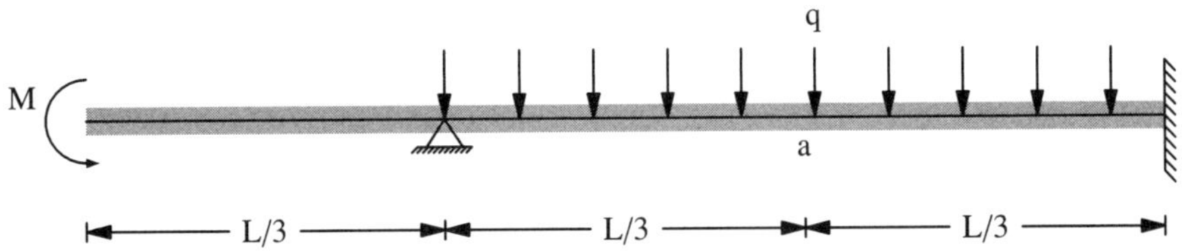

Figure 4.71.

4.27 Using the simplest possible finite element model, determine the maximum bending and shear stress in a uniform beam simply supported at the left end and spring supported at the right end, as shown in Figure 4.72. Assume the following numerical data:

$$a = 5\text{ kN/m};\qquad L = 5\text{ m};\qquad E = 200\text{ GPa};\qquad I = 10^6\text{ mm}^4;$$

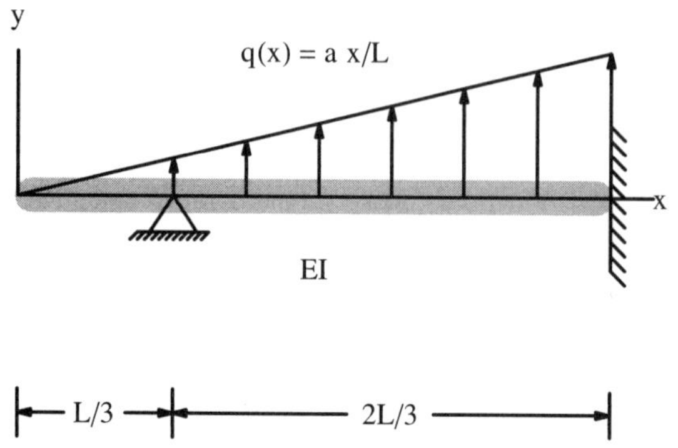

Figure 4.72. Uniform beam subjected to a linearly increasing load

4.28 Two overhanging beams are joined together through a simple pin connection as shown in Figure 4.73. Determine displacement, shear force, and bending moments. Use the following numerical data:

$$L = 8 \text{ m}; \qquad E = 200 \text{ GPa}; \qquad q = 3 \text{ kN/m}$$

Beam cross section: Standard I-shape with $I = 125.3 \times 10^{-6}$ m^4 and dimensions (in mm) as shown in the figure.

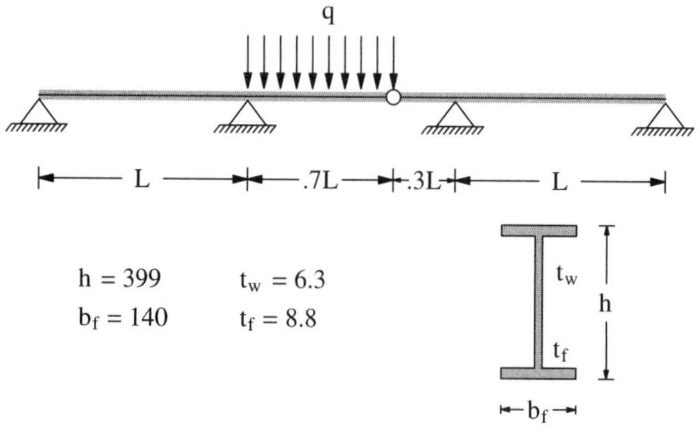

Figure 4.73.

Plane Frames

4.29 A 300-mm-wide and 100-mm-thick bar is supported and loaded as shown in Figure 4.74. Determine displacement, shear force, and bending moments. Assume E = 10 GPa.

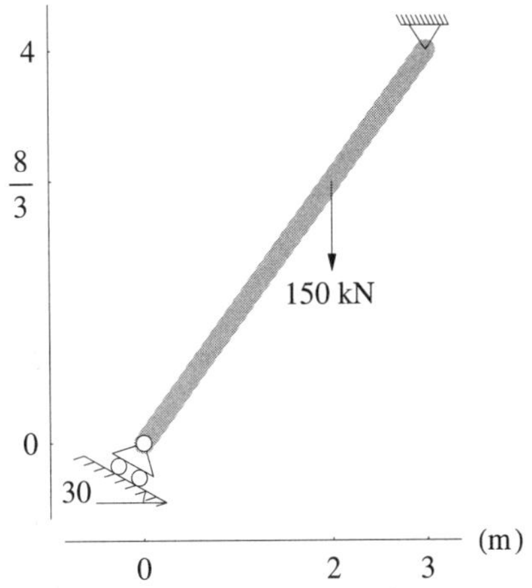

Figure 4.74.

4.30 Determine displacements, bending moments, and shear forces in the plane frame shown in Figure 4.75. Draw free-body diagrams for each element clearly showing all element end forces and moments. Assume $q = 10\,\text{kN/m}$, $L = 2\,\text{m}$, $E = 210\,\text{GPa}$, $A = 4 \times 10^{-2}\,\text{m}^2$, and $I = 4 \times 10^{-4}\,\text{m}^4$. Take advantage of symmetry.

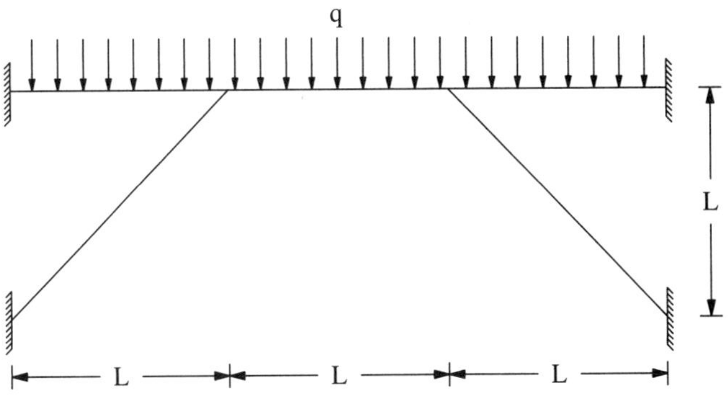

Figure 4.75.

4.31 Determine displacements, bending moments, and shear forces in the plane frame shown in Figure 4.76. Draw free-body diagrams for each element clearly showing all element end forces and moments. Assume $q = 10\,\text{kN/m}$, $L = 2\,\text{m}$, $E = 210\,\text{GPa}$, $A = 4 \times 10^{-2}\,\text{m}^2$, and $I = 4 \times 10^{-4}\,\text{m}^4$. Take advantage of symmetry. Use $\text{kN} \cdot \text{m}$ units.

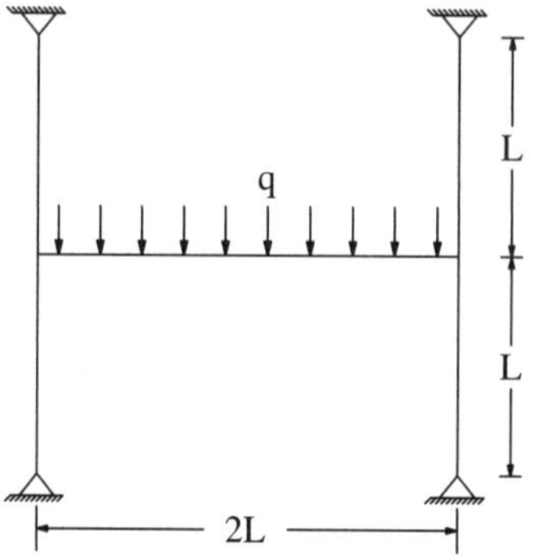

Figure 4.76.

4.32 The element stiffness matrices for vertical and horizontal elements of a plane frame have already been computed and assembled into a 9×9 global matrix as follows:

$$\begin{pmatrix}
1875. & 0 & -3750. & -1875. & 0 & -3750. & 0 & 0 & 0 \\
0 & 25000. & 0 & 0 & -25000. & 0 & 0 & 0 & 0 \\
-3750. & 0 & 10000. & 3750. & 0 & 5000. & 0 & 0 & 0 \\
-1875. & 0 & 3750. & 51875. & 0 & 3750. & -50000. & 0 & 0 \\
0 & -25000. & 0 & 0 & 40000. & 15000. & 0 & -15000. & 15000. \\
-3750. & 0 & 5000. & 3750. & 15000. & 30000. & 0 & -15000. & 10000. \\
0 & 0 & 0 & -50000. & 0 & 0 & 50000. & 0 & 0 \\
0 & 0 & 0 & 0 & -15000. & -15000. & 0 & 15000. & -15000. \\
0 & 0 & 0 & 0 & 15000. & 10000. & 0 & -15000. & 20000.
\end{pmatrix}$$

It is decided to brace the frame by an inclined truss member as shown in Figure 4.77. Write down the stiffness matrix for the inclined truss member. Assemble it into the global matrix. Determine nodal displacements due to a downward concentrated load

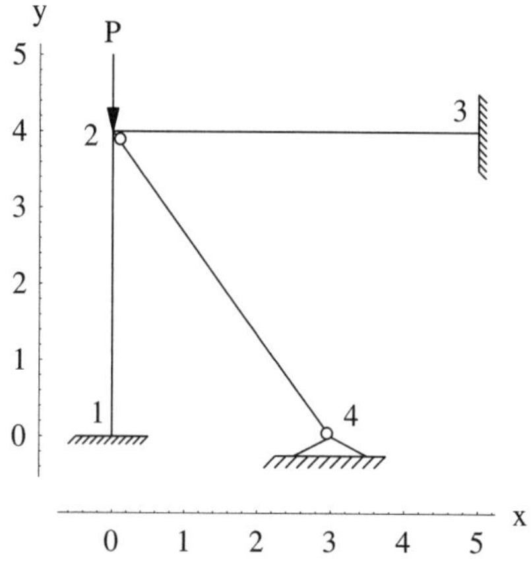

Figure 4.77.

of $P = 50$ kN applied at the corner. For all members, $E = 210$ GPa, $I = 10^{-4}\ \text{m}^4$, and $A = 10^{-2}\ \text{m}^2$.

4.33 Determine displacement, shear force, and bending moments in the plane frame shown in Figure 4.78. Use the following numerical data:

$$P = 12\,\text{kN}; \qquad E = 200\,\text{GPa}; \qquad q = 3\,\text{kN/m}$$

Cross-section: Standard I-shape with $A = 4.95 \times 10^{-3}\ \text{m}^2$ and $I = 125.3 \times 10^{-6}\ \text{m}^4$.

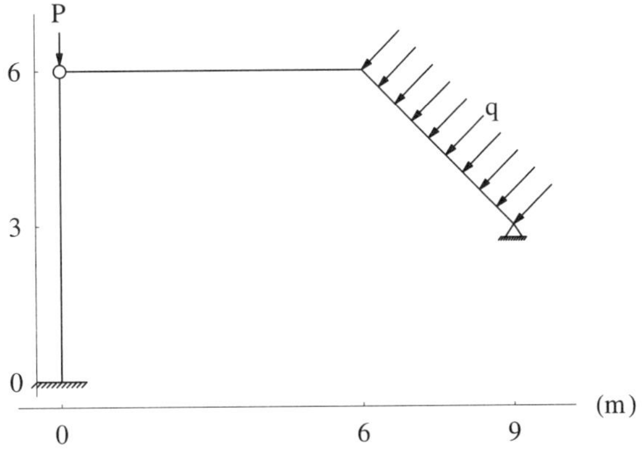

Figure 4.78.

4.34 Two 200 mm × 200 mm square bars are joined through a pin and loaded as shown in Figure 4.79. Determine displacement, shear force, and bending moments. Assume $E = 10$ GPa and $P = 20$ kN.

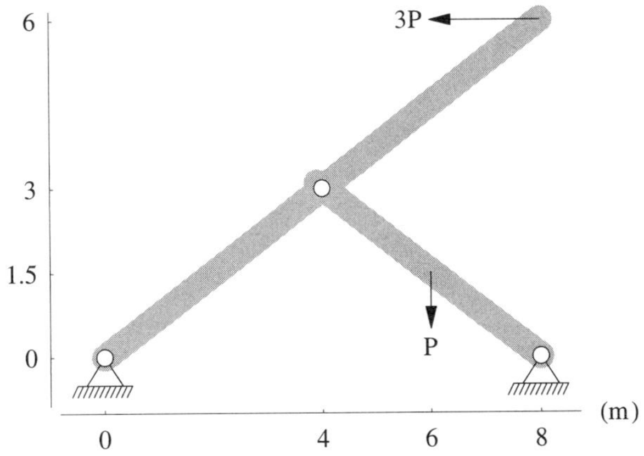

Figure 4.79.

4.35 The lower portion of an aluminum step bracket, shown in Figure 4.80, is subjected to a uniform pressure of 20 N/mm^2. The left end is fixed and the right end is free. The dimensions of the bracket are thickness = 3 mm, L = 150 mm, b = 10 mm, and h = 24 mm. The material properties are E = 70,000 N/mm^2 and ν = 0.3. For a preliminary analysis determine stresses in the bracket using a simple model consisting of three plane frame elements.

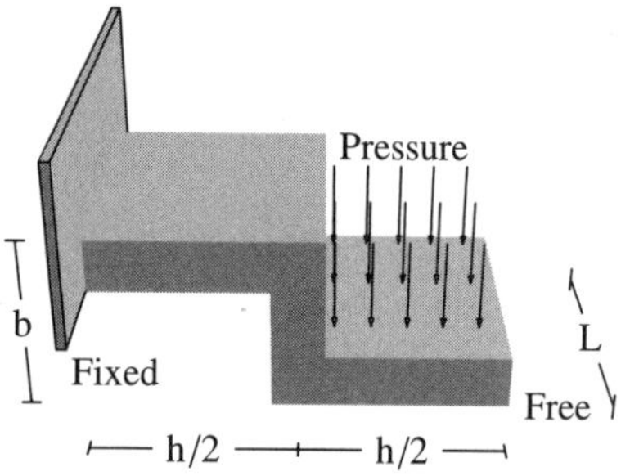

Figure 4.80. Step bracket

4.36 The top surface of an S-shaped aluminum block, shown in Figure 4.81, is subjected to a uniform pressure of 20 N/mm^2. The bottom is fixed. The dimensions of the block are t = 3 mm, L = 15 mm, b = 10 mm, and h = 24 mm. The material properties are E = 70,000 N/mm^2 and ν = 0.3. For a preliminary analysis determine stresses in the block using a simple model consisting of four plane frame elements.

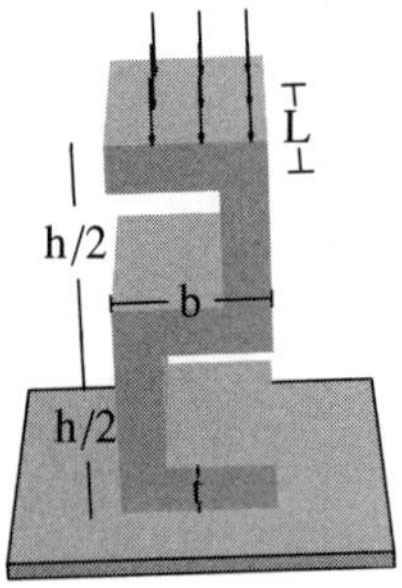

Figure 4.81. S-shaped aluminum block

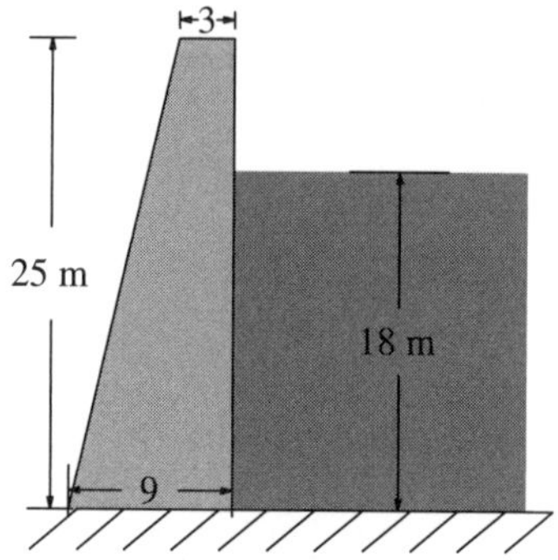

Figure 4.82. Concrete dam

4.37 The cross section of a concrete dam is shown in Figure 4.82. Using a simple plane frame model, estimate stresses in the dam due to self-weight of the dam and the water pressure. Use a reasonable number of elements and assign average section properties to each element. Assume unit thickness perpendicular to the plane. The depth of the water behind the dam is 18 m. The density of concrete is 2400 kg/m^3 and that of water is 1000 kg/m^3. The modulus of elasticity of concrete is 30 GPa.

4.38 The side view of a pry bar is shown in Figure 4.83. The cross section of the bar is rectangular with thickness = $\frac{1}{2}$ in and width (perpendicular to the plane of paper) = 1.2 in. The other dimensions (in inches) are shown in the figure. The material properties are $E = 29 \times 10^6$ psi and $\nu = 0.3$. A load of $P = 200$ lb is applied at the center of the handle. Using a suitable number of plane frame elements, determine maximum deflection and stress in the bar.

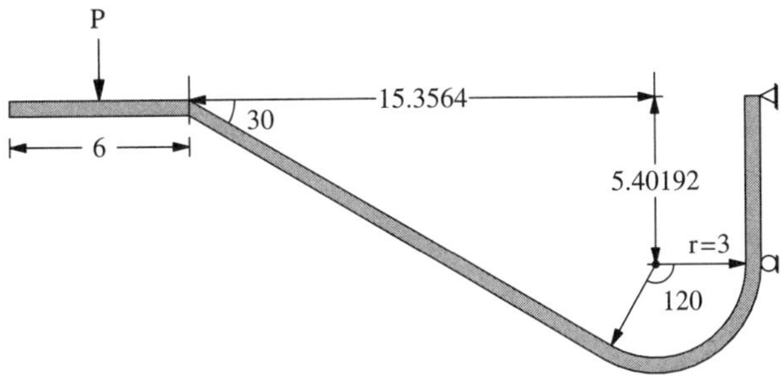

Figure 4.83. Pry bar

Space Frames

4.39 A cantilever beam is supported by a beam fixed at both ends, as shown in Figure 4.84. Analyze the system when it is subjected to loading as shown. Use the following numerical data:

$L = 4$ m; $b = 1$ m; $q = 100$ N/m; $E = 200$ GPa; $G = 100$ GPa

Beams, W360 × 44.6: $A = 5.71 \times 10^{-3}$ m^2; $J = 0.1729 \times 10^{-6}$ m^4; $I_{max} = 121.1 \times 10^{-6}$ m^4; $I_{min} = 8.16 \times 10^{-6}$ m^4

Beams, W310 × 38.7: $A = 4.94 \times 10^{-3}$ m^2; $J = 0.1421 \times 10^{-6}$ m^4; $I_{max} = 84.9 \times 10^{-6}$ m^4; $I_{min} = 7.2 \times 10^{-6}$ m^4

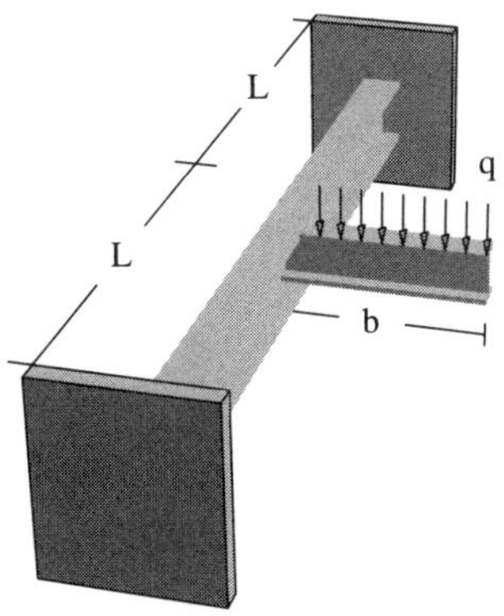

Figure 4.84.

4.40 Two beam cantilever out of an H-frame, as shown in Figure 4.85. The columns are fixed at the top and the bottom. Analyze the frame when it is subjected to loading as shown. Use the following numerical data.

$L = 20$ ft; $h = 15$ ft; $b = 20$ ft; $P = 30$ kips; $q = 2$ kips/ft; $w = 0.5$ kip/ft

Beams, W16 × 40: $A = 11.8\ \text{in}^2$; $J = 0.851\ \text{in}^4$;

$I_{max} = 518\ \text{in}^4$; $I_{min} = 28.9\ \text{in}^4$

Columns, W12 × 72: $A = 21.1\ \text{in}^2$; $J = 3.062\ \text{in}^4$;

$I_{max} = 597\ \text{in}^4$; $I_{min} = 195\ \text{in}^4$

$E = 29{,}000$ ksi; $G = 11{,}200$ ksi

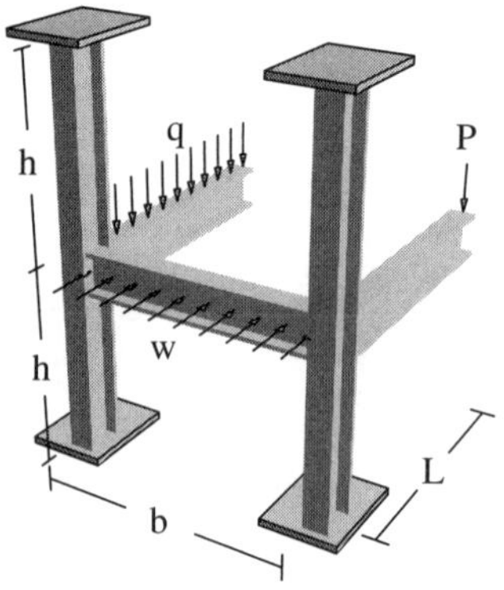

Figure 4.85.

CHAPTER FIVE

TWO-DIMENSIONAL ELEMENTS

In the previous chapters the basic finite element concepts were demonstrated through simple one-dimensional elements. For most of those problems either the exact solution is known or there are several other numerical methods available that may be more convenient to use than the finite element method. This is not the case for two- and three-dimensional problems. The exact solutions are rare for these problems. Most other numerical methods are not as widely applicable as the finite element method. One of the key advantages of the finite element method is that the fundamental concepts extend easily from one-dimensional to higher dimensional problems. The only complication is that the required integration and differentiation become more difficult, but these are calculus-related problems and have nothing to do with the finite element concepts. A review of vector calculus, especially integration over two- and three-dimensional domains, will make the transition from one-dimensional problems to two- and three-dimensional problems a lot smoother.

In this chapter the basic finite element concepts are illustrated with reference to the following partial differential equation defined over an arbitrary two-dimensional region, such as the one shown in Figure 5.1:

$$\frac{\partial}{\partial x}\left(k_x \frac{\partial u}{\partial x}\right) + \frac{\partial}{\partial y}\left(k_y \frac{\partial u}{\partial y}\right) + pu + q = 0$$

where $k_x(x, y)$, $k_y(x, y)$, $p(x, y)$, and $q(x, y)$ are known functions defined over the area. The unknown solution is $u(x, y)$. The equation can easily be recognized as a generalization of the one-dimensional boundary value problem considered in Chapter 3. Steady-state heat flow, variety of fluid flow, and torsion of planar sections are some of the common engineering applications that are governed by the differential equations which are special cases of this general boundary value problem.

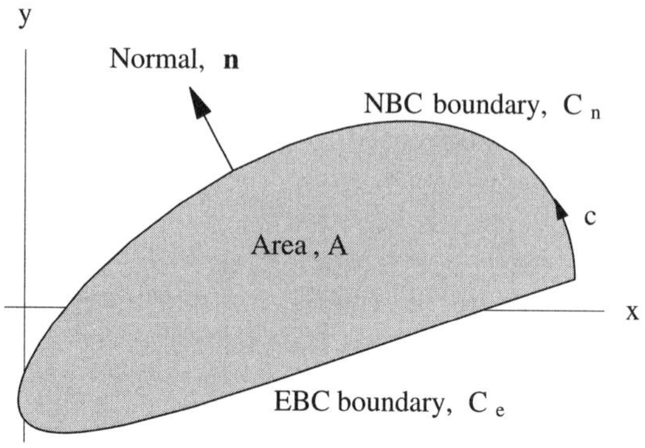

Figure 5.1. Two-dimensional solution domain

The differential equation is second order and therefore, based on the discussion in Chapter 2, the boundary conditions involving u are essential and those involving its derivatives are natural boundary conditions. The area of the solution domain is denoted by A and its boundary by C. The boundary is defined in terms of a coordinate c that runs along the boundary and an outer unit normal to the boundary $\boldsymbol{n}$, as shown in Figure 5.1. The x and y components of the normal vector are denoted by n_x and n_y:

$$\boldsymbol{n} = \begin{pmatrix} n_x \\ n_y \end{pmatrix}; \qquad |\boldsymbol{n}| = \sqrt{n_x^2 + n_y^2} = 1$$

The formulation presented in this chapter can handle boundary conditions of the following types:

(i) Essential boundary condition specified on portion of the boundary indicated by C_e:

$$u(c) \text{ specified on } C_e$$

where c is a coordinate along the boundary.

(ii) Natural boundary condition specified on portion of the boundary indicated by C_n:

$$k_x \frac{\partial u}{\partial x} n_x + k_y \frac{\partial u}{\partial y} n_y = \alpha(c)u(c) + \beta(c) \text{ on } C_n$$

where $\alpha(c)$ and $\beta(c)$ are specified parameters along the boundary. When $k_x = k_y \equiv k$, the left-hand side is the derivative of u in the direction of the outer normal to the boundary, and the boundary condition is expressed as

$$k\left(\frac{\partial u}{\partial x} n_x + \frac{\partial u}{\partial y} n_y\right) \equiv k\frac{\partial u}{\partial n} = \alpha u + \beta$$

Observe carefully that this expression does not allow specification of *any* arbitrary first derivative of u. Only the *normal derivative* can be specified. For example, on a boundary that is perpendicular to the x axis, the components of the unit normal to the boundary will be $n_x = 1$ and $n_y = 0$ and thus only $\partial u/\partial x$ can be specified by this form. The reason for choosing the normal derivative is because it gives rise to a convenient weak form. Furthermore, for most physical problems it is natural to specify the normal derivative or it is possible to use mathematical manipulations to express the derivative in this form. Thus it is not a severe limitation for practical problems.

A few selected applications that are governed by the differential equation of this form are presented in the first section. The second section presents a general finite element formulation for the problem. Rectangular and triangular elements for solution of this differential equation are presented in the last two sections in this chapter. These elements are used to obtain approximate solutions for a variety of problems.

5.1 SELECTED APPLICATIONS OF THE 2D BVP

5.1.1 Two-Dimensional Potential Flow

The problem of two-dimensional frictionless and incompressible fluid flow, known as potential flow, is governed by the following two equations:

1. Continuity equation: $\frac{\partial u}{\partial x} + \frac{\partial v}{\partial y} = 0$
2. Irrotational flow condition: $\frac{\partial u}{\partial y} - \frac{\partial v}{\partial x} = 0$

where $u(x, y)$ and $v(x, y)$ are the fluid velocities in the x and y directions. Two different formulations are used to find solutions of these equations and are known as stream function and potential function formulations.

In the *stream function* formulation it is assumed that a scalar function, called the stream function $\psi(x, y)$, exists that is related to the fluid velocities in the x and y directions as follows:

$$u = \frac{\partial \psi}{\partial y}; \qquad v = -\frac{\partial \psi}{\partial x}$$

With this assumption the continuity equation is satisfied exactly while the irrotational flow condition yields the following second-order partial differential equation:

$$\frac{\partial^2 \psi}{\partial x^2} + \frac{\partial^2 \psi}{\partial y^2} = 0$$

In the *potential function* formulation it is assumed that a scalar function, called the potential function $\phi(x, y)$, exists that is related to the fluid velocities in the x and y directions as follows:

$$u = \frac{\partial \phi}{\partial x}; \qquad v = \frac{\partial \phi}{\partial y}$$

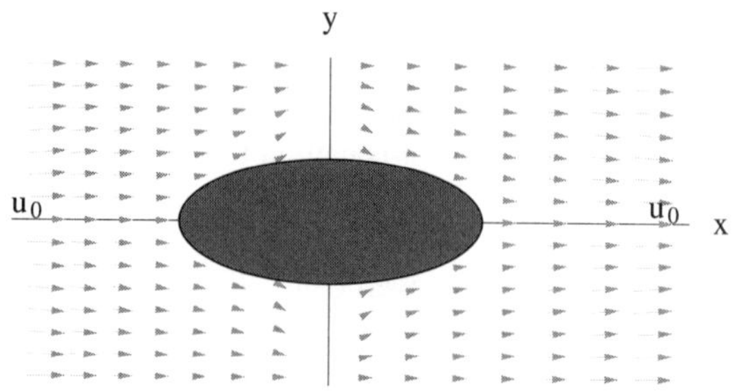

Figure 5.2. Flow around an object

With this assumption the irrotational flow condition is satisfied exactly while the continuity equation yields the following second-order partial differential equation:

$$\frac{\partial^2 \phi}{\partial x^2} + \frac{\partial^2 \phi}{\partial y^2} = 0$$

Thus both formulations are governed by the differential equation of the same type with $k_x = k_y = 1$ and $p = q = 0$. However, the two formulations differ in the way the boundary conditions are imposed. To illustrate this difference, consider a typical potential flow problem to determine fluid flow around a symmetric object, as shown in Figure 5.2. Away from the object the fluid is flowing in the x direction at a known velocity u_0. In the vicinity of the object the velocities change and are unknown. At a sufficient distance downstream from the object, once again we have a constant flow in the x direction.

The first task is to establish the boundaries of the solution domain. On all sides of the object, we must extend the solution domain far enough so that the assumption of uniform flow in the x direction is valid (indicated by horizontal streamlines). Since the computational effort will depend on the size of the solution domain, extending the domain too far would be inefficient. However, if the domain is not large enough, the results will be inaccurate. Thus few trials may be necessary before settling on a proper size for the solution domain. It is generally recommended to extend the computational domain in all directions to about four times the size of the object.

The next task is to define appropriate boundary conditions on all sides. Taking advantage of symmetry, we need to model only a quarter of the solution domain, as shown in Figure 5.3. For the stream function formulation ψ, we can take any arbitrary reference value for the base because the velocities depend only on the derivatives of ψ. We choose $\psi = 0$ along $x = 0$ and come up with the following boundary conditions:

Boundary Conditions for Stream Function Formulation

1. On the bottom and along the obstruction

$$\psi = 0$$

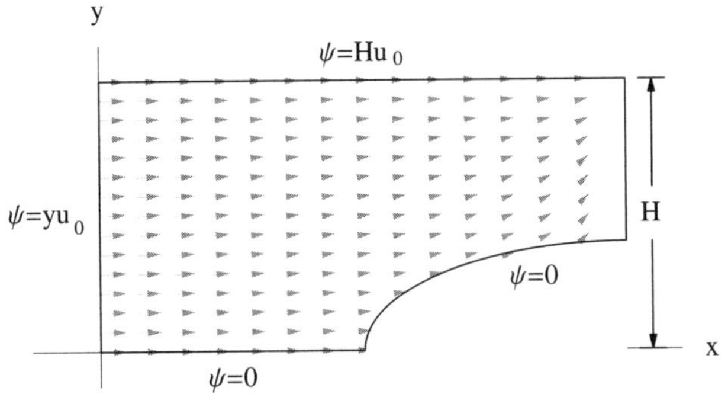

Figure 5.3. Boundary conditions in terms of stream function for flow around an object

2. On the left side

$$u = u_0 \Longrightarrow \frac{\partial \psi}{\partial y} = u_0$$

Integrating this expression, we have

$$\psi = y u_0$$

Thus ψ varies linearly with height on this side.

3. On the top: From linear variation of ψ on the left side with $y = H$, we have

$$\psi = H u_0$$

4. On the right side, because of symmetry,

$$\frac{\partial \psi}{\partial x} = 0$$

Note that the original form of the boundary condition $\partial\psi/\partial y = u_0$ on the left side is not in the form that can be handled by the natural boundary condition defined for the differential equation since the normal derivative for this side is

$$n_x = -1; \qquad n_y = 0 \qquad \Longrightarrow \frac{\partial \psi}{\partial n} = -\frac{\partial \psi}{\partial x}$$

Thus we can specify $\partial\psi/\partial x$ on the ends but not $\partial\psi/\partial y$. On the right side $\partial\psi/\partial n \equiv \partial\psi/\partial x$, and thus we can use this boundary condition directly.

With the potential function formulation the boundary conditions on a quarter of the solution domain are as shown in Figure 5.4. On the bottom, top, and left side we have natural boundary conditions. The right vertical symmetry line must be a potential line and therefore we must specify that ϕ = constant. The actual value does not matter.

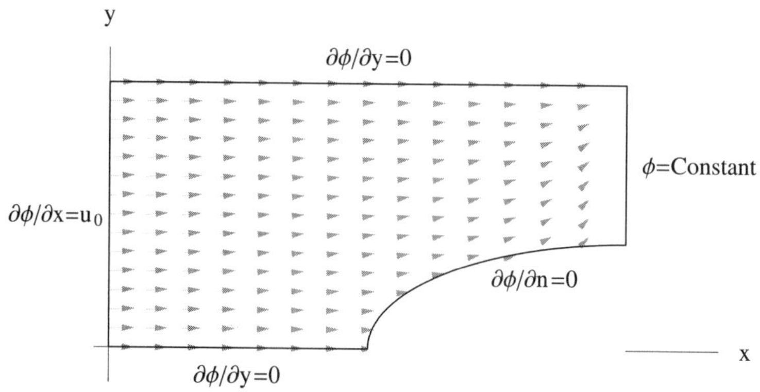

Figure 5.4. Boundary conditions in terms of potential function for flow around an object

Boundary Conditions for Potential Function Formulation

1. On the top and bottom sides

$$\frac{\partial \phi}{\partial y} = 0$$

2. Along the obstruction

$$\frac{\partial \phi}{\partial n} \equiv \frac{\partial \phi}{\partial x} n_x + \frac{\partial \phi}{\partial y} n_y = 0$$

3. On the left end

$$u = u_0 \Longrightarrow \frac{\partial \phi}{\partial x} = u_0$$

4. On the right end

$$\phi = \text{constant}$$

5.1.2 Steady-State Heat Flow

Consider the problem of finding the steady-state temperature distribution $T(x, y)$ in long chimneylike structures. Assuming no temperature gradient in the longitudinal direction, we can take a unit slice of such a structure and model it as a two-dimensional problem. Using conservation of energy on a differential volume, the following governing differential equation can easily be established:

$$\frac{\partial}{\partial x}\left(k_x \frac{\partial T}{\partial x}\right) + \frac{\partial}{\partial y}\left(k_y \frac{\partial T}{\partial y}\right) + Q = 0$$

where k_x and k_y are thermal conductivities in the x and y directions and $Q(x, y)$ is specified heat generation per unit volume. Typical units for k are W/m · °C or Btu/hr · ft · °F and those for Q are W/m^3 or Btu/hr · ft^3. The possible boundary conditions are as follows:

(i) Known temperature along a boundary:

$$T(c) = T_0 \text{ specified}$$

(ii) Specified heat flux along a boundary:

$$-k\frac{\partial T}{\partial n} \equiv -\left(k_x\frac{\partial T}{\partial x}n_x + k_y\frac{\partial T}{\partial y}n_y\right) = q_0 \text{ specified}$$

The sign convention for heat flow is that heat flowing into a body is positive and that out of the body is negative. Comparing with the general form this boundary condition implies $\alpha = 0$ and $\beta = -q_0$. On an insulated boundary or across a line of symmetry there is no heat flow and thus $q_0 = 0$.

(iii) Convection heat loss along a boundary:

$$-k\frac{\partial T}{\partial n} = h(T(c) - T_\infty)$$

where h is the convection coefficient, $T(c)$ is the unknown temperature at the boundary, and T_∞ is the temperature of the surrounding fluid. Typical units for h are $W/m^2 \cdot {}^\circ C$ and $Btu/hr \cdot ft^2 \cdot {}^\circ F$. Comparing with the general form, this boundary condition implies $\alpha = -h$ and $\beta = hT_\infty$.

5.1.3 Bars Subjected to Torsion

The problem of determining stress distribution in bars of arbitrary cross section subjected to twist can be formulated in terms of a stress function ϕ defined as follows:

$$\tau_{zx} = \frac{\partial \phi}{\partial y}; \qquad \tau_{zy} = -\frac{\partial \phi}{\partial x}$$

The total shear stress can be computed by taking the vector sum as follows:

$$\tau = \sqrt{\tau_{zx}^2 + \tau_{zy}^2}$$

All other stress components on the cross section are assumed to be zero. (See Chapter 7 for a review of basic stress and strain concepts and derivation of stress equilibrium equations.) With this assumption the stress equilibrium equations are automatically satisfied. The strain compatibility conditions lead to the following differential equation:

$$\frac{\partial^2 \phi}{\partial x^2} + \frac{\partial^2 \phi}{\partial y^2} + 2G\theta = 0$$

where G is the shear modulus and θ is the angle of twist per unit length. The boundary condition is that ϕ must be constant on the boundary. Usually $\phi = 0$ is chosen on the boundary.

The total torque is related to the stress function as follows:

$$T = 2 \iint_A \phi \, dA$$

Using these equations, the finite element formulation can be employed to analyze the following two situations.

Determining Stresses in a Bar Subjected to a Given Torque T The governing differential equation needs θ, the angle of twist per unit length. The torque does not show up in the governing differential equation and thus we cannot proceed directly as in other common stress analysis situations. Recognizing that the stress function ϕ depends linearly on $G\theta$, we use the following procedure:

(i) Assume any convenient value of $G\theta_a$. To clearly identify that this value is assumed, we have used the subscript a.

(ii) Find the stress function solution ϕ_a from the governing differential equation corresponding to $G\theta_a$. Compute the total torque T_a by integrating over the bar cross section using

$$T_a = 2 \iint_A \phi_a \, dA$$

(iii) Use ratios and proportions to determine the actual values of the angle of twist and ϕ as follows:

$$\frac{G\theta}{T} = \frac{G\theta_a}{T_a} \Longrightarrow G\theta = \frac{T}{T_a} G\theta_a$$

$$\frac{\phi}{T} = \frac{\phi_a}{T_a} \Longrightarrow \phi = \frac{T}{T_a} \phi_a$$

Once ϕ is known, the two nonzero stress components can be determined by simple differentiation. All other stress components are zero.

Determining Torsional Constant An important use of the stress function approach is to determine the torsional constant J for a given section. The torsional constant is a property of a cross section and is related to the torque as follows:

$$T = JG\theta$$

Thus, if we set $G\theta = 1$, then the total torque computed is the torsional constant of the cross section:

Figure 5.5. Rectangular waveguide

(i) Set $G\theta = 1$.

(ii) Find the stress function solution ϕ by solving the governing differential equation. Compute the total torque T by integrating over the bar cross section using

$$T \equiv J = 2 \iint_A \phi \, dA$$

Recall from Chapter 4 that J values are required for space frame members. The equations given there for some simple cross sections are based on the closed-form solutions of ϕ available for those sections. For complicated sections, closed-form solutions are not possible. However, one can use the finite element formulation presented here to determine a numerical value of J for virtually any cross section.

5.1.4 Waveguides in Electromagnetics

Waveguides are hollow conductors that are employed to efficiently transmit electrical signals at microwave frequencies (in the GHz range). A waveguide with a rectangular cross section is shown in Figure 5.5. Electromagnetic waves inside a waveguide experience multiple reflections from the enclosing walls and produce discrete modes that depend on the shape and size of the waveguide, the medium within the waveguide, and the operating frequency. The two types of modes supported by a waveguide are identified as the *transverse magnetic* (TM) and the *transverse electric* (TE). These modes can propagate only when the operating frequency is higher than a certain frequency known as the *cutoff frequency*. Thus determination of the cutoff frequency is a critical parameter in the design of waveguides.

The governing differential equation for determination of the cutoff frequency is the Helmholtz equation, written in terms of a scalar potential ψ as follows:

$$\frac{\partial^2 \psi}{\partial x^2} + \frac{\partial^2 \psi}{\partial y^2} + k_c^2 \psi = 0$$

where k_c is the unknown cutoff frequency. The potential is related to the transverse electric field components as follows:

$$\text{For TM modes:} \quad E_x = -Z_0 \frac{\partial \psi}{\partial x}; \qquad E_y = -Z_0 \frac{\partial \psi}{\partial y}$$

$$\text{For TE modes:} \quad E_x = -\frac{\partial \psi}{\partial y}; \qquad E_y = \frac{\partial \psi}{\partial x}$$

where Z_0 is the characteristic wave impedance for TM modes. For the TM modes ψ is set to 0 along the boundary. For the TE modes the normal derivative of the potential $(\partial\psi/\partial n)$ is 0 on the boundary.

This differential equation is of the same form as the general BVP with $k_x = k_y = 1$, $p = k_c^2$, and $q = 0$. However, there is one key difference in this application, and that is the cutoff frequency k_c is not known. The situation is similar to the buckling problem considered in Chapter 3. Proceeding with the standard finite element formulation in the usual manner will give rise to an eigenvalue problem. The eigenvalues of the system will correspond to the cutoff frequencies and the corresponding eigenvectors will be the wave propagation modes.

5.2 INTEGRATION BY PARTS IN HIGHER DIMENSIONS

In order to derive the weak form for two- and three-dimensional boundary value problems, we need a formula equivalent to the integration by parts for one-dimensional problems. This is accomplished by using the well-known Gauss and Green-Gauss theorems from vector calculus. These theorems are presented here in the context of a two-dimensional problem. Their extension to the three-dimensional boundary value problems is obvious.

Gauss's Divergence Theorem Consider a vector of functions $F = (F_1(x, y), F_2(x, y))^T$. The divergence of $\boldsymbol{F}$ is defined as follows:

$$\operatorname{div} \boldsymbol{F} = \frac{\partial F_1}{\partial x} + \frac{\partial F_2}{\partial y}$$

According to Gauss's divergence theorem, the area integral of div $\boldsymbol{F}$ is equal to line integral of its dot product with the unit outer normal $\boldsymbol{n}$. That is,

$$\iint_A \operatorname{div} \boldsymbol{F}\, dA = \int_c \boldsymbol{F}^T \boldsymbol{n}\, dc$$

The $dA = dx\,dy$ is the differential area. All terms in the boundary integral must be expressed in terms of the boundary coordinate c. The dc in the boundary integral is the length of a differential segment on the boundary:

$$dc = \sqrt{dx^2 + dy^2}$$

Written explicitly,

$$\iint_A \left(\frac{\partial F_1}{\partial x} + \frac{\partial F_2}{\partial y} \right) dA = \int_c (F_1 n_x + F_2 n_y)\, dc$$

where n_x and n_y are the components of the unit normal to the boundary.

Green-Gauss Theorem Given another function $g(x, y)$, the integral of the divergence of the product $g\boldsymbol{F}$ can be written as follows:

$$\iint_A \operatorname{div}(g\boldsymbol{F})\, dA = \int_c g\boldsymbol{F}^T \boldsymbol{n}\, dc$$

Using the product rule, the divergence of the product $g\boldsymbol{F}$ can be written as follows:

$$\operatorname{div}(g\boldsymbol{F}) = \frac{\partial (gF_1)}{\partial x} + \frac{\partial (gF_2)}{\partial y} = g\frac{\partial F_1}{\partial x} + F_1 \frac{\partial g}{\partial x} + g\frac{\partial F_2}{\partial y} + F_2 \frac{\partial g}{\partial y}$$

Rearranging terms,

$$\operatorname{div}(g\boldsymbol{F}) = g\left(\frac{\partial F_1}{\partial x} + \frac{\partial F_2}{\partial y} \right) + \begin{pmatrix} \frac{\partial g}{\partial x} & \frac{\partial g}{\partial y} \end{pmatrix} \begin{pmatrix} F_1 \\ F_2 \end{pmatrix}$$

Noting that the row vector in the second term is the transpose of the gradient vector of g,

$$\nabla g \equiv \begin{pmatrix} \frac{\partial g}{\partial x} & \frac{\partial g}{\partial y} \end{pmatrix}^T$$

we have

$$\operatorname{div}(g\boldsymbol{F}) = g(\operatorname{div} \boldsymbol{F}) + (\nabla g)^T \boldsymbol{F}$$

Thus the integral of $\operatorname{div}(g\boldsymbol{F})$ can be expressed as follows:

$$\iint_A g(\operatorname{div} \boldsymbol{F})\, dA + \iint_A (\nabla g)^T \boldsymbol{F}\, dA = \int_c g\boldsymbol{F}^T \boldsymbol{n}\, dc$$

Rearranging terms, we get the following relationship, commonly known as the Green-Gauss theorem:

$$\iint_A g(\operatorname{div} \boldsymbol{F})\, dA = \int_c g\boldsymbol{F}^T n\, dc - \iint (\nabla g)^T \boldsymbol{F}\, dA$$

Green-Gauss Theorem as Integration by Parts in Two Dimensions The Green-Gauss theorem can be written in a form very similar to the usual one-dimensional integration by parts by taking the vector of functions with $F_1 = f$ and $F_2 = 0$, i.e., $\boldsymbol{F} = (f(x, y), 0)^T$. The divergence of this function is

$$\operatorname{div} \boldsymbol{F} = \frac{\partial f}{\partial x}$$

and thus applying the Green-Gauss theorem, we have

$$\iint_A g\left(\frac{\partial f}{\partial x}\right) dA = \int_c gfn_x\, dc - \iint_A \frac{\partial g}{\partial x} f\, dA$$

Similarly, by taking the vector of functions with $F_1 = 0$ and $F_2 = f$, we can see that

$$\iint_A g\left(\frac{\partial f}{\partial y}\right) dA = \int_c gfn_y\, dc - \iint_A \frac{\partial g}{\partial y} f\, dA$$

Note that, if the derivative on function $f(x, y)$ is with respect to x, then the boundary integral involves n_x and vice versa. Similar to the one-dimensional case, this form of the Green-Gauss theorem gives an area integral in which the derivative is switched from one function to the other. We also get a boundary integral involving the product of the functions and the component of the unit normal. These forms, and their obvious extensions to three dimensions, will be used in the following section and the remaining chapters to obtain weak forms for given boundary value problems.

Example 5.1 A two-dimensional boundary value problem is stated as follows:

$$2\frac{\partial^2 u}{\partial x^2} - 3\frac{\partial^2 u}{\partial y^2} + (x + y)u + \frac{x}{y} = 0; \qquad 1 < x < 3; \qquad 1 < y < 2$$

$$u(x, 1) = \sin(x); \qquad u(1, y) = \sin(y); \qquad \frac{\partial u}{\partial y}(x, 2) = x^3; \qquad \frac{\partial u}{\partial x}(3, y) = u(3, y) + 1$$

Comparing the problem with the general form, identify the terms $k_x, k_y, p, \ldots, \beta$ for this problem. Classify the boundary conditions into essential and natural types. Obtain a weak form for the problem.

The solution domain is a rectangle, as shown in Figure 5.6. The coefficients in the differential equation are as follows:

$$k_x = 2; \qquad k_y = -3; \qquad p = x + y; \qquad q = \frac{x}{y}$$

The boundary conditions are as follows:

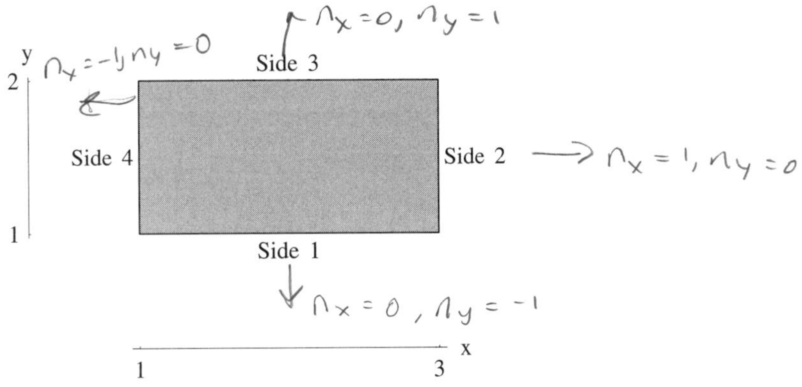

Figure 5.6.

Essential: On side 1, $u = \sin(x)$; on side 4, $u = \sin(y)$.

Natural: On side 3

$$\frac{\partial u}{\partial y} = x^3; \qquad n_x = 0; \qquad n_y = 1$$

$$k_y \frac{\partial u}{\partial y} n_y = -3\frac{\partial u}{\partial y} \Longrightarrow k_y \frac{\partial u}{\partial y} n_y = -3x^3 \Longrightarrow \alpha = 0 \quad \text{and} \quad \beta = -3x^3$$

Natural: On side 2

$$\frac{\partial u}{\partial x} = u(3, y) + 1; \qquad n_x = 1, \qquad n_y = 0$$

$$k_x \frac{\partial u}{\partial x} n_x = 2\frac{\partial u}{\partial x} \Longrightarrow k_x \frac{\partial u}{\partial x} n_x = 2u(3, y) + 2 \Longrightarrow \alpha = 2 \quad \text{and} \quad \beta = 2$$

Multiplying the differential equation by the weighting function N_i, the Galerkin weighted residual is as follows:

$$\iint_A \left(2\frac{\partial^2 u}{\partial x^2} - 3\frac{\partial^2 u}{\partial y^2} + (x + y)u + \frac{x}{y}\right) N_i \, dA = 0; \qquad i = 1, 2, \ldots$$

Using integration by parts on the first two terms,

$$\iint_A 2\frac{\partial^2 u}{\partial x^2} N_i \, dA = \int_C 2\frac{\partial u}{\partial x} N_i n_x \, dc - \iint_A 2\frac{\partial u}{\partial x}\frac{\partial N_i}{\partial x} \, dA$$

$$\iint_A -3\frac{\partial^2 u}{\partial y^2} N_i \, dA = \int_C -3\frac{\partial u}{\partial y} N_i n_y \, dc + \iint_A 3\frac{\partial u}{\partial y}\frac{\partial N_i}{\partial y} \, dA$$

For the rectangular domain for this example, by substituting the normal vector components, the expressions can be written more explicitly:

First term:

$$\int_{x=1}^{3}\int_{y=1}^{2} 2\frac{\partial^2 u}{\partial x^2}N_i\,dx\,dy = \int_{x=1}^{3} 2\frac{\partial u}{\partial x}N_i(x,1)\times(0)\,dx + \int_{y=1}^{2} 2\frac{\partial u}{\partial x}N_i(3,y)\times(1)\,dy$$
$$+ \int_{x=1}^{3} 2\frac{\partial u}{\partial x}N_i(x,2)\times(0)\,dx + \int_{y=1}^{2} 2\frac{\partial u}{\partial x}N_i(1,y)\times(-1)dy$$
$$- \int_{x=1}^{3}\int_{y=1}^{2} 2\frac{\partial u}{\partial x}\frac{\partial N_i}{\partial x}\,dx\,dy$$

$$\int_{x=1}^{3}\int_{y=1}^{2} 2\frac{\partial^2 u}{\partial x^2}N_i\,dx\,dy = \int_{y=1}^{2} 2\frac{\partial u}{\partial x}N_i(3,y)\,dy$$
$$- \int_{y=1}^{2} 2\frac{\partial u}{\partial x}N_i(1,y)\,dy - \int_{x=1}^{3}\int_{y=1}^{2} 2\frac{\partial u}{\partial x}\frac{\partial N_i}{\partial x}\,dx\,dy$$

Second term:

$$\int_{x=1}^{3}\int_{y=1}^{2} -3\frac{\partial^2 u}{\partial y^2}N_i\,dx\,dy = \int_{x=1}^{3} -3\frac{\partial u}{\partial y}N_i(x,1)\times(-1)\,dx$$
$$+ \int_{y=1}^{2} -3\frac{\partial u}{\partial y}N_i(3,y)\times(0)\,dy + \int_{x=1}^{3} -3\frac{\partial u}{\partial y}N_i(x,2)\times(1)\,dx$$
$$+ \int_{y=1}^{2} -3\frac{\partial u}{\partial y}N_i(1,y)\times(0)dy + \int_{x=1}^{3}\int_{y=1}^{2} 3\frac{\partial u}{\partial x}\frac{\partial N_i}{\partial x}\,dx\,dy$$

$$\int_{x=1}^{3}\int_{y=1}^{2} -3\frac{\partial^2 u}{\partial y^2}N_i\,dx\,dy = \int_{x=1}^{3} 3\frac{\partial u}{\partial y}N_i(x,1)\,dx - \int_{x=1}^{3} 3\frac{\partial u}{\partial y}N_i(x,2)\,dx$$
$$+ \int_{x=1}^{3}\int_{y=1}^{2} 3\frac{\partial u}{\partial x}\frac{\partial N_i}{\partial x}\,dx\,dy$$

Thus the weighted residual is

$$\int_{y=1}^{2} 2\frac{\partial u}{\partial x}N_i(3,y)\,dy - \int_{y=1}^{2} 2\frac{\partial u}{\partial x}N_i(1,y)dy + \int_{x=1}^{3} 3\frac{\partial u}{\partial y}N_i(x,1)dx - \int_{x=1}^{3} 3\frac{\partial u}{\partial y}N_i(x,2)\,dx$$
$$+ \int_{x=1}^{3}\int_{y=1}^{2}\left(-2\frac{\partial u}{\partial x}\frac{\partial N_i}{\partial x} + 3\frac{\partial u}{\partial y}\frac{\partial N_i}{\partial y} + \left((x+y)u + \frac{x}{y}\right)N_i\right)dx\,dy = 0$$

Consider boundary terms:

First term:

$$\int_{y=1}^{2} 2\frac{\partial u}{\partial x}N_i(3,y)\,dy$$

Side 2: $\dfrac{\partial u}{\partial x} = u(3, y) + 1$; Natural

$$\Longrightarrow \int_{y=1}^{2} 2(u(3, y) + 1)N_i(3, y)\, dy$$

Second term:

$$-\int_{y=1}^{2} 2\frac{\partial u}{\partial x} N_i(1, y)\, dy$$

Side 4: $u = \sin(y)$; Essential

Must satisfy explicitly: $-\displaystyle\int_{y=1}^{2} 2\frac{\partial u}{\partial x} N_i(1, y)\, dy = 0$

Third term:

$$\int_{x=1}^{3} 3\frac{\partial u}{\partial y} N_i(x, 1)\, dx$$

Side 1: $u = \sin(x)$; Essential

Must satisfy explicitly: $\displaystyle\int_{x=1}^{3} 3\frac{\partial u}{\partial y} N_i(x, 1)\, dx = 0$

Fourth term:

$$-\int_{x=1}^{3} 3\frac{\partial u}{\partial y} N_i(x, 2)\, dx$$

Side 3: $\dfrac{\partial u}{\partial y} = x^3$; Natural

$$\Longrightarrow \int_{y=1}^{2} -3x^3 N_i(x, 2)\, dx$$

Thus the weak form is

$$\int_{y=1}^{2} 2(u(3, y) + 1)N_i(3, y)\, dy + \int_{y=1}^{2} -3x^3 N_i(x, 2)dx$$
$$+ \int_{x=1}^{3}\int_{y=1}^{2} \left(-2\frac{\partial u}{\partial x}\frac{\partial N_i}{\partial x} + 3\frac{\partial u}{\partial y}\frac{\partial N_i}{\partial y} + \left((x + y)u + \frac{x}{y}\right)N_i\right) dx\, dy = 0$$

Admissible solutions must satisfy the EBC on sides 1 and 4.

5.3 FINITE ELEMENT EQUATIONS USING THE GALERKIN METHOD

The solution domain is discretized into elements. For simplicity of notation we do not introduce any super- or subscripts and assume that A and C now refer to an arbitrary element area and its boundary. For the general derivation considered in this section, no specific shape of the element is assumed. Explicit equations for rectangular and triangular are developed in later sections.

The general form of finite element equations for the problem can be written using exactly the same steps used for the one-dimensional problems. Recall from Chapter 2 that for one-dimensional problems a weak form is written by using standard steps of writing the weighted residual, integration by parts, and incorporating the natural boundary conditions. Furthermore, in the finite element form the weighting functions are the same as the interpolation functions N_i used to define an assumed solution over a finite element. The same general procedure works for two-dimensional problems.

The Galerkin weighted residual for the two-dimensional boundary value problem is

$$\iint_A \left(\frac{\partial}{\partial x}\left(k_x \frac{\partial u}{\partial x}\right) + \frac{\partial}{\partial y}\left(k_y \frac{\partial u}{\partial y}\right) + pu + q \right) N_i \, dA = 0; \qquad i = 1, 2, \ldots$$

Using the Green-Gauss theorem on the first two terms in the weighted residual, we have

$$\iint_A \frac{\partial}{\partial x}\left(k_x \frac{\partial u}{\partial x}\right) N_i \, dA = \int_C k_x \frac{\partial u}{\partial x} N_i n_x \, dc - \iint k_x \frac{\partial u}{\partial x}\frac{\partial N_i}{\partial x} \, dA$$

$$\iint_A \frac{\partial}{\partial y}\left(k_y \frac{\partial u}{\partial y}\right) N_i \, dA = \int_C k_y \frac{\partial u}{\partial y} N_i n_y \, dc - \iint_A k_y \frac{\partial u}{\partial y}\frac{\partial N_i}{\partial y} \, dA$$

Thus the weighted residual is

$$\int_C \left(k_x \frac{\partial u}{\partial x} n_x + k_y \frac{\partial u}{\partial y} n_y\right) N_i \, dc + \iint_A \left(-k_x \frac{\partial u}{\partial x}\frac{\partial N_i}{\partial x} - k_y \frac{\partial u}{\partial y}\frac{\partial N_i}{\partial y} + puN_i + qN_i\right) dA = 0$$

Splitting the boundary integral into two parts, the one over which the essential boundary condition is specified (C_e) and the other where the natural boundary condition is specified (C_n), and substituting the natural boundary condition, we get

$$\int_{C_e} \left(k_x \frac{\partial u}{\partial x} n_x + k_y \frac{\partial u}{\partial y} n_y\right) N_i \, dc + \int_{C_n} (\alpha u + \beta) N_i \, dc$$
$$+ \iint_A \left(-k_x \frac{\partial u}{\partial x}\frac{\partial N_i}{\partial x} - k_y \frac{\partial u}{\partial y}\frac{\partial N_i}{\partial y} + puN_i + qN_i\right) dA = 0$$

As was the case for one-dimensional problems, requiring the assumed solution to satisfy the essential boundary conditions results in weighting functions that are zero over the C_e boundary. Thus with the admissible assumed solutions the weak-form equivalent to the given boundary value problem is as follows:

$$\int_{C_n} (\alpha u + \beta) N_i \, dc + \iint_A \left(-k_x \frac{\partial u}{\partial x}\frac{\partial N_i}{\partial x} - k_y \frac{\partial u}{\partial y}\frac{\partial N_i}{\partial y} + puN_i + qN_i\right) dA = 0; \qquad i = 1, 2, \ldots$$

The assumed solution over an element is written as follows:

$$u(x,y) = \begin{pmatrix} N_1(x,y) & N_2(x,y) & \cdots & N_n(x,y) \end{pmatrix} \begin{pmatrix} u_1 \\ u_2 \\ \vdots \\ u_n \end{pmatrix} \equiv \boldsymbol{N}^T \boldsymbol{d}$$

$$\frac{\partial u}{\partial x} = \begin{pmatrix} \frac{\partial N_1}{\partial x} & \frac{\partial N_2}{\partial x} & \cdots & \frac{\partial N_n}{\partial x} \end{pmatrix} \begin{pmatrix} u_1 \\ u_2 \\ \vdots \\ u_n \end{pmatrix} \equiv \boldsymbol{B}_x^T \boldsymbol{d}$$

$$\frac{\partial u}{\partial y} = \begin{pmatrix} \frac{\partial N_1}{\partial y} & \frac{\partial N_2}{\partial y} & \cdots & \frac{\partial N_n}{\partial y} \end{pmatrix} \begin{pmatrix} u_1 \\ u_2 \\ \vdots \\ u_n \end{pmatrix} \equiv \boldsymbol{B}_y^T \boldsymbol{d}$$

where $u_1, u_2, \ldots$ are the unknown solutions at the element nodes. Over the boundary of the element the assumed solution must be written in terms of boundary coordinate c in the following form:

$$u(c) = \begin{pmatrix} N_1(c) & N_2(c) & \cdots & N_n(c) \end{pmatrix} \begin{pmatrix} u_1 \\ u_2 \\ \vdots \\ u_n \end{pmatrix} = \boldsymbol{N}_c^T \boldsymbol{d}$$

Substituting these into the weak form, we have

$$\int_{C_n} (\alpha N_i(c) \boldsymbol{N}_c^T \boldsymbol{d} + \beta N_i) dc + \iint_A \left(-k_x \frac{\partial N_i}{\partial x} \boldsymbol{B}_x^T \boldsymbol{d} - k_y \frac{\partial N_i}{\partial y} \boldsymbol{B}_y^T \boldsymbol{d} + p N_i \boldsymbol{N}^T \boldsymbol{d} + q N_i \right) dA = 0; \qquad i = 1, 2, \ldots$$

Writing all n equations, with N_i, $i = 1, 2, \ldots, n$, together in a matrix form yields

$$\int_{C_n} \left(\alpha \begin{pmatrix} N_1(c) \\ N_2(c) \\ \vdots \\ N_n(c) \end{pmatrix} \boldsymbol{N}_c^T \boldsymbol{d} + \beta \begin{pmatrix} N_1(c) \\ N_2(c) \\ \vdots \\ N_n(c) \end{pmatrix} \right) dc + \iint_A \left(-k_x \begin{pmatrix} \partial N_1/\partial x \\ \partial N_2/\partial x \\ \vdots \\ \partial N_2/\partial x \end{pmatrix} \boldsymbol{B}_x^T \boldsymbol{d} - \cdots \right) dA = \begin{pmatrix} 0 \\ 0 \\ \vdots \\ 0 \end{pmatrix}$$

or

$$\int_{C_n} (\alpha \boldsymbol{N}_c \boldsymbol{N}_c^T \boldsymbol{d} + \beta \boldsymbol{N}_c)\, dc + \iint_A (-k_x \boldsymbol{B}_x \boldsymbol{B}_x^T \boldsymbol{d} - k_y \boldsymbol{B}_y \boldsymbol{B}_y^T \boldsymbol{d} + p \boldsymbol{N} \boldsymbol{N}^T \boldsymbol{d} + q \boldsymbol{N})\, dA = 0$$

Rearranging terms by keeping quantities involving $\boldsymbol{d}$ on the left-hand side, we get the element equations as follows:

$$\left(-\int_{C_n} \alpha \boldsymbol{N}_c \boldsymbol{N}_c^T\, dc + \iint_A (k_x \boldsymbol{B}_x \boldsymbol{B}_x^T + k_y \boldsymbol{B}_y \boldsymbol{B}_y^T - p\boldsymbol{N}\boldsymbol{N}^T) dA\right)\boldsymbol{d} = \int_{C_n} \beta \boldsymbol{N}_c\, dc + \iint_A q\boldsymbol{N}\, dA$$

The first two terms inside the area integral can be arranged in a more compact form by using matrices as follows:

$$k_x \boldsymbol{B}_x \boldsymbol{B}_x^T + k_y \boldsymbol{B}_y \boldsymbol{B}_y^T = (\boldsymbol{B}_x \quad \boldsymbol{B}_y)\begin{pmatrix} k_x & 0 \\ 0 & k_y \end{pmatrix}\begin{pmatrix} \boldsymbol{B}_x^T \\ \boldsymbol{B}_y^T \end{pmatrix} \equiv \boldsymbol{BCB}^T$$

where

$$\boldsymbol{B}^T = \begin{pmatrix} \frac{\partial N_1}{\partial x} & \frac{\partial N_2}{\partial x} & \cdots & \frac{\partial N_n}{\partial x} \\ \frac{\partial N_1}{\partial y} & \frac{\partial N_2}{\partial y} & \cdots & \frac{\partial N_n}{\partial y} \end{pmatrix} \quad \text{and} \quad \boldsymbol{C} = \begin{pmatrix} k_x & 0 \\ 0 & k_y \end{pmatrix}$$

Thus the finite element equations are as follows:

$$\left(-\int_{C_n} \alpha \boldsymbol{N}_c \boldsymbol{N}_c^T\, dc + \iint_A (\boldsymbol{BCB}^T - p\boldsymbol{N}\boldsymbol{N}^T)\, dA\right)\boldsymbol{d} = \int_{C_n} \beta \boldsymbol{N}_c\, dc + \iint_A q\boldsymbol{N}\, dA$$

Define $n \times n$ matrices

$$\boldsymbol{k}_k = \iint_A \boldsymbol{BCB}^T\, dA; \qquad \boldsymbol{k}_p = -\iint_A p\boldsymbol{N}\boldsymbol{N}^T\, dA; \qquad \boldsymbol{k}_\alpha = -\int_{C_n} \alpha \boldsymbol{N}_c \boldsymbol{N}_c^T\, dc$$

Define $n \times 1$ vectors

$$\boldsymbol{r}_q = \iint_A q\boldsymbol{N}\, dA; \qquad \boldsymbol{r}_\beta = \int_{C_n} \beta \boldsymbol{N}_c\, dc$$

Thus the element equations are

$$(\boldsymbol{k}_k + \boldsymbol{k}_p + \boldsymbol{k}_\alpha)\boldsymbol{d} = \boldsymbol{r}_q + \boldsymbol{r}_\beta$$

The quantities $\boldsymbol{k}_\alpha$ and $\boldsymbol{r}_\beta$ are a result of applied natural boundary conditions and will affect only those elements that have a boundary on which an NBC is specified. For the interior elements these terms will be zero.

In order to actually evaluate the integrals defining the element equations, we need to assume an element shape and develop appropriate interpolation functions. Any element shape can be created as long as it is useful in modeling the geometry of practical solution domains and it is possible to carry out required integrations and differentiations. Some typical elements are presented in the remaining sections and the following chapter.

5.4 RECTANGULAR FINITE ELEMENTS

A rectangular element is the simplest element that can be developed for two-dimensional problems. The assumed solutions can easily be written by using the Lagrange interpolation formula in the x and y directions. The area and boundary integrals are also straightforward to evaluate over rectangular domains. However, from a practical point of view a rectangular element is not very useful since it is difficult to model irregular geometries using rectangular elements alone. The rectangular elements are developed here primarily to show the procedure without integration issues complicating the discussion.

5.4.1 Four-Node Rectangular Element

A typical rectangular element with dimensions $2a \times 2b$ is shown in Figure 5.7. The boundary is defined by four sides of the element. The outer normals and boundary coordinate c for each side are also shown in the figure. The computations are easier if we choose a local coordinate system (s, t) with the origin at the element center. For each element these local coordinates are related to the global coordinates by a simple translation. Denoting the global coordinates of the element centroid by (x_c, y_c), we have the following relationship between the local and the global coordinates:

$$s = x - x_c; \qquad t = y - y_c$$

Note that with this change of variables

$$ds = dx; \qquad dt = dy$$

and therefore the derivatives with respect to x and y are exactly the same as those with respect to s and t. In the local coordinate system, the nodal coordinates are $(-a, -b)$, $(a, -b)$, (a, b), and $(-a, b)$. To evaluate boundary terms, for each side a boundary coordinate c is defined with the origin at the center of each side. The positive directions of these boundary coordinates are consistent with the requirement of moving counterclockwise around the boundary of the element.

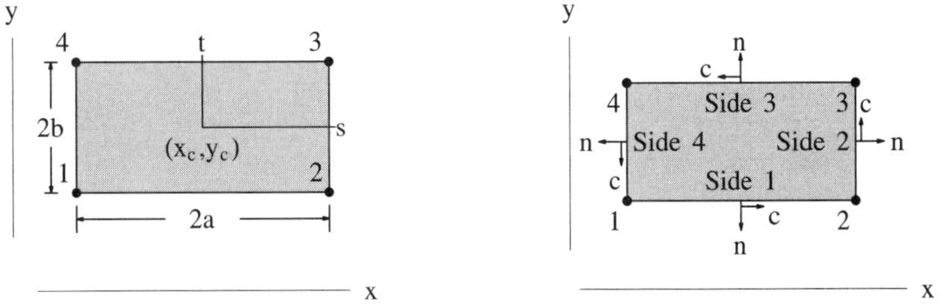

Figure 5.7. Four-node rectangular element

Assumed Solution Assuming unknown solutions at the four corners as nodal degrees of freedom, we have a total of four degrees of freedom. The assumed finite element solutions are needed in the following form:

$$\text{Over A:} \qquad u(s,t) = \begin{pmatrix} N_1(s,t) & N_2(s,t) & N_3(s,t) & N_4(s,t) \end{pmatrix} \begin{pmatrix} u_1 \\ u_2 \\ u_3 \\ u_4 \end{pmatrix} = \boldsymbol{N}^T \boldsymbol{d}$$

$$\text{Over C:} \qquad u(c) = \begin{pmatrix} N_1(c) & N_2(c) & N_3(c) & N_4(c) \end{pmatrix} \begin{pmatrix} u_1 \\ u_2 \\ u_3 \\ u_4 \end{pmatrix} = \boldsymbol{N}_c^T \boldsymbol{d}$$

Over the area the assumed solution can be written starting from a polynomial in two dimensions with four coefficients. Following the discussion in Chapter 2, we must use complete polynomials as much as possible. A complete linear polynomial in two dimensions has three terms $c_0 + c_1 s + c_2 t$. For the fourth term we must choose from one of the three quadratic terms $c_3 st + c_4 s^2 + c_5 t^2$. For an element that is appropriate for any general application the assumed solution should be symmetric; otherwise the element will give different results depending upon its orientation. Based on this consideration, we choose the st term as the fourth term, and thus a suitable polynomial solution for a rectangular element is as follows:

$$u(s,t) = c_0 + c_1 s + c_2 t + c_3 st = \begin{pmatrix} 1 & s & t & st \end{pmatrix} \begin{pmatrix} c_0 \\ c_1 \\ c_2 \\ c_3 \end{pmatrix}$$

The coefficients $c_0, c_1, \ldots$ can be expressed in terms of nodal degrees of freedom by evaluating the polynomial at the nodes as follows:

$$\begin{pmatrix} u_1 \\ u_2 \\ u_3 \\ u_4 \end{pmatrix} = \begin{pmatrix} 1 & -a & -b & ab \\ 1 & a & -b & -ab \\ 1 & a & b & ab \\ 1 & -a & b & -ab \end{pmatrix} \begin{pmatrix} c_0 \\ c_1 \\ c_2 \\ c_3 \end{pmatrix}$$

Inverting the 4×4 matrix, we get

$$\begin{pmatrix} c_0 \\ c_1 \\ c_2 \\ c_3 \end{pmatrix} = \begin{pmatrix} \frac{1}{4} & \frac{1}{4} & \frac{1}{4} & \frac{1}{4} \\ -\frac{1}{4a} & \frac{1}{4a} & \frac{1}{4a} & -\frac{1}{4a} \\ -\frac{1}{4b} & -\frac{1}{4b} & \frac{1}{4b} & \frac{1}{4b} \\ \frac{1}{4ab} & -\frac{1}{4ab} & \frac{1}{4ab} & -\frac{1}{4ab} \end{pmatrix} \begin{pmatrix} u_1 \\ u_2 \\ u_3 \\ u_4 \end{pmatrix}$$

Substituting into the polynomial, we have

$$u(s,t) = \begin{pmatrix} 1 & s & t & st \end{pmatrix} \begin{pmatrix} \frac{1}{4} & \frac{1}{4} & \frac{1}{4} & \frac{1}{4} \\ -\frac{1}{4a} & \frac{1}{4a} & \frac{1}{4a} & -\frac{1}{4a} \\ -\frac{1}{4b} & -\frac{1}{4b} & \frac{1}{4b} & \frac{1}{4b} \\ \frac{1}{4ab} & -\frac{1}{4ab} & \frac{1}{4ab} & -\frac{1}{4ab} \end{pmatrix} \begin{pmatrix} u_1 \\ u_2 \\ u_3 \\ u_4 \end{pmatrix}$$

Carrying out matrix multiplication,

$$u(s,t) = \begin{pmatrix} \frac{(a-s)(b-t)}{4ab} & \frac{(a+s)(b-t)}{4ab} & \frac{(a+s)(b+t)}{4ab} & \frac{(a-s)(b+t)}{4ab} \end{pmatrix} \begin{pmatrix} u_1 \\ u_2 \\ u_3 \\ u_4 \end{pmatrix}$$

$$= \begin{pmatrix} N_1 & N_2 & N_3 & N_4 \end{pmatrix} \begin{pmatrix} u_1 \\ u_2 \\ u_3 \\ u_4 \end{pmatrix} = \boldsymbol{N}^T \boldsymbol{d}$$

$$N_1 = \frac{(a-s)(b-t)}{4ab}; \qquad N_2 = \frac{(a+s)(b-t)}{4ab};$$

$$N_3 = \frac{(a+s)(b+t)}{4ab}; \qquad N_4 = \frac{(a-s)(b+t)}{4ab}$$

where $N_1, \ldots, N_4$ are the required interpolation functions. Note that each of the interpolation functions N_i is 1 at the ith node of the element and 0 at every other node.

The solutions along the element sides are written by substituting appropriate values of s and t for each side. Expressing boundary interpolation functions in terms of c for each side, we have

For side 1: $\quad s = c, t = -b \Longrightarrow \boldsymbol{N}_{c,1-2}^T = \begin{pmatrix} \frac{a-c}{2a} & \frac{a+c}{2a} & 0 & 0 \end{pmatrix}; \quad -a \le c \le a$

For side 2: $\quad s = a, t = c \Longrightarrow \boldsymbol{N}_{c,2-3}^T = \begin{pmatrix} 0 & \frac{b-c}{2b} & \frac{b+c}{2b} & 0 \end{pmatrix}; \quad -b \le c \le b$

For side 3: $\quad s = -c, t = b \Longrightarrow \boldsymbol{N}_{c,3-4}^T = \begin{pmatrix} 0 & 0 & \frac{a-c}{2a} & \frac{a+c}{2a} \end{pmatrix}; \quad -a \le c \le a$

For side 4: $\quad s = -a, t = -c \Longrightarrow \boldsymbol{N}_{c,4-1}^T = \begin{pmatrix} \frac{b+c}{2b} & 0 & 0 & \frac{b-c}{2b} \end{pmatrix}; \quad -b \le c \le b$

Element Equations The element equations involve the following matrices and vectors:

$$\boldsymbol{k}_k = \iint_A \boldsymbol{BCB}^T \, dA; \qquad \boldsymbol{k}_p = -\iint_A p\boldsymbol{NN}^T \, dA; \qquad \boldsymbol{k}_\alpha = -\int_{C_n} \alpha \boldsymbol{N}_c \boldsymbol{N}_c^T \, dc$$

$$\boldsymbol{r}_q = \iint_A q\boldsymbol{N} \, dA; \qquad \boldsymbol{r}_\beta = \int_{C_n} \beta \boldsymbol{N}_c \, dc$$

$$B^T = \begin{pmatrix} \frac{\partial N_1}{\partial s} & \frac{\partial N_2}{\partial s} & \cdots & \frac{\partial N_n}{\partial s} \\ \frac{\partial N_1}{\partial t} & \frac{\partial N_2}{\partial t} & \cdots & \frac{\partial N_n}{\partial t} \end{pmatrix} \quad \text{and} \quad C = \begin{pmatrix} k_x & 0 \\ 0 & k_y \end{pmatrix}$$

For simplicity in integrations, it will be assumed that k_x, k_y, p, q, α, and β are all constant over an element. Thus these terms can be taken out of the integrals. Substituting the assumed solution and carrying out integrations, the matrices and vectors needed for element equations can easily be written:

$$N^T = \begin{pmatrix} \frac{(a-s)(b-t)}{4ab} & \frac{(a+s)(b-t)}{4ab} & \frac{(a+s)(b+t)}{4ab} & \frac{(a-s)(b+t)}{4ab} \end{pmatrix}$$

$$B^T = \begin{pmatrix} \frac{t-b}{4ab} & \frac{b-t}{4ab} & \frac{b+t}{4ab} & -\frac{b+t}{4ab} \\ \frac{s-a}{4ab} & -\frac{a+s}{4ab} & \frac{a+s}{4ab} & \frac{a-s}{4ab} \end{pmatrix}$$

$$BCB^T = \begin{pmatrix} \frac{k_y(a-s)^2+k_x(b-t)^2}{16a^2b^2} & \frac{k_y(a^2-s^2)-k_x(b-t)^2}{16a^2b^2} & \frac{k_y(s^2-a^2)+k_x(t^2-b^2)}{16a^2b^2} & \frac{k_x(b^2-t^2)-k_y(a-s)^2}{16a^2b^2} \\ \frac{k_y(a^2-s^2)-k_x(b-t)^2}{16a^2b^2} & \frac{k_y(a+s)^2+k_x(b-t)^2}{16a^2b^2} & \frac{k_x(b^2-t^2)-k_y(a+s)^2}{16a^2b^2} & \frac{k_y(s^2-a^2)+k_x(t^2-b^2)}{16a^2b^2} \\ \frac{k_y(s^2-a^2)+k_x(t^2-b^2)}{16a^2b^2} & \frac{k_x(b^2-t^2)-k_y(a+s)^2}{16a^2b^2} & \frac{k_y(a+s)^2+k_x(b+t)^2}{16a^2b^2} & \frac{k_y(a^2-s^2)-k_x(b+t)^2}{16a^2b^2} \\ \frac{k_x(b^2-t^2)-k_y(a-s)^2}{16a^2b^2} & \frac{k_y(s^2-a^2)+k_x(t^2-b^2)}{16a^2b^2} & \frac{k_y(a^2-s^2)-k_x(b+t)^2}{16a^2b^2} & \frac{k_y(a-s)^2+k_x(b+t)^2}{16a^2b^2} \end{pmatrix}$$

$$k_k = \int_{-a}^{a}\int_{-b}^{b} BCB^T\,ds\,dt = \begin{pmatrix} \frac{k_ya}{3b}+\frac{k_xb}{3a} & \frac{k_ya}{6b}-\frac{k_xb}{3a} & -\frac{k_ya}{6b}-\frac{k_xb}{6a} & \frac{k_xb}{6a}-\frac{k_ya}{3b} \\ \frac{k_ya}{6b}-\frac{k_xb}{3a} & \frac{k_ya}{3b}+\frac{k_xb}{3a} & \frac{k_xb}{6a}-\frac{k_ya}{3b} & -\frac{k_ya}{6b}-\frac{k_xb}{6a} \\ -\frac{k_ya}{6b}-\frac{k_xb}{6a} & \frac{k_xb}{6a}-\frac{k_ya}{3b} & \frac{k_ya}{3b}+\frac{k_xb}{3a} & \frac{k_ya}{6b}-\frac{k_xb}{3a} \\ \frac{k_xb}{6a}-\frac{k_ya}{3b} & -\frac{k_ya}{6b}-\frac{k_xb}{6a} & \frac{k_ya}{6b}-\frac{k_xb}{3a} & \frac{k_ya}{3b}+\frac{k_xb}{3a} \end{pmatrix}$$

$$NN^T = \begin{pmatrix} \frac{(a-s)^2(b-t)^2}{16a^2b^2} & \frac{(a-s)(a+s)(b-t)^2}{16a^2b^2} & \frac{(a-s)(a+s)(b-t)(b+t)}{16a^2b^2} & \frac{(a-s)^2(b-t)(b+t)}{16a^2b^2} \\ \frac{(a-s)(a+s)(b-t)^2}{16a^2b^2} & \frac{(a+s)^2(b-t)^2}{16a^2b^2} & \frac{(a+s)^2(b-t)(b+t)}{16a^2b^2} & \frac{(a^2-s^2)(b^2-t^2)}{16a^2b^2} \\ \frac{(a-s)(a+s)(b-t)(b+t)}{16a^2b^2} & \frac{(a+s)^2(b-t)(b+t)}{16a^2b^2} & \frac{(a+s)^2(b+t)^2}{16a^2b^2} & \frac{(a-s)(a+s)(b+t)^2}{16a^2b^2} \\ \frac{(a-s)^2(b-t)(b+t)}{16a^2b^2} & \frac{(a^2-s^2)(b^2-t^2)}{16a^2b^2} & \frac{(a-s)(a+s)(b+t)^2}{16a^2b^2} & \frac{(a-s)^2(b+t)^2}{16a^2b^2} \end{pmatrix}$$

$$k_p = \int_{-a}^{a}\int_{-b}^{b} -pNN^T\,ds\,dt = \begin{pmatrix} -\frac{4}{9}abp & -\frac{2}{9}abp & -\frac{1}{9}abp & -\frac{2}{9}abp \\ -\frac{2}{9}abp & -\frac{4}{9}abp & -\frac{2}{9}abp & -\frac{1}{9}abp \\ -\frac{1}{9}abp & -\frac{2}{9}abp & -\frac{4}{9}abp & -\frac{2}{9}abp \\ -\frac{2}{9}abp & -\frac{1}{9}abp & -\frac{2}{9}abp & -\frac{4}{9}abp \end{pmatrix}$$

$$r_q^T = \int_{-a}^{a}\int_{-b}^{b} qN^T\,ds\,dt = (abq \quad abq \quad abq \quad abq)$$

For the NBC on side 1

$$\boldsymbol{N}_c^T = \left(\tfrac{1}{2} - \tfrac{c}{2a} \quad \tfrac{c}{2a} + \tfrac{1}{2} \quad 0 \quad 0\right)$$

$$\boldsymbol{k}_\alpha = \int_{-a}^{a} -\alpha \boldsymbol{N}_c \boldsymbol{N}_c^T \, dc = \begin{pmatrix} -\frac{2a\alpha}{3} & -\frac{a\alpha}{3} & 0 & 0 \\ -\frac{a\alpha}{3} & -\frac{2a\alpha}{3} & 0 & 0 \\ 0 & 0 & 0 & 0 \\ 0 & 0 & 0 & 0 \end{pmatrix}$$

$$\boldsymbol{r}_\beta^T = \int_{-a}^{a} \beta \boldsymbol{N}_c^T \, dc = (a\beta \quad a\beta \quad 0 \quad 0)$$

For the NBC on side 2

$$\boldsymbol{N}_c^T = \left(0 \quad \tfrac{1}{2} - \tfrac{c}{2b} \quad \tfrac{c}{2b} + \tfrac{1}{2} \quad 0\right)$$

$$\boldsymbol{k}_\alpha = \int_{-b}^{b} -\alpha \boldsymbol{N}_c \boldsymbol{N}_c^T \, dc = \begin{pmatrix} 0 & 0 & 0 & 0 \\ 0 & -\frac{2b\alpha}{3} & -\frac{b\alpha}{3} & 0 \\ 0 & -\frac{b\alpha}{3} & -\frac{2b\alpha}{3} & 0 \\ 0 & 0 & 0 & 0 \end{pmatrix}$$

$$\boldsymbol{r}_\beta^T = \int_{-b}^{b} \beta \boldsymbol{N}_c^T \, dc = (0 \quad b\beta \quad b\beta \quad 0)$$

For the NBC on side 3

$$\boldsymbol{N}_c^T = \left(0 \quad 0 \quad \tfrac{1}{2} - \tfrac{c}{2a} \quad \tfrac{c}{2a} + \tfrac{1}{2}\right)$$

$$\boldsymbol{k}_\alpha = \int_{-a}^{a} -\alpha \boldsymbol{N}_c \boldsymbol{N}_c^T \, dc = \begin{pmatrix} 0 & 0 & 0 & 0 \\ 0 & 0 & 0 & 0 \\ 0 & 0 & -\frac{2a\alpha}{3} & -\frac{a\alpha}{3} \\ 0 & 0 & -\frac{a\alpha}{3} & -\frac{2a\alpha}{3} \end{pmatrix}$$

$$\boldsymbol{r}_\beta^T = \int_{-a}^{a} \beta \boldsymbol{N}_c^T \, dc = (0 \quad 0 \quad a\beta \quad a\beta)$$

For the NBC on side 4

$$\boldsymbol{N}_c^T = \left(\tfrac{c}{2b} + \tfrac{1}{2} \quad 0 \quad 0 \quad \tfrac{1}{2} - \tfrac{c}{2b}\right)$$

$$\boldsymbol{k}_\alpha = \int_{-b}^{b} -\alpha \boldsymbol{N}_c \boldsymbol{N}_c^T \, dc = \begin{pmatrix} -\frac{2b\alpha}{3} & 0 & 0 & -\frac{b\alpha}{3} \\ 0 & 0 & 0 & 0 \\ 0 & 0 & 0 & 0 \\ -\frac{b\alpha}{3} & 0 & 0 & -\frac{2b\alpha}{3} \end{pmatrix}$$

$$r_\beta^T = \int_{-b}^{b} \beta N_c^T \, dc = (\, b\beta \quad 0 \quad 0 \quad b\beta)$$

All quantities have now been defined in the element equations. The complete element equations are

$$(\boldsymbol{k}_k + \boldsymbol{k}_p + \boldsymbol{k}_\alpha)\boldsymbol{d} = \boldsymbol{r}_q + \boldsymbol{r}_\beta$$

Using these element equations and following the usual finite element discretization and assembly procedures, approximate numerical solutions to a variety of practical problems that are governed by the differential equation of the type considered here can be obtained.

Example 5.2 Laplace Equation over a Square Domain Consider solution of the Laplace equation over a square domain:

$$\frac{\partial^2 u}{\partial x^2} + \frac{\partial^2 u}{\partial y^2} = 0; \qquad 0 < x < 1; \qquad 0 < y < 1$$

with the boundary conditions

$$u(0, y) = 0; \qquad u(1, y) = 0$$

$$u(x, 0) = x(1 - x); \qquad u(x, 1) = 0$$

Comparing this problem to the general form of the 2D BVP, we can see that here we have $k_x = k_y = 1$ and $p = q = 0$. The boundary conditions on all four sides are of the essential type.

An exact solution of the problem is known and is given by the sum of the following infinite series:

$$\text{Exact } u(x, y) = \sum_{n=1}^{\infty} -\frac{4\sin(n\pi x)((-1)^n - 1)\sinh(n\pi(1 - y))}{\sinh(n\pi)n^3\pi^3}$$

We obtain a finite element solution of the problem using four square elements as shown in Figure 5.8. There are nine nodes and thus we have a total of nine degrees of freedom.

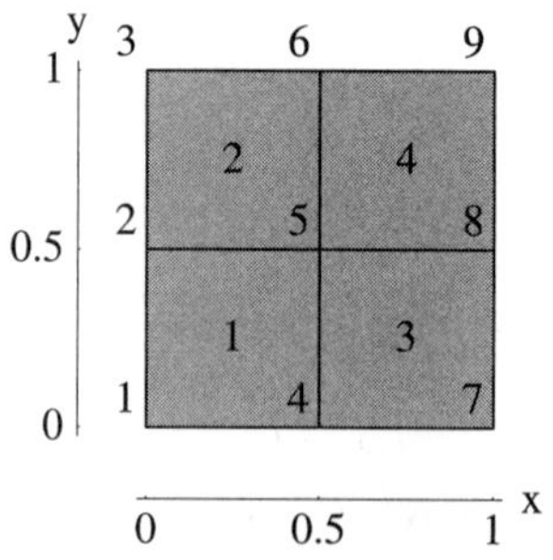

Figure 5.8. Four-element model

From the given essential boundary conditions the following nodal values are known:

$$\{\text{node, value}\}: (\{1,0\} \quad \{2,0\} \quad \{3,0\} \quad \{4,\tfrac{1}{4}\} \quad \{6,0\} \quad \{7,0\} \quad \{8,0\} \quad \{9,0\})$$

Equations for element 1 are as follows:

$$\text{Element dimensions:} \qquad a=\tfrac{1}{4}; \qquad b=\tfrac{1}{4}$$

$$k_x = 1; \qquad k_y = 1; \qquad p = 0; \qquad q = 0$$

$$\begin{pmatrix} \frac{2}{3} & -\frac{1}{6} & -\frac{1}{3} & -\frac{1}{6} \\ -\frac{1}{6} & \frac{2}{3} & -\frac{1}{6} & -\frac{1}{3} \\ -\frac{1}{3} & -\frac{1}{6} & \frac{2}{3} & -\frac{1}{6} \\ -\frac{1}{6} & -\frac{1}{3} & -\frac{1}{6} & \frac{2}{3} \end{pmatrix} \begin{pmatrix} u_1 \\ u_4 \\ u_5 \\ u_2 \end{pmatrix} = \begin{pmatrix} 0 \\ 0 \\ 0 \\ 0 \end{pmatrix}$$

Equations for element 2 are

$$\begin{pmatrix} \frac{2}{3} & -\frac{1}{6} & -\frac{1}{3} & -\frac{1}{6} \\ -\frac{1}{6} & \frac{2}{3} & -\frac{1}{6} & -\frac{1}{3} \\ -\frac{1}{3} & -\frac{1}{6} & \frac{2}{3} & -\frac{1}{6} \\ -\frac{1}{6} & -\frac{1}{3} & -\frac{1}{6} & \frac{2}{3} \end{pmatrix} \begin{pmatrix} u_2 \\ u_5 \\ u_6 \\ u_3 \end{pmatrix} = \begin{pmatrix} 0 \\ 0 \\ 0 \\ 0 \end{pmatrix}$$

Processing the remaining elements in exactly the same manner, we get the following global equations:

$$\begin{pmatrix} \frac{2}{3} & -\frac{1}{6} & 0 & -\frac{1}{6} & -\frac{1}{3} & 0 & 0 & 0 & 0 \\ -\frac{1}{6} & \frac{4}{3} & -\frac{1}{6} & -\frac{1}{3} & -\frac{1}{3} & -\frac{1}{3} & 0 & 0 & 0 \\ 0 & -\frac{1}{6} & \frac{2}{3} & 0 & -\frac{1}{3} & -\frac{1}{6} & 0 & 0 & 0 \\ -\frac{1}{6} & -\frac{1}{3} & 0 & \frac{4}{3} & -\frac{1}{3} & 0 & -\frac{1}{6} & -\frac{1}{3} & 0 \\ -\frac{1}{3} & -\frac{1}{3} & -\frac{1}{3} & -\frac{1}{3} & \frac{8}{3} & -\frac{1}{3} & -\frac{1}{3} & -\frac{1}{3} & -\frac{1}{3} \\ 0 & -\frac{1}{3} & -\frac{1}{6} & 0 & -\frac{1}{3} & \frac{4}{3} & 0 & -\frac{1}{3} & -\frac{1}{6} \\ 0 & 0 & 0 & -\frac{1}{6} & -\frac{1}{3} & 0 & \frac{2}{3} & -\frac{1}{6} & 0 \\ 0 & 0 & 0 & -\frac{1}{3} & -\frac{1}{3} & -\frac{1}{3} & -\frac{1}{6} & \frac{4}{3} & -\frac{1}{6} \\ 0 & 0 & 0 & 0 & -\frac{1}{3} & -\frac{1}{6} & 0 & -\frac{1}{6} & \frac{2}{3} \end{pmatrix} \begin{pmatrix} u_1 \\ u_2 \\ u_3 \\ u_4 \\ u_5 \\ u_6 \\ u_7 \\ u_8 \\ u_9 \end{pmatrix} = \begin{pmatrix} 0 \\ 0 \\ 0 \\ 0 \\ 0 \\ 0 \\ 0 \\ 0 \\ 0 \end{pmatrix}$$

Essential boundary conditions are as follows:

Node	dof	Value
1	u_1	0
2	u_2	0
3	u_3	0
4	u_4	$\frac{1}{4}$
6	u_6	0
7	u_7	0
8	u_8	0
9	u_9	0

Delete equations {1, 2, 3, 4, 6, 7, 8, 9}:

$$\begin{pmatrix} -\frac{1}{3} & -\frac{1}{3} & -\frac{1}{3} & -\frac{1}{3} & \frac{8}{3} & -\frac{1}{3} & -\frac{1}{3} & -\frac{1}{3} & -\frac{1}{3} \end{pmatrix} \begin{pmatrix} 0 \\ 0 \\ 0 \\ \frac{1}{4} \\ u_5 \\ 0 \\ 0 \\ 0 \\ 0 \end{pmatrix} = (0)$$

$$\left(\tfrac{8}{3}\right)(u_5) = \left(\tfrac{1}{12}\right)$$

Solving the final system of global equations, we get

$$\left\{u_5 = \tfrac{1}{32}\right\}$$

The solution for element 1 is as follows:

Coordinates of element center: $x_c = \frac{1}{4}$; $y_c = \frac{1}{4}$
Element dimensions: $a = \frac{1}{4}$; $b = \frac{1}{4}$
Interpolation functions in local element coordinates:

$$\boldsymbol{N}^T = \left\{4ts - s - t + \tfrac{1}{4}, -4ts + s - t + \tfrac{1}{4}, 4ts + s + t + \tfrac{1}{4}, -4ts - s + t + \tfrac{1}{4}\right\}$$

Shift for global coordinates: $s = x - \frac{1}{4}$; $t = y - \frac{1}{4}$
Interpolation functions in global coordinates:

$$\boldsymbol{N}^T = \{4yx - 2x - 2y + 1, 2x - 4xy, 4xy, 2y - 4xy\}$$

Nodal values, $\boldsymbol{d}^T = \left\{0, \frac{1}{4}, \frac{1}{32}, 0\right\}$

$$u(x, y) = \boldsymbol{N}^T\boldsymbol{d} = \frac{x}{2} - \frac{7xy}{8}$$

$$\frac{\partial u}{\partial x} = \frac{1}{2} - \frac{7y}{8}; \ \frac{\partial u}{\partial y} = -\frac{7x}{8}$$

Solutions over the remaining elements can be determined in exactly the same manner. The resulting solution and their derivatives at the element centers are as follows:

	x	y	u	$\partial u/\partial x$	$\partial u/\partial y$
1	0.25	0.25	0.0703125	0.28125	−0.21875
2	0.25	0.75	0.0078125	0.03125	−0.03125
3	0.75	0.25	0.0703125	−0.28125	−0.21875
4	0.75	0.75	0.0078125	−0.03125	−0.03125

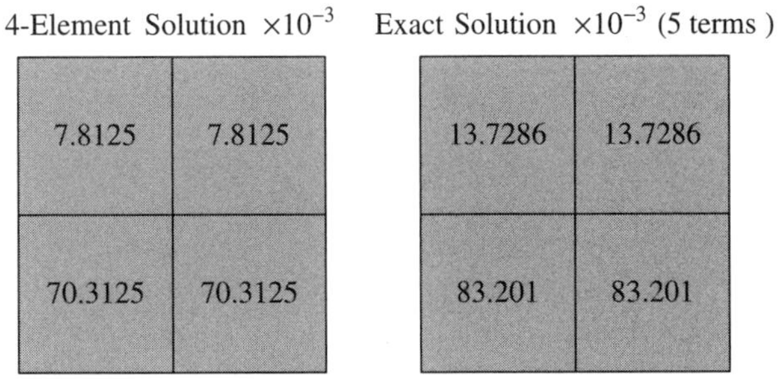

Figure 5.9. Four-element solution and the exact solution

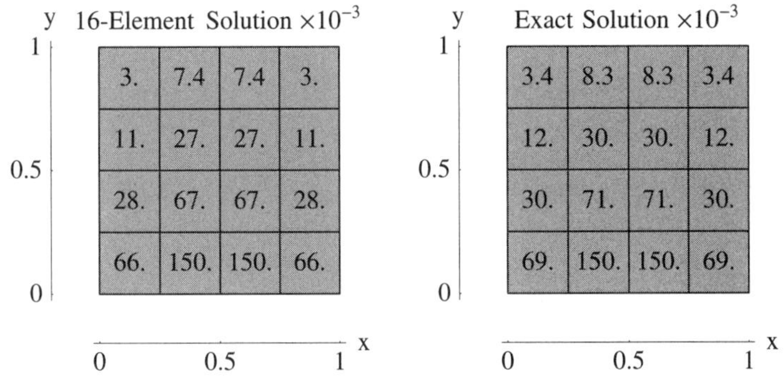

Figure 5.10. Finite element solution using 16 elements and the exact solution

In Figure 5.9 the finite element solution at the element centers is compared with the exact solution with five terms in the sum. Even with this coarse mesh the solution at the center is not too bad. To get a feel for the convergence of the solution, the problem is solved using 16 elements. The exact and the finite element solutions are compared in Figure 5.10. The 16-element solution is much closer to the exact solution. The solution can obviously be improved even further by using more elements.

It should be noted that the solution is symmetric with respect to the center line in the x direction. Thus we could have taken advantage of this and modeled only the right or left half of the solution domain. This would reduce the computational effort considerably and is especially advantageous when using fine meshes.

Example 5.3 Heat Flow in an L-Shaped Body Consider two-dimensional heat flow over an L-shaped body with thermal conductivity $k = 45$ W/m · °C shown in Figure 5.11. The bottom is maintained at $T_0 = 110$°C. Convection heat loss takes place on the top where the ambient air temperature is 20°C and the convection heat transfer coefficient $h = 55$ W/m^2 · °C. The right side is insulated. The left side is subjected to heat flux at a uniform rate $q_0 = 8000$ W/m^2. Heat is generated in the body at a rate $Q = 5 \times 10^6$ W/m^3. Determine the temperature distribution in the body.

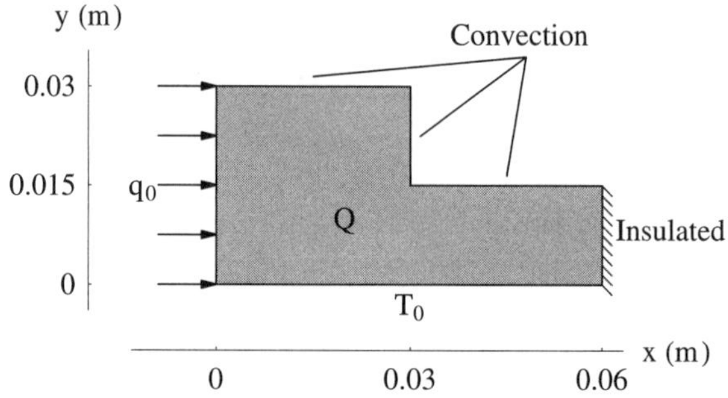

Figure 5.11. L-shaped body

As shown earlier, the governing differential equation for a heat flow problem is a special case of the general form. With the numerical values given for this example

$$k_x = k_y = 45; \qquad p = 0; \qquad q = 5 \times 10^6$$

The boundary conditions are as follows:

For all nodes on the bottom side, $T = 110$

On the left side ($n_x = -1, n_y = 0$):

$$-k\frac{\partial T}{\partial n} = k\frac{\partial T}{\partial x} = q_0 \Longrightarrow \alpha = 0; \qquad \beta = 8000$$

On the right side, $\alpha = 0; \beta = 0$

For convection on the horizontal portion of the top side ($n_x = 0, n_y = 1$):

$$-k\frac{\partial T}{\partial n} = -k\frac{\partial T}{\partial y} = h(T - T_\infty)$$

$$\Longrightarrow \alpha = -h = -55; \qquad \beta = hT_\infty = 55 \times 20 = 1100$$

For convection on the vertical portion of the top side ($n_x = 1, n_y = 0$):

$$-k\frac{\partial T}{\partial n} = -k\frac{\partial T}{\partial x} = h(T - T_\infty)$$

$$\Longrightarrow \alpha = -h = -55; \qquad \beta = hT_\infty = 55 \times 20 = 1100$$

We obtain a finite element solution of the problem using six square elements as shown in Figure 5.12. There are 13 nodes and thus we have a total of 13 degrees of freedom.

The complete finite element solution is as follows:

Equations for element 1:

Element dimensions: $a = 0.0075$; $b = 0.0075$

$k_x = 45$; $k_y = 45$; $p = 0$; $q = 5000000$

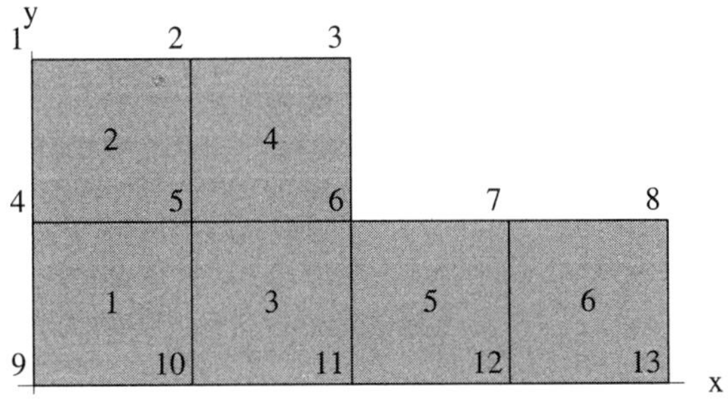

Figure 5.12. Six-element model

$$k_k = \begin{pmatrix} 30. & -7.5 & -15. & -7.5 \\ -7.5 & 30. & -7.5 & -15. \\ -15. & -7.5 & 30. & -7.5 \\ -7.5 & -15. & -7.5 & 30. \end{pmatrix}; k_p = \begin{pmatrix} 0 & 0 & 0 & 0 \\ 0 & 0 & 0 & 0 \\ 0 & 0 & 0 & 0 \\ 0 & 0 & 0 & 0 \end{pmatrix}; r_q = \begin{pmatrix} 281.25 \\ 281.25 \\ 281.25 \\ 281.25 \end{pmatrix}$$

NBC on side 4:

$$L = 0.015; \qquad \alpha = 0; \qquad \beta = 8000$$

$$k_\alpha = \begin{pmatrix} 0 & 0 & 0 & 0 \\ 0 & 0 & 0 & 0 \\ 0 & 0 & 0 & 0 \\ 0 & 0 & 0 & 0 \end{pmatrix}; \qquad r_\beta = \begin{pmatrix} 60. \\ 0 \\ 0 \\ 60. \end{pmatrix}$$

Complete element equations:

$$\begin{pmatrix} 30. & -7.5 & -15. & -7.5 \\ -7.5 & 30. & -7.5 & -15. \\ -15. & -7.5 & 30. & -7.5 \\ -7.5 & -15. & -7.5 & 30. \end{pmatrix} \begin{pmatrix} T_9 \\ T_{10} \\ T_5 \\ T_4 \end{pmatrix} = \begin{pmatrix} 341.25 \\ 281.25 \\ 281.25 \\ 341.25 \end{pmatrix}$$

Equations for element 2:

Element dimensions: $a = 0.0075$; $b = 0.0075$

$k_x = 45$; $k_y = 45$; $p = 0$; $q = 5000000$

$$k_k = \begin{pmatrix} 30. & -7.5 & -15. & -7.5 \\ -7.5 & 30. & -7.5 & -15. \\ -15. & -7.5 & 30. & -7.5 \\ -7.5 & -15. & -7.5 & 30. \end{pmatrix}; k_p = \begin{pmatrix} 0 & 0 & 0 & 0 \\ 0 & 0 & 0 & 0 \\ 0 & 0 & 0 & 0 \\ 0 & 0 & 0 & 0 \end{pmatrix}; r_q = \begin{pmatrix} 281.25 \\ 281.25 \\ 281.25 \\ 281.25 \end{pmatrix}$$

NBC on side 3:

$$L = 0.015; \qquad \alpha = -55; \qquad \beta = 1100$$

$$k_\alpha = \begin{pmatrix} 0 & 0 & 0 & 0 \\ 0 & 0 & 0 & 0 \\ 0 & 0 & 0.275 & 0.1375 \\ 0 & 0 & 0.1375 & 0.275 \end{pmatrix}; \qquad r_\beta = \begin{pmatrix} 0 \\ 0 \\ 8.25 \\ 8.25 \end{pmatrix}$$

NBC on side 4:

$$L = 0.015; \qquad \alpha = 0; \qquad \beta = 8000$$

$$\boldsymbol{k}_\alpha = \begin{pmatrix} 0 & 0 & 0 & 0 \\ 0 & 0 & 0 & 0 \\ 0 & 0 & 0 & 0 \\ 0 & 0 & 0 & 0 \end{pmatrix}; \qquad \boldsymbol{r}_\beta = \begin{pmatrix} 60. \\ 0 \\ 0 \\ 60. \end{pmatrix}$$

Complete element equations:

$$\begin{pmatrix} 30. & -7.5 & -15. & -7.5 \\ -7.5 & 30. & -7.5 & -15. \\ -15. & -7.5 & 30.275 & -7.3625 \\ -7.5 & -15. & -7.3625 & 30.275 \end{pmatrix} \begin{pmatrix} T_4 \\ T_5 \\ T_2 \\ T_1 \end{pmatrix} = \begin{pmatrix} 341.25 \\ 281.25 \\ 289.5 \\ 349.5 \end{pmatrix}$$

Processing the remaining elements in exactly the same manner, we get the following global equations:

$$\begin{pmatrix}
30.275 & -7.3625 & 0 & -7.5 & -15. & 0 & 0 & 0 & 0 & 0 & 0 & 0 & 0 \\
-7.3625 & 60.55 & -7.3625 & -15. & -15. & 15. & 0 & 0 & 0 & 0 & 0 & 0 & 0 \\
0 & -7.3625 & 30.55 & 0 & -15. & -7.3625 & 0 & 0 & 0 & 0 & 0 & 0 & 0 \\
-7.5 & -15. & 0 & 60. & -15. & 0 & 0 & 0 & -7.5 & -15. & 0 & 0 & 0 \\
-15. & -15. & -15. & -15. & 120. & 15. & 0 & 0 & -15. & -15. & -15. & 0 & 0 \\
0 & -15. & -7.3625 & 0 & -15. & 90.55 & -7.3625 & 0 & 0 & -15. & -15. & -15. & 0 \\
0 & 0 & 0 & 0 & 0 & -7.3625 & 60.55 & -7.3625 & 0 & 0 & -15. & -15. & -15. \\
0 & 0 & 0 & 0 & 0 & 0 & -7.3625 & 30.275 & 0 & 0 & 0 & -15. & -7.5 \\
0 & 0 & 0 & -7.5 & -15. & 0 & 0 & 0 & 30. & -7.5 & 0 & 0 & 0 \\
0 & 0 & 0 & -15. & -15. & -15. & 0 & 0 & -7.5 & 60. & -7.5 & 0 & 0 \\
0 & 0 & 0 & 0 & -15. & -15. & -15. & 0 & 0 & -7.5 & 60. & -7.5 & 0 \\
0 & 0 & 0 & 0 & 0 & -15. & -15. & -15. & 0 & 0 & -7.5 & 60. & -7.5 \\
0 & 0 & 0 & 0 & 0 & 0 & -15. & -7.5 & 0 & 0 & 0 & -7.5 & 30.
\end{pmatrix}
\begin{pmatrix} T_1 \\ T_2 \\ T_3 \\ T_4 \\ T_5 \\ T_6 \\ T_7 \\ T_8 \\ T_9 \\ T_{10} \\ T_{11} \\ T_{12} \\ T_{13} \end{pmatrix} =
\begin{pmatrix} 349.5 \\ 579. \\ 297.75 \\ 682.5 \\ 1125. \\ 860.25 \\ 579. \\ 289.5 \\ 341.25 \\ 562.5 \\ 562.5 \\ 562.5 \\ 281.25 \end{pmatrix}$$

Essential boundary conditions:

Node	dof	Value
9	T_9	110
10	T_{10}	110
11	T_{11}	110
12	T_{12}	110
13	T_{13}	110

Delete equations {9, 10, 11, 12, 13}:

$$\begin{pmatrix}
30.275 & -7.3625 & 0 & -7.5 & -15. & 0 & 0 & 0 & 0 & 0 & 0 & 0 & 0 \\
-7.3625 & 60.55 & -7.3625 & -15. & -15. & -15. & 0 & 0 & 0 & 0 & 0 & 0 & 0 \\
0 & -7.3625 & 30.55 & 0 & -15. & -7.3625 & 0 & 0 & 0 & 0 & 0 & 0 & 0 \\
-7.5 & -15. & 0 & 60. & -15. & 0 & 0 & 0 & -7.5 & -15. & 0 & 0 & 0 \\
-15. & -15. & -15. & -15. & 120. & -15. & 0 & 0 & -15. & -15. & -15. & 0 & 0 \\
0 & -15. & -7.3625 & 0 & -15. & 90.55 & -7.3625 & 0 & 0 & -15. & -15. & -15. & 0 \\
0 & 0 & 0 & 0 & 0 & -7.3625 & 60.55 & -7.3625 & 0 & 0 & -15. & -15. & -15. \\
0 & 0 & 0 & 0 & 0 & 0 & -7.3625 & 30.275 & 0 & 0 & 0 & -15. & -7.5
\end{pmatrix}
\begin{pmatrix} T_1 \\ T_2 \\ T_3 \\ T_4 \\ T_5 \\ T_6 \\ T_7 \\ T_8 \\ 110 \\ 110 \\ 110 \\ 110 \\ 110 \end{pmatrix} =
\begin{pmatrix} 349.5 \\ 579. \\ 297.75 \\ 682.5 \\ 1125. \\ 860.25 \\ 579. \\ 289.5 \end{pmatrix}$$

Extract columns {9, 10, 11, 12, 13}.

Multiply each column by its respective known value {110, 110, 110, 110, 110}.

Move all resulting vectors to the right-hand side.

After adjusting for essential boundary conditions we have

$$\begin{pmatrix} 30.275 & -7.3625 & 0 & -7.5 & -15. & 0 & 0 & 0 \\ -7.3625 & 60.55 & -7.3625 & -15. & -15. & -15. & 0 & 0 \\ 0 & -7.3625 & 30.55 & 0 & -15. & -7.3625 & 0 & 0 \\ -7.5 & -15. & 0 & 60. & -15. & 0 & 0 & 0 \\ -15. & -15. & -15. & -15. & 120. & -15. & 0 & 0 \\ 0 & -15. & -7.3625 & 0 & -15. & 90.55 & -7.3625 & 0 \\ 0 & 0 & 0 & 0 & 0 & -7.3625 & 60.55 & -7.3625 \\ 0 & 0 & 0 & 0 & 0 & 0 & -7.3625 & 30.275 \end{pmatrix} \begin{pmatrix} T_1 \\ T_2 \\ T_3 \\ T_4 \\ T_5 \\ T_6 \\ T_7 \\ T_8 \end{pmatrix} = \begin{pmatrix} 349.5 \\ 579. \\ 297.75 \\ 3157.5 \\ 6075. \\ 5810.25 \\ 5529. \\ 2764.5 \end{pmatrix}$$

Solving the final system of global equations, we get

$$\{T_1 = 154.962, T_2 = 151.228, T_3 = 148.673, T_4 = 145.433,$$
$$T_5 = 142.521, T_6 = 134.871, T_7 = 122.436, T_8 = 121.088\}$$

Solution for element 1:

Interpolation functions in local element coordinates:

$$\begin{aligned} N^T = \{&4444.44ts - 33.3333s - 33.3333t + 0.25, \\ &-4444.44ts + 33.3333s - 33.3333t + 0.25, \\ &4444.44ts + 33.3333s + 33.3333t + 0.25, \\ &-4444.44ts - 33.3333s + 33.3333t + 0.25\} \end{aligned}$$

Shift for global coordinates: $s = x - 0.0075$; $t = y - 0.0075$

Interpolation functions in global coordinates:

$$\begin{aligned} N^T = \{&4444.44yx - 66.6667x - 66.6667y + 1., 66.6667x - 4444.44xy, \\ &4444.44xy, 66.6667y - 4444.44xy\} \end{aligned}$$

Nodal values, $\boldsymbol{d}^T = \{110, 110, 142.521, 145.433\}$

$T(x, y) = \boldsymbol{N}^T \boldsymbol{d} = -12940.9xy + 2362.17y + 110.$

$\partial T/\partial x = -12940.9y$; $\partial T/\partial y = 2362.17 - 12940.9x$

Solutions over the remaining elements can be determined in exactly the same manner. The solution derivatives at the element centers are shown in Figure 5.13. These values differ considerably over adjacent elements, indicating that the mesh is coarse. A finer mesh is needed for more reliable results.

The problem is solved with a finer mesh using ANSYS (AnsysFiles\Chap5\LShape-Thermal.inp on the book web site). The resulting solution in the form nodal temperature contours and vectors of thermal gradient ($\partial T/\partial x$, $\partial T/\partial y$) is shown in Figure 5.14.

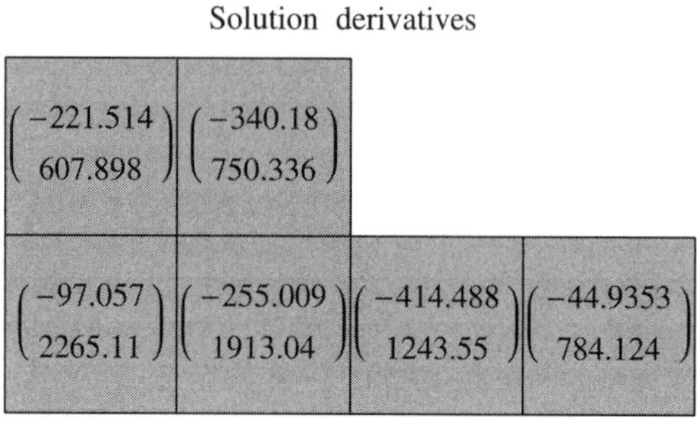

Figure 5.13. Solution derivatives ($\partial T/\partial x$, $\partial T/\partial y$) for the six-element model

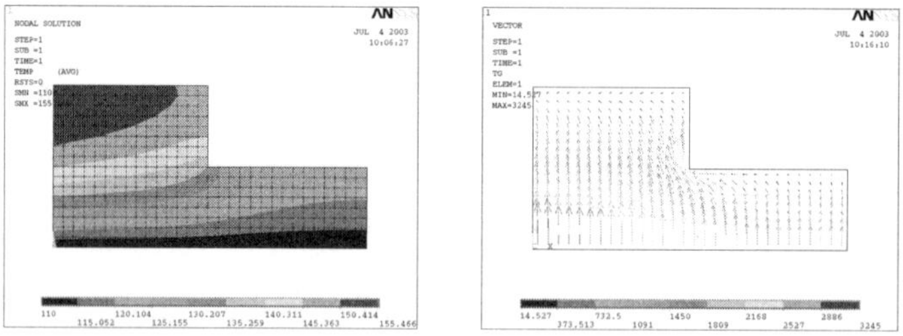

Figure 5.14. Nodal temperature contours and thermal gradient vectors

◆ *Mathematica*/MATLAB Implementation 5.1 on the Book Web Site: *Four-node rectangular element for 2D BVP*

Example 5.4 Torsion of a Rectangular Shaft Find stresses developed in a 4-cm × 8-cm rectangular shaft when it is subjected to a torque of 500 N · m. The shaft is 1 m long and $G = 76.9$ GPa. A quarter of the domain needs to be modeled because of symmetry. As shown in Figure 5.15, a coarse four-element mesh is used.

The governing differential equation for the problem is

$$\frac{\partial^2 \phi}{\partial x^2} + \frac{\partial^2 \phi}{\partial y^2} + 2G\theta = 0$$

where G is the shear modulus and θ is the angle of twist per unit length. The boundary

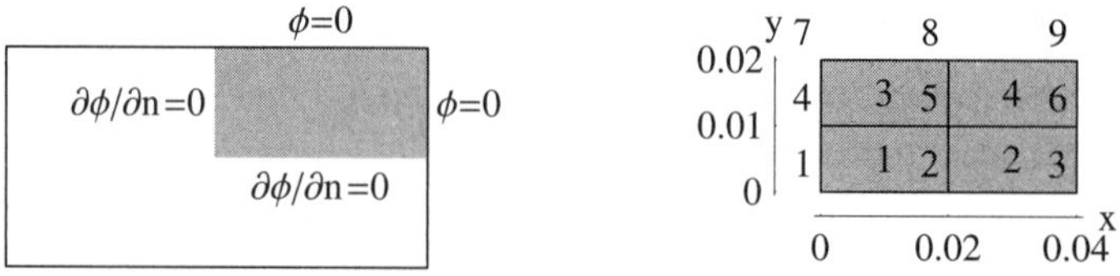

Figure 5.15. Rectangular shaft subjected to torque

condition is $\phi = 0$ on the boundary. As a result of the essential boundary condition, $\phi = 0$ at nodes {3, 6, 9, 8, 7}. There are no nonzero natural boundary conditions.

Since θ is unknown, we start by arbitrarily assuming $G\theta = 1$. After performing the analysis, we compute the total torque T_a. This torque corresponds to the assumed value of $G\theta$. Since the relationship between the torque and the angle of twist is linear, the actual value of θ can then be computed using the given value of torque T as follows:

$$\theta = \frac{T}{GT_a}$$

The actual ϕ values are obtained by multiplying the computed values by the actual $G\theta$ value. The complete finite element solution, using newtons and meters, is as follows:

Equations for element 1:

Element dimensions: $a = 0.01$; $b = 0.005$

$k_x = 1$; $k_y = 1$; $p = 0$; $q = 2$

$$\boldsymbol{k}_k = \begin{pmatrix} 0.833333 & 0.166667 & -0.416667 & -0.583333 \\ 0.166667 & 0.833333 & -0.583333 & -0.416667 \\ -0.416667 & -0.583333 & 0.833333 & 0.166667 \\ -0.583333 & -0.416667 & 0.166667 & 0.833333 \end{pmatrix};$$

$$\boldsymbol{k}_p = \begin{pmatrix} 0 & 0 & 0 & 0 \\ 0 & 0 & 0 & 0 \\ 0 & 0 & 0 & 0 \\ 0 & 0 & 0 & 0 \end{pmatrix}; \boldsymbol{r}_q = \begin{pmatrix} 0.0001 \\ 0.0001 \\ 0.0001 \\ 0.0001 \end{pmatrix}$$

Complete element equations:

$$\begin{pmatrix} 0.833333 & 0.166667 & -0.416667 & -0.583333 \\ 0.166667 & 0.833333 & -0.583333 & -0.416667 \\ -0.416667 & -0.583333 & 0.833333 & 0.166667 \\ -0.583333 & -0.416667 & 0.166667 & 0.833333 \end{pmatrix} \begin{pmatrix} \phi_1 \\ \phi_2 \\ \phi_5 \\ \phi_4 \end{pmatrix} = \begin{pmatrix} 0.0001 \\ 0.0001 \\ 0.0001 \\ 0.0001 \end{pmatrix}$$

Processing the remaining elements in exactly the same manner, we get the following global equations:

$$\begin{pmatrix} 0.833333 & 0.166667 & 0 & -0.583333 & -0.416667 & 0 & 0 & 0 & 0 \\ 0.166667 & 1.66667 & 0.166667 & -0.416667 & -1.16667 & -0.416667 & 0 & 0 & 0 \\ 0 & 0.166667 & 0.833333 & 0 & -0.416667 & -0.583333 & 0 & 0 & 0 \\ -0.583333 & -0.416667 & 0 & 1.66667 & 0.333333 & 0 & -0.583333 & -0.416667 & 0 \\ -0.416667 & -1.16667 & -0.416667 & 0.333333 & 3.33333 & 0.333333 & -0.416667 & -1.16667 & -0.416667 \\ 0 & -0.416667 & -0.583333 & 0 & 0.333333 & 1.66667 & 0 & -0.416667 & -0.583333 \\ 0 & 0 & 0 & -0.583333 & -0.416667 & 0 & 0.833333 & 0.16667 & 0 \\ 0 & 0 & 0 & -0.416667 & -1.16667 & -0.416667 & 0.16667 & 1.66667 & 0.166667 \\ 0 & 0 & 0 & 0 & -0.416667 & -0.583333 & 0 & 0.166667 & 0.833333 \end{pmatrix} \begin{pmatrix} \phi_1 \\ \phi_2 \\ \phi_3 \\ \phi_4 \\ \phi_5 \\ \phi_6 \\ \phi_7 \\ \phi_8 \\ \phi_9 \end{pmatrix} = \begin{pmatrix} 0.0001 \\ 0.0002 \\ 0.0001 \\ 0.0002 \\ 0.0004 \\ 0.0002 \\ 0.0001 \\ 0.0002 \\ 0.0001 \end{pmatrix}$$

Essential boundary conditions:

Node	dof	Value
3	ϕ_3	0
6	ϕ_6	0
7	ϕ_7	0
8	ϕ_8	0
9	ϕ_9	0

Remove {3, 6, 7, 8, 9} rows and columns.

After adjusting for essential boundary conditions, we have

$$\begin{pmatrix} 0.833333 & 0.166667 & -0.583333 & -0.416667 \\ 0.166667 & 1.66667 & -0.416667 & -1.16667 \\ -0.583333 & -0.416667 & 1.66667 & 0.333333 \\ -0.416667 & -1.16667 & 0.333333 & 3.33333 \end{pmatrix} \begin{pmatrix} \phi_1 \\ \phi_2 \\ \phi_4 \\ \phi_5 \end{pmatrix} = \begin{pmatrix} 0.0001 \\ 0.0002 \\ 0.0002 \\ 0.0004 \end{pmatrix}$$

Solving the final system of global equations, we get

$$\{\phi_1 = 0.000380919, \phi_2 = 0.000331898, \phi_4 = 0.000285245, \phi_5 = 0.000255255\}$$

Solution for element 1:

Interpolation functions in local element coordinates:

$$N^T = \{5000.ts - 25.s - 50.t + 0.25, -5000.ts + 25.s - 50.t + 0.25,$$
$$5000.ts + 25.s + 50.t + 0.25, -5000.ts - 25.s + 50.t + 0.25\}$$

Shift for global coordinates: $s = x - 0.01$; $t = y - 0.005$

Interpolation functions in global coordinates:

$$\boldsymbol{N}^T = \{5000.yx - 50.x - 100.y + 1., 50.x - 5000.xy, 5000.xy, 100.y - 5000.xy\}$$

Nodal values, $\boldsymbol{d}^T = \{0.000380919, 0.000331898, 0.000255255, 0.000285245\}$

$\phi(x, y) = \boldsymbol{N}^T\boldsymbol{d} = 0.0951558yx - 0.0024511x - 0.00956742y + 0.000380919$

$\partial\phi/\partial x = 0.0951558y - 0.0024511$; $\partial\phi/\partial y = 0.0951558x - 0.00956742$

Solutions over the remaining elements can be determined in exactly the same manner. A solution summary and the integral of ϕ over each element is summarized as follows:

	ϕ_a	$\iint \phi_a \, dA$
1	$0.0951558yx - 0.0024511x - 0.00956742y + 0.000380919$	6.26658×10^{-8}
2	$0.383215yx - 0.0165949x - 0.0153286y + 0.000663795$	2.93576×10^{-8}
3	$0.149954yx - 0.00299907x - 0.0285245y + 0.000570491$	2.7025×10^{-8}
4	$1.27627yx - 0.0255255x - 0.0510509y + 0.00102102$	1.27627×10^{-8}

The total torque is given by

$$T = 2\iint_A \phi\, dA$$

Summing $\iint \phi\, dA$ contributions from all elements and multiplying by 2 give the total torque. Since we are modeling one-fourth of the shape, the torque for the entire section is

$$T_a = 4 \times 2 \times \sum\left(\iint \phi_a\, dA\right) = 1.05449 \times 10^{-6}\ \text{N}\cdot\text{m}$$

Since the actual torque is 500 N · m, the actual value of the angle of twist is

$$G\theta = \frac{500}{T_a} = 4.74163 \times 10^{8}; \qquad \theta = 0.00616597\ \text{rad/m}$$

The ϕ values are simply scaled by this value of $G\theta$, and thus the solution corresponding to a torque of 500 N · m is as follows:

$$\phi = \left(\frac{500}{T_a}\right)\phi_a = 4.74163 \times 10^{8}\phi_a$$

	ϕ ($\times 10^6$)	$\tau_{yz} = -\partial\phi/\partial x$ (MPa)	$\tau_{xz} = \partial\phi/\partial y$ (MPa)
1	$45.1194yx - 1.16222x - 4.53652y + 0.180618$	$1.16222 - 45.1194y$	$45.1194x - 4.53652$
2	$181.706yx - 7.86868x - 7.26826y + 0.314747$	$7.86868 - 181.706y$	$181.706x - 7.26826$
3	$71.1025yx - 1.42205x - 13.5253y + 0.270506$	$1.42205 - 71.1025y$	$71.1025x - 13.5253$
4	$605.162yx - 12.1032x - 24.2065y + 0.484129$	$12.1032 - 605.162y$	$605.162x - 24.2065$

Stresses at element centroids are as follows:

	τ_{yz} (MPa)	τ_{xz} (MPa)
1	0.936622	−4.08532
2	6.96015	−1.81706
3	0.355513	−12.8143
4	3.02581	−6.05162

The maximum shear stress occurs at the midpoint of the long side (node 7), which from element 3 is

$$\text{Stresses at node 7: } \tau_{yz} = 2.22045 \times 10^{-16}\ \text{MPa}; \qquad \tau_{xz} = -13.5253\ \text{MPa}$$

$$\tau_{max} = 13.5253\ \text{MPa}$$

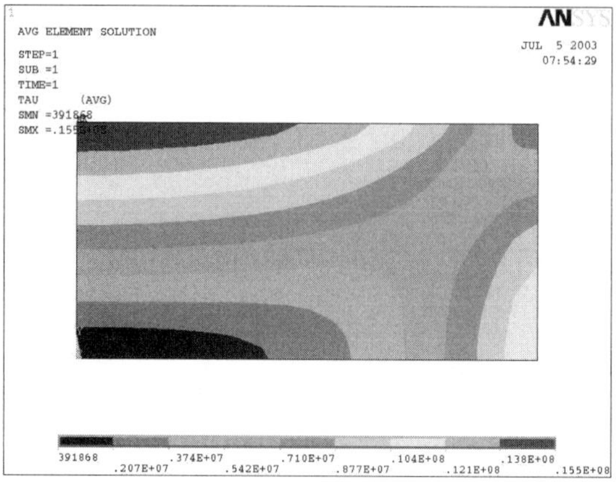

Figure 5.16. Shear stress contours using ANSYS

An exact solution for the problem is available as follows (W. C. Young and R. G. Budynas, *Roark's Formulas for Stress and Strain*, seventh edition, p. 401, McGraw-Hill, 2002).

$$\tau_{\max} = \frac{3T}{8ab^2}\left(1 + 0.6095\frac{b}{a} + 0.8865\left(\frac{b}{a}\right)^2 - 1.8023\left(\frac{b}{a}\right)^3 + 0.91\left(\frac{b}{a}\right)^4\right)$$

where $2a$ is the longer dimension of the section and $2b$ is the shorter dimension:

$$\text{Exact solution: } \tau_{\max} = 15.9136\,\text{MPa}$$

The problem is solved with 20×20 mesh using ANSYS (AnsysFiles\Chap5\Rect-Torsion.inp on the book web site). The resulting shear stresses are shown in Figure 5.16. The maximum shear stress is seen as 15.5 MPa, which is fairly close to the exact value.

5.4.2 Eight-Node Rectangular Element

The interpolation functions for the four-node rectangular element presented in the previous section were based on the following assumed solution:

$$u(x, y) = c_0 + c_1 s + c_2 t + c_3 st$$

We can add more terms to the polynomial and use the same process as before to derive interpolation functions for higher order elements. For an eight-node rectangular element $2a \times 2b$ as shown in Figure 5.17 we use the following polynomial consisting of a complete

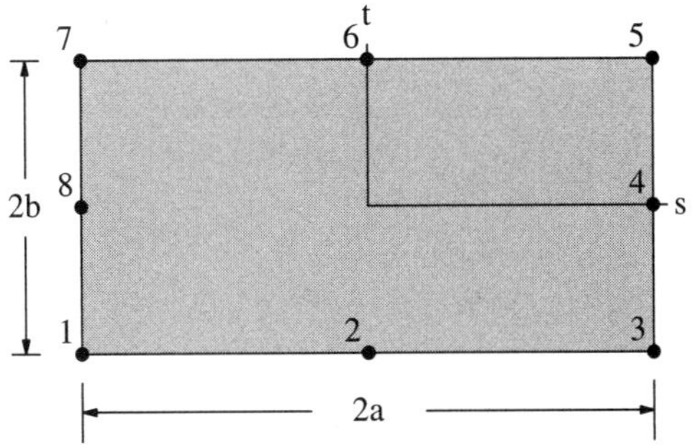

Figure 5.17. Eight-node rectangular element

quadratic (six terms) with two additional cubic terms:

$$u(s,t) = \begin{pmatrix} 1 & s & t & s^2 & st & t^2 & s^2t & st^2 \end{pmatrix} \begin{pmatrix} c_0 \\ c_1 \\ c_2 \\ c_3 \\ c_4 \\ c_5 \\ c_6 \\ c_7 \end{pmatrix}$$

By evaluating u at the eight nodes and solving the resulting system of equations, we can evaluate the coefficients in terms of nodal degrees of freedom, resulting in the following interpolation functions:

$$\text{Interpolation functions, } \boldsymbol{N} = \begin{pmatrix} -\frac{(a-s)(b-t)(bs+a(b+t))}{4a^2b^2} \\ \frac{(a^2-s^2)(b-t)}{2a^2b} \\ -\frac{(a+s)(b-t)(a(b+t)-bs)}{4a^2b^2} \\ \frac{(a+s)(b^2-t^2)}{2ab^2} \\ -\frac{(a+s)(a(b-t)-bs)(b+t)}{4a^2b^2} \\ \frac{(a^2-s^2)(b+t)}{2a^2b} \\ -\frac{(a-s)(bs+a(b-t))(b+t)}{4a^2b^2} \\ \frac{(a-s)(b^2-t^2)}{2ab^2} \end{pmatrix}$$

These interpolation functions are known as serendipity functions and will be used extensively in the following chapters. Note that each of the interpolation functions N_i is 1 at the ith node and 0 at every other node. Using these interpolation functions, the following element matrices are obtained:

$$\boldsymbol{N}^T = \begin{pmatrix} -\frac{(a-s)(b-t)(bs+a(b+t))}{4a^2b^2} & \frac{(a^2-s^2)(b-t)}{2a^2b} & -\frac{(a+s)(b-t)(a(b+t)-bs)}{4a^2b^2} & \frac{(a+s)(b^2-t^2)}{2ab^2} & -\frac{(a+s)(a(b-t)-bs)(b+t)}{4a^2b^2} & \frac{(a^2-s^2)(b+t)}{2a^2b} & -\frac{(a-s)(bs+a(b-t))(b+t)}{4a^2b^2} & \frac{(a-s)(b^2-t^2)}{2ab^2} \end{pmatrix}$$

$$\boldsymbol{B}^T = \begin{pmatrix} \frac{(b-t)(2bs+at)}{4a^2b^2} & \frac{s(t-b)}{a^2b} & -\frac{(b-t)(at-2bs)}{4a^2b^2} & \frac{b^2-t^2}{2ab^2} & \frac{(b+t)(2bs+at)}{4a^2b^2} & -\frac{s(b+t)}{a^2b} & -\frac{(b+t)(at-2bs)}{4a^2b^2} & \frac{t^2-b^2}{2ab^2} \\ \frac{(a-s)(bs+2at)}{4a^2b^2} & \frac{s^2-a^2}{2a^2b} & \frac{(a+s)(2at-bs)}{4a^2b^2} & -\frac{(a+s)t}{ab^2} & \frac{(a+s)(bs+2at)}{4a^2b^2} & \frac{a^2-s^2}{2a^2b} & \frac{(a-s)(2at-bs)}{4a^2b^2} & \frac{(s-a)t}{ab^2} \end{pmatrix}$$

$$\boldsymbol{k}_k = \int_{-a}^{a}\int_{-b}^{b} \boldsymbol{BCB}^T \, ds\, dt$$

$$= \begin{pmatrix}
\frac{26(k_ya^2+k_xb^2)}{45ab} & \frac{k_ya}{15b}-\frac{8k_xb}{9a} & \frac{17k_ya}{90b}+\frac{14k_xb}{45a} & -\frac{4k_ya}{9b}-\frac{k_xb}{15a} & \frac{23(k_ya^2+k_xb^2)}{90ab} & -\frac{k_ya}{15b}-\frac{4k_xb}{9a} & \frac{14k_ya}{45b}+\frac{17k_xb}{90a} & \frac{k_xb}{15a}-\frac{8k_ya}{9b} \\
\frac{k_ya}{15b}-\frac{8k_xb}{9a} & \frac{8k_ya}{15b}+\frac{16k_xb}{9a} & \frac{k_ya}{15b}-\frac{8k_xb}{9a} & 0 & -\frac{k_ya}{15b}-\frac{4k_xb}{9a} & \frac{8k_xb}{9a}-\frac{8k_ya}{15b} & -\frac{k_ya}{15b}-\frac{4k_xb}{9a} & 0 \\
\frac{17k_ya}{90b}+\frac{14k_xb}{45a} & \frac{k_ya}{15b}-\frac{8k_xb}{9a} & \frac{26(k_ya^2+k_xb^2)}{45ab} & \frac{k_xb}{15a}-\frac{8k_ya}{9b} & \frac{14k_ya}{45b}+\frac{17k_xb}{90a} & -\frac{k_ya}{15b}-\frac{4k_xb}{9a} & \frac{23(k_ya^2+k_xb^2)}{90ab} & -\frac{4k_ya}{9b}-\frac{k_xb}{15a} \\
-\frac{4k_ya}{9b}-\frac{k_xb}{15a} & 0 & \frac{k_xb}{15a}-\frac{8k_ya}{9b} & \frac{16k_ya}{9b}+\frac{8k_xb}{15a} & \frac{k_xb}{15a}-\frac{8k_ya}{9b} & 0 & -\frac{4k_ya}{9b}-\frac{k_xb}{15a} & \frac{8k_ya}{9b}-\frac{8k_xb}{15a} \\
\frac{23(k_ya^2+k_xb^2)}{90ab} & -\frac{k_ya}{15b}-\frac{4k_xb}{9a} & \frac{14k_ya}{45b}+\frac{17k_xb}{90a} & \frac{k_xb}{15a}-\frac{8k_ya}{9b} & \frac{26(k_ya^2+k_xb^2)}{45ab} & \frac{k_ya}{15b}-\frac{8k_xb}{9a} & \frac{17k_ya}{90b}+\frac{14k_xb}{45a} & -\frac{4k_ya}{9b}-\frac{k_xb}{15a} \\
-\frac{k_ya}{15b}-\frac{4k_xb}{9a} & \frac{8k_xb}{9a}-\frac{8k_ya}{15b} & -\frac{k_ya}{15b}-\frac{4k_xb}{9a} & 0 & \frac{k_ya}{15b}-\frac{8k_xb}{9a} & \frac{8k_ya}{15b}+\frac{16k_xb}{9a} & \frac{k_ya}{15b}-\frac{8k_xb}{9a} & 0 \\
\frac{14k_ya}{45b}+\frac{17k_xb}{90a} & -\frac{k_ya}{15b}-\frac{4k_xb}{9a} & \frac{23(k_ya^2+k_xb^2)}{90ab} & -\frac{4k_ya}{9b}-\frac{k_xb}{15a} & \frac{17k_ya}{90b}+\frac{14k_xb}{45a} & \frac{k_ya}{15b}-\frac{8k_xb}{9a} & \frac{26(k_ya^2+k_xb^2)}{45ab} & \frac{k_xb}{15a}-\frac{8k_ya}{9b} \\
\frac{k_xb}{15a}-\frac{8k_ya}{9b} & 0 & -\frac{4k_ya}{9b}-\frac{k_xb}{15a} & \frac{8k_ya}{9b}-\frac{8k_xb}{15a} & -\frac{4k_ya}{9b}-\frac{k_xb}{15a} & 0 & \frac{k_xb}{15a}-\frac{8k_ya}{9b} & \frac{16k_ya}{9b}+\frac{8k_xb}{15a}
\end{pmatrix}$$

$$\boldsymbol{k}_p = \int_{-a}^{a}\int_{-b}^{b} -p\boldsymbol{NN}^T \, ds\, dt = \begin{pmatrix}
-\frac{2}{15}abp & \frac{2abp}{15} & -\frac{2}{45}abp & \frac{8abp}{45} & -\frac{1}{15}abp & \frac{8abp}{45} & -\frac{2}{45}abp & \frac{2abp}{15} \\
\frac{2abp}{15} & -\frac{32}{45}abp & \frac{2abp}{15} & -\frac{4}{9}abp & \frac{8abp}{45} & -\frac{16}{45}abp & \frac{8abp}{45} & -\frac{4}{9}abp \\
-\frac{2}{45}abp & \frac{2abp}{15} & -\frac{2}{15}abp & \frac{2abp}{15} & -\frac{2}{45}abp & \frac{8abp}{45} & -\frac{1}{15}abp & \frac{8abp}{45} \\
\frac{8abp}{45} & -\frac{4}{9}abp & \frac{2abp}{15} & -\frac{32}{45}abp & \frac{2abp}{15} & -\frac{4}{9}abp & \frac{8abp}{45} & -\frac{16}{45}abp \\
-\frac{1}{15}abp & \frac{8abp}{45} & -\frac{2}{45}abp & \frac{2abp}{15} & -\frac{2}{15}abp & \frac{2abp}{15} & -\frac{2}{45}abp & \frac{8abp}{45} \\
\frac{8abp}{45} & -\frac{16}{45}abp & \frac{8abp}{45} & -\frac{4}{9}abp & \frac{2abp}{15} & -\frac{32}{45}abp & \frac{2abp}{15} & -\frac{4}{9}abp \\
-\frac{2}{45}abp & \frac{8abp}{45} & -\frac{1}{15}abp & \frac{8abp}{45} & -\frac{2}{45}abp & \frac{2abp}{15} & -\frac{2}{15}abp & \frac{2abp}{15} \\
\frac{2abp}{15} & -\frac{4}{9}abp & \frac{8abp}{45} & -\frac{16}{45}abp & \frac{8abp}{45} & -\frac{4}{9}abp & \frac{2abp}{15} & -\frac{32}{45}abp
\end{pmatrix}$$

$$\boldsymbol{r}_q^T = \int_{-a}^{a}\int_{-b}^{b} q\boldsymbol{N}^T \, ds\, dt = \begin{pmatrix} -\frac{1}{3}abq & \frac{4abq}{3} & -\frac{1}{3}abq & \frac{4abq}{3} & -\frac{1}{3}abq & \frac{4abq}{3} & -\frac{1}{3}abq & \frac{4abq}{3} \end{pmatrix}$$

For the NBC on side 1

$$\boldsymbol{N}_c^T = \begin{pmatrix} \frac{c^2}{2a^2}-\frac{c}{2a} & 1-\frac{c^2}{a^2} & \frac{c^2}{2a^2}+\frac{c}{2a} & 0 & 0 & 0 & 0 & 0 \end{pmatrix}$$

$$\boldsymbol{k}_\alpha = \int_{-a}^{a} -\alpha \boldsymbol{N}_c\boldsymbol{N}_c^T \, dc = \begin{pmatrix}
-\frac{4a\alpha}{15} & -\frac{2a\alpha}{15} & \frac{a\alpha}{15} & 0 & 0 & 0 & 0 & 0 \\
-\frac{2a\alpha}{15} & -\frac{16a\alpha}{15} & -\frac{2a\alpha}{15} & 0 & 0 & 0 & 0 & 0 \\
\frac{a\alpha}{15} & -\frac{2a\alpha}{15} & -\frac{4a\alpha}{15} & 0 & 0 & 0 & 0 & 0 \\
0 & 0 & 0 & 0 & 0 & 0 & 0 & 0 \\
0 & 0 & 0 & 0 & 0 & 0 & 0 & 0 \\
0 & 0 & 0 & 0 & 0 & 0 & 0 & 0 \\
0 & 0 & 0 & 0 & 0 & 0 & 0 & 0 \\
0 & 0 & 0 & 0 & 0 & 0 & 0 & 0
\end{pmatrix}$$

$$r_\beta^T = \int_{-a}^{a} \beta N_c^T\, dc = \begin{pmatrix} \frac{a\beta}{3} & \frac{4a\beta}{3} & \frac{a\beta}{3} & 0 & 0 & 0 & 0 & 0 \end{pmatrix}$$

For the NBC on side 2

$$N_c^T = \begin{pmatrix} 0 & 0 & \frac{c^2}{2b^2} - \frac{c}{2b} & 1 - \frac{c^2}{b^2} & \frac{c^2}{2b^2} + \frac{c}{2b} & 0 & 0 & 0 \end{pmatrix}$$

$$k_\alpha = \int_{-b}^{b} -\alpha N_c N_c^T\, dc = \begin{pmatrix} 0 & 0 & 0 & 0 & 0 & 0 & 0 & 0 \\ 0 & 0 & 0 & 0 & 0 & 0 & 0 & 0 \\ 0 & 0 & -\frac{4b\alpha}{15} & -\frac{2b\alpha}{15} & \frac{b\alpha}{15} & 0 & 0 & 0 \\ 0 & 0 & -\frac{2b\alpha}{15} & -\frac{16b\alpha}{15} & -\frac{2b\alpha}{15} & 0 & 0 & 0 \\ 0 & 0 & \frac{b\alpha}{15} & -\frac{2b\alpha}{15} & -\frac{4b\alpha}{15} & 0 & 0 & 0 \\ 0 & 0 & 0 & 0 & 0 & 0 & 0 & 0 \\ 0 & 0 & 0 & 0 & 0 & 0 & 0 & 0 \\ 0 & 0 & 0 & 0 & 0 & 0 & 0 & 0 \end{pmatrix}$$

$$r_\beta^T = \int_{-b}^{b} \beta N_c^T\, dc = \begin{pmatrix} 0 & 0 & \frac{b\beta}{3} & \frac{4b\beta}{3} & \frac{b\beta}{3} & 0 & 0 & 0 \end{pmatrix}$$

For the NBC on side 3

$$N_c^T = \begin{pmatrix} 0 & 0 & 0 & 0 & \frac{c^2}{2a^2} - \frac{c}{2a} & 1 - \frac{c^2}{a^2} & \frac{c^2}{2a^2} + \frac{c}{2a} & 0 \end{pmatrix}$$

$$k_\alpha = \int_{-a}^{a} -\alpha N_c N_c^T\, dc = \begin{pmatrix} 0 & 0 & 0 & 0 & 0 & 0 & 0 & 0 \\ 0 & 0 & 0 & 0 & 0 & 0 & 0 & 0 \\ 0 & 0 & 0 & 0 & 0 & 0 & 0 & 0 \\ 0 & 0 & 0 & 0 & 0 & 0 & 0 & 0 \\ 0 & 0 & 0 & 0 & -\frac{4a\alpha}{15} & -\frac{2a\alpha}{15} & \frac{a\alpha}{15} & 0 \\ 0 & 0 & 0 & 0 & -\frac{2a\alpha}{15} & -\frac{16a\alpha}{15} & -\frac{2a\alpha}{15} & 0 \\ 0 & 0 & 0 & 0 & \frac{a\alpha}{15} & -\frac{2a\alpha}{15} & -\frac{4a\alpha}{15} & 0 \\ 0 & 0 & 0 & 0 & 0 & 0 & 0 & 0 \end{pmatrix}$$

$$r_\beta^T = \int_{-a}^{a} \beta N_c^T\, dc = \begin{pmatrix} 0 & 0 & 0 & 0 & \frac{a\beta}{3} & \frac{4a\beta}{3} & \frac{a\beta}{3} & 0 \end{pmatrix}$$

For the NBC on side 4

$$N_c^T = \begin{pmatrix} \frac{c^2}{2b^2} + \frac{c}{2b} & 0 & 0 & 0 & 0 & 0 & \frac{c^2}{2b^2} - \frac{c}{2b} & 1 - \frac{c^2}{b^2} \end{pmatrix}$$

$$k_\alpha = \int_{-b}^{b} -\alpha N_c N_c^T \, dc = \begin{pmatrix} -\frac{4b\alpha}{15} & 0 & 0 & 0 & 0 & 0 & \frac{b\alpha}{15} & -\frac{2b\alpha}{15} \\ 0 & 0 & 0 & 0 & 0 & 0 & 0 & 0 \\ 0 & 0 & 0 & 0 & 0 & 0 & 0 & 0 \\ 0 & 0 & 0 & 0 & 0 & 0 & 0 & 0 \\ 0 & 0 & 0 & 0 & 0 & 0 & 0 & 0 \\ 0 & 0 & 0 & 0 & 0 & 0 & 0 & 0 \\ \frac{b\alpha}{15} & 0 & 0 & 0 & 0 & 0 & -\frac{4b\alpha}{15} & -\frac{2b\alpha}{15} \\ -\frac{2b\alpha}{15} & 0 & 0 & 0 & 0 & 0 & -\frac{2b\alpha}{15} & -\frac{16b\alpha}{15} \end{pmatrix}$$

$$r_\beta^T = \int_{-b}^{b} \beta N_c^T \, dc = \begin{pmatrix} \frac{b\beta}{3} & 0 & 0 & 0 & 0 & 0 & \frac{b\beta}{3} & \frac{4b\beta}{3} \end{pmatrix}$$

◆ ***Mathematica*/MATLAB Implementation 5.2 on the Book Web Site:**
Eight-node rectangular element for 2D BVP

5.4.3 Lagrange Interpolation for Rectangular Elements

The Lagrange interpolation formula was presented in Chapter 2 as a convenient way to write interpolation functions for higher order one-dimensional elements for second-order problems. We can use products of appropriate Lagrange interpolation formulas in the x and y directions to write interpolation functions for rectangular elements. The procedure is justified in detail for a four-node element followed by several examples of product Lagrange interpolation functions for higher order rectangular elements.

Four-Node Rectangular Element For a four-node rectangular element with dimensions $2a \times 2b$ as shown in Figure 5.6, we note the following:

(a) Along side 1–2, $t = -b$ is constant and therefore the interpolation functions must be functions only of s. Thus

$$u(s, -b) = \begin{pmatrix} n_1(s) & n_2(s) \end{pmatrix} \begin{pmatrix} u_1 \\ u_2 \end{pmatrix} \equiv \begin{pmatrix} -\frac{s-a}{2a} & \frac{a+s}{2a} \end{pmatrix} \begin{pmatrix} u_1 \\ u_2 \end{pmatrix}$$

where the interpolation functions n_1 and n_2 are written using the one-dimensional Lagrange interpolation formula.

(b) Similarly along side 4–3, $t = b$ and therefore the interpolation functions are

$$u(s, b) = \begin{pmatrix} n_4(s) & n_3(s) \end{pmatrix} \begin{pmatrix} u_4 \\ u_3 \end{pmatrix} \equiv \begin{pmatrix} -\frac{s-a}{2a} & \frac{a+s}{2a} \end{pmatrix} \begin{pmatrix} u_4 \\ u_3 \end{pmatrix}$$

Expanding to include all four degrees of freedom, the interpolations for $u(s, -b)$ and $u(s, b)$ can be written as follows:

$$u(s,-b) = \begin{pmatrix} -\frac{s-a}{2a} & \frac{a+s}{2a} & 0 & 0 \end{pmatrix} \begin{pmatrix} u_1 \\ u_2 \\ u_3 \\ u_4 \end{pmatrix}$$

$$u(s,b) = \begin{pmatrix} 0 & 0 & \frac{a+s}{2a} & -\frac{s-a}{2a} \end{pmatrix} \begin{pmatrix} u_1 \\ u_2 \\ u_3 \\ u_4 \end{pmatrix}$$

(c) In the t direction we can consider linear interpolation between $u(s,-b)$ and $u(s,b)$. Using the Lagrange interpolation formula in the t direction, we have

$$u(s,t) = \begin{pmatrix} -\frac{t-b}{2b} & \frac{b+t}{2b} \end{pmatrix} \begin{pmatrix} u(s,-b) \\ u(s,b) \end{pmatrix}$$

Thus the interpolation for $u(s,t)$ can be written as follows:

$$u(s,t) = \begin{pmatrix} -\frac{t-b}{2b} & \frac{b+t}{2b} \end{pmatrix} \begin{pmatrix} -\frac{s-a}{2a} & \frac{a+s}{2a} & 0 & 0 \\ 0 & 0 & \frac{a+s}{2a} & -\frac{s-a}{2a} \end{pmatrix} \begin{pmatrix} u_1 \\ u_2 \\ u_3 \\ u_4 \end{pmatrix}$$

Multiplying the two matrices, we get the same interpolation functions as the ones obtained earlier by writing a four-term polynomial:

$$u(s,t) = \begin{pmatrix} \frac{(a-s)(b-t)}{4ab} & \frac{(a+s)(b-t)}{4ab} & \frac{(a+s)(b+t)}{4ab} & \frac{(a-s)(b+t)}{4ab} \end{pmatrix} \begin{pmatrix} u_1 \\ u_2 \\ u_3 \\ u_4 \end{pmatrix}$$

Each interpolation function can be seen as simply a product of the linear Lagrange interpolation functions in the s and t directions.

Six-Node Rectangular Element Consider a six-node rectangular element $2a \times 2b$ as shown in Figure 5.18. Along the bottom and the top in the s direction there are three nodes. Thus we can define quadratic interpolation functions along these two sides. In the t direction, we still have only two values and thus we can use linear interpolation. Therefore interpolation functions for the element are as follows:

$$u(s,t) = \begin{pmatrix} -\frac{t-b}{2b} & \frac{b+t}{2b} \end{pmatrix} \begin{pmatrix} \frac{s(s-a)}{2a^2} & -\frac{(s-a)(a+s)}{a^2} & \frac{s(a+s)}{2a^2} & 0 & 0 & 0 \\ 0 & 0 & 0 & \frac{s(s-a)}{2a^2} & -\frac{(s-a)(a+s)}{a^2} & \frac{s(a+s)}{2a^2} \end{pmatrix} \begin{pmatrix} u_1 \\ u_2 \\ u_3 \\ u_4 \\ u_5 \\ u_6 \end{pmatrix}$$

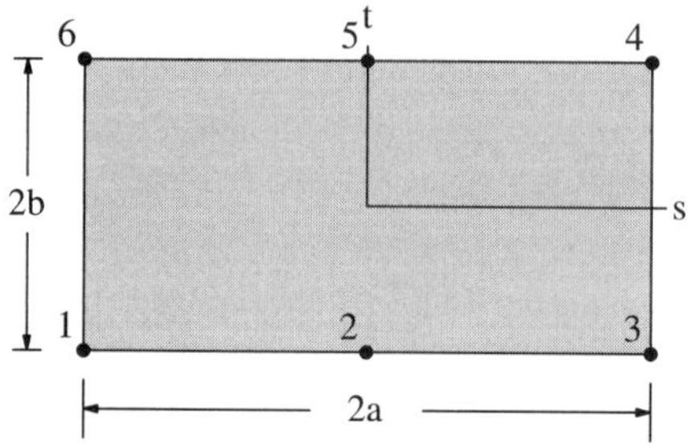

Figure 5.18. Six-node rectangular element

$$
\text{Interpolation functions} = \begin{pmatrix} -\frac{s(s-a)(t-b)}{4a^2 b} \\ \frac{(s-a)(a+s)(t-b)}{2a^2 b} \\ -\frac{s(a+s)(t-b)}{4a^2 b} \\ \frac{s(a+s)(b+t)}{4a^2 b} \\ -\frac{(s-a)(a+s)(b+t)}{2a^2 b} \\ \frac{s(s-a)(b+t)}{4a^2 b} \end{pmatrix}
$$

Nine-Node Rectangular Element In order to use quadratic interpolation in both the s and t directions, we need three nodes along each line. Thus we have a nine-node rectangular element $2a \times 2b$ as shown in Figure 5.19. Along the bottom, middle, and top in the s direction we can define quadratic interpolation functions. Now in the t direction we also can use quadratic interpolation. Thus interpolation functions for the element are as follows:

$$
u(s,t) = \left(\frac{t(t-b)}{2b^2} \quad -\frac{(t-b)(b+t)}{b^2} \quad \frac{t(b+t)}{2b^2} \right) \begin{pmatrix} \frac{s(s-a)}{2a^2} & -\frac{(s-a)(a+s)}{a^2} & \frac{s(a+s)}{2a^2} & \cdots & 0 \\ 0 & 0 & 0 & \cdots & -\frac{(s-a)(a+s)}{a^2} \\ 0 & 0 & 0 & \cdots & 0 \end{pmatrix} \begin{pmatrix} u_1 \\ u_2 \\ \vdots \\ u_9 \end{pmatrix}
$$

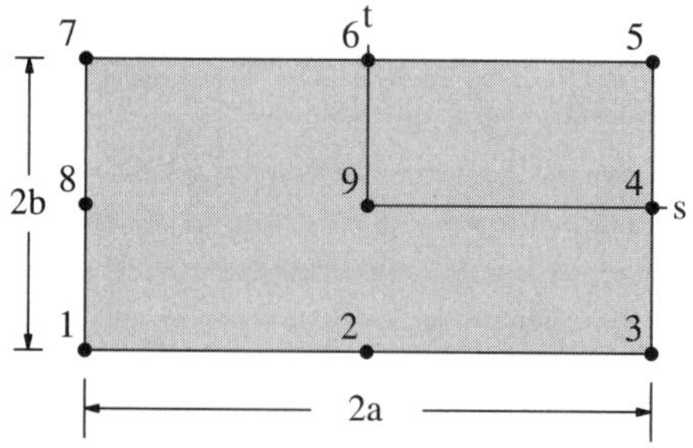

Figure 5.19. Nine-node rectangular element

$$
\text{Interpolation functions} = \begin{pmatrix} \frac{s(s-a)t(t-b)}{4a^2b^2} \\ -\frac{(s-a)(a+s)t(t-b)}{2a^2b^2} \\ \frac{s(a+s)t(t-b)}{4a^2b^2} \\ -\frac{s(a+s)(t-b)(b+t)}{2a^2b^2} \\ \frac{s(a+s)t(b+t)}{4a^2b^2} \\ -\frac{(s-a)(a+s)t(b+t)}{2a^2b^2} \\ \frac{s(s-a)t(b+t)}{4a^2b^2} \\ -\frac{s(s-a)(t-b)(b+t)}{2a^2b^2} \\ \frac{(s-a)(a+s)(t-b)(b+t)}{a^2b^2} \end{pmatrix}
$$

Example 5.5 TM Modes for Waveguides Using Nine-Node Lagrange Elements As discussed earlier, the TM modes for electromagnetic waveguides are obtained from solving the following eigenvalue problem:

$$
\frac{\partial^2 \psi}{\partial x^2} + \frac{\partial^2 \psi}{\partial y^2} + k_c^2 \psi = 0
$$

$$
\psi = 0 \text{ on the boundary}
$$

where k_c is the unknown cutoff frequency. Comparing this equation to the general form, the finite element equations for this problem are as follows:

$$
(\boldsymbol{k}_k + k_c^2 \boldsymbol{k}_p)\boldsymbol{d} = \boldsymbol{0}
$$

where

$$
\boldsymbol{k}_k = \iint_A \boldsymbol{B}\boldsymbol{B}^T \, dA; \qquad \boldsymbol{k}_p = -\iint_A \boldsymbol{N}\boldsymbol{N}^T \, dA
$$

$$
\boldsymbol{B}^T = \begin{pmatrix} \frac{\partial N_1}{\partial x} & \frac{\partial N_2}{\partial x} & \cdots & \frac{\partial N_n}{\partial x} \\ \frac{\partial N_1}{\partial y} & \frac{\partial N_2}{\partial y} & \cdots & \frac{\partial N_n}{\partial y} \end{pmatrix}
$$

Using the interpolation functions of the nine-node rectangular element and carrying out required differentiations and integrations, the following explicit expressions for the element matrices can be derived:

$$
\boldsymbol{k}_k = \begin{pmatrix}
\frac{14(a^2+b^2)}{45ab} & \frac{7a}{45b}-\frac{16b}{45a} & \frac{2b}{45a}-\frac{7a}{90b} & \frac{4a}{45b}+\frac{b}{45a} & -\frac{a^2+b^2}{90ab} & \frac{a}{45b}+\frac{4b}{45a} & \frac{2a}{45b}-\frac{7b}{90a} & \frac{7b}{45a}-\frac{16a}{45b} & -\frac{8(a^2+b^2)}{45ab} \\
\frac{7a}{45b}-\frac{16b}{45a} & \frac{56a}{45b}+\frac{32b}{45a} & \frac{7a}{45b}-\frac{16b}{45a} & -\frac{8(a^2+b^2)}{45ab} & \frac{a}{45b}+\frac{4b}{45a} & \frac{8a}{45b}-\frac{8b}{45a} & \frac{a}{45b}+\frac{4b}{45a} & -\frac{8(a^2+b^2)}{45ab} & \frac{16b}{45a}-\frac{64a}{45b} \\
\frac{2b}{45a}-\frac{7a}{90b} & \frac{7a}{45b}-\frac{16b}{45a} & \frac{14(a^2+b^2)}{45ab} & \frac{7b}{45a}-\frac{16a}{45b} & \frac{2a}{45b}-\frac{7b}{90a} & \frac{a}{45b}+\frac{4b}{45a} & -\frac{a^2+b^2}{90ab} & \frac{4a}{45b}+\frac{b}{45a} & -\frac{8(a^2+b^2)}{45ab} \\
\frac{4a}{45b}+\frac{b}{45a} & -\frac{8(a^2+b^2)}{45ab} & \frac{7b}{45a}-\frac{16a}{45b} & \frac{32a}{45b}+\frac{56b}{45a} & \frac{7b}{45a}-\frac{16a}{45b} & -\frac{8(a^2+b^2)}{45ab} & \frac{4a}{45b}+\frac{b}{45a} & \frac{8b}{45a}-\frac{8a}{45b} & \frac{16(a^2-4b^2)}{45ab} \\
-\frac{a^2+b^2}{90ab} & \frac{a}{45b}+\frac{4b}{45a} & \frac{2a}{45b}-\frac{7b}{90a} & \frac{7b}{45a}-\frac{16a}{45b} & \frac{14(a^2+b^2)}{45ab} & \frac{7a}{45b}-\frac{16b}{45a} & \frac{2b}{45a}-\frac{7a}{90b} & \frac{4a}{45b}+\frac{b}{45a} & -\frac{8(a^2+b^2)}{45ab} \\
\frac{a}{45b}+\frac{4b}{45a} & \frac{8a}{45b}-\frac{8b}{45a} & \frac{a}{45b}+\frac{4b}{45a} & -\frac{8(a^2+b^2)}{45ab} & \frac{7a}{45b}-\frac{16b}{45a} & \frac{56a}{45b}+\frac{32b}{45a} & \frac{7a}{45b}-\frac{16b}{45a} & -\frac{8(a^2+b^2)}{45ab} & \frac{16b}{45a}-\frac{64a}{45b} \\
\frac{2a}{45b}-\frac{7b}{90a} & \frac{a}{45b}+\frac{4b}{45a} & -\frac{a^2+b^2}{90ab} & \frac{4a}{45b}+\frac{b}{45a} & \frac{2b}{45a}-\frac{7a}{90b} & \frac{7a}{45b}-\frac{16b}{45a} & \frac{14(a^2+b^2)}{45ab} & \frac{7b}{45a}-\frac{16a}{45b} & -\frac{8(a^2+b^2)}{45ab} \\
\frac{7b}{45a}-\frac{16a}{45b} & -\frac{8(a^2+b^2)}{45ab} & \frac{4a}{45b}+\frac{b}{45a} & \frac{8b}{45a}-\frac{8a}{45b} & \frac{4a}{45b}+\frac{b}{45a} & -\frac{8(a^2+b^2)}{45ab} & \frac{7b}{45a}-\frac{16a}{45b} & \frac{32a}{45b}+\frac{56b}{45a} & \frac{16(a^2-4b^2)}{45ab} \\
-\frac{8(a^2+b^2)}{45ab} & \frac{16b}{45a}-\frac{64a}{45b} & -\frac{8(a^2+b^2)}{45ab} & \frac{16(a^2-4b^2)}{45ab} & -\frac{8(a^2+b^2)}{45ab} & \frac{16b}{45a}-\frac{64a}{45b} & -\frac{8(a^2+b^2)}{45ab} & \frac{16(a^2-4b^2)}{45ab} & \frac{128(a^2+b^2)}{45ab}
\end{pmatrix}
$$

$$
\boldsymbol{k}_p = \frac{ab}{225}\begin{pmatrix}
-16 & -8 & 4 & 2 & -1 & 2 & 4 & -8 & -4 \\
-8 & -64 & -8 & -4 & 2 & 16 & 2 & -4 & -32 \\
4 & -8 & -16 & -8 & 4 & 2 & -1 & 2 & -4 \\
2 & -4 & -8 & -64 & -8 & -4 & 2 & 16 & -32 \\
-1 & 2 & 4 & -8 & -16 & -8 & 4 & 2 & -4 \\
2 & 16 & 2 & -4 & -8 & -64 & -8 & -4 & -32 \\
4 & 2 & -1 & 2 & 4 & -8 & -16 & -8 & -4 \\
-8 & -4 & 2 & 16 & 2 & -4 & -8 & -64 & -32 \\
-4 & -32 & -4 & -32 & -4 & -32 & -4 & -32 & -256
\end{pmatrix}
$$

As was the case in the buckling problem considered in Chapter 3, for each element in the model, we need to write matrices $\boldsymbol{k}_k$ and $\boldsymbol{k}_p$ and then assemble them in the usual manner to obtain corresponding global matrices. The resulting system of global equations is then of the following form, which is recognized as a generalized eigenvalue problem:

$$(\boldsymbol{K}_k + k_c^2\boldsymbol{K}_p)\boldsymbol{d} = \boldsymbol{0} \Longrightarrow \boldsymbol{K}_k\boldsymbol{d} = \lambda(-\boldsymbol{K}_p)\boldsymbol{d}$$

where $\lambda = k_c^2$. After adjusting for the essential boundary conditions, the eigenvalues and eigenvectors of the system are computed. The square roots of the eigenvalues are the cutoff frequencies and the corresponding eigenvectors are the wave propagation modes.

As a specific example, consider a rectangular waveguide with width 2 units and height 1 unit. Using four elements the model is as shown in Figure 5.20. Note that even though the geometry is symmetric we must model the entire domain when computing modes. This is because some of the wave propagation modes are nonsymmetric even for a symmetric system. Thus, if we model only half or a quarter of the domain, we will not be able to capture the nonsymmetric modes.

Since all four elements have the same size, the finite element matrices for all elements are same. Substituting $a = \frac{1}{2}$ and $b = \frac{1}{4}$, the $\boldsymbol{k}_k$ and $\boldsymbol{k}_p$ matrices are as follows:

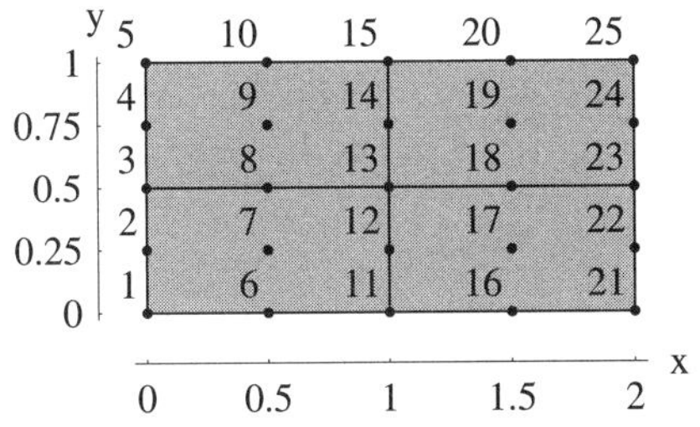

Figure 5.20. Four-element model for rectangular waveguide

```
kk = {{7/9, 2/15, -2/15, 17/90, -1/36, 4/45, 1/20, -19/30, -4/9},
  {2/15, 128/45, 2/15, -4/9, 4/45, 4/15, 4/45, -4/9, -8/3},
  {-2/15, 2/15, 7/9, -19/30, 1/20, 4/45, -1/36, 17/90, -4/9},
  {17/90, -4/9, -19/30, 92/45, -19/30, -4/9, 17/90, -4/15, 0},
  {-1/36, 4/45, 1/20, -19/30, 7/9, 2/15, -2/15, 17/90, -4/9},
  {4/45, 4/15, 4/45, -4/9, 2/15, 128/45, 2/15, -4/9, -8/3},
  {1/20, 4/45, -1/36, 17/90, -2/15, 2/15, 7/9, -19/30, -4/9},
  {-19/30, -4/9, 17/90, -4/15, 17/90, -4/9, -19/30, 92/45, 0},
  {-4/9, -8/3, -4/9, 0, -4/9, -8/3, -4/9, 0, 64/9}};
```

```
kp = {{-2/225, -1/225, 1/450, 1/900, -1/1800, 1/900, 1/450, -1/225, -1/450},
  {-1/225, -8/225, -1/225, -1/450, 1/900, 2/225, 1/900, -1/450, -4/225},
  {1/450, -1/225, -2/225, -1/225, 1/450, 1/900, -1/1800, 1/900, -1/450},
  {1/900, -1/450, -1/225, -8/225, -1/225, -1/450, 1/900, 2/225, -4/225},
  {-1/1800, 1/900, 1/450, -1/225, -2/225, -1/225, 1/450, 1/900, -1/450},
  {1/900, 2/225, 1/900, -1/450, -1/225, -8/225, -1/225, -1/450, -4/225},
  {1/450, 1/900, -1/1800, 1/900, 1/450, -1/225, -2/225, -1/225, -1/450},
  {-1/225, -1/450, 1/900, 2/225, 1/900, -1/450, -1/225, -8/225, -4/225},
  {-1/450, -4/225, -1/450, -4/225, -1/450, -4/225, -1/450, -4/225, -32/225}};
```

These matrices can be assembled using the standard assembly process to form the global matrices as follows:

```
lm = {{1, 6, 11, 12, 13, 8, 3, 2, 7}, {3, 8, 13, 14, 15, 10, 5, 4, 9},
  {11, 16, 21, 22, 23, 18, 13, 12, 17}, {13, 18, 23, 24, 25, 20, 15, 14, 19}};
Kk = Kp = Table[0, {25}, {25}];
Do[Kk[[lm[[i]], lm[[i]]]] += kk; Kp[[lm[[i]], lm[[i]]]] += kp, {i, 1, 4}];
```

Incorporating the essential boundary conditions, the reduced system of matrices is as follows:

```
debc = {1, 2, 3, 4, 5, 6, 11, 16, 21, 22, 23, 24, 10, 15, 20, 25};
ebcVals = Table[0, {Length[debc]}];
df = Complement[Range[25], debc];

Kkf = Kk[[df, df]];

Kpf = Kp[[df, df]];
```

The eigenvalues and the corresponding modes can be computed by solving the generalized eigenvalue problem. The five lowest eigenvalues and modes are computed as follows:

```
{evals, evecs} = Eigensystem[{Kkf, −Kpf}//N, −5]; evals
```

{50., 42.486, 42.1246, 19.9438, 12.4298}

The cutoff frequencies are square roots of the eigenvalues:

```
Sqrt[evals]
```

{7.07107, 6.51813, 6.49034, 4.46585, 3.52559}

An analytical solution for a rectangular waveguide is well known. This formula gives the lowest cutoff frequency as follows:

```
Sqrt[(π/2)^2 + (π/1)^2]//N
```

3.51241

Even with this coarse mesh the computed cutoff frequency compares very well with the exact analytical value. The electric field corresponding to the lowest TM mode can be computed by first determining the solution over each element and then taking its derivatives:

$$\text{For TM modes (with } Z_0 = 1\text{):} \qquad E_x = -\frac{\partial \psi}{\partial x}; \qquad E_y = -\frac{\partial \psi}{\partial y}$$

The complete vector of nodal values in the lowest mode, in the original order established by the node numbering, is obtained by combining these values with those specified as essential boundary conditions as follows:

```
d = Table[0, {25}];
d[[debc]] = ebcVals//N;
d[[df]] = evecs[[5]];
d
```

{0., 0., 0., 0., 0., 0., 0.249883, 0.353553, 0.249883, 0., 0., 0.353553, 0.500234, 0.353553, 0., 0., 0.249883, 0.353553, 0.249883, 0., 0., 0., 0., 0., 0.}

The following function is created to compute the electric field at several points on each element:

```
EVPRect9Results[a_, b_, {xc_, yc_}, d_] := Module[{n, electricField, s, t},
    n = {(s * (s − a) * t * (t − b))/(4 * a^2 * b^2),
        −(((s − a) * (a + s) * t * (t − b))/(2 * a^2 * b^2)),
        (s * (a + s) * t * (t − b))/(4 * a^2 * b^2),
        −((s * (a + s) * (t − b) * (b + t))/(2 * a^2 * b^2)),
        (s * (a + s) * t * (b + t))/(4 * a^2 * b^2),
        −(((s − a) * (a + s) * t * (b + t))/(2 * a^2 * b^2)),
        (s * (s − a) * t * (b + t))/(4 * a^2 * b^2),
```

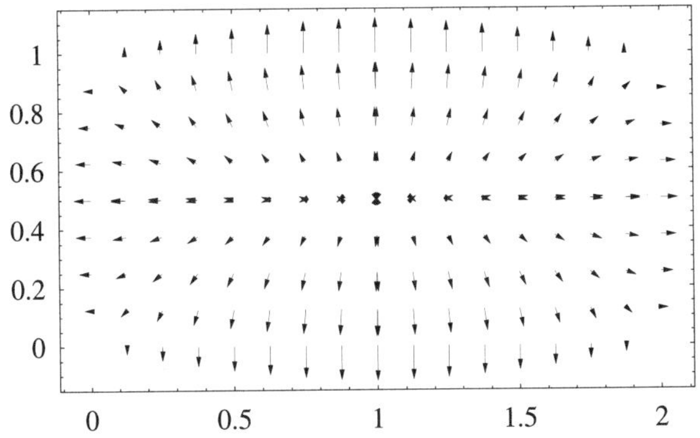

Figure 5.21. Electric field distribution in the lowest TM mode

```
        -((s * (s - a) * (t - b) * (b + t))/(2 * a^2 * b^2)),
        ((s - a) * (a + s) * (t - b) * (b + t))/(a^2 * b^2)};
    electricField = {-D[n, s].d, -D[n, t].d};
    Map[{# + {xc, yc}, electricField/.Thread[{s, t} → #]}&,
      Flatten[Table[{s, t}, {s, -a, a, 1/8}, {t, -b, b, 1/8}], 1]]
];

centers = {{1/2, 1/4}, {1/2, 3/4}, {1.5, 1/4}, {1.5, 3/4}};
esol = Flatten[Table[EVPRect9Results[1/2, 1/4, centers[[i]],
  d[[lm[[i]]]]], {i, 1, 4}], 1];
```

The electric field is shown in the form of a vector plot in Figure 5.21.

```
Needs["Graphics`PlotField`"];
ListPlotVectorField[esol, ScaleFactor → 0.12, Frame → True,
    HeadLength → 0.015];
```

5.5 TRIANGULAR FINITE ELEMENTS

A triangular element is simple yet versatile for two-dimensional problems. Almost any two-dimensional shape can be discretized into triangular elements. The only errors introduced in the geometry of the model are those resulting from approximating curved boundaries by straight lines representing edges of triangular elements. These errors can be reduced by increasing the number of elements. Closed-form formulas are available for evaluating integrals over triangular domains. The evaluation of boundary integrals is only slightly more complicated as compared to the rectangular domains.

Equations for a simple three-node triangular element are developed in detail in the following section. The procedure for developing higher order triangular elements is outlined in the last section.

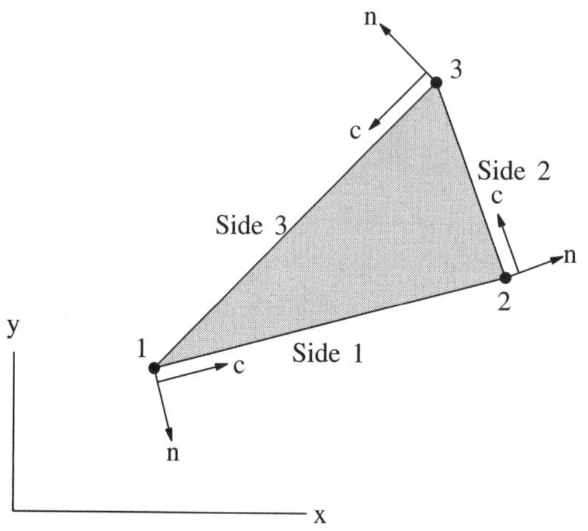

Figure 5.22. Three-node triangular element

5.5.1 Three-Node Triangular Element

A typical three-node triangular element is shown in Figure 5.22. The nodal coordinates are (x_1, y_1), (x_2, y_2), and (x_3, y_3). The directions of the outer normal and the boundary coordinates c for each side are also shown in the figure. The origin of the boundary coordinate c is the first node of each side. Thus for a given side c goes from 0 to the length of that side.

Assumed Solution Assuming unknown solutions at the three corners as nodal degrees of freedom, we have a total of three degrees of freedom, u_1, u_2, and u_3. The assumed finite element solutions are needed in the following form:

$$\text{Over A:} \qquad u(x, y) = \begin{pmatrix} N_1(x, y) & N_2(x, y) & N_3(x, y) \end{pmatrix} \begin{pmatrix} u_1 \\ u_2 \\ u_3 \end{pmatrix} = \boldsymbol{N}^T \boldsymbol{d}$$

$$\text{Over C:} \qquad u(c) = \begin{pmatrix} N_1(c) & N_2(c) & N_3(c) \end{pmatrix} \begin{pmatrix} u_1 \\ u_2 \\ u_3 \end{pmatrix} = \boldsymbol{N}_c^T \boldsymbol{d}$$

A complete linear polynomial in two dimensions has three terms and thus a finite element solution for the element is based on the following polynomial:

$$u(x, y) = c_0 + c_1 x + c_2 y = \begin{pmatrix} 1 & x & y \end{pmatrix} \begin{pmatrix} c_0 \\ c_1 \\ c_2 \end{pmatrix}$$

The coefficients c_0, c_1, and c_2 can be expressed in terms of nodal degrees of freedom by evaluating the polynomial at the nodes as follows:

$$\begin{pmatrix} u_1 \\ u_2 \\ u_3 \end{pmatrix} = \begin{pmatrix} 1 & x_1 & y_1 \\ 1 & x_2 & y_2 \\ 1 & x_3 & y_3 \end{pmatrix} \begin{pmatrix} c_0 \\ c_1 \\ c_2 \end{pmatrix}$$

Inverting the 3×3 matrix, we get

$$\begin{pmatrix} c_0 \\ c_1 \\ c_2 \end{pmatrix} = \frac{1}{2A} \begin{pmatrix} f_1 & f_2 & f_3 \\ b_1 & b_2 & b_3 \\ c_1 & c_2 & c_3 \end{pmatrix} \begin{pmatrix} u_1 \\ u_2 \\ u_3 \end{pmatrix}$$

where A is the area of the triangle and we have introduced the notation

$$f_1 = x_2 y_3 - x_3 y_2; \qquad f_2 = x_3 y_1 - x_1 y_3; \qquad f_3 = x_1 y_2 - x_2 y_1;$$

$$b_1 = y_2 - y_3; \qquad b_2 = y_3 - y_1; \qquad b_3 = y_1 - y_2;$$

$$c_1 = x_3 - x_2; \qquad c_2 = x_1 - x_3; \qquad c_3 = x_2 - x_1;$$

$$A = \tfrac{1}{2}(-x_2 y_1 + x_3 y_1 + x_1 y_2 - x_3 y_2 - x_1 y_3 + x_2 y_3) \equiv \tfrac{1}{2}\left(f_1 + f_2 + f_3\right)$$

Substituting the coefficients into the polynomial, we have

$$u(x, y) = (\,1 \quad x \quad y\,) \frac{1}{2A} \begin{pmatrix} f_1 & f_2 & f_3 \\ b_1 & b_2 & b_3 \\ c_1 & c_2 & c_3 \end{pmatrix} \begin{pmatrix} u_1 \\ u_2 \\ u_3 \end{pmatrix}$$

Carrying out multiplication, we get the finite element assumed solution over the element as follows:

$$u(x, y) = (\,N_1 \quad N_2 \quad N_3\,) \begin{pmatrix} u_1 \\ u_2 \\ u_3 \end{pmatrix} = \boldsymbol{N}^T \boldsymbol{d}$$

$$N_1 = \frac{1}{2A}(x_3(y - y_2) + x(y_2 - y_3) + x_2(-y + y_3)) \equiv \frac{1}{2A}(x b_1 + y c_1 + f_1)$$

$$N_2 = \frac{1}{2A}(x_3(-y + y_1) + x_1(y - y_3) + x(-y_1 + y_3)) \equiv \frac{1}{2A}(x b_2 + y c_2 + f_2)$$

$$N_3 = \frac{1}{2A}(x_2(y - y_1) + x(y_1 - y_2) + x_1(-y + y_2)) \equiv \frac{1}{2A}(x b_3 + y c_3 + f_3)$$

Note that each of the interpolation functions N_i is 1 at the ith node and 0 at every other node.

The boundary is defined by three sides of the element. For each side the solution is linear in terms of coordinate c for that side and the two degrees of freedom at the ends of that side. We can write this linear solution easily using the Lagrange interpolation formula:

For side 1

$$u(c) = \begin{pmatrix} -\frac{c - L_{12}}{L_{12}} & \frac{c}{L_{12}} & 0 \end{pmatrix} \begin{pmatrix} u_1 \\ u_2 \\ u_3 \end{pmatrix}; \qquad 0 \le c \le L_{12}$$

where $L_{12} = \sqrt{(x_2 - x_1)^2 + (y_2 - y_1)^2}$ is the length of side 1–2.

For side 2

$$u(c) = \begin{pmatrix} 0 & -\frac{c - L_{23}}{L_{23}} & \frac{c}{L_{23}} \end{pmatrix} \begin{pmatrix} u_1 \\ u_2 \\ u_3 \end{pmatrix}; \qquad 0 \le c \le L_{23}$$

where $L_{23} = \sqrt{(x_3 - x_2)^2 + (y_3 - y_2)^2}$ is the length of side 2–3.

For side 3

$$u(c) = \begin{pmatrix} \frac{c}{L_{31}} & 0 & -\frac{c - L_{31}}{L_{31}} \end{pmatrix} \begin{pmatrix} u_1 \\ u_2 \\ u_3 \end{pmatrix}; \qquad 0 \le c \le L_{31}$$

where $L_{31} = \sqrt{(x_3 - x_1)^2 + (y_3 - y_1)^2}$ is the length of side 3–1.

Element Equations Substituting the assumed solution and carrying out integrations, the matrices and vectors needed for element equations can easily be written. For simplicity in integrations, it will be assumed that k_x, k_y, p, q, α, and β are all constant over an element. Thus these terms can be taken out of the integrals:

$$\boldsymbol{k}_k = \iint_A \boldsymbol{BCB}^T\, dA; \qquad \boldsymbol{k}_p = -\iint_A p\boldsymbol{NN}^T\, dA; \qquad \boldsymbol{k}_\alpha = -\int_{C_n} \alpha \boldsymbol{N}_c \boldsymbol{N}_c^T\, dc$$

$$\boldsymbol{r}_q = \iint_A q\boldsymbol{N}\, dA; \qquad \boldsymbol{r}_\beta = \int_{C_n} \beta \boldsymbol{N}_c\, dc$$

$$\boldsymbol{N}^T = \frac{1}{2A}\begin{pmatrix} xb_1 + yc_1 + f_1 & xb_2 + yc_2 + f_2 & xb_3 + yc_3 + f_3 \end{pmatrix}$$

$$\boldsymbol{B}^T = \begin{pmatrix} \frac{\partial N_1}{\partial x} & \frac{\partial N_2}{\partial x} & \frac{\partial N_3}{\partial x} \\ \frac{\partial N_1}{\partial y} & \frac{\partial N_2}{\partial y} & \frac{\partial N_3}{\partial y} \end{pmatrix} = \frac{1}{2A}\begin{pmatrix} b_1 & b_2 & b_3 \\ c_1 & c_2 & c_3 \end{pmatrix}$$

$$\boldsymbol{C} = \begin{pmatrix} k_x & 0 \\ 0 & k_y \end{pmatrix}$$

Since all terms in the $\boldsymbol{B}$ and $\boldsymbol{C}$ matrices are constants, the integrations to obtain $\boldsymbol{k}_k$ matrix are trivial:

$$\boldsymbol{k}_k = \iint_A \boldsymbol{BCB}^T\, dA = A\boldsymbol{BCB}^T = \frac{1}{4A}\begin{pmatrix} k_x b_1^2 + k_y c_1^2 & k_x b_1 b_2 + k_y c_1 c_2 & k_x b_1 b_3 + k_y c_1 c_3 \\ k_x b_1 b_2 + k_y c_1 c_2 & k_x b_2^2 + k_y c_2^2 & k_x b_2 b_3 + k_y c_2 c_3 \\ k_x b_1 b_3 + k_y c_1 c_3 & k_x b_2 b_3 + k_y c_2 c_3 & k_x b_3^2 + k_y c_3^2 \end{pmatrix}$$

The matrix $\boldsymbol{k}_p$ involves the $\boldsymbol{NN}^T$ matrix in which the terms are functions of x and y:

$$\boldsymbol{NN}^T = \frac{1}{4A^2}\begin{pmatrix} (xb_1 + yc_1 + f_1)^2 & (xb_1 + yc_1 + f_1)(xb_2 + yc_2 + f_2) & (xb_1 + yc_1 + f_1)(xb_3 + yc_3 + f_3) \\ (xb_1 + yc_1 + f_1)(xb_2 + yc_2 + f_2) & (xb_2 + yc_2 + f_2)^2 & (xb_2 + yc_2 + f_2)(xb_3 + yc_3 + f_3) \\ (xb_1 + yc_1 + f_1)(xb_3 + yc_3 + f_3) & (xb_2 + yc_2 + f_2)(xb_3 + yc_3 + f_3) & (xb_3 + yc_3 + f_3)^2 \end{pmatrix}$$

Each term must be integrated over a triangular region. In terms of x and y coordinates this integration is not an easy task. It is possible to derive a simple closed-form integration formula; however, this needs special coordinates for a triangle called *area coordinates*. Here we outline the procedure for performing integration directly in terms of x and y coordinates.

As shown in Figure 5.23, the integrations can be performed by dividing the triangle into two parts, one with the x limits from x_1 to x_3 and the other from x_3 to x_2. In the y direction the integration limits are established by writing equations for lines defining triangle boundaries:

Line 1–2: $$y_{12} = y_1 + \frac{(x - x_1)(y_2 - y_1)}{x_2 - x_1}$$

Line 1–3: $$y_{13} = y_1 + \frac{(x - x_1)(y_3 - y_1)}{x_3 - x_1}$$

Line 3–2: $$y_{32} = y_3 + \frac{(x - x_3)(y_2 - y_3)}{x_2 - x_3}$$

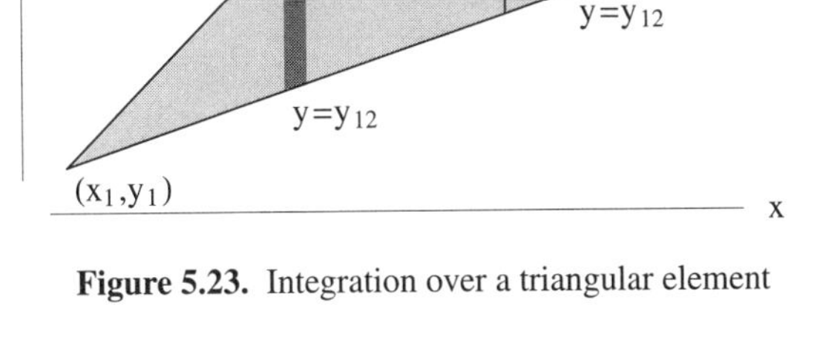

Figure 5.23. Integration over a triangular element

The integration of a function $f(x, y)$ over the triangular element is thus performed as follows:

$$\iint_A f(x,y)\,dA = \int_{x=x_1}^{x_3}\left(\int_{y=y_{12}}^{y_{13}} f(x,y)\,dy\right)dx + \int_{x=x_3}^{x_2}\left(\int_{y=y_{12}}^{y_{32}} f(x,y)\,dy\right)dx$$

Because of the complicated integration limits on y, the algebra obviously is quite messy. The computations can easily be performed using a computer algebra system, such as *Mathematica*. Using this procedure and integrating each term in the $\boldsymbol{k}_p$ matrix and $\boldsymbol{r}_q$ vector, we get

$$\boldsymbol{k}_p = -\frac{pA}{12}\begin{pmatrix} 2 & 1 & 1 \\ 1 & 2 & 1 \\ 1 & 1 & 2 \end{pmatrix}; \qquad \boldsymbol{r}_q = \frac{qA}{3}\begin{pmatrix} 1 \\ 1 \\ 1 \end{pmatrix}$$

The boundary integrals are evaluated separately for each side. For the natural boundary condition on side 1

$$\boldsymbol{N}_c^T = \begin{pmatrix} -\frac{c-L_{12}}{L_{12}} & \frac{c}{L_{12}} & 0 \end{pmatrix}; \qquad 0 \le c \le L_{12}$$

$$L_{12} = \sqrt{(x_2-x_1)^2 + (y_2-y_1)^2}$$

$$\boldsymbol{N}_c\boldsymbol{N}_c^T = \begin{pmatrix} \frac{(c-L_{12})^2}{L_{12}^2} & -\frac{c(c-L_{12})}{L_{12}^2} & 0 \\ -\frac{c(c-L_{12})}{L_{12}^2} & \frac{c^2}{L_{12}^2} & 0 \\ 0 & 0 & 0 \end{pmatrix}$$

$$\boldsymbol{k}_\alpha = -\int_{C_n} \alpha \boldsymbol{N}_c\boldsymbol{N}_c^T\,dc = -\alpha\int_{c=0}^{L_{12}} \boldsymbol{N}_c\boldsymbol{N}_c^T\,dc = -\frac{\alpha L_{12}}{6}\begin{pmatrix} 2 & 1 & 0 \\ 1 & 2 & 0 \\ 0 & 0 & 0 \end{pmatrix}$$

Similarly,

$$\boldsymbol{r}_\beta = \int_{C_n} \beta \boldsymbol{N}_c\,ds = \frac{\beta L_{12}}{2}\begin{pmatrix} 1 \\ 1 \\ 0 \end{pmatrix}$$

For the natural boundary condition on side 2,

$$\boldsymbol{k}_\alpha = -\frac{\alpha L_{23}}{6}\begin{pmatrix} 0 & 0 & 0 \\ 0 & 2 & 1 \\ 0 & 1 & 2 \end{pmatrix}; \qquad \boldsymbol{r}_\beta = \frac{\beta L_{23}}{2}\begin{pmatrix} 0 \\ 1 \\ 1 \end{pmatrix}$$

For the natural boundary condition on side 3,

$$\boldsymbol{k}_\alpha = -\frac{\alpha L_{31}}{6}\begin{pmatrix} 2 & 0 & 1 \\ 0 & 0 & 0 \\ 1 & 0 & 2 \end{pmatrix}; \qquad \boldsymbol{r}_\beta = \frac{\beta L_{31}}{2}\begin{pmatrix} 1 \\ 0 \\ 1 \end{pmatrix}$$

All quantities have now been defined in the element equations. The complete element equations are

$$(\boldsymbol{k}_k + \boldsymbol{k}_p + \boldsymbol{k}_\alpha)\boldsymbol{d} = \boldsymbol{r}_q + \boldsymbol{r}_\beta$$

Example 5.6 Stream Function Formulation for Fluid Flow around a Cylinder Consider fluid flow in the direction perpendicular to a long cylinder as shown in Figure 5.24. The cylinder diameter is 4.5 cm. At a distance of about 3 times the diameter of the cylinder, both upstream and downstream, the flow can be considered uniform with a velocity $u_0 = 5$ cm/s in the x direction. Determine the flow velocity near the cylinder.

We choose a computational domain that extends 3 times the cylinder diameter upstream and downstream and 1.5 times the diameter above and below the cylinder. Taking advantage of symmetry, we need to model only a quarter of the solution domain as shown in Figure 5.25. The applicable boundary conditions are also shown in the figure.

The governing differential equation in terms of stream function $\psi(x, y)$ is as follows:

$$\frac{\partial^2 \psi}{\partial x^2} + \frac{\partial^2 \psi}{\partial y^2} = 0$$

Compared to the general form, $k_x = k_y = 1$ and $p = q = 0$. The fluid velocity is related to stream function as follows:

$$u = \frac{\partial \psi}{\partial y}; \qquad v = -\frac{\partial \psi}{\partial x}$$

We consider a very coarse model consisting of only eight elements as shown in Figure 5.26. Note the large error in modeling the geometry near the cylinder. With only two element

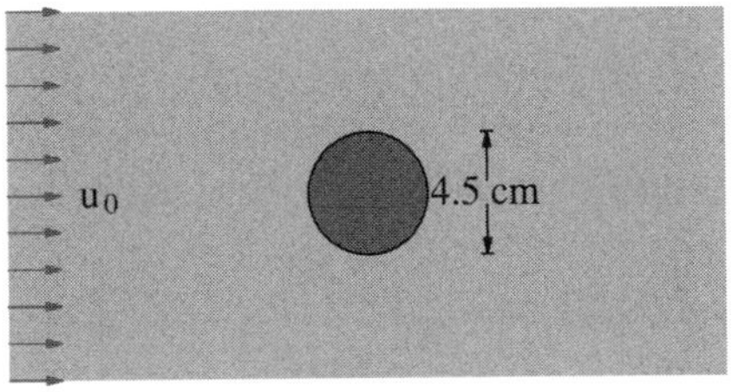

Figure 5.24. Two-dimensional flow around a cylinder

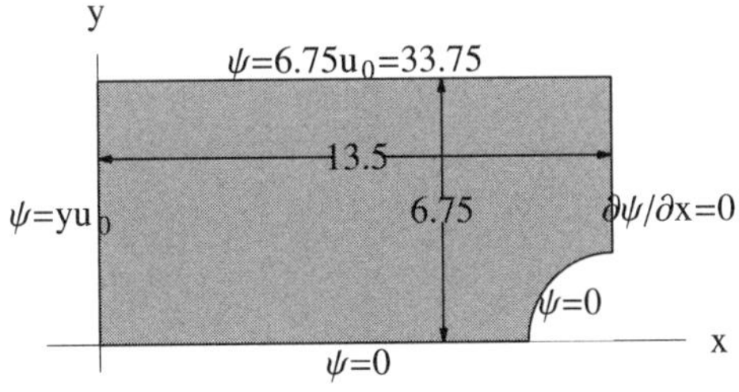

Figure 5.25. One-fourth of the solution domain for flow around a cylinder

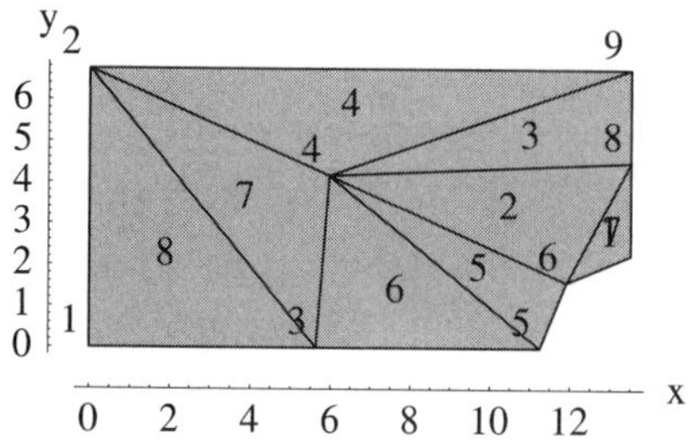

Figure 5.26. Eight-element model for flow around a cylinder

sides representing a quarter circle, instead of a cylinder, computationally we are actually computing flow around an eight-sided polygon. The results clearly will not be very accurate. The coarse mesh is used here to show all calculations.

The complete finite element solution is as follows:

Equations for element 1:

$k_x = 1;\ k_y = 1;\ p = 0;\ q = 0$

Nodal coordinates:

Element Node	Global Node Number	x	y
1	6	11.909	1.59099
2	7	13.5	2.25
3	8	13.5	4.5

Using these values, we get

$$b_1 = -2.25; \quad b_2 = 2.90901; \quad b_3 = -0.65901$$

$$c_1 = 0.; \quad c_2 = -1.59099; \quad c_3 = 1.59099$$

Element area, $A = 1.78986$

$$\boldsymbol{B}^T = \begin{pmatrix} -2.25 & 2.90901 & -0.65901 \\ 0. & -1.59099 & 1.59099 \end{pmatrix}$$

$$\boldsymbol{k}_k = \begin{pmatrix} 0.707107 & -0.914214 & 0.207107 \\ -0.914214 & 1.53553 & -0.62132 \\ 0.207107 & -0.62132 & 0.414214 \end{pmatrix}; \quad \boldsymbol{k}_p = \begin{pmatrix} 0 & 0 & 0 \\ 0 & 0 & 0 \\ 0 & 0 & 0 \end{pmatrix}; \quad \boldsymbol{r}_q = \begin{pmatrix} 0 \\ 0 \\ 0 \end{pmatrix}$$

Complete element equations:

$$\begin{pmatrix} 0.707107 & -0.914214 & 0.207107 \\ -0.914214 & 1.53553 & -0.62132 \\ 0.207107 & -0.62132 & 0.414214 \end{pmatrix} \begin{pmatrix} \psi_6 \\ \psi_7 \\ \psi_8 \end{pmatrix} = \begin{pmatrix} 0 \\ 0 \\ 0 \end{pmatrix}$$

Processing the remaining elements in exactly the same manner, we get the following global equations:

$$\begin{pmatrix} 1.01667 & -0.416667 & -0.6 & 0 & 0 & 0 & 0 & 0 & 0 \\ -0.416667 & 1.6704 & 0.171236 & -1.97454 & 0 & 0 & 0 & 0 & 0.549572 \\ -0.6 & 0.171236 & 2.38819 & -1.62591 & -0.33352 & 0 & 0 & 0 & 0 \\ 0 & -1.97454 & -1.62591 & 5.3325 & -0.0314544 & -0.443273 & 0 & -0.124963 & -1.13236 \\ 0 & 0 & -0.33352 & -0.0314544 & 2.2574 & -1.89242 & 0 & 0 & 0 \\ 0 & 0 & 0 & -0.443273 & -1.89242 & 4.07147 & -0.914214 & -0.821559 & 0 \\ 0 & 0 & 0 & 0 & 0 & -0.914214 & 1.53553 & -0.62132 & 0 \\ 0 & 0 & 0 & -0.124963 & 0 & -0.821559 & -0.62132 & 3.26965 & -1.70181 \\ 0 & 0.549572 & 0 & -1.13236 & 0 & 0 & 0 & -1.70181 & 2.2846 \end{pmatrix} \begin{pmatrix} \psi_1 \\ \psi_2 \\ \psi_3 \\ \psi_4 \\ \psi_5 \\ \psi_6 \\ \psi_7 \\ \psi_8 \\ \psi_9 \end{pmatrix} = \begin{pmatrix} 0 \\ 0 \\ 0 \\ 0 \\ 0 \\ 0 \\ 0 \\ 0 \\ 0 \end{pmatrix}$$

Essential boundary conditions:

Node	dof	Value
1	ψ_1	0
2	ψ_2	33.75
3	ψ_3	0
4	ψ_4	0
5	ψ_5	0
6	ψ_6	0
7	ψ_7	0
9	ψ_2	33.75

Delete equations {1, 2, 3, 5, 6, 7, 9}:

$$\begin{pmatrix} 0 & -1.97454 & -1.62591 & 5.3325 & -0.0314544 & -0.443273 & 0 & -0.124963 & -1.13236 \\ 0 & 0 & 0 & -0.124963 & 0 & -0.821559 & -0.62132 & 3.26965 & -1.70181 \end{pmatrix} \begin{pmatrix} 0 \\ 33.75 \\ 0 \\ \psi_4 \\ 0 \\ 0 \\ 0 \\ \psi_8 \\ 33.75 \end{pmatrix} = \begin{pmatrix} 0 \\ 0 \end{pmatrix}$$

After adjusting for essential boundary conditions, we have

$$\begin{pmatrix} 5.3325 & -0.124963 \\ -0.124963 & 3.26965 \end{pmatrix} \begin{pmatrix} \psi_4 \\ \psi_8 \end{pmatrix} = \begin{pmatrix} 104.858 \\ 57.436 \end{pmatrix}$$

Solving the final system of global equations, we get

$$\{\psi_4 = 20.0936, \psi_8 = 18.3344\}$$

Solution for element 1:

$x_1 = 11.909; x_2 = 13.5; x_3 = 13.5$

$y_1 = 1.59099; y_2 = 2.25; y_3 = 4.5$

$b_1 = -2.25; b_2 = 2.90901; b_3 = -0.65901$

$c_1 = 0.; c_2 = -1.59099; c_3 = 1.59099$
$f_1 = 30.375; f_2 = -32.1122; f_3 = 5.3169$
Element area, $A = 1.78986$
Substituting these into the formulas for triangle interpolation functions, we get
Interpolation functions,

$$N^T = \{8.48528 - 0.628539x, 0.812634x - 0.444444y - 8.97056,$$
$$-0.184095x + 0.444444y + 1.48528\}$$

Nodal values, $\boldsymbol{d}^T = \{0, 0, 18.3344\}$
$\psi(x, y) = \boldsymbol{N}^T\boldsymbol{d} = -3.37526x + 8.14861y + 27.2317$
$\partial\psi/\partial x = -3.37526; \partial\psi/\partial y = 8.14861$

Solutions over the remaining elements can be determined in exactly the same manner. A solution summary for all elements is as follows:

Solution at Element Centroids

	x Coordinate	y Coordinate	ψ	$\partial\psi/\partial x$	$\partial\psi/\partial y$
1	12.9697	2.78033	6.11146	−3.37526	8.14861
2	10.4545	3.4205	12.8093	−0.520816	6.58746
3	10.9848	5.14017	24.0593	−0.532342	6.85139
4	6.48483	5.89017	29.1979	0	5.2942
5	9.7045	1.9205	6.69786	−2.86112	1.18512
6	7.60983	1.39017	6.69786	0	4.81803
7	3.85983	3.64017	17.9479	−0.199449	4.83379
8	1.875	2.25	11.25	0	5.

In order to get a better solution, we use a 120-element model as shown in Figure 5.27. The following table shows partial results for the stream function values and the velocities in the x and y directions obtained at the centroids of the elements. Using the u and v values, the velocity vectors shown in Figure 5.28 are obtained. The velocity vectors are tangent to the stream lines and show that the finite element solution is reasonable.

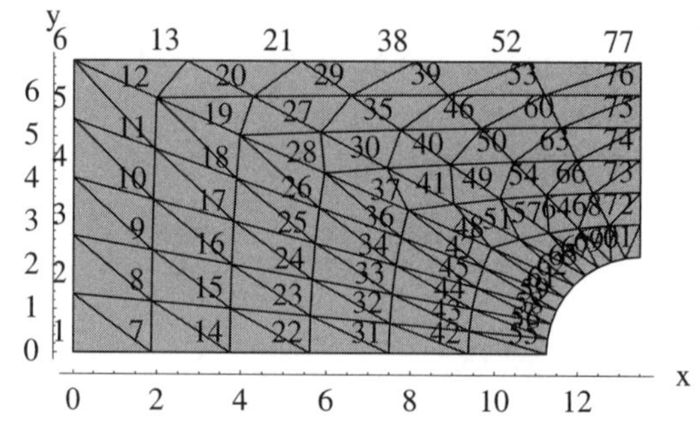

Figure 5.27. A 120-element model for flow around a cylinder

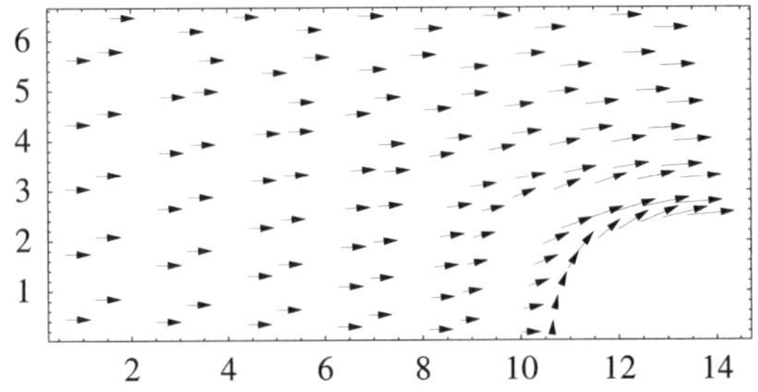

Figure 5.28. Velocity vectors for flow around a cylinder

x	y	ψ	$u = \partial\psi/\partial y$	$v = -\partial\psi/\partial x$
11.5662	2.02003	3.12636	4.35271	3.67236
10.8125	2.22896	6.34974	6.18323	2.19341
9.85069	2.85397	11.5733	4.41314	1.68798
8.96234	3.07584	13.6316	5.32864	0.65605
8.13516	3.68791	17.319	4.99405	0.608097
7.11216	3.92273	18.9254	5.17981	0.24295
6.41963	4.52185	22.17	5.09408	0.236236
5.26198	4.76962	23.6071	5.10947	0.0807255
4.7041	5.35579	26.6377	5.07742	0.0794096
3.41179	5.6165	28.0187	5.05675	0.0197334
2.98857	6.18973	30.9208	5.03959	0.0194431
1.56161	6.46339	32.3102	5.02342	0
11.9899	2.20921	2.9261	5.79379	3.55624
11.3639	2.42207	6.05247	6.87944	2.36017
10.676	3.00993	11.1313	5.19344	1.99278
9.91652	3.23292	13.4288	5.6888	1.0255
9.36214	3.81065	17.1543	5.26698	0.978779
8.46913	4.04377	18.9862	5.31546	0.441183
8.04823	4.61137	22.1837	5.29723	0.440073
7.02174	4.85463	23.761	5.20322	0.179558

Example 5.7 Torsion Constant of a C Shape Find torsional constant J for the standard C12 × 30 section shown in Figure 5.29. The section dimensions are as follows:

$$h = 12\text{ in}; \qquad t_w = 0.51\text{ in}; \qquad b_f = 3.17\text{ in}; \qquad t_f = 0.501\text{ in}$$

Taking advantage of symmetry, we model only half the cross section. Since ϕ is assigned a zero value on all the outside boundary nodes, the mesh must have some interior nodes. Thus the simplest possible model with triangular elements is the eight-element model shown in Figure 5.30. With such a coarse mesh, we do not expect a very accurate solution. The mesh is used to show most computations explicitly.

As a result of essential boundary conditions, $\phi = 0$ at nodes $\{1, 2, 5, 6, 7, 8, 9\}$. Setting $G\theta = 1$, we can obtain the finite element solution using usual steps:

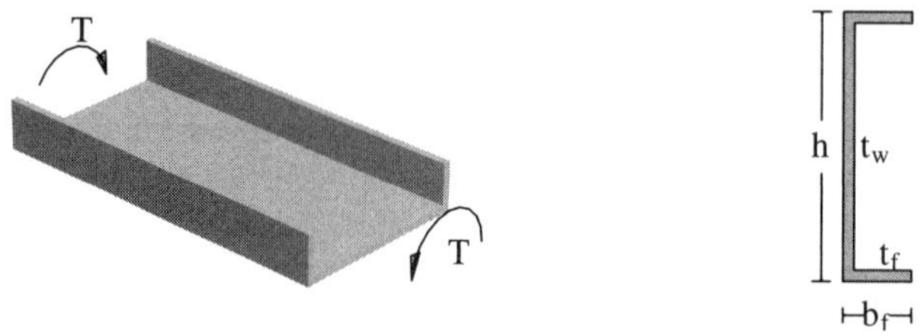

Figure 5.29. C shape subjected to torque

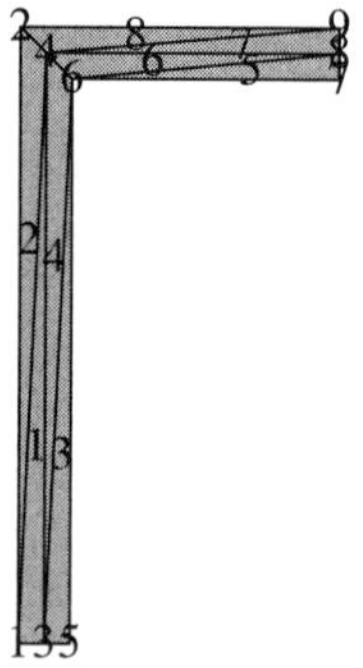

Figure 5.30. Eight-triangular-element model for half of the C shape

Equations for element 1:

$k_x = 1; k_y = 1; p = 0; q = 2$

Nodal coordinates:

Element Node	Global Node Number	x	y
1	1	0.	0.
2	3	0.255	0.
3	4	0.255	5.7495

$x_1 = 0.; x_2 = 0.255; x_3 = 0.255$

$y_1 = 0.; y_2 = 0.; y_3 = 5.7495$

Using these values, we get

$$b_1 = -5.7495; \quad b_2 = 5.7495; \quad b_3 = 0.$$
$$c_1 = 0.; \quad c_2 = -0.255; \quad c_3 = 0.255$$

Element area, $A = 0.733061$

$$\boldsymbol{B}^T = \begin{pmatrix} -5.7495 & 5.7495 & 0. \\ 0. & -0.255 & 0.255 \end{pmatrix}$$

$$\boldsymbol{k}_k = \begin{pmatrix} 11.2735 & -11.2735 & 0. \\ -11.2735 & 11.2957 & -0.0221758 \\ 0. & -0.0221758 & 0.0221758 \end{pmatrix}; \boldsymbol{k}_p = \begin{pmatrix} 0 & 0 & 0 \\ 0 & 0 & 0 \\ 0 & 0 & 0 \end{pmatrix}; \boldsymbol{r}_q = \begin{pmatrix} 0.488707 \\ 0.488707 \\ 0.488707 \end{pmatrix}$$

Complete element equations:

$$\begin{pmatrix} 11.2735 & -11.2735 & 0 \\ -11.2735 & 11.2957 & -0.0221758 \\ 0 & -0.0221758 & 0.0221758 \end{pmatrix} \begin{pmatrix} \phi_1 \\ \phi_3 \\ \phi_4 \end{pmatrix} = \begin{pmatrix} 0.488707 \\ 0.488707 \\ 0.488707 \end{pmatrix}$$

Processing the remaining elements in exactly the same manner, we get the following global equations:

$$\begin{pmatrix} 11.3153 & 0.44942 & -11.2735 & -0.491176 & 0 & 0 & 0 & 0 & 0 \\ 0.44942 & 16.2139 & 0 & -17.0919 & 0 & 0 & 0 & 0 & 0.428528 \\ -11.2735 & 0 & 22.1216 & 0.425425 & -10.7824 & -0.491176 & 0 & 0 & 0 \\ -0.491176 & -17.0919 & 0.425425 & 33.3798 & 0 & -16.0917 & 0 & 0.378522 & -0.508982 \\ 0 & 0 & -10.7824 & 0 & 10.8055 & -0.023186 & 0 & 0 & 0 \\ 0 & 0 & -0.491176 & -16.0917 & -0.023186 & 17.1622 & -0.0470865 & -0.508982 & 0 \\ 0 & 0 & 0 & 0 & 0 & -0.0470865 & 5.35647 & -5.30938 & 0 \\ 0 & 0 & 0 & 0.378522 & 0 & -0.508982 & -5.30938 & 11.2582 & -5.81836 \\ 0 & 0.428528 & 0 & -0.508982 & 0 & 0 & 0 & -5.81836 & 5.89882 \end{pmatrix} \begin{pmatrix} \phi_1 \\ \phi_2 \\ \phi_3 \\ \phi_4 \\ \phi_5 \\ \phi_6 \\ \phi_7 \\ \phi_8 \\ \phi_9 \end{pmatrix} = \begin{pmatrix} 0.998707 \\ 0.774695 \\ 1.44483 \\ 2.23892 \\ 0.467415 \\ 1.42164 \\ 0.22211 \\ 0.708915 \\ 0.508098 \end{pmatrix}$$

Essential boundary conditions:

Node	dof	Value
1	ϕ_1	0
2	ϕ_2	0
5	ϕ_5	0
6	ϕ_6	0
7	ϕ_7	0
8	ϕ_8	0
9	ϕ_9	0

Remove {1, 2, 5, 6, 7, 8, 9} rows and columns:

$$\begin{pmatrix} 22.1216 & 0.425425 \\ 0.425425 & 33.3798 \end{pmatrix} \begin{pmatrix} \phi_3 \\ \phi_4 \end{pmatrix} = \begin{pmatrix} 1.44483 \\ 2.23892 \end{pmatrix}$$

Solving the final system of global equations, we get

$$\{\phi_3 = 0.0640388, \phi_4 = 0.0662577\}$$

Solution for element 1:

$x_1 = 0.; x_2 = 0.255; x_3 = 0.255$

$y_1 = 0.; y_2 = 0.; y_3 = 5.7495$

$b_1 = -5.7495; b_2 = 5.7495; b_3 = 0.$

$c_1 = 0.; c_2 = -0.255; c_3 = 0.255$

$f_1 = 1.46612; f_2 = 0.; f_3 = 0.$

Element area, $A = 0.733061$

Substituting these into the formulas for triangle interpolation functions, we get

Interpolation functions, $\boldsymbol{N}^T = \{1. - 3.92157x, 3.92157x - 0.173928y, 0.173928y\}$

Nodal values, $\boldsymbol{d}^T = \{0, 0.0640388, 0.0662577\}$

$\phi(x, y) = \boldsymbol{N}^T\boldsymbol{d} = 0.251132x + 0.000385934y$

$\partial\phi/\partial x = 0.251132;\ \partial\phi/\partial y = 0.000385934$

Solutions over the remaining elements can be determined in exactly the same manner.

The total torque is given by

$$T = 2\iint_A \phi\, dA$$

The integral of ϕ over each element can be evaluated as described earlier. Since ϕ is a linear function over each element, using the procedure for integration over a triangle discussed earlier, it can be shown that the integral over each element is

$$\iint_{A^{(e)}} \phi^{(e)}\, dA = \frac{A^{(e)}}{3}(\phi_1 + \phi_2 + \phi_3)$$

where $A^{(e)}$ is the area of the element and ϕ_1, ϕ_2, ϕ_3 are the values at its nodes. Using this formula, the integral of ϕ over each element is evaluated and the results are summarized as follows:

	ϕ	$\iint \phi\, dA$
1	$0.251132x + 0.000385934y$	0.0318384
2	$0.259834x$	0.0168957
3	$0.128078 - 0.251132x$	0.0149663
4	$-0.259455x + 0.000385934y + 0.1302$	0.0318384
5	0	0
6	$-0.0227299x + 0.241364y - 1.31567$	0.00806364
7	$0.0720538 - 0.0227299x$	0.00806364
8	$1.58701 - 0.264502y$	0.00876904

Summing $\iint \phi\, dA$ contributions from all elements and multiplying by 2 give the total torque. Since we are modeling half of the C shape, the torque for the entire section is twice this value. The total torque is

$$T = 2 \times 2 \times \sum\left(\iint \phi\, dA\right) = 0.481741$$

Since $T = JG\theta$ and we have used $G\theta = 1$, the computations show that the torsional constant J for the section is 0.48 in^4. The J value tabulated in the steel design handbook for this section is 0.87 in^4. As expected, the computed value has a large error of almost 45%. Solution improves considerably if we use a finer mesh involving 64 triangular elements:

Using 8 element: $J = 0.481741$; Error $= 45.\%$

Using 64 elements: $J = 0.754029$; Error $= 13.\%$

◆ ***Mathematica*/MATLAB Implementation 5.3 on the Book Web Site:**
Triangular element for 2D BVP

5.5.2 Higher Order Triangular Elements

Higher order triangular elements are difficult to develop by using the same procedure as the one used for a three-node triangular element. For example, to develop a quadratic triangular element, we can start with the following polynomial:

$$u(x, y) = c_0 + c_1 x + c_2 y + c_3 x^2 + c_4 xy + c_5 y^2$$

The coefficients $c_0, c_1, \ldots$ can be expressed in terms of nodal degrees of freedom by evaluating the polynomial at the nodes. Since there are six coefficients, the element must have six nodes. Placing three nodes along each side, the quadratic triangular element is as shown in Figure 5.31. The nodal coordinates are $(x_1, y_1), (x_2, y_2), \ldots$.

Evaluating the polynomial at the nodes, we get

$$\begin{pmatrix} u_1 \\ u_2 \\ u_3 \\ u_4 \\ u_5 \\ u_6 \end{pmatrix} = \begin{pmatrix} 1 & x_1 & y_1 & x_1^2 & x_1 y_1 & y_1^2 \\ 1 & x_2 & y_2 & x_2^2 & x_2 y_2 & y_2^2 \\ 1 & x_3 & y_3 & x_3^2 & x_3 y_3 & y_3^2 \\ 1 & x_4 & y_4 & x_4^2 & x_4 y_4 & y_4^2 \\ 1 & x_5 & y_5 & x_5^2 & x_5 y_5 & y_5^2 \\ 1 & x_6 & y_6 & x_6^2 & x_6 y_6 & y_6^2 \end{pmatrix} \begin{pmatrix} c_0 \\ c_1 \\ c_2 \\ c_3 \\ c_4 \\ c_5 \end{pmatrix} \Longrightarrow \boldsymbol{d} = \boldsymbol{A}\boldsymbol{c}$$

The explicit inverse of matrix A in symbolic form is not possible. However, substituting numerical values of nodal coordinates for each element, it is possible to invert the matrix and write interpolation functions for each element as

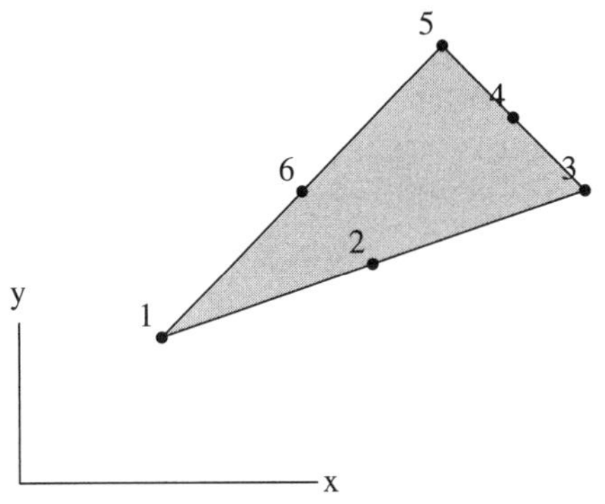

Figure 5.31. Six-node triangular element

$$u(x, y) = (1 \quad x \quad y \quad x^2 \quad xy \quad y^2)\begin{pmatrix} c_0 \\ c_1 \\ c_2 \\ c_3 \\ c_4 \\ c_5 \end{pmatrix} = (1 \quad x \quad y \quad x^2 \quad xy \quad y^2)\boldsymbol{A}^{-1}\boldsymbol{d} \equiv \boldsymbol{N}^T\boldsymbol{d}$$

$$\boldsymbol{N}^T = (1 \quad x \quad y \quad x^2 \quad xy \quad y^2)\boldsymbol{A}^{-1}$$

These interpolation functions must be established for each element in the model. This is to be followed by computing derivatives and evaluating integrals for each element. The computations are not entirely numerical in nature and thus are difficult to program in a numerically oriented language such as C or Fortran. A procedure based on area coordinates for triangles is more suitable for numerical treatment. Details of this procedure can be found in several finite element books, for example, see Gallagher [22].

It is also possible to form triangles by *collapsing* one side of a quadrilateral defined by using the mapping concept discussed in the following chapter. This procedure is very convenient from a programming point of view and is the one implemented in most commercial computer programs.

PROBLEMS

General Form of Two-Dimensional Boundary Value Problem

5.1 A two-dimensional boundary value problem is stated as follows:

$$-\frac{\partial^2 u}{\partial x^2} - \frac{\partial^2 u}{\partial y^2} - 2x - 2y + 4 = 0; \qquad 0 < x < 1; \qquad 0 < y < 1$$

$$u(x, 1) = x^2; \qquad u(0, y) = y^2; \qquad \frac{\partial u}{\partial y}(x, 0) = 2 - 2x - x^3; \qquad \frac{\partial u}{\partial x}(1, y) = 1 - 3y$$

(a) Draw a sketch of the solution domain and show the applicable conditions on the boundary.

(b) Comparing the problem with the general form, identify the terms k_x, k_y, $p, \ldots, \beta$ for this problem.

(c) Classify the boundary conditions into essential and natural types.

(d) Obtain a weak form for the problem.

5.2 Consider the following boundary value problem defined over a rectangular domain shown in Figure 5.32:

$$-\frac{\partial^2 \psi}{\partial x^2} - \frac{\partial^2 \psi}{\partial y^2} + 1 = 0; \qquad 0 < x < 1; \qquad 0 < y < \tfrac{1}{2}$$

$$\frac{\partial \psi}{\partial y}(x, 0) = 0; \qquad \psi(1, y) = 0; \qquad \psi(x, \tfrac{1}{2}) = 0; \qquad \frac{\partial \psi}{\partial x}(0, y) = 0$$

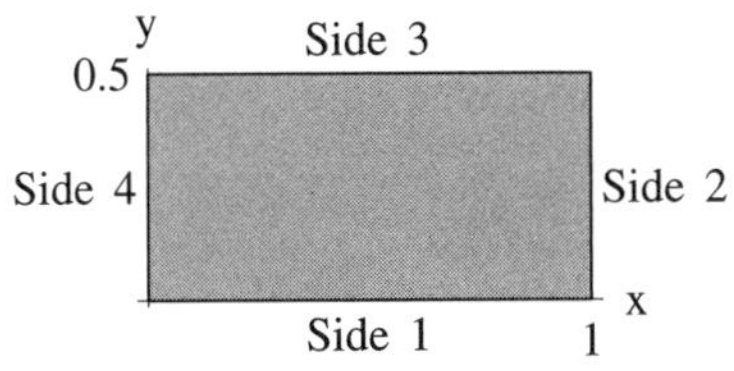

Figure 5.32.

(a) Obtain a weak form for the problem.

(b) Show that an approximate solution of the following form is an admissible solution for the problem:

$$\psi(x, y) = (x - 1)\left(y - \tfrac{1}{2}\right)(a_1 + a_2 x + a_3 y + \cdots)$$

(c) Starting with a solution of the form given in (b), with three unknown coefficients, use the Galerkin method to obtain an approximate solution of the problem.

Rectangular Finite Elements

5.3 Develop the $\boldsymbol{r}_q$ vector for the four-node rectangular element in Figure 5.33 if $q(x, y)$ varies linearly over the element. Use the interpolation functions to express q in terms of its values at the four nodes of the element $q_1, \ldots, q_4$. Thus $q(x, y) = N_1(x, y)q_1 + \cdots + N_4(x, y)q_4$.

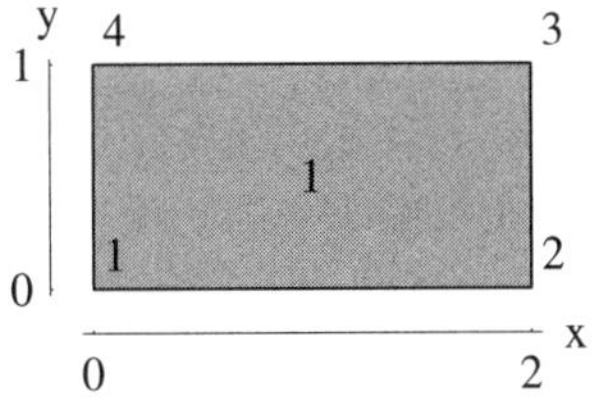

Figure 5.33.

5.4 A two-dimensional boundary value problem is stated as follows:

$$\frac{\partial^2 u}{\partial x^2} + 2\frac{\partial^2 u}{\partial y^2} = 0; \qquad 0 < x < 2; \qquad 0 < y < 1$$

$$u = 1 \text{ on side 2–3}; \qquad \frac{\partial u}{\partial y} + u = 2 \text{ on side 1–2}$$

$$\frac{\partial u}{\partial n} = 0 \text{ on the remaining two sides}$$

Using only one finite element shown in Figure 5.33, obtain an approximate solution of the problem.

5.5 Use the two square elements in Figure 5.34 to find an approximate solution of the following two-dimensional boundary value problem. Report ψ and its x and y derivatives at the element centroids:

$$\frac{\partial^2\psi}{\partial x^2} + \frac{\partial^2\psi}{\partial y^2} + 1 = 0; \qquad 0 < x < 1; \qquad 0 < y < \tfrac{1}{2}$$

$$\psi(x, \tfrac{1}{2}) = 1 - x; \qquad \frac{\partial\psi}{\partial y}(x, 0) = 0; \qquad \frac{\partial\psi}{\partial x}(0, y) = 0; \qquad \frac{\partial\psi}{\partial x}(1, y) = 0$$

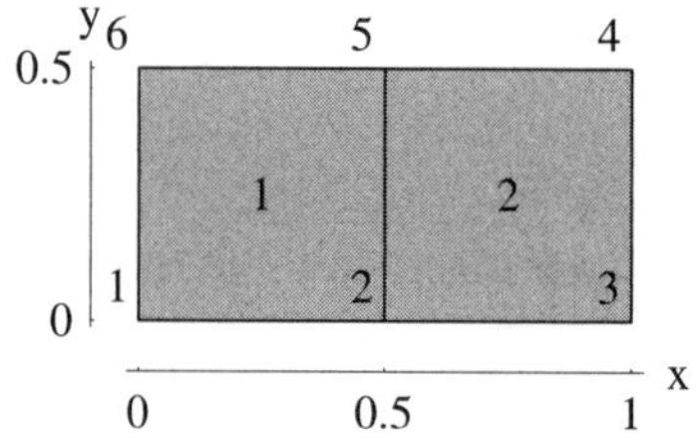

Figure 5.34.

5.6 Determine the temperature distribution in a rectangular body with a thermal conductivity of 10 W/m · °C, as shown in Figure 5.35. The bottom is maintained at 300°C. The top and the right side are insulated. The left side is subjected to heat loss by convection with h = 30 W/m^2 · °C and T_∞ = 30°C. For simplicity use only one element.

$$k_x\frac{\partial^2 T}{\partial x^2} + k_y\frac{\partial^2 T}{\partial y^2} = 0; \qquad 0 < x < 2; \qquad 0 < y < 1$$

$$T(x, 0) = 300°\text{C}; \quad -k\frac{\partial T}{\partial n}(0, y) = h(T - T_\infty); \quad \frac{\partial T}{\partial x}(2, y) = 0; \quad \frac{\partial T}{\partial y}(x, 1) = 0$$

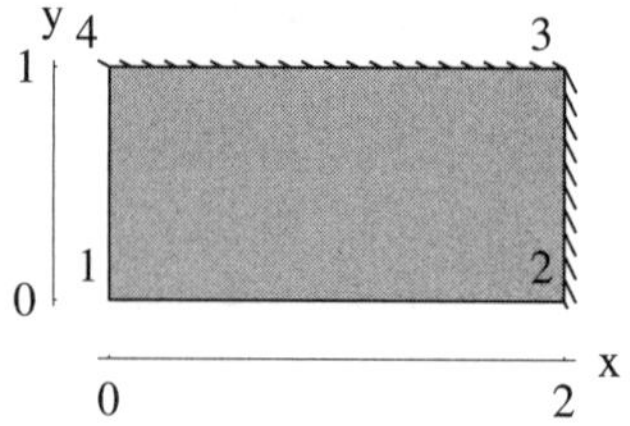

Figure 5.35.

5.7 Find stresses developed in the 1 cm × 1 cm square shaft shown in Figure 5.36 when it is subjected to a torque of 175 N · cm. The shaft is 1 m long and $G = 8 \times 10^6$ N/cm^2. Take advantage of symmetry and model a quarter of the domain using only one element.

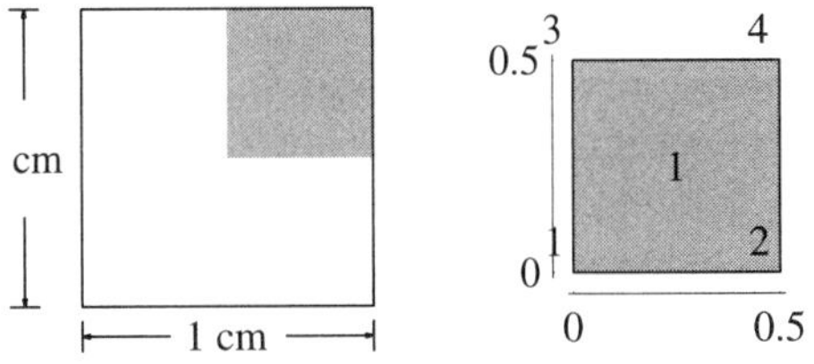

Figure 5.36. Rectangular shaft subjected to torque

5.8 Find stresses developed in the 12 cm × 12 cm cross-shaped shaft shown in Figure 5.37 when it is subjected to a torque of 500 N · m. The shaft is 1 m long and G = 76.9 GPa. Take advantage of symmetry and model a quarter of the domain using three elements.

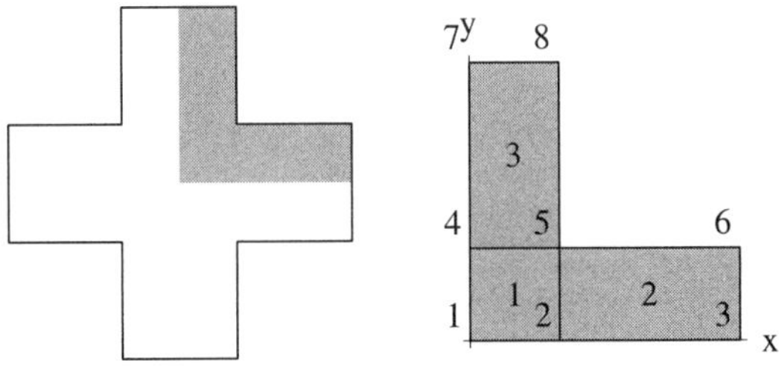

Figure 5.37. Cross-shaped shaft subjected to torque

5.9 Compute the first five TE modes for a rectangular waveguide with width = 2 units and height = 1 unit. Use four-node rectangular elements as shown in Figure 5.38. Note that, even though the geometry is symmetric, we must model the entire domain when computing modes.

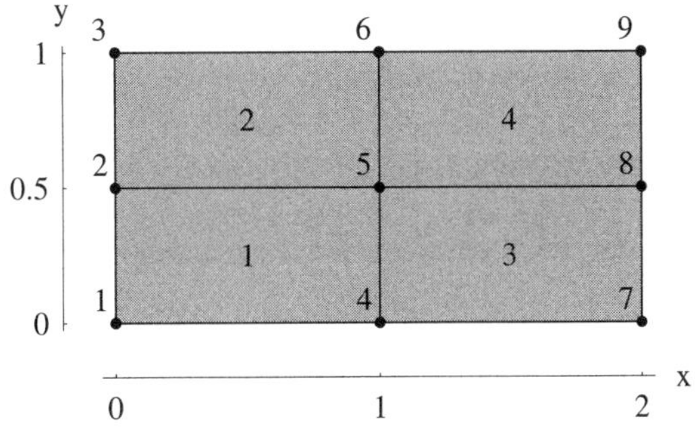

Figure 5.38. Four-element model for rectangular waveguide

5.10 Compute the first five TM modes for a cross-shaped waveguide. Use four-node rectangular elements as shown in Figure 5.39. Note that, even though the geometry is symmetric, we must model the entire domain when computing modes.

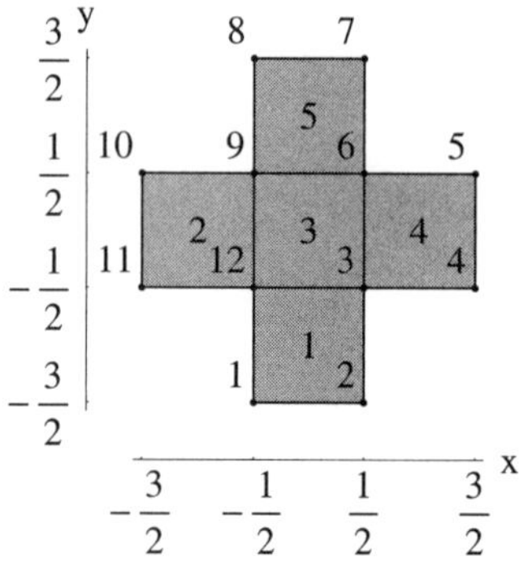

Figure 5.39. Five-element model for cross-shaped waveguide

5.11 Use Lagrange interpolation to develop interpolation functions for the rectangular element shown in Figure 5.40. The midside nodes in the x direction are placed at the third points.

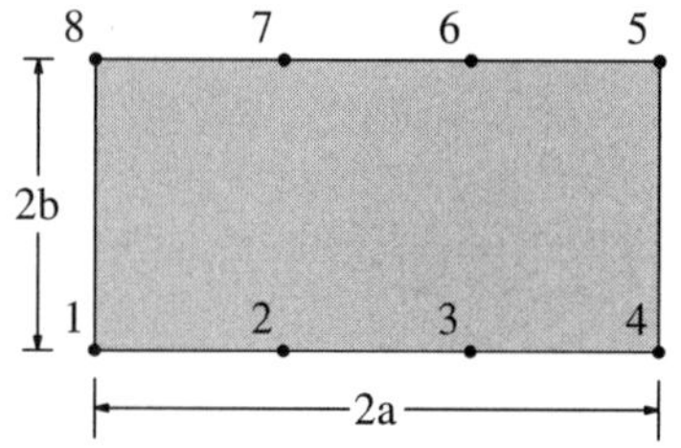

Figure 5.40.

Triangular Finite Elements

5.12 Evaluate the following integral over the triangle shown in Figure 5.41.

$$\iint_A 4356 N_1 N_2 \, dA$$

where N_1 and N_2 are the triangle interpolation functions.
Answer = 11979

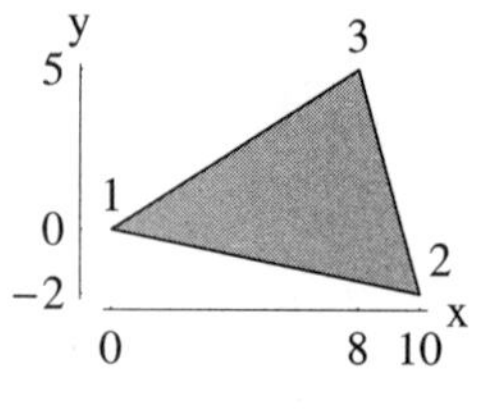

Figure 5.41.

5.13 Evaluate the following area integral over the triangular element shown in Figure 5.42.

$$\iint_A \frac{(x^3 - 4.3)y}{10000}\, dA$$

Answer = 62.9638

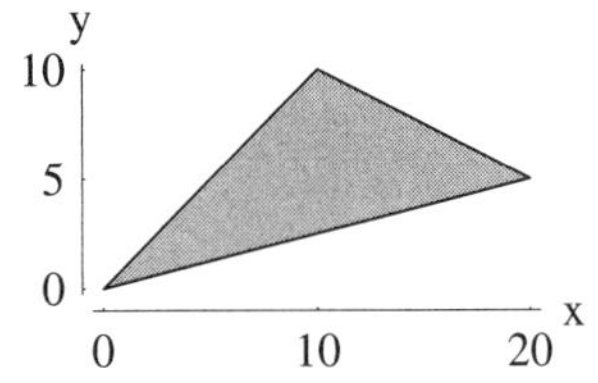

Figure 5.42.

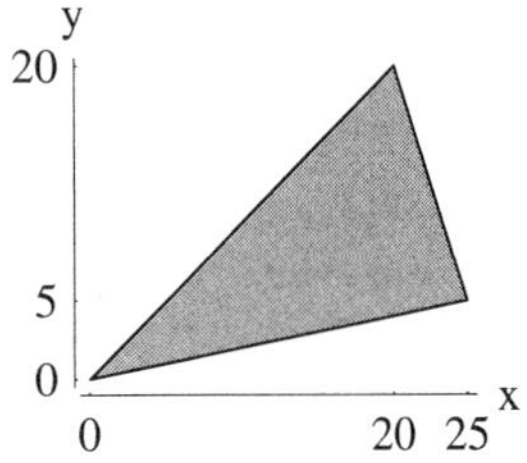

Figure 5.43.

5.14 Evaluate the following area integral over the triangular element shown in Figure 5.43.

$$\iint_A (x + y^3)\, dA$$

Answer = 215500

5.15 Consider fluid flow in the transition region shown in Figure 5.44. The numerical data are as follows:

$H_u = 32$ cm; $H_d = 12$ cm; $L = 50$ cm;

Upstream velocity $V_u = 36$ cm/s

Downstream velocity $V_d = \dfrac{H_u V_u}{H_d} = 96$ cm/s

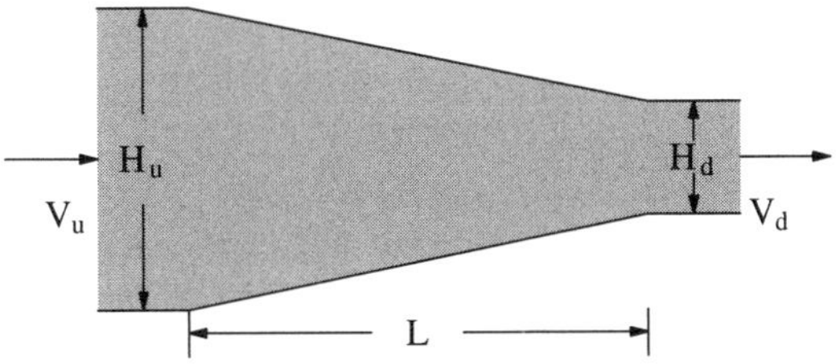

Figure 5.44.

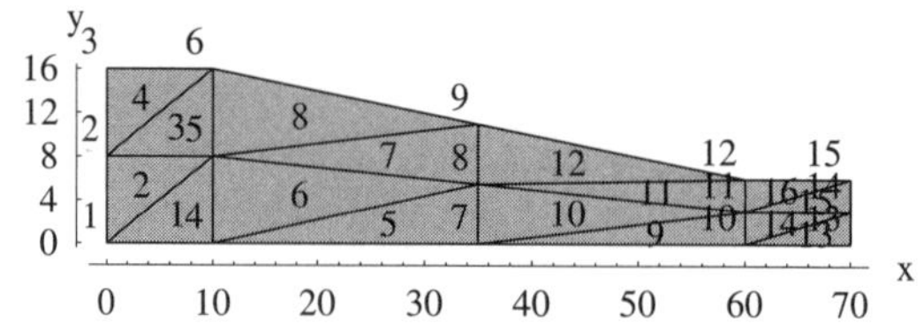

Figure 5.45.

Taking advantage of symmetry, a finite element model of half of the domain is as shown in Figure 5.45. Using the stream function formulation, equations for the first 15 elements are written and assembled to obtain the following global equations. Include the last element in the system and complete the solution for nodal stream function values. What are the computed fluid velocities at $x = 30$ cm and $y = 3$ cm?

$$\begin{pmatrix}
\frac{41}{40} & -\frac{5}{8} & 0 & -\frac{2}{5} & 0 & 0 & 0 & 0 & 0 & 0 & 0 & 0 & 0 & 0 & 0 \\
-\frac{5}{8} & \frac{41}{20} & -\frac{5}{8} & 0 & -\frac{4}{5} & 0 & 0 & 0 & 0 & 0 & 0 & 0 & 0 & 0 & 0 \\
0 & -\frac{5}{8} & \frac{41}{40} & 0 & 0 & -\frac{2}{5} & 0 & 0 & 0 & 0 & 0 & 0 & 0 & 0 & 0 \\
-\frac{2}{5} & 0 & 0 & \frac{4341}{1600} & -\frac{689}{320} & 0 & -\frac{11}{100} & -\frac{1}{20} & 0 & 0 & 0 & 0 & 0 & 0 & 0 \\
0 & -\frac{4}{5} & 0 & -\frac{689}{320} & \frac{8677}{1600} & -\frac{43}{20} & 0 & -\frac{17}{100} & -\frac{3}{20} & 0 & 0 & 0 & 0 & 0 & 0 \\
0 & 0 & -\frac{2}{5} & 0 & -\frac{43}{20} & \frac{261}{100} & 0 & 0 & -\frac{3}{50} & 0 & 0 & 0 & 0 & 0 & 0 \\
0 & 0 & 0 & -\frac{11}{100} & 0 & 0 & \frac{1303}{275} & -\frac{497}{110} & 0 & -\frac{3}{50} & -\frac{1}{20} & 0 & 0 & 0 & 0 \\
0 & 0 & 0 & -\frac{1}{20} & -\frac{17}{100} & 0 & -\frac{497}{110} & \frac{5207}{550} & -\frac{248}{55} & 0 & -\frac{7}{100} & -\frac{3}{20} & 0 & 0 & 0 \\
0 & 0 & 0 & 0 & -\frac{3}{20} & -\frac{3}{50} & 0 & -\frac{248}{55} & \frac{2601}{550} & 0 & 0 & -\frac{1}{100} & 0 & 0 & 0 \\
0 & 0 & 0 & 0 & 0 & 0 & -\frac{3}{50} & 0 & 0 & \frac{1813}{300} & -\frac{35}{6} & 0 & -\frac{3}{20} & 0 & 0 \\
0 & 0 & 0 & 0 & 0 & 0 & -\frac{1}{20} & -\frac{7}{100} & 0 & -\frac{35}{6} & \frac{6247}{600} & -\frac{499}{120} & 0 & -\frac{3}{10} & 0 \\
0 & 0 & 0 & 0 & 0 & 0 & 0 & -\frac{3}{20} & -\frac{1}{100} & 0 & -\frac{499}{120} & \frac{2591}{600} & 0 & 0 & 0 \\
0 & 0 & 0 & 0 & 0 & 0 & 0 & 0 & 0 & -\frac{3}{20} & 0 & 0 & \frac{109}{60} & -\frac{5}{3} & 0 \\
0 & 0 & 0 & 0 & 0 & 0 & 0 & 0 & 0 & 0 & -\frac{3}{10} & 0 & -\frac{5}{3} & \frac{109}{30} & -\frac{5}{3} \\
0 & 0 & 0 & 0 & 0 & 0 & 0 & 0 & 0 & 0 & 0 & 0 & 0 & -\frac{5}{3} & \frac{5}{3}
\end{pmatrix}
\begin{pmatrix} \psi_1 \\ \psi_2 \\ \psi_3 \\ \psi_4 \\ \psi_5 \\ \psi_6 \\ \psi_7 \\ \psi_8 \\ \psi_9 \\ \psi_{10} \\ \psi_{11} \\ \psi_{12} \\ \psi_{13} \\ \psi_{14} \\ \psi_{15} \end{pmatrix}
=
\begin{pmatrix} 0 \\ 0 \\ 0 \\ 0 \\ 0 \\ 0 \\ 0 \\ 0 \\ 0 \\ 0 \\ 0 \\ 0 \\ 0 \\ 0 \\ 0 \end{pmatrix}$$

5.16 Use the potential function formulation to determine fluid flow in the direction perpendicular to a long diamond-shaped bar, as shown in Figure 5.46. At a distance of about 3 times the size of the bar, on both the upstream and the downstream, the flow can be considered uniform with a velocity of $u_0 = 7$ cm/s in the x direction. Determine the flow velocity near the bar.

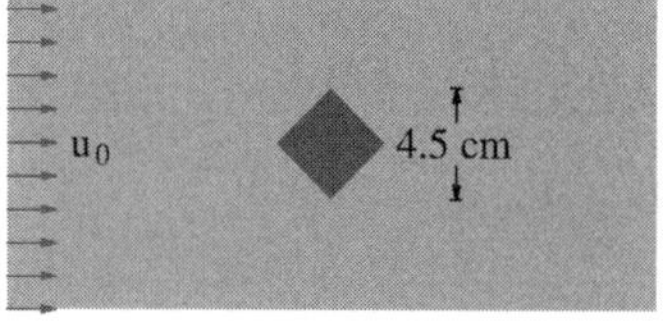

Figure 5.46.

5.17 An equilateral triangular shaped bar is subjected to a torque $T = 500\,\text{N}\cdot\text{m}$. Assume $h = 5$ cm, $E = 70$ GPa, and $\nu = 0.33$.

(a) Verify that the following $\phi(x, y)$ satisfies the differential equation and the boundary conditions and hence represents the exact solution. Note the origin of the coordinate system is at the centroid of the triangle, as shown in Figure 5.47.

$$\phi(x, y) = -G\left(-\frac{2h^2}{27} + \frac{1}{2}(x^2 + y^2) - \frac{x^3 - 3xy^2}{2h}\right)\theta$$

(b) From the exact solution compute the exact values of the torsional constant J, angle of twist θ per unit length, and maximum shear stress τ_{max}. The maximum shear stress occurs at midsides $x = h/6$, $y = \pm h/2\sqrt{3}$ and is equal to $\tau_{max} = G\theta h/2$.

(c) Take advantage of symmetry and model half of the cross section using five triangular elements. From the finite element solution compute the approximate values of J, θ, and τ_{max}. Compare the finite element values with the exact solution.

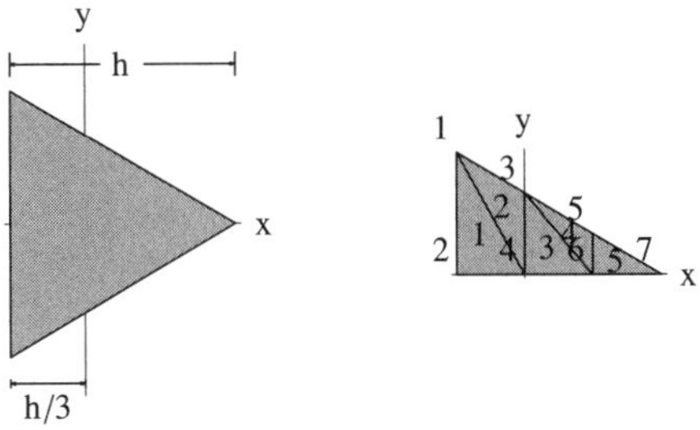

Figure 5.47.

5.18 Find stresses developed in a 4 cm × 8 cm rectangular shaft when it is subjected to a torque of 500 N · m. The shaft is 1 m long and $G = 76.9$ GPa. An eighth of the domain needs to be modeled because of symmetry. As shown in Figure 5.48, a coarse four-element mesh is used.

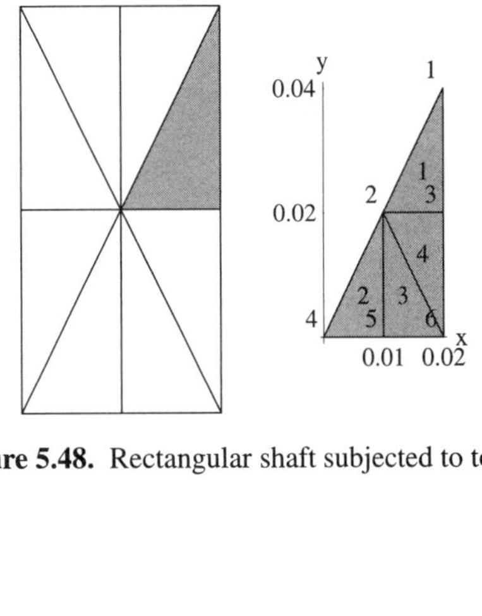

Figure 5.48. Rectangular shaft subjected to torque

5.19 Compute the first five TM modes for a rectangular waveguide with width = 2 units and height = 1 unit. Use triangular elements as shown in Figure 5.49. Note that, even though the geometry is symmetric, we must model the entire domain when computing modes.

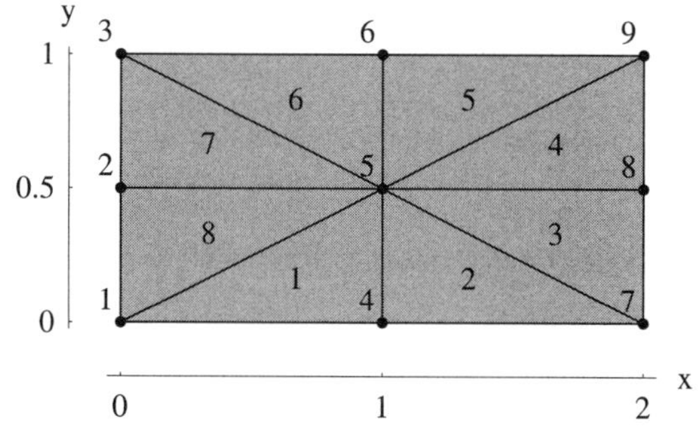

Figure 5.49. Eight-element model for rectangular waveguide

5.20 Compute the first five TE modes for a cross-shaped waveguide. Use four-node rectangular elements as shown in Figure 5.50. Note that, even though the geometry is symmetric, we must model the entire domain when computing modes.

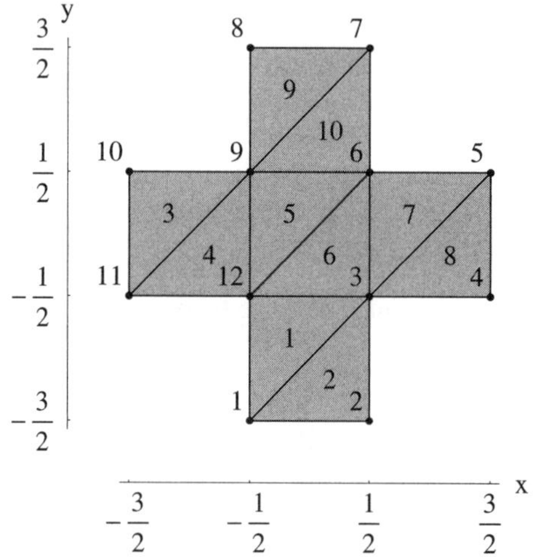

Figure 5.50. Five-element model for cross-shaped waveguide

CHAPTER SIX

MAPPED ELEMENTS

In Chapter 5, rectangular and triangular finite elements were developed for solution of two-dimensional boundary value problems. As pointed out there, the rectangular elements are not very useful for modeling complicated shapes. Triangular elements are more versatile, but since the elements have straight sides, a large number of such elements must be used to model curved geometries. From a theoretical point of view there is no restriction on an element's shape. However, practical requirements of being able to easily define an element geometry and to carry out required integrations and differentiations dictated simple element geometries considered in Chapter 5. Quadrilateral elements, with straight or curved sides, are much more useful in accurately modeling arbitrary shapes. Successful development of these elements is based on the key concept of mapping to facilitate integration. A complicated shape is mapped into a simpler one through an appropriate transformation of coordinates. The integration is then carried out over the mapped shape.

Mapping is presented as a change of variables in textbooks on calculus and is reviewed briefly in the first section. In the second section the Lagrange interpolation is used to develop appropriate mapping from arbitrary quadrilateral shapes to square shapes. With this mapping, even though the integration limits are easy to establish, the integrands become quite complicated, and it is usually not possible to find closed-form expressions for the integrals. Numerical integration therefore becomes necessary to evaluate these integrals. The Gaussian quadrature is the preferred method for evaluating finite element integrals. A brief description of the Gaussian quadrature is presented in the third section. Four- and eight-node quadrilateral elements are presented as illustrations of practical elements based on the mapping concept and the use of numerical integration.

6.1 INTEGRATION USING CHANGE OF VARIABLES

Frequently a suitable change of variables simplifies integration considerably. The concept is straightforward for one-dimensional integrals; however, for two- and three-dimensional integrals a change of variables is associated with mapping from a given to another computational domain. A simple one-dimensional example is presented first. Two- and three-dimensional cases are discussed in later subsections.

6.1.1 One-Dimensional Integrals

Consider evaluation of the following integral:

$$\int_{x=a}^{b} f(x)\,dx$$

Assume a change of variables from x to s in the form of a function $x(s)$ as follows:

$$x = x(s) \Longrightarrow dx = \frac{dx}{ds}\,ds$$

The integrand can be expressed in terms of s by substituting for x:

$$f(x)\,dx = f(x(s))\frac{dx}{ds}\,ds$$

We also need to establish new integration limits in terms of s. The lower limit for s, denoted by s_a, is obtained by solving the equation that is obtained by substituting the change of variables into the lower limit:

$$x = a \Longrightarrow x(s_a) = a; \qquad \text{solve for } s_a$$

Similarly, at the upper limit

$$x = b \Longrightarrow x(s_b) = b; \qquad \text{solve for } s_b$$

Solving these equations, we get s_a and s_b, and thus the integral becomes

$$\int_{x=a}^{b} f(x)\,dx \equiv \int_{s=s_a}^{s_b} f(x(s))\frac{dx}{ds}\,ds$$

Example 6.1 Evaluate the following integral:

$$I = \int_{x=0}^{r} \sqrt{r^2 - x^2}\,dx$$

Direct evaluation of the integral is cumbersome; however, the following change of variable simplifies the integral considerably:

$$x = r\sin(s) \Longrightarrow dx = r\cos(s)\,ds$$

$$\text{At } x = 0: \qquad r\sin(s) = 0 \Longrightarrow s_a = 0$$

$$\text{At } x = r: \qquad r\sin(s) = r; \qquad \sin(s) = 1 \Longrightarrow s_b = \pi/2$$

With these substitutions, the integral is

$$\int_{x=0}^{r} \sqrt{r^2 - x^2}\,dx = \int_{s=0}^{\pi/2} \sqrt{r^2 - r^2\sin^2(s)}r\cos(s)\,ds = \int_{t=0}^{\pi/2} r\sqrt{1 - \sin^2(s)}r\cos(s)\,ds$$

Noting that $\sqrt{1 - \sin^2(s)} = \sqrt{\cos^2(s)} = \cos(s)$, we have

$$I = r^2 \int_{s=0}^{\pi/2} \cos^2(s)\,ds$$

The integral of $\cos^2(s)$ is $s/2 + \frac{1}{4}\sin(2s)$ and thus we have

$$r^2 \int_{s=0}^{\pi/2} \cos^2(s)\,ds = r^2\left[\frac{s}{2} + \frac{1}{4}\sin(2s)\right]_{s=0}^{\pi/2} = \frac{\pi r^2}{4}$$

6.1.2 Two-Dimensional Area Integrals

Suppose we have a function $f(x, y)$ of two variables x and y that needs to be integrated over a two-dimensional region

$$\iint_{A_{xy}} f(x, y)\,dA_{xy}$$

The subscript xy is used on A to indicate that the integration is over a region defined in terms of x and y. Consider a change of variables to s, t given by mapping functions as follows:

$$x = x(s, t); \qquad y = y(s, t)$$

Before proceeding, we must define the relationship between the differential area dA_{xy} in the x, y domain and the dA_{st} in the transformed s, t domain. It can be shown that (the derivation is given after the example)

$$dA_{xy} = \det \boldsymbol{J}\,ds\,dt \Longrightarrow dA_{st}$$

where $\det \boldsymbol{J}$ is called the Jacobian and is the determinant of the following 2×2 matrix of derivatives of the mapping functions:

$$\boldsymbol{J} = \begin{pmatrix} \frac{\partial x}{\partial s} & \frac{\partial x}{\partial t} \\ \frac{\partial y}{\partial s} & \frac{\partial y}{\partial t} \end{pmatrix}; \qquad \det \boldsymbol{J} = \left| \frac{\partial x}{\partial s}\frac{\partial y}{\partial t} - \frac{\partial x}{\partial t}\frac{\partial y}{\partial s} \right|$$

In terms of new variables, the integral is written as follows:

$$\iint_{A_{st}} f(x(s,t), y(s,t)) \det \boldsymbol{J}\, ds\, dt$$

Example 6.2 Evaluate the following integral over a four-sided region bounded by the curves $C_1 : xy = 2$, $C_2 : y^2 = x$, $C_3 : xy = 4$, and $C_4 : y^2 = 3x$:

$$\iint_{A_{xy}} \left(\frac{x^6}{y^3} \right) dA_{xy}$$

In terms of x and y the area is quite complicated. However, the area gets mapped to a square, as shown in Figure 6.1, if we introduce the following change of variables:

$$xy = s; \qquad y^2 = tx \Longrightarrow t = \frac{y^2}{x}$$

Substituting these, the bounding curves are

$$C_1 : s = 2; \qquad C_2 : y^2 = tx = x \Longrightarrow t = 1; \qquad C_3 : s = 4; \qquad C_4 : t = 3$$

Thus, in terms of (s, t) the integration region is a square, as shown in the figure.

From the given change of variables we can solve for x and y to get

$$x = \frac{s^{2/3}}{\sqrt[3]{t}}; \qquad y = \sqrt[3]{s}\sqrt[3]{t}$$

The Jacobian matrix and the Jacobian of the transformation are as follows.

$$\boldsymbol{J} = \begin{pmatrix} \frac{\partial x}{\partial s} & \frac{\partial x}{\partial t} \\ \frac{\partial y}{\partial s} & \frac{\partial y}{\partial t} \end{pmatrix} = \begin{pmatrix} \frac{2}{3\sqrt[3]{s}\sqrt[3]{t}} & -\frac{s^{2/3}}{3t^{4/3}} \\ \frac{\sqrt[3]{t}}{3s^{2/3}} & \frac{\sqrt[3]{s}}{3t^{2/3}} \end{pmatrix}; \qquad \det \boldsymbol{J} = \frac{1}{3t}$$

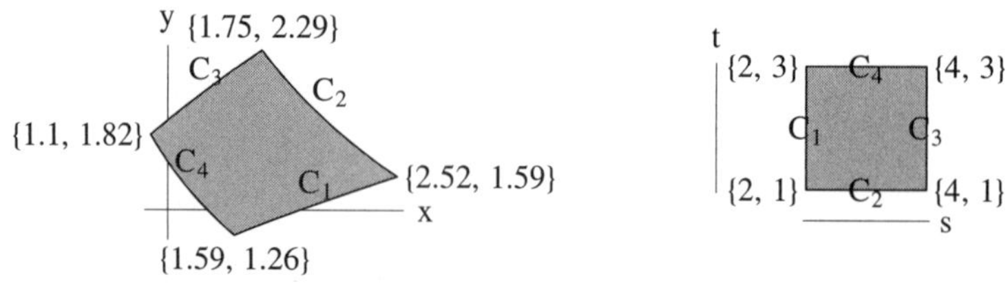

Figure 6.1. Two-dimensional area in original x-y and transformed s-t coordinates

Thus the integral can be evaluated as follows:

$$\iint_{A_{xy}} \left(\frac{x^6}{y^3}\right) dA_{xy} = \iint_{A_{st}} \frac{(s^{2/3}/\sqrt[3]{t})^6}{(\sqrt[3]{s}\,\sqrt[3]{t})^3} \det \boldsymbol{J}\, ds\, dt = \iint_{A_{st}} \frac{s^3}{3t^4}\, ds\, dt$$

$$= \int_{s=2}^{4} \left(\int_{t=1}^{3} \frac{s^3}{3t^4}\, dt \right) ds = \int_{s=2}^{4} \frac{26 s^3}{243}\, ds = \frac{520}{81}$$

Derivation of $dA_{xy} = \det \boldsymbol{J}\, ds\ dt$ An important relationship used in the computation with mapping is that $dA_{xy} = \det \boldsymbol{J}\, ds\, dt$. For the derivation of this result, assume that in the s, t coordinates the differential area dA_{st} is a small rectangle, as shown in Figure 6.2. The points P_i, $i = 0, \ldots, 3$, denote the vertices of this rectangle. With the lengths of the sides denoted by ds and dt, the coordinates of the vertices are expressed as follows:

$$P_0(s_0, t_0); \qquad P_1(s_0 + ds, t_0); \qquad P_2(s_0 + ds, t_0 + dt); \qquad P_3(s_0, t_0 + dt)$$

The corresponding points in the x, y domain are denoted by Q_i, $i = 0, \ldots, 3$, and are obtained from the mapping giving the differential area dA_{xy} as shown in Figure 6.2. The coordinates of point Q_0 are determined as follows:

$$Q_0(x_0, y_0) = \big(x(s_0, t_0), y(s_0, t_0)\big)$$

Similarly, the coordinates of point Q_1 are determined as follows:

$$Q_1(x_1, y_1) = \big(x(s_0 + ds, t_0), y(s_0 + ds, t_0)\big)$$

In general, the coordinates x_1, y_1 are complicated functions of ds. However, since we are dealing with small differential lengths, we can write these coordinates using a Taylor series as follows:

$$x_1 = x_0 + \frac{\partial x(s_0, t_0)}{\partial s}\, ds; \qquad y_1 = y_0 + \frac{\partial y(s_0, t_0)}{\partial s}\, ds$$

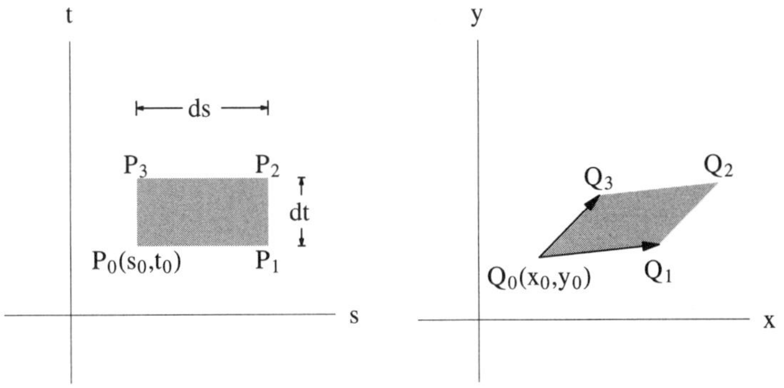

Figure 6.2. Differential area in mapped and original coordinates

In a similar manner the coordinates of the other two points can be expressed in terms of derivatives of the mapping functions. Thus the coordinates of points Q_i, $i = 0, \ldots, 3$, can be written as follows:

$$Q_0(x_0, y_0) = \left(x(s_0, t_0), y(s_0, t_0)\right)$$

$$Q_1(x_1, y_1) = \left(x_0 + \frac{\partial x(s_0, t_0)}{\partial s} ds, y_0 + \frac{\partial y(s_0, t_0)}{\partial s} ds\right)$$

$$Q_2(x_2, y_2) = \left(x_0 + \frac{\partial x(s_0, t_0)}{\partial s} ds + \frac{\partial x(s_0, t_0)}{\partial t} dt, y_0 + \frac{\partial y(s_0, t_0)}{\partial s} ds + \frac{\partial y(s_0, t_0)}{\partial t} dt\right)$$

$$Q_3(x_3, y_3) = \left(x_0 + \frac{\partial x(s_0, t_0)}{\partial t} dt, y_0 + \frac{\partial y(s_0, t_0)}{\partial t} dt\right)$$

These four points form a parallelogram whose area can be determined by taking the cross product of vectors Q_0Q_1 and Q_0Q_3 as follows:

$$\text{Vector } Q_0Q_1 = Q_1 - Q_0 = \left(\frac{\partial x}{\partial s} ds, \frac{\partial y}{\partial s} ds\right)$$

$$\text{Vector } Q_0Q_3 = Q_3 - Q_0 = \left(\frac{\partial x}{\partial t} dt, \frac{\partial y}{\partial t} dt\right)$$

$$dA_{xy} = Q_0Q_3 \times Q_0Q_1 = \frac{\partial x}{\partial s} ds \frac{\partial y}{\partial t} dt - \frac{\partial y}{\partial s} ds \frac{\partial x}{\partial t} dt \equiv \det \boldsymbol{J}\, ds\, dt$$

where

$$\det \boldsymbol{J} = \left| \frac{\partial x}{\partial s}\frac{\partial y}{\partial t} - \frac{\partial x}{\partial t}\frac{\partial y}{\partial s} \right|$$

is the Jacobian as defined earlier. Thus, with the change of variables, the area integral is transformed as follows:

$$\iint_{A_{xy}} f(x, y) dA_{xy} = \iint_{A_{st}} f\left(x(s, t), y(s, t)\right) \det \boldsymbol{J}\, ds\, dt$$

Note that, since $ds\, dt$ is the area dA_{st}, the ratio of the areas A_{xy}/A_{st} is equal to the absolute value of the Jacobian.

6.1.3 Three-Dimensional Volume Integrals

Consider a function $f(x, y, z)$ of three variables x, y, and z that needs to be integrated over a three-dimensional region

$$\iiint_{V_{xyz}} f(x, y, z)\, dV_{xyz}$$

The subscript xyz is used on V to indicate that the integration is over a volume defined in terms of x, y, and z. Consider a change of variables to r, s, t given by the mapping functions as follows:

$$x = x(r, s, t); \qquad y = y(r, s, t); \qquad z = z(r, s, t)$$

Following the same reasoning as for the two-dimensional case, it can be shown that the relationship between the differential volume dV_{xyz} in the x, y, z domain and dV_{rst} in the transformed r, s, t domain is given by

$$dV_{xyz} = \det \boldsymbol{J}\, dr\, ds\, dt \equiv dV_{rst}$$

where $\det \boldsymbol{J}$ is the Jacobian and is the determinant of the following 3×3 matrix of derivatives of the mapping functions:

$$J = \begin{pmatrix} \frac{\partial x}{\partial r} & \frac{\partial x}{\partial s} & \frac{\partial x}{\partial t} \\ \frac{\partial y}{\partial r} & \frac{\partial y}{\partial s} & \frac{\partial y}{\partial t} \\ \frac{\partial z}{\partial r} & \frac{\partial z}{\partial s} & \frac{\partial z}{\partial t} \end{pmatrix}$$

In terms of these new variables, the integral is written as follows:

$$\iiint_{V_{xyz}} f(x, y, z)\, dV_{xyz} = \iiint_{V_{rst}} f\big(x(r, s, t), y(r, s, t), z(r, s, t)\big) \det \boldsymbol{J}\, dr\, ds\, dt$$

6.2 MAPPING QUADRILATERALS USING INTERPOLATION FUNCTIONS

Recall that the Lagrange interpolation formula, introduced in Chapter 2 for one-dimensional problems and extended to rectangular elements in Chapter 5, is a convenient formula for defining a polynomial that passes through a given set of data. So far we have used this formula to derive interpolation functions that pass through the nodal degrees of freedom. In this section we will use the formula to develop appropriate functions that map a given quadrilateral to a square. The key idea is to define suitable elements, called parent or master elements, and use the nodal coordinates of the actual element as data for interpolation over the master element. The idea is first explained for mapping lines and then is extended to the quadrilaterals. The same idea also works for three-dimensional solid elements as well.

6.2.1 Mapping Lines

Mapping a Straight Line Consider two straight lines as shown in Figure 6.3. The one called the master is two units long and is lying along the s coordinate in an s, t coordinate system. The other is of an arbitrary length and orientation in an x, y coordinate system.

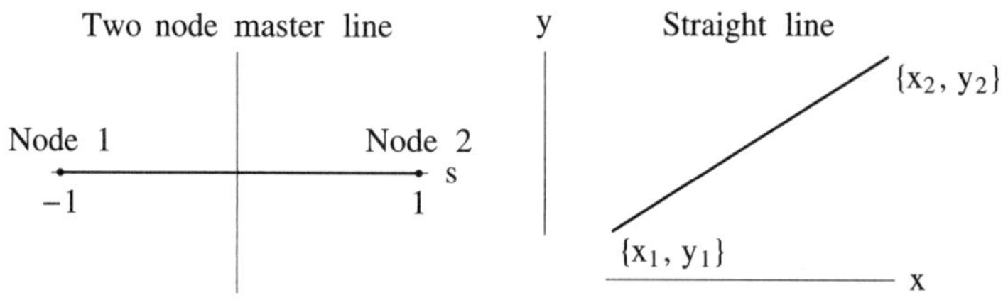

Figure 6.3. Master line in s-t and a straight line in the x-y coordinates

If some data are given at the two ends of the master line, we can use the Lagrange interpolation to describe that data as a function of s. For example, suppose that the following temperature data are given:

s	Temperature
−1	T_1
1	T_2

Then the temperature as a function of s can be written as follows:

$$T(s) = N_1T_1 + N_2T_2 = \left(\frac{1-s}{2}\right)T_1 + \left(\frac{1+s}{2}\right)T_2$$

where N_1 and N_2 are written using the Lagrange interpolation formula presented in Chapter 2.

In an analogous manner we can look at the x, y coordinates of the given straight line as two sets of data points for the master line and write linear interpolation between them. That is, for the x coordinates the data are as follows:

s	x coordinate
−1	x_1
1	x_2

The linear interpolation is

$$x(s) = \tfrac{1}{2}(1-s)x_1 + \tfrac{1}{2}(s+1)x_2$$

Similarly, for the y coordinate we can write

$$y(s) = \tfrac{1}{2}(1-s)y_1 + \tfrac{1}{2}(1+s)y_2$$

We can clearly see that $x(s)$ and $y(s)$ represent the mapping of the straight line to the master line by evaluating the coordinates of the line at selected s values as follows:

At $s = -1$ $\quad x(-1) = x_1;\quad y(-1) = y_1 \Longrightarrow$ Starting point of line

At $s = 1$ $\quad x(1) = x_2;\quad y(1) = y_2 \Longrightarrow$ Ending point of line

At $s = 0$ $\quad x(0) = (x_1 + x_2)/2;\quad y(1) = (y_1 + y_2)/2 \Longrightarrow$ Midpoint of line

Mapping a Curve The same idea can be used to write a mapping between a space curve and a two-unit-long line in the s coordinate. If a curve is defined in terms of n points $(x_1, y_1, z_1), (x_2, y_2, z_2), \ldots$, then we choose a two-unit-long master line element with n equally spaced nodes. For the master element we write the $(n-1)$-order Lagrange interpolation functions $N_i(s)$, $i = 1, 2, \ldots, n$. The mapping of x and y coordinates is then as follows:

$$x(s) = N_1(s)x_1 + N_2(s)x_2 + \cdots + N_n(s)x_n$$
$$y(s) = N_1(s)y_1 + N_2(s)y_2 + \cdots + N_n(s)y_n$$
$$z(s) = N_1(s)z_1 + N_2(s)z_2 + \cdots + N_n(s)z_n$$

As an example, consider a quadratic curve in the x, y plane described in terms of coordinates at three points as shown in Figure 6.4. The master element is defined with nodes at $s = -1, 0, 1$. For interpolating between the three data points, we write the quadratic Lagrange interpolation functions for the master line as follows:

$$N_1 = \tfrac{1}{2}(-1+s)s; \qquad N_2 = 1 - s^2; \qquad N_3 = \tfrac{1}{2}s(1+s)$$

The mapping of x and y coordinates is then as follows:

$$x(s) = \tfrac{1}{2}(-1+s)sx_1 + (1-s^2)x_2 + \tfrac{1}{2}s(1+s)x_3$$
$$y(s) = \tfrac{1}{2}(-1+s)sy_1 + (1-s^2)y_2 + \tfrac{1}{2}s(1+s)y_3$$

We can clearly see that $x(s)$ and $y(s)$ map the three given points on the curve to the three nodes on the master line by evaluating the coordinates of the curve at the corresponding s values as follows:

At $s = -1$: $\quad x(-1) = x_1; \quad y(-1) = y_1 \Longrightarrow$ Starting point of line

At $s = 0$: $\quad x(0) = x_2; \quad y(0) = y_2 \Longrightarrow$ Second point on line

At $s = 1$: $\quad x(1) = x_3; \quad y(1) = y_3 \Longrightarrow$ Ending point of line

In order to see that the intermediate points are mapped correctly as well, we consider the following numerical example.

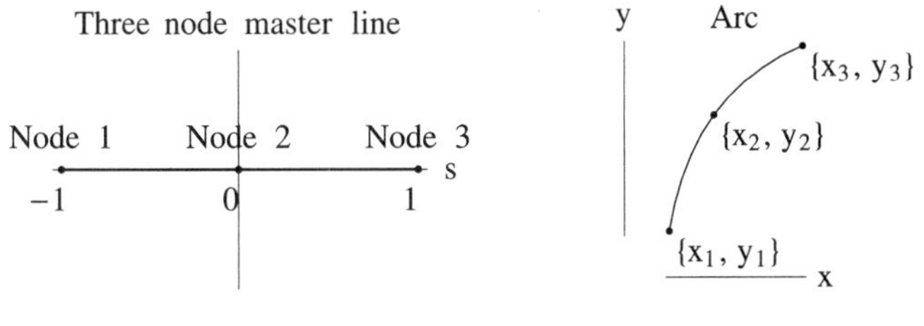

Figure 6.4. Master line in s-t and a quadratic curve line in the x-y coordinates

Example 6.3 Given a quadratic curve that passes through points {1, 1}, {2, 7/2}, {4, 5}, develop an appropriate mapping to map this curve to a straight line two units long. Numerically demonstrate that all points on the curve are mapped uniquely to the points on the straight line.

In this example we have

$$x_1 = 1; \quad y_1 = 1; \quad x_2 = 2; \quad y_2 = \tfrac{7}{2}; \quad x_3 = 4; \quad y_3 = 5$$

Substituting the given coordinates into the mapping of x and y coordinates, we have

$$x(s) = 1\left(\tfrac{1}{2}(s-1)s\right) + 2\left(-(s-1)(s+1)\right) + 4\left(\tfrac{1}{2}s(s+1)\right) = \tfrac{1}{2}s^2 + \tfrac{3}{2}s + 2$$
$$y(s) = 1\left(\tfrac{1}{2}(s-1)s\right) + \tfrac{7}{2}\left(-(s-1)(s+1)\right) + 5\left(\tfrac{1}{2}s(s+1)\right) = -\tfrac{1}{2}s^2 + 2s + \tfrac{7}{2}$$

Using this mapping, we evaluate x and y coordinates for several values of s. The computations are summarized in the following table:

s	$x(s)$	$y(s)$
−1	1	1
−0.75	1.15625	1.71875
−0.5	1.375	2.375
−0.25	1.65625	2.96875
0.	2.	3.5
0.25	2.40625	3.96875
0.5	2.875	4.375
0.75	3.40625	4.71875
1.	4.	5.

The table clearly demonstrates that the given points are mapped into the nodes of the master element and all points in between are mapped appropriately.

Restrictions on Mapping of Lines For the mapping to be useful in finite element applications, it must be one to one. That is, each point on the master element must map into a unique point on the given curve. Unfortunately, the mapping based on the Lagrange interpolation does not always satisfy this requirement. As an example, consider a line passing through the following three data points. These points are the same as those used in Example 6.3 except for the x coordinate of the second point.

$$\text{Coordinates: } \{1, 1\}, \{\tfrac{5}{4}, \tfrac{7}{2}\}, \{4, 5\}$$

Using these coordinates with the quadratic Lagrange interpolation functions, we get the following mapping:

$$x(s) = 1\left(\tfrac{1}{2}(s-1)s\right) + \tfrac{5}{4}\left(-(s-1)(s+1)\right) + 4\left(\tfrac{1}{2}s(s+1)\right) = \tfrac{5}{4}s^2 + \tfrac{3}{2}s + \tfrac{5}{4}$$
$$y(s) = 1\left(\tfrac{1}{2}(s-1)s\right) + \tfrac{7}{2}\left(-(s-1)(s+1)\right) + 5\left(\tfrac{1}{2}s(s+1)\right) = -\tfrac{1}{2}s^2 + 2s + \tfrac{7}{2}$$

The following table shows several x, y points evaluated from this mapping for different s values. The y values are all reasonable. However, some of the x values between $s = -1, 0$ are not reasonable. We would expect all these values to be between 1 and $\frac{5}{4} = 1.25$. However, the computed values for $s = -0.8, -0.6, -0.4$ are all less than 1. Also different values of s are getting mapped to the same point in x. For example, $x = 1$ is obtained from both $s = -1$ and $s = -0.2$. This shows that the mapping is not one to one.

s	$x(s)$	$y(s)$
−1	1	1
−0.8	0.85	1.58
−0.6	0.8	2.12
−0.4	0.85	2.62
−0.2	1.	3.08
0	1.25	3.5
0.2	1.6	3.88
0.4	2.05	4.22
0.6	2.6	4.52
0.8	3.25	4.78
1.	4.	5.

From this example it is clear that for mapping to be one to one the functions $x(s)$ and $y(s)$ should be either increasing or decreasing functions over the entire range $-1 \leq s < 1$. Assuming that the coordinates are given in increasing order, for our present situation the mapping should always be increasing. Mathematically this requirement means that the derivative of the mapping functions must be positive. Thus

$$\frac{dx}{ds} > 0 \quad \text{and} \quad \frac{dy}{ds} > 0 \quad \text{for all } -1 \leq s \leq 1$$

For the numerical example we have

$$\frac{dx}{ds} = \frac{5s}{2} + \frac{3}{2}; \qquad \frac{dy}{ds} = 2 - s$$

These derivatives are plotted in Figure 6.5. We can clearly see that dx/ds is negative from $s = -1$ to $s = -0.6$. Therefore the mapping for the x coordinate is not good over this range. The mapping for the y coordinate is good for the entire range.

Using the positive-derivative requirement for the mapping function, it is possible to write a general requirement for placement of the middle point for a quadratic curve. Differentiating the general formula for mapping of a quadratic curve, we have

$$\frac{dx}{ds} = \frac{1}{2}\left((-1 + 2s)x_1 - 4sx_2 + (1 + 2s)x_3\right)$$

Setting $L_x = x_3 - x_1$, where L_x is the horizontal length of the line, we have

$$\frac{L_x}{2} + L_x s + 2sx_1 - 2sx_2 > 0$$

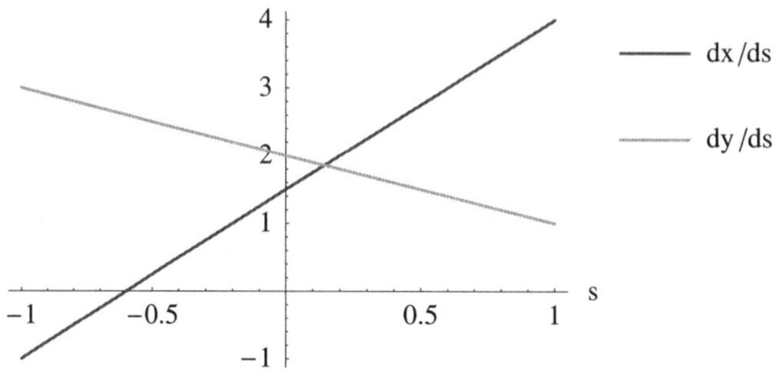

Figure 6.5. Derivatives for the mapping

At $s = -1$

$$-\frac{L_x}{2} - 2x_1 + 2x_2 > 0 \Longrightarrow x_2 > x_1 + \frac{L_x}{4}$$

At $s = 1$

$$\frac{3L_x}{2} + 2x_1 - 2x_2 > 0 \Longrightarrow x_2 < x_1 + \frac{3L_x}{4}$$

The derivative will be positive over the entire range for s between -1 and 1 if

$$x_1 + \frac{L_x}{4} < x_2 < x_1 + \frac{3L_x}{4}$$

In other words, the mapping of a quadratic curve is fine as long as the second point is chosen anywhere over the middle half of the curve.

◆ *Mathematica*/MATLAB Implementation 6.1 on the Book Web Site: *Mapping lines*

6.2.2 Mapping Quadrilateral Areas

For mapping two-dimensional areas the master area is a 2×2 square in the s, t coordinates. The interpolation functions are written by substituting numerical values of the nodal coordinates of the master area into the expressions given for rectangular elements in Chapter 5. The mapping between x and y coordinates is obtained by treating the coordinates of the actual element as the data for Lagrange interpolation. The idea is first explained by defining mapping for a quadrilateral with straight sides. It is then generalized to quadrilaterals with curved sides.

Quadrilaterals with Straight Sides Consider a 2×2 square master area and an arbitrary quadrilateral area as shown in Figure 6.6. The interpolation functions for the master area can easily be written by taking products of linear Lagrange interpolation functions

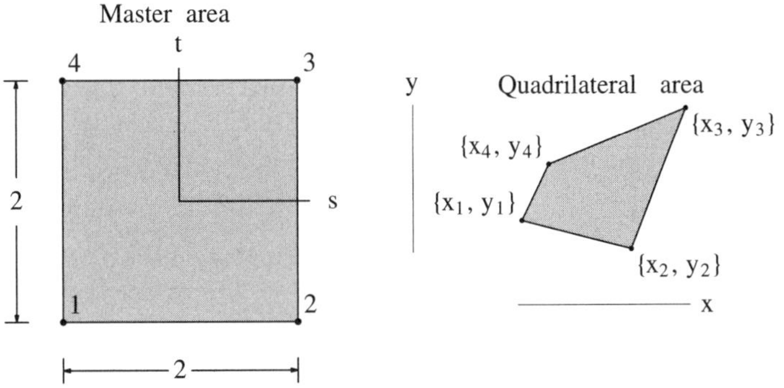

Figure 6.6. Four-node master area and actual quadrilateral area

in the s and t directions as discussed in Chapter 5. Using the numerical values of the coordinates at the four nodes of the master area, the following interpolation functions are obtained:

$$N = \begin{pmatrix} \frac{1}{4}(1-s)(1-t) \\ \frac{1}{4}(s+1)(1-t) \\ \frac{1}{4}(s+1)(t+1) \\ \frac{1}{4}(1-s)(t+1) \end{pmatrix}$$

Multiplying these interpolation functions with nodal coordinates of the quadrilateral, the mapping for the quadrilateral is then as follows:

$$x(s,t) = \tfrac{1}{4}(1-s)(1-t)x_1 + \tfrac{1}{4}(s+1)(1-t)x_2 + \tfrac{1}{4}(s+1)(t+1)x_3 + \tfrac{1}{4}(1-s)(t+1)x_4$$
$$y(s,t) = \tfrac{1}{4}(1-s)(1-t)y_1 + \tfrac{1}{4}(s+1)(1-t)y_2 + \tfrac{1}{4}(s+1)(t+1)y_3 + \tfrac{1}{4}(1-s)(t+1)y_4$$

To see that $x(s,t)$ and $y(s,t)$ represent the appropriate mapping, consider the sides of the quadrilateral. The side (x_1, y_1) to (x_2, y_2) of the quadrilateral should be mapped to the master area side 1–2. To verify this, we evaluate the coordinates of the quadrilateral at $t = -1$ and selected s values as follows:

At $(s,t) = (-1,-1)$: $x(-1,-1) = x_1$; $y(-1,-1) = y_1 \Longrightarrow$ Starting point of side 1–2

At $(s,t) = (0,-1)$: $x = \frac{1}{2}(x_1 + x_2)$; $y = \frac{1}{2}(y_1 + y_2) \Longrightarrow$ Midpoint of side 1–2

At $(s,t) = (1,-1)$: $x = x_2$; $y = y_2 \Longrightarrow$ Endpoint of side 1–2

Similarly we can verify that each point on the master area maps to a unique point on the quadrilateral. The origin of the area (0, 0) corresponds to the point with x, y coordinates as the average of the coordinates of the four corners:

$$(s,t) = (0,0) \Longrightarrow \{\tfrac{1}{4}(x_1 + x_2 + x_3 + x_4), \tfrac{1}{4}(y_1 + y_2 + y_3 + y_4)\}$$

Restrictions on Mapping of Areas The mapping must be one to one to be useful in the finite element computations. For mapping of lines it was shown that this requirement is fulfilled if the derivative of the mapping is positive over the range of the master line. For the mapping of areas, a similar criteria can be developed. The mapping now is a function of two variables; therefore, using the chain rule, we write its total derivatives as follows:

$$dx = \frac{\partial x}{\partial s} ds + \frac{\partial x}{\partial t} dt$$

$$dy = \frac{\partial y}{\partial s} ds + \frac{\partial y}{\partial t} dt$$

Writing the two differentials in a matrix form, we have

$$\begin{pmatrix} dx \\ dy \end{pmatrix} = \begin{pmatrix} \frac{\partial x}{\partial s} & \frac{\partial x}{\partial t} \\ \frac{\partial y}{\partial s} & \frac{\partial y}{\partial t} \end{pmatrix} \begin{pmatrix} ds \\ dt \end{pmatrix}$$

For given ds and dt values this equation determines a corresponding value of dx and dy. For mapping to be one to one, we should be able get unique values of ds and dt for any given values of dx and dy. That is, we should be able to get an inverse relationship

$$\begin{pmatrix} ds \\ dt \end{pmatrix} = \begin{pmatrix} \frac{\partial x}{\partial s} & \frac{\partial x}{\partial t} \\ \frac{\partial y}{\partial s} & \frac{\partial y}{\partial t} \end{pmatrix}^{-1} \begin{pmatrix} dx \\ dy \end{pmatrix} \equiv \frac{1}{\det \boldsymbol{J}} \begin{pmatrix} \frac{\partial y}{\partial t} & -\frac{\partial x}{\partial t} \\ -\frac{\partial y}{\partial s} & \frac{\partial x}{\partial s} \end{pmatrix} \begin{pmatrix} dx \\ dy \end{pmatrix}$$

where $\boldsymbol{J}$ is the 2×2 Jacobian matrix and $\det \boldsymbol{J}$ is its determinant and is simply called the Jacobian:

$$\boldsymbol{J} = \begin{pmatrix} \frac{\partial x}{\partial s} & \frac{\partial x}{\partial t} \\ \frac{\partial y}{\partial s} & \frac{\partial y}{\partial t} \end{pmatrix}; \qquad \det \boldsymbol{J} = \frac{\partial x}{\partial s}\frac{\partial y}{\partial t} - \frac{\partial x}{\partial t}\frac{\partial y}{\partial s}$$

It is evident that $\det \boldsymbol{J}$ must not be zero anywhere over $-1 \le s, t \le 1$ for the mapping to be invertible. This requirement is satisfied as long as $\det \boldsymbol{J}$ is either positive or negative over the entire master element. By adopting a convention of defining areas by moving in a counterclockwise direction around the boundaries, $\det \boldsymbol{J}$ should always be positive. Thus, for the mapping to be valid, we have the following criteria:

$$\text{Mapping is valid if } \det \boldsymbol{J} > 0 \text{ for all } -1 \le s, t \le 1$$

Example 6.4 Good Mapping Determine the mapping to a 2×2 square for a quadrilateral with the following corner coordinates:

	x	y
1	0.5	0
2	2.25	1.3
3	3.1	2.7
4	1.1	2

Substituting these coordinates into the equations for mapping, we get

$$x(s,t) = \tfrac{1}{4}(1-s)(1-t)0.5 + \tfrac{1}{4}(s+1)(1-t)2.25 + \tfrac{1}{4}(s+1)(t+1)3.1 + \tfrac{1}{4}(1-s)(t+1)1.1$$
$$y(s,t) = \tfrac{1}{4}(1-s)(1-t)0 + \tfrac{1}{4}(s+1)(1-t)1.3 + \tfrac{1}{4}(s+1)(t+1)2.7 + \tfrac{1}{4}(1-s)(t+1)2$$

These expressions simplify to

$$x(s,t) = 0.0625ts + 0.9375s + 0.3625t + 1.7375$$
$$y(s,t) = -0.15ts + 0.5s + 0.85t + 1.5$$

Differentiating these expressions, the Jacobian matrix and the Jacobian of the mapping are as follows:

$$\boldsymbol{J} = \begin{pmatrix} 0.0625t + 0.9375 & 0.0625s + 0.3625 \\ 0.5 - 0.15t & 0.85 - 0.15s \end{pmatrix}$$
$$\det \boldsymbol{J} = -0.171875s + 0.1075t + 0.615625$$

By solving the equation $\det \boldsymbol{J} = 0$ for s, we find the line over which the Jacobian is zero as follows:

$$s = -5.81818(-0.1075t - 0.615625)$$

Figure 6.7 shows the line along which the Jacobian is zero. This line is clearly outside of the master area. Thus $\det \boldsymbol{J} \neq 0$ over $-1 \leq s, t \leq 1$ and hence the mapping is good.

Since $\det \boldsymbol{J}$ is a simple function for this example, it is easy to set $\det \boldsymbol{J} = 0$ and solve for s to draw the line along which $\det \boldsymbol{J}$ is zero. When $\det \boldsymbol{J}$ is a complicated function, this will not be possible. Thus, to generate a zero Jacobian line, we need a tool to draw contour plot of functions of two variables. Both MATLAB and *Mathematica* have functions for drawing contour plots. The following *Mathematica* commands are used to draw Figure 6.7.

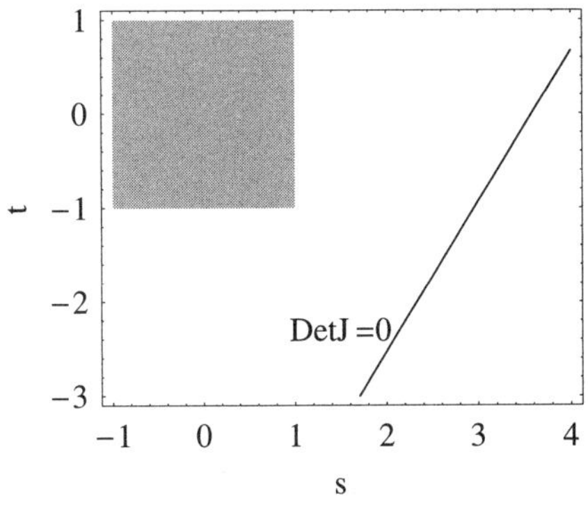

Figure 6.7. Zero Jacobian line and the master area

```
detJ = −0.171875s + 0.1075t + 0.615625;
Block[{$DisplayFunction = Identity},
    zeroContour = ContourPlot[detJ, {s, −4, 4}, {t, −3, 3}, Contours → {0},
    ContourShading → False, FrameLabel → {"s", "t"},
    AspectRatio → Automatic]];
Show[{zeroContour, Graphics[{GrayLevel[0.7], Rectangle[{−1, −1}, {1, 1}],
    GrayLevel[0], Text["DetJ = 0", {1.5, −2.3}]}]}];
```

Figure 6.8 provides a graphical illustration of the mapping. A regular grid of points is chosen on the master area. For each pair of s, t values, the corresponding x, y values are computed from the mapping. All points are clearly mapped properly.

s	t	$x(s, t)$	$y(s, t)$
−1	−1	0.5	0.
−1	−0.5	0.65	0.5
−1	0.	0.8	1.
−1	0.5	0.95	1.5
−1	1.	1.1	2.
−0.5	−1	0.9375	0.325
−0.5	−0.5	1.10313	0.7875
−0.5	0.	1.26875	1.25
−0.5	0.5	1.43438	1.7125
−0.5	1.	1.6	2.175
0.	−1	1.375	0.65
0.	−0.5	1.55625	1.075
0.	0.	1.7375	1.5
0.	0.5	1.91875	1.925
0.	1.	2.1	2.35

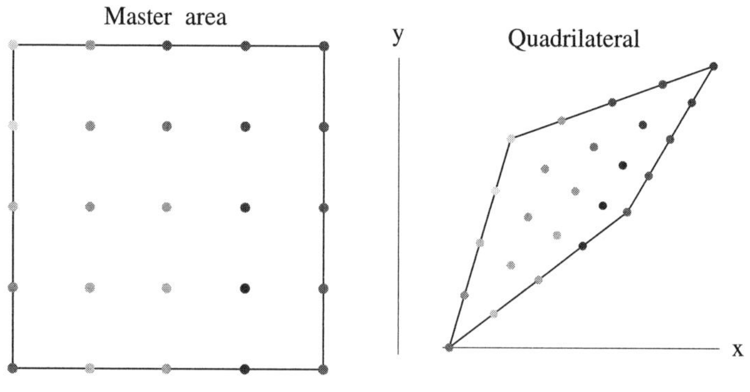

Figure 6.8. Four-node master area and actual quadrilateral area

s	t	$x(s,t)$	$y(s,t)$
0.5	−1	1.8125	0.975
0.5	−0.5	2.00938	1.3625
0.5	0.	2.20625	1.75
0.5	0.5	2.40312	2.1375
0.5	1.	2.6	2.525
1.	−1	2.25	1.3
1.	−0.5	2.4625	1.65
1.	0.	2.675	2.
1.	0.5	2.8875	2.35
1.	1.	3.1	2.7

Example 6.5 Bad Mapping Determine the mapping to a 2×2 square for a quadrilateral with the following corner coordinates:

	x	y
1	0.5	0
2	1.2	1.3
3	3.1	2.7
4	1.1	2

Substituting these coordinates into the equations for mapping, we get

$$x(s,t) = 0.325ts + 0.675s + 0.625t + 1.475$$

$$y(s,t) = -0.15ts + 0.5s + 0.85t + 1.5$$

The Jacobian matrix and the Jacobian of the mapping are as follows:

$$\boldsymbol{J} = \begin{pmatrix} 0.325t + 0.675 & 0.325s + 0.625 \\ 0.5 - 0.15t & 0.85 - 0.15s \end{pmatrix}; \qquad \det \boldsymbol{J} = -0.26375s + 0.37t + 0.26125$$

By solving the equation $\det \boldsymbol{J} = 0$ for s, we find the line over which the Jacobian is zero as follows:

$$s = -3.79147(-0.37t - 0.26125)$$

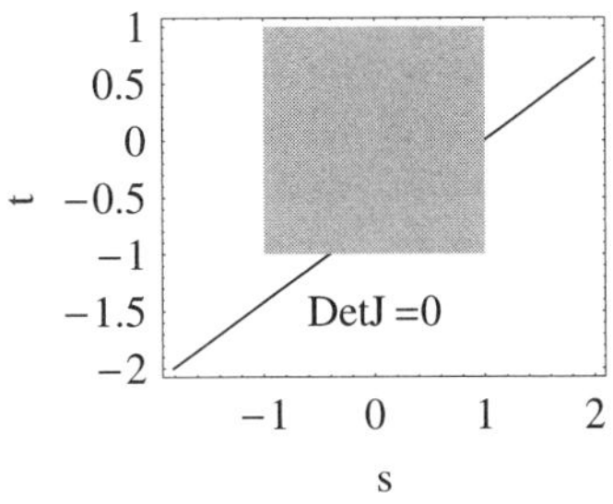

Figure 6.9. Zero Jacobian line and the master area

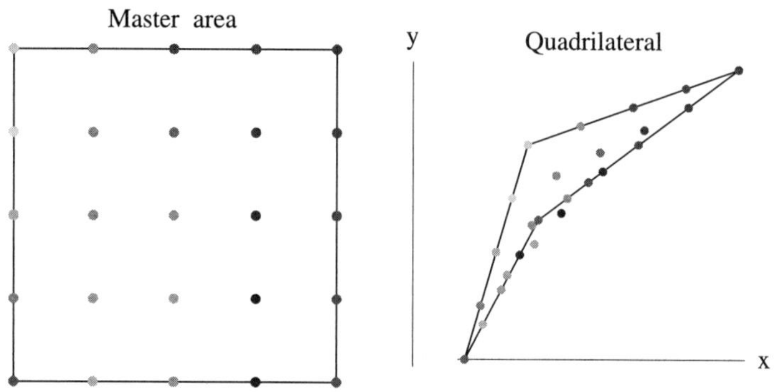

Figure 6.10. Four-node master area and actual quadrilateral area

Figure 6.9 shows the line along which the Jacobian is zero. This line overlaps the master area and hence the mapping is not one to one. Figure 6.10 provides a graphical illustration of the mapping. A regular grid of points is chosen on the master area. For each pair of s, t values, the corresponding x, y values are computed from the mapping. It is clear from the figure that some points are mapped outside of the quadrilateral.

s	t	$x(s,t)$	$y(s,t)$
−1	−1	0.5	0.
−1	−0.5	0.65	0.5
−1	0.	0.8	1.
−1	0.5	0.95	1.5
−1	1.	1.1	2.
−0.5	−1	0.675	0.325
−0.5	−0.5	0.90625	0.7875
−0.5	0.	1.1375	1.25
−0.5	0.5	1.36875	1.7125
−0.5	1.	1.6	2.175
0.	−1	0.85	0.65
0.	−0.5	1.1625	1.075
0.	0.	1.475	1.5
0.	0.5	1.7875	1.925
0.	1.	2.1	2.35
0.5	−1	1.025	0.975
0.5	−0.5	1.41875	1.3625
0.5	0.	1.8125	1.75
0.5	0.5	2.20625	2.1375
0.5	1.	2.6	2.525
1.	−1	1.2	1.3
1.	−0.5	1.675	1.65
1.	0.	2.15	2.
1.	0.5	2.625	2.35
1.	1.	3.1	2.7

Quadrilaterals with Curved Sides Using appropriate interpolation functions, it is possible to map quadrilaterals with one or more sides curved. The master element side that corresponds to a curve will have as many equally spaced nodes as those needed to define the curve on the actual area. For example, for a side represented by a quadratic curve, there will be three nodes on the master element placed at −1, 0, and 1. Thus, if all sides of an element are quadratic curves, then there will be a total of eight nodes. The interpolation functions for this eight-node master element are the serendipity functions written for rectangular elements in Chapter 5. If the two opposite sides are curves, then the interpolation functions are written using the product Lagrange formula. If one, two, or three adjacent sides of an element are curves, it is possible to use the procedures discussed in Chapter 5 to derive the following sets of interpolation functions:

(i) Interpolation functions when all four sides are quadratic curves—eight nodes (Figure 6.11):

$$N = \begin{pmatrix} -\frac{1}{4}(-1+s)(-1+t)(1+s+t) \\ \frac{1}{2}(-1+s^2)(-1+t) \\ \frac{1}{4}(-1+t)(1-s^2+t+st) \\ -\frac{1}{2}(1+s)(-1+t^2) \\ \frac{1}{4}(1+s)(1+t)(-1+s+t) \\ -\frac{1}{2}(-1+s^2)(1+t) \\ \frac{1}{4}(-1+s)(1+s-t)(1+t) \\ \frac{1}{2}(-1+s)(-1+t^2) \end{pmatrix}$$

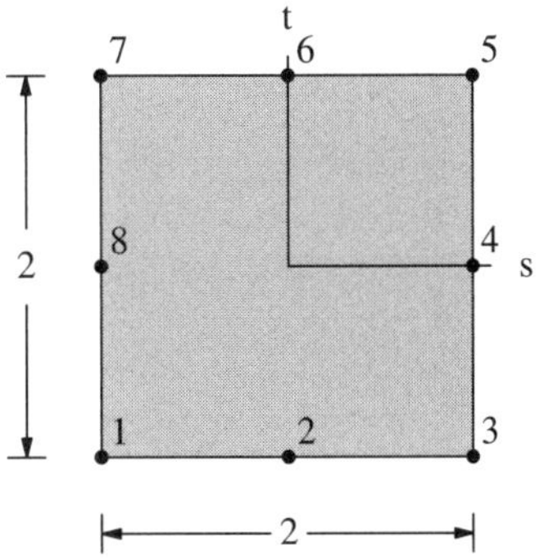

Figure 6.11.

(ii) Interpolation functions when first three sides are quadratic curves—seven nodes (Figure 6.12):

$$N = \begin{pmatrix} -\frac{1}{4}(-1+s)\,s\,(-1+t) \\ \frac{1}{2}(-1+s^2)(-1+t) \\ \frac{1}{4}(-1+t)(1-s^2+t+st) \\ -\frac{1}{2}(1+s)(-1+t^2) \\ \frac{1}{4}(1+s)(1+t)(-1+s+t) \\ -\frac{1}{2}(-1+s^2)(1+t) \\ \frac{1}{4}(-1+s)\,s\,(1+t) \end{pmatrix}$$

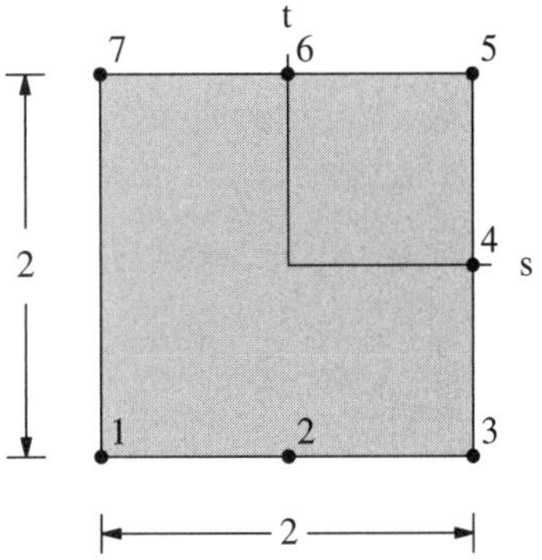

Figure 6.12.

(iii) Interpolation functions when the opposite sides are quadratic curves—six nodes (Figure 6.13):

$$N = \begin{pmatrix} \frac{1}{4}(-1+s)\,s\,(1-t) \\ -\frac{1}{2}(-1+s)(1+s)(1-t) \\ \frac{1}{4}s\,(1+s)(1-t) \\ \frac{1}{4}s\,(1+s)(1+t) \\ -\frac{1}{2}(-1+s)(1+s)(1+t) \\ \frac{1}{4}(-1+s)\,s\,(1+t) \end{pmatrix}$$

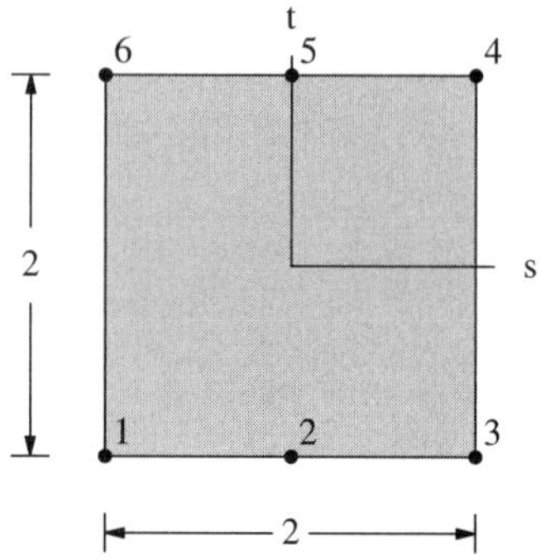

Figure 6.13.

(iv) Interpolation functions when the two adjacent sides are quadratic curves—six nodes (Figure 6.14):

$$N = \begin{pmatrix} -\frac{1}{4}(-1+s)\,s\,(-1+t) \\ \frac{1}{2}(-1+s^2)(-1+t) \\ \frac{1}{4}(-1+t)(1-s^2+t+s\,t) \\ -\frac{1}{2}(1+s)(-1+t^2) \\ \frac{1}{4}(1+s)\,t\,(1+t) \\ -\frac{1}{4}(-1+s)(1+t) \end{pmatrix}$$

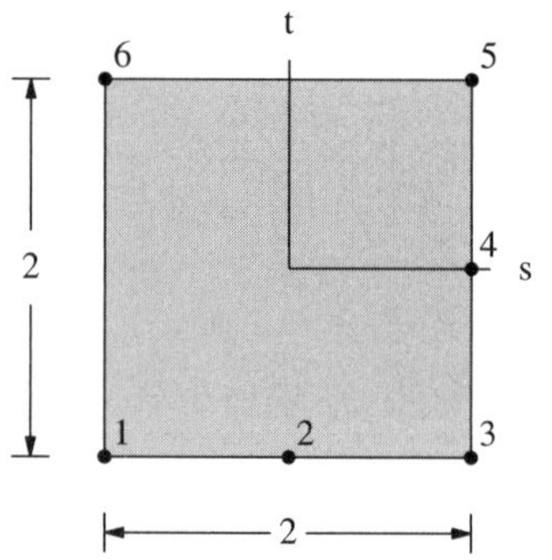

Figure 6.14.

(v) Interpolation functions when the first side is a quadratic curve—five nodes (Figure 6.15):

$$N = \begin{pmatrix} -\frac{1}{4}(-1+s)\,s\,(-1+t) \\ \frac{1}{2}(-1+s^2)(-1+t) \\ -\frac{1}{4}s\,(1+s)(-1+t) \\ \frac{1}{4}(1+s)(1+t) \\ -\frac{1}{4}(-1+s)(1+t) \end{pmatrix}$$

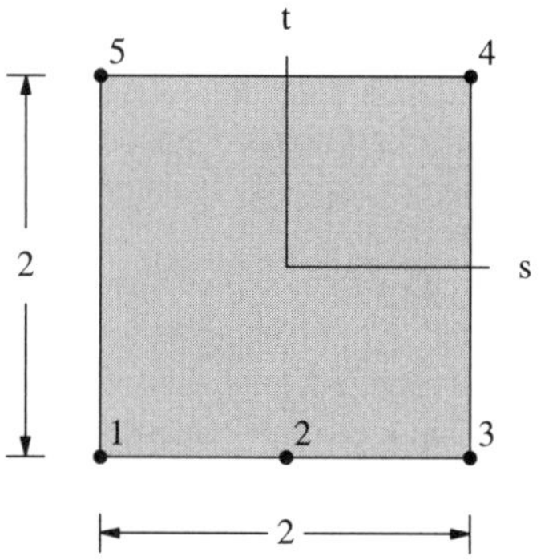

Figure 6.15.

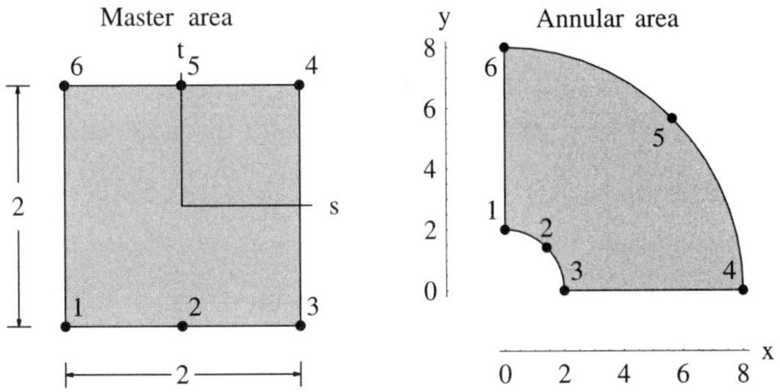

Figure 6.16. Six-node master area and actual annular area

The following examples illustrate the procedure of mapping arbitrary quadrilaterals with one or more curved sides using these interpolation functions.

Example 6.6 Annular Area Develop a mapping to a 2×2 square for the annular area shown in Figure 6.16. The inner radius is 2 units and the outer radius is 8 units. The circular arcs are defined by three points. The coordinates of all six key points are as follows:

	x	y
1	0	2
2	$\sqrt{2}$	$\sqrt{2}$
3	2	0
4	8	0
5	$4\sqrt{2}$	$4\sqrt{2}$
6	0	8

The node numbering is chosen in such a way that three nodes are placed on the first and third sides of the master element. The interpolation functions for this six-node element are given in case (iii). Multiplying these interpolation functions by the x and y coordinates of the six key points of the annular area, we get the following mapping:

$$\begin{aligned}
x(s,t) &= \tfrac{1}{4}(s-1)s(1-t)x_1 - \tfrac{1}{2}(s-1)(s+1)(1-t)x_2 + \tfrac{1}{4}s(s+1)(1-t)x_3 \\
&\quad + \tfrac{1}{4}s(s+1)(t+1)x_4 - \tfrac{1}{2}(s-1)(s+1)(t+1)x_5 + \tfrac{1}{4}(s-1)s(t+1)x_6 \\
&= -\frac{3ts^2}{\sqrt{2}} + \frac{3ts^2}{2} - \frac{5s^2}{\sqrt{2}} + \frac{5s^2}{2} + \frac{3ts}{2} + \frac{5s}{2} + \frac{3t}{\sqrt{2}} + \frac{5}{\sqrt{2}} \\
y(s,t) &= \tfrac{1}{4}(s-1)s(1-t)y_1 - \tfrac{1}{2}(s-1)(s+1)(1-t)y_2 + \tfrac{1}{4}s(s+1)(1-t)y_3 \\
&\quad + \tfrac{1}{4}s(s+1)(t+1)y_4 - \tfrac{1}{2}(s-1)(s+1)(t+1)y_5 + \tfrac{1}{4}(s-1)s(t+1)y_6 \\
&= -\frac{3ts^2}{\sqrt{2}} + \frac{3ts^2}{2} - \frac{5s^2}{\sqrt{2}} + \frac{5s^2}{2} - \frac{3ts}{2} - \frac{5s}{2} + \frac{3t}{\sqrt{2}} + \frac{5}{\sqrt{2}}
\end{aligned}$$

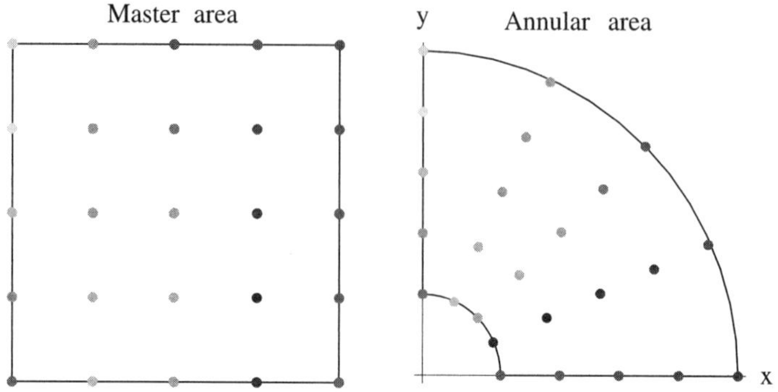

Figure 6.17. Mapping from the master area to the annular area

The Jacobian matrix and the Jacobian of the mapping are as follows:

$$J = \begin{pmatrix} -3\sqrt{2}ts + 3ts - 5\sqrt{2}s + 5s + \frac{3t}{2} + \frac{5}{2} & -\frac{3s^2}{\sqrt{2}} + \frac{3s^2}{2} + \frac{3s}{2} + \frac{3}{\sqrt{2}} \\ -3\sqrt{2}ts + 3ts - 5\sqrt{2}s + 5s - \frac{3t}{2} - \frac{5}{2} & -\frac{3s^2}{\sqrt{2}} + \frac{3s^2}{2} - \frac{3s}{2} + \frac{3}{\sqrt{2}} \end{pmatrix}$$

$$\det J = \frac{9ts^2}{\sqrt{2}} - \frac{9ts^2}{2} + \frac{15s^2}{\sqrt{2}} - \frac{15s^2}{2} + \frac{9t}{\sqrt{2}} + \frac{15}{\sqrt{2}}$$

Setting $\det J = 0$ and solving for t, we see that the Jacobian is zero when

$$t = -\frac{5}{3}$$

This is clearly outside of the master area and hence the mapping is good. Figure 6.17 provides a graphical illustration of the mapping. A regular grid of points is chosen on the master area. For each pair of s, t values, the corresponding x, y values are computed from the mapping. All points are clearly mapped properly.

Example 6.7 Quadrilateral with Curved Sides Develop a mapping to a 2×2 square for the quadrilateral area shown in Figure 6.18. All four sides are curved and are defined by three points on each side for a total of eight points. The master element has eight nodes as shown in the figure. The coordinates of the key points are as follows:

	x	y
1	0.5	0.5
2	1.1	1.6
3	1.7	2.1
4	1.5	2.5
5	1.1	2.8
6	0.75	2.5
7	0.5	2.5
8	0.25	1.5

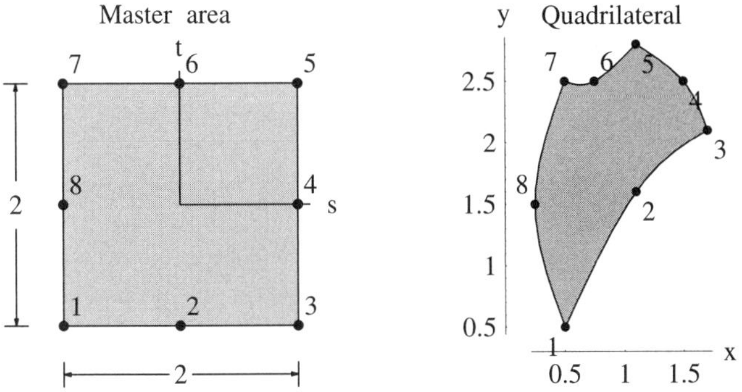

Figure 6.18. Eight-node master area and actual quadrilateral

The interpolation functions for the master element are the standard serendipity shape functions given in case (i). Multiplying these interpolation functions by the x and y coordinates of the eight key points of the annular area we get the following mapping:

$$x(s,t) = -\tfrac{1}{4}(s-1)(t-1)(s+t+1)x_1 + \cdots + \tfrac{1}{2}(s-1)(t^2-1)x_8$$
$$= 0.025ts^2 + 0.025s^2 - 0.175t^2 s - 0.15ts + 0.625s + 0.075t^2 - 0.175t + 0.85$$
$$y(s,t) = -\tfrac{1}{4}(s-1)(t-1)(s+t+1)y_1 + \cdots + \tfrac{1}{2}(s-1)(t^2-1)y_8$$
$$= 0.225ts^2 - 0.075s^2 - 0.025t^2 s - 0.325ts + 0.5s - 0.025t^2 + 0.45t + 2.075$$

The Jacobian matrix and the Jacobian of the mapping are as follows:

$$\boldsymbol{J} = \begin{pmatrix} -0.175t^2 + 0.05st - 0.15t + 0.05s + 0.625 & 0.025s^2 - 0.35ts - 0.15s + 0.15t - 0.175 \\ -0.025t^2 + 0.45st - 0.325t - 0.15s + 0.5 & 0.225s^2 - 0.05ts - 0.325s - 0.05t + 0.45 \end{pmatrix}$$

$$\det \boldsymbol{J} = 0.015s^3 + 0.11625t^2 s^2 - 0.029375ts^2 + 0.089375s^2 - 0.123125t^2 s$$
$$+ 0.265ts - 0.131875s + 0.0125t^3 - 0.026875t^2 - 0.230625t + 0.36875$$

Figure 6.19 shows the contour line along which the Jacobian is zero. The line is clearly outside of the master area and hence the mapping is good. Figure 6.20 provides a graphical illustration of the mapping. A regular grid of points is chosen on the master area. For each pair of s, t values, the corresponding x, y values are computed from the mapping. All points are clearly mapped properly.

Guidelines for Mapped Element Shapes From the examples presented in this section, it is clear that when using mapped elements the finite element discretization must be done in such a way that no element has a zero or near-zero Jacobian. Elements that are close to square will obviously be the best. Elements that are excessively distorted may introduce significant numerical errors. Figure 6.21 presents some practical guidelines for desired element shapes. Most commercial finite element computer programs will issue a warning if the finite element mesh used does not meet these guidelines.

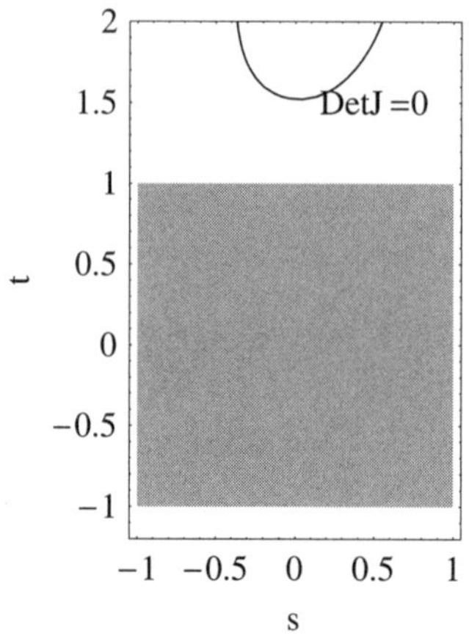

Figure 6.19. Contour for zero Jacobian

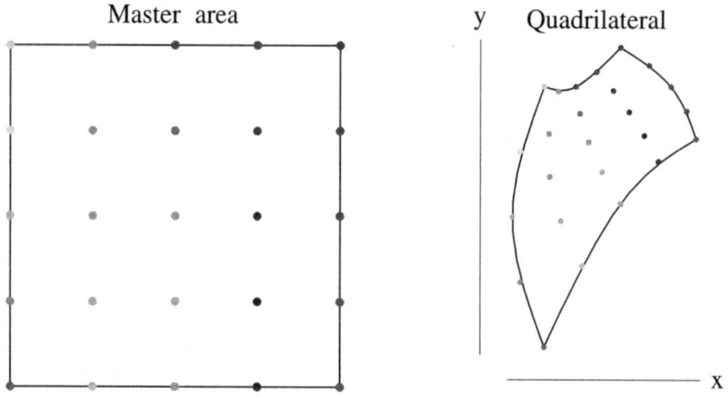

Figure 6.20. Mapping from the master area to the quadrilateral

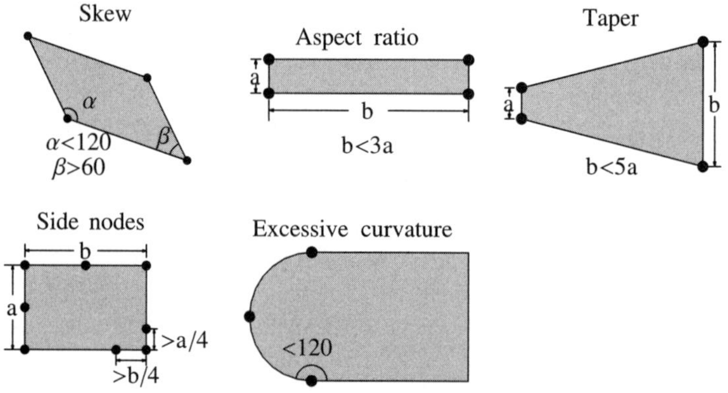

Figure 6.21. Guidelines for mapped element shapes

◆ *Mathematica*/MATLAB Implementation 6.2 on the Book Web Site: *Mapping areas*

6.2.3 Mapped Mesh Generation

The mapping of areas provides a simple method for finite element mesh generation. The method can be used for any geometry that can be divided into four-sided quadrilaterals. The quadrilaterals can have curved sides. The procedure is as follows:

1. Divide the given geometry into four-sided regions and get coordinates of key points defining the quadrilaterals.
2. Decide on the number of nodes to be created on the two adjacent sides of the quadrilaterals. The opposite sides will automatically be divided into that many nodes. Make sure to choose the same number of nodes for the sides that are common between the two quadrilaterals.
3. Map each quadrilateral into a 2×2 square master area and generate the mapping $x(s, t)$ and $y(s, t)$.
4. Divide the master area into a regular grid based on the number of nodes to be created on the s and t sides of the master areas.
5. Substitute the s, t coordinates of the master area grid into the mapping functions to obtain the x, y coordinates of the nodes.
6. Assign node numbers in any convenient order and use the pattern to define element connectivity.
7. Since each quadrilateral area is processed independently, the common sides will have two different sets of nodes. Eliminate the duplicate nodes at the same location and redefine the connectivity of elements affected by this operation.

The following example illustrates the procedure.

Example 6.8 Mapping is used to generate a finite element mesh consisting of triangular elements for the two-dimensional model of a culvert shown in Figure 6.22. Taking advantage of symmetry, only the right half of the culvert is modeled. The area can be divided into two four-sided quadrilaterals as shown in the figure. For each area the curved side is defined by three key points and the straight sides by two corner points. The key point locations are as follows:

	x	y
1	0.	3.
2	1.14805	2.77164
3	2.12132	2.12132
4	2.77164	1.14805
5	3.	0.
6	6	0
7	3.2	5.196
8	0	5.196

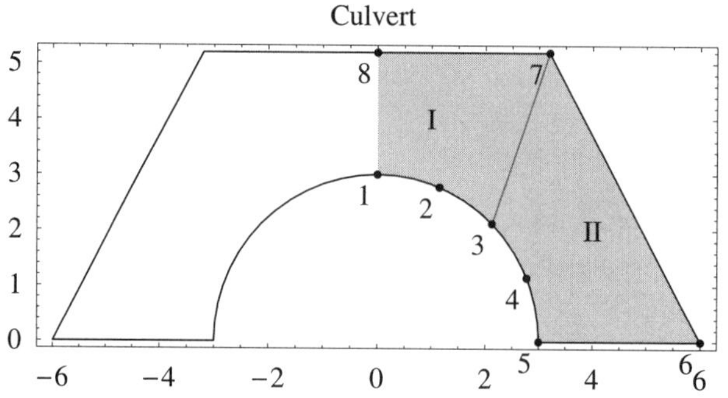

Figure 6.22. Two-dimensional model of a culvert

Each area has one curved side and therefore the appropriate master area interpolation functions are those given in case (v). Defining the area I by using the key point order as {1, 2, 3, 7, 8}, the mapping for area I is as follows:

$$\begin{aligned} x(s,t) &= -\tfrac{1}{4}(s-1)s(t-1)x_1 + \tfrac{1}{2}(s^2-1)(t-1)x_2 - \tfrac{1}{4}s(s+1)(t-1)x_3 \\ &\quad + \tfrac{1}{4}(s+1)(t+1)x_7 - \tfrac{1}{4}(s-1)(t+1)x_8 \\ &= 0.0436951ts^2 - 0.0436951s^2 + 0.26967ts + 1.33033s + 0.225975t + 1.37403 \\ y(s,t) &= -\tfrac{1}{4}(s-1)s(t-1)y_1 + \tfrac{1}{2}(s^2-1)(t-1)y_2 - \tfrac{1}{4}s(s+1)(t-1)y_3 \\ &\quad + \tfrac{1}{4}(s+1)(t+1)y_7 - \tfrac{1}{4}(s-1)(t+1)y_8 \\ &= 0.105489ts^2 - 0.105489s^2 + 0.21967ts - 0.21967s + 1.21218t + 3.98382 \end{aligned}$$

The mapped area is now a 2×2 square and can be subdivided into uniform elements. As an example, we create a mesh with five nodes in the s direction (along sides 1-2-3 and 7-8) and 3 in the t direction (sides 3-7 and 8-1). In the s direction the nodes are placed at $s = -1$, $-\frac{1}{2}$, 0, $\frac{1}{2}$, 1 and in the t coordinates at $t = -1, 0, 1$. Substituting these s, t locations into the mapping gives the corresponding x, y coordinates of the nodes as follows:

	s	t	x	y
1	−1	−1	0	3.
2	−1	0	0	4.098
3	−1	1	0	5.196
4	$-\frac{1}{2}$	−1	0.595873	2.93856
5	$-\frac{1}{2}$	0	0.697936	4.06728
6	$-\frac{1}{2}$	1	0.8	5.196
7	0	−1	1.14805	2.77164
8	0	0	1.37403	3.98382

	s	t	x	y
9	0	1	1.6	5.196
10	$\frac{1}{2}$	−1	1.65653	2.49922
11	$\frac{1}{2}$	0	2.02827	3.84761
12	$\frac{1}{2}$	1	2.4	5.196
13	1	−1	2.12132	2.12132
14	1	0	2.66066	3.65866
15	1	1	3.2	5.196

Similarly, defining the area II using the key point order as {3, 4, 5, 6, 7}, the mapping for area II is as follows:

$$\begin{aligned}x(s,t) &= -\tfrac{1}{4}(s-1)s(t-1)x_3 + \tfrac{1}{2}(s^2-1)(t-1)x_4 - \tfrac{1}{4}s(s+1)(t-1)x_5 \\ &\quad + \tfrac{1}{4}(s+1)(t+1)x_6 - \tfrac{1}{4}(s-1)(t+1)x_7 \\ &= 0.105489ts^2 - 0.105489s^2 + 0.48033ts + 0.91967s + 0.914181t + 3.68582 \\ y(s,t) &= -\tfrac{1}{4}(s-1)s(t-1)y_3 + \tfrac{1}{2}(s^2-1)(t-1)y_4 - \tfrac{1}{4}s(s+1)(t-1)y_5 \\ &\quad + \tfrac{1}{4}(s+1)(t+1)y_6 - \tfrac{1}{4}(s-1)(t+1)y_7 \\ &= 0.0436951ts^2 - 0.0436951s^2 - 0.76867ts - 1.82933s + 0.724975t + 1.87303\end{aligned}$$

Since this area meets with area I along side 3-7, we must create only three nodes in the t direction (sides 5-6 and 3-7) for this area. In the s direction (sides 3-4-5 and 6-7) we can create as many nodes as we like. For illustration purposes we create three nodes in the s direction as well. The locations of the nodes are determined from the mapping as follows:

	s	t	x	y
1	−1	−1	2.12132	2.12132
2	−1	0	2.66066	3.65866
3	−1	1	3.2	5.196
4	0	−1	2.77164	1.14805
5	0	0	3.68582	1.87303
6	0	1	4.6	2.598
7	1	−1	3.	0
8	1	0	4.5	0
9	1	1	6.	0

The nodes and elements created in each area are shown in Figure 6.23. For each area the triangular elements are defined between adjacent columns of nodes. The only remaining task is to merge the nodes from the two areas, eliminate the duplicate nodes along the common edge, and get the final finite element mesh shown in Figure 6.24.

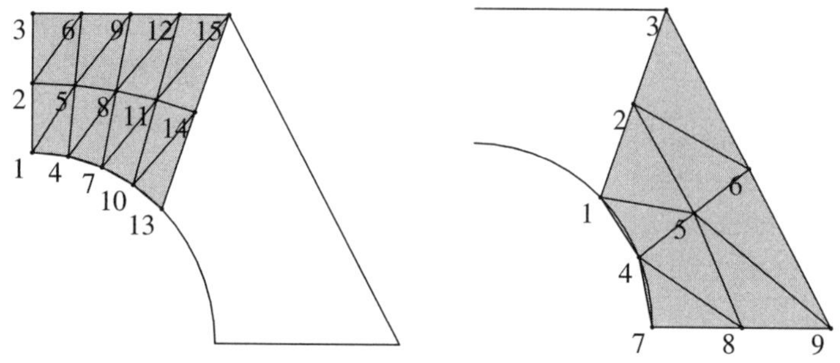

Figure 6.23. Triangular element mesh in the two areas

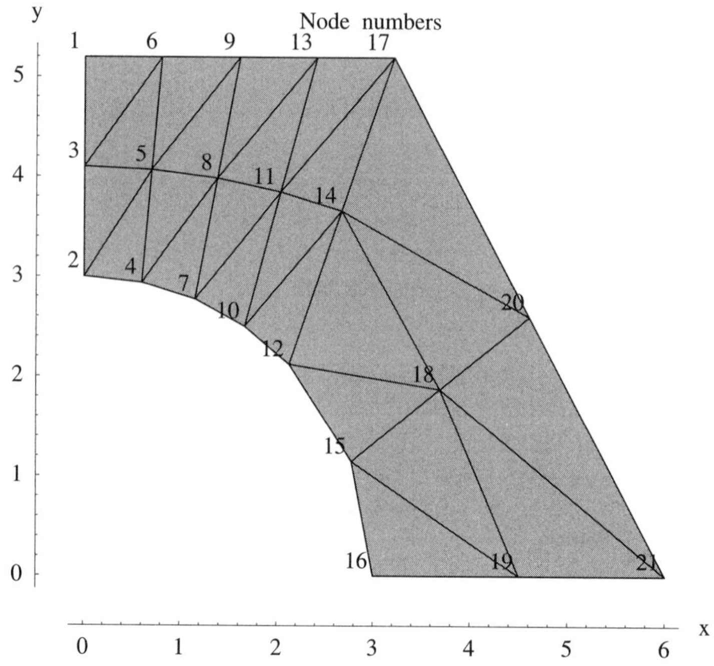

Figure 6.24. Complete mesh for the culvert

6.3 NUMERICAL INTEGRATION USING GAUSS QUADRATURE

When mapping is used to develop finite elements, especially higher order elements, the integrands become quite complicated and it is not possible to evaluate the required integrals in a closed form. We must use numerical integration in these situations. Simple numerical integration techniques such as the trapezoidal rule and Simpson's formula do not work that well in finite element computations. The Gauss quadrature, also called the Gauss-Legendre quadrature, is more suitable for finite element applications and has become the standard tool for developing mapped elements.

The basic Gauss quadrature formulas are derived for one-dimensional integrals in the following section. These formulas extend easily to two- and three-dimensional integrals.

6.3.1 Gauss Quadrature for One-Dimensional Integrals

In the Gauss quadrature, the basic derivation assumes that the integral of a function $f(s)$ is to be evaluated over an interval $-1 \leq s \leq 1$. The main idea is to represent the integral in the following form:

$$I = \int_{-1}^{1} f(s)\,ds \approx w_1 f(s_1) + w_2 f(s_2) + \cdots + w_n f(s_n)$$

The n points $(s_1, s_2, \ldots)$ are known as Gauss points and the corresponding coefficients $(w_1, w_2, \ldots)$ are known as weights. The locations of Gauss points and the appropriate weights are determined by requiring that the above equation give exact integrals for constant, linear, quadratic, etc., terms in a polynomial.

One-Point Formula To start with the simplest case, consider the derivation of a one-point Gauss quadrature. The integral is represented in the following form:

$$I = \int_{-1}^{1} f(s)\,ds \approx w_1 f(s_1)$$

The two unknowns (w_1 and s_1) are determined by making this formula give exact integrals for the first two terms in a polynomial:

Constant term:	$f(s) = 1;$	$\int_{-1}^{1} 1\,ds = 2 \Longrightarrow w_1 = 2$
Linear term:	$f(s) = s;$	$\int_{-1}^{1} s\,ds = 0 \Longrightarrow w_1 s_1 = 0$

The first equation gives $w_1 = 2$ and the second equation gives $s_1 = 0$. Thus the one-point Gauss quadrature formula is

$$I = \int_{-1}^{1} f(s)\,ds \approx 2f(0)$$

Clearly the one-point formula will give exact integrals for linear polynomials. For all other functions the results will be approximate.

Two-Point Formula Now consider derivation of a two-point Gauss quadrature. The integral is represented by the following form:

$$I = \int_{-1}^{1} f(s)\,ds \approx w_1 f(s_1) + w_2 f(s_2)$$

The four unknowns (w_1, s_1, w_2, and s_2) are determined by making this formula give exact integrals for the first four terms in a polynomial:

Constant term:	$f(s) = 1;$	$\int_{-1}^{1} 1\, ds = 2 \Longrightarrow w_1 + w_2 = 2$
Linear term:	$f(s) = s;$	$\int_{-1}^{1} s\, ds = 0 \Longrightarrow w_1 s_1 + w_2 s_2 = 0$
Quadratic term:	$f(s) = s^2;$	$\int_{-1}^{1} s^2\, ds = \frac{2}{3} \Longrightarrow w_1 s_1^2 + w_2 s_2^2 = \frac{2}{3}$
Cubic term:	$f(s) = s^3;$	$\int_{-1}^{1} s^3\, ds = 0 \Longrightarrow w_1 s_1^3 + w_2 s_2^3 = 0$

Solving these four equations simultaneously, we get

$$w_1 = w_2 = 1; \qquad s_1 = -\frac{1}{\sqrt{3}}; \qquad s_2 = \frac{1}{\sqrt{3}}$$

Thus the two-point Gauss quadrature formula is

$$I = \int_{-1}^{1} f(s)\, ds \approx f\left(-\frac{1}{\sqrt{3}}\right) + f\left(\frac{1}{\sqrt{3}}\right)$$

Clearly the two-point formula will give exact integrals for polynomials up to the third order. For other functions the results will be approximate.

Three-Point Formula For a three-point Gauss quadrature the integral is represented in the following form:

$$I = \int_{-1}^{1} f(s)\, ds \approx w_1 f(s_1) + w_2 f(s_2) + w_3 f(s_3)$$

The six unknowns ($w_1, s_1, \ldots$) are determined from the following six conditions:

$$f(s) = 1; \qquad \int_{-1}^{1} 1\, ds = 2 \Longrightarrow w_1 + w_2 + w_3 = 2$$

$$f(s) = s; \qquad \int_{-1}^{1} s\, ds = 0 \Longrightarrow w_1 s_1 + w_2 s_2 + w_3 s_3 = 0$$

$$f(s) = s^2; \qquad \int_{-1}^{1} s^2\, ds = \frac{2}{3} \Longrightarrow w_1 s_1^2 + w_2 s_2^2 + w_3 s_3^2 = \frac{2}{3}$$

$$f(s) = s^3; \qquad \int_{-1}^{1} s^3\, ds = 0 \Longrightarrow w_1 s_1^3 + w_2 s_2^3 + w_3 s_3^3 = 0$$

$$f(s) = s^4; \qquad \int_{-1}^{1} s^4\, ds = \frac{2}{5} \Longrightarrow w_1 s_1^4 + w_2 s_2^4 + w_3 s_3^4 = \frac{2}{5}$$

$$f(s) = s^5; \qquad \int_{-1}^{1} s^5\, ds = 0 \Longrightarrow w_1 s_1^5 + w_2 s_2^5 + w_3 s_3^5 = 0$$

Solving these equations simultaneously, we get

$$w_1 = \frac{5}{9}; \qquad w_2 = \frac{8}{9}; \qquad w_3 = \frac{5}{9}$$

$$s_1 = -\sqrt{\frac{3}{5}}; \qquad s_2 = 0; \qquad s_3 = \sqrt{\frac{3}{5}}$$

Thus the three-point Gauss quadrature formula is

$$I = \int_{-1}^{1} f(s)\,ds \approx \frac{5}{9} f\left(-\sqrt{\frac{3}{5}}\right) + \frac{8}{9} f(0) + \frac{5}{9} f\left(\sqrt{\frac{3}{5}}\right)$$

Clearly the three-point formula will give exact integrals for polynomials up to the fifth order. For other functions the results will be approximate.

Table of Gauss Points and Weights Using the same procedure, one can determine Gauss points and corresponding weights for a formula with any number of points. The following table lists Gauss point locations and weights for formulas up to six points. The table also indicates the order of the polynomial that is integrated exactly by the corresponding formula. Note the given locations and weights must be used only when evaluating integrals from −1 to 1. For different integration limits an appropriate change of variables must be introduced prior to using these values.

Points	Locations	Weights	Order
1	0.	2.	1
2	−0.57735 02691 89626	1.	3
	0.57735 02691 89626	1.	
3	−0.77459 66692 41483	0.55555 55555 55556	5
	0.	0.88888 88888 88889	
	0.77459 66692 41483	0.55555 55555 55556	
4	−0.86113 63115 94053	0.34785 48451 37454	7
	−0.33998 10435 84856	065214 51548 62546	
	0.33998 10435 84856	0.65214 51548 62546	
	0.86113 63115 94053	0.34785 48451 37454	
5	−0.90617 98459 38664	0.23692 68850 56189	9
	−0.53846 93101 05683	0.47862 86704 99366	
	0.	0.56888 88888 88889	
	0.53846 93101 05683	0.47862 86704 99366	
	0.90617 98459 38664	0.23692 68850 56189	
6	−0.93246 95142 03152	0.17132 44923 7917	11
	−0.66120 93864 66265	0.36076 15730 48139	
	−0.23861 91860 83197	0.46791 39345 72691	
	0.23861 91860 83197	0.46791 39345 72691	
	0.66120 93864 66265	0.36076 15730 48139	
	0.93246 95142 03152	0.17132 44923 7917	

Example 6.9 Evaluate the following integral using two-, three-, and four-point Gauss quadrature:

$$I = \int_{-1}^{1} (0.2 + 25s - 200s^2 + 675s^3 - 900s^4 + 400s^5)\,ds$$

$$f(s) = 400s^5 - 900s^4 + 675s^3 - 200s^2 + 25s + 0.2$$

Using the two-point formula, the integral is evaluated as follows:

i	s_i	$f(s_i)$	w_i	$w_i f(s_i)$
1	−0.57735	−336.464	1.	−336.464
2	0.57735	3.53091	1.	3.53091
				$I \approx -332.933$

Using three-point formula the integral is evaluated as follows.

i	s_i	$f(s_i)$	w_i	$w_i f(s_i)$
1	−0.774597	−888.418	0.555556	−493.566
2	0.	0.2	0.888889	0.177778
3	0.774597	0.818488	0.555556	0.454716
				$I \approx -492.933$

Since the given function f is a fifth-order polynomial, the three-point formula should give us the exact integral. A direct evaluation of the integral confirms this. As the following computation with four points demonstrates, using any formula with $n \geq 3$ will give the exact integral for this function:

i	s_i	$f(s_i)$	w_i	$w_i f(s_i)$
1	−0.861136	−1285.01	0.347855	−446.998
2	−0.339981	−71.784	0.652145	−46.8136
3	0.339981	1.90044	0.652145	1.23936
4	0.861136	−1.03804	0.347855	−0.361088
				$I \approx -492.933$

Example 6.10 Evaluate the following integral using two- through six-point Gauss quadrature:

$$I = \int_{-1}^{1} \left(\frac{\exp[s] \sin[s]}{1 + s^2} \right) ds$$

$$f(s) = \frac{e^s \sin(s)}{s^2 + 1}$$

Using the two-point formula the integral is evaluated as follows:

i	s_i	$f(s_i)$	w_i	$w_i f(s_i)$
1	−0.57735	−0.229805	1.	−0.229805
2	0.57735	0.729188	1.	0.729188
				$I \approx 0.499383$

Using the three-point formula the integral is evaluated as follows:

i	s_i	$f(s_i)$	w_i	$w_i f(s_i)$
1	−0.774597	−0.201474	0.555556	−0.11193
2	0.	0.	0.888889	0.
3	0.774597	0.948475	0.555556	0.526931
				$I \approx 0.415$

Using the four-point formula, the integral is evaluated as follows:

i	s_i	$f(s_i)$	w_i	$w_i f(s_i)$
1	−0.861136	−0.184111	0.347855	−0.0640438
2	−0.339981	−0.212765	0.652145	−0.138754
3	0.339981	0.419956	0.652145	0.273873
4	0.861136	1.03051	0.347855	0.358468
				$I \approx 0.429543$

Using the five-point formula, the integral is evaluated as follows:

i	s_i	$f(s_i)$	w_i	$w_i f(s_i)$
1	−0.90618	−0.174647	0.236927	−0.0413786
2	−0.538469	−0.232028	0.478629	−0.111055
3	0.	0.	0.568889	0.
4	0.538469	0.681159	0.478629	0.326022
5	0.90618	1.06969	0.236927	0.253439
				$I \approx 0.427028$

Using the six-point formula, the integral is evaluated as follows:

i	s_i	$f(s_i)$	w_i	$w_i f(s_i)$
1	−0.93247	−0.169073	0.171324	−0.0289664
2	−0.661209	−0.220568	0.360762	−0.0795726
3	−0.238619	−0.176155	0.467914	−0.0824254
4	0.238619	0.283895	0.467914	0.132838
5	0.661209	0.827679	0.360762	0.298595
6	0.93247	1.09146	0.171324	0.186994
				$I \approx 0.427462$

The first three digits of the integral obtained by using five and six points are identical. Thus the integral has essentially converged.

◆ *Mathematica*/MATLAB Implementation 6.3 on the Book Web Site:
One-dimensional numerical integration

6.3.2 Gauss Quadrature for Area Integrals

The one-dimensional Gauss quadrature formulas extend easily to two-dimensional integrals as long as the integration region is a 2×2 square with the origin at the center as shown in Figure 6.25. Thus we consider evaluating the following integral:

$$I = \int_{-1}^{1}\int_{-1}^{1} f(s,t)\,ds\,dt$$

Considering a vertical strip, shown in dark shade in the figure, the integral in the t direction for a given s can be evaluated using an n-point one-dimensional Gauss quadrature formula as follows:

$$I \approx \int_{-1}^{1}\left(\sum_{j=1}^{n} w_j f(s,t_j)\right)ds$$

The resulting s integrals can be evaluated by using m points in the s direction, giving

$$I \approx \sum_{i=1}^{m} w_i\left(\sum_{j=1}^{n} w_j f(s_i,t_j)\right) = \sum_{i=1}^{m}\sum_{j=1}^{n} w_i w_j f(s_i,t_j)$$

Note that the total number of points is $m \times n$. Usually the same number of points is used in both directions, and therefore we get $1 \times 1 = 1, 2 \times 2 = 4, 3 \times 3 = 9, \ldots$ point formulas. The following table shows the locations and weights of some of these formulas. To save space, only six significant figures are shown in the table. Actual numerical computations should use more precision.

Quadrature	Points	s_i	t_j	$w_i \times w_j$
1×1	1	0.	0.	4.
2×2	1	−0.57735	−0.57735	1.
	2	−0.57735	0.57735	1.
	3	0.57735	−0.57735	1.
	4	0.57735	0.57735	1.

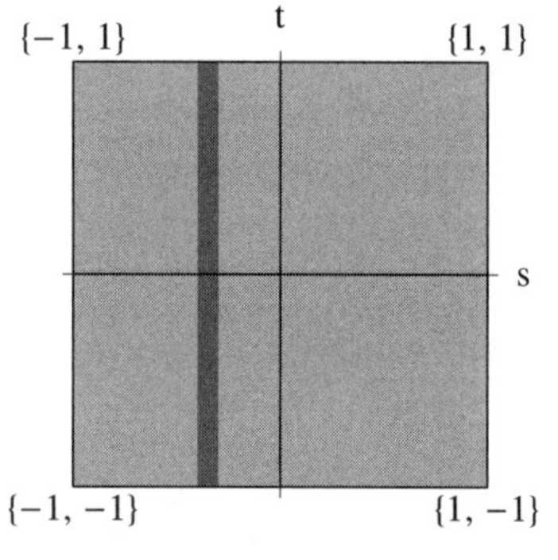

Figure 6.25. 2×2 Square area for two-dimensional Gauss quadrature

Quadrature	Points	s_i	t_j	$w_i \times w_j$
3 × 3	1	−0.774597	−0.774597	0.308642
	2	−0.774597	0.	0.493827
	3	−0.774597	0.774597	0.308642
	4	0.	−0.774597	0.493827
	5	0.	0.	0.790123
	6	0.	0.774597	0.493827
	7	0.774597	−0.774597	0.308642
	8	0.774597	0.	0.493827
	9	0.774597	0.774597	0.308642
4 × 4	1	−0.861136	−0.861136	0.121003
	2	−0.861136	−0.339981	0.226852
	3	−0.861136	0.339981	0.226852
	4	−0.861136	0.861136	0.121003
	5	−0.339981	−0.861136	0.226852
	6	−0.339981	−0.339981	0.425293
	7	−0.339981	0.339981	0.425293
	8	−0.339981	0.861136	0.226852
	9	0.339981	−0.861136	0.226852
	10	0.339981	−0.339981	0.425293
	11	0.339981	0.339981	0.425293
	12	0.339981	0.861136	0.226852
	13	0.861136	−0.861136	0.121003
	14	0.861136	−0.339981	0.226852
	15	0.861136	0.339981	0.226852
	16	0.861136	0.861136	0.121003

Example 6.11 Evaluate the following integral using 2 × 2-, 3 × 2-, and 3 × 3-point Gauss quadrature:

$$I = \int_{-1}^{1}\int_{-1}^{1}(400s^5 + 675s^3 + 25s - 900s^2t^6 - 200t^2 + 0.2)\,ds\,dt$$

$$f(s,t) = -900s^2t^6 - 200t^2 + 400s^5 + 675s^3 + 25s + 0.2$$

Using a 2 × 2 Gauss quadrature, the integral is computed as follows:

Point	s_i, t_j	$f(s_i, t_j)$	$w_i \times w_j$	$w_i \times w_j f(s_i, t_j)$
1	−0.57735	−247.575	1.	−247.575
	−0.57735	−247.575	1.	−247.575
2	−0.57735	−247.575	1.	−247.575
	0.57735	−247.575	1.	−247.575
3	0.57735	92.4198	1.	92.4198
	−0.57735	92.4198	1.	92.4198
4	0.57735	92.4198	1.	92.4198
	0.57735	92.4198	1.	92.4198
				$I \approx -310.311$

With 3 × 2 Gauss quadrature, there are three points in the s direction and two points in the t direction as follows:

$$s \text{ direction points}, s_i = \{-0.774597, 0., 0.774597\}$$
$$\text{weights}, w_i = \{0.555556, 0.888889, 0.555556\}$$
$$t \text{ direction points}, t_j = \{-0.57735, 0.57735\}$$
$$\text{weights}, w_j = \{1., 1.\}$$

Thus the integral is computed as follows:

Point	s_i, t_j	$f(s_i, t_j)$	$w_i \times w_j$	$w_i \times w_j f(s_i, t_j)$
1	−0.774597, −0.57735	−531.085	0.555556	−295.047
2	−0.774597, 0.57735	−531.085	0.555556	−295.047
3	0., −0.57735	−66.4667	0.888889	−59.0815
4	0., 0.57735	−66.4667	0.888889	−59.0815
5	0.774597, −0.57735	358.152	0.555556	198.973
6	0.774597, 0.57735	358.152	0.555556	198.973
				$I \approx -310.311$

Using a 3 × 3 Gauss quadrature, the integral is computed as follows:

Point	s_i, t_j	$f(s_i, t_j)$	$w_i \times w_j$	$w_i \times w_j f(s_i, t_j)$
1	−0.774597, −0.774597	−681.058	0.308642	−210.203
2	−0.774597, 0.	−444.418	0.493827	−219.466
3	−0.774597, 0.774597	−681.058	0.308642	−210.203
4	0., −0.774597	−119.8	0.493827	−59.1605
5	0., 0.	0.2	0.790123	0.158025
6	0., 0.774597	−119.8	0.493827	−59.1605
7	0.774597, −0.774597	208.178	0.308642	64.2526

Point	s_i, t_j	$f(s_i, t_j)$	$w_i \times w_j$	$w_i \times w_j f(s_i, t_j)$
8	0.774597 0.	444.818	0.493827	219.663
9	0.774597 0.774597	208.178	0.308642	64.2526
				$I \approx -409.867$

By direct evaluation, it can easily be verified that the exact value of the integral is as follows:

```
Integrate[f, {s, −1, 1}, {t, −1, 1}]
```

−437.295

Thus we need to further increase the number of integration points to get a more accurate solution.

◆ *Mathematica*/MATLAB Implementation 6.4 on the Book Web Site: *Two-dimensional numerical integration*

6.3.3 Gauss Quadrature for Volume Integrals

Extension of the one-dimensional Gauss quadrature to three-dimensional integrals follows the same line of reasoning as for the two-dimensional case. The integration region now must be a $2 \times 2 \times 2$ cube with the origin at the center as shown in Figure 6.26. Thus we consider evaluating the following integral:

$$I = \int_{-1}^{1} \int_{-1}^{1} \int_{-1}^{1} f(r, s, t)\, dr\, ds\, dt$$

Taking m points in the r direction, n points in the s direction, and p points in the t direction,

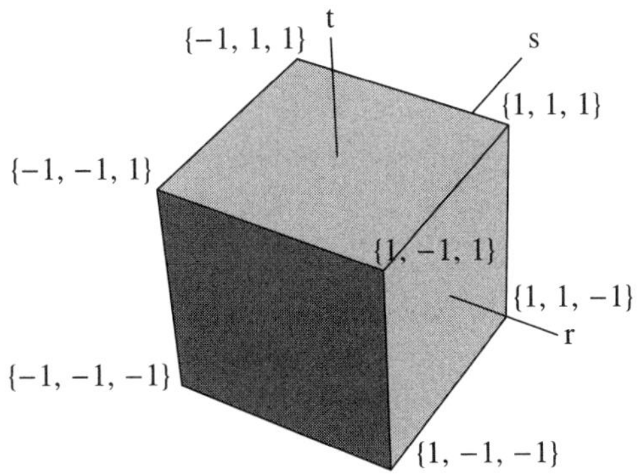

Figure 6.26. $2 \times 2 \times 2$ cube for three-dimensional Gauss quadrature

the $m \times n \times p$ Gauss quadrature for volume integration is as follows:

$$I \approx \sum_{i=1}^{m} \sum_{j=1}^{n} \sum_{k=1}^{p} w_i w_j w_k f(r_i, s_j, t_k)$$

Usually the same number of points are used in each direction. The following table shows the locations and weights of some of these formulas:

Quadrature	Points	r_i	s_j	t_k	$w_i \times w_j \times w_k$
$1 \times 1 \times 1$	1	0.	0.	0.	8.
$2 \times 2 \times 2$	1	−0.57735	−0.57735	−0.57735	1.
	2 −0.57735	−0.57735	0.57735	1.	
	3	−0.57735	0.57735	−0.57735	1.
	4	−0.57735	0.57735	0.57735	1.
	5	0.57735	−0.57735	−0.57735	1.
	6	0.57735	−0.57735	0.57735	1.
	7	0.57735	0.57735	−0.57735	1.
	8	0.57735	0.57735	0.57735	1.
$3 \times 3 \times 3$	1	−0.774597	−0.774597	−0.774597	0.171468
	2	−0.774597	−0.774597	0.	0.274348
	3	−0.774597	−0.774597	0.774597	0.171468
	4	−0.774597	0.	−0.774597	0.274348
	5	−0.774597	0.	0.	0.438957
	6	−0.774597	0.	0.774597	0.274348
	7	−0.774597	0.774597	−0.774597	0.171468
	8	0.774597	0.774597	0.	0.274348
	9	0.774597	0.774597	0.774597	0.171468
	10	0.	−0.774597	0.774597	0.274348
	11	0.	−0.774597	0.	0.438957
	12	0.	−0.774597	0.774597	0.274348
	13	0.	0.	−0.774597	0.438957
	14	0.	0.	0.	0.702332
	15	0.	0.	0.774597	0.438957
	16	0.	0.774597	−0.774597	0.274348
	17	0.	0.774597	0.	0.438957
	18	0.	0.774597	0.774597	0.274348
	19	0.774597	−0.774597	0.774597	0.171468
	20	0.774597	−0.774597	0.	0.274348
	21	0.774597	−0.774597	0.774597	0.171468
	22	0.774597	0.	−0.774597	0.274348
	23	0.774597	0.	0.	0.438957
	24	0.774597	0.	0.774597	0.274348
	25	0.774597	0.774597	−0.774597	0.171468
	26	0.774597	0.774597	0.	0.274348
	27	0.774597	0.774597	0.774597	0.171468

Example 6.12 Evaluate the following integral using (1 × 2 × 3)- and (3 × 3 × 3)-point Gauss quadrature:

$$I = \int_{-1}^{1}\int_{-1}^{1}\int_{-1}^{1}(400t^5 + 675t^3 - 900s^4 - 200s^2 + 25r + 0.2)\,dr\,ds\,dt$$

$$f(r, s, t) = 400t^5 + 675t^3 - 900s^4 - 200s^2 + 25r + 0.2$$

With 1 × 2 × 3 Gauss quadrature, there is one point in the r direction, two points in the s directions, and three points in the t direction as follows:

$$r \text{ direction points, } r_i = \{0.\}$$
$$\text{weights, } w_i = \{2.\}$$
$$s \text{ direction points, } s_j = \{-0.57735, 0.57735\}$$
$$\text{weights, } w_j = \{1., 1.\}$$
$$t \text{ direction points, } t_k = \{-0.774597, 0., 0.774597\}$$
$$\text{weights, } w_k = \{0.555556, 0.888889, 0.555556\}$$

Thus the integral is computed as follows:

Point	r_i, s_j, t_k	$f(r_i, s_j, t_k)$	$w_i \times w_j \times w_k$	$w_i \times w_j \times w_k f(r_i, s_j, t_k)$
1	0., −0.57735, −0.774597	−591.72	1.11111	−657.467
2	0., −0.57735, 0.	−166.467	1.77778	−295.941
3	0., −0.57735, 0.774597	258.787	1.11111	287.541
4	0., −0.57735, 0.774597	−591.72	1.11111	−657.467
5	0., 0.57735, 0.	−166.467	1.77778	−295.941
6	0., 0.57735, 0.774597	258.787	1.11111	287.541

$$I \approx -1331.73$$

It can be verified that using a $3 \times 3 \times 3$ Gauss quadrature (27 points) the integral comes out to be -1971.73, which is the exact value of the integral.

◆ ***Mathematica*/MATLAB Implementation 6.5 on the Book Web Site:** *Three-dimensional numerical integration*

6.4 FINITE ELEMENT COMPUTATIONS INVOLVING MAPPED ELEMENTS

The finite element equations for a second-order boundary value problem were derived in Chapter 5. For any arbitrary shaped element in the x, y coordinate system, the element equations are as follows:

$$(\boldsymbol{k}_k + \boldsymbol{k}_p + \boldsymbol{k}_\alpha)\boldsymbol{d} = \boldsymbol{r}_q + \boldsymbol{r}_\beta$$

where

$$\boldsymbol{k}_k = \iint_A \boldsymbol{BCB}^T \, dA; \qquad \boldsymbol{k}_p = -\iint_A p\boldsymbol{NN}^T \, dA; \qquad \boldsymbol{k}_\alpha = -\int_{C_n} \alpha \boldsymbol{N}_c \boldsymbol{N}_c^T \, dc$$

$$\boldsymbol{r}_q = \iint_A q\boldsymbol{N} \, dA; \qquad \boldsymbol{r}_\beta = \int_{C_n} \beta \boldsymbol{N}_c \, dc$$

$$u(x, y) = \begin{pmatrix} N_1(x, y) & N_2(x, y) & \cdots & N_n(x, y) \end{pmatrix} \begin{pmatrix} u_1 \\ u_2 \\ \vdots \\ u_n \end{pmatrix} = \boldsymbol{N}^T \boldsymbol{d}$$

$$\boldsymbol{B}^T = \begin{pmatrix} \frac{\partial N_1}{\partial x} & \frac{\partial N_2}{\partial x} & \cdots & \frac{\partial N_n}{\partial x} \\ \frac{\partial N_1}{\partial y} & \frac{\partial N_2}{\partial y} & \cdots & \frac{\partial N_n}{\partial y} \end{pmatrix} \quad \text{and} \quad \boldsymbol{C} = \begin{pmatrix} k_x & 0 \\ 0 & k_y \end{pmatrix}$$

$$u(c) = \begin{pmatrix} N_1(c) & N_2(c) & \cdots & N_n(c) \end{pmatrix} \begin{pmatrix} u_1 \\ u_2 \\ \vdots \\ u_n \end{pmatrix} = \boldsymbol{N}_c^T \boldsymbol{d}$$

These equations involve selecting a suitable element shape, developing an appropriate assumed solution over the element area and its boundary in terms of interpolation functions $\boldsymbol{N}$ and $\boldsymbol{N}_c$, computing derivatives of the interpolation functions with respect to x and y, and carrying out integrations over the element area and over the boundary of the element.

The mapping concept makes these computations possible for arbitrary shaped elements. However, it also adds an additional layer of complexity that must be dealt with when performing necessary computations. Procedures for carrying out these computations are explained in this section. It is assumed that the chosen quadrilateral element has been mapped to a 2 × 2 square element. Recall that the mapping between the actual element coordinates and the master element coordinates is written as follows:

$$x(s,t) = N_1 x_1 + N_2 x_2 + \cdots + N_n x_n$$
$$y(s,t) = N_1 y_1 + N_2 y_2 + \cdots + N_n y_n$$

where N_i are interpolation functions written over suitable master elements.

6.4.1 Assumed Solution

There is no simple formula for writing an assumed solution in terms of interpolation functions for an arbitrary shaped quadrilateral. One can use the basic method of starting with a suitable polynomial and evaluating the coefficients of this polynomial in terms of nodal degrees of freedom, as was done for a four-node rectangular element in Chapter 5. However, unless the shape is very simple, the procedure will not give useful closed-form expressions for the interpolation functions. The only possibility would be to employ the procedure in a numerical scheme that evaluates these interpolation functions and their derivatives for each element using the numerical values of the nodal coordinates. Beside being expensive computationally, the procedure is numerically unstable and hence not practical for general applications.

The practical approach is to write the assumed solution in terms of the mapped element coordinates s, t. Since the mapped element is a 2 × 2 square, with the origin at the center, the interpolation functions are exactly the same as those used for mapping. As examples, the assumed solutions for four- and eight-node elements are as follows:

Assumed solution for a four-node element (Figure 6.27):

$$u(s,t) = N_1 u_1 + N_2 u_2 + N_3 u_3 + N_4 u_4 \equiv \boldsymbol{N}^T \boldsymbol{d}$$

$$\boldsymbol{N} = \begin{pmatrix} \frac{1}{4}(-1+s)(-1+t) \\ -\frac{1}{4}(1+s)(-1+t) \\ \frac{1}{4}(1+s)(1+t) \\ -\frac{1}{4}(-1+s)(1+t) \end{pmatrix}$$

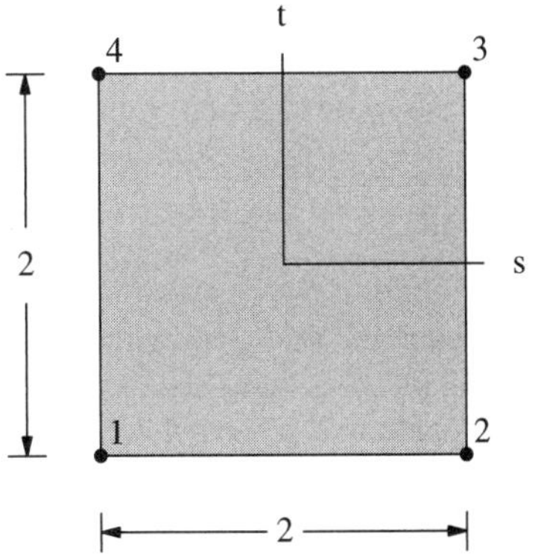

Figure 6.27.

Assumed solution for an eight-node element (Figure 6.28):

$$u(s,t) = N_1 u_1 + N_2 u_2 + \cdots + N_8 u_8 \equiv \boldsymbol{N}^T \boldsymbol{d}$$

$$\boldsymbol{N} = \begin{pmatrix} -\frac{1}{4}(-1+s)(-1+t)(1+s+t) \\ \frac{1}{2}(-1+s^2)(-1+t) \\ \frac{1}{4}(-1+t)(1-s^2+t+st) \\ -\frac{1}{2}(1+s)(-1+t^2) \\ \frac{1}{4}(1+s)(1+t)(-1+s+t) \\ -\frac{1}{2}(-1+s^2)(1+t) \\ \frac{1}{4}(-1+s)(1+s-t)(1+t) \\ \frac{1}{2}(-1+s)(-1+t^2) \end{pmatrix}$$

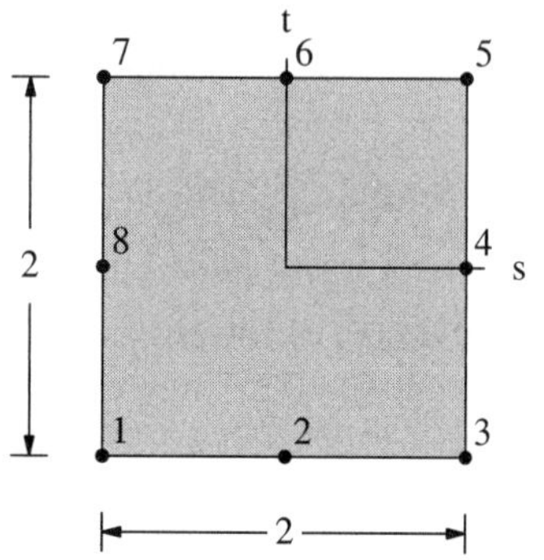

Figure 6.28.

The solutions along a boundary of an element can be written simply by setting the applicable s or t to ± 1. For example, for an eight-node element along side 1 ($t = -1$), the solution is as follows:

$$u(s) = N_1(s,-1)u_1 + N_2(s,-1)u_2 + N_3(s,-1)u_3 + 0u_4 + \cdots + 0u_8$$

or

$$u(s) = \tfrac{1}{2}(s-1)su_1 + (1-s^2)u_2 + \tfrac{1}{2}(s^2+s)u_3 + 0u_4 + \cdots + 0u_8 \equiv \boldsymbol{N}_c^T \boldsymbol{d}$$

Thus for side 1 of an eight-node element,

$$\boldsymbol{N}_c^T = \left(\tfrac{1}{2}(s-1)s \quad 1-s^2 \quad \tfrac{1}{2}(s^2+s) \quad 0 \quad 0 \quad 0 \quad 0 \quad 0 \right)$$

6.4.2 Derivatives of the Assumed Solution

The interpolation functions are written over the master element in terms of s, t. However the element equations need x and y derivatives of these interpolation functions. Using the mapping $x(s,t)$ and $y(s,t)$ and employing the chain rule of differentiation, the derivatives of the ith interpolation can be written as follows:

$$\frac{\partial N_i}{\partial s} = \frac{\partial N_i}{\partial x}\frac{\partial x}{\partial s} + \frac{\partial N_i}{\partial y}\frac{\partial y}{\partial s}$$

$$\frac{\partial N_i}{\partial t} = \frac{\partial N_i}{\partial x}\frac{\partial x}{\partial t} + \frac{\partial N_i}{\partial y}\frac{\partial y}{\partial t}$$

Writing the two equations together in a matrix form, we have

$$\begin{pmatrix} \frac{\partial N_i}{\partial s} \\ \frac{\partial N_i}{\partial t} \end{pmatrix} = \begin{pmatrix} \frac{\partial x}{\partial s} & \frac{\partial y}{\partial s} \\ \frac{\partial x}{\partial t} & \frac{\partial y}{\partial t} \end{pmatrix} \begin{pmatrix} \frac{\partial N_i}{\partial x} \\ \frac{\partial N_i}{\partial y} \end{pmatrix}$$

Here we notice that the 2×2 matrix is the transpose of the Jacobian matrix defined earlier in Section 6.1:

$$\boldsymbol{J} = \begin{pmatrix} \frac{\partial x}{\partial s} & \frac{\partial x}{\partial t} \\ \frac{\partial y}{\partial s} & \frac{\partial y}{\partial t} \end{pmatrix} \equiv \begin{pmatrix} J_{11} & J_{12} \\ J_{21} & J_{22} \end{pmatrix}$$

The x and y derivatives of the ith interpolation function can thus be computed from its s and t derivatives and the inverse of the $\boldsymbol{J}^T$ matrix as follows:

$$\begin{pmatrix} \frac{\partial N_i}{\partial x} \\ \frac{\partial N_i}{\partial y} \end{pmatrix} = \boldsymbol{J}^{-T} \begin{pmatrix} \frac{\partial N_i}{\partial s} \\ \frac{\partial N_i}{\partial t} \end{pmatrix} = \frac{1}{\det \boldsymbol{J}} \begin{pmatrix} \frac{\partial y}{\partial t} & -\frac{\partial y}{\partial s} \\ -\frac{\partial x}{\partial t} & \frac{\partial x}{\partial s} \end{pmatrix} \begin{pmatrix} \frac{\partial N_i}{\partial s} \\ \frac{\partial N_i}{\partial t} \end{pmatrix} \equiv \frac{1}{\det \boldsymbol{J}} \begin{pmatrix} J_{22} & -J_{21} \\ -J_{12} & J_{11} \end{pmatrix} \begin{pmatrix} \frac{\partial N_i}{\partial s} \\ \frac{\partial N_i}{\partial t} \end{pmatrix}$$

where $\det \boldsymbol{J}$ is the determinant of the Jacobian matrix,

$$\det \boldsymbol{J} = \frac{\partial x}{\partial s}\frac{\partial y}{\partial t} - \frac{\partial x}{\partial t}\frac{\partial y}{\partial s}$$

Separating the x and y derivatives, we have

$$\frac{\partial N_i}{\partial x} = \frac{1}{\det \boldsymbol{J}} \left(J_{22}\frac{\partial N_i}{\partial s} - J_{21}\frac{\partial N_i}{\partial t} \right)$$

$$\frac{\partial N_i}{\partial y} = \frac{1}{\det \boldsymbol{J}} \left(-J_{12}\frac{\partial N_i}{\partial s} + J_{11}\frac{\partial N_i}{\partial t} \right)$$

The complete vectors of derivatives of all interpolation functions $\boldsymbol{B}_x$ and $\boldsymbol{B}_y$ can be written as follows:

$$\boldsymbol{B}_x = \begin{pmatrix} \frac{\partial N_1}{\partial x} \\ \frac{\partial N_2}{\partial x} \\ \vdots \end{pmatrix} = \frac{J_{22}}{\det \boldsymbol{J}} \begin{pmatrix} \frac{\partial N_1}{\partial s} \\ \frac{\partial N_2}{\partial s} \\ \vdots \end{pmatrix} - \frac{J_{21}}{\det \boldsymbol{J}} \begin{pmatrix} \frac{\partial N_1}{\partial t} \\ \frac{\partial N_2}{\partial t} \\ \vdots \end{pmatrix}$$

$$\boldsymbol{B}_y = \begin{pmatrix} \frac{\partial N_1}{\partial y} \\ \frac{\partial N_2}{\partial y} \\ \vdots \end{pmatrix} = -\frac{J_{12}}{\det \boldsymbol{J}} \begin{pmatrix} \frac{\partial N_1}{\partial s} \\ \frac{\partial N_2}{\partial s} \\ \vdots \end{pmatrix} + \frac{J_{11}}{\det \boldsymbol{J}} \begin{pmatrix} \frac{\partial N_1}{\partial t} \\ \frac{\partial N_2}{\partial t} \\ \vdots \end{pmatrix}$$

The matrix of derivatives of the interpolation functions $\boldsymbol{B}$ can be written as follows:

$$\boldsymbol{B}^T = \begin{pmatrix} \frac{\partial N_1}{\partial x} & \frac{\partial N_2}{\partial x} & \cdots & \frac{\partial N_n}{\partial x} \\ \frac{\partial N_1}{\partial y} & \frac{\partial N_2}{\partial y} & \cdots & \frac{\partial N_n}{\partial y} \end{pmatrix}$$

$$\boldsymbol{B}^T = \frac{1}{\det \boldsymbol{J}} \begin{pmatrix} J_{22}\frac{\partial N_1}{\partial s} - J_{21}\frac{\partial N_1}{\partial t} & J_{22}\frac{\partial N_2}{\partial s} - J_{21}\frac{\partial N_2}{\partial t} & \cdots & J_{22}\frac{\partial N_n}{\partial s} - J_{21}\frac{\partial N_n}{\partial t} \\ -J_{12}\frac{\partial N_1}{\partial s} + J_{11}\frac{\partial N_1}{\partial t} & -J_{12}\frac{\partial N_2}{\partial s} + J_{11}\frac{\partial N_2}{\partial t} & \cdots & -J_{12}\frac{\partial N_n}{\partial s} + J_{11}\frac{\partial N_n}{\partial t} \end{pmatrix}$$

It should be noted that the derivatives are being computed with respect to x and y but they are still expressed in terms of s and t. Using the inverse of the mapping, it is possible to express the derivatives in terms of x and y. However, it is not necessary because the integration will also be performed in terms of s and t. Furthermore, the explicit expression for the inverse mapping is possible only in simple mapping situations. In general, it is difficult to write explicit expressions for inverse mapping for quadrilaterals with curved sides.

Example 6.13 Rectangular Element The purpose of this example is to demonstrate that the derivatives computed using the mapping equations are correct. To do this, we consider a 2×1 rectangular element and evaluate $\partial f/\partial x$ and $\partial f/\partial y$, where $f(x, y) = x^3 + y^2$. Since $f(x, y)$ is an explicit function of x, y, we know that the partial derivatives should be

$$\frac{\partial f}{\partial x} = 3x^2; \qquad \frac{\partial f}{\partial y} = 2y$$

To demonstrate that we get the same derivatives using the equations derived in this section, we map the element to a 2×2 square as shown in Figure 6.29. Using the interpolation functions for the four-node master element and multiplying with the coordinates of the rectangular element, the mapping is as follows:

$$x(s, t) = \tfrac{1}{2}(s + 1)(1 - t) + \tfrac{1}{2}(s + 1)(t + 1) = s + 1$$
$$y(s, t) = \tfrac{1}{4}(1 - s)(t + 1) + \tfrac{1}{4}(s + 1)(t + 1) = \tfrac{1}{2}t + \tfrac{1}{2}$$

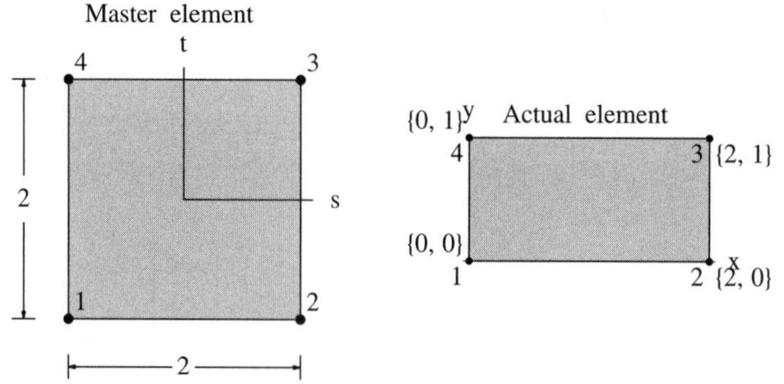

Figure 6.29. Four-node master element and actual rectangular element

The Jacobian matrix and the Jacobian of the mapping are as follows:

$$J = \begin{pmatrix} \frac{\partial x}{\partial s} & \frac{\partial x}{\partial t} \\ \frac{\partial y}{\partial s} & \frac{\partial y}{\partial t} \end{pmatrix} = \begin{pmatrix} 1 & 0 \\ 0 & \frac{1}{2} \end{pmatrix}; \qquad \det J = \tfrac{1}{2}$$

Using the mapping, the given function $f(x, y)$ can be written as follows:

$$f(x, y) = x^3 + y^2 \Longrightarrow f(s, t) = (1 + s)^3 + \left(\tfrac{1}{2} + \tfrac{t}{2}\right)^2$$

The x, y derivatives of $f(s, t)$ can now be computed as follows:

$$\begin{pmatrix} \frac{\partial f}{\partial x} \\ \frac{\partial f}{\partial y} \end{pmatrix} = \frac{1}{\det J}\begin{pmatrix} \frac{\partial y}{\partial t} & -\frac{\partial y}{\partial s} \\ -\frac{\partial x}{\partial t} & \frac{\partial x}{\partial s} \end{pmatrix}\begin{pmatrix} \frac{\partial f}{\partial s} \\ \frac{\partial f}{\partial t} \end{pmatrix} = \frac{1}{1/2}\begin{pmatrix} \frac{1}{2} & 0 \\ 0 & 1 \end{pmatrix}\begin{pmatrix} 3(1+s)^2 \\ (\frac{1}{2} + \frac{t}{2}) \end{pmatrix} = \begin{pmatrix} 3(1+s)^2 \\ 1+t \end{pmatrix}$$

Since the mapping is very simple, we can easily write its inverse, giving

$$s = x - 1; \qquad t = 2y - 1$$

Substituting these, we see that the computed derivatives are exactly what we expected:

$$\frac{\partial f}{\partial x} = 3(1 + s)^2 = 3(1 + x - 1)^2 = 3x^2$$
$$\frac{\partial f}{\partial y} = 1 + t = 1 + 2y - 1 = 2y$$

Example 6.14 Four-Node Quadrilateral Element For the four-node quadrilateral element shown in Figure 6.30, evaluate the $\boldsymbol{B}_x$ vector at node 3. The interpolation functions and their derivatives are as follows:

$$N^T = \left\{\frac{1}{4}(1-s)(1-t), \frac{1}{4}(s+1)(1-t), \frac{1}{4}(s+1)(t+1), \frac{1}{4}(1-s)(t+1)\right\}$$
$$\frac{\partial N^T}{\partial s} = \left\{\frac{t-1}{4}, \frac{1-t}{4}, \frac{t+1}{4}, \frac{1}{4}(-t-1)\right\}$$
$$\frac{\partial N^T}{\partial t} = \left\{\frac{s-1}{4}, \frac{1}{4}(-s-1), \frac{s+1}{4}, \frac{1-s}{4}\right\}$$

Multiplying these interpolation functions with the coordinates of the actual element, the mapping is as follows:

$$x(s, t) = \frac{ts}{2} + \frac{3s}{2} + \frac{t}{2} + \frac{3}{2}; \qquad y(s, t) = \frac{ts}{4} + \frac{s}{4} + \frac{3t}{4} + \frac{3}{4}$$

$$J = \begin{pmatrix} \frac{t}{2} + \frac{3}{2} & \frac{s}{2} + \frac{1}{2} \\ \frac{t}{4} + \frac{1}{4} & \frac{s}{4} + \frac{3}{4} \end{pmatrix}; \qquad \det J = \frac{s}{4} + \frac{t}{4} + 1$$

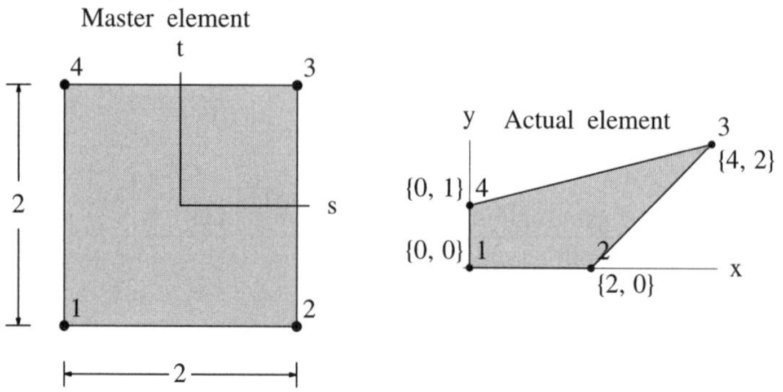

Figure 6.30. Four-node master element and actual element

The required $\boldsymbol{B}_x$ vector is as follows:

$$\boldsymbol{B}_x = \begin{pmatrix} \frac{\partial N_1}{\partial x} \\ \frac{\partial N_2}{\partial x} \\ \vdots \end{pmatrix} = \frac{1}{\det \boldsymbol{J}} J_{22} \begin{pmatrix} \frac{\partial N_1}{\partial s} \\ \frac{\partial N_2}{\partial s} \\ \vdots \end{pmatrix} - \frac{1}{\det \boldsymbol{J}} J_{21} \begin{pmatrix} \frac{\partial N_1}{\partial t} \\ \frac{\partial N_2}{\partial t} \\ \vdots \end{pmatrix}$$

All quantities needed to evaluate this matrix are available as functions of s, t. A symbolic evaluation will obviously be tedious. However, numerical evaluation at a given point is not too difficult. As an example we evaluate this vector at node 3 of the element. At this point $s = t = 1$; therefore we get the following numerical values:

$$\boldsymbol{J} = \begin{pmatrix} 2 & 1 \\ \frac{1}{2} & 1 \end{pmatrix}; \qquad \det \boldsymbol{J} = \tfrac{3}{2}$$

$$\boldsymbol{N}^T = (0 \quad 0 \quad 1 \quad 0)$$

$$\frac{\partial \boldsymbol{N}^T}{\partial s} = (0 \quad 0 \quad \tfrac{1}{2} \quad -\tfrac{1}{2})$$

$$\frac{\partial \boldsymbol{N}^T}{\partial t} = (0 \quad -\tfrac{1}{2} \quad \tfrac{1}{2} \quad 0)$$

The $\boldsymbol{B}_x$ vector at (1, 1) can now be evaluated as follows:

$$\boldsymbol{B}_x(1,1) = \begin{pmatrix} \frac{\partial N_1}{\partial x}(1,1) \\ \frac{\partial N_2}{\partial x}(1,1) \\ \frac{\partial N_3}{\partial x}(1,1) \\ \frac{\partial N_4}{\partial x}(1,1) \end{pmatrix} = \frac{1}{3/2} 1 \begin{pmatrix} 0 \\ 0 \\ \frac{1}{2} \\ -\frac{1}{2} \end{pmatrix} - \frac{1}{3/2} \frac{1}{2} \begin{pmatrix} 0 \\ -\frac{1}{2} \\ \frac{1}{2} \\ 0 \end{pmatrix} = \begin{pmatrix} 0 \\ \frac{1}{6} \\ \frac{1}{6} \\ -\frac{1}{3} \end{pmatrix}$$

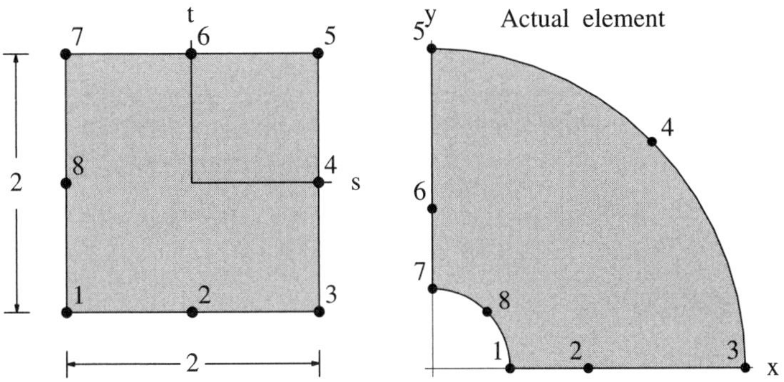

Figure 6.31. Eight-node master element and actual element

Example 6.15 Eight-Node Quadrilateral Element Evaluate the $\boldsymbol{B}_y$ vector at $s = \frac{1}{2}$ and $t = -\frac{1}{4}$ for the eight-node element. The master and the actual element are shown in Figure 6.31. The nodal point coordinates are as follows:

	x	y
1	2.	0.
2	4.	0.
3	8.	0.
4	5.65685	5.65685
5	0.	8.
6	0.	4.
7	0.	2.
8	1.41421	1.41421

The interpolation functions and their derivatives for the master element are as follows:

$$\begin{aligned} N^T = \Big\{ & -\tfrac{1}{4}(s-1)(t-1)(s+t+1), \tfrac{1}{2}(s^2-1)(t-1), \tfrac{1}{4}(t-1)(-s^2+ts+t+1), \\ & -\tfrac{1}{2}(s+1)(t^2-1), \tfrac{1}{4}(s+1)(t+1)(s+t-1), -\tfrac{1}{2}(s^2-1)(t+1), \\ & \tfrac{1}{4}(s-1)(s-t+1)(t+1), \tfrac{1}{2}(s-1)(t^2-1)\Big\} \end{aligned}$$

$$\begin{aligned} \frac{\partial N^T}{\partial s} = \Big\{ & -\tfrac{1}{4}(t-1)(2s+t), s(t-1), -\tfrac{1}{4}(2s-t)(t-1), \tfrac{1}{2}(1-t^2), \\ & \tfrac{1}{4}(t+1)(2s+t), -s(t+1), \tfrac{1}{4}(2s-t)(t+1), \tfrac{1}{2}(t^2-1)\Big\} \end{aligned}$$

$$\begin{aligned} \frac{\partial N^T}{\partial t} = \Big\{ & -\tfrac{1}{4}(s-1)(s+2t), \tfrac{1}{2}(s^2-1), -\tfrac{1}{4}(s+1)(s-2t), -(s+1)t, \\ & \tfrac{1}{4}(s+1)(s+2t), \tfrac{1}{2}(1-s^2), \tfrac{1}{4}(s-1)(s-2t), (s-1)t\Big\} \end{aligned}$$

Multiplying these interpolation functions with the coordinates of the actual element, the mapping is as follows:

$$x(s,t) = -0.5ts^2 + 0.5s^2 - 0.62132t^2s - 1.5ts + 2.12132s - 1.03553t^2 - 2.t + 3.03553$$

$$y(s,t) = 0.5ts^2 + 0.5s^2 - 0.62132t^2s + 1.5ts + 2.12132s - 1.03553t^2 + 2.t + 3.03553$$

$$J = \begin{pmatrix} -0.62132t^2 - 1.st - 1.5t + 1.s + 2.12132 & -0.5s^2 - 1.24264ts - 1.5s - 2.07107t - 2. \\ -0.62132t^2 + 1.st + 1.5t + 1.s + 2.12132 & 0.5s^2 - 1.24264ts + 1.5s - 2.07107t + 2. \end{pmatrix}$$

$$\det J = 1.s^3 + 1.86396t^2s^2 + 5.12132s^2 + 6.0061t^2s + 10.364s + 3.72792t^2 + 8.48528$$

At the given point $s = \frac{1}{2}$ and $t = -\frac{1}{4}$, the numerical values are as follows:

$$J = \begin{pmatrix} 3.08249 & -2.2019 \\ 2.08249 & 3.5481 \end{pmatrix}; \qquad \det J = 15.5224$$

$$N^T = (-0.195313 \quad 0.46875 \quad -0.117188 \quad 0.703125 \quad -0.210938 \quad 0.28125 \quad -0.164063 \quad 0.234375)$$

$$\frac{\partial N^T}{\partial s} = (0.234375 \quad -0.625 \quad 0.390625 \quad 0.46875 \quad 0.140625 \quad -0.375 \quad 0.234375 \quad -0.46875)$$

$$\frac{\partial N^T}{\partial t} = (0. \quad -0.375 \quad -0.375 \quad 0.375 \quad 0. \quad 0.375 \quad -0.125 \quad 0.125)$$

The B_y vector at the given point can now be evaluated as follows:

$$B_y = \begin{pmatrix} \frac{\partial N_1}{\partial y} \\ \frac{\partial N_2}{\partial y} \\ \vdots \end{pmatrix} = -\frac{1}{\det J} J_{12} \begin{pmatrix} \frac{\partial N_1}{\partial s} \\ \frac{\partial N_2}{\partial s} \\ \vdots \end{pmatrix} + \frac{1}{\det J} J_{11} \begin{pmatrix} \frac{\partial N_1}{\partial t} \\ \frac{\partial N_2}{\partial t} \\ \vdots \end{pmatrix}$$

$$B_y\left(\tfrac{1}{2}, -\tfrac{1}{4}\right) = -\begin{pmatrix} -0.0332469 \\ 0.0886583 \\ -0.0554114 \\ -0.0664937 \\ -0.0199481 \\ 0.053195 \\ -0.0332469 \\ 0.0664937 \end{pmatrix} + \begin{pmatrix} 0 \\ -0.0744687 \\ -0.0744687 \\ 0.0744687 \\ 0 \\ 0.0744687 \\ -0.0248229 \\ 0.0248229 \end{pmatrix} = \begin{pmatrix} 0.0332469 \\ -0.163127 \\ -0.0190573 \\ 0.140962 \\ 0.0199481 \\ 0.0212737 \\ 0.00842396 \\ -0.0416708 \end{pmatrix}$$

6.4.3 Evaluation of Area Integrals

In developing element equations the matrices k_k, k_p and vector r_q all involve integrations over the area of the element:

$$k_k = \iint_A BCB^T \, dA; \qquad k_p = -\iint_A pNN^T \, dA; \qquad r_q = \iint_A qN \, dA$$

The arbitrary element area in the x, y coordinate is already mapped into a 2×2 square, and therefore with the change of variables these integrals can be written as follows:

$$k_k = \int_{-1}^{1}\int_{-1}^{1} \boldsymbol{BCB}^T \det \boldsymbol{J}\, ds\, dt$$

$$k_p = -\int_{-1}^{1}\int_{-1}^{1} p\boldsymbol{NN}^T \det \boldsymbol{J}\, ds\, dt$$

$$r_q = \int_{-1}^{1}\int_{-1}^{1} q\boldsymbol{N} \det \boldsymbol{J}\, ds\, dt$$

where $\det \boldsymbol{J}$ is the Jacobian of the mapping. The interpolation functions and their derivatives are already expressed in terms of s, t and are thus in correct form for this integration. The given functions k_x, k_y, p, and q are in terms of x, y coordinates and must be converted into the s, t form by using the mapping functions. Frequently it is assumed that these quantities are constant over an element in which case they can simply be taken out of the integral sign.

All integrands are expressed in terms of s, t, and therefore the above element matrices can be established by evaluating integrals of the following form:

$$\int_{-1}^{1}\int_{-1}^{1} f(s,t)\, ds\, dt$$

A simple closed-form integration is, however, impossible because of the $\det \boldsymbol{J}$ term in the integrals, which usually is a messy expression involving nodal coordinates. We must resort to numerical integration for evaluation of these matrices. Using an $m \times n$ Gauss quadrature formula, the element matrices are then evaluated numerically as follows:

$$k_k = \int_{-1}^{1}\int_{-1}^{1} \boldsymbol{BCB}^T \det \boldsymbol{J}\, ds\, dt = \sum_{i=1}^{m}\sum_{j=1}^{n} w_i w_j \boldsymbol{B}(s_i,t_j)\boldsymbol{C}(s_i,t_j)\boldsymbol{B}^T(s_i,t_j) \det \boldsymbol{J}(s_i,t_j)$$

$$k_p = -\int_{-1}^{1}\int_{-1}^{1} p\boldsymbol{NN}^T \det \boldsymbol{J}\, ds\, dt = -\sum_{i=1}^{m}\sum_{j=1}^{n} w_i w_j p(s_i,t_j)\boldsymbol{N}(s_i,t_j)\boldsymbol{N}^T(s_i,t_j) \det \boldsymbol{J}(s_i,t_j)$$

$$r_q = \int_{-1}^{1}\int_{-1}^{1} q\boldsymbol{N} \det \boldsymbol{J}\, ds\, dt = \sum_{i=1}^{m}\sum_{j=1}^{n} w_i w_j q(s_i,t_j)\boldsymbol{N}(s_i,t_j) \det \boldsymbol{J}(s_i,t_j)$$

where (s_i, t_j) are the Gauss points and w_i and w_j are the corresponding weights.

The number of integration points used in evaluating these matrices depends on the order of terms present in the integrands. In most practical applications we assume k_x, k_y, p, and q are constant over an element. The terms in $\boldsymbol{N}$ are all known. The $\boldsymbol{B}$ matrix involves not only known derivatives of $\boldsymbol{N}$ with respect to s and t but also terms from the Jacobian matrix $\boldsymbol{J}$ and its determinant. Thus the complexity of the terms in the integrand depends on the mapping. If the actual elements are rectangles or squares, then $\boldsymbol{J}$ involves constant terms, and it is fairly easy to determine the number of points necessary for evaluating these integrals exactly. However, for a general quadrilateral, $\boldsymbol{J}$ is not constant, and thus the integrands in these matrices are not simple polynomials. Therefore, in general, integration over mapped elements is approximate and does introduce another source of error in the finite element results. As long as the elements are not too distorted, the following integration rules give reasonable results:

Four-node quadrilaterals: use 2×2 integration (4 points)
Eight-node quadrilaterals: use 3×3 integration (9 points)

Example 6.16 Rectangular Element The purpose of this example is to demonstrate that the integrals computed using the mapping equations are correct. To do this, we consider a 2×1 rectangular element and evaluate the following integral:

$$\iint_A f(x, y)\, dx\, dy$$

where $f(x, y) = 3x^2 + 2y$.

The master and the actual element are shown in Figure 6.32. Since $f(x, y)$ is an explicit function of x, y and the area is rectangular, the integral can directly be evaluated as follows:

$$\int_{y=0}^{1} \int_{x=0}^{2} (3x^2 + 2y)\, dx\, dy = \int_0^1 [x^3 + 2xy]_{x=0}^{2}\, dy = \int_0^1 (8 + 4y)\, dy = [8y + 2y^2]_{y=0}^{1} = 10$$

To demonstrate that we get the same integrals using the equations derived in this section, we map the element to a 2×2 square. Using the interpolation functions for the four-node master element and multiplying with the coordinates of the rectangular element, the mapping is as follows:

$$x(s, t) = s + 1; \qquad y(s, t) = \frac{t}{2} + \frac{1}{2}$$

$$\boldsymbol{J} = \begin{pmatrix} \frac{\partial x}{\partial s} & \frac{\partial x}{\partial t} \\ \frac{\partial y}{\partial s} & \frac{\partial y}{\partial t} \end{pmatrix} = \begin{pmatrix} 1 & 0 \\ 0 & \frac{1}{2} \end{pmatrix}; \qquad \det \boldsymbol{J} = \frac{1}{2}$$

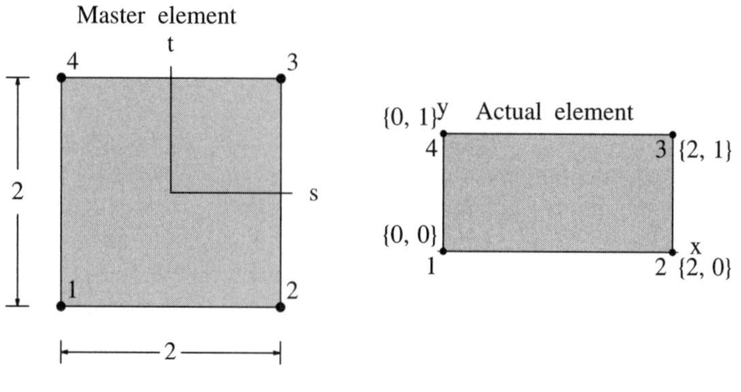

Figure 6.32. Four-node master element and actual rectangular element

The given function $f(x, y)$ can be written as follows:

$$f(x, y) = 3x^2 + 2y \Longrightarrow f(s, t) = 3(1 + s)^2 + 2\left(\tfrac{1}{2} + \tfrac{t}{2}\right)$$

The given integral can now be computed as follows:

$$\iint_A f(x, y)\, dx\, dy = \int_{t=-1}^{1} \int_{s=-1}^{1} f(s, t) \det \boldsymbol{J}\, ds\, dt$$

$$= \tfrac{1}{2} \int_{t=-1}^{1} \int_{s=-1}^{1} (3(1 + s)^2 + 1 + t)\, ds\, dt$$

$$= \tfrac{1}{2} \int_{t=-1}^{1} [(1 + s)^3 + s(1 + t)]_{s=-1}^{1}\, dt = \tfrac{1}{2} \int_{t=-1}^{1} (10 + 2t)\, dt = 10$$

This value agrees with the one computed directly.

Example 6.17 Four-Node Quadrilateral Element Evaluate matrix $\boldsymbol{k}_k$ for the four-node quadrilateral element shown in Figure 6.33. Assume $k_x = 2$ and $k_y = 1$ over the element. The interpolation functions and their derivatives are as follows:

$$\boldsymbol{N}^T = \left\{\tfrac{1}{4}(s - 1)(t - 1), -\tfrac{1}{4}(s + 1)(t - 1), \tfrac{1}{4}(s + 1)(t + 1), -\tfrac{1}{4}(s - 1)(t + 1)\right\}$$

$$\frac{\partial \boldsymbol{N}^T}{\partial s} = \left\{\frac{t - 1}{4}, \frac{1 - t}{4}, \frac{t + 1}{4}, \frac{1}{4}(-t - 1)\right\}$$

$$\frac{\partial \boldsymbol{N}^T}{\partial t} = \left\{\frac{s - 1}{4}, \frac{1}{4}(-s - 1), \frac{s + 1}{4}, \frac{1 - s}{4}\right\}$$

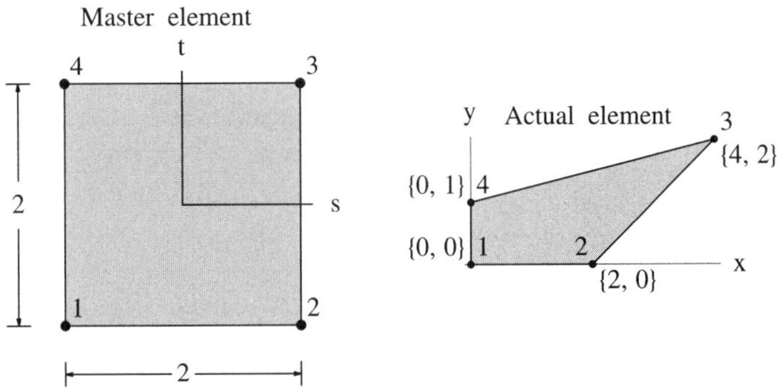

Figure 6.33. Four-node master element and actual element

Mapping to the master element,

$$x(s,t) = \frac{ts}{2} + \frac{3s}{2} + \frac{t}{2} + \frac{3}{2}; \qquad y(s,t) = \frac{ts}{4} + \frac{s}{4} + \frac{3t}{4} + \frac{3}{4}$$

$$\boldsymbol{J} = \begin{pmatrix} \frac{t}{2} + \frac{3}{2} & \frac{s}{2} + \frac{1}{2} \\ \frac{t}{4} + \frac{1}{4} & \frac{s}{4} + \frac{3}{4} \end{pmatrix}; \qquad \det \boldsymbol{J} = \frac{s}{4} + \frac{t}{4} + 1$$

The matrix $\boldsymbol{k}_k$ is to be evaluated as follows:

$$\boldsymbol{k}_k = \int_{-1}^{1}\int_{-1}^{1} \boldsymbol{BCB}^T \det \boldsymbol{J}\, ds\, dt = \sum_{i=1}^{m}\sum_{j=1}^{n} w_i w_j \boldsymbol{B}(s_i,t_j)\boldsymbol{C}(s_i,t_j)\boldsymbol{B}^T(s_i,t_j) \det \boldsymbol{J}(s_i,t_j)$$

Using 2×2 Gauss quadrature we need to evaluate the integrand at the four Gauss points and then sum the contributions from each Gauss point to get the integral:

Gauss Quadrature Points and Weights

	Point		
	s	t	Weight
1	−0.57735	−0.57735	1.
2	−0.57735	0.57735	1.
3	0.57735	−0.57735	1.
4	0.57735	0.57735	1.

Computation of element matrices at $\{s \to -0.57735, t \to -0.57735\}$ with weight = 1:

$$\boldsymbol{J} = \begin{pmatrix} 1.21132 & 0.211325 \\ 0.105662 & 0.605662 \end{pmatrix}; \qquad \det \boldsymbol{J} = 0.711325$$

$$\boldsymbol{N}^T = (0.622008 \quad 0.166667 \quad 0.0446582 \quad 0.166667)$$

$$\frac{\partial \boldsymbol{N}^T}{\partial s} = (-0.394338 \quad 0.394338 \quad 0.105662 \quad -0.105662)$$

$$\frac{\partial \boldsymbol{N}^T}{\partial t} = (-0.394338 \quad -0.105662 \quad 0.105662 \quad 0.394338)$$

$$\boldsymbol{B}^T = \begin{pmatrix} -0.277185 & 0.351457 & 0.0742716 & -0.148543 \\ -0.554371 & -0.297086 & 0.148543 & 0.702914 \end{pmatrix}$$

$$\boldsymbol{C} = \begin{pmatrix} 2 & 0 \\ 0 & 1 \end{pmatrix}$$

$$\boldsymbol{k}_k = \begin{pmatrix} 0.327914 & -0.0214404 & -0.0878642 & -0.218609 \\ -0.0214404 & 0.23851 & 0.00574493 & -0.222815 \\ -0.0878642 & 0.00574493 & 0.0235431 & 0.0585761 \\ -0.218609 & -0.222815 & 0.0585761 & 0.382848 \end{pmatrix}$$

Computation of element matrices at $\{s \to -0.57735, t \to 0.57735\}$ with weight = 1.,

$$\boldsymbol{J} = \begin{pmatrix} 1.78868 & 0.211325 \\ 0.394338 & 0.605662 \end{pmatrix}; \qquad \det \boldsymbol{J} = 1.$$

$$\boldsymbol{N}^T = (0.166667 \quad 0.0446582 \quad 0.166667 \quad 0.622008)$$

$$\frac{\partial \boldsymbol{N}^T}{\partial s} = (-0.105662 \quad 0.105662 \quad 0.394338 \quad -0.394338)$$

$$\frac{\partial \boldsymbol{N}^T}{\partial t} = (-0.394338 \quad -0.105662 \quad 0.105662 \quad 0.394338)$$

$$\boldsymbol{B}^T = \begin{pmatrix} 0.0915064 & 0.105662 & 0.197169 & -0.394338 \\ -0.683013 & -0.211325 & 0.105662 & 0.788675 \end{pmatrix}$$

$$\boldsymbol{C} = \begin{pmatrix} 2 & 0 \\ 0 & 1 \end{pmatrix}$$

$$\boldsymbol{k}_k = \begin{pmatrix} 0.483253 & 0.163675 & -0.0360844 & -0.610844 \\ 0.163675 & 0.0669873 & 0.0193376 & -0.25 \\ -0.0360844 & 0.0193376 & 0.0889156 & -0.0721688 \\ -0.610844 & -0.25 & -0.0721688 & 0.933013 \end{pmatrix}$$

Computation of element matrices at $\{s \to 0.57735, t \to -0.57735\}$ with weight = 1.,

$$\boldsymbol{J} = \begin{pmatrix} 1.21132 & 0.788675 \\ 0.105662 & 0.894338 \end{pmatrix}; \qquad \det \boldsymbol{J} = 1.$$

$$\boldsymbol{N}^T = (0.166667 \quad 0.622008 \quad 0.166667 \quad 0.0446582)$$

$$\frac{\partial \boldsymbol{N}^T}{\partial s} = (-0.394338 \quad 0.394338 \quad 0.105662 \quad -0.105662)$$

$$\frac{\partial \boldsymbol{N}^T}{\partial t} = (-0.105662 \quad -0.394338 \quad 0.394338 \quad 0.105662)$$

$$\boldsymbol{B}^T = \begin{pmatrix} -0.341506 & 0.394338 & 0.0528312 & -0.105662 \\ 0.183013 & -0.788675 & 0.394338 & 0.211325 \end{pmatrix}$$

$$\boldsymbol{C} = \begin{pmatrix} 2 & 0 \\ 0 & 1 \end{pmatrix}$$

$$\boldsymbol{k}_k = \begin{pmatrix} 0.266747 & -0.413675 & 0.0360844 & 0.110844 \\ -0.413675 & 0.933013 & -0.269338 & -0.25 \\ 0.0360844 & -0.269338 & 0.161084 & 0.0721688 \\ 0.110844 & -0.25 & 0.0721688 & 0.0669873 \end{pmatrix}$$

Computation of element matrices at $\{s \to 0.57735, t \to 0.57735\}$ with weight = 1.,

$$J = \begin{pmatrix} 1.78868 & 0.788675 \\ 0.394338 & 0.894338 \end{pmatrix}; \qquad \det J = 1.28868$$

$$N^T = (\,0.0446582 \quad 0.166667 \quad 0.622008 \quad 0.166667\,)$$

$$\frac{\partial N^T}{\partial s} = (\,-0.105662 \quad 0.105662 \quad 0.394338 \quad -0.394338\,)$$

$$\frac{\partial N^T}{\partial t} = (\,-0.105662 \quad -0.394338 \quad 0.394338 \quad 0.105662\,)$$

$$B^T = \begin{pmatrix} -0.0409965 & 0.193998 & 0.153001 & -0.306002 \\ -0.0819931 & -0.612005 & 0.306002 & 0.387995 \end{pmatrix}$$

$$C = \begin{pmatrix} 2 & 0 \\ 0 & 1 \end{pmatrix}$$

$$k_k = \begin{pmatrix} 0.0129954 & 0.0441676 & -0.0484994 & -0.00866359 \\ 0.0441676 & 0.579672 & -0.164836 & -0.459003 \\ -0.0484994 & -0.164836 & 0.181002 & 0.0323329 \\ -0.00866359 & -0.459003 & 0.0323329 & 0.435334 \end{pmatrix}$$

Summing contributions from all points, we get

$$k_k = \begin{pmatrix} 1.09091 & -0.227273 & -0.136364 & -0.727273 \\ -0.227273 & 1.81818 & -0.409091 & -1.18182 \\ -0.136364 & -0.409091 & 0.454545 & 0.0909091 \\ -0.727273 & -1.18182 & 0.0909091 & 1.81818 \end{pmatrix}$$

Example 6.18 Eight-Node Quadrilateral Element Evaluate matrix k_p for the eight-node element shown in Figure 6.34. Assume $p = 5$ over the element.

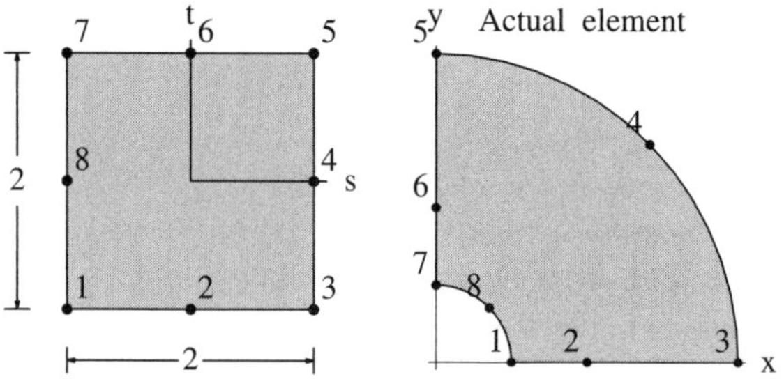

Figure 6.34. Eight-node master element and actual element

	x	y
1	2.	0.
2	4.	0.
3	8.	0.
4	5.65685	5.65685
5	0.	8.
6	0.	4.
7	0.	2.
8	1.41421	1.41421

Mapping to the master element,

$$x(s,t) = -0.5ts^2 + 0.5s^2 - 0.62132t^2 s - 1.5ts + 2.12132s - 1.03553t^2 - 2.t + 3.03553$$

$$y(s,t) = 0.5ts^2 + 0.5s^2 - 0.62132t^2 s + 1.5ts + 2.12132s - 1.03553t^2 + 2.t + 3.03553$$

$$\boldsymbol{J} = \begin{pmatrix} -0.62132t^2 - 1.st - 1.5t + 1.s + 2.12132 & -0.5s^2 - 1.24264ts - 1.5s - 2.07107t - 2. \\ -0.62132t^2 + 1.st + 1.5t + 1.s + 2.12132 & 0.5s^2 - 1.24264ts + 1.5s - 2.07107t + 2. \end{pmatrix}$$

$$\det \boldsymbol{J} = 1.s^3 + 1.86396t^2 s^2 + 5.12132s^2 + 6.0061t^2 s + 10.364s + 3.72792t^2 + 8.48528$$

The matrix $\boldsymbol{k}_p$ is to be evaluated as follows:

$$\boldsymbol{k}_p = -\int_{-1}^{1}\int_{-1}^{1} p\boldsymbol{N}\boldsymbol{N}^T \det\boldsymbol{J}\, ds\, dt = -\sum_{i=1}^{m}\sum_{j=1}^{n} w_i w_j p(s_i, t_j)\boldsymbol{N}(s_i, t_j)\boldsymbol{N}^T(s_i, t_j)\det\boldsymbol{J}(s_i, t_j)$$

Using 3 × 3 Gauss quadrature, we need to evaluate the integrand at the nine Gauss points and then sum the contributions from each Gauss point to get the integral:

	Point		
	s	t	Weight = $w_i \times w_j$
1	−0.774597	−0.774597	0.308642
2	−0.774597	0	0.493827
3	−0.774597	0.774597	0.308642
4	0	−0.774597	0.493827
5	0	0	0.790123
6	0	0.774597	0.493827
7	0.774597	−0.774597	0.308642
8	0.774597	0	0.493827
9	0.774597	0.774597	0.308642

Computation of the $\boldsymbol{k}_p$ matrix at point 1 is as follows:

Point = $\{s \to -0.774597, t \to -0.774597\}$; Weight = 0.308642

$$\boldsymbol{J} = \begin{pmatrix} 1.53583 & -0.279447 \\ 0.412036 & 1.99676 \end{pmatrix}; \det\boldsymbol{J} = 3.18182$$

$$N^T = (0.432379 \quad 0.354919 \quad -0.1 \quad 0.0450807 \quad -0.032379 \quad 0.0450807 \quad -0.1 \quad 0.354919)$$

$$k_p = \begin{pmatrix} 0.917974 & 0.753521 & -0.212308 & 0.0957097 & -0.0687431 & 0.0957097 & -0.212308 & 0.753521 \\ 0.753521 & 0.618529 & -0.174273 & 0.0785635 & -0.0564279 & 0.0785635 & -0.174273 & 0.618529 \\ -0.212308 & -0.174273 & 0.0491022 & -0.0221356 & 0.0158988 & -0.0221356 & 0.0491022 & -0.174273 \\ 0.0957097 & 0.0785635 & -0.0221356 & 0.00997888 & -0.00716729 & 0.00997888 & -0.0221356 & 0.0785635 \\ -0.0687431 & -0.0564279 & 0.0158988 & -0.00716729 & 0.00514787 & -0.00716729 & 0.0158988 & -0.0564279 \\ 0.0957097 & 0.0785635 & -0.0221356 & 0.00997888 & -0.00716729 & 0.00997888 & -0.0221356 & 0.0785635 \\ -0.212308 & -0.174273 & 0.0491022 & -0.0221356 & 0.0158988 & -0.0221356 & 0.0491022 & -0.174273 \\ 0.753521 & 0.618529 & -0.174273 & 0.0785635 & -0.0564279 & 0.0785635 & -0.174273 & 0.618529 \end{pmatrix}$$

Computation of the k_p matrix at point 2 is as follows:

Point = $\{s \to -0.774597, t \to 0\}$; Weight = 0.493827

$$J = \begin{pmatrix} 1.34672 & -1.1381 \\ 1.34672 & 1.1381 \end{pmatrix}; \ \det J = 3.06543$$

$$N^T = (-0.1 \quad 0.2 \quad -0.1 \quad 0.112702 \quad -0.1 \quad 0.2 \quad -0.1 \quad 0.887298)$$

$$k_p = \begin{pmatrix} 0.0756895 & -0.151379 & 0.0756895 & -0.0853034 & 0.0756895 & -0.151379 & 0.0756895 & -0.671592 \\ -0.151379 & 0.302758 & -0.151379 & 0.170607 & -0.151379 & 0.302758 & -0.151379 & 1.34318 \\ 0.0756895 & -0.151379 & 0.0756895 & -0.0853034 & 0.0756895 & -0.151379 & 0.0756895 & -0.671592 \\ -0.0853034 & 0.170607 & -0.0853034 & 0.0961383 & -0.0853034 & 0.170607 & -0.0853034 & 0.756895 \\ 0.0756895 & -0.151379 & 0.0756895 & -0.0853034 & 0.0756895 & -0.151379 & 0.0756895 & -0.671592 \\ -0.151379 & 0.302758 & -0.151379 & 0.170607 & -0.151379 & 0.302758 & -0.151379 & 1.34318 \\ 0.0756895 & -0.151379 & 0.0756895 & -0.0853034 & 0.0756895 & -0.151379 & 0.0756895 & -0.671592 \\ -0.671592 & 1.34318 & -0.671592 & 0.756895 & -0.671592 & 1.34318 & -0.671592 & 5.95902 \end{pmatrix}$$

After computing and summing contributions from all nine points, the k_p matrix for the element is as follows:

$$k_p = \begin{pmatrix} 4.60149 & -8.71619 & 1.40948 & -11.5208 & 3.01679 & -8.69358 & 3.0281 & -6.13668 \\ -8.71619 & 37.5578 & -2.2417 & 28.3616 & -8.69358 & 17.3872 & -8.69358 & 17.5481 \\ 1.40948 & -2.2417 & 11.076 & -5.06893 & -0.209149 & -8.69358 & 3.01679 & -6.11406 \\ -11.5208 & 28.3616 & -5.06893 & 59.6034 & -5.06893 & 28.3616 & -11.5208 & 17.6349 \\ 3.01679 & -8.69358 & -0.209149 & -5.06893 & 11.076 & -2.2417 & 1.40948 & -6.11406 \\ -8.69358 & 17.3872 & -8.69358 & 28.3616 & -2.2417 & 37.5578 & -8.71619 & 17.5481 \\ 3.0281 & -8.69358 & 3.01679 & -11.5208 & 1.40948 & -8.71619 & 4.60149 & -6.13668 \\ -6.13668 & 17.5481 & -6.11406 & 17.6349 & -6.11406 & 17.5481 & -6.13668 & 18.4849 \end{pmatrix}$$

6.4.4 Evaluation of Boundary Integrals

Elements involving a boundary where a nonzero natural boundary condition is specified require computation of the k_α matrix and the r_β vector:

$$k_\alpha = -\int_{C_n} \alpha N_c N_c^T \, dc; \qquad r_\beta = \int_{C_n} \beta N_c \, dc$$

where c is a coordinate along the boundary and the integration is along the boundary of the element. In general, the element boundary is an arbitrary curve in the x, y coordinate

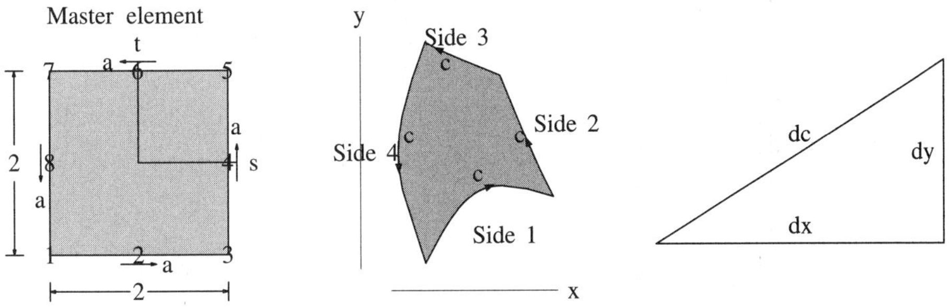

Figure 6.35. Boundary coordinates on the master and the actual element

system. Once mapped, each boundary curve is either a horizontal or a vertical line representing a side of the master square element. The boundary coordinate c for each side is mapped to a coordinate $-1 \leq a \leq 1$ for each side, as shown in Figure 6.35. The mapping for each side $x(a)$ and $y(a)$ can be obtained from the element mapping as follows:

For side 1:	$a = s;$	$t = -1$
For side 2:	$s = 1;$	$a = t$
For side 3:	$a = -s;$	$t = 1$
For side 4:	$s = -1;$	$a = -t$

To write interpolation functions N_c along a side, we substitute the appropriate s, t values in the interpolation functions $N(s, t)$. The only remaining task is to relate the differential length dc along an element side to the differential line segment da along the side of the master element. This can be done easily by noting that

$$dc = \sqrt{dx^2 + dy^2}$$

Dividing both sides by da, we have

$$\frac{dc}{da} = \sqrt{\left(\frac{dx}{da}\right)^2 + \left(\frac{dy}{da}\right)^2} \Longrightarrow dc = J_c\, da$$

where J_c is called the Jacobian of a side,

$$J_c = \sqrt{\left(\frac{dx}{da}\right)^2 + \left(\frac{dy}{da}\right)^2}$$

The boundary integrals along the side of the element can now be written as follows:

$$\boldsymbol{k}_\alpha = -\int_{-1}^{1} \alpha \boldsymbol{N}_c \boldsymbol{N}_c^T J_c\, da; \qquad \boldsymbol{r}_\beta = \int_{-1}^{1} \beta \boldsymbol{N}_c J_c\, da$$

Of course, the known quantities α and β, if not constant, must be expressed in terms of a using the mapping for the side. Now these integrals can easily be evaluated using one-dimensional Gauss quadrature formulas:

$$\boldsymbol{k}_\alpha = -\int_{-1}^{1} \alpha \boldsymbol{N}_c \boldsymbol{N}_c^T J_c \, da = -\sum_{i=1}^{n} w_i \alpha(a_i) \boldsymbol{N}_c(a_i) \boldsymbol{N}_c^T(a_i) J_c(a_i)$$

$$\boldsymbol{r}_\beta = \int_{-1}^{1} \beta \boldsymbol{N}_c J_c \, da = \sum_{i=1}^{n} w_i \beta(a_i) \boldsymbol{N}_c(a_i) J_c(a_i)$$

where a_i are the one-dimensional Gauss points and w_i are the corresponding weights.

Example 6.19 Four-Node Quadrilateral Element Evaluate the matrix $\boldsymbol{k}_\alpha$ for the four-node quadrilateral element shown in Figure 6.36. Assume α is equal to 1.35 over side 3-4 of the element. The interpolation functions are

$$\boldsymbol{N}^T = \left\{\tfrac{1}{4}(s-1)(t-1), -\tfrac{1}{4}(s+1)(t-1), \tfrac{1}{4}(s+1)(t+1), -\tfrac{1}{4}(s-1)(t+1)\right\}$$

Mapping to the master element,

$$x(s,t) = \frac{ts}{2} + \frac{3s}{2} + \frac{t}{2} + \frac{3}{2}$$

$$y(s,t) = \frac{ts}{4} + \frac{s}{4} + \frac{3t}{4} + \frac{3}{4}$$

For the NBC on side 3, $a = -s$ and $t = 1$. Thus

$$\boldsymbol{N}_c^T = \begin{pmatrix} 0 & 0 & \dfrac{1-a}{2} & \dfrac{a+1}{2} \end{pmatrix}$$

$$x(a) = 2 - 2a; \qquad y(a) = \frac{3}{2} - \frac{a}{2}$$

$$\frac{dx}{da} = -2; \qquad \frac{dy}{da} = -\frac{1}{2}; \qquad J_c = \frac{\sqrt{17}}{2}$$

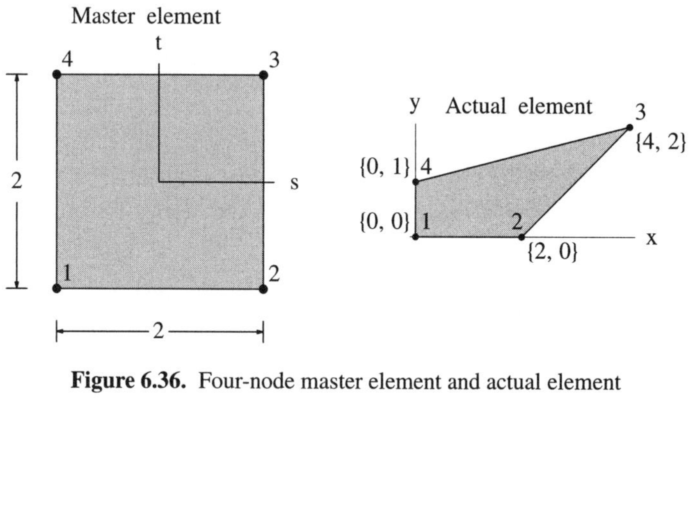

Figure 6.36. Four-node master element and actual element

Using two Gauss points, the matrix $\boldsymbol{k}_\alpha$ for the side is established as follows:

Gauss point = −0.57735; Weight = 1.; $J_c = \sqrt{17}/2$

$$N_c^T = (0 \quad 0 \quad 0.788675 \quad 0.211325)$$

$$\boldsymbol{k}_\alpha = \begin{pmatrix} 0 & 0 & 0 & 0 \\ 0 & 0 & 0 & 0 \\ 0 & 0 & -1.73111 & -0.463849 \\ 0 & 0 & -0.463849 & -0.124288 \end{pmatrix}$$

Gauss point = 0.57735; Weight = 1.; $J_c = \dfrac{\sqrt{17}}{2}$

$$N_c^T = (0 \quad 0 \quad 0.211325 \quad 0.788675)$$

$$\boldsymbol{k}_\alpha = \begin{pmatrix} 0 & 0 & 0 & 0 \\ 0 & 0 & 0 & 0 \\ 0 & 0 & -0.124288 & -0.463849 \\ 0 & 0 & -0.463849 & -1.73111 \end{pmatrix}$$

Summing contributions from all Gauss points,

$$\boldsymbol{k}_\alpha = \begin{pmatrix} 0 & 0 & 0 & 0 \\ 0 & 0 & 0 & 0 \\ 0 & 0 & -1.8554 & -0.927699 \\ 0 & 0 & -0.927699 & -1.8554 \end{pmatrix}$$

Example 6.20 Eight-Node Quadrilateral Element Compute vector $\boldsymbol{r}_\beta$ for the eight-node element shown in Figure 6.37. Assume β is equal to 23.4 over side 7-8-1 of the element.

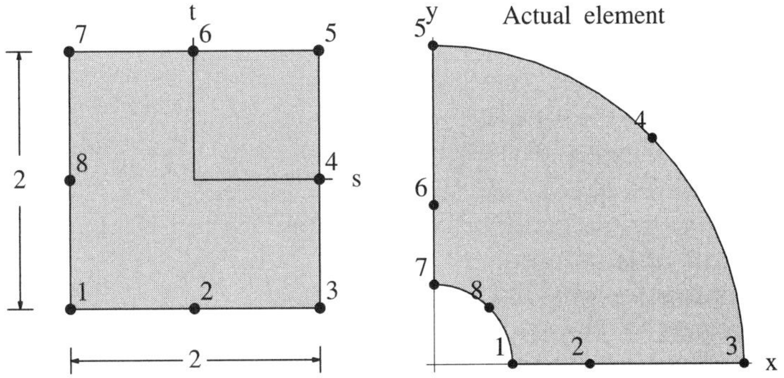

Figure 6.37. Eight-node master element and actual element

	x	y
1	2.	0.
2	4.	0.
3	8.	0.
4	5.65685	5.65685
5	0.	8.
6	0.	4.
7	0.	2.
8	1.41421	1.41421

Interpolation functions and their derivatives are

$$N^T = \Big\{ -\tfrac{1}{4}(s-1)(t-1)(s+t+1), \tfrac{1}{2}(s^2-1)(t-1), \tfrac{1}{4}(t-1)(-s^2+ts+t+1),$$
$$-\tfrac{1}{2}(s+1)(t^2-1), \tfrac{1}{4}(s+1)(t+1)(s+t-1), -\tfrac{1}{2}(s^2-1)(t+1),$$
$$\tfrac{1}{4}(s-1)(s-t+1)(t+1), \tfrac{1}{2}(s-1)(t^2-1)\Big\}$$

Mapping to the master element,

$$x(s,t) = -0.5ts^2 + 0.5s^2 - 0.62132t^2s - 1.5ts + 2.12132s - 1.03553t^2 - 2.t + 3.03553$$
$$y(s,t) = 0.5ts^2 + 0.5s^2 - 0.62132t^2s + 1.5ts + 2.12132s - 1.03553t^2 + 2.t + 3.03553$$

We now compute $\boldsymbol{r}_\beta$ resulting from the NBC for side 4 with $\beta = 23.4$. For this side $a = -t$ and $s = -1$. Thus

$$N_c^T = \left(\frac{a+1}{2} + \frac{1}{2}(a^2-1) \quad 0 \quad 0 \quad 0 \quad 0 \quad 0 \quad \frac{1-a}{2} + \frac{1}{2}(a^2-1) \quad 1-a^2 \right)$$
$$x(a) = -0.414214a^2 + 1.a + 1.41421; \qquad y(a) = -0.414214a^2 - 1.a + 1.41421$$
$$\frac{dx}{da} = 1. - 0.828427a$$
$$\frac{dy}{da} = -0.828427a - 1.$$
$$J_c = \sqrt{(-0.828427a - 1.)^2 + (1. - 0.828427a)^2}$$

Using three Gauss points, the vector $\boldsymbol{r}_\beta$ for the side is established as follows:

Gauss point = −0.774597; weight = 0.555556; $J_c = 1.68034$:

$$N_c^T = (-0.0872983 \quad 0 \quad 0 \quad 0 \quad 0 \quad 0 \quad 0.687298 \quad 0.4)$$
$$\boldsymbol{r}_\beta^T = (-1.90698 \quad 0 \quad 0 \quad 0 \quad 0 \quad 0 \quad 15.0137 \quad 8.73778)$$

Gauss point = 0; weight = 0.888889; $J_c = 1.41421$:

$$N_c^T = (0 \quad 0 \quad 0 \quad 0 \quad 0 \quad 0 \quad 0 \quad 1)$$

$$r_\beta^T = (0 \quad 0 \quad 0 \quad 0 \quad 0 \quad 0 \quad 0 \quad 29.4156)$$

Gauss point = 0.774597; weight = 0.555556; $J_c = 1.68034$:

$$N_c^T = (0.687298 \quad 0 \quad 0 \quad 0 \quad 0 \quad 0 \quad -0.0872983 \quad 0.4)$$

$$r_\beta^T = (15.0137 \quad 0 \quad 0 \quad 0 \quad 0 \quad 0 \quad -1.90698 \quad 8.73778)$$

Summing contributions from all Gauss points,

$$r_\beta^T = (13.1067 \quad 0 \quad 0 \quad 0 \quad 0 \quad 0 \quad 13.1067 \quad 46.8912)$$

6.5 COMPLETE *MATHEMATICA* AND MATLAB SOLUTIONS OF 2D BVP INVOLVING MAPPED ELEMENTS

As seen from the previous sections, for any arbitrary shaped element in the x, y coordinate system, the element equations are developed by first creating a mapping as follows.

$$x(s,t) = N_1 x_1 + N_2 x_2 + \cdots + N_n x_n$$
$$y(s,t) = N_1 y_1 + N_2 y_2 + \cdots + N_n y_n$$

where N_i are the interpolation functions for the appropriate master element and x_i and y_i are the nodal coordinates. The Jacobian matrix of the mapping is as follows:

$$\boldsymbol{J} = \begin{pmatrix} \frac{\partial x}{\partial s} & \frac{\partial x}{\partial t} \\ \frac{\partial y}{\partial s} & \frac{\partial y}{\partial t} \end{pmatrix} = \begin{pmatrix} J_{11} & J_{12} \\ J_{21} & J_{22} \end{pmatrix}$$

The actual finite element equations are

$$(\boldsymbol{k}_k + \boldsymbol{k}_p + \boldsymbol{k}_\alpha)\boldsymbol{d} = \boldsymbol{r}_q + \boldsymbol{r}_\beta$$

where using mapping the three terms involving area integrals are evaluated as follows:

$$\boldsymbol{k}_k = \iint_A \boldsymbol{BCB}^T \, dA = \int_{-1}^{1}\int_{-1}^{1} \boldsymbol{BCB}^T \det \boldsymbol{J} \, ds \, dt$$

$$= \sum_{i=1}^{m}\sum_{j=1}^{n} w_i w_j \boldsymbol{B}(s_i, t_j)\boldsymbol{C}(s_i, t_j)\boldsymbol{B}^T(s_i, t_j) \det \boldsymbol{J}(s_i, t_j)$$

$$\boldsymbol{B}^T = \frac{1}{\det \boldsymbol{J}} \begin{pmatrix} J_{22}\frac{\partial N_1}{\partial s} - J_{21}\frac{\partial N_1}{\partial t} & J_{22}\frac{\partial N_2}{\partial s} - J_{21}\frac{\partial N_2}{\partial t} & \cdots & J_{22}\frac{\partial N_n}{\partial s} - J_{21}\frac{\partial N_n}{\partial t} \\ -J_{12}\frac{\partial N_1}{\partial s} + J_{11}\frac{\partial N_1}{\partial t} & -J_{12}\frac{\partial N_2}{\partial s} + J_{11}\frac{\partial N_2}{\partial t} & \cdots & -J_{12}\frac{\partial N_n}{\partial s} + J_{11}\frac{\partial N_n}{\partial t} \end{pmatrix}$$

$$C = \begin{pmatrix} k_x & 0 \\ 0 & k_y \end{pmatrix}$$

$$k_p = -\iint_A pNN^T\, dA = -\int_{-1}^{1}\int_{-1}^{1} pNN^T \det J\, ds\, dt$$

$$= -\sum_{i=1}^{m}\sum_{j=1}^{n} w_i w_j p(s_i, t_j) N(s_i, t_j) N^T(s_i, t_j) \det J(s_i, t_j)$$

$$r_q = \iint_A qN\, dA = \int_{-1}^{1}\int_{-1}^{1} qN \det J\, ds\, dt = \sum_{i=1}^{m}\sum_{j=1}^{n} w_i w_j q(s_i, t_j) N(s_i, t_j) \det J(s_i, t_j)$$

where s_i, t_j are the Gauss points and w_i, w_j are the corresponding weights.

The two natural boundary condition terms involve line integrals:

$$k_\alpha = -\int_{C_n} \alpha N_c N_c^T\, dc = -\int_{-1}^{1} \alpha N_c N_c^T J_c\, da = -\sum_{i=1}^{n} w_i \alpha(a_i) N_c(a_i) N_c^T(a_i) J_c(a_i)$$

$$r_\beta = \int_{C_n} \beta N_c\, dc = \int_{-1}^{1} \beta N_c J_c\, da = \sum_{i=1}^{n} w_i \beta(a_i) N_c(a_i) J_c(a_i)$$

where J_c is the Jacobian of the side,

$$J_c = \sqrt{\left(\frac{dx}{da}\right)^2 + \left(\frac{dy}{da}\right)^2}$$

where a_i are the Gauss points and w_i are the corresponding weights.

Calculations involving mapped elements are clearly very tedious to perform by hand. However, it is not too difficult to create MATLAB and *Mathematica* functions to develop the element equations numerically. The following implementations are for four-node and eight-node quadrilateral elements. They can be used as templates to create functions for other mapped elements.

MATLAB Implementation: Four-Node Quadrilateral Element for 2D BVP The analysis of two-dimensional boundary value problems using mapped elements can be performed conveniently by writing three MATLAB functions, one for defining the element $k_k + k_p$ and r_q vectors, one for evaluating natural boundary terms, and the third for computing the element solution. The BVPQuad4Element employs 2×2 integration. The BVPQuad4NBCTerm employs two-point integration. The functions can be modified easily to use any other integration formula. The BVPQuad4Results function computes results at the point of the element that corresponds to the origin of the master element ($s = t = 0$). The function returns the location of this point (in terms of x, y coordinates) and the solution and its x and y derivatives at this point. The function can easily be modified to compute results at any other point.

MatlabFiles\Chap6\BVPQuad4Element.m

```
function [ke, rq] = BVPQuad4Element(kx, ky, p, q, coord)
% [ke, rq] = BVPQuad4Element(kx, ky, p, q, coord)
% Generates for a 4 node quadrilateral element for 2d BVP
% kx, ky, p, q = parameters defining the BVP
% coord = coordinates at the element ends

% Use 2x2 integration. Gauss point locations and weights
pt=1/sqrt(3);
gpLocs = [-pt,-pt; -pt,pt; pt,-pt; pt,pt];
gpWts = [1,1,1,1];
kk=zeros(4); kp=zeros(4); rq=zeros(4,1);
for i=1:length(gpWts)
  s = gpLocs(i, 1); t = gpLocs(i, 2); w = gpWts(i);
  n = [(1/4)*(1 - s)*(1 - t), (1/4)*(s + 1)*(1 -t), ...
      (1/4)*(s + 1)*(t + 1), (1/4)*(1 - s)*(t + 1)];
  dns=[(-1 + t)/4, (1 - t)/4, (1 + t)/4, (-1 - t)/4];
  dnt=[(-1 + s)/4, (-1 - s)/4, (1 + s)/4, (1 - s)/4];
  x = n*coord(:,1); y = n*coord(:,2);
  dxs = dns*coord(:,1); dxt = dnt*coord(:,1);
  dys = dns*coord(:,2); dyt = dnt*coord(:,2);
  J = [dxs, dxt; dys, dyt]; detJ = det(J);
  bx = (J(2, 2)*dns - J(2, 1)*dnt)/detJ;
  by = (-J(1, 2)*dns + J(1, 1)*dnt)/detJ;
  b = [bx; by];
  c = [kx, 0; 0, ky];
  kk = kk + detJ*w* b'*c*b;
  kp = kp - detJ*w*p * n'*n;
  rq = rq + detJ*w*q * n';
end
ke=kk+kp;
```

MatlabFiles\Chap6\BVPQuad4NBCTerm.m

```
function [ka, rb] = BVPQuad4NBCTerm(side, alpha, beta, coord)
% [ka, rb] = BVPQuad4NBCTerm(side, alpha, beta, coord)
% Generates kalpha and rbeta when NBC is specified along a side
% side = side over which the NBC is specified
% alpha and beta = coefficients specifying the NBC
% coord = coordinates at the element ends

% Use 2 point integration. Gauss point locations and weights
pt=-1/sqrt(3);
gpLocs = [-pt, pt];
gpWts = [1,1];
ka=zeros(4); rb=zeros(4,1);
```

```
for i=1:length(gpWts)
  a = gpLocs(i); w = gpWts(i);
  switch (side)
  case 1
    n = [(1 - a)/2, (1 + a)/2, 0, 0];
    dna = [-1/2, 1/2, 0, 0];
  case 2
    n = [0, (1 - a)/2, (1 + a)/2, 0];
    dna = [0, -1/2, 1/2, 0];
  case 3
    n = [0, 0, (1 - a)/2, (1 + a)/2];
    dna = [0, 0, -1/2, 1/2];
  case 4
    n = [(1 + a)/2, 0, 0, (1 - a)/2];
    dna = [1/2, 0, 0, -1/2];
  end
  dxa = dna*coord(:,1); dya = dna*coord(:,2);
  Jc=sqrt(dxa^2 + dya^2);
  ka = ka - alpha*Jc*w*n'*n;
  rb = rb + beta*Jc*w*n';
end
```

MatlabFiles\Chap6\BVPQuad4Results.m

```
function results = BVPQuad4Results(coord, dn)
% results = BVPQuad4Results(coord, dn)
% Computes element solution for a quadrilateral element for 2D BVP
% coord = nodal coordinates
% dn = nodal solution
% The solution is computed at the element center
% The output variables are loc, u and its x and y derivatives
s = 0; t = 0;
n = [(1/4)*(1 - s)*(1 - t), (1/4)*(s + 1)*(1 -t), ...
    (1/4)*(s + 1)*(t + 1), (1/4)*(1 - s)*(t + 1)];
dns=[(-1 + t)/4, (1 - t)/4, (1 + t)/4, (-1 - t)/4];
dnt=[(-1 + s)/4, (-1 - s)/4, (1 + s)/4, (1 - s)/4];
x = n*coord(:,1); y = n*coord(:,2);
dxs = dns*coord(:,1); dxt = dnt*coord(:,1);
dys = dns*coord(:,2); dyt = dnt*coord(:,2);
J = [dxs, dxt; dys, dyt]; detJ = det(J);
bx = (J(2, 2)*dns - J(2, 1)*dnt)/detJ;
by = (-J(1, 2)*dns + J(1, 1)*dnt)/detJ;
b = [bx; by];
results=[x, y, n*dn, bx*dn, by*dn];
```

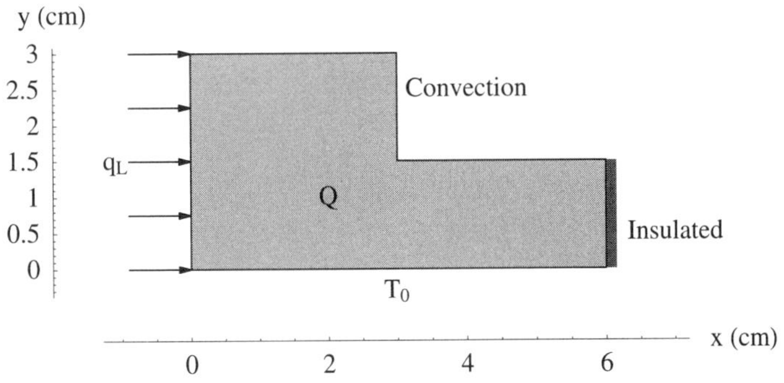

Figure 6.38. L-shaped body

Example 6.21 Heat Flow in an L-Shaped Body Using Quad4 Elements Consider two-dimensional heat flow over an L-shaped body with thermal conductivity k = 45 W/m · °C shown in Figure 6.38. The bottom is maintained at T_0 = 110 °C. Convection heat loss takes place on the top where the ambient air temperature is 20°C and the convection heat transfer coefficient is h = 55 W/m^2 · °C. The right side is insulated. The left side is subjected to heat flux at a uniform rate of q_L = 8000 W/m^2. Heat is generated in the body at a rate of $Q = 5 \times 10^6$ W/m^3. Determine the temperature distribution in the body.

To demonstrate the use of the functions presented in this section, we obtain a finite element solution of the problem using only two quadrilateral elements, as shown in Figure 6.39. Obviously we do not expect very good results. The mesh is chosen to simplify calculations. The steps are exactly those used in other MATLAB implementations.

MatlabFiles\Chap6\LHeatQuad4Ex.m

```
% Heat flow through an L-shaped body using quad4 elements
h = 55; tf = 20; htf = h*tf;
kx = 45;  ky = 45; Q = 5*10^6; ql = 8000; t0 = 110;
```

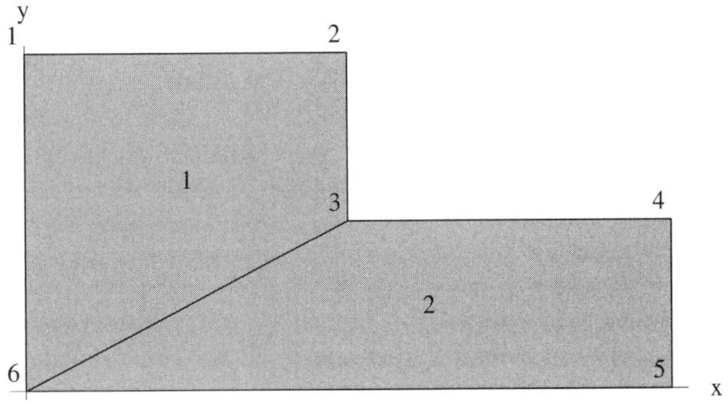

Figure 6.39. Two four-node quadrilateral element model

```
nodes = 1.5/100*[0, 2; 2, 2; 2, 1; 4, 1; 4, 0; 0, 0];
lmm = [6, 3, 2, 1; 6, 5, 4, 3];
debc = [5:6]; ebcVals=t0*ones(length(debc),1);
dof=length(nodes); elems=size(lmm,1);
K=zeros(dof); R = zeros(dof,1);
% Generate equations for each element and assemble them.
for i=1:elems
  lm = lmm(i,:);
  [k, r] = BVPQuad4Element(kx, ky, 0, Q, nodes(lm,:));
  K(lm, lm) = K(lm, lm) + k;
  R(lm) = R(lm) + r;
end
% Compute and assemble NBC contributions
lm = lmm(1,:);
[k, r] = BVPQuad4NBCTerm(4, 0, ql, nodes(lm,:));
K(lm, lm) = K(lm, lm) + k;
R(lm) = R(lm) + r;
[k, r] = BVPQuad4NBCTerm(2, -h, htf, nodes(lm,:));
K(lm, lm) = K(lm, lm) + k;
R(lm) = R(lm) + r;
[k, r] = BVPQuad4NBCTerm(3, -h, htf, nodes(lm,:));
K(lm, lm) = K(lm, lm) + k;
R(lm) = R(lm) + r;

lm = lmm(2,:);
[k, r] = BVPQuad4NBCTerm(3, -h, htf, nodes(lm,:));
K(lm, lm) = K(lm, lm) + k;
R(lm) = R(lm) + r;

% Nodal solution
d = NodalSoln(K, R, debc, ebcVals)
results=[];
for i=1:elems
  results = [results; BVPQuad4Results(nodes(lmm(i,:),:), ...
      d(lmm(i,:)))];
end
format short g
results
```

```
>> LHeatQuad4Ex

d =

  153.3936
  142.9067
  132.8533
  124.5394
```

```
  110.0000
  110.0000

results =

       0.015      0.01875      134.79      -90.82      1187.7
      0.0375       0.0075      119.35     -92.377      1338.8
```

Mathematica Implementation: Eight-Node Quadrilateral Element for 2D BVP

The functions for eight-node quadrilateral elements are almost identical to those for the four node quadrilaterals. The only changes are the dimensions of the matrices and that we use three points for integration and eight node serendipity interpolation functions.

```
Needs["NumericalMath`GaussianQuadrature`"];

BVPQuad8Element[kx_, ky_, p_, q_, {{x1_, y1_}, {x2_, y2_}, {x3_, y3_}, {x4_, y4_},
      {x5_, y5_}, {x6_, y6_}, {x7_, y7_}, {x8_, y8_}}] :=
  Module[{sloc, swts, tloc, twts, gpLocs, gpWts, n, s, t, dns, dnt, xst, yst,
      J, npt, Jpt, dnsPt, dntPt, detJ, bx, by, bt, b, c, kk, kp, rq},
    {sloc, swts} = Transpose[GaussianQuadratureWeights[3, −1, 1]];
    {tloc, twts} = Transpose[GaussianQuadratureWeights[3, −1, 1]];
    gpLocs = Flatten[Outer[List, sloc, tloc], 1];
    gpWts = Flatten[Outer[Times, swts, twts], 1];
    n =
      {(1/4) * (1 − s) * (1 − t) − (1/4) * (1 − s^2) * (1 − t)
      −(1/4) * (1 − s) * (1 − t^2), (1/2) * (1 − s^2) * (1 − t), (1/4) * (s + 1) * (1 − t)
      −(1/4) * (1 − s^2) * (1 − t) − (1/4) * (s + 1) * (1 − t^2),
      (1/2) * (s + 1) * (1 − t^2), (1/4) * (s + 1) * (t + 1) − (1/4) * (1 − s^2) * (t + 1)
      −(1/4) * (s + 1) * (1 − t^2), (1/2) * (1 − s^2) * (t + 1), (1/4) * (1 − s) * (t + 1)
      −(1/4) * (1 − s^2) * (t + 1) − (1/4) * (1 − s) * (1 − t^2),
      (1/2) * (1 − s) * (1 − t^2)};
    {dns, dnt} = {D[n, s], D[n, t]};
    xst = n.{x1, x2, x3, x4, x5, x6, x7, x8};
    yst = n.{y1, y2, y3, y4, y5, y6, y7, y8};
    J = {{D[xst, s], D[xst, t]}, {D[yst, s], D[yst, t]}};
    kk = kp = Table[0, {8}, {8}];
    rq = Table[0, {8}];
    Do[pt = {s → gpLocs[[i, 1]], t → gpLocs[[i, 2]]};
      w = gpWts[[i]];
      {npt, Jpt, dnsPt, dntPt} = {n, J, dns, dnt}/.pt;
      detJ = det[Jpt];
      bx = (Jpt[[2, 2]]dnsPt − Jpt[[2, 1]]dntPt)/detJ;
      by = (−Jpt[[1, 2]]dnsPt + Jpt[[1, 1]]dntPt)/detJ;
      bt = {bx, by}; b = Transpose[bt];
      c = {{kx, 0}, {0, ky}};
      kk+ = detJ wb.c.bt;
      kp+ = −detJ w p Outer[Times, npt, npt];
```

```
        rq+ = detJ w q npt, {i, 1, Length[gpWts]}
      ];
      {kk + kp, rq}
    ]

BVPQuad8NBCTerm[side_, α_, β_, {{x1_, y1_}, {x2_, y2_}, {x3_, y3_}, {x4_, y4_},
        {x5_, y5_}, {x6_, y6_}, {x7_, y7_}, {x8_, y8_}}] :=
      Module[{k, r, pts, wts, n, s, t, a, xa, ya, Jc, Jcpt},
      k = Table[0, {8}, {8}]; r = Table[0, {8}];
      {pts, wts} = Transpose[GaussianQuadratureWeights[2, -1, 1]];
      n =
        {(1/4) * (1 - s) * (1 - t) - (1/4) * (1 - s^2) * (1 - t)
        -(1/4) * (1 - s) * (1 - t^2), (1/2) * (1 - s^2) * (1 - t),
          (1/4) * (s + 1) * (1 - t) - (1/4) * (1 - s^2) * (1 - t)
        -(1/4) * (s + 1) * (1 - t^2), (1/2) * (s + 1) * (1 - t^2),
          (1/4) * (s + 1) * (t + 1) - (1/4) * (1 - s^2) * (t + 1)
        -(1/4) * (s + 1) * (1 - t^2), (1/2) * (1 - s^2) * (t + 1),
          (1/4) * (1 - s) * (t + 1) - (1/4) * (1 - s^2) * (t + 1)
        -(1/4) * (1 - s) * (1 - t^2), (1/2) * (1 - s) * (1 - t^2)};
      n = Switch[side, 1, n/.{s → a, t → -1},
            2, n/.{s → 1, t → a},
            3, n/.{s → -a, t → 1},
            4, n/.{s → -1, t → -a}
        ];
      xa = n.{x1, x2, x3, x4, x5, x6, x7, x8};
      ya = n.{y1, y2, y3, y4, y5, y6, y7, y8};
      Jc = Sqrt[D[xa, a]^2 + D[ya, a]^2];
      Do[{npt, Jcpt} = {n, Jc}/.a → pts[[i]];
        k+ = -α * Jcpt * wts[[i]] Outer[Times, npt, npt];
        r+ = β * Jcpt * wts[[i]]npt,
        {i, 1, Length[pts]}
      ];
      {k, r}
];

BVPQuad8Results[{{x1_, y1_}, {x2_, y2_}, {x3_, y3_}, {x4_, y4_},
        {x5_, y5_}, {x6_, y6_}, {x7_, y7_}, {x8_, y8_}}, d_] :=
  Module[{pt, n, s, t, dns, dnt, bx, by, xst, yst, J, npt, Jpt, dnsPt, dntPt, detJ, loc},
  pt = {s → 0, t → 0};
    n =
      {(1/4) * (1 - s) * (1 - t) - (1/4) * (1 - s^2) * (1 - t)
        -(1/4) * (1 - s) * (1 - t^2), (1/2) * (1 - s^2) * (1 - t),
        (1/4) * (s + 1) * (1 - t) - (1/4) * (1 - s^2) * (1 - t)
        -(1/4) * (s + 1) * (1 - t^2), (1/2) * (s + 1) * (1 - t^2),
        (1/4) * (s + 1) * (t + 1) - (1/4) * (1 - s^2) * (t + 1)
        -(1/4) * (s + 1) * (1 - t^2), (1/2) * (1 - s^2) * (t + 1),
```

```
        (1/4) * (1 - s) * (t + 1) - (1/4) * (1 - s^2) * (t + 1)
        -(1/4) * (1 - s) * (1 - t^2), (1/2) * (1 - s) * (1 - t^2)};
    {dns, dnt} = {D[n, s], D[n, t]};
    xst = n.{x1, x2, x3, x4, x5, x6, x7, x8};
    yst = n.{y1, y2, y3, y4, y5, y6, y7, y8};
    J = {{D[xst, s], D[xst, t]}, {D[yst, s], D[yst, t]}};
        {loc, npt, Jpt, dnsPt, dntPt} =
    detJ = det[Jpt];
    bx = (Jpt[[2, 2]] dnsPt - Jpt[[2, 1]] dntPt)/detJ;
    by = (-Jpt[[1, 2]] dnsPt + Jpt[[1, 1]] dntPt)/detJ;
    {loc, npt.d, bx.d, by.d}
]
```

Example 6.22 Heat Flow in an L-Shaped Body Using Quad8 Elements To demonstrate the use of the functions presented in this section, we obtain a finite element solution of the problem using only two quadrilateral elements, as shown in Figure 6.40.

```
Clear[k, r];
h = 55; tf = 20; htf = h * tf;
kx = ky = 45; Q = 5 * 10^6; ql = 8000; t0 = 110;
nodes = 1.5/100{{0, 2}, {1, 2}, {2, 2}, {2, 1.5}, {2, 1}, {3, 1}, {4, 1},
        {4, 0.5}, {4, 0}, {2, 0}, {0, 0}, {0, 1}, {1, 0.5}};
lm[1] = {11, 13, 5, 4, 3, 2, 1, 12};
lm[2] = {11, 10, 9, 8, 7, 6, 5, 13};
Do[{k[i], r[i]} = BVPQuad8Element[kx, ky, 0, Q, nodes[[lm[i]]]], {i, 1, 2}];
{k[1], r[1]}+ = BVPQuad8NBCTerm[4, 0, ql, nodes[[lm[1]]]];
{k[1], r[1]}+ = BVPQuad8NBCTerm[2, -h, htf, nodes[[lm[1]]]];
{k[1], r[1]}+ = BVPQuad8NBCTerm[3, -h, htf, nodes[[lm[1]]]];
{k[2], r[2]}+ = BVPQuad8NBCTerm[3, -h, htf, nodes[[lm[2]]]];
R = Table[0, {13}];
```

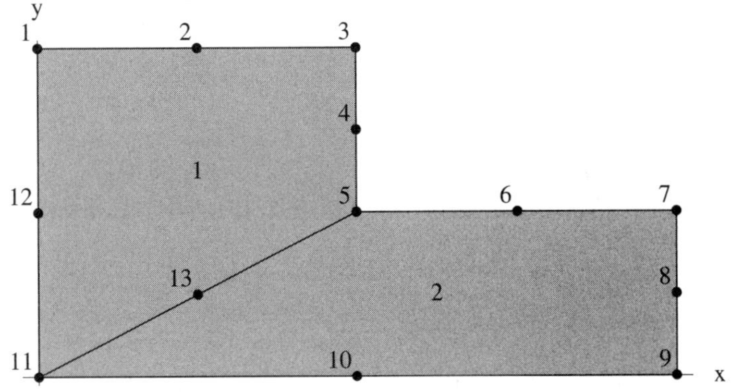

Figure 6.40. Two eight-node quadrilateral element model

```
K = Table[0, {13}, {13}];
Do[K[[lm[i], lm[i]]]+ = k[i]; R[[lm[i]]]+ = r[i], {i, 1, 2}];
```

Essential boundary conditions are as follows:

```
debc = Range[9, 11];
ebcVals = Table[t0, {3}];
```

```
df = Complement[Range[13], debc]
```

{1, 2, 3, 4, 5, 6, 7, 8, 12, 13}

```
Kf = K[[df, df]]
```

$$\begin{pmatrix} 57.3302 & -33.3087 & 23.1028 & -36.1508 & 30.4563 & 0 & 0 & 0 & -46.6746 & -16.5079 \\ -33.3087 & 95.8127 & -20.4516 & 3.65079 & -28.0159 & 0 & 0 & 0 & 3.1746 & -5.07937 \\ 23.1028 & -20.4516 & 58.5885 & -68.0829 & 33.5653 & 0 & 0 & 0 & -28.4127 & -19.3651 \\ -36.1508 & 3.65079 & -68.0829 & 156.24 & -82.6067 & 0 & 0 & 0 & 42.0635 & 16.3492 \\ 30.4563 & -28.0159 & 33.5653 & -82.6067 & 257.914 & -85.3722 & 51.9123 & -90.6349 & -33.6508 & -156.349 \\ 0 & 0 & 0 & 0 & -85.3722 & 133.749 & -27.277 & 14.6032 & 0 & 65.3968 \\ 0 & 0 & 0 & 0 & 51.9123 & -27.277 & 70.6397 & -103.968 & 0 & -56.0317 \\ 0 & 0 & 0 & 0 & -90.6349 & 14.6032 & -103.968 & 258.889 & 0 & 101.111 \\ -46.6746 & 3.1746 & -28.4127 & 42.0635 & -33.6508 & 0 & 0 & 0 & 125.968 & -23.1746 \\ -16.5079 & -5.07937 & -19.3651 & 16.3492 & -156.349 & 65.3968 & -56.0317 & 101.111 & -23.1746 & 353.968 \end{pmatrix}$$

```
Rf = R[[df]] − K[[df, debc]].ebcVals
```

{−2627.56, 2762.08, −2665.76, 4411.4, −11968.1, 12022., −7457., 20925., 5732.3, 30884.9}

The solution for nodal unknowns can now be obtained by using the LinearSolve function as follows:

```
dfVals = LinearSolve[Kf, Rf]
```

{156.445, 150.755, 149.201, 144.228, 133.843, 123.998, 121.754, 119.151, 144.677, 129.134}

The complete vector of nodal values, in the original order established by the node numbering, is as obtained by combining these values with those specified as essential boundary conditions as follows:

```
d = Table[0, {13}];
d[[debc]] = ebcVals;
d[[df]] = dfVals;
d
```

{156.445, 150.755, 149.201, 144.228, 133.843, 123.998, 121.754, 119.151,
110, 110, 110, 144.677, 129.134}

Using the BVPQuad8Results function, the solution and its x and y derivatives for each element can now be computed as follows:

```
BVPQuad8Results[nodes[[lm[1]]], d[[lm[1]]]]
```

{{0.015, 0.01875}, 147.025, −255.209, 960.97}

```
BVPQuad8Results[nodes[[lm[2]]], d[[lm[2]]]]
```

{{0.0375, 0.0075}, 122.242, −221.837, 1155.06}

6.6 TRIANGULAR ELEMENTS BY COLLAPSING QUADRILATERALS

By *collapsing* one side of a quadrilateral, it is possible to map triangular areas into 2 × 2 square areas. As seen in Figure 6.41, this is accomplished by considering all nodes on side 4 of the master area to be physically located at the same point. Thus for a straight-sided triangle nodes 1 and 4 are coincident. The mapping in terms of its three physical nodes is as follows:

$$x(s,t) = \left(\tfrac{1}{4}(s-1)(t-1) + \tfrac{1}{4}(1-s)(t+1)\right)x_1 + \tfrac{1}{4}(-s-1)(t-1)x_2 + \tfrac{1}{4}(s+1)(t+1)x_3$$

$$y(s,t) = \left(\tfrac{1}{4}(s-1)(t-1) + \tfrac{1}{4}(1-s)(t+1)\right)y_1 + \tfrac{1}{4}(-s-1)(t-1)y_2 + \tfrac{1}{4}(s+1)(t+1)y_3$$

For a triangle with all three sides curved nodes 1, 7, and 8 are coincident. The mapping in terms of its six physical nodes is as follows:

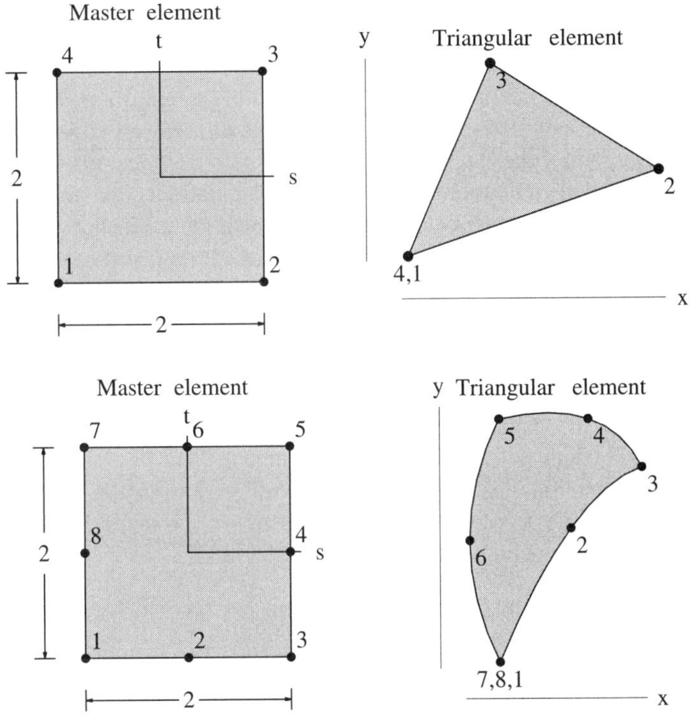

Figure 6.41. Mapping triangular elements into quadrilaterals

$$
\begin{aligned}
x(s,t) = {} & \left(\tfrac{1}{4}(s-1)(s-t+1)(t+1) + \tfrac{1}{4}(1-s)(t-1)(s+t+1) + \tfrac{1}{2}(s-1)(t^2-1)\right)x_1 \\
& + \tfrac{1}{2}(s^2-1)(t-1)x_2 + \tfrac{1}{4}(t-1)(-s^2+ts+t+1)x_3 + \tfrac{1}{2}(-s-1)(t^2-1)x_4 \\
& + \tfrac{1}{4}(s+1)(t+1)(s+t-1)x_5 + \tfrac{1}{2}(1-s^2)(t+1)x_6 \\
y(s,t) = {} & \left(\tfrac{1}{4}(s-1)(s-t+1)(t+1) + \tfrac{1}{4}(1-s)(t-1)(s+t+1) + \tfrac{1}{2}(s-1)(t^2-1)\right)y_1 \\
& + \tfrac{1}{2}(s^2-1)(t-1)y_2 + \tfrac{1}{4}(t-1)(-s^2+ts+t+1)y_3 + \tfrac{1}{2}(-s-1)(t^2-1)y_4 \\
& + \tfrac{1}{4}(s+1)(t+1)(s+t-1)y_5 + \tfrac{1}{2}(1-s^2)(t+1)y_6
\end{aligned}
$$

One key advantage of this procedure is that we do not need to develop separate interpolation functions and computer programs for triangular elements. Any computer implementation of quadrilateral elements can be used to handle triangular elements as well simply by repeating the nodes that define the last side of the quadrilateral.

6.7 INFINITE ELEMENTS

Many practical problems involve computational domains that are not well defined and are essentially infinite in one or more directions. One such situation was encountered in Chapter 5 when modeling fluid flow around an object. There are no definite boundaries of the flow region. It was suggested to simply extend the boundaries *far enough* from the object so that assumption of uniform flow is valid. Another common situation that involves infinite boundaries is analysis of structures supported on ground. In most such cases part of the ground must be included in the finite element model. However, usually there is no rigid boundary that defines the extent of ground that should be included in the model. In all these situations the results will not be reliable if the computational domain is too small. On the other hand, making the computational domain very large wastes computational effort. The so-called *infinite elements,* presented in this section, are designed to be used as the last layer of elements in the direction where the boundary is not well defined.

6.7.1 One-Dimensional BVP

The key idea in developing an infinite element is to create an appropriate mapping such that an infinite element is mapped to a standard master element. Consider mapping of an infinite line to a standard three-node master line as shown in Figure 6.42. We need a mapping so that the point $s = -1$ on the master line is mapped to x_1, $s = 0$ is mapped to x_2, and $s = 1$ is mapped to $x = \infty$. An appropriate mapping can be developed by starting with a relationship in the following form:

$$x(s) = a_0 + a_1 s + a_2 f(s)$$

where a_0, a_1, a_2, and $f(s)$ are to be determined to give the desired mapping. The coefficients a_0 and a_1 can be established by the requirement of mapping x_1 and x_2 to $s = -1$ and $s = 0$,

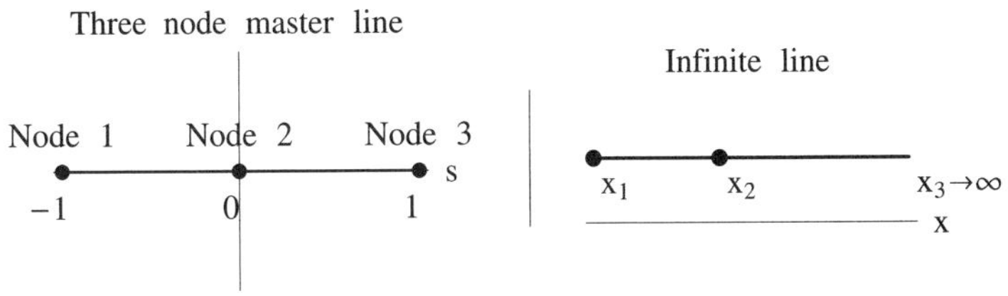

Figure 6.42. Master line for mapping an actual infinite line

respectively, as follows.

$$\text{At } s = -1: \quad x_1 = a_0 - a_1 + a_2 f(-1)$$
$$\text{At } s = 0: \quad x_2 = a_0 + a_2 f(0)$$

Solving the two equations for a_0 and a_1, we get

$$a_0 = x_2 - a_2 f(0); \qquad a_1 = -x_1 + x_2 + a_2 f(-1) - a_2 f(0)$$

Substituting these, the mapping takes the following form:

$$x(s) = x_2 - a_2 f(0) + [-x_1 + x_2 + a_2 f(-1) - a_2 f(0)]s + a_2 f(s)$$

Rearranging terms,

$$x(s) = x_2 + s(x_2 - x_1) + a_2[s(f(-1) - f(0)) - f(0) + f(s)]$$

The next task is to choose $f(s)$ and a_2 so that $x = \infty$ when $s = 1$. Clearly, $f(s) = 1/(1-s)$ will do the job, giving

$$x(s) = a_2\left(-\frac{s}{2} + \frac{1}{1-s} - 1\right) + x_2 + s(x_2 - x_1)$$

The coefficient a_2 can be chosen arbitrarily. A common choice is to set a_2 equal to twice the distance between the two nodes, i.e., $a_2 = 2(x_2 - x_1)$. Using this value of a_2, the mapping takes the following simple form:

$$x(s) = \frac{2s}{s-1}x_1 - \frac{(s+1)}{s-1}x_2$$

Note the element has three nodes. However, only the first two, which are at finite locations, are used in the mapping.

The assumed solution over the mapped element is written using the standard quadratic interpolation functions:

$$u(s) = \begin{pmatrix} \frac{s^2}{2} - \frac{s}{2} & 1 - s^2 & \frac{s^2}{2} + \frac{s}{2} \end{pmatrix} \begin{pmatrix} u_1 \\ u_2 \\ u_3 \end{pmatrix} \equiv \boldsymbol{N}^T\boldsymbol{d}$$

The derivative of this solution with respect to x is calculated using the chain rule as usual for the mapped elements:

$$\frac{du}{ds} = \frac{du}{dx}\frac{dx}{ds} \Longrightarrow \frac{du}{dx} = \frac{1}{J}\frac{du}{ds} = \frac{1}{J}\begin{pmatrix} s - \frac{1}{2} & -2s & s + \frac{1}{2} \end{pmatrix}\begin{pmatrix} u_1 \\ u_2 \\ u_3 \end{pmatrix} \equiv \boldsymbol{B}^T\boldsymbol{d}$$

where

$$J = \frac{dx}{ds} = -\frac{2(x_1 - x_2)}{(s-1)^2}$$

The element equations for a one-dimensional element for the 1D BVP considered in Chapter 3 can be developed in a usual manner:

$$\boldsymbol{k}_k = \int_{x_1}^{x_3} k\boldsymbol{B}\boldsymbol{B}^T\, dx = \int_{-1}^{1} k\boldsymbol{B}\boldsymbol{B}^T J\, ds$$

$$\boldsymbol{k}_p = \int_{x_1}^{x_3} -p\boldsymbol{N}\boldsymbol{N}^T\, dx = \int_{-1}^{1} -p\boldsymbol{N}\boldsymbol{N}^T J\, ds$$

$$\boldsymbol{r}_q = \int_{x_1}^{x_3} q\boldsymbol{N}\, dx = \int_{-1}^{1} q\boldsymbol{N} J\, ds$$

Because of singularity in J, the integrals cannot be evaluated in the usual sense. However, since the intention of these elements is to approximate the behavior of a region that is far away from the actual region of interest, this fact is ignored. In numerical implementations integrations are typically carried out using a two-point Gauss quadrature formula. Thus the element equations are

$$(\boldsymbol{k}_k + \boldsymbol{k}_p)\boldsymbol{d} = \boldsymbol{r}_q \Longrightarrow \boldsymbol{k}\boldsymbol{d} = \boldsymbol{r}$$

Mathematica Functions for Solution of One-Dimensional Boundary Value Problem Involving Infinite Domain The following functions are developed for performing computations for both the standard and the infinite elements for the one-dimensional boundary value problem. The functions are a little more general than those presented in Chapter 3. The same function can handle a linear, quadratic, or infinite element. The coefficients k, p, and q can be any arbitrary functions of x. The integrations are performed using the two-point Gauss quadrature.

```
BVP1DElement[k_, p_, q_, nodes_, type_ : Linear] :=
  Module[{n, nm, s, xs, dxs, ks, ps, qs, kk, kp, ke, re, gpw},
    Switch[type,
      Linear, n = nm = {(1 - s)/2, (1 + s)/2},
      Quadratic, n = nm = {-s/2 + s^2/2, 1 - s^2, s/2 + s^2/2},
      Infinite, n = {-s/2 + s^2/2, 1 - s^2, s/2 + s^2/2};
        nm = {(2 * s)/(-1 + s), -(1 + s)/(-1 + s)}
```

```
    ];
    gpw = {{s- > -1/Sqrt[3], s- > 1/Sqrt[3]}, {1, 1}}//N;
    xs = nm.nodes; J = D[xs, s];
    {ks, ps, qs} = {k, p, q}/.x → xs;
    b = D[n, s]/J;
    kk = ks Outer[Times, b, b]J;
    kk = Apply[Plus, MapThread[#2 * kk/.#1&, gpw]];
    kp = -ps Outer[Times, n, n]J; v
    kp = Apply[Plus, MapThread[#2 * kp/.#1&, gpw]];
    ke = kk + kp;
    re = qs n J;
    re = Apply[Plus, MapThread[#2 * re/.#1&, gpw]];
    {ke, re}]
Clear[x]; BVP1DElementSolution[nodes_, nodalSol_, type_ : Linear] :=
  Module[{n, nm, s, xs, u, x1, x2},
    x1 = nodes[[1]];
    Switch[type,
      Linear, n = nm = {(1 - s)/2, (1 + s)/2};
          x2 = nodes[[2]],
      Quadratic, n = nm = {-s/2 + s^2/2, 1 - s^2, s/2 + s^2/2};
          x2 = nodes[[3]],
      Infinite, n = {-s/2 + s^2/2, 1 - s^2, s/2 + s^2/2};
          nm = {(2 * s)/(-1 + s), -(1 + s)/(-1 + s)};
          x2 = ∞
    ];
    xs = nm.nodes;
    n = n/.Solve[xs == x, s][[1]];
    u = Simplify[n.nodalSol];
    {x1 < x < x2, u}]
```

Example 6.23 1D BVP with Infinite Domain Consider solution of the following one-dimensional boundary value problem:

$$\frac{d^2u}{dx^2} = \frac{2}{x^3}; \qquad 1 < x < \infty$$

$$u(1) = 1; \qquad u(\infty) = 0$$

It can easily be verified that the following is the exact solution for this problem:

$$u_{\text{exact}} = \frac{1}{x}$$

This problem is of the type considered in Chapter 3 with

$$k = 1; \qquad p = 0; \qquad q = -\frac{2}{x^3}$$

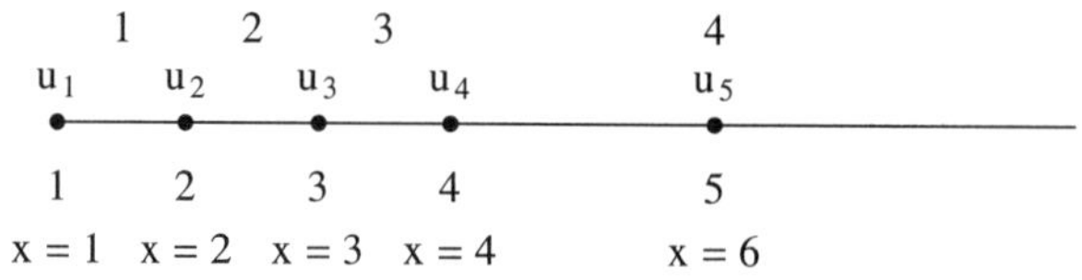

Figure 6.43. Three linear and one infinite element model

We consider the finite element solution of this problem using three standard linear elements and one infinite element. The finite element model is as shown in Figure 6.43. The model has six degrees of freedom. Only five are shown on the figure. The degree of freedom u_6 is at $x = \infty$.

```
k = 1; p = 0; q = −2/x^3;
nodes = {1, 2, 3, 4, 6};
```

```
Do[lm[i] = {i, i + 1};
    {ke[i], re[i]} = BVP1DElement[k, p, q, nodes[[lm[i]]], Linear], {i, 1, 3}];
lm[4] = {4, 5, 6};
{ke[4], re[4]} = BVP1DElement[k, p, q, nodes[[{4, 5}]], Infinite];
K = Table[0, {6}, {6}]; R = Table[0, {6}];
Do[K[[lm[i], lm[i]]]+ = ke[i];
R[[lm[i]]]+ = re[i], {i, 1, 4}];
```

The essential boundary conditions are as follows:

```
debc = {1, 6};
ebcVals = {1, 0};
```

```
df = Complement[Range[6], debc]
```

{2, 3, 4, 5}

```
Kf = K[[df, df]]
```

$$\begin{pmatrix} 2. & -1. & 0 & 0 \\ -1. & 2. & -1. & 0 \\ 0 & -1. & 1.72222 & -0.777778 \\ 0 & 0 & -0.777778 & 0.888889 \end{pmatrix}$$

```
Rf = R[[df]] − K[[df, debc]].ebcVals
```

{0.660606, −0.0836104, −0.0329453, −0.0427856}

The solution for nodal unknowns can now be obtained by using the LinearSolve function as follows:

```
dfVals = LinearSolve[Kf, Rf]//N
```

{0.49616, 0.331713, 0.250877, 0.171384}

The complete vector of nodal values, in the original order established by the node numbering, is as obtained by combining these values with those specified as essential boundary conditions as follows:

```
d = Table[0, {6}];
d[[debc]] = ebcVals;
d[[df]] = dfVals;
d
```

{1, 0.49616, 0.331713, 0.250877, 0.171384, 0}

The complete solution for each element can be computed as follows:

```
res[1] = BVP1DElementSolution[nodes[[lm[1]]], d[[lm[1]]], Linear]
```

$\{1 < x < 2, 1.50384 - 0.50384x\}$

```
res[2] = BVP1DElementSolution[nodes[[lm[2]]], d[[lm[2]]], Linear]
```

$\{2 < x < 3, 0.825052 - 0.164446x\}$

```
res[3] = BVP1DElementSolution[nodes[[lm[3]]], d[[lm[3]]], Linear]
```

$\{3 < x < 4, 0.574221 - 0.080836x\}$

```
res[4] = BVP1DElementSolution[nodes[[{4, 5}]], d[[lm[4]]], Infinite]
```

$$\left\{4 < x < \infty, \frac{0.869317x - 2.47376}{(x - 2.)^2}\right\}$$

In Figure 6.44 the finite element solution is compared with the exact solution. This plot is generated as follows:

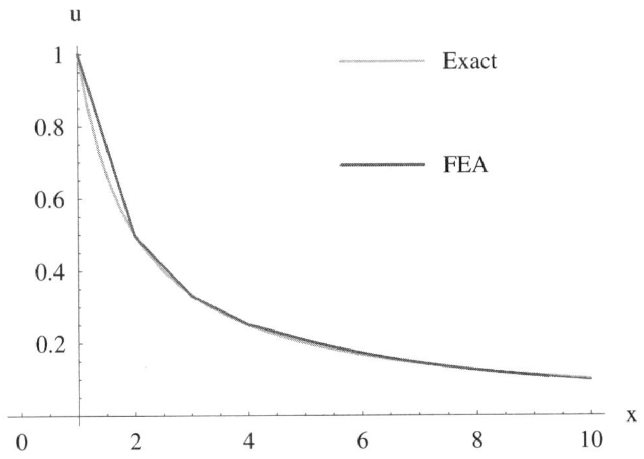

Figure 6.44.

```
esol = Apply[Which, Flatten[Table[res[i], {i, 1, 4}]]];
Plot[{1/x, Evaluate[esol]}, {x, 1, 10}, PlotStyle → {Hue[0.5], Hue[0.9]},
    AxesLabel → {"x", "u"}, AxesOrigin → {1, 0},
    PlotLegend → {"Exact", "FEA"}, LegendShadow → None,
    LegendPosition → {0, 0}];
```

6.7.2 Two-Dimensional BVP

The ideas presented for the one-dimensional problem can be easily extended to two-dimensional problems. The simplest infinite element for a two-dimensional boundary value problem is a six-node element shown in Figure 6.45. In the infinite direction each side has three nodes, with nodes 5 and 6 located at infinity. The master element is a usual 2×2 square element used with other mapped elements. Note that, since the nodes at infinity are not used explicitly in the mapping functions, the node-numbering scheme is different than our usual convention of moving counterclockwise around the element. From the figure it is clear that in the t direction we can use the standard linear Lagrange interpolation function:

$$\text{Mapping functions in the } t \text{ direction:} \qquad \frac{1-t}{2}; \qquad \frac{1+t}{2}$$

In the s direction we use the mapping functions developed for the one-dimensional infinite element:

$$\text{Mapping functions in the } s \text{ direction:} \qquad \frac{2s}{s-1}; \qquad -\frac{(s+1)}{s-1}$$

Taking the product of these functions, we have the following mapping of the infinite element to the 2×2 master element:

$$x(s,t) = \frac{s(1-t)}{s-1}x_1 + \frac{s(t+1)}{s-1}x_2 - \frac{(s+1)(1-t)}{2(s-1)}x_3 - \frac{(s+1)(t+1)}{2(s-1)}x_4$$

and

$$y(s,t) = \frac{s(1-t)}{s-1}y_1 + \frac{s(t+1)}{s-1}y_2 - \frac{(s+1)(1-t)}{2(s-1)}y_3 - \frac{(s+1)(t+1)}{2(s-1)}y_4$$

The assumed solution over the master element is written using the product of quadratic Lagrange interpolation functions in the s direction and the linear function in the t direction:

$$u = N_1 u_1 + \cdots + N_6 u_6 \equiv \boldsymbol{N}^T \boldsymbol{d}$$

$$\boldsymbol{N} = \begin{pmatrix} \frac{1}{4}(s-1)s(1-t) \\ \frac{1}{4}(s-1)s(t+1) \\ -\frac{1}{2}(s-1)(s+1)(1-t) \\ -\frac{1}{2}(s-1)(s+1)(t+1) \\ \frac{1}{4}s(s+1)(1-t) \\ \frac{1}{4}s(s+1)(t+1) \end{pmatrix}$$

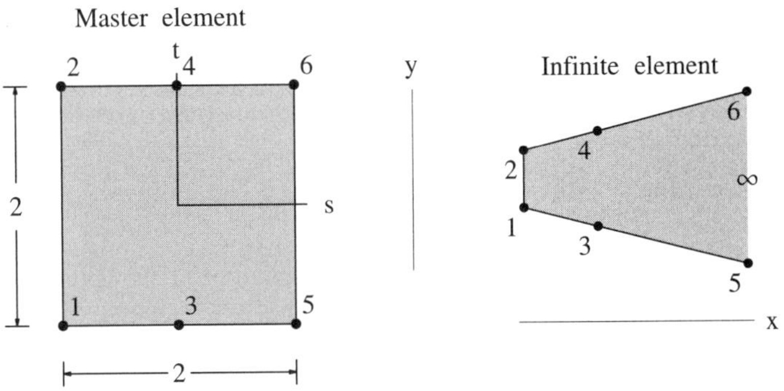

Figure 6.45. Mapped infinite element

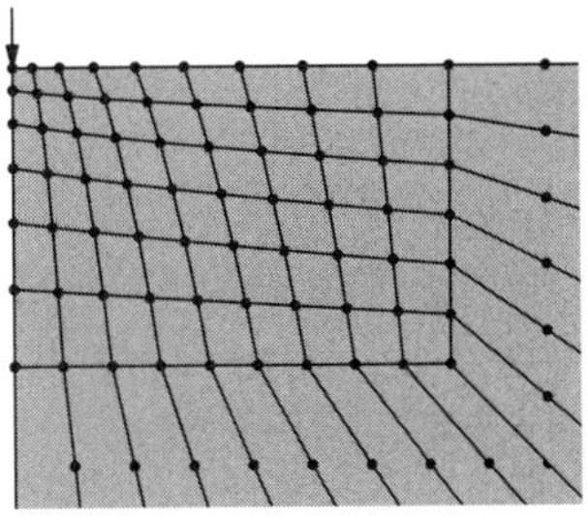

Figure 6.46. Finite element mesh terminated with infinite elements on the left and the bottom

Using these interpolation functions, the element equations can be derived in the usual manner. As mentioned already, the integrals in the element equations are singular. However, since only crude approximations are desired, in actual implementations integrations are typically carried out using a 2×2 Gauss quadrature. A typical finite element mesh involving infinite elements is as shown in Figure 6.46. All other elements are the standard quadrilateral elements. Only the last layer of elements in the x and y directions are the infinite elements.

PROBLEMS

Integration Using Change of Variables

6.1 Show that the following integral over a four-sided region bounded by the curves shown in Figure 6.47, evaluates to the value given:

$$\iint_{A_{xy}} (x - y)\, dA_{xy} = -\frac{23}{20}$$

$$C_1:\ y = \frac{2x^2}{15} + \frac{7x}{15}; \quad C_2:\ y = 7 - 2x; \quad C_3:\ y = \frac{x}{2} + 2; \quad C_4:\ y = \frac{7x}{2} - x^2$$

Use the following change of variables:

$$x(s,t) = \frac{ts^2}{8} - \frac{s^2}{8} - \frac{3ts}{8} + \frac{7s}{8} + \frac{3}{2}$$

$$y(s,t) = \frac{st^2}{8} - \frac{t^2}{8} - \frac{3st}{8} + \frac{7t}{8} + \frac{s}{2} + 2$$

Verify that with this change of variables, in terms of (s, t), the bounding curves are

$$C_1:\ t = -1; \qquad C_2:\ s = 1; \qquad C_3:\ t = 1; \qquad C_4:\ s = -1$$

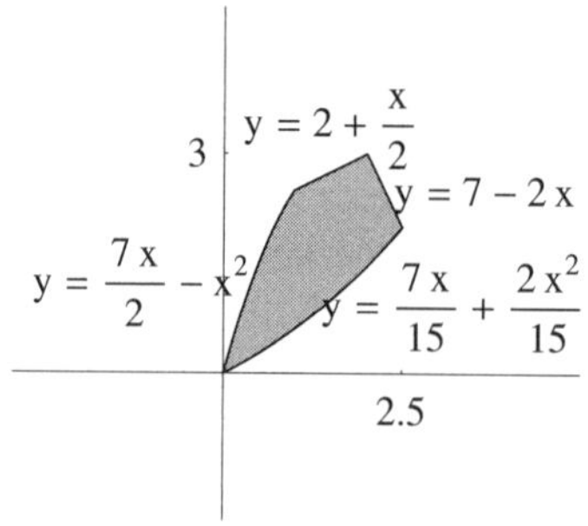

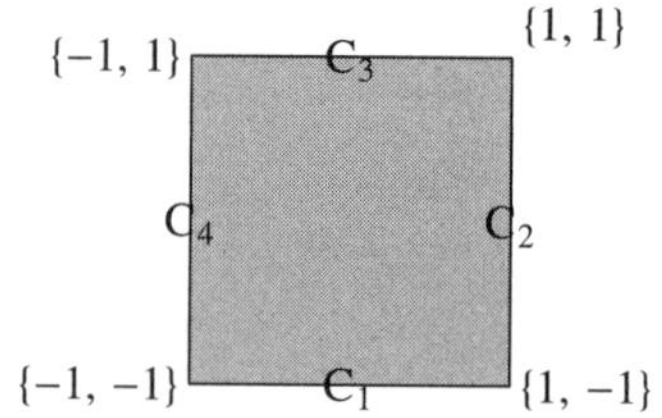

Figure 6.47.

6.2 Show that the following integral over a four-sided region bounded by the curves shown in Figure 6.48 evaluates to the value given:

$$\iint_{A_{xy}} (xy) dA_{xy} = \frac{151}{12}$$

$$C_1:\ y = \frac{x}{3}; \qquad C_2:\ y = 7 - 2x; \qquad C_3:\ y = \frac{11}{3} - \frac{x}{3}; \qquad C_4:\ y = -4x$$

Use the following change of variables:

$$x(s,t) = \frac{3s}{2} - \frac{t}{2} + 1; \qquad y(s,t) = -\frac{st}{2} + \frac{3t}{2} + 2$$

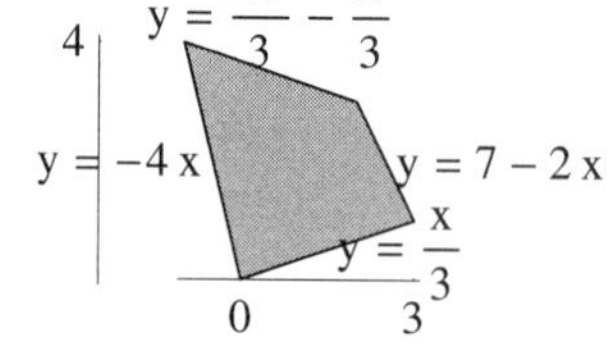

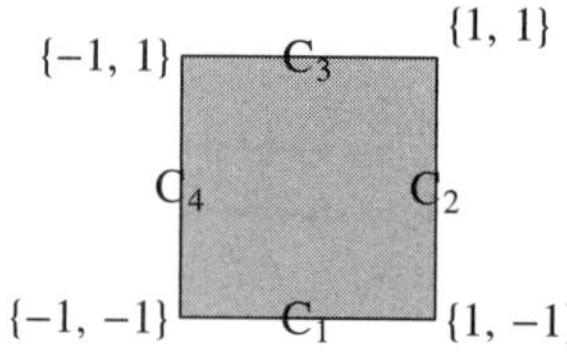

Figure 6.48.

Verify that with this change of variables, in terms of (s, t), the bounding curves are

$$C_1:\ t = -1; \qquad C_2:\ s = 1; \qquad C_3:\ t = 1; \qquad C_4:\ s = -1$$

and therefore the given quadrilateral is mapped to the 2×2 square as shown.

6.3 A function $f = x^2 y$ is to be integrated over a quadrilateral shown in Figure 6.49. The sides of the quadrilateral are defined by the following equations:

$$C_1:\ y = -\frac{x}{3}; \qquad C_2:\ y = 14 - 5x; \qquad C_3:\ y = 4; \qquad C_4:\ y = -4x$$

The following change of variables is proposed for evaluation of the integral:

$$x(s,t) = \frac{3s}{2} - \frac{t}{2} + 1; \qquad y(s,t) = \frac{ts}{4} - \frac{s}{4} + \frac{9t}{4} + \frac{7}{4}$$

$$\boldsymbol{J} = \begin{pmatrix} \frac{3}{2} & -\frac{1}{2} \\ \frac{t}{4} - \frac{1}{4} & \frac{s}{4} + \frac{9}{4} \end{pmatrix}; \qquad \det \boldsymbol{J} = \frac{3s}{8} + \frac{t}{8} + \frac{13}{4}$$

(a) Verify that with this change of variables, in terms of (s, t), the bounding curves are

$$C_1:\ t = -1; \qquad C_2:\ s = 1; \qquad C_3:\ t = 1; \qquad C_4:\ s = -1$$

and therefore the given quadrilateral is mapped to the 2×2 square as shown.

(b) Show that the integral is

$$\iint_{A_{xy}} (x^2 y)\, dA_{xy} = \frac{911}{30} = 30.3667$$

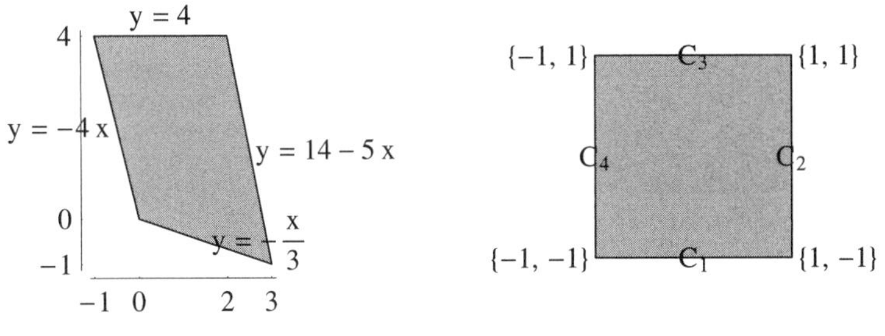

Figure 6.49.

Numerical Integration

6.4 Use numerical integration to evaluate the integral in Problem 6.2. Use 1×1 and 2×2 Gaussian integration. Show all calculations.

6.5 Use numerical integration to evaluate the integral in Problem 6.3. Use 1×1 and 2×2 Gaussian integration. Show all calculations.

Mapping Quadrilaterals

6.6 Develop an appropriate mapping to map the quadrilateral shown in Figure 6.50 to the 2×2 square master area. Compute the determinant of the Jacobian matrix and use it to demonstrate that the mapping is good. The coordinates of the key points are as follows:

	x	y
1	8.	0.
2	5.65685	5.65685
3	0.	8.
4	0	3
5	2	0

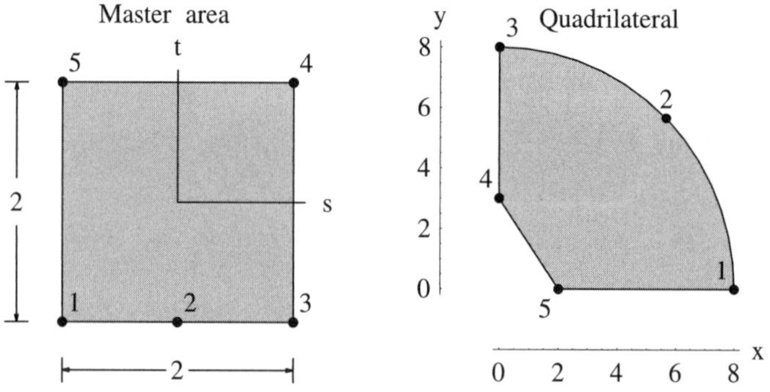

Figure 6.50.

6.7 Develop an appropriate mapping to map the quadrilateral shown in Figure 6.51 to the 2×2 square master area. Compute the determinant of the Jacobian matrix and use it to demonstrate that the mapping is good.

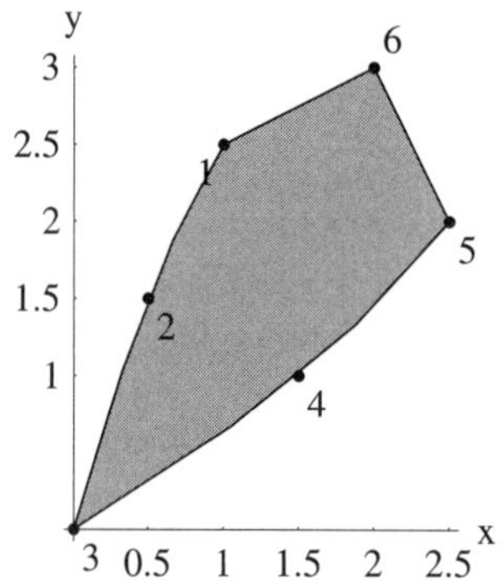

Figure 6.51.

6.8 A quadrilateral element with one curved side and the other three sides straight is mapped to a 2×2 square master area using five-node serendipity interpolation functions. The mapping is computed as follows:

$$x(s, t) = 0.828427ts^2 - 0.828427s^2 + 1.5ts - 2.5s - 2.32843t + 3.32843$$
$$y(s, t) = 0.828427ts^2 - 0.828427s^2 - 1.25ts + 2.75s - 2.07843t + 3.57843$$

Assume that the solution over the element is also written using the same five-node serendipity interpolation functions. Compute x and y derivatives of the interpolation functions at $x = y = 4$.

6.9 Use 2×2 Gauss quadrature to integrate $f = x^2y$ over the quadrilateral shown in Figure 6.52.

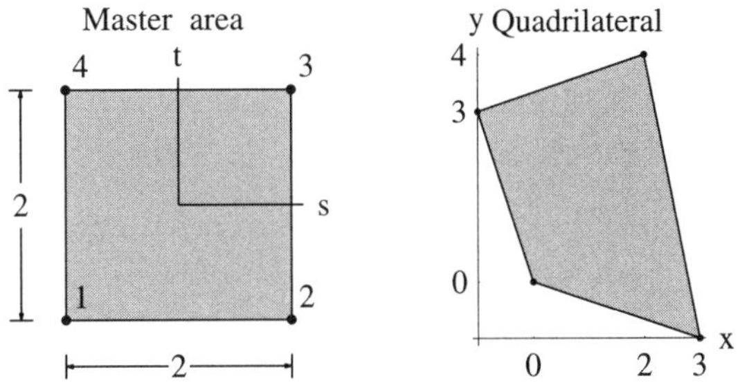

Figure 6.52.

6.10 A four-node quadrilateral element is shown in Figure 6.53.

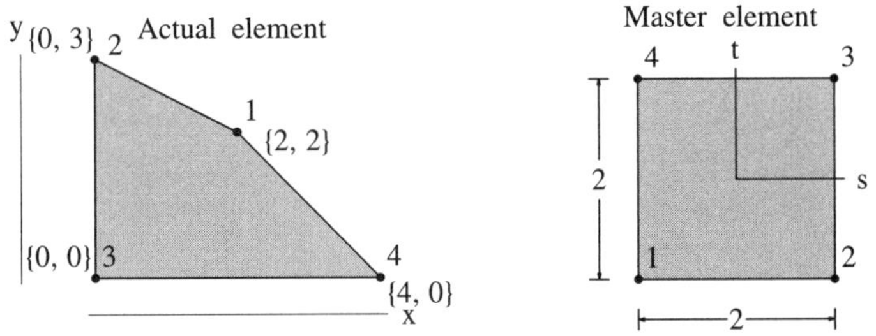

Figure 6.53.

(a) Develop an appropriate mapping to map the actual element into the 2×2 master element shown in the figure.

(b) Verify that the mapping is good.

(c) Compute $\partial N_3/\partial y$, where $N_3(s, t)$ is the third interpolation function for the master element. Give the value of the derivative at node 3 of the element.

(d) Compute $\iint_A 64 N_2 N_3 \, dA$, where $N_2(s,t)$ and $N_3(s,t)$ are the second and the third interpolation functions for the master element and A is the area of the actual element. Use numerical integration with a 1×1 Gauss quadrature formula. No credit will be given if any other integration method is used, even if the integral is correct.

(e) Compute $\int_c 4 N_{3c} \, dc$, where c is the line 3-4 of the element and N_{3c} is the third interpolation function for the master element expressed in a coordinate that runs along this line. Use numerical integration with a two-point Gauss quadrature formula. No credit will be given if any other integration method is used, even if the integral is correct.

6.11 A 2D boundary value problem is being solved by using eight-node serendipity elements. An element near a boundary is as shown in the Figure 6.54. The nodal coordinates are as follows:

	x	y
1	6.	0.
2	4.24264	4.24264
3	0.	6.
4	0.	3.
5	0	0
6	$\frac{3}{2}$	−2
7	3	−4
8	4.5	−2.

(a) Determine the vector $\boldsymbol{r}_\beta$ as a result of the natural boundary condition $\partial T/\partial n = 100$ on side 5-6-7 of the element. For simplicity use only a one-point integration.

For the following parts use the following mapping:

$$x = 2.87 + 2.25s - 0.62s^2 + 1.37t + 0.75st - 0.62s^2 t$$

$$y = 1.12 - 2.5s - 0.62s^2 + 3.12t - 0.5st - 0.62s^2 t$$

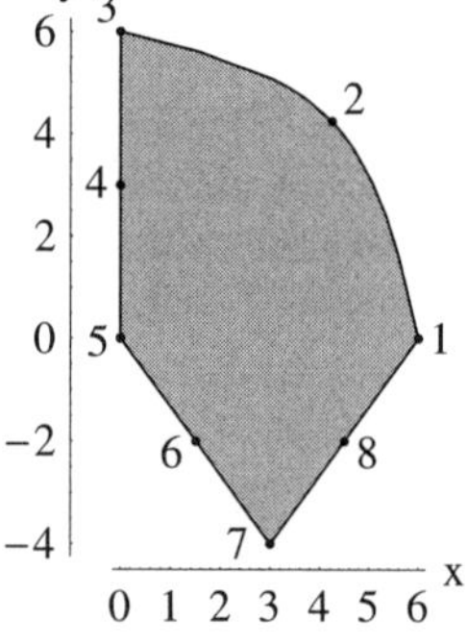

Figure 6.54.

Also assume that after complete assembly and considerations of boundary conditions the following nodal solution is obtained:

$$\boldsymbol{d} = (0 \quad 0 \quad 0 \quad 0 \quad 1 \quad 0 \quad 0 \quad 0)$$

(b) Determine the solution T at the centroid of the element.

(c) Determine the x derivative of the solution $\partial T/\partial x$ at node 5 of the element.

Finite Element Computations over Mapped Elements

6.12 Evaluate matrix $\boldsymbol{k}_k$ for the four-node quadrilateral element shown in Figure 6.55. Assume $k_x = 0.5$ and $k_y = 1.3$ over the element. Use 2×2 Gaussian quadrature.

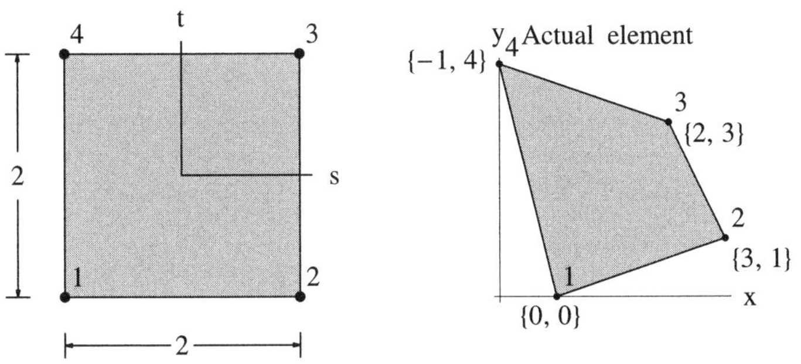

Figure 6.55.

6.13 Compute matrix $\boldsymbol{k}_\alpha$ for the eight-node element shown in Figure 6.56. Assume α is equal to 2 over side 1-2-3 of the element.

	1	2	3	4	5	6	7	8
x	1	$\frac{1}{2}$	0	$\frac{3}{2}$	$\frac{5}{2}$	$\frac{9}{4}$	2	$\frac{3}{2}$
y	$\frac{5}{2}$	$\frac{3}{2}$	0	1	2	$\frac{5}{2}$	3	$\frac{11}{4}$

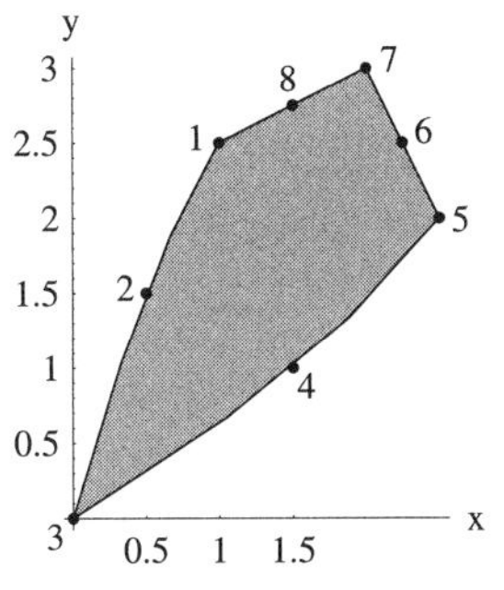

Figure 6.56.

6.14 The square duct shown in Figure 6.57 carries hot gases at a temperature of 300°C. The duct is insulated by a layer of square fiberglass that has a thermal conductivity of 0.04 W/m · °C. The outside surface of the fiberglass is subjected to convection with $h = 15$ W/m^2 · °C and $T_\infty = 30$°C. Take advantage of symmetry and model only one-eighth of the section, shown in dark shade. Assume that the temperature on the inside surface is same as that of the hot gases. Using only one four-node quadrilateral finite element, determine the nodal temperatures.

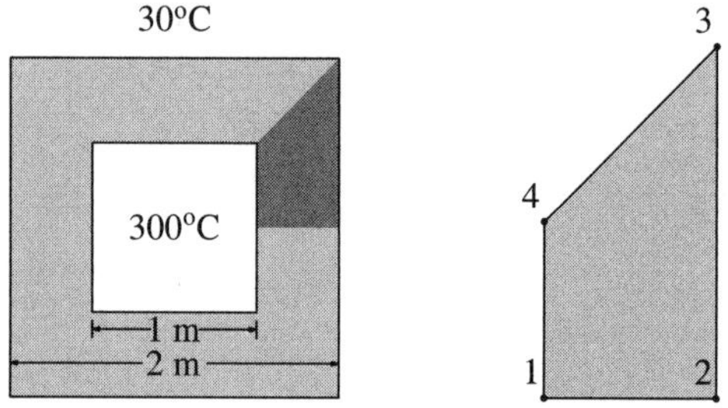

Figure 6.57.

6.15 The square duct shown in Figure 6.58 carries hot gases at a temperature of 300°C. The duct is insulated by a layer of circular fiberglass that has a thermal conductivity of 0.04 W/m · °C. The outside surface of the fiberglass is subjected to convection with $h = 15$ W/m^2 · °C and $T_\infty = 20$°C. Determine heat loss through the insulation. Take advantage of symmetry and model only one-eighth of the section, shown in dark shade. Use only one finite element.

	1	2	3	4	5	6	7	8
x	1.	0.92388	0.707107	0.603553	0.5	0.5	0.5	0.75
y	0.	0.382683	0.707107	0.603553	0.5	0.25	0.	0.

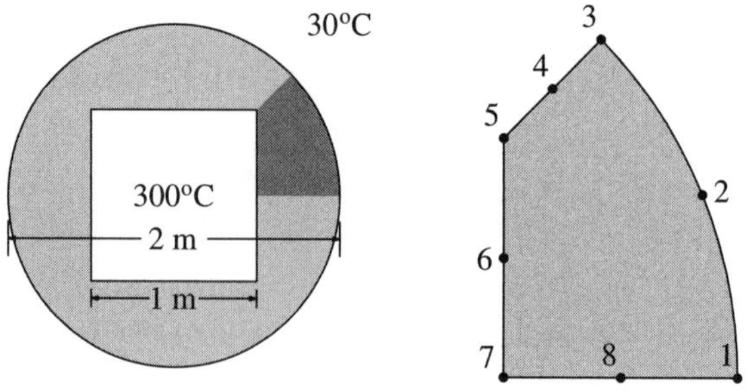

Figure 6.58.

CHAPTER SEVEN

ANALYSIS OF ELASTIC SOLIDS

The problem of determining stresses and strains in elastic solids subjected to loading and temperature changes is considered in this chapter. In addition to externally applied loading, the solid may be subjected to body forces that are distributed over its volume. Inertia forces and self-weight are the two common examples of the body forces. The fundamental concepts from elasticity are reviewed in Section 7.1. Using these concepts, the governing differential equations in terms of stresses and displacements are derived in Section 7.2. These equations are much more complicated than the one considered in Chapters 5 and 6. However, since the equations are still of the second order, all the fundamental concepts introduced in the earlier chapters are still applicable. Section 7.3 presents the general form of finite element equations for analysis of elastic solids. Specific elements for analysis of plane stress and plane strain problems are presented in Section 7.4.

7.1 FUNDAMENTAL CONCEPTS IN ELASTICITY

7.1.1 Stresses

Internal forces are produced in a body when it is subjected to external loading. These internal forces are characterized in terms of stresses. The stress at a point P is defined by passing a plane section through that point and considering force per unit area on the differential element. Denoting the area of this differential element by ΔA and the force on it by $\Delta \boldsymbol{F}$, the *stress vector* at point P for the plane whose unit normal is $\boldsymbol{n}$ is written as follows:

$$\boldsymbol{t}_n(P) = \lim_{\Delta A \to 0} \frac{\Delta \boldsymbol{F}}{\Delta \mathrm{A}}$$

The normal and shear stresses are written by considering components of the force in the normal (ΔF_n) and tangential (ΔF_t) directions of the plane as follows:

$$\text{Normal stress:} \qquad \sigma_n(P) = \lim_{\Delta A \to 0} \frac{\Delta F_n}{\Delta A}$$

$$\text{Shear stress:} \qquad \tau_n(P) = \lim \frac{\Delta F_t}{\Delta A}$$

Note that the definition of stress is intimately tied to the plane section that is characterized in terms of its unit normal $\boldsymbol{n}$. Since an infinite number of planes can be passed through a given point P, the stress state at the point must clearly state the plane section being considered. The usual practice is to define stresses in terms of planes whose normals are along the coordinate directions. For example, in a cartesian coordinate system, the stresses on a plane whose normal is along the x axis (the y-z plane) are as follows.

$$\text{Normal stress:} \qquad \sigma_x(P) = \lim_{\Delta A \to 0} \frac{\Delta F_x}{\Delta A}$$

$$\text{Shear stress:} \qquad \tau_x(P) = \lim_{\Delta A \to 0} \frac{\Delta F_t}{\Delta A}$$

The shear stress is further expressed in terms of the two in-plane force components in the y and z directions:

$$\text{Shear stress components:} \qquad \tau_{xy}(P) = \lim_{\Delta A \to 0} \frac{\Delta F_y}{\Delta A}; \qquad \tau_{xz}(P) = \lim_{\Delta A \to 0} \frac{\Delta F_z}{\Delta A}$$

Similarly, we can define the stresses on the planes whose normals are along the y axis (the x-z plane) and along the z-axis (the x-y plane). The complete stress state at point P can then be described in terms of nine stress components:

$$\text{Stress vector on plane with normal along } x \text{ axis:} \qquad \boldsymbol{t}_x = (\sigma_x \quad \tau_{xy} \quad \tau_{xz})^T$$

$$\text{Stress vector on plane with normal along } y \text{ axis:} \qquad \boldsymbol{t}_y = (\tau_{yx} \quad \sigma_y \quad \tau_{yz})^T$$

$$\text{Stress vector on plane with normal along } z \text{ axis:} \qquad \boldsymbol{t}_z = (\tau_{zx} \quad \tau_{zy} \quad \sigma_z)^T$$

These stress components are shown in Figure 7.1 on a differential block. On the three hidden faces of the block the outer normals are in the opposite directions; thus the directions of the stresses will be opposite to the ones shown in the figure. Writing the stress vectors as rows in a 3×3 matrix, the so-called *stress tensor* is defined as follows:

$$\boldsymbol{S} = \begin{pmatrix} \sigma_x & \tau_{xy} & \tau_{xz} \\ \tau_{yx} & \sigma_y & \tau_{yz} \\ \tau_{zx} & \tau_{zy} & \sigma_z \end{pmatrix}$$

Using the moment equilibrium condition, it can be shown that the following shear stress components are equal:

$$\tau_{xy} = \tau_{yx}; \qquad \tau_{yz} = \tau_{zy}; \qquad \tau_{zx} = \tau_{xz}$$

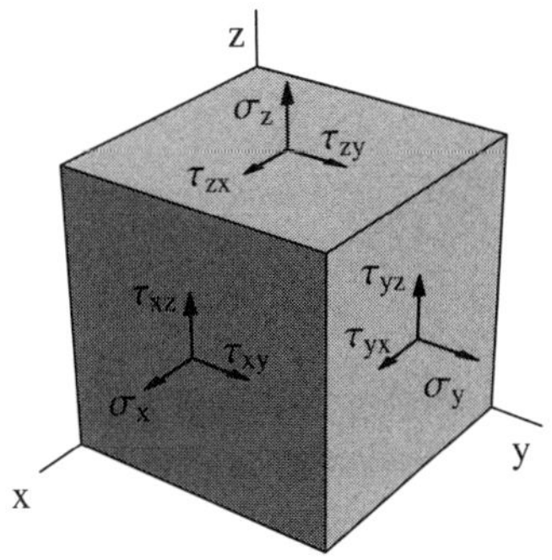

Figure 7.1. Notation for stress components in rectangular coordinates

Thus the stress tensor $\boldsymbol{S}$ is a symmetric matrix and there are six unique stress components. In the finite element formulation, it is convenient to arrange the six stress components in a vector form as follows:

$$\boldsymbol{\sigma} = (\sigma_x \quad \sigma_y \quad \sigma_z \quad \tau_{xy} \quad \tau_{yz} \quad \tau_{zx})^T$$

Stresses on an Inclined Plane It is possible to relate stresses on any inclined plane to those on the planes normal to the coordinate axes. To derive this relationship, consider an infinitesimal wedge (tetrahedron), shown in Figure 7.2. The stress vectors on the three vertical planes are $-\boldsymbol{t}_x$, $-\boldsymbol{t}_y$, and $-\boldsymbol{t}_z$. The stress vector on the inclined plane with unit normal $\boldsymbol{n} = (n_x \quad n_y \quad n_z)^T$ is denoted by $\boldsymbol{t}_n$. If the area of the inclined plane is dA, then from geometry the areas of the wedge perpendicular to the coordinate axes are, respectively, $\boldsymbol{n}_x\, dA$, $\boldsymbol{n}_y\, dA$, and $\boldsymbol{n}_z\, dA$. The equilibrium of forces acting on the tetrahedron gives

$$\boldsymbol{t}_n\, dA = \boldsymbol{t}_x n_x\, dA + \boldsymbol{t}_y n_y\, dA + \boldsymbol{t}_z n_z\, dA$$

or

$$\boldsymbol{t}_n = \begin{pmatrix} \sigma_x \\ \tau_{xy} \\ \tau_{xz} \end{pmatrix} n_x + \begin{pmatrix} \tau_{yx} \\ \sigma_y \\ \tau_{yz} \end{pmatrix} n_y + \begin{pmatrix} \tau_{zx} \\ \tau_{zy} \\ \sigma_z \end{pmatrix} n_z = \begin{pmatrix} \sigma_x & \tau_{xy} & \tau_{xz} \\ \tau_{xy} & \sigma_y & \tau_{yz} \\ \tau_{zx} & \tau_{yz} & \sigma_z \end{pmatrix} \begin{pmatrix} n_x \\ n_y \\ n_z \end{pmatrix}$$

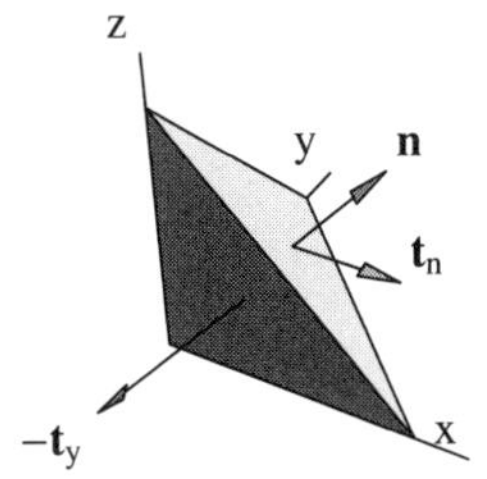

Figure 7.2. Stresses on an inclined plane

Thus the stress vector on any inclined plane can be obtained by multiplying the stress tensor $\boldsymbol{S}$ with the normal vector for the plane:

$$\boldsymbol{t}_n = \boldsymbol{S}\boldsymbol{n}$$

This is an important result because it shows that the six component stresses $\sigma_x, \ldots, \tau_{zx}$ are enough to uniquely determine the stress state with respect to any plane passing through that point. The components of the stress vector $\boldsymbol{t}_n$ in the direction of the normal to the plane is by definition the normal stress on the plane, and hence

$$\sigma_n = \boldsymbol{t}_n^T \boldsymbol{n} \equiv \boldsymbol{n}^T \boldsymbol{t}_n$$

Finally, using the vector sum, the shear stress on the plane can be written as follows:

$$\tau_n = \sqrt{\boldsymbol{t}_n^T \boldsymbol{t}_n - \sigma_n^2}$$

Example 7.1 Stress components at a point in a solid are given as follows:

$$\sigma_x = 9; \qquad \sigma_y = -3; \qquad \sigma_z = 3; \qquad \tau_{xy} = -2; \qquad \tau_{xz} = 0; \qquad \tau_{yz} = 2\ \text{N/m}^2$$

Compute the stress vector, normal stress, and shear stress on a plane that passes through (1, 0, 0), (0, 2, 0), and (0, 0, 1).

The first task is to compute the unit normal for the plane that passes through the given points. This can be done by defining two vectors, one from point 1 to 2 and the other from point 1 to 3. The normal vector is then the cross product of the two vectors. Finally the unit normal vector $\boldsymbol{n}$ is obtained by dividing the normal vector by its length:

Vector from point 1 to 2: $\boldsymbol{v}_{12} = \{0, 2, 0\} - \{1, 0, 0\} = \{-1, 2, 0\}$

Vector from point 1 to 3: $\boldsymbol{v}_{13} = \{0, 0, 1\} - \{1, 0, 0\} = \{-1, 0, 1\}$

Normal vector: $\boldsymbol{v}_n = \boldsymbol{v}_{12} \times \boldsymbol{v}_{13} = \left\{ \begin{vmatrix} 2 & 0 \\ 0 & 1 \end{vmatrix}, - \begin{vmatrix} -1 & 0 \\ -1 & 1 \end{vmatrix}, \begin{vmatrix} -1 & 2 \\ -1 & 0 \end{vmatrix} \right\} = \{2, 1, 2\}$

$$\text{Length} = \sqrt{2^2 + 1^2 + 2^2} = 3$$

Unit normal vector: $\boldsymbol{n} = \left\{ \frac{2}{3}, \frac{1}{3}, \frac{2}{3} \right\}$

From the given stress components, the stress tensor is

$$\boldsymbol{S} = \begin{pmatrix} 9 & -2 & 0 \\ -2 & -3 & 2 \\ 0 & 2 & 3 \end{pmatrix} \text{N/m}^2$$

The stress vector on the plane with unit normal $\boldsymbol{n}$ is

$$\boldsymbol{t}_n = \boldsymbol{S}\boldsymbol{n} = \left\{ \frac{16}{3}, -1, \frac{8}{3} \right\}$$

The normal and shear stresses on the plane can now be computed to get

$$\text{Normal stress on the plane:} \qquad \sigma_n = \boldsymbol{n}^T \boldsymbol{t}_n = 5$$

$$\text{Shear stress on the plane:} \qquad \tau_n = \frac{2\sqrt{26}}{3}$$

Applied Surface Forces The relationship between the usual stress components on planes perpendicular to the coordinate axes and those acting on an inclined plane was derived by considering an arbitrarily placed tetrahedron. If the tetrahedron is taken near the boundary of a solid such that the inclined plane is along the boundary, then the components of the stress vector $\boldsymbol{t}_n$ on the inclined plane must be equal to the applied forces on the boundary. Denoting the components of the applied forces on the boundary by q_x, q_y, and q_z, the applied surface forces are related to stresses as follows:

$$q_x = \sigma_x n_x + \tau_{xy} n_y + \tau_{xz} n_z$$
$$q_y = \tau_{yx} n_x + \sigma_y n_y + \tau_{yz} n_z$$
$$q_z = \tau_{zx} n_x + \tau_{zy} n_y + \sigma_z n_z$$

Principal Directions and Principal Stresses Principal directions are the unit normals for the planes over which there are no shear stresses and thus the normal stresses are maximum. Since the off-diagonal terms in the stress tensor $\boldsymbol{S}$ are the shear stresses, it follows that the principal planes are those that make the $\boldsymbol{S}$ matrix diagonal. Thus principal stresses and directions can be determined by solving the following eigenvalue problem:

$$(\boldsymbol{S} - \sigma \boldsymbol{I})\boldsymbol{n}_p = 0$$

where σ is a principal stress, $\boldsymbol{I}$ is a 3×3 identity matrix, and $\boldsymbol{n}_p$ is a normal vector for the principal plane. The eigenvalues are determined by setting $\det(\boldsymbol{S} - \sigma \boldsymbol{I}) = 0$. For each eigenvalue the corresponding eigenvector is determined by solving the above equation for $\boldsymbol{n}_p$.

The principal stresses are ordered according to their algebraic values:

$$\text{Maximum: } \sigma_1; \qquad \text{Intermediate: } \sigma_2; \qquad \text{Minimum: } \sigma_3$$

It can be shown that the maximum shear stress on any plane is given as follows:

$$\tau_{\max} = \frac{\sigma_1 - \sigma_3}{2}$$

where σ_1 is the maximum principal stress and σ_3 is the minimum principal stress. This $\tau_{\max}$ acts on a plane inclined at 45° to the directions of both σ_1 and σ_3.

Example 7.2 Stress components at a point in a solid are given as follows:

$$\sigma_x = 10; \quad \sigma_y = -7; \quad \sigma_z = 5; \quad \tau_{xy} = -3; \quad \tau_{xz} = 0; \quad \tau_{yz} = 2\ \text{N/m}^2$$

Compute principal stresses and their directions.

From the given stress components, the stress tensor is

$$S = \begin{pmatrix} 10 & -3 & 0 \\ -3 & -7 & 2 \\ 0 & 2 & 5 \end{pmatrix} \text{N/m}^2$$

Mathematica's Eigensystem command is used to determine the eigenvalues and corresponding eigenvectors of a matrix:

```
Eigensystem[{{10., -3, 0}, {-3, -7, 2}, {0, 2, 5}}]
```

$$\begin{pmatrix} 10.5353 & -7.81721 & 5.2819 \\ \{-0.982474, 0.175309, 0.0633422\} & \{-0.164108, -0.974648, 0.152084\} & \{-0.0883982, -0.139024, -0.986336\} \end{pmatrix}$$

Ordering the eigenvalues from highest to lowest values, we have the following principal stresses and unit normals for the principal planes:

	Principal Stresses	Normal Vectors for Principal Planes
1	10.5353	−0.982474 0.175309 0.0633422
2	5.2819	−0.0883982 −0.139024 −0.986336
3	−7.81721	−0.164108 −0.974648 0.152084

7.1.2 Stress Failure Criteria

The ultimate and yield strengths of engineering materials are determined from the tests conducted on specimens loaded in one direction. In order to use this information for general three-dimensional solids, we must convert the three-dimensional stress state into an equivalent value that is representative of failure stress for that body. The three most commonly used stress failure criteria are as follows.

Maximum Shear Stress Failure—Tresca Criterion According to this criterion, the material failure occurs when the maximum shear stress at any plane in the material reaches the value of shear stress as determined by a uniaxial tension test. The maximum shear stress on any plane is

$$\tau_{max} = \frac{\sigma_1 - \sigma_3}{2}$$

where σ_1 is the maximum principal stress and σ_3 is the minimum principal stress. In a uniaxial test σ_1 is the applied axial stress and $\sigma_3 = 0$; therefore the maximum shear stress is $\sigma_f/2$, where σ_f is the failure stress of the material in a uniaxial test. Thus, according to the maximum shear stress failure criterion, the stress failure occurs when

$$\tau_{max} = \frac{\sigma_1 - \sigma_3}{2} = \frac{\sigma_f}{2}$$

Using this criteria, the factor of safety can be defined as follows:

$$\text{Factor of safety} = \frac{\sigma_f/2}{\tau_{max}} \equiv \frac{\sigma_f}{\sigma_1 - \sigma_3}$$

Example 7.3 If a material has yield strength $\sigma_f = 50$ MPa, what is the factor of safety using the maximum shear stress failure criterion if the state of stress at a point in a solid is given as follows:

$$\sigma_x = 5; \quad \sigma_y = -18; \quad \sigma_z = 0; \quad \tau_{xy} = 15; \quad \tau_{xz} = 0; \quad \tau_{yz} = 0\ \text{MPa}$$

Component stresses = {5., −18., 0., 15., 0., 0.}

Principal stresses = {12.4011, 0., −25.4011}; $\tau_{max} = 18.9011$

$$\text{Factor of safety} = \frac{\sigma_f/2}{\tau_{max}} = 1.32268$$

Von Mises Failure Criterion The von Mises criterion is the most commonly used stress failure criterion for metals. It assumes that the stress failure occurs when the octahedral shear stress value is equal to the octahedral shear stress at yielding in a uniaxial tensile test.

An octahedral plane is a plane that makes equal angles with the principal stress directions. The shear stress on the octahedral plane is known as the octahedral shear stress. Using stress transformation for an inclined plane, it can be shown that on an octahedral plane, the stresses are as follows:

Octahedral normal stress: $$\sigma_{oct} = \frac{\sigma_1 + \sigma_2 + \sigma_3}{3}$$

Octahedral shear stress: $$\tau_{oct} = \tfrac{1}{3}\sqrt{(\sigma_1 - \sigma_2)^2 + (\sigma_2 - \sigma_3)^2 + (\sigma_3 - \sigma_1)^2}$$

where $\sigma_1 > \sigma_2 > \sigma_3$ are the principal stresses. The octahedral shear stress can also be expressed in terms of normal and shear stress components as follows:

$$\tau_{oct} = \tfrac{1}{3}\sqrt{(\sigma_x - \sigma_y)^2 + (\sigma_y - \sigma_z)^2 + (\sigma_z - \sigma_x)^2 + 6(\tau_{xy}^2 + \tau_{yz}^2 + \tau_{zx}^2)}$$

Using this equation, the octahedral shear stress at yielding in a uniaxial tensile test ($\sigma_x = \sigma_f, \sigma_y = \cdots = \tau_{zx} = 0$) is

$$\tau_{octUni} = \tfrac{1}{3}\sqrt{2\sigma_f^2} = \tfrac{\sqrt{2}}{3}\sigma_f$$

Thus, according to the von Mises criterion, the stress failure occurs when

$$\tfrac{1}{3}\sqrt{(\sigma_x-\sigma_y)^2+(\sigma_y-\sigma_z)^2+(\sigma_z-\sigma_x)^2+6(\tau_{xy}^2+\tau_{yz}^2+\tau_{zx}^2)}=\tfrac{\sqrt{2}}{3}\sigma_f$$

or

$$\tfrac{1}{\sqrt{2}}\sqrt{(\sigma_x-\sigma_y)^2+(\sigma_y-\sigma_z)^2+(\sigma_z-\sigma_x)^2+6(\tau_{xy}^2+\tau_{yz}^2+\tau_{zx}^2)}=\sigma_f$$

The quantity on the left-hand side is usually referred to as the *effective stress* or *von Mises stress* and is denoted by σ_e:

$$\sigma_e=\tfrac{1}{\sqrt{2}}\sqrt{(\sigma_x-\sigma_y)^2+(\sigma_y-\sigma_z)^2+(\sigma_z-\sigma_x)^2+6(\tau_{xy}^2+\tau_{yz}^2+\tau_{zx}^2)}$$

Using this criterion, the factor of safety can be defined as follows:

$$\text{Factor of safety}=\frac{\sigma_f}{\sigma_e}$$

Example 7.4 If a material has yield strength $\sigma_f = 50\,\text{MPa}$, what is the factor of safety using the von Mises failure criterion if the state of stress at a point in a solid is given as follows:

$$\sigma_x = 5; \quad \sigma_y = -18; \quad \sigma_z = 0; \quad \tau_{xy} = 15; \quad \tau_{xz} = 0; \quad \tau_{yz} = 0\,\text{MPa}$$

Component stresses = {5., −18., 0., 15., 0., 0.}

$$\sigma_e = 33.3766; \qquad \text{Factor of safety} = \frac{\sigma_f}{\sigma_e} = 1.49805$$

Mohr-Coulomb Failure Criterion In brittle materials the failure strengths in tension and compression tests are usually different. According to the Mohr-Coulomb criterion, the failure of brittle materials is predicted when

$$\frac{\sigma_1}{\sigma_{tf}}-\frac{\sigma_3}{\sigma_{cf}}=1$$

where σ_1 is the maximum principal stress, σ_3 is the minimum principal stress, σ_{tf} is failure stress in the uniaxial tension test, and σ_{cf} is failure stress in uniaxial compression. Note that proper sign of σ_1 and σ_3 must be used in this equation while σ_{tf} and σ_{cf} are always positive.

Using this criterion, the factor of safety can be defined as follows:

$$\text{Factor of safety}=\frac{1}{\sigma_1/\sigma_{tf}-\sigma_3/\sigma_{cf}}$$

Example 7.5 If a material has failure strength in uniaxial tension $\sigma_{tf} = 15\text{MPa}$ and that in uniaxial compression $\sigma_{cf} = 100\,\text{MPa}$, what is the factor of safety using the Mohr-Coulomb failure criterion if the state of stress at a point in a solid is given as follows:

$$\sigma_x = 5; \quad \sigma_y = -18; \quad \sigma_z = 0; \quad \tau_{xy} = 15; \quad \tau_{xz} = 0; \quad \tau_{yz} = 0\ \text{MPa}$$

Principal stresses = {12.4011, 0., −25.4011}; Factor of safety = 0.925285

7.1.3 Strains

When subjected to loading, a point originally at (x, y, z) displaces to $(x + u, y + v, z + w)$. The changes (u, v, w) are the displacement components in the coordinate directions and are collected together in a displacement vector $\boldsymbol{u} = (u \quad v \quad w)^T$. The displacement of a point does not indicate the actual deformation of the solid at that point since the displacement could simply be a rigid-body motion of the solid. To correctly characterize deformations in solids, we consider an infinitesimal cube in the original configuration and determine its change in shape and size in the displaced position. The change in size is characterized by changes in the lengths of the sides of the cube. The change in shape is determined by considering the change in the angle between two intersecting lines from the original 90°.

To characterize changes in length, consider a differential line segment AB parallel to the x axis in the original configuration. After displacement the line moves to a position indicated by $A'B'$, as shown in Figure 7.3. The displacement of point A is $\boldsymbol{u}$ and that of the neighboring point B is $\boldsymbol{u} + d\boldsymbol{u}$. Assume point A is at (x, y, z). Since the line originally is assumed to be parallel to the x axis, point B is at $(x + dx, y, z)$. Due to displacement, the coordinates of point A' are $(x + u, y + v, z + w)$ and that of point B are $(x + dx + u + du, y + v + dv, z + w + dw)$. From these coordinates the original and the deformed lengths can be computed as follows:

$$L_{AB} = dx$$

$$L_{A'B'} = \sqrt{dv^2 + dw^2 + (-du - dx)^2}$$

Since line AB is parallel to the x axis, the increments in displacements are functions only

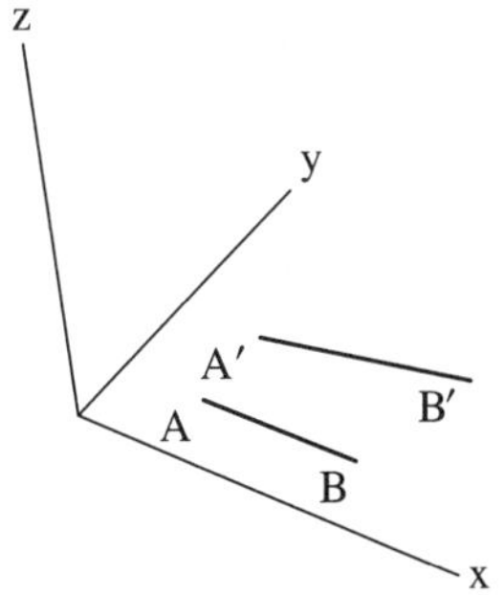

Figure 7.3. Deformation of a differential line segment originally parallel to the x axis

of x, and therefore

$$du = \frac{\partial u}{\partial x}\,dx; \qquad dv = \frac{\partial v}{\partial x}\,dx; \qquad dw = \frac{\partial w}{\partial x}\,dx$$

Substituting these expressions, the deformed length $L_{A'B'}$ can be written as follows:

$$L_{A'B'} = \sqrt{\left(\frac{\partial v}{\partial x}\right)^2 dx^2 + \frac{\partial w}{\partial x}^2 dx^2 + \left(-\frac{\partial u}{\partial x}\,dx - dx\right)^2}$$

or

$$L_{A'B'} = dx\sqrt{\left(\frac{\partial v}{\partial x}\right)^2 + \left(\frac{\partial w}{\partial x}\right)^2 + \left(1 + \frac{\partial u}{\partial x}\right)^2}$$

The normal strain in the x direction is then defined as the change in length of line AB divided by its original length:

$$\epsilon_x = \frac{L_{A'B'} - L_{AB}}{L_{AB}} = \sqrt{\left(\frac{\partial v}{\partial x}\right)^2 + \left(\frac{\partial w}{\partial x}\right)^2 + \left(1 + \frac{\partial u}{\partial x}\right)^2} - 1$$

So far this expression is exact. To simplify further, we assume small displacements and that the displacement derivatives are much smaller as compared to 1. Under this assumption

$$\left(\frac{\partial v}{\partial x}\right)^2 + \left(\frac{\partial w}{\partial x}\right)^2 \ll \left(1 + \frac{\partial u}{\partial x}\right)^2$$

and thus neglecting the first two terms under the radical sign, we get

$$\epsilon_x \approx \frac{\partial u}{\partial x}$$

Proceeding in a similar manner, with lines parallel to y and z axes, the normal strains in the y and z directions can be written as follows:

$$\epsilon_y \approx \frac{\partial v}{\partial y}; \qquad \epsilon_z \approx \frac{\partial w}{\partial z}$$

The normal strains quantify changes in the lengths of a differential cube. The changes in its angles, which are originally 90°, are quantified in terms of shear strains. Consider two lines AB and AC, originally lying in the x-y plane, meeting at 90° at point A. After displacement the lines move to $A'B'$ and $A'C'$, as shown in Figure 7.4. The point A at (x, y, z) is displaced to point A' at $(x + u, y + v, z)$. Since we are interested in determining change in angles alone, and we have already made the assumption that the displacements are small, the change in lengths is ignored. Thus the point B' moves up by increment in displacement dv relative to

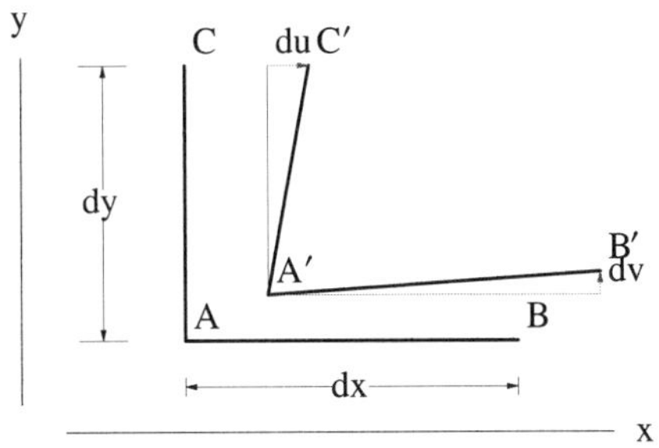

Figure 7.4. Change in the angle between two differential line segments in the x-y plane

point A' and point C' moves to the right by du. The total change in the original right angle is thus

$$\text{Angle change} = \frac{du}{dy} + \frac{dv}{dx}$$

Along AB, $dy = dz = 0$, and therefore

$$dv = \frac{\partial v}{\partial x} dx$$

Similarly, along AC, $dx = dz = 0$, and therefore

$$du = \frac{\partial u}{\partial y} dy$$

Using these expressions and defining the angle change as the shear strain γ_{xy}, we have

$$\gamma_{xy} = \frac{\partial u}{\partial y} + \frac{\partial v}{\partial x}$$

Proceeding in a similar fashion by taking lines in the y-z and z-x planes, the following shear strains can be defined:

$$\gamma_{yz} = \frac{\partial v}{\partial z} + \frac{\partial w}{\partial y}; \qquad \gamma_{zx} = \frac{\partial u}{\partial z} + \frac{\partial w}{\partial x}$$

The complete vector of strain-displacement relationships is written as follows:

$$\epsilon = \begin{pmatrix} \epsilon_x \\ \epsilon_y \\ \epsilon_z \\ \gamma_{xy} \\ \gamma_{yz} \\ \gamma_{zx} \end{pmatrix} = \begin{pmatrix} \frac{\partial u}{\partial x} \\ \frac{\partial v}{\partial y} \\ \frac{\partial w}{\partial z} \\ \frac{\partial u}{\partial y} + \frac{\partial v}{\partial x} \\ \frac{\partial v}{\partial z} + \frac{\partial w}{\partial y} \\ \frac{\partial u}{\partial z} + \frac{\partial w}{\partial x} \end{pmatrix}$$

7.1.4 Constitutive Equations

The stresses and strains are related through constitutive equations. The simplest form of the constitutive equations is the linear elasticity, which is a generalization of the familiar Hooke's law. Assuming stresses are proportional to strains, the most general form of linear elastic constitutive equations is as follows:

$$\begin{pmatrix} \sigma_x \\ \sigma_y \\ \sigma_z \\ \tau_{xy} \\ \tau_{yz} \\ \tau_{zx} \end{pmatrix} = \begin{pmatrix} C_{11} & C_{12} & C_{13} & C_{14} & C_{15} & C_{16} \\ C_{21} & C_{22} & C_{23} & C_{24} & C_{25} & C_{26} \\ C_{31} & C_{32} & C_{33} & C_{34} & C_{35} & C_{36} \\ C_{41} & C_{42} & C_{43} & C_{44} & C_{45} & C_{46} \\ C_{51} & C_{52} & C_{53} & C_{54} & C_{55} & C_{56} \\ C_{61} & C_{62} & C_{63} & C_{64} & C_{65} & C_{66} \end{pmatrix} \begin{pmatrix} \epsilon_x \\ \epsilon_y \\ \epsilon_z \\ \gamma_{xy} \\ \gamma_{yz} \\ \gamma_{zx} \end{pmatrix}$$

or

$$\sigma = \boldsymbol{C}\epsilon$$

The 36 coefficients C_{ij} are called elastic constants and must be established somehow for each material. Several assumptions are introduced to reduce the number of elastic constants in the constitutive equations. It is almost universally assumed in linear elasticity that the constitutive matrix $\boldsymbol{C}$ is symmetric. With this assumption the number of constants is reduced to 21. This number can be reduced further if the material properties are symmetric about some known planes. For example, if the material properties are symmetric about the xy plane, then it can be shown that the $\boldsymbol{C}$ matrix has the following form with 13 independent coefficients:

$$\begin{pmatrix} \sigma_x \\ \sigma_y \\ \sigma_z \\ \tau_{xy} \\ \tau_{yz} \\ \tau_{zx} \end{pmatrix} = \begin{pmatrix} C_{11} & C_{12} & C_{13} & C_{14} & 0 & 0 \\ C_{12} & C_{22} & C_{23} & C_{24} & 0 & 0 \\ C_{13} & C_{23} & C_{33} & C_{34} & 0 & 0 \\ C_{14} & C_{24} & C_{34} & C_{44} & 0 & 0 \\ 0 & 0 & 0 & 0 & C_{55} & C_{56} \\ 0 & 0 & 0 & 0 & C_{56} & C_{66} \end{pmatrix} \begin{pmatrix} \epsilon_x \\ \epsilon_y \\ \epsilon_z \\ \gamma_{xy} \\ \gamma_{yz} \\ \gamma_{zx} \end{pmatrix}$$

If the material properties are symmetric about all three coordinate planes, then the number of constants reduces to 9 and the material is called *orthotropic*. Wood is commonly treated as an orthotropic material. The constitutive equations for an orthotropic material are usually expressed as follows:

$$\begin{pmatrix} \sigma_x \\ \sigma_y \\ \sigma_z \\ \tau_{xy} \\ \tau_{yz} \\ \tau_{zx} \end{pmatrix} = \begin{pmatrix} C_{11} & C_{12} & C_{13} & 0 & 0 & 0 \\ C_{12} & C_{22} & C_{23} & 0 & 0 & 0 \\ C_{13} & C_{23} & C_{33} & 0 & 0 & 0 \\ 0 & 0 & 0 & C_{44} & 0 & 0 \\ 0 & 0 & 0 & 0 & C_{55} & 0 \\ 0 & 0 & 0 & 0 & 0 & C_{66} \end{pmatrix} \begin{pmatrix} \epsilon_x \\ \epsilon_y \\ \epsilon_z \\ \gamma_{xy} \\ \gamma_{yz} \\ \gamma_{zx} \end{pmatrix}$$

Finally, if the material properties are symmetric about every plane, then the material is called *isotropic* and only two independent constants are needed to completely define the constitutive equations. The constants are identified as E, the Young's modulus of elasticity, and ν, the Poisson's ratio. The constitutive equations for an isotropic material are as follows:

$$\begin{pmatrix} \sigma_x \\ \sigma_y \\ \sigma_z \\ \tau_{xy} \\ \tau_{yz} \\ \tau_{zx} \end{pmatrix} = \frac{E}{(1+\nu)(1-2\nu)} \begin{pmatrix} 1-\nu & \nu & \nu & 0 & 0 & 0 \\ \nu & 1-\nu & \nu & 0 & 0 & 0 \\ \nu & \nu & 1-\nu & 0 & 0 & 0 \\ 0 & 0 & 0 & \frac{1-2\nu}{2} & 0 & 0 \\ 0 & 0 & 0 & 0 & \frac{1-2\nu}{2} & 0 \\ 0 & 0 & 0 & 0 & 0 & \frac{1-2\nu}{2} \end{pmatrix} \begin{pmatrix} \epsilon_x \\ \epsilon_y \\ \epsilon_z \\ \gamma_{xy} \\ \gamma_{yz} \\ \gamma_{zx} \end{pmatrix}$$

In the inverse form the relationship can be written as follows:

$$\epsilon = \boldsymbol{D}\sigma$$

For an isotropic material,

$$\boldsymbol{D} = \begin{pmatrix} \frac{1}{E} & -\frac{\nu}{E} & -\frac{\nu}{E} & 0 & 0 & 0 \\ -\frac{\nu}{E} & \frac{1}{E} & -\frac{\nu}{E} & 0 & 0 & 0 \\ -\frac{\nu}{E} & -\frac{\nu}{E} & \frac{1}{E} & 0 & 0 & 0 \\ 0 & 0 & 0 & \frac{1}{G} & 0 & 0 \\ 0 & 0 & 0 & 0 & \frac{1}{G} & 0 \\ 0 & 0 & 0 & 0 & 0 & \frac{1}{G} \end{pmatrix}$$

where G is known as the shear modulus and is related to E and ν as follows:

$$G = \frac{E}{2(1+\nu)}$$

For an orthotropic material,

$$\boldsymbol{D} = \begin{pmatrix} \frac{1}{E_x} & -\frac{\nu_{xy}}{E_y} & -\frac{\nu_{xz}}{E_z} & 0 & 0 & 0 \\ -\frac{\nu_{yx}}{E_x} & \frac{1}{E_y} & -\frac{\nu_{yz}}{E_z} & 0 & 0 & 0 \\ -\frac{\nu_{zx}}{E_x} & -\frac{\nu_{zy}}{E_y} & \frac{1}{E_z} & 0 & 0 & 0 \\ 0 & 0 & 0 & \frac{1}{G_{xy}} & 0 & 0 \\ 0 & 0 & 0 & 0 & \frac{1}{G_{yz}} & 0 \\ 0 & 0 & 0 & 0 & 0 & \frac{1}{G_{xz}} \end{pmatrix}$$

where E_x is Young's modulus in the x direction, ν_{xy} is Poisson's ratio relating ϵ_x to σ_y/E_y, etc. The $\boldsymbol{D}$ matrix is symmetric and therefore

$$\frac{\nu_{xy}}{E_y} = \frac{\nu_{yx}}{E_x} \qquad \frac{\nu_{xz}}{E_z} = \frac{\nu_{zx}}{E_x} \cdot \qquad \frac{\nu_{yz}}{E_z} = \frac{\nu_{zy}}{E_y}$$

The constitutive equations give a relationship between stress and strain at a point. If the constitutive equations do not change from one point to another point in the material, then the material is known as homogeneous; otherwise it is inhomogeneous. In practice, for most linear elasticity problems it is assumed that the material is homogeneous and isotropic. Thus the material behavior is usually described in terms of E and ν values that are constant for the entire solid.

7.1.5 Temperature Effects and Initial Strains

In general, it is assumed that prior to loading there are no stresses and strains in the body. However, there are situations when there are initial strains that must be taken into consideration. The most common situations causing initial strains in solids are the temperature and moisture changes.

A temperature rise of ΔT results in a uniform strain that depends on the coefficient of thermal expansion α of the material. The temperature change does not cause shear strains. Thus the initial strain vector due to temperature change is as follows:

$$\epsilon_0 = \begin{pmatrix} \alpha\Delta T \\ \alpha\Delta T \\ \alpha\Delta T \\ 0 \\ 0 \\ 0 \end{pmatrix}$$

The stresses are caused by the difference in the total strains and the initial strains:

$$\boldsymbol{\sigma} = \boldsymbol{C}(\boldsymbol{\epsilon} - \boldsymbol{\epsilon}_0)$$

where ϵ_0 is the vector of initial strains. In the inverse form

$$\boldsymbol{\epsilon} = \boldsymbol{D}\boldsymbol{\sigma} + \boldsymbol{\epsilon}_0$$

7.2 GOVERNING DIFFERENTIAL EQUATIONS

Consider an arbitrary solid supported in a stable manner under the influence of externally applied forces on its surface such as pressure and concentrated loads. In addition, the solid may be subjected to applied forces distributed over the entire volume of the body. These forces are known as *body forces*. Typical examples of body forces are self-weight of the object and centrifugal forces developed in rotating objects. The governing differential equations can be developed easily by considering equilibrium of forces acting on an isolated part of the object. This approach gives three equilibrium equations in terms of stresses. These equations are derived in the first section. No constitutive equations are used in deriving these equations and therefore they are applicable to all materials.

The most common finite element formulations are based on displacements as primary unknowns. To use this formulation, the stress equilibrium equations must be expressed in terms of displacements using the strain-displacement and constitutive equations. The displacement equilibrium equations presented in this section assume linear strain-displacement relations and linear elastic stress-strain behavior. Thus the finite element formulation based on these equations, presented in the following sections, is applicable only to small-displacement problems that are restricted to linear elastic behavior.

7.2.1 Stress Equilibrium Equations

If a body is in equilibrium, then any arbitrary part of the body must also be in equilibrium. The surface S of an arbitrary part isolated from a solid is subjected to stress vector $\boldsymbol{t}_n = (t_{nx} \quad t_{ny} \quad t_{nz})^T$. The normal to the surface is $\boldsymbol{n} = (n_x \quad n_y \quad n_z)^T$. The body forces $\boldsymbol{b} = (b_x \quad b_y \quad b_z)^T$ are distributed over the entire volume V of the part. The equilibrium of the isolated part requires that

$$\iint_S t_{nx}\, dS + \iiint_V b_x\, dV = 0$$

$$\iint_S t_{ny}\, dS + \iiint_V b_y\, dV = 0$$

$$\iint_S t_{ny}\, dS + \iiint_V b_y\, dV = 0$$

Since $t_{nx} = \sigma_x n_x + \tau_{xy} n_y + \tau_{xz} n_z$ the first equation can be written as follows:

$$\iint_S \sigma_x n_x\, dS + \iint_S \tau_{xy} n_y\, dS + \iint_S \tau_{xz} n_z\, dS + \iiint_V b_x\, dV = 0$$

We can use the divergence theorem to convert surface integrals to volume integrals. The divergence theorem is a special case of the Green-Gauss (integration-by-parts) theorem:

$$\iiint_V f \frac{\partial g}{\partial x}\, dV = \iint_S f g n_x\, dS - \iiint_V \frac{\partial f}{\partial x} g\, dV$$

If $g = 1$, then $\partial g/\partial x = 0$, and we have

$$0 = \iint_S f n_x\, dS - \iiint_V \frac{\partial f}{\partial x}\, dV \Longrightarrow \iint_S f n_x\, dS = \iiint_V \frac{\partial f}{\partial x}\, dV$$

Similarly,

$$\iint_S f n_y\, dS = \iiint_V \frac{\partial f}{\partial y}\, dV \qquad \text{and} \qquad \iint_S f n_z\, dS = \iiint_V \frac{\partial f}{\partial z}\, dV$$

Using the divergence theorem, the surface integrals can be written in terms of volume integrals, giving

$$\iiint_V \left(\frac{\partial \sigma_x}{\partial x} + \frac{\partial \tau_{xy}}{\partial y} + \frac{\partial \tau_{xz}}{\partial z} + b_x \right) dV = 0$$

Since this holds for any arbitrary region, the integrand must be zero:

$$\frac{\partial \sigma_x}{\partial x} + \frac{\partial \tau_{xy}}{\partial y} + \frac{\partial \tau_{xz}}{\partial z} + b_x = 0$$

Proceeding in the same manner with the other two equations, we get the following equations in terms of stresses:

$$\frac{\partial \tau_{yx}}{\partial x} + \frac{\partial \sigma_y}{\partial y} + \frac{\partial \tau_{yz}}{\partial z} + b_y = 0$$

$$\frac{\partial \tau_{zx}}{\partial x} + \frac{\partial \tau_{zy}}{\partial y} + \frac{\partial \sigma_z}{\partial z} + b_z = 0$$

7.2.2 Governing Differential Equations in Terms of Displacements

The governing differential equations can be expressed in terms of displacements by using the constitutive equations and the strain-displacement relationships. For a linear elastic isotropic material we have

$$\begin{pmatrix} \sigma_x \\ \sigma_y \\ \sigma_z \\ \tau_{xy} \\ \tau_{yz} \\ \tau_{zx} \end{pmatrix} = \frac{E}{(1+\nu)(1-2\nu)} \begin{pmatrix} 1-\nu & \nu & \nu & 0 & 0 & 0 \\ \nu & 1-\nu & \nu & 0 & 0 & 0 \\ \nu & \nu & 1-\nu & 0 & 0 & 0 \\ 0 & 0 & 0 & \frac{1-2\nu}{2} & 0 & 0 \\ 0 & 0 & 0 & 0 & \frac{1-2\nu}{2} & 0 \\ 0 & 0 & 0 & 0 & 0 & \frac{1-2\nu}{2} \end{pmatrix} \begin{pmatrix} \epsilon_x \\ \epsilon_y \\ \epsilon_z \\ \gamma_{xy} \\ \gamma_{yz} \\ \gamma_{zx} \end{pmatrix}$$

Substituting strain-displacement equations $\epsilon_x = \partial u/\partial x, \ldots$ into the constitutive equations and carrying out matrix multiplication, we get

$$\sigma_x = \frac{E}{(1+\nu)(1-2\nu)} \left((1-\nu)\frac{\partial u}{\partial x} + \nu \frac{\partial v}{\partial y} + \nu \frac{\partial w}{\partial z} \right)$$

$$\sigma_y = \frac{E}{(1+\nu)(1-2\nu)} \left(\nu \frac{\partial u}{\partial x} + (1-\nu)\frac{\partial v}{\partial y} + \nu \frac{\partial w}{\partial z} \right)$$

$$\sigma_z = \frac{E}{(1+\nu)(1-2\nu)} \left(\nu \frac{\partial u}{\partial x} + \nu \frac{\partial v}{\partial y} + (1-\nu)\frac{\partial w}{\partial z} \right)$$

$$\tau_{xy} = G\left(\frac{\partial u}{\partial y} + \frac{\partial v}{\partial x} \right); \qquad \tau_{yz} = G\left(\frac{\partial v}{\partial z} + \frac{\partial w}{\partial y} \right); \quad \tau_{zx} = G\left(\frac{\partial u}{\partial z} + \frac{\partial w}{\partial x} \right)$$

Substituting the stresses in terms of displacement derivatives into the first stress equilibrium equation, we have

$$\frac{\partial \sigma_x}{\partial x} + \frac{\partial \tau_{xy}}{\partial y} + \frac{\partial \tau_{xz}}{\partial z} + b_x = 0$$

$$\frac{E}{(1+\nu)(1-2\nu)}\left((1-\nu)\frac{\partial^2 u}{\partial x^2} + \nu\frac{\partial^2 v}{\partial x \partial y} + \nu\frac{\partial^2 w}{\partial x \partial z}\right)$$
$$+ G\left(\frac{\partial^2 u}{\partial y^2} + \frac{\partial^2 v}{\partial x \partial y}\right) + G\left(\frac{\partial^2 u}{\partial z^2} + \frac{\partial^2 w}{\partial x \partial z}\right) + b_x = 0$$

Noting that $E/(1+\nu) = 2G$ and adding and subtracting $\nu\,\partial^2 u/\partial x^2$ in the first term, we have

$$\frac{2G}{1-2\nu}\left((1-2\nu)\frac{\partial^2 u}{\partial x^2} + \nu\frac{\partial^2 u}{\partial x^2} + \nu\frac{\partial^2 v}{\partial x \partial y} + \nu\frac{\partial^2 w}{\partial x \partial z}\right) + G\left(\frac{\partial^2 u}{\partial y^2} + \frac{\partial^2 v}{\partial x \partial y}\right)$$
$$+ G\left(\frac{\partial^2 u}{\partial z^2} + \frac{\partial^2 w}{\partial x \partial z}\right) + b_x = 0$$

The terms can now be grouped as follows:

$$G\left(\frac{\partial^2 u}{\partial x^2} + \frac{\partial^2 u}{\partial y^2} + \frac{\partial^2 u}{\partial z^2}\right) + \left(G + \frac{2G\nu}{1-2\nu}\right)\frac{\partial^2 u}{\partial x^2} + \frac{2G\nu}{1-2\nu}\left(\frac{\partial^2 v}{\partial x \partial y} + \frac{\partial^2 w}{\partial x \partial z}\right)$$
$$+ G\left(\frac{\partial^2 v}{\partial x \partial y} + \frac{\partial^2 w}{\partial x \partial z}\right) + b_x = 0$$

Simplifying this, the final form of the governing differential equation can be written as follows:

$$G\left(\frac{\partial^2 u}{\partial x^2} + \frac{\partial^2 u}{\partial y^2} + \frac{\partial^2 u}{\partial z^2}\right) + \frac{G}{1-2\nu}\frac{\partial}{\partial x}\left(\frac{\partial u}{\partial x} + \frac{\partial v}{\partial y} + \frac{\partial w}{\partial z}\right) + b_x = 0$$

Proceeding in exactly the same manner, the other two equilibrium equations give

$$G\left(\frac{\partial^2 v}{\partial x^2} + \frac{\partial^2 v}{\partial y^2} + \frac{\partial^2 v}{\partial z^2}\right) + \frac{G}{1-2\nu}\frac{\partial}{\partial y}\left(\frac{\partial u}{\partial x} + \frac{\partial v}{\partial y} + \frac{\partial w}{\partial z}\right) + b_y = 0$$

$$G\left(\frac{\partial^2 w}{\partial x^2} + \frac{\partial^2 w}{\partial y^2} + \frac{\partial^2 w}{\partial z^2}\right) + \frac{G}{1-2\nu}\frac{\partial}{\partial z}\left(\frac{\partial u}{\partial x} + \frac{\partial v}{\partial y} + \frac{\partial w}{\partial z}\right) + b_z = 0$$

In order to find displacements $u(x, y, z)$, $v(x, y, z)$, and $w(x, y, z)$, we must solve these three partial differential equations simultaneously. Since the equations are of the second order, the specified displacement boundary conditions are of the essential type. The derivatives of displacements are related to strains, which in turn are related to stresses; thus the applied surface forces are the natural boundary conditions.

7.3 GENERAL FORM OF FINITE ELEMENT EQUATIONS

7.3.1 Potential Energy Functional

It can be shown that a solution of the elasticity problem minimizes the following potential energy functional:

$$\Pi_p(u, v, w) = U - W$$

where U = strain energy and W = work done by applied forces.

The strain energy is defined as follows:

$$U = \frac{1}{2} \iiint_V \epsilon^T \sigma \, dV$$

where

$$\epsilon = (\epsilon_x \quad \epsilon_y \quad \epsilon_z \quad \gamma_{xy} \quad \gamma_{yz} \quad \gamma_{zx})^T$$

is the strain vector,

$$\sigma = (\sigma_x \quad \sigma_y \quad \sigma_z \quad \tau_{xy} \quad \tau_{yz} \quad \tau_{zx})^T$$

is the stress vector, and V is the volume of the solid.

The work-done term W incorporates work done by all applied forces, including body forces, distributed surface forces, and concentrated forces. The work done by the body forces is computed by integrating over the volume and is written as follows:

$$W_b = \iiint_V (b_x u + b_y v + b_z w) \, dV$$

The work done by the distributed surface forces is computed by integrating over the surface on which these forces are applied. Denoting the applied distributed forces by $\boldsymbol{q} = (q_x \quad q_y \quad q_z)^T$, this work done is as follows:

$$W_q = \iint_S (q_x u + q_y v + q_z w) \, dS$$

where S is the part of the surface over which these forces are applied.

If there are any concentrated forces applied, then the work done by these forces is simply the components of forces multiplied by the displacement components at the point where the forces are applied.

$$W_f = \sum_i (F_{xi} u_i + F_{yi} v_i + F_{zi} w_i)$$

The total work done W is the sum of these three terms.

7.3.2 Weak Form

Even though the finite element equations can be derived using the potential energy, it is instructive to derive the weak form. As will be seen in this section, the weak form for an elasticity problem takes the form of the familiar principle of virtual displacements. For obtaining the weak form, instead of using the differential equations in terms of displacements, it is more convenient to start with the stress equilibrium equations. The equations in terms of stresses are simpler and therefore the algebraic manipulations in deriving the weak form are easier in this form than in the displacement form. Furthermore, since the constitutive equations are not used, the final form is applicable to a general elasticity problem. Thus we construct the weak form corresponding to the following differential equations:

$$\frac{\partial \sigma_x}{\partial x} + \frac{\partial \tau_{xy}}{\partial y} + \frac{\partial \tau_{xz}}{\partial z} + b_x = 0$$

$$\frac{\partial \tau_{yx}}{\partial x} + \frac{\partial \sigma_y}{\partial y} + \frac{\partial \tau_{yz}}{\partial z} + b_y = 0$$

$$\frac{\partial \tau_{zx}}{\partial x} + \frac{\partial \tau_{zy}}{\partial y} + \frac{\partial \sigma_z}{\partial z} + b_z = 0$$

Since there are three coupled differential equations, we need three different symbols for the weighting functions. Denoting the weighting functions by $\overline{u}$, $\overline{v}$, and $\overline{w}$, multiplying each equation by its weighting function, integrating over the volume, and adding all three terms, the total weighted residual is as follows.

$$\iiint_V \left(\frac{\partial \sigma_x}{\partial x} + \frac{\partial \tau_{xy}}{\partial y} + \frac{\partial \tau_{xz}}{\partial z} + b_x\right)\overline{u} + \left(\frac{\partial \tau_{yx}}{\partial x} + \frac{\partial \sigma_y}{\partial y} + \frac{\partial \tau_{yz}}{\partial z} + b_y\right)\overline{v}$$
$$+ \left(\frac{\partial \tau_{zx}}{\partial x} + \frac{\partial \tau_{zy}}{\partial y} + \frac{\partial \sigma_z}{\partial z} + b_z\right)\overline{w}\, dV = 0$$

Using the Green-Gauss theorem on each of the nine terms involving the stress derivatives, we get

$$\iint_S (\sigma_x n_x + \tau_{xy} n_y + \tau_{xz} n_z)\overline{u} + (\tau_{yx} n_x + \sigma_x n_y + \tau_{yz} n_z)\overline{v} + (\tau_{zx} n_x + \tau_{zy} n_y + \sigma_z n_z)\overline{w}\, dS$$
$$- \iiint_V \left(\sigma_x \frac{\partial \overline{u}}{\partial x} + \tau_{xy} \frac{\partial \overline{u}}{\partial y} + \tau_{xz} \frac{\partial \overline{u}}{\partial z}\right) + \left(\tau_{yx} \frac{\partial \overline{v}}{\partial x} + \sigma_y \frac{\partial \overline{v}}{\partial y} + \tau_{yz} \frac{\partial \overline{v}}{\partial z}\right)$$
$$+ \left(\tau_{zx} \frac{\partial \overline{w}}{\partial x} + \tau_{zy} \frac{\partial \overline{w}}{\partial y} + \sigma_z \frac{\partial \overline{w}}{\partial z}\right) dV$$
$$+ \iiint_V (b_x \overline{u} + b_y \overline{v} + b_z \overline{w})\, dV = 0$$

On the surface the applied forces are $q_x = \sigma_x n_x + \tau_{xy} n_y + \tau_{xz} n_z$, etc. Substituting these and rearranging terms, we get the following weak form:

$$\iiint_V \sigma_x \frac{\partial \overline{u}}{\partial x} + \sigma_y \frac{\partial \overline{v}}{\partial y} + \sigma_z \frac{\partial \overline{w}}{\partial z} + \tau_{xy}\left(\frac{\partial \overline{u}}{\partial y} + \frac{\partial \overline{v}}{\partial x}\right) + \tau_{yz}\left(\frac{\partial \overline{v}}{\partial z} + \frac{\partial \overline{w}}{\partial y}\right) + \tau_{xz}\left(\frac{\partial \overline{u}}{\partial z} + \frac{\partial \overline{w}}{\partial x}\right) dV$$

$$= \iint_S (q_x \overline{u} + q_y \overline{v} + q_z \overline{w})\, dS + \iiint_V (b_x \overline{u} + b_y \overline{v} + b_z \overline{w})\, dV$$

If we interpret the weighting functions as virtual displacements, then their derivatives can be interpreted as virtual strains and defined as follows:

$$\frac{\partial \overline{u}}{\partial x} = (\overline{\epsilon})_x; \qquad \frac{\partial \overline{v}}{\partial y} = (\overline{\epsilon})_y; \qquad \frac{\partial \overline{w}}{\partial z} = (\overline{\epsilon})_z$$

$$\frac{\partial \overline{u}}{\partial y} + \frac{\partial \overline{v}}{\partial x} = (\overline{\gamma})_{xy}; \qquad \frac{\partial \overline{v}}{\partial z} + \frac{\partial \overline{w}}{\partial y} = (\overline{\gamma})_{yz}; \qquad \frac{\partial \overline{u}}{\partial z} + \frac{\partial \overline{w}}{\partial x} = (\overline{\gamma})_{zx}$$

Substituting these, the weak form is

$$\iiint_V (\sigma_x (\overline{\epsilon})_x + \sigma_y (\overline{\epsilon})_y + \sigma_z (\overline{\epsilon})_z + \tau_{xy} (\overline{\gamma})_{xy} + \tau_{yz} (\overline{\gamma})_{yz} + \tau_{xz} (\overline{\gamma})_{zx})\, dV$$

$$= \iint_S (q_x \overline{u} + q_y \overline{v} + q_z \overline{w})\, dS + \iiint_V (b_x \overline{u} + b_y \overline{v} + b_z \overline{w})\, dV$$

The integral on the left-hand side is interpreted as the virtual work done by the internal stresses. Those on the right-hand side represent the work done by the applied forces. Thus the above weak form is simply a statement of the principle of virtual displacement.

7.3.3 Finite Element Equations

The finite element equations can be derived using either the weak form or the potential energy functional. The volume and the surface integrals in these forms now are evaluated over an assumed element. In order to actually carry out these integrations, we will have to assume a suitable shape for the element. However, the general form of the finite element equations can be written that is applicable to any element shape.

Since there are three unknown displacements, we need three separate interpolations, one each for $u(x, y, z)$, $v(x, y, z)$, and $w(x, y, z)$. In principle, we could use different interpolations for different displacements. However, in practice, we use the same interpolation functions for all three displacements. Thus

$$u(x, y, z) = N_1 u_1 + N_2 u_2 + \cdots$$
$$v(x, y, z) = N_1 v_1 + N_2 v_2 + \cdots$$
$$w(x, y, z) = N_1 w_1 + N_2 w_2 + \cdots$$

where $u_1, v_1, w_1, u_2, \ldots$ are the nodal degrees of freedom and $N_i(x, y, z)$ are suitable interpolation functions. Arranging the terms so that all three nodal degrees of freedom at a node are grouped together, the three equations can be written as follows:

$$\boldsymbol{u}(x, y, z) \equiv \begin{pmatrix} u \\ v \\ w \end{pmatrix} = \begin{pmatrix} N_1 & 0 & 0 & N_2 & 0 & \cdots \\ 0 & N_1 & 0 & 0 & N_2 & \cdots \\ 0 & 0 & N_1 & 0 & 0 & \cdots \end{pmatrix} \begin{pmatrix} u_1 \\ v_1 \\ w_1 \\ u_2 \\ \vdots \end{pmatrix} \equiv \boldsymbol{N}^T \boldsymbol{d}$$

From the assumed solution the element strain vector can be computed by appropriate differentiation as follows:

$$\boldsymbol{\epsilon} = \begin{pmatrix} \epsilon_x \\ \epsilon_y \\ \epsilon_z \\ \gamma_{xy} \\ \gamma_{yz} \\ \gamma_{zx} \end{pmatrix} = \begin{pmatrix} \frac{\partial u}{\partial x} \\ \frac{\partial v}{\partial y} \\ \frac{\partial w}{\partial z} \\ \frac{\partial u}{\partial y} + \frac{\partial v}{\partial x} \\ \frac{\partial v}{\partial z} + \frac{\partial w}{\partial y} \\ \frac{\partial u}{\partial z} + \frac{\partial w}{\partial x} \end{pmatrix} = \begin{pmatrix} \frac{\partial N_1}{\partial x} & 0 & 0 & \frac{\partial N_2}{\partial x} & 0 & 0 & \cdots \\ 0 & \frac{\partial N_1}{\partial y} & 0 & 0 & \frac{\partial N_2}{\partial y} & 0 & \cdots \\ 0 & 0 & \frac{\partial N_1}{\partial z} & 0 & 0 & \frac{\partial N_2}{\partial z} & \cdots \\ \frac{\partial N_1}{\partial y} & \frac{\partial N_1}{\partial x} & 0 & \frac{\partial N_2}{\partial y} & \frac{\partial N_2}{\partial x} & 0 & \cdots \\ 0 & \frac{\partial N_1}{\partial z} & \frac{\partial N_1}{\partial y} & 0 & \frac{\partial N_2}{\partial z} & \frac{\partial N_2}{\partial y} & \cdots \\ \frac{\partial N_1}{\partial z} & 0 & \frac{\partial N_1}{\partial x} & \frac{\partial N_2}{\partial z} & 0 & \frac{\partial N_2}{\partial x} & \cdots \end{pmatrix} \begin{pmatrix} u_1 \\ v_1 \\ w_1 \\ u_2 \\ \vdots \end{pmatrix} \equiv \boldsymbol{B}^T \boldsymbol{d}$$

Using the constitutive matrix appropriate for the material, the element stress vector can be written as follows:

$$\boldsymbol{\sigma} = \boldsymbol{C\epsilon} = \boldsymbol{CB}^T \boldsymbol{d}$$

The strain energy over an element can now be written as follows:

$$U = \tfrac{1}{2} \iiint_V \boldsymbol{\epsilon}^T \boldsymbol{\sigma}\, dV = \tfrac{1}{2} \iiint_V (\boldsymbol{B}^T \boldsymbol{d})^T \boldsymbol{CB}^T \boldsymbol{d}\, dV$$

$$= \tfrac{1}{2} \boldsymbol{d}^T \iiint_V \boldsymbol{BCB}^T\, dV\, \boldsymbol{d} = \tfrac{1}{2} \boldsymbol{d}^T \boldsymbol{kd}$$

where $\boldsymbol{k}$ is known as the element stiffness matrix

$$\boldsymbol{k} = \iiint_V \boldsymbol{BCB}^T\, dV$$

The work done by the body forces can be evaluated as follows:

$$W_b = \iiint_V (b_x u + b_y v + b_z w)\, dV = \iiint_V \begin{pmatrix} b_x & b_y & b_z \end{pmatrix} \begin{pmatrix} u \\ v \\ w \end{pmatrix} dV$$

Substituting the assumed solution, we have

$$W_b = \iiint_V (b_x \quad b_y \quad b_z) \boldsymbol{N}^T \, dV \, \boldsymbol{d} \equiv \boldsymbol{r}_b^T \boldsymbol{d}$$

where $\boldsymbol{r}_b^T$ is the transpose of the equivalent nodal load vector due to body forces,

$$\boldsymbol{r}_b = \iiint_V \boldsymbol{N} \begin{pmatrix} b_x \\ b_y \\ b_z \end{pmatrix} dV = \begin{pmatrix} \iiint_V N_1 b_x \, dV \\ \iiint_V N_1 b_y \, dV \\ \iiint_V N_1 b_z \, dV \\ \iiint_V N_2 b_x \, dV \\ \vdots \end{pmatrix}$$

Assuming concentrated forces to be applied only at the nodes, the work done by the concentrated nodal forces can be evaluated as follows:

$$W_f = \sum_i (F_{xi} u_i + F_{yi} v_i + F_{zi} w_i) = (F_{x1} \quad F_{y1} \quad F_{z1} \quad F_{x2} \quad \cdots) \begin{pmatrix} u_1 \\ v_1 \\ w_1 \\ u_2 \\ \vdots \end{pmatrix} \equiv \boldsymbol{r}_f^T \boldsymbol{d}$$

where $\boldsymbol{r}_f^T$ is the transpose of the applied nodal load vector:

$$\boldsymbol{r}_f = (F_{x1} \quad F_{y1} \quad F_{z1} \quad F_{x2} \quad \cdots)^T$$

The work done by the applied surface forces is a little more difficult to write because we must express the interpolation functions in terms of surface parameters over which these forces are applied. The details will be presented when considering specific elements. For the general discussion, here we assume that the appropriate interpolation functions specific to the surface S are denoted by vector $\boldsymbol{N}_s$. Then the work done by the surface forces can be written as follows:

$$W_q = \iint_S (q_x u + q_y v + q_z w) \, dS = \iint_S (q_x \quad q_y \quad q_z) \begin{pmatrix} u \\ v \\ w \end{pmatrix} dS$$

$$= \iint_S (q_x \quad q_y \quad q_z) \boldsymbol{N}_s^T \, dS \, \boldsymbol{d} \equiv \boldsymbol{r}_q^T \boldsymbol{d}$$

where $\boldsymbol{r}_q^T$ is the transpose of the equivalent nodal load vector due to surface forces and $\boldsymbol{N}_s^T$ are the displacement interpolation functions expressed in terms of coordinates along the surface S:

$$\boldsymbol{r}_q = \iint_S \boldsymbol{N}_s \begin{pmatrix} q_x \\ q_y \\ q_z \end{pmatrix} dS = \begin{pmatrix} \iint_S N_{s1} q_x \, dS \\ \iint_S N_{s1} q_y \, dS \\ \iint_S N_{s1} q_z \, dS \\ \iint_S N_{s2} q_x \, dS \\ \vdots \end{pmatrix}$$

All terms in the potential energy have now been written in terms of nodal unknowns. Thus the potential energy for a finite element is as follows:

$$\Pi_p(u, v, w) = U - W = \tfrac{1}{2}\boldsymbol{d}^T \boldsymbol{k} \boldsymbol{d} - (\boldsymbol{r}_q^T + \boldsymbol{r}_b^T + \boldsymbol{r}_f^T)\boldsymbol{d}$$

Using the necessary conditions for the minimum of potential energy as $\partial \Pi_p / \partial d_i = 0$, we get the element equations as follows:

$$\boldsymbol{kd} = \boldsymbol{r}_q + \boldsymbol{r}_b + \boldsymbol{r}_f$$

The concentrated forces applied at nodes can be assembled directly into the global load vector during assembly. Thus they are usually not considered as part of the element equations and the complete element equations are written as follows:

$$\boldsymbol{kd} = \boldsymbol{r}_q + \boldsymbol{r}_b$$

7.3.4 Finite Element Equations in the Presence of Initial Strains

Recall that a temperature rise of ΔT results in a uniform strain that depends on the coefficient of thermal expansion α of the material. The temperature change does not cause shear strains. Thus the initial strain vector due to temperature change is as follows:

$$\epsilon_0 = \begin{pmatrix} \alpha \Delta T \\ \alpha \Delta T \\ \alpha \Delta T \\ 0 \\ 0 \\ 0 \end{pmatrix}$$

In the presence of initial strain ϵ_0 due to temperature change or another similar cause, the constitutive equations are written as follows:

$$\sigma = \boldsymbol{C}(\epsilon - \epsilon_0)$$

The strain energy over an element can now be written as follows:

$$U = \frac{1}{2} \iiint\limits_V (\epsilon - \epsilon_0)^T C(\epsilon - \epsilon_0)\, dV$$

Expanding the product,

$$U = \frac{1}{2} \iiint\limits_V \epsilon^T C\epsilon\, dV - \frac{1}{2} \iiint\limits_V \epsilon_0^T C\epsilon\, dV - \frac{1}{2} \iiint\limits_V \epsilon^T C\epsilon_0\, dV + \frac{1}{2} \iiint\limits_V \epsilon_0^T C\epsilon_0\, dV$$

The last term involves all known quantities that do not depend on the assumed solution. This term will drop out when writing the necessary conditions for the minimum of the potential energy. Thus the last term can be ignored. Since $\boldsymbol{C}$ is a symmetric matrix, the transpose of the third term is the same as the second term and hence the second and the third terms can be combined together. The effective strain energy can therefore be expressed as follows:

$$U_e = \frac{1}{2} \iiint\limits_V \epsilon^T C\epsilon\, dV - \iiint\limits_V \epsilon_0^T C\epsilon\, dV$$

Substituting the strains in terms of the assumed solution ($\epsilon = \boldsymbol{B}^T \boldsymbol{d}$), we get

$$U_e = \frac{1}{2} d^T \iiint\limits_V \boldsymbol{BCB}^T\, dV\, \boldsymbol{d} - \iiint\limits_V \epsilon_0^T \boldsymbol{CB}^T\, dV\, \boldsymbol{d} = \frac{1}{2}\boldsymbol{d}^T \boldsymbol{kd} - \boldsymbol{r}_\epsilon^T \boldsymbol{d}$$

where $\boldsymbol{k}$ is the usual element stiffness matrix

$$\boldsymbol{k} = \iint\limits_V \boldsymbol{BCB}^T\, dV$$

where $\boldsymbol{r}_\epsilon^T$ is the transpose of the equivalent nodal load vector due to initial strains ϵ_0,

$$\boldsymbol{r}_\epsilon = \iiint\limits_V \boldsymbol{BC}\epsilon_0\, dV$$

Thus the effect of initial strain is to add another vector to the right-hand side of the element equations. The element stiffness matrix remains unaffected.

7.4 PLANE STRESS AND PLANE STRAIN

In principle, any stress analysis situation can be handled by the general finite element formulation given in the previous section. However, the computational cost will be very high because of the need to evaluate three-dimensional volume and surface integrals for each element. Furthermore, with three degrees of freedom at each node, the resulting system of finite element equations could be very large, requiring specialized computer hardware and software for their solution. Therefore for common everyday situations it is desirable to introduce simplifications in order to reduce the problem size and still be able to get reasonably accurate solutions.

The axial deformation problem, considered in Chapter 2, assumes that bodies are long and slender and are loaded in the axial direction only. Denoting the axial direction by x, the key assumption is that σ_x is the only nonzero stress component. This reduces the problem to only a single differential equation in terms of axial displacement u and results in a simple finite element formulation. The formulation clearly is effective for truss-type structures, as demonstrated in Chapter 4. Just imagine the effort that it would take to analyze even the simplest truss structure by the three-dimensional formulation outlined in the previous section.

The beam bending formulation, presented in Chapter 4, is another specialized formulation based on assumptions on strain components that reduce the problem to a single differential equation.

The plane stress and plane strain represent the next level of approximate behavior in an attempt to reduce the analysis problem to a more manageable form. This section presents this formulation in detail and considers several practical applications.

It is very important to clearly understand all approximations introduced in a particular formulation and to take advantage of these situations when appropriate. It is the analyst's responsibility to recognize when a particular simplified model is appropriate. As a general rule, it is good practice to start with simplified models first and gradually move toward more sophisticated models. For many stress analysis situations, it is possible to create simple truss and beam and frame models to predict the overall behavior of the structure. To get a more accurate picture of an actual stress pattern, a plane stress or plane strain model can then be employed, either for the entire structure or only for critical areas. A few most critical areas can finally be analyzed using the full three-dimensional model. The results from the lower order models are useful in defining appropriate loading and boundary conditions for isolated portions of the structure using the higher dimensional models.

Another caution is in order here when we are dealing with an approximate solution technique such as the finite element method. When describing solution accuracy using a particular element type, the reference generally is to the accuracy with respect to the solution of the governing differential equation. Thus an accurate finite element solution of an axial deformation problem does not mean that the stresses predicted by the solution are exactly what the component being analyzed actually sees. All it means is that we have an accurate solution to the mathematical model. How well the solution of the mathematical model actually represents the true stresses and strains in the structure depends on the suitability of the assumptions inherent in the model. This is an important point to realize when interpreting results produced by some of today's sophisticated commercial finite element software packages. Some of these programs give solution accuracy as a standard part of their output. A near-zero percent error in stresses may not mean much if you are using a plane stress formulation to analyze a physical situation that violates the plane stress assumptions.

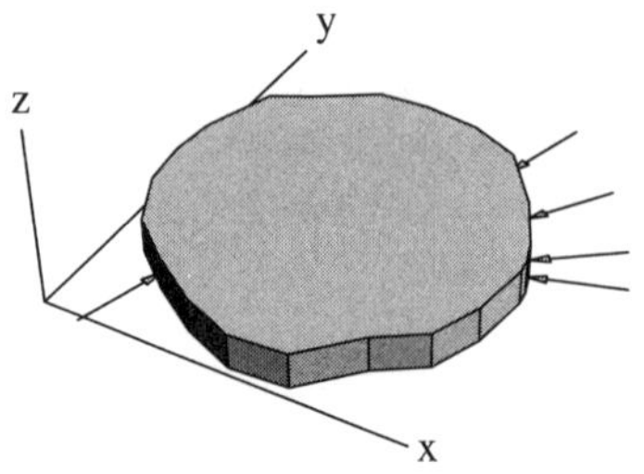

Figure 7.5. A thin solid subjected to in-plane loading suitable for plane stress model

7.4.1 Plane Stress Problem

Consider a body that is much thinner in one direction as compared to the other two. Assume that the thin direction is the z direction and thus the body is lying in the xy plane, as shown in Figure 7.5. The situation may be considered a *plane stress* problem if all applied forces are acting in the xy plane as well. Analysis of beams, brackets, hooks, etc., fall into this category. Under these circumstances it generally is reasonable to assume that the following *stress* components are zero:

$$\text{Assumed zero stresses: } \{\sigma_z, \tau_{yz}, \tau_{zx}\}$$

Substituting these zero-stress components in the inverse stress-strain relationship, we get

$$\begin{pmatrix} \epsilon_x \\ \epsilon_y \\ \epsilon_z \\ \gamma_{xy} \\ \gamma_{yz} \\ \gamma_{zx} \end{pmatrix} = \begin{pmatrix} \frac{1}{E} & -\frac{\nu}{E} & -\frac{\nu}{E} & 0 & 0 & 0 \\ -\frac{\nu}{E} & \frac{1}{E} & -\frac{\nu}{E} & 0 & 0 & 0 \\ -\frac{\nu}{E} & -\frac{\nu}{E} & \frac{1}{E} & 0 & 0 & 0 \\ 0 & 0 & 0 & \frac{1}{G} & 0 & 0 \\ 0 & 0 & 0 & 0 & \frac{1}{G} & 0 \\ 0 & 0 & 0 & 0 & 0 & \frac{1}{G} \end{pmatrix} \begin{pmatrix} \sigma_x \\ \sigma_y \\ 0 \\ \tau_{xy} \\ 0 \\ 0 \end{pmatrix} = \begin{pmatrix} \frac{\sigma_x - \nu\sigma_y}{E} \\ \frac{\sigma_y - \nu\sigma_x}{E} \\ -\frac{\nu(\sigma_x + \sigma_y)}{E} \\ \frac{\tau_{xy}}{G} \\ 0 \\ 0 \end{pmatrix}$$

These relationships can be written as follows:

$$\begin{pmatrix} \epsilon_x \\ \epsilon_y \\ \gamma_{xy} \end{pmatrix} = \begin{pmatrix} \frac{1}{E} & -\frac{\nu}{E} & 0 \\ -\frac{\nu}{E} & \frac{1}{E} & 0 \\ 0 & 0 & \frac{1}{G} \end{pmatrix} \begin{pmatrix} \sigma_x \\ \sigma_y \\ \tau_{xy} \end{pmatrix}$$

$$\epsilon_z = -\frac{\nu(\sigma_x + \sigma_y)}{E}; \qquad \gamma_{yz} = 0; \qquad \gamma_{zx} = 0$$

Note that ϵ_z is not zero in the plane stress formulation. However, knowing σ_x and σ_y, it can be determined from using the above equation. Thus the primary unknowns are the stresses and strains in the xy plane. The plane stress problem can therefore be expressed in terms of displacements along the two coordinate directions u, v and the following stress and strain components:

$$\text{Stresses:} \quad \{\sigma_x, \sigma_y, \tau_{xy}\}; \qquad \text{Strains:} \quad \{\epsilon_x, \epsilon_y, \gamma_{xy}\}$$

Strain-displacement relationships (assuming small displacements) can be written as

$$\epsilon_x = \frac{\partial u}{\partial x}; \qquad \epsilon_y = \frac{\partial v}{\partial y}; \qquad \gamma_{xy} = \frac{\partial u}{\partial y} + \frac{\partial v}{\partial x}$$

and the stress-strain relationship as

$$\begin{pmatrix} \epsilon_x \\ \epsilon_y \\ \gamma_{xy} \end{pmatrix} = \begin{pmatrix} \frac{1}{E} & -\frac{\nu}{E} & 0 \\ -\frac{\nu}{E} & \frac{1}{E} & 0 \\ 0 & 0 & \frac{1}{G} \end{pmatrix} \begin{pmatrix} \sigma_x \\ \sigma_y \\ \tau_{xy} \end{pmatrix}$$

Inverting this matrix equation, we can express stresses as functions of strains as follows:

$$\begin{pmatrix} \sigma_x \\ \sigma_y \\ \tau_{xy} \end{pmatrix} = \frac{E}{1-\nu^2} \begin{pmatrix} 1 & \nu & 0 \\ \nu & 1 & 0 \\ 0 & 0 & \frac{1-\nu}{2} \end{pmatrix} \begin{pmatrix} \epsilon_x \\ \epsilon_y \\ \gamma_{xy} \end{pmatrix}; \qquad \boldsymbol{\sigma} = \boldsymbol{C}\boldsymbol{\epsilon}$$

If there is a temperature change of ΔT, the initial strains vector for plane stress is

$$\epsilon_0 = (\alpha\Delta T \quad \alpha\Delta T \quad 0)^T \qquad \text{and} \qquad \epsilon_z = -\frac{\nu(\sigma_x + \sigma_y)}{E} + \alpha\Delta T$$

$$\boldsymbol{\sigma} = \boldsymbol{C}(\boldsymbol{\epsilon} - \boldsymbol{\epsilon}_0)$$

7.4.2 Plane Strain Problem

Consider a body that is much larger in one direction as compared to the other two. Assume that the long direction is the z direction and consider a unit slice (thickness $h = 1$) of the body lying in the xy plane as shown in Figure 7.6. The situation may be considered a

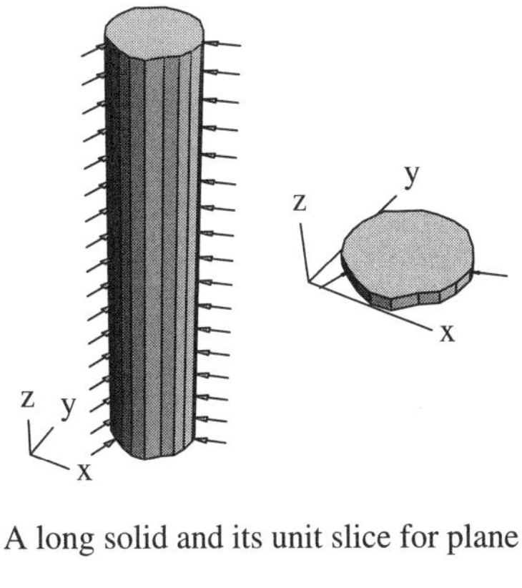

Figure 7.6. A long solid and its unit slice for plane strain model

plane strain problem if all applied forces are acting in the xy plane as well. Analysis of many dams and shafts falls into this category. Under these circumstances it generally is reasonable to assume that the following *strain* components are zero:

$$\text{Assumed zero strains:} \qquad \{\epsilon_z, \gamma_{yz}, \gamma_{zx}\}$$

Introducing these zero strains into the stress-strain relationship, we get

$$\begin{pmatrix} \sigma_x \\ \sigma_y \\ \sigma_z \\ \tau_{xy} \\ \tau_{yz} \\ \tau_{zx} \end{pmatrix} = \frac{E}{(1+\nu)(1-2\nu)} \begin{pmatrix} 1-\nu & \nu & \nu & 0 & 0 & 0 \\ \nu & 1-\nu & \nu & 0 & 0 & 0 \\ \nu & \nu & 1-\nu & 0 & 0 & 0 \\ 0 & 0 & 0 & \frac{1-2\nu}{2} & 0 & 0 \\ 0 & 0 & 0 & 0 & \frac{1-2\nu}{2} & 0 \\ 0 & 0 & 0 & 0 & 0 & \frac{1-2\nu}{2} \end{pmatrix} \begin{pmatrix} \epsilon_x \\ \epsilon_y \\ 0 \\ \gamma_{xy} \\ 0 \\ 0 \end{pmatrix}$$

$$= \frac{E}{(1+\nu)(1-2\nu)} \begin{pmatrix} (1-\nu)\epsilon_x - \nu\epsilon_y \\ (1-\nu)\epsilon_y - \nu\epsilon_x \\ \nu(\epsilon_x + \epsilon_y) \\ \frac{(1-2\nu)\gamma_{xy}}{2} \\ 0 \\ 0 \end{pmatrix}$$

These relationships can be written as follows:

$$\begin{pmatrix} \sigma_x \\ \sigma_y \\ \tau_{xy} \end{pmatrix} = \frac{E}{(1+\nu)(1-2\nu)} \begin{pmatrix} 1-\nu & \nu & 0 \\ \nu & 1-\nu & 0 \\ 0 & 0 & \frac{1-2\nu}{2} \end{pmatrix} \begin{pmatrix} \epsilon_x \\ \epsilon_y \\ \gamma_{xy} \end{pmatrix}$$

$$\sigma_z = \frac{E\nu(\epsilon_x + \epsilon_y)}{(1+\nu)(1-2\nu)}; \qquad \tau_{yz} = 0; \qquad \tau_{zx} = 0$$

Note that σ_z is not zero in the plane strain formulation. However, knowing ϵ_x and ϵ_y, it can be determined from using the above equation. Thus the primary unknowns are the stresses and strains in the xy plane. Similar to the plane stress case, the plane strain problem can also be expressed in terms of displacements along two coordinate directions: u, v. In fact, the only difference between the plane stress and the plane strain problems is in the constitutive matrices:

$$\text{Stresses:} \quad \{\sigma_x, \sigma_y, \tau_{xy}\}; \qquad \text{Strains:} \quad \{\epsilon_x, \epsilon_y, \gamma_{xy}\}$$

Strain-displacement relationships (assuming small displacements) can be written as

$$\epsilon_x = \frac{\partial u}{\partial x}; \qquad \epsilon_y = \frac{\partial v}{\partial y}; \qquad \gamma_{xy} = \frac{\partial u}{\partial y} + \frac{\partial v}{\partial x}$$

and the stress-strain relationship as

$$\sigma = C\epsilon$$

$$\begin{pmatrix} \sigma_x \\ \sigma_y \\ \tau_{xy} \end{pmatrix} = \frac{E}{(1+\nu)(1-2\nu)} \begin{pmatrix} 1-\nu & \nu & 0 \\ \nu & 1-\nu & 0 \\ 0 & 0 & \frac{1-2\nu}{2} \end{pmatrix} \begin{pmatrix} \epsilon_x \\ \epsilon_y \\ \gamma_{xy} \end{pmatrix}$$

If there is a temperature change of ΔT, the initial strain vector for plane strain is established as follows:

$$\begin{pmatrix} \sigma_x \\ \sigma_y \\ \sigma_z \\ \tau_{xy} \\ \tau_{yz} \\ \tau_{zx} \end{pmatrix} = \frac{E}{(1+\nu)(1-2\nu)} \begin{pmatrix} 1-\nu & \nu & \nu & 0 & 0 & 0 \\ \nu & 1-\nu & \nu & 0 & 0 & 0 \\ \nu & \nu & 1-\nu & 0 & 0 & 0 \\ 0 & 0 & 0 & \frac{1-2\nu}{2} & 0 & 0 \\ 0 & 0 & 0 & 0 & \frac{1-2\nu}{2} & 0 \\ 0 & 0 & 0 & 0 & 0 & \frac{1-2\nu}{2} \end{pmatrix} \begin{pmatrix} \epsilon_x - \alpha\Delta T \\ \epsilon_y - \alpha\Delta T \\ -\alpha\Delta T \\ \gamma_{xy} \\ 0 \\ 0 \end{pmatrix}$$

or

$$\begin{pmatrix} \sigma_x \\ \sigma_y \\ \sigma_z \\ \tau_{xy} \\ \tau_{yz} \\ \tau_{zx} \end{pmatrix} = \frac{E}{(1+\nu)(1-2\nu)} \begin{pmatrix} (1-\nu)\epsilon_x - \nu\epsilon_y - (1+\nu)\alpha\Delta T \\ (1-\nu)\epsilon_y - \nu\epsilon_x - (1+\nu)\alpha\Delta T \\ \nu(\epsilon_x + \epsilon_y) - (1+\nu)\alpha\Delta T \\ \frac{(1-2\nu)\gamma_{xy}}{2} \\ 0 \\ 0 \end{pmatrix}$$

$$\begin{pmatrix} \sigma_x \\ \sigma_y \\ \tau_{xy} \end{pmatrix} = \frac{E}{(1+\nu)(1-2\nu)} \begin{pmatrix} 1-\nu & \nu & 0 \\ \nu & 1-\nu & 0 \\ 0 & 0 & \frac{1-2\nu}{2} \end{pmatrix} \begin{pmatrix} \epsilon_x - (1+\nu)\alpha\Delta T \\ \epsilon_y - (1+\nu)\alpha\Delta T \\ \gamma_{xy} \end{pmatrix}$$

$$\sigma_z = \frac{E(\nu(\epsilon_x + \epsilon_y) - (1+\nu)\alpha\Delta T)}{(1+\nu)(1-2\nu)}; \qquad \tau_{yz} = 0; \qquad \tau_{zx} = 0$$

Thus

$$\sigma = C(\epsilon - \epsilon_0)$$

with

$$\epsilon_0 = (1+\nu)(\alpha\Delta T \quad \alpha\Delta T \quad 0)^T$$

7.4.3 Finite Element Equations

The primary unknowns in both the plane stress and the plane strain formulations are the in-plane displacements u and v. In each formulation the in-plane stresses and strains are determined directly from these displacement components. The out-of-plane stress or strain can be determined from the in-plane components. The only difference in the two formulations is that they use slightly different constitutive equations. Thus both formulations can be handled by essentially the same finite element formulation. For a plane stress problem the model thickness is denoted by h. By assumption of a unit slice, for a plane strain problem $h = 1$.

For a general three-dimensional elasticity problem the finite element equations were derived in Section 7.3. For a plane problem, assuming a constant thickness over an element, the volume integrals reduce to integrals over the element area A and the surface integrals reduce to line integrals over the element sides c. The applied loads and body forces have components only in the xy plane. Thus the element equations for a plane stress or strain problem are as follows:

Assumed solution:

$$\boldsymbol{u}(x, y) \equiv \begin{pmatrix} u \\ v \end{pmatrix} = \begin{pmatrix} N_1 & 0 & N_2 & 0 & \cdots \\ 0 & N_1 & 0 & N_2 & \cdots \end{pmatrix} \begin{pmatrix} u_1 \\ v_1 \\ u_2 \\ \vdots \end{pmatrix} \equiv \boldsymbol{N}^T \boldsymbol{d}$$

Strain-displacement:

$$\boldsymbol{\epsilon} = \begin{pmatrix} \epsilon_x \\ \epsilon_y \\ \gamma_{xy} \end{pmatrix} = \begin{pmatrix} \frac{\partial u}{\partial x} \\ \frac{\partial v}{\partial y} \\ \frac{\partial u}{\partial y} + \frac{\partial v}{\partial x} \end{pmatrix} = \begin{pmatrix} \frac{\partial N_1}{\partial x} & 0 & \frac{\partial N_2}{\partial x} & 0 & \cdots \\ 0 & \frac{\partial N_1}{\partial y} & 0 & \frac{\partial N_2}{\partial y} & \cdots \\ \frac{\partial N_1}{\partial y} & \frac{\partial N_1}{\partial x} & \frac{\partial N_2}{\partial y} & \frac{\partial N_2}{\partial x} & \cdots \end{pmatrix} \begin{pmatrix} u_1 \\ v_1 \\ u_2 \\ \vdots \end{pmatrix} \equiv \boldsymbol{B}^T \boldsymbol{d}$$

Constitutive relationship:

$$\boldsymbol{\sigma} = \boldsymbol{C}(\boldsymbol{\epsilon} - \boldsymbol{\epsilon}_0)$$

Plane stress: $$\begin{pmatrix} \sigma_x \\ \sigma_y \\ \tau_{xy} \end{pmatrix} = \frac{E}{1 - \nu^2} \begin{pmatrix} 1 & \nu & 0 \\ \nu & 1 & 0 \\ 0 & 0 & \frac{1-\nu}{2} \end{pmatrix} \left(\begin{pmatrix} \epsilon_x \\ \epsilon_y \\ \gamma_{xy} \end{pmatrix} - \begin{pmatrix} \alpha\Delta T \\ \alpha\Delta T \\ 0 \end{pmatrix} \right)$$

Plane strain: $$\begin{pmatrix} \sigma_x \\ \sigma_y \\ \tau_{xy} \end{pmatrix} = \frac{E}{(1 + \nu)(1 - 2\nu)} \begin{pmatrix} 1 - \nu & \nu & 0 \\ \nu & 1 - \nu & 0 \\ 0 & 0 & \frac{1-2\nu}{2} \end{pmatrix} \left(\begin{pmatrix} \epsilon_x \\ \epsilon_y \\ \gamma_{xy} \end{pmatrix} - \begin{pmatrix} (1 + \nu)\alpha\Delta T \\ (1 + \nu)\alpha\Delta T \\ 0 \end{pmatrix} \right)$$

Element equations:

$$\boldsymbol{kd} = \boldsymbol{r}_q + \boldsymbol{r}_b + \boldsymbol{r}_\epsilon$$

Stiffness matrix:

$$\boldsymbol{k} = \iint_A \boldsymbol{BCB}^T \, dA$$

Equivalent load vector due to distributed loads:

$$\boldsymbol{r}_q = h \int_c \boldsymbol{N}_c \begin{pmatrix} q_x \\ q_y \end{pmatrix} dc$$

where $\boldsymbol{N}_c$ are the displacement interpolation functions expressed in terms of coordinates along the boundary c.

Equivalent load vector due to body forces:

$$\boldsymbol{r}_b = h \iint_A \boldsymbol{N} \begin{pmatrix} b_x \\ b_y \end{pmatrix} dA$$

Equivalent load vector from initial strains due to temperature change:

$$\boldsymbol{r}_\epsilon = h \iint_A \boldsymbol{B} \boldsymbol{C} \epsilon_0 \, dA$$

For plane stress: $\epsilon_0 = (\alpha \Delta T \quad \alpha \Delta T \quad 0)^T$

For plane strain: $\epsilon_0 = (1 + \nu)(\alpha \Delta T \quad \alpha \Delta T \quad 0)^T$

7.4.4 Three-Node Triangular Element

A typical three-node triangular element is shown in Figure 7.7. The nodal coordinates are (x_1, y_1), (x_2, y_2), and (x_3, y_3). With the unknown displacements at the three corners as nodal degrees of freedom, we have a total of six degrees of freedom, $u_1, v_1, u_2, \ldots, v_3$. The directions of outer normal and the boundary coordinates c for each side are also shown in the figure. For simplicity in integration it is assumed that any applied body or surface forces are constant over an element.

The interpolation functions for a three-node triangular element were derived in Chapter 5. Using, these the assumed finite element solution is as follows:

$$\begin{pmatrix} u \\ v \end{pmatrix} = \begin{pmatrix} N_1 & 0 & N_2 & 0 & N_3 & 0 \\ 0 & N_1 & 0 & N_2 & 0 & N_3 \end{pmatrix} \begin{pmatrix} u_1 \\ v_1 \\ u_2 \\ v_2 \\ u_3 \\ v_3 \end{pmatrix} \equiv \boldsymbol{N}^T \boldsymbol{d}$$

$$N_1 = \frac{1}{2A}\left(x_3(y - y_2) + x(y_2 - y_3) + x_2(-y + y_3)\right) \equiv \frac{1}{2A}(xb_1 + yc_1 + f_1)$$

$$N_2 = \frac{1}{2A}\left(x_3(-y + y_1) + x_1(y - y_3) + x(-y_1 + y_3)\right) \equiv \frac{1}{2A}(xb_2 + yc_2 + f_2)$$

$$N_3 = \frac{1}{2A}\left(x_2(y - y_1) + x(y_1 - y_2) + x_1(-y + y_2)\right) \equiv \frac{1}{2A}(xb_3 + yc_3 + f_3)$$

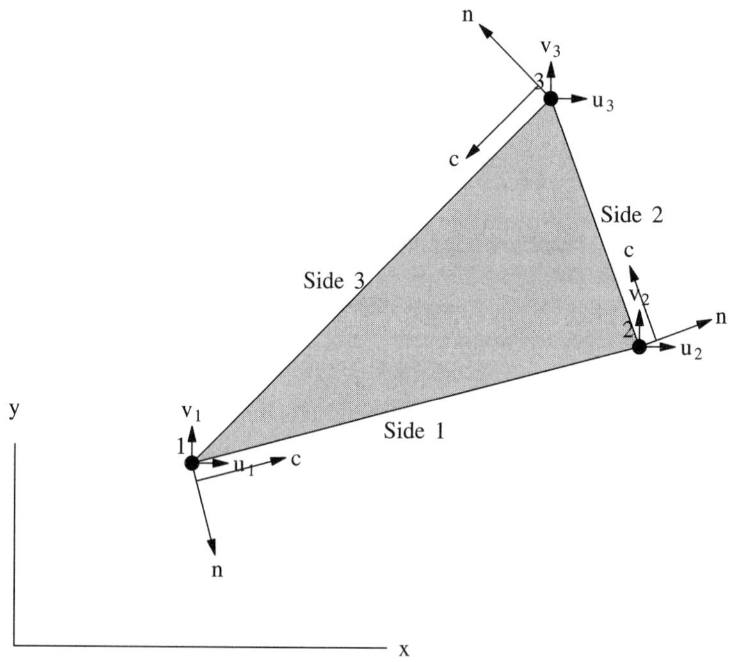

Figure 7.7. Three-node triangular element

$$f_1 = x_2y_3 - x_3y_2; \qquad f_2 = x_3y_1 - x_1y_3; \qquad f_3 = x_1y_2 - x_2y_1$$
$$b_1 = y_2 - y_3; \qquad b_2 = y_3 - y_1; \qquad b_3 = y_1 - y_2$$
$$c_1 = x_3 - x_2; \qquad c_2 = x_1 - x_3; \qquad c_3 = x_2 - x_1;$$
$$A = \text{area of triangle} = \tfrac{1}{2}(f_1 + f_2 + f_3)$$

By differentiating the displacements, the strain-displacement matrix $\boldsymbol{B}$ can be established as follows:

$$\begin{pmatrix} \epsilon_x \\ \epsilon_y \\ \gamma_{xy} \end{pmatrix} = \begin{pmatrix} \frac{\partial u}{\partial x} \\ \frac{\partial v}{\partial y} \\ \frac{\partial u}{\partial y} + \frac{\partial v}{\partial x} \end{pmatrix} = \frac{1}{2A}\begin{pmatrix} b_1 & 0 & b_2 & 0 & b_3 & 0 \\ 0 & c_1 & 0 & c_2 & 0 & c_3 \\ c_1 & b_1 & c_2 & b_2 & c_3 & b_3 \end{pmatrix}\begin{pmatrix} u_1 \\ v_1 \\ u_2 \\ v_2 \\ u_3 \\ v_3 \end{pmatrix} \equiv \boldsymbol{B}^T\boldsymbol{d}$$

Since none of the terms in the $\boldsymbol{B}$ matrix is a function of x or y, the integration to obtain the element stiffness matrix is trivial. The element stiffness matrix is therefore given by

$$\boldsymbol{k} = h\iint_A \boldsymbol{BCB}^T\, dA = hA\boldsymbol{BCB}^T$$

where $\boldsymbol{C}$ is the appropriate constitutive matrix:

$$\text{Plane stress:} \qquad \boldsymbol{C} = \frac{E}{1-\nu^2}\begin{pmatrix} 1 & \nu & 0 \\ \nu & 1 & 0 \\ 0 & 0 & \frac{1-\nu}{2} \end{pmatrix}$$

$$\text{Plane strain:} \qquad \boldsymbol{C} = \frac{E}{(1+\nu)(1-2\nu)}\begin{pmatrix} 1-\nu & \nu & 0 \\ \nu & 1-\nu & 0 \\ 0 & 0 & \frac{1-2\nu}{2} \end{pmatrix}$$

The equivalent load vector due to body forces is

$$\boldsymbol{r}_b = h \iint_A \boldsymbol{N}\begin{pmatrix} b_x \\ b_y \end{pmatrix} dA = \begin{pmatrix} h \iint_A N_1 b_x \, dA \\ h \iint_A N_1 b_y \, dA \\ h \iint_A N_2 b_x \, dA \\ \vdots \end{pmatrix}$$

The integration over a triangle can be carried out as explained in Chapter 5. With the body force components assumed constant over an element, the integration yields the following $\boldsymbol{r}_b$ vector:

$$\boldsymbol{r}_b = \frac{hA}{3}(b_x \quad b_y \quad b_x \quad b_y \quad b_x \quad b_y)^T$$

The equivalent load vector from initial strains due to temperature change is

$$\boldsymbol{r}_\epsilon = h \iint_A \boldsymbol{B}\boldsymbol{C}\epsilon_0 \, dA = hA\boldsymbol{B}\boldsymbol{C}\epsilon_0$$

$$\text{For plane stress:} \qquad \epsilon_0 = (\alpha\Delta T \quad \alpha\Delta T \quad 0)^T$$

$$\text{For plane strain:} \qquad \epsilon_0 = (1+\nu)(\alpha\Delta T \quad \alpha\Delta T \quad 0)^T$$

The distributed loads, if any, are applied along one or more sides of the element. The equivalent load vector due to distributed loads is

$$\boldsymbol{r}_q = h \int_c \boldsymbol{N}_c \begin{pmatrix} q_x \\ q_y \end{pmatrix} dc$$

where c is the part of the boundary over which the forces are applied. Generally, the applied surface forces are known in terms of components that are in the normal and tangential direction to the surface. For example, the applied pressure in a pressure vessel is always

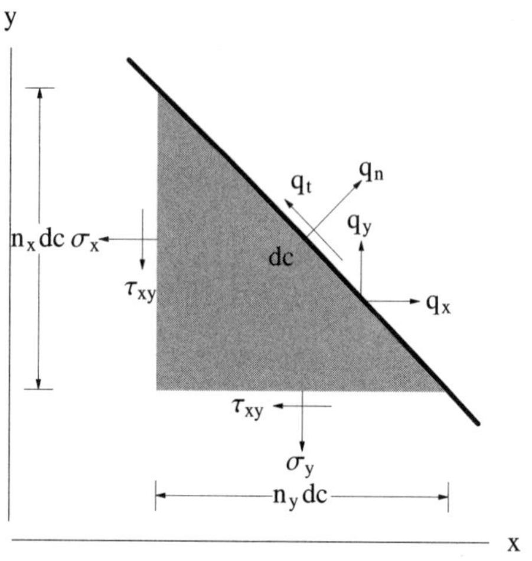

Figure 7.8. Equilibrium of a differential element on the boundary

normal to the surface of the vessel. Denoting the normal and the tangential components of the applied loading by q_n and q_t, with reference to Figure 7.8, these forces can be related to the components in the coordinate directions as follows:

$$\begin{pmatrix} q_x \\ q_y \end{pmatrix} = \begin{pmatrix} n_x & -n_y \\ n_y & n_x \end{pmatrix} \begin{pmatrix} q_n \\ q_t \end{pmatrix}$$

where n_x and n_y are the components of the unit normal to the boundary on which the forces are applied.

The boundary of the element is defined by its three sides. For each side the solution is linear in terms of coordinate c for that side and the degrees of freedom at the ends of that side. For side 1:

$$\begin{pmatrix} u \\ v \end{pmatrix} = \begin{pmatrix} -\frac{c-L_{12}}{L_{12}} & 0 & \frac{c}{L_{12}} & 0 & 0 & 0 \\ 0 & -\frac{c-L_{12}}{L_{12}} & 0 & \frac{c}{L_{12}} & 0 & 0 \end{pmatrix} \begin{pmatrix} u_1 \\ v_1 \\ u_2 \\ v_2 \\ u_3 \\ v_3 \end{pmatrix} \equiv N_c^T d; \qquad 0 \le c \le L_{12}$$

where $L_{12} = \sqrt{(x_2 - x_1)^2 + (y_2 - y_1)^2}$ is the length of side 1. The components of the unit normal to the side can be computed as follows:

$$n_x = \frac{y_2 - y_1}{L_{12}}; \qquad n_y = -\frac{x_2 - x_1}{L_{12}}$$

The equivalent load vector for applied loading on side 1-2 of the element therefore is

$$\boldsymbol{r}_q = h\int_c \boldsymbol{N}_c \begin{pmatrix} q_x \\ q_y \end{pmatrix} dc = \begin{pmatrix} h\int_0^{L_{12}} -\frac{c-L_{12}}{L_{12}} q_x\, dc \\ h\int_0^{L_{12}} -\frac{c-L_{12}}{L_{12}} q_y\, dc \\ h\int_0^{L_{12}} \frac{c}{L_{12}} q_x\, dc \\ h\int_0^{L_{12}} \frac{c}{L_{12}} q_y\, dc \\ 0 \\ 0 \end{pmatrix}$$

Carrying out integration,

$$\boldsymbol{r}_q = \frac{hL_{12}}{2}(q_x \quad q_y \quad q_x \quad q_y \quad 0 \quad 0)^T$$

In terms of specified normal and tangential components,

$$\boldsymbol{r}_q = \frac{hL_{12}}{2}(n_x q_n - n_y q_t \quad n_y q_n + n_x q_t \quad n_x q_n - n_y q_t \quad n_y q_n + n_x q_t \quad 0 \quad 0)^T$$

Proceeding in a similar manner, the equivalent load vector for normal and tangential pressures applied on sides 2 and 3 can be written. For side 2

$$\boldsymbol{r}_q = \frac{hL_{23}}{2}(0 \quad 0 \quad n_x q_n - n_y q_t \quad n_y q_n + n_x q_t \quad n_x q_n - n_y q_t \quad n_y q_n + n_x q_t)^T$$

where $L_{23} = \sqrt{(x_3 - x_2)^2 + (y_3 - y_2)^2}$ is the length of side 2-3 and

$$n_x = \frac{y_3 - y_2}{L_{23}}; \qquad n_y = -\frac{x_3 - x_2}{L_{23}}$$

For side 3

$$\boldsymbol{r}_q = \frac{hL_{31}}{2}(n_x q_n - n_y q_t \quad n_y q_n + n_x q_t \quad 0 \quad 0 \quad n_x q_n - n_y q_t \quad n_y q_n + n_x q_t)^T$$

where $L_{31} = \sqrt{(x_3 - x_1)^2 + (y_3 - y_1)^2}$ is the length of side 3-1 and

$$n_x = \frac{y_1 - y_3}{L_{31}}; \qquad n_y = -\frac{x_1 - x_3}{L_{31}}$$

The element equations are

$$\boldsymbol{kd} = \boldsymbol{r}_q + \boldsymbol{r}_b + \boldsymbol{r}_\epsilon$$

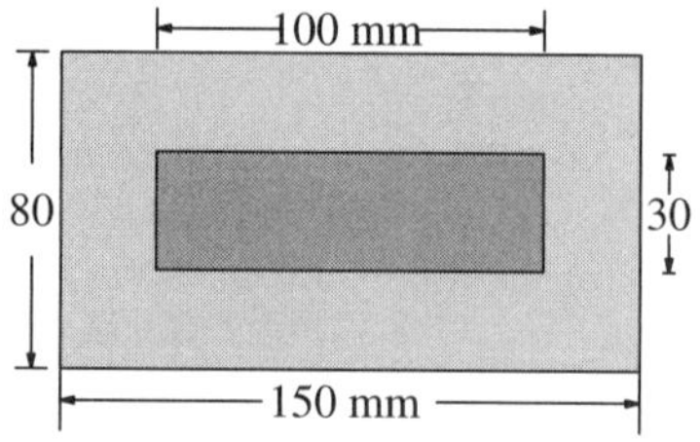

Figure 7.9. Steel and aluminum assembly subjected to temperature change

Example 7.6 Thermal Stresses A 5-mm-thick symmetric assembly of steel and aluminum plates, shown in Figure 7.9, is created at room temperature. Determine stresses and deformed shape if the temperature of the assembly is increased by 70°C above room temperature. Assume a perfect bond between the two materials. Use the following data:

Steel plate: 80×150 mm; $E = 200$ GPa; $\nu = 0.3$; $\alpha = 12 \times 10^{-6}$/°C

Aluminum plate: 30×100 mm; $E = 70$ GPa; $\nu = 0.33$; $\alpha = 23 \times 10^{-6}$/°C

Since the thickness of the assembly is much smaller than the other dimensions and there are no out-of-plane loads, the problem can be treated as a plane stress situation. Using symmetry, a quarter of the assembly is modeled as shown in Figure 7.10. The first two elements are in the aluminum plate and the remaining six in the steel plate. A coarse mesh is used to show all calculations. Due to symmetry nodes 2 and 3 can displace in the x direction only while nodes 4 and 7 can displace in the y direction alone. Node 1, being on both axes of symmetry, cannot displace in either direction. Note that in addition to reducing the model size the use of symmetry provides enough boundary conditions so that there is no rigid-body motion in the model. Since no support conditions are given for the assembly, analysis of a full finite element model would not be possible without introducing artificial supports.

The complete finite element calculations are as follows. The numerical values are in the newton-millimeters units. The displacements will be in millimeters and the stresses in megapascals.

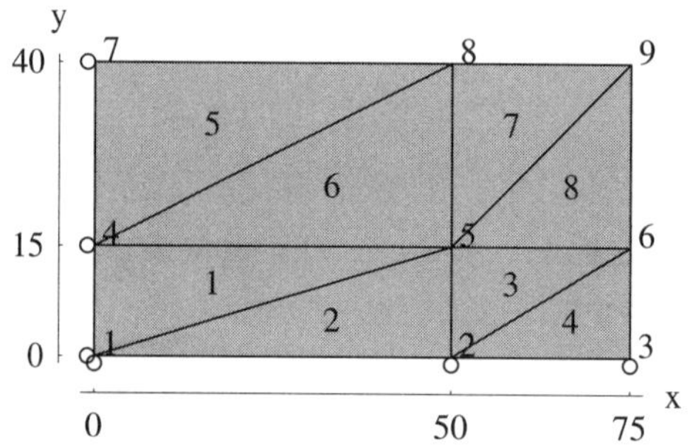

Figure 7.10. Finite element model of one-fourth of the assembly

Essential boundary conditions:

Node	dof	Value
1	u_1	0
	v_1	0
2	v_2	0
3	v_3	0
4	u_4	0
7	u_7	0

Equations for element 1:

$h = 5; E = 70000; \nu = 0.33$

Plane stress constitutive matrix, $\boldsymbol{C} = \begin{pmatrix} 78554.6 & 25923. & 0 \\ 25923. & 78554.6 & 0 \\ 0 & 0 & 26315.8 \end{pmatrix}$

Nodal coordinates:

Element node	Global node number	x	y
1	1	0	0
2	5	50	15
3	4	0	15

$x_1 = 0; x_2 = 50; x_3 = 0$

$y_1 = 0; y_2 = 15; y_3 = 15$

Using these values, we get

$$b_1 = 0; \quad b_2 = 15; \quad b_3 = -15$$

$$c_1 = -50; \quad c_2 = 0; \quad c_3 = 50$$

$$f_1 = 750; \quad f_2 = 0; \quad f_3 = 0$$

Element area, $A = 375$

$$\boldsymbol{B}^T = \begin{pmatrix} 0 & 0 & \frac{1}{50} & 0 & -\frac{1}{50} & 0 \\ 0 & -\frac{1}{15} & 0 & 0 & 0 & \frac{1}{15} \\ -\frac{1}{15} & 0 & 0 & \frac{1}{50} & \frac{1}{15} & -\frac{1}{50} \end{pmatrix}$$

Thus the element stiffness matrix is

$$\boldsymbol{k} = hA\boldsymbol{B}\boldsymbol{C}\boldsymbol{B}^T = 10^6 \begin{pmatrix} 0.219298 & 0 & 0 & -0.0657895 & -0.219298 & 0.0657895 \\ 0 & 0.654622 & -0.0648075 & 0 & 0.0648075 & -0.654622 \\ 0 & -0.0648075 & 0.0589159 & 0 & -0.0589159 & 0.0648075 \\ -0.0657895 & 0 & 0 & 0.0197368 & 0.0657895 & -0.0197368 \\ -0.219298 & 0.0648075 & -0.0589159 & 0.0657895 & 0.278214 & -0.130597 \\ 0.0657895 & -0.654622 & 0.0648075 & -0.0197368 & -0.130597 & 0.674358 \end{pmatrix}$$

Load vector due to temperature change:

$$\alpha = \frac{23}{1000000}; \qquad \Delta T = 70; \qquad \epsilon_0^T = \begin{pmatrix} \frac{161}{100000} & \frac{161}{100000} & 0 \end{pmatrix}$$

$$\boldsymbol{r}_\epsilon^T = (0. \quad -21026.1 \quad 6307.84 \quad 0. \quad -6307.84 \quad 21026.1)$$

Complete equations for element 1:

$$10^6 \begin{pmatrix} 0.219298 & 0 & 0 & -0.0657895 & -0.219298 & 0.0657895 \\ 0 & 0.654622 & -0.0648075 & 0 & 0.0648075 & -0.654622 \\ 0 & -0.0648075 & 0.0589159 & 0 & -0.0589159 & 0.0648075 \\ -0.0657895 & 0 & 0 & 0.0197368 & 0.0657895 & -0.0197368 \\ -0.219298 & 0.0648075 & -0.0589159 & 0.0657895 & 0.278214 & -0.130597 \\ 0.0657895 & -0.654622 & 0.0648075 & -0.0197368 & -0.130597 & 0.674358 \end{pmatrix} \begin{pmatrix} u_1 \\ v_1 \\ u_5 \\ v_5 \\ u_4 \\ v_4 \end{pmatrix} = \begin{pmatrix} 0. \\ -21026.1 \\ 6307.84 \\ 0. \\ -6307.84 \\ 21026.1 \end{pmatrix}$$

Processing the remaining elements in the same manner, the global equations after assembly and essential boundary conditions are as follows:

$$10^6 \begin{pmatrix} 0.928397 & -0.32967 & 0 & -0.539811 & 0.257115 & 0 & -0.357143 & 0 & 0 & 0 & 0 & 0 \\ -0.32967 & 0.650183 & 0 & 0 & 0 & -0.320513 & 0.164835 & 0 & 0 & 0 & 0 & 0 \\ 0 & 0 & 1.86941 & 0.257115 & -0.115891 & 0 & 0 & -1.0989 & -0.357143 & 0 & 0 & 0 \\ -0.539811 & 0 & 0.257115 & 2.3295 & -0.714286 & -0.879121 & 0.357143 & 0 & -0.576923 & 0.357143 & 0 & -0.357143 \\ 0.257115 & 0 & -0.115891 & -0.714286 & 3.64231 & 0.357143 & -0.307692 & 0 & 0.357143 & -1.64835 & -0.357143 & 0 \\ 0 & -0.320513 & 0 & -0.879121 & 0.357143 & 1.39194 & -0.357143 & 0 & 0 & 0 & -0.192308 & 0.164835 \\ -0.357143 & 0.164835 & 0 & 0.357143 & -0.307692 & -0.357143 & 1.77289 & 0 & 0 & 0 & 0.192308 & -0.549451 \\ 0 & 0 & -1.0989 & 0 & 0 & 0 & 0 & 1.19505 & 0.164835 & -0.0961538 & 0 & 0 \\ 0 & 0 & -0.357143 & -0.576923 & 0.357143 & 0 & 0 & 0.164835 & 1.4011 & -0.357143 & -0.549451 & 0.192308 \\ 0 & 0 & 0 & 0.357143 & -1.64835 & 0 & 0 & -0.0961538 & -0.357143 & 1.93681 & 0.164835 & -0.192308 \\ 0 & 0 & 0 & 0 & -0.357143 & -0.192308 & 0.192308 & 0 & -0.549451 & 0.164835 & 0.741758 & 0 \\ 0 & 0 & 0 & -0.357143 & 0 & 0.164835 & -0.549451 & 0 & 0.192308 & -0.192308 & 0 & 0.741758 \end{pmatrix} \begin{pmatrix} u_2 \\ u_3 \\ v_4 \\ u_5 \\ v_5 \\ u_6 \\ v_6 \\ v_7 \\ u_8 \\ v_8 \\ u_9 \\ v_9 \end{pmatrix} = \begin{pmatrix} -2692.16 \\ 9000. \\ -8973.88 \\ -2692.16 \\ -8973.88 \\ 24000. \\ 0 \\ 30000. \\ 0 \\ 45000. \\ 15000. \\ 15000. \end{pmatrix}$$

Solving the final system of global equations, we get

$$\{u_2 = 0.0513447, u_3 = 0.0703132, v_4 = 0.0252714, u_5 = 0.0495551, v_5 = 0.0186102,$$
$$u_6 = 0.069253, v_6 = 0.0146016, v_7 = 0.0445986, u_8 = 0.0498168, v_8 = 0.0388815,$$
$$u_9 = 0.0716126, v_9 = 0.0366734\}$$

Solution for element 1:

$x_1 = 0; x_2 = 50; x_3 = 0$

$y_1 = 0; y_2 = 15; y_3 = 15$

Using these values, we get

$$b_1 = 0; \qquad b_2 = 15; \qquad b_3 = -15$$

$$c_1 = -50; \qquad c_2 = 0; \qquad c_3 = 50$$

$$f_1 = 750; \qquad f_2 = 0; \qquad f_3 = 0$$

Element area, $A = 375$

$$\boldsymbol{B}^T = \begin{pmatrix} 0 & 0 & \frac{1}{50} & 0 & -\frac{1}{50} & 0 \\ 0 & -\frac{1}{15} & 0 & 0 & 0 & \frac{1}{15} \\ -\frac{1}{15} & 0 & 0 & \frac{1}{50} & \frac{1}{15} & -\frac{1}{50} \end{pmatrix}$$

Substituting these into the formulas for triangle interpolation functions, we get

$$\text{Interpolation functions, } \left\{1 - \frac{y}{15}, \frac{x}{50}, \frac{y}{15} - \frac{x}{50}\right\}$$

$$N^T = \begin{pmatrix} 1 - \frac{y}{15} & 0 & \frac{x}{50} & 0 & \frac{y}{15} - \frac{x}{50} & 0 \\ 0 & 1 - \frac{y}{15} & 0 & \frac{x}{50} & 0 & \frac{y}{15} - \frac{x}{50} \end{pmatrix}$$

From the global solution the displacements at the element nodes are (displacements at nodes {1, 5, 4}):

$$d^T = \{0, 0, 0.0495551, 0.0186102, 0, 0.0252714\}$$

The displacement distribution over the element is

$$\begin{pmatrix} u(x, y) \\ v(x, y) \end{pmatrix} = N^T d = \begin{pmatrix} 0.000991101x \\ 0.00168476y - 0.000133224x \end{pmatrix}$$

In-plane strain components, $\epsilon = B^T d = (0.000991101 \quad 0.00168476 \quad -0.000133224)$
Initial strains: $\epsilon_0^T = \left(\frac{161}{100000} \quad \frac{161}{100000} \quad 0\right)$
In-plane stress components, $\sigma = C(\epsilon - \epsilon_0) = (-46.6793 \quad -10.1709 \quad -3.50591)$
Computing out-of-plane strain and stress components using appropriate formulas, the complete strain and stress vectors are as follows:

$$\epsilon^T = (0.000991101 \quad 0.00168476 \quad 0.00187801 \quad -0.000133224 \quad 0 \quad 0)$$

$$\sigma^T = (-46.6793 \quad -10.1709 \quad 0 \quad -3.50591 \quad 0 \quad 0)$$

Substituting these stress components into appropriate formulas,

Principal stresses = $(0 \quad -9.83725 \quad -47.0129)$

Effective stress (von Mises) = 42.9477

Solutions over the remaining elements can be computed in a similar manner. A summary of the solution at element centers is as follows:

	Coordinates	Displacements	Stresses	Principal Stresses	Effective Stress
1	$\left\{\begin{matrix} \frac{50}{3} \\ 10 \end{matrix}\right.$	0.0165184 0.0146272	$\begin{Bmatrix} -46.6793 \\ -10.1709 \\ 0 \\ -3.50591 \\ 0 \\ 0 \end{Bmatrix}$	$\begin{Bmatrix} 0 \\ -9.83725 \\ -47.0129 \end{Bmatrix}$	42.9477

	Coordinates	Displacements	Stresses	Principal Stresses	Effective Stress
2	$\begin{cases} \frac{100}{3} \\ 5 \end{cases}$	0.0336333 0.0062034	$\begin{Bmatrix} -55.3796 \\ -44.1277 \\ 0 \\ -3.13967 \\ 0 \\ 0 \end{Bmatrix}$	$\begin{Bmatrix} 0 \\ -43.3109 \\ -56.1964 \end{Bmatrix}$	50.9897
3	$\begin{cases} \frac{175}{3} \\ 10 \end{cases}$	0.0567176 0.0110706	$\begin{Bmatrix} 14.9716 \\ 84.6275 \\ 0 \\ -21.5116 \\ 0 \\ 0 \end{Bmatrix}$	$\begin{Bmatrix} 90.7353 \\ 8.86375 \\ 0 \end{Bmatrix}$	86.6441
4	$\begin{cases} \frac{200}{5} \\ 0.0636369 \end{cases}$	0.0048672 0.0048672	$\begin{Bmatrix} -9.06129 \\ 23.9696 \\ 0 \\ -5.43678 \\ 0 \\ 0 \end{Bmatrix}$	$\begin{Bmatrix} 24.8415 \\ 0 \\ -9.93316 \end{Bmatrix}$	31.0245
5	$\begin{cases} \frac{50}{30} \\ \frac{95}{3} \end{cases}$	0.0166056 0.0362505	$\begin{Bmatrix} 29.9479 \\ -4.39778 \\ 0 \\ -8.79555 \\ 0 \\ 0 \end{Bmatrix}$	$\begin{Bmatrix} 32.0694 \\ 0 \\ -6.51919 \end{Bmatrix}$	35.7772
6	$\begin{cases} \frac{100}{3} \\ \frac{70}{3} \end{cases}$	0.033124 0.0275877	$\begin{Bmatrix} 31.2874 \\ 3.55695 \\ 0 \\ -9.44265 \\ 0 \\ 0 \end{Bmatrix}$	$\begin{Bmatrix} 34.1974 \\ 0.646946 \\ 0 \end{Bmatrix}$	33.8785
7	$\begin{cases} \frac{175}{3} \\ \frac{95}{3} \end{cases}$	0.0569948 0.0313884	$\begin{Bmatrix} 5.07389 \\ -4.3071 \\ 0 \\ -5.98876 \\ 0 \\ 0 \end{Bmatrix}$	$\begin{Bmatrix} 7.99036 \\ 0 \\ -7.22357 \end{Bmatrix}$	13.1812
8	$\begin{cases} \frac{200}{3} \\ \frac{70}{3} \end{cases}$	0.0634735 0.0232951	$\begin{Bmatrix} -8.62005 \\ 5.98876 \\ 0 \\ -5.07389 \\ 0 \\ 0 \end{Bmatrix}$	$\begin{Bmatrix} 7.57809 \\ 0 \\ -10.2094 \end{Bmatrix}$	15.4605

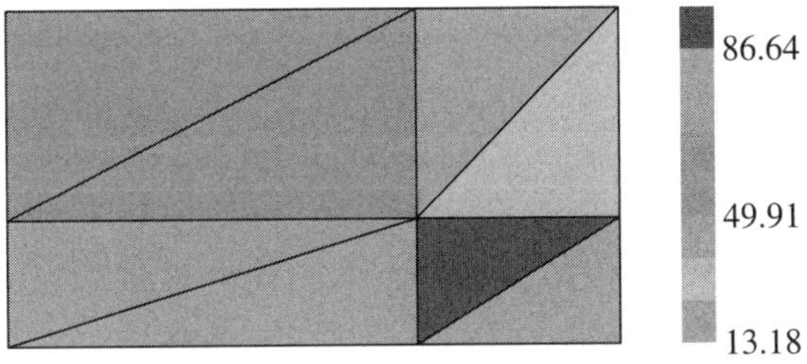

Figure 7.11. Equivalent stress plot

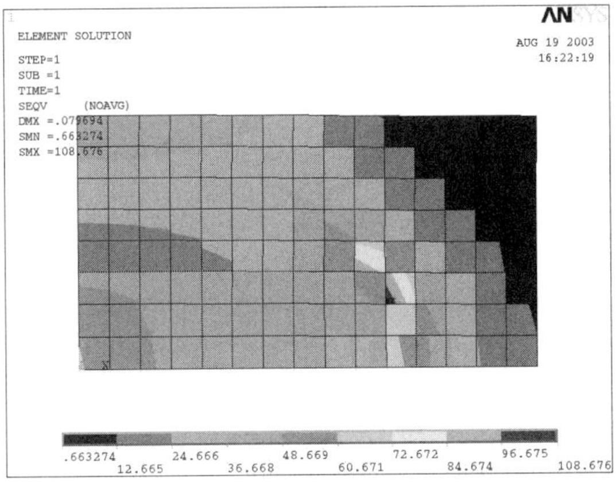

Figure 7.12. Equivalent stress plot using uniform mesh

Figure 7.11 shows equivalent von Mises stresses in different elements. There are large differences in stresses among neighboring elements, indicating that the solution is not reliable and mesh must be refined.

The problem is solved using ANSYS (AnsysFiles\Chap7\ThermalStress.inp on the book web site) with a uniform mesh with an element size of approximately 5 mm. The computed equivalent stresses from elements are shown in Figure 7.12. The results show considerable improvement. However, there are still large differences in stresses among neighboring elements near the interface between the two materials. This suggests using a finer mesh in the interface region. Results from a mesh that is refined only in the regions of high stress gradient are shown in Figure 7.13. Overall the results appear satisfactory. The only exceptions are the few elements at the sharp corner at the interface between the two materials. A sharp corner represents stress singularity, and thus further refinement of the mesh near the corner will only be a wasteful attempt at capturing this singularity. In actual design situations, the corners are rounded to avoid high stress gradients. Thus any further refinement of the model should include the actual rounded profile of the corner to get realistic results.

◆ *Mathematica*/MATLAB Implementation 7.1 on the Book Web Site: *Triangular element for plane stress and plane strain*

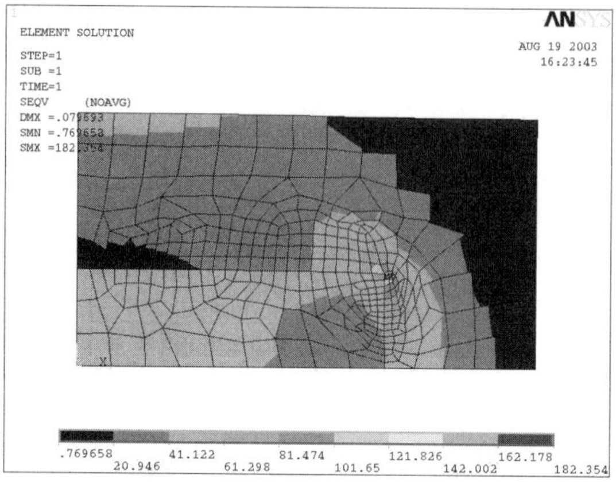

Figure 7.13. Equivalent stress plot using finer mesh in high stress gradient region

7.4.5 Mapped Quadrilateral Elements

The quadrilateral elements for the elasticity problem can be developed in exactly the same manner as those for the two-dimensional boundary value problem in Chapter 6. The actual element is mapped to a master element using appropriate interpolation functions:

$$x = \boldsymbol{N}(s,t)^T \boldsymbol{x}_n; \qquad y = \boldsymbol{N}^T \boldsymbol{y}_n$$

$$\boldsymbol{J} = \begin{pmatrix} \frac{\partial x}{\partial s} & \frac{\partial x}{\partial t} \\ \frac{\partial y}{\partial s} & \frac{\partial y}{\partial t} \end{pmatrix} = \begin{pmatrix} J_{11} & J_{12} \\ J_{21} & J_{22} \end{pmatrix}; \qquad \det \boldsymbol{J} = \left| \frac{\partial x}{\partial s}\frac{\partial y}{\partial t} - \frac{\partial x}{\partial t}\frac{\partial y}{\partial s} \right|$$

The assumed solution over the master element is written as follows:

$$u(s,t) = \boldsymbol{N}(s,t)^T \boldsymbol{d}$$

The element strain vector is

$$\boldsymbol{\epsilon} = \begin{pmatrix} \epsilon_x \\ \epsilon_y \\ \gamma_{xy} \end{pmatrix} = \begin{pmatrix} \frac{\partial u}{\partial x} \\ \frac{\partial v}{\partial y} \\ \frac{\partial u}{\partial y} + \frac{\partial v}{\partial x} \end{pmatrix} = \begin{pmatrix} \frac{\partial N_1}{\partial x} & 0 & \frac{\partial N_2}{\partial x} & 0 & \cdots \\ 0 & \frac{\partial N_1}{\partial y} & 0 & \frac{\partial N_2}{\partial y} & \cdots \\ \frac{\partial N_1}{\partial y} & \frac{\partial N_1}{\partial x} & \frac{\partial N_2}{\partial y} & \frac{\partial N_2}{\partial x} & \cdots \end{pmatrix} \begin{pmatrix} u_1 \\ v_1 \\ u_2 \\ v_2 \\ \vdots \end{pmatrix} = \boldsymbol{B}^T \boldsymbol{d}$$

As explained in Chapter 6, using mapping, the derivatives of the interpolation functions with respect to x and y are computed as follows:

$$\frac{\partial N_i}{\partial x} = \frac{1}{\det \boldsymbol{J}} \left(J_{22}\frac{\partial N_i}{\partial s} - J_{21}\frac{\partial N_i}{\partial t} \right); \qquad \frac{\partial N_i}{\partial y} = \frac{1}{\det \boldsymbol{J}} \left(-J_{12}\frac{\partial N_i}{\partial s} + J_{11}\frac{\partial N_i}{\partial t} \right)$$

$$\boldsymbol{B}^T = \frac{1}{\det \boldsymbol{J}} \begin{pmatrix} J_{22}\frac{\partial N_1}{\partial s} - J_{21}\frac{\partial N_1}{\partial t} & 0 & J_{22}\frac{\partial N_2}{\partial s} - J_{21}\frac{\partial N_2}{\partial t} & \cdots \\ 0 & -J_{12}\frac{\partial N_1}{\partial s} + J_{11}\frac{\partial N_1}{\partial t} & 0 & \cdots \\ -J_{12}\frac{\partial N_1}{\partial s} + J_{11}\frac{\partial N_1}{\partial t} & J_{22}\frac{\partial N_1}{\partial s} - J_{21}\frac{\partial N_1}{\partial t} & -J_{12}\frac{\partial N_2}{\partial s} + J_{11}\frac{\partial N_2}{\partial t} & \cdots \end{pmatrix}$$

The element equations are obtained using numerical integration, as discussed in Chapter 6:

Element stiffness matrix:

$$\boldsymbol{k} = h \iint_A \boldsymbol{BCB}^T\, dA = h \int_{-1}^{1}\int_{-1}^{1} \boldsymbol{BCB}^T \det \boldsymbol{J}\, ds\, dt$$

$$= h \sum_{i=1}^{m}\sum_{j=1}^{n} w_i w_j \boldsymbol{B}(s_i, t_j)\boldsymbol{CB}^T(s_i, t_j) \det \boldsymbol{J}(s_i, t_j)$$

Equivalent load vector due to body forces:

$$\boldsymbol{r}_b = h \iint_A \boldsymbol{N}\begin{pmatrix} b_x \\ b_y \end{pmatrix} dA = h \int_{-1}^{1}\int_{-1}^{1} \boldsymbol{N}\begin{pmatrix} b_x \\ b_y \end{pmatrix} \det \boldsymbol{J}\, ds\, dt$$

$$= h \sum_{i=1}^{m}\sum_{j=1}^{n} w_i w_j \boldsymbol{N}(s_i, t_j)\begin{pmatrix} b_x \\ b_y \end{pmatrix} \det \boldsymbol{J}(s_i, t_j)$$

Equivalent load vector due to initial strains:

$$\boldsymbol{r}_\epsilon = h \iint_A \boldsymbol{BC}\epsilon_0\, dA = h \int_{-1}^{1}\int_{-1}^{1} \boldsymbol{BC}\epsilon_0 \det \boldsymbol{J}\, ds\, dt$$

$$= h \sum_{i=1}^{m}\sum_{j=1}^{n} w_i w_j \boldsymbol{B}(s_i, t_j)\boldsymbol{C}\epsilon_0 \det \boldsymbol{J}(s_i, t_j)$$

Equivalent load vector due to distributed loads:

$$\boldsymbol{r}_q = h \int_C \boldsymbol{N}_c \begin{pmatrix} q_x \\ q_y \end{pmatrix} dc = h \int_{-1}^{1} \boldsymbol{N}_c \begin{pmatrix} q_x \\ q_y \end{pmatrix} J_c\, da = h \sum_{i=1}^{n} w_i \boldsymbol{N}_c(a_i) \begin{pmatrix} q_x \\ q_y \end{pmatrix} J_c(a_i)$$

where c is the part of the boundary over which the forces are applied.

As seen in Chapter 6, to write interpolation functions $\boldsymbol{N}_c(a)$ along each side, we substitute the appropriate s, t values in the interpolation functions $\boldsymbol{N}(s, t)$. For example, for side 1, $s = a$ and $t = -1$. The differential length dc along an element side is related to the differential line segment da along the side of the master element as follows:

$$dc = \sqrt{dx^2 + dy^2}$$

Dividing both sides by da, we have

$$\frac{dc}{da} = \sqrt{\left(\frac{dx}{da}\right)^2 + \left(\frac{dy}{da}\right)^2} \Longrightarrow dc = J_c\, da$$

where J_c is the Jacobian of the side:

$$J_c = \sqrt{\left(\frac{dx}{da}\right)^2 + \left(\frac{dy}{da}\right)^2}$$

Generally the applied surface forces are known in terms of components that are in the normal and tangential directions to the surface. Denoting the normal and the tangential components of the applied loading by q_n and q_t, with reference to Figure 7.8, these forces can be related to the components in the coordinate directions as follows:

$$\begin{pmatrix} q_x \\ q_y \end{pmatrix} = \begin{pmatrix} n_x & -n_y \\ n_y & n_x \end{pmatrix} \begin{pmatrix} q_n \\ q_t \end{pmatrix}$$

where n_x and n_y are the components of the unit normal to the boundary on which the forces are applied. For mapped elements, the components of the unit normal are as follows:

$$n_x = \frac{dy/da}{J_c}; \qquad n_y = -\frac{dx/da}{J_c}$$

Example 7.7 Notched Beam Find stresses in the notched beam of rectangular cross section shown in Figure 7.14. The following numerical values are used:

$a = 48\,\text{in}$; $b = 12\,\text{in}$; $c = 5\,\text{in}$; $d = 12\,\text{in}$; Thickness $= 4\,\text{in}$;

$E = 3 \times 10^6\,\text{lb/in}^2$; $\nu = 0.2$; $q = 50\,\text{lb/in}^2$

Since the thickness of the beam is much smaller than the other dimensions and there are no out-of-plane loads, the problem can be treated as a plane stress situation. Using symmetry half of the beam is modeled. To show all calculations, a very coarse model involving only three elements is used as shown in Figure 7.15.

The essential boundary conditions are as follows:

Node	dof	Value
1	u_1	0
2	u_2	0
7	u_7	0
	v_7	0
8	u_8	0
	v_8	0

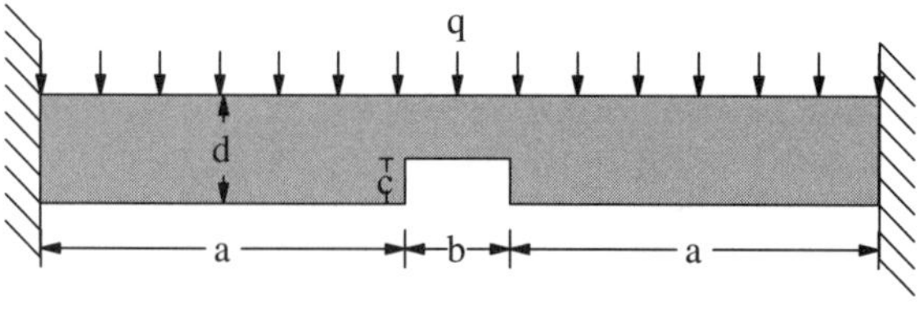

Figure 7.14. Notched beam

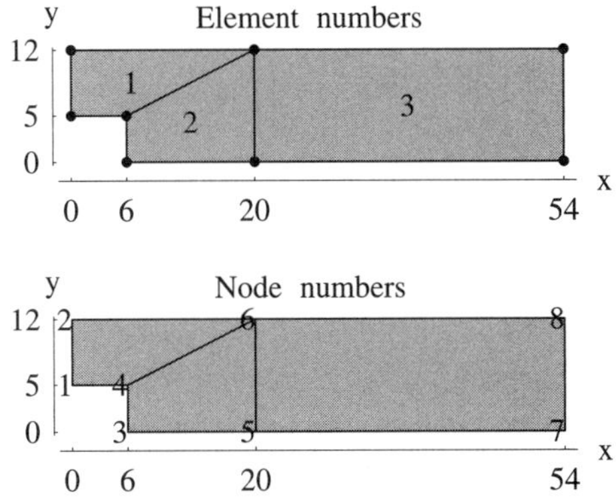

Figure 7.15. Three Quad4 element model for notched beam

Equations for element 1 are as follows:

$E = 3000000$; $\nu = 0.2$; $h = 4$

Nodal coordinates:

Element Node	Global Node Number	x	y
1	1	0.	5.
2	4	6.	5.
3	6	20.	12.
4	2	0.	12.

Mapping to the master element:

$$x(s,t) = \boldsymbol{N}^T \boldsymbol{x}_n = 3.5ts + 6.5s + 3.5t + 6.5; \qquad y(s,t) = \boldsymbol{N}^T \boldsymbol{y}_n = 3.5t + 8.5$$

$$\boldsymbol{J} = \begin{pmatrix} 3.5t + 6.5 & 3.5s + 3.5 \\ 0 & 3.5 \end{pmatrix}; \qquad \det \boldsymbol{J} = 12.25t + 22.75$$

$$\text{Plane stress } \boldsymbol{C} = \begin{pmatrix} 3.125 \times 10^6 & 625000. & 0 \\ 625000. & 3.125 \times 10^6 & 0 \\ 0 & 0 & 1.25 \times 10^6 \end{pmatrix}$$

For numerical integration the Gauss quadrature points and weights are as follows:

	s	t	Weight
1	−0.57735	−0.57735	1.
2	−0.57735	0.57735	1.
3	0.57735	−0.57735	1.
4	0.57735	0.57735	1.

Computation of element matrices at {−0.57735, −0.57735} with weight = 1:

$$\boldsymbol{J} = \begin{pmatrix} 4.47927 & 1.47927 \\ 0 & 3.5 \end{pmatrix}; \qquad \det \boldsymbol{J} = 15.6775$$

To save space numerical values of only four interpolation functions N_1, N_2, N_3, and N_4 are shown below. For element NBC calculations these values must obviously be arranged into full 2×8 matrices as shown in the text.

$$\boldsymbol{N}^T = (0.622008 \quad 0.166667 \quad 0.0446582 \quad 0.166667)$$

$$\frac{\partial \boldsymbol{N}^T}{\partial s} = (-0.394338 \quad 0.394338 \quad 0.105662 \quad -0.105662)$$

$$\frac{\partial \boldsymbol{N}^T}{\partial t} = (-0.394338 \quad -0.105662 \quad 0.105662 \quad 0.394338)$$

$$\frac{\partial \boldsymbol{N}^T}{\partial x} = (-0.088036 \quad 0.088036 \quad 0.0235892 \quad -0.0235892)$$

$$\frac{\partial \boldsymbol{N}^T}{\partial y} = (-0.0754595 \quad -0.0673977 \quad 0.0202193 \quad 0.122638)$$

$$\boldsymbol{B}^T = \begin{pmatrix} -0.088036 & 0 & 0.088036 & 0 & 0.0235892 & 0 & -0.0235892 & 0 \\ 0 & -0.0754595 & 0 & -0.0673977 & 0 & 0.0202193 & 0 & 0.122638 \\ -0.0754595 & -0.088036 & -0.0673977 & 0.088036 & 0.0202193 & 0.0235892 & 0.122638 & -0.0235892 \end{pmatrix}$$

$$\boldsymbol{k} = 10^6 \begin{pmatrix} 1.96517 & 0.781108 & -1.12016 & -0.288186 & -0.526565 & -0.209297 & -0.318444 & -0.283625 \\ 0.781108 & 1.7234 & 0.204736 & 0.389125 & -0.209297 & -0.461783 & -0.776547 & -1.65074 \\ -1.12016 & 0.204736 & 1.87489 & -0.697658 & 0.300146 & -0.0548588 & -1.05488 & 0.547781 \\ -0.288186 & 0.389125 & -0.697658 & 1.4977 & 0.0772193 & -0.104266 & 0.908625 & -1.78256 \\ -0.526565 & -0.209297 & 0.300146 & 0.0772193 & 0.141093 & 0.056081 & 0.0853267 & 0.075997 \\ -0.209297 & -0.461783 & -0.0548588 & -0.104266 & 0.056081 & 0.123734 & 0.208075 & 0.442314 \\ -0.318444 & -0.776547 & -1.05488 & 0.908625 & 0.0853267 & 0.208075 & 1.28799 & -0.340153 \\ -0.283625 & -1.65074 & 0.547781 & -1.78256 & 0.075997 & 0.442314 & -0.340153 & 2.99099 \end{pmatrix}$$

Similarly, evaluating at the remaining Gauss points and adding contributions result in the following:

$$\boldsymbol{k} = 10^6 \begin{pmatrix} 6.26209 & -0.0163755 & -4.11923 & 1.26638 & -0.951731 & -1.12991 & -1.19113 & -0.120087 \\ -0.0163755 & 9.0354 & 2.51638 & -3.67826 & -1.12991 & 0.228478 & -1.37009 & -5.58562 \\ -4.11923 & 2.51638 & 11.2621 & -3.76638 & -1.19113 & -1.37009 & -5.95173 & 2.62009 \\ 1.26638 & -3.67826 & -3.76638 & 21.5354 & -0.120087 & -5.58562 & 2.62009 & -12.2715 \\ -0.951731 & -1.12991 & -1.19113 & -0.120087 & 2.54484 & 0.786026 & -0.401981 & 0.463974 \\ -1.12991 & 0.228478 & -1.37009 & -5.58562 & 0.786026 & 2.55069 & 1.71397 & 2.80646 \\ -1.19113 & -1.37009 & -5.95173 & 2.62009 & -0.401981 & 1.71397 & 7.54484 & -2.96397 \\ -0.120087 & -5.58562 & 2.62009 & -12.2715 & 0.463974 & 2.80646 & -2.96397 & 15.0507 \end{pmatrix}$$

$$\boldsymbol{r}^T = (0 \quad 0 \quad 0 \quad 0 \quad 0 \quad 0 \quad 0 \quad 0)$$

Computation of element matrices resulting from the NBC:

NBC on side 3 with $q = \{-50, 0\}$

$$\boldsymbol{N}_c^T = \begin{pmatrix} 0 & 0 & \frac{1-a}{2} & \frac{a+1}{2} \end{pmatrix}$$

$x(a) = 10. - 10.a$; $y(a) = 12.$

$dx/da = -10.; dy/da = 0.$

$\boldsymbol{J}_c = 10.$

Gauss point = −0.57735; Weight = 1.; $\boldsymbol{J}_c = 10.$

$\boldsymbol{N}_c^T = (0 \quad 0 \quad 0.788675 \quad 0.211325)$

$\boldsymbol{r}_q^T = (0 \quad 0 \quad 0 \quad 0 \quad 0 \quad -1577.35 \quad 0 \quad -422.65)$

Gauss point = 0.57735; Weight = 1.; $\boldsymbol{J}_c = 10.$

$\boldsymbol{N}_c^T = (0 \quad 0 \quad 0.211325 \quad 0.788675)$

$\boldsymbol{r}_q^T = (0 \quad 0 \quad 0 \quad 0 \quad 0 \quad -422.65 \quad 0 \quad -1577.35)$

Summing contributions from all Gauss points:

$$\boldsymbol{r}_q^T = (0 \quad 0 \quad 0 \quad 0 \quad 0 \quad -2000. \quad 0 \quad -2000.)$$

Complete element equations for element 1:

$$10^6 \begin{pmatrix} 6.26209 & -0.0163755 & -4.11923 & 1.26638 & -0.951731 & -1.12991 & -1.19113 & -0.120087 \\ -0.0163755 & 9.0354 & 2.51638 & -3.67826 & -1.12991 & 0.228478 & -1.37009 & -5.58562 \\ -4.11923 & 2.51638 & 11.2621 & -3.76638 & -1.19113 & -1.37009 & -5.95173 & 2.62009 \\ 1.26638 & -3.67826 & -3.76638 & 21.5354 & -0.120087 & -5.58562 & v2.62009 & -12.2715 \\ -0.951731 & -1.12991 & -1.19113 & -0.120087 & 2.54484 & 0.786026 & -0.401981 & 0.463974 \\ -1.12991 & 0.228478 & -1.37009 & -5.58562 & 0.786026 & 2.55069 & 1.71397 & 2.80646 \\ -1.19113 & -1.37009 & -5.95173 & 2.62009 & -0.401981 & 1.71397 & 7.54484 & -2.96397 \\ -0.120087 & -5.58562 & 2.62009 & -12.2715 & 0.463974 & 2.80646 & -2.96397 & 15.0507 \end{pmatrix} \begin{pmatrix} u_1 \\ v_1 \\ u_4 \\ v_4 \\ u_6 \\ v_6 \\ u_2 \\ v_2 \end{pmatrix} = \begin{pmatrix} 0. \\ 0. \\ 0. \\ 0. \\ 0. \\ -2000. \\ 0. \\ -2000. \end{pmatrix}$$

The equations for the remaining elements are developed in exactly the same manner. After assembly of all elements and incorporating essential boundary conditions, the global system of equations is as follows:

$$10^6 \begin{pmatrix} 9.0354 & -5.58562 & 0 & 0 & 2.51638 & -3.67826 & 0 & 0 & -1.12991 & 0.228478 \\ -5.58562 & 15.0507 & 0 & 0 & 2.62009 & -12.2715 & 0 & 0 & 0.463974 & 2.80646 \\ 0 & 0 & 5.7276 & 0.559291 & -3.49546 & 1.94071 & -0.771918 & -1.17054 & -1.46023 & -1.32946 \\ 0 & 0 & 0.559291 & 9.65901 & 0.690709 & -8.76615 & 0.0794621 & 2.74625 & -1.32946 & -3.6391 \\ 2.51638 & 2.62009 & -3.49546 & 0.690709 & 20.1147 & -6.95708 & -4.58523 & 2.42054 & -1.96304 & -1.29063 \\ -3.67826 & -12.2715 & 1.94071 & -8.76615 & -6.95708 & 32.4444 & 2.42054 & -4.8891 & -1.29063 & -2.83938 \\ 0 & 0 & -0.771918 & 0.0794621 & -4.58523 & 2.42054 & 12.5561 & -0.866443 & -4.99308 & 0.866443 \\ 0 & 0 & -1.17054 & 2.74625 & 2.42054 & -4.8891 & -0.866443 & 19.7912 & 0.866443 & -16.766 \\ -1.12991 & 0.463974 & -1.46023 & -1.32946 & -1.96304 & -1.29063 & -4.99308 & 0.866443 & 11.9759 & -0.0804163 \\ 0.228478 & 2.80646 & -1.32946 & -3.6391 & -1.29063 & -2.83938 & 0.866443 & -16.766 & -0.0804163 & 21.0919 \end{pmatrix} \begin{pmatrix} v_1 \\ v_2 \\ u_3 \\ v_3 \\ u_4 \\ v_4 \\ u_5 \\ v_5 \\ u_6 \\ v_6 \end{pmatrix} = \begin{pmatrix} 0 \\ -2000. \\ 0 \\ 0 \\ 0 \\ 0 \\ 0 \\ 0 \\ 0 \\ -5400. \end{pmatrix}$$

Solving the final system of global equations, we get

$$\{v_1 = -0.0183155, v_2 = -0.0183204, u_3 = 0.00275915, v_3 = -0.0166486,$$
$$u_4 = 0.00114552, v_4 = -0.0164634, u_5 = 0.00305003, v_5 = -0.0113566,$$
$$u_6 = -0.00210128, v_6 = -0.0116254\}$$

The solution for element 1 is as follows:

Element nodal displacements:

Element node	Global node number	u	v
1	1	0	−0.0183155
2	4	0.00114552	−0.0164634
3	6	−0.00210128	−0.0116254
4	2	0	−0.0183204

$$d^T = (0 \quad -0.0183155 \quad 0.00114552 \quad -0.0164634 \quad -0.00210128 \quad -0.0116254 \quad 0 \quad -0.0183204)$$

Using appropriate s and t values, the solution at any point over the element can be computed. For example, consider the solution at the element centroid:

Element solution at $\{s, t\} = \{0, 0\} \Longrightarrow \{x, y\} = \{6.5, 8.5\}$:

$$N^T = \left\{\tfrac{1}{4}, \tfrac{1}{4}, \tfrac{1}{4}, \tfrac{1}{4}\right\}$$

Displacements $= N^T d = \{-0.00023894, -0.0161812\}$

$$J = \begin{pmatrix} 6.5 & 3.5 \\ 0 & 3.5 \end{pmatrix}; \qquad \det J = 22.75$$

$$\frac{\partial N^T}{\partial s} = \left(-\tfrac{1}{4} \quad \tfrac{1}{4} \quad \tfrac{1}{4} \quad -\tfrac{1}{4}\right)$$

$$\frac{\partial N^T}{\partial t} = \left(-\tfrac{1}{4} \quad -\tfrac{1}{4} \quad \tfrac{1}{4} \quad \tfrac{1}{4}\right)$$

$$\frac{\partial N^T}{\partial x} = (-0.0384615 \quad 0.0384615 \quad 0.0384615 \quad -0.0384615)$$

$$\frac{\partial N^T}{\partial y} = (-0.032967 \quad -0.10989 \quad 0.032967 \quad 0.10989)$$

$$B^T = \begin{pmatrix} -0.0384615 & 0 & 0.0384615 & 0 & 0.0384615 & 0 & -0.0384615 & 0 \\ 0 & -0.032967 & 0 & -0.10989 & 0 & 0.032967 & 0 & 0.10989 \\ -0.032967 & -0.0384615 & -0.10989 & 0.0384615 & 0.032967 & 0.0384615 & 0.10989 & -0.0384615 \end{pmatrix}$$

In-plane strain components,

$$\epsilon = B^T d = (-0.00003676 \quad 0.0000164861 \quad 0.000133582)$$

In-plane stress components, $\sigma = C\epsilon = (-104.571 \quad 28.544 \quad 166.978)$

Computing out-of-plane strain and stress components using appropriate formulas, the complete strain and stress vectors are as follows:

$$\epsilon^T = (-0.00003676 \quad 0.0000164861 \quad 5.06848 \times 10^{-6} \quad 0.000133582 \quad 0 \quad 0)$$

$$\sigma^T = (-104.571 \quad 28.544 \quad 0 \quad 166.978 \quad 0 \quad 0)$$

Substituting these stress components into appropriate formulas,

Principal stresses = (141.741   0.   −217.768)

Effective stress (von Mises) = 313.656

As another example, consider the solution at node 1 of the element:

Element solution at $\{s, t\} = \{-1, -1\} \Longrightarrow \{x, y\} = \{0., 5.\}$:

$$N^T = \{1, 0, 0, 0\}$$

Displacements $= N^T d = \{0., -0.0183155\}$

$$J = \begin{pmatrix} 3. & 0 \\ 0 & 3.5 \end{pmatrix} \det J = 10.5$$

$$\frac{\partial N^T}{\partial s} = \left(-\tfrac{1}{2} \quad \tfrac{1}{2} \quad 0 \quad 0\right)$$

$$\frac{\partial N^T}{\partial t} = \left(-\tfrac{1}{2} \quad 0 \quad 0 \quad \tfrac{1}{2}\right)$$

$$\frac{\partial N^T}{\partial x} = (-0.166667 \quad 0.166667 \quad 0 \quad 0.)$$

$$\frac{\partial N^T}{\partial y} = (-0.142857 \quad 0. \quad 0 \quad 0.142857)$$

$$B^T = \begin{pmatrix} -0.166667 & 0 & 0.166667 & 0 & 0 & 0 & 0 & 0 \\ 0 & -0.142857 & 0 & 0 & 0 & 0 & 0 & 0.142857 \\ -0.142857 & -0.166667 & 0 & 0.166667 & 0 & 0 & 0.142857 & 0 \end{pmatrix}$$

In-plane strain components,

$$\epsilon = B^T d = (0.00019092 \quad -6.97447 \times 10^{-7} \quad 0.000308689)$$

In-plane stress components, $\sigma = C\epsilon = (596.188 \quad 117.145 \quad 385.861)$

Computing out-of-plane strain and stress components using appropriate formulas, the complete strain and stress vectors are as follows:

$$\epsilon^T = (0.00019092 \quad -6.97447 \times 10^{-7} \quad -0.0000475555 \quad 0.000308689 \quad 0 \quad 0)$$

$$\sigma^T = (596.188 \quad 117.145 \quad 0 \quad 385.861 \quad 0 \quad 0)$$

Substituting these stress components into appropriate formulas,

Principal stresses = (810.824 0. −97.4913)

Effective stress (von Mises) = 863.706

Similarly, solutions at selected points for all other elements can be computed. A summary of solutions at element centroids is as follows:

Solution at Selected Points on Elements

	Coordinates	Displacements	Stresses	Principal Stresses	Effective Stress
1	{6.5, 8.5}	−0.00023894 −0.0161812	(−104.571, 28.544, 0, 166.978, 0, 0)	{141.741, 0, −217.768}	313.656
2	{13., 4.25}	0.00121336 −0.0140235	(−22.0848, −19.1666, 0, −43.6555, 0, 0)	{23.0542, 0., −64.3056}	78.4169

	Coordinates	Displacements	Stresses	Principal Stresses	Effective Stress
3	$\begin{Bmatrix} 37. \\ 6. \end{Bmatrix}$	0.000237189 −0.00574551	$\begin{Bmatrix} -50.6003 \\ -43.7172 \\ 0 \\ 154.167 \\ 0 \\ 0 \end{Bmatrix}$	$\begin{Bmatrix} 107.046 \\ 0. \\ -201.364 \end{Bmatrix}$	271.222

◆ *Mathematica*/MATLAB Implementation 7.2 on the Book Web Site: *Four-node quadrilateral element for plane stress and plane strain*

Example 7.8 Notched Beam Using Transition from Eight- to Four-Node Elements Many practical problems can be analyzed efficiently by using higher order elements in the region of high stress gradients and low-order elements elsewhere. Appropriate order elements must be used in the transition region between the high- to low-order elements. To demonstrate this, consider analysis of the notched beam of Figure 7.14. To capture stress concentration in the vicinity of the notch, we employ eight-node elements. Away from the notch the stresses do not change that rapidly and thus we could use four-node elements. To maintain the compatibility of the displacement field over the entire mesh it is necessary to use five-node elements in the transition region from four- to eight-node elements. Taking advantage of symmetry, the right half of the beam is modeled as shown in Figure 7.16. The first 12 elements are the eight-node elements. The elements 16 through 21 are four-node elements. The elements 13, 14, and 15 in the transition region must have a quadratic

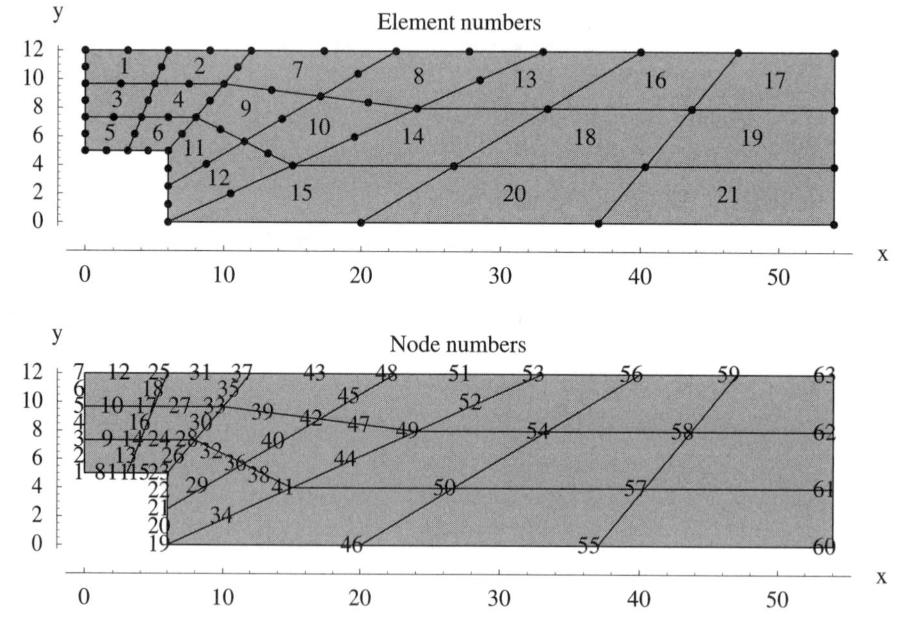

Figure 7.16. Finite element model of the notched beam

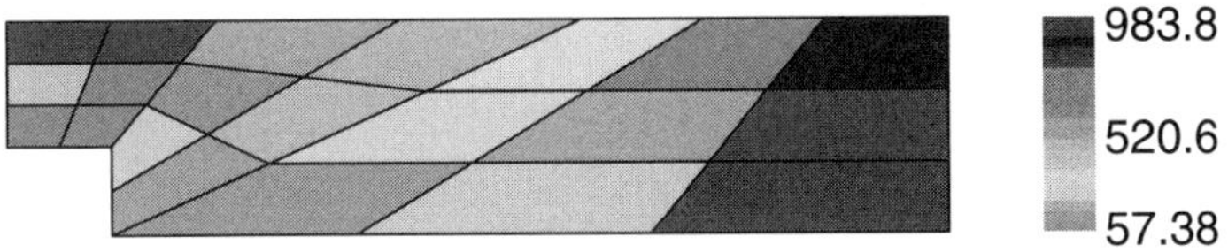

Figure 7.17. Equivalent stresses at element centers

displacement field along their left edges and linear along the right sides. Thus five-node elements are used. Due to symmetry, nodes 1 through 7 are restrained in the x direction. Both horizontal and vertical displacements are zero at nodes 60 through 63 because of the fixed support.

The effective stresses at element centers are used to create an element stress plot as shown in Figure 7.17. There are significant jumps in stresses across element boundaries, and thus the model needs to be refined further. Results obtained from refined models using ANSYS and ABAQUS are presented in Appendix A.

7.5 PLANAR FINITE ELEMENT MODELS

Several practical finite element applications that can be modeled effectively by using the plane stress or the plane strain assumptions are presented in this section.

7.5.1 Pressure Vessels

A thick cylinder subjected to internal or external pressure is a classic example of a plane strain situation. However, pressure vessels of various cross-sectional shapes can also be effectively modeled as plane strain problems. As long as the main body is long, so that the end conditions can be ignored, the model should produce reliable results. Here we consider a simple cylindrical pressure vessel as shown in Figure 7.18. The exact analytical solution of this plane strain problem is well known. The tangential and radial stresses are given by the following formulas:

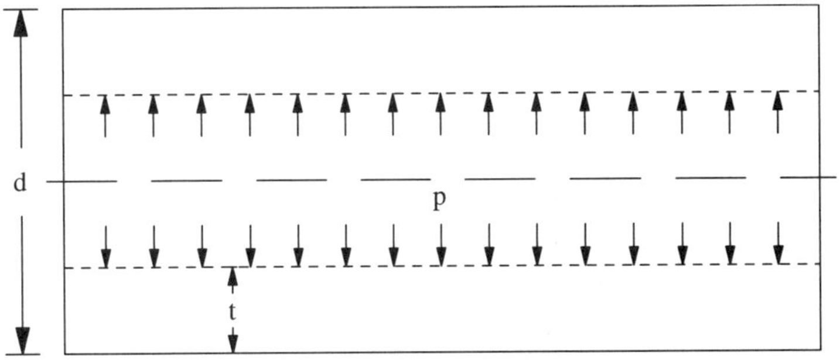

Figure 7.18. Long thick cylinder subjected to internal pressure

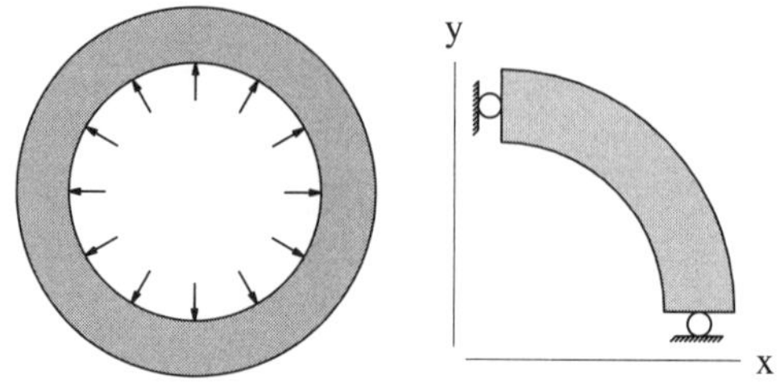

Figure 7.19. Plane strain model of a long thick cylinder subjected to internal pressure

$$\sigma_t = \frac{p_i r_i^2}{r_0^2 - r_i^2}\left(1 + \frac{r_0^2}{r^2}\right); \qquad \sigma_r = \frac{p_i r_i^2}{r_0^2 - r_i^2}\left(1 - \frac{r_0^2}{r^2}\right)$$

where p_i is the internal pressure on the cylinder, r_i and r_0 are the inner and the outer radii, and r is the radius of the point where the stress is to be computed. Note that these stresses are independent of any material properties. For the numerical results presented in this section we use the following data:

$$p_i = 20\,\text{ksi}; \qquad r_i = 5\,\text{in}; \qquad r_0 = 15\,\text{in}$$

With these values the exact solution predicts the following tangential (hoop) stresses in the cylinder:

$$\text{At } r = 5: \quad \sigma_t = 25\,\text{ksi}; \qquad \text{At } r = 15: \quad \sigma_t = 5\,\text{ksi}$$

For a finite element solution with the plane strain assumption we need to model a cross section of the cylinder. We can take advantage of symmetry and model one-fourth of the region as shown in Figure 7.19. All nodes along the horizontal plane of symmetry can move only in the horizontal direction, and those on the vertical plane of symmetry can move only in the vertical direction.

Example 7.9 One-Element Solution To show all calculations, only a single eight-node element is used as shown in Figure 7.20. Because of symmetry, the following boundary conditions are imposed:

At nodes 3, 4, 5: y displacement = 0

At nodes 7, 8, 1: x displacement = 0

Using kip-inches, the calculations follow.

$$E = 30000; \qquad \nu = 0.3; \qquad h = 1$$

Mapping to the master element,

$$x(s,t) = N^T x_n = -1.03553ts^2 - 2.07107s^2 + 2.5ts + 5.s + 3.53553t + 7.07107$$

$$y(s,t) = N^T y_n = -1.03553ts^2 - 2.07107s^2 - 2.5ts - 5.s + 3.53553t + 7.07107$$

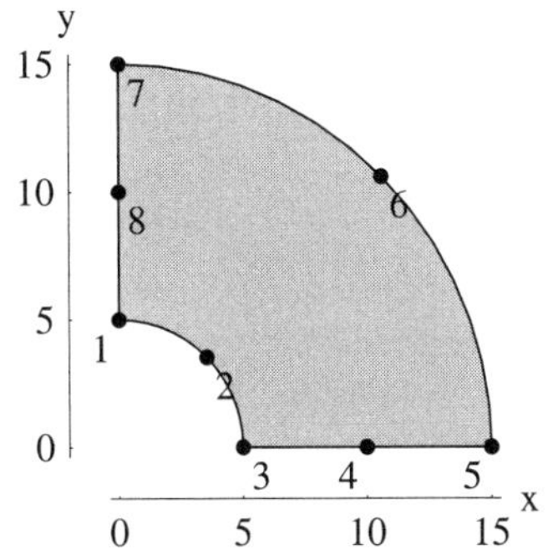

Figure 7.20. One-element plane strain model

$$J = \begin{pmatrix} -2.07107ts - 4.14214s + 2.5t + 5. & -1.03553s^2 + 2.5s + 3.53553 \\ -2.07107ts - 4.14214s - 2.5t - 5. & -1.03553s^2 - 2.5s + 3.53553 \end{pmatrix}$$

$$\det J = 5.17767ts^2 + 10.3553s^2 + 17.6777t + 35.3553$$

$$\text{Plane strain } C = \begin{pmatrix} 40384.6 & 17307.7 & 0 \\ 17307.7 & 40384.6 & 0 \\ 0 & 0 & 11538.5 \end{pmatrix}$$

For numerical integration the Gauss quadrature points and weights are as follows:

	s	t	Weight
1	−0.774597	−0.774597	0.308642
2	−0.774597	0.	0.493827
3	−0.774597	0.774597	0.308642
4	0.	−0.774597	0.493827
5	0.	0.	0.790123
6	0.	0.774597	0.493827
7	0.774597	−0.774597	0.308642
8	0.774597	0.	0.493827
9	0.774597	0.774597	0.308642

Computation of element matrices at $\{-0.774597, -0.774597\}$ with weight $= 0.308642$ is as follows:

$$J = \begin{pmatrix} 5.02935 & 0.977722 \\ -1.09766 & 4.85071 \end{pmatrix}; \qquad \det J = 25.4691$$

$$N^T = (0.432379 \quad 0.354919 \quad -0.1 \quad 0.0450807 \quad -0.032379 \quad 0.0450807 \quad -0.1 \quad 0.354919)$$

$$\frac{\partial N^T}{\partial s} = (-1.03095 \quad 1.3746 \quad -0.343649 \quad 0.2 \quad -0.130948 \quad 0.174597 \quad -0.0436492 \quad -0.2)$$

$$\frac{\partial N^T}{\partial t} = (-1.03095 \quad -0.2 \quad -0.0436492 \quad 0.174597 \\ -0.130948 \quad 0.2 \quad -0.343649 \quad 1.3746)$$

$$\frac{\partial N^T}{\partial x} = (-0.24078 \quad 0.253178 \quad -0.0673307 \quad 0.0456156 \\ -0.0305831 \quad 0.0418723 \quad -0.0231237 \quad 0.0211513)$$

$$\frac{\partial N^T}{\partial y} = (-0.164003 \quad -0.0922625 \quad 0.00457284 \quad 0.0267997 \\ -0.0208311 \quad 0.0327912 \quad -0.0661843 \quad 0.279117)$$

$$\boldsymbol{B} = \begin{pmatrix} -0.24078 & 0 & -0.164003 \\ 0 & -0.164003 & -0.24078 \\ 0.253178 & 0 & -0.0922625 \\ 0 & -0.0922625 & 0.253178 \\ -0.0673307 & 0 & 0.00457284 \\ 0 & 0.00457284 & -0.0673307 \\ 0.0456156 & 0 & 0.0267997 \\ 0 & 0.0267997 & 0.0456156 \\ -0.0305831 & 0 & -0.0208311 \\ 0 & -0.0208311 & -0.0305831 \\ 0.0418723 & 0 & 0.0327912 \\ 0 & 0.0327912 & 0.0418723 \\ -0.0231237 & 0 & -0.0661843 \\ 0 & -0.0661843 & -0.0231237 \\ 0.0211513 & 0 & 0.279117 \\ 0 & 0.279117 & 0.0211513 \end{pmatrix}$$

$$\boldsymbol{k} = \begin{pmatrix} 20844.2 & 8954.26 & -17979.8 & -743.723 & 5078.55 & 851.771 & -3885.39 & -1556.48 & 2647.56 & 1137.34 & -3688.39 & -1697.07 & 2752.03 & 2512.1 & -5768.74 & -9458.2 \\ 8954.26 & 13797.1 & -3634.26 & -725.672 & 1402.49 & 1232.37 & -1603.11 & -2391.51 & 1137.34 & 1752.46 & -1650.44 & -2621.7 & 1961.38 & 3950.82 & -6567.66 & -14993.9 \\ -17979.8 & -3634.26 & 21120.9 & -5296.74 & -5449.85 & 720.964 & 3442.01 & 541.403 & -2283.74 & -461.612 & 3091. & 779.112 & -1304.67 & -2086.26 & -635.765 & 9437.39 \\ -743.723 & -725.672 & -5296.74 & 8516.24 & 950.184 & -1680.1 & 42.8277 & 262.562 & -94.4652 & -92.1724 & 227.405 & 1.11193 & -1229.58 & 1407.49 & 6144.09 & -7689.46 \\ 5078.55 & 1402.49 & -5449.85 & 950.184 & 1441.06 & -69.8162 & -963.9 & -226.58 & 645.061 & 178.14 & -881.402 & -283.018 & 466.809 & 596.693 & -336.332 & -2548.09 \\ 851.771 & 1232.37 & 720.964 & -1680.1 & -69.8162 & 417.829 & -135.287 & -239.671 & 108.189 & 156.532 & -174.206 & -208.113 & 389.803 & 45.1387 & -1691.42 & 276.018 \\ -3885.39 & -1603.11 & 3442.01 & 42.8277 & -963.9 & -135.287 & 725.704 & 277.204 & -493.51 & -203.622 & 686.061 & 305.289 & -495.734 & -466.958 & 984.766 & 1783.66 \\ -1556.48 & -2391.51 & 541.403 & 262.562 & -226.58 & -239.671 & 277.204 & 416.736 & -197.699 & -303.761 & 288.345 & 452.223 & -358.146 & -658.752 & 1231.95 & 2462.17 \\ 2647.56 & 1137.34 & -2283.74 & -94.4652 & 645.061 & 108.189 & -493.51 & -197.699 & 336.284 & 144.461 & -468.487 & -215.556 & 349.554 & 319.078 & -732.726 & -1201.35 \\ 1137.34 & 1752.46 & -461.612 & -92.1724 & 178.14 & 156.532 & -203.622 & -303.761 & 144.461 & 222.592 & -209.633 & -333. & 249.128 & 501.82 & -834.203 & -1904.47 \\ -3688.39 & -1650.44 & 3091. & 227.405 & -881.402 & -174.206 & 686.061 & 288.345 & -468.487 & -209.633 & 654.122 & 311.344 & -504.223 & -445.817 & 1111.32 & 1653. \\ -1697.07 & -2621.7 & 779.112 & 1.11193 & -283.018 & -208.113 & 305.289 & 452.223 & -215.556 & -333. & 311.344 & 500.377 & -354.524 & -776.787 & 1154.42 & 2985.89 \\ 2752.03 & 1961.38 & -1304.67 & -1229.58 & 466.809 & 389.803 & -495.734 & -358.146 & 349.554 & 249.128 & -504.223 & -354.524 & 567.054 & 347.032 & -1830.82 & -1005.09 \\ 2512.1 & 3950.82 & -2086.26 & 1407.49 & 596.693 & 45.1387 & -466.958 & -658.752 & 319.078 & 501.82 & -445.817 & -776.787 & 347.032 & 1439.08 & -775.87 & -5908.81 \\ -5768.74 & -6567.66 & -635.765 & 6144.09 & -336.332 & -1691.42 & 984.766 & 1231.95 & -732.726 & -834.203 & 1111.32 & 1154.42 & -1830.82 & -775.87 & 7208.3 & 1338.69 \\ -9458.2 & -14993.9 & 9437.39 & -7689.46 & -2548.09 & 276.018 & 1783.66 & 2462.17 & -1201.35 & -1904.47 & 1653. & 2985.89 & -1005.09 & -5908.81 & 1338.69 & 24772.6 \end{pmatrix}$$

We must perform similar computations at the remaining eight Gauss points. Summing contributions from all points, we get

$$\boldsymbol{k} = \begin{pmatrix} 36733.5 & 12876.3 & -27675.8 & -81.2534 & 13375.3 & 6621.25 & -25336.9 & -10596.2 & 18861.9 & 3094.06 & -6756.94 & 862.763 & 12055.6 & 4640.68 & -21256.5 & -17417.6 \\ 12876.3 & 29235.1 & -3927.41 & -13841. & 7582.79 & 13375.3 & -10596.2 & -13639.6 & 3094.06 & 9989.62 & 862.763 & -4928.03 & 3679.14 & 19081.9 & -13571.5 & -39273.3 \\ -27675.8 & -3927.41 & 58331.4 & -7381.3 & -13841. & -81.2534 & 7797.13 & 11355.9 & -19175.3 & 874.112 & -5843.92 & -13070.1 & -9681.35 & 874.112 & 10088.9 & 11355.9 \\ -81.2534 & -13841. & -7381.3 & 58331.4 & -3927.41 & -27675.8 & 11355.9 & 10088.9 & 874.112 & -9681.35 & -13070.1 & -5843.92 & 874.112 & -19175.3 & 11355.9 & 7797.13 \\ 13375.3 & 7582.79 & -13841. & -3927.41 & 29235.1 & 12876.3 & -39273.3 & -13571.5 & 19081.9 & 3679.14 & -4928.03 & 862.763 & 9989.62 & 3094.06 & -13639.6 & -10596.2 \\ 6621.25 & 13375.3 & -81.2534 & -27675.8 & 12876.3 & 36733.5 & -17417.6 & -21256.5 & 4640.68 & 12055.6 & 862.763 & -6756.94 & 3094.06 & 18861.9 & -10596.2 & -25336.9 \\ -25336.9 & -10596.2 & 7797.13 & 11355.9 & -39273.3 & -17417.6 & 108701. & 20689.5 & -62430.1 & -7762.13 & 1719.55 & -11355.9 & -16853.6 & -3516.93 & 25676. & 18603.3 \\ -10596.2 & -13639.6 & 11355.9 & 10088.9 & -13571.5 & -21256.5 & 20689.5 & 57600.7 & -11608.3 & -19829.7 & -11355.9 & -19605.6 & -3516.93 & -19034.2 & 18603.3 & 25676. \\ 18861.9 & 3094.06 & -19175.3 & 874.112 & 19081.9 & 4640.68 & -62430.1 & -11608.3 & 53178.9 & -3329.31 & -1545.83 & 11827.3 & 11062.8 & -1981.59 & -19034.2 & -3516.93 \\ 3094.06 & 9989.62 & 874.112 & -9681.35 & 3679.14 & 12055.6 & -7762.13 & -19829.7 & -3329.31 & 19206.8 & 7981.11 & -5950.17 & -1020.06 & 11062.8 & -3516.93 & -16853.6 \\ -6756.94 & 862.763 & -5843.92 & -13070.1 & -4928.03 & 862.763 & 1719.55 & -11355.9 & -1545.83 & 7981.11 & 42911. & 14248. & -5950.17 & 11827.3 & -19605.6 & -11355.9 \\ 862.763 & -4928.03 & -13070.1 & -5843.92 & 862.763 & -6756.94 & -11355.9 & -19605.6 & 11827.3 & -5950.17 & 14248. & 42911. & 7981.11 & -1545.83 & -11355.9 & 1719.55 \\ 12055.6 & 3679.14 & -9681.35 & 874.112 & 9989.62 & 3094.06 & -16853.6 & -3516.93 & 11062.8 & -1020.06 & -5950.17 & 7981.11 & 19206.8 & -3329.31 & -19829.7 & -7762.13 \\ 4640.68 & 19081.9 & 874.112 & -19175.3 & 3094.06 & 18861.9 & -3516.93 & -19034.2 & -1981.59 & 11062.8 & 11827.3 & -1545.83 & -3329.31 & 53178.9 & -11608.3 & -62430.1 \\ -21256.5 & -13571.5 & 10088.9 & 11355.9 & -13639.6 & -10596.2 & 25676. & 18603.3 & -19034.2 & -3516.93 & -19605.6 & -11355.9 & -19829.7 & -11608.3 & 57600.7 & 20689.5 \\ -17417.6 & -39273.3 & 11355.9 & 7797.13 & -10596.2 & -25336.9 & 18603.3 & 25676. & -3516.93 & -16853.6 & -11355.9 & 1719.55 & -7762.13 & -62430.1 & 20689.5 & 108701. \end{pmatrix}$$

Computation of element matrices resulting from the NBC is as follows:

NBC on side 1 with $q = \{-20, 0\}$

$$N_c^T = \begin{pmatrix} \frac{1-a}{2} + \frac{1}{2}(a^2-1) & 1-a^2 & \frac{a+1}{2} + \frac{1}{2}(a^2-1) & 0 & 0 & 0 & 0 & 0 \end{pmatrix}$$

$$x(a) = -1.03553a^2 + 2.5a + 3.53553; y(a) = -1.03553a^2 - 2.5a + 3.53553$$

$$\frac{dx}{da} = 2.5 - 2.07107a; \frac{dy}{da} = -2.07107a - 2.5$$

$$J_c = \sqrt{(-2.07107a - 2.5)^2 + (2.07107a - 2.5)^2}$$

Gauss point $= -0.774597$; Weight $= 0.555556$; $J_c = 4.20086$

$$N_c^T = (0.687298 \quad 0.4 \quad -0.0872983 \quad 0 \quad 0 \quad 0 \quad 0 \quad 0)$$

$$r_q^T = (6.84059 \quad 31.3427 \quad 3.98115 \quad 18.2411 \quad -0.868868 \quad -3.98104 \quad 0 \quad 0 \quad 0 \quad 0 \quad 0 \quad 0 \quad 0 \quad 0 \quad 0 \quad 0)$$

Gauss point $= 0.$; Weight $= 0.888889$; $J_c = 3.53553$

$$N_c^T = (0. \quad 1. \quad 0. \quad 0 \quad 0 \quad 0 \quad 0 \quad 0)$$

$$r_q^T = (0 \quad 0 \quad 44.4444 \quad 44.4444 \quad 0 \quad 0 \quad 0 \quad 0 \quad 0 \quad 0 \quad 0 \quad 0 \quad 0 \quad 0 \quad 0 \quad 0)$$

Gauss point $= 0.774597$; Weight $= 0.555556$; $J_c = 4.20086$

$$N_c^T = (-0.0872983 \quad 0.4 \quad 0.687298 \quad 0 \quad 0 \quad 0 \quad 0 \quad 0)$$

$$r_q^T = (-3.98104 \quad -0.868868 \quad 18.2411 \quad 3.98115 \quad 31.3427 \quad 6.84059 \quad 0 \quad 0 \quad 0 \quad 0 \quad 0 \quad 0 \quad 0 \quad 0 \quad 0 \quad 0)$$

Summing contributions from all Gauss points,

$$r_q^T = (2.85955 \quad 30.4738 \quad 66.6667 \quad 66.6667 \quad 30.4738 \quad 2.85955 \quad 0 \quad 0 \quad 0 \quad 0 \quad 0 \quad 0 \quad 0 \quad 0 \quad 0 \quad 0)$$

The complete element equations for element 1 are

$$\begin{pmatrix}
36733.5 & 12876.3 & -27675.8 & -81.2534 & 13375.3 & 6621.25 & -25336.9 & -10596.2 & 18861.9 & 3094.06 & -6756.94 & 862.763 & 12055.6 & 4640.68 & -21256.5 & -17417.6 \\
12876.3 & 29235.1 & -3927.41 & -13841. & 7582.79 & 13375.3 & -10596.2 & -13639.6 & 3094.06 & 9989.62 & 862.763 & -4928.03 & 3679.14 & 19081.9 & -13571.5 & -39273.3 \\
-27675.8 & -3927.41 & 58331.4 & -7381.3 & -13841. & -81.2534 & 7797.13 & 11355.9 & -19175.3 & 874.112 & -5843.92 & -13070.1 & -9681.35 & 874.112 & 10088.9 & 11355.9 \\
-81.2534 & -13841. & -7381.3 & 58331.4 & -3927.41 & -27675.8 & 11355.9 & 10088.9 & 874.112 & -9681.35 & -13070.1 & -5843.92 & 874.112 & -19175.3 & 11355.9 & 7797.13 \\
13375.3 & 7582.79 & -13841. & -3927.41 & 29235.1 & 12876.3 & -39273.3 & -13571.5 & 19081.9 & 3679.14 & -4928.03 & 862.763 & 9989.62 & 3094.06 & -13639.6 & -10596.2 \\
6621.25 & 13375.3 & -81.2534 & -27675.8 & 12876.3 & 36733.5 & -17417.6 & -21256.5 & 4640.68 & 12055.6 & 862.763 & -6756.94 & 3094.06 & 18861.9 & -10596.2 & -25336.9 \\
-25336.9 & -10596.2 & 7797.13 & 11355.9 & -39273.3 & -17417.6 & 108701. & 20689.5 & -62430.1 & -7762.13 & 1719.55 & -11355.9 & -16853.6 & -3516.93 & 25676. & 18603.3 \\
-10596.2 & -13639.6 & 11355.9 & 10088.9 & -13571.5 & -21256.5 & 20689.5 & 57600.7 & -11608.3 & -19829.7 & -11355.9 & -19605.6 & -3516.93 & -19034.2 & 18603.3 & 25676. \\
18861.9 & 3094.06 & -19175.3 & 874.112 & 19081.9 & 4640.68 & -62430.1 & -11608.3 & 53178.9 & -3329.31 & -1545.83 & 11827.3 & 11062.8 & -1981.59 & -19034.2 & -3516.93 \\
3094.06 & 9989.62 & 874.112 & -9681.35 & 3679.14 & 12055.6 & -7762.13 & -19829.7 & -3329.31 & 19206.8 & 7981.11 & -5950.17 & -1020.06 & 11062.8 & -3516.93 & -16853.6 \\
-6756.94 & 862.763 & -5843.92 & -13070.1 & -4928.03 & 862.763 & 1719.55 & -11355.9 & -1545.83 & 7981.11 & 42911. & 14248. & -5950.17 & 11827.3 & -19605.6 & -11355.9 \\
862.763 & -4928.03 & -13070.1 & -5843.92 & 862.763 & -6756.94 & -11355.9 & -19605.6 & 11827.3 & -5950.17 & 14248. & 42911. & 7981.11 & -1545.83 & -11355.9 & 1719.55 \\
12055.6 & 3679.14 & -9681.35 & 874.112 & 9989.62 & 3094.06 & -16853.6 & -3516.93 & 11062.8 & -1020.06 & -5950.17 & 7981.11 & 19206.8 & -3329.31 & -19829.7 & -7762.13 \\
4640.68 & 19081.9 & 874.112 & -19175.3 & 3094.06 & 18861.9 & -3516.93 & -19034.2 & -1981.59 & 11062.8 & 11827.3 & -1545.83 & -3329.31 & 53178.9 & -11608.3 & -62430.1 \\
-21256.5 & -13571.5 & 10088.9 & 11355.9 & -13639.6 & -10596.2 & 25676. & 18603.3 & -19034.2 & -3516.93 & -19605.6 & -11355.9 & -19829.7 & -11608.3 & 57600.7 & 20689.5 \\
-17417.6 & -39273.3 & 11355.9 & 7797.13 & -10596.2 & -25336.9 & 18603.3 & 25676. & -3516.93 & -16853.6 & -11355.9 & 1719.55 & -7762.13 & -62430.1 & 20689.5 & 108701.
\end{pmatrix}
\begin{pmatrix} u_1 \\ v_1 \\ u_2 \\ v_2 \\ u_3 \\ v_3 \\ u_4 \\ v_4 \\ u_5 \\ v_5 \\ u_6 \\ v_6 \\ u_7 \\ v_7 \\ u_8 \\ v_8 \end{pmatrix}
=
\begin{pmatrix} 2.85955 \\ 30.4738 \\ 66.6667 \\ 66.6667 \\ 30.4738 \\ 2.85955 \\ 0. \\ 0. \\ 0. \\ 0. \\ 0. \\ 0. \\ 0. \\ 0. \\ 0. \\ 0. \end{pmatrix}$$

Since there is only one element, the global equations are the same as the element equations:

Essential boundary conditions:

Node	dof	Value
1	u_1	0
3	v_3	0
4	v_4	0
5	v_5	0
7	u_7	0
8	u_8	0

Remove {1, 6, 8, 10, 13, 15} rows and columns.

After adjusting for essential boundary conditions, we have

$$\begin{pmatrix} 29235.1 & -3927.41 & -13841. & 7582.79 & -10596.2 & 3094.06 & 862.763 & -4928.03 & 19081.9 & -39273.3 \\ -3927.41 & 58331.4 & -7381.3 & -13841. & 7797.13 & -19175.3 & -5843.92 & -13070.1 & 874.112 & 11355.9 \\ -13841. & -7381.3 & 58331.4 & -3927.41 & 11355.9 & 874.112 & -13070.1 & -5843.92 & -19175.3 & 7797.13 \\ 7582.79 & -13841. & -3927.41 & 29235.1 & -39273.3 & 19081.9 & -4928.03 & 862.763 & 3094.06 & -10596.2 \\ -10596.2 & 7797.13 & 11355.9 & -39273.3 & 108701. & -62430.1 & 1719.55 & -11355.9 & -3516.93 & 18603.3 \\ 3094.06 & -19175.3 & 874.112 & 19081.9 & -62430.1 & 53178.9 & -1545.83 & 11827.3 & -1981.59 & -3516.93 \\ 862.763 & -5843.92 & -13070.1 & -4928.03 & 1719.55 & -1545.83 & 42911. & 14248. & 11827.3 & -11355.9 \\ -4928.03 & -13070.1 & -5843.92 & 862.763 & -11355.9 & 11827.3 & 14248. & 42911. & -1545.83 & 1719.55 \\ 19081.9 & 874.112 & -19175.3 & 3094.06 & -3516.93 & -1981.59 & 11827.3 & -1545.83 & 53178.9 & -62430.1 \\ -39273.3 & 11355.9 & 7797.13 & -10596.2 & 18603.3 & -3516.93 & -11355.9 & 1719.55 & -62430.1 & 108701. \end{pmatrix} \begin{pmatrix} v_1 \\ u_2 \\ v_2 \\ u_3 \\ u_4 \\ u_5 \\ u_6 \\ v_6 \\ v_7 \\ v_8 \end{pmatrix} = \begin{pmatrix} 30.4738 \\ 66.6667 \\ 66.6667 \\ 30.4738 \\ 0 \\ 0 \\ 0 \\ 0 \\ 0 \\ 0 \end{pmatrix}$$

Solving the final system of global equations, we get

$$\{v_1 = 0.00494694, u_2 = 0.0033909, v_2 = 0.0033909, u_3 = 0.00494694, u_4 = 0.00271213,$$
$$u_5 = 0.00225656, u_6 = 0.00152523, v_6 = 0.00152523, v_7 = 0.00225656, v_8 = 0.00271213\}$$

The element solution is as follows:

$$\boldsymbol{d}^T = (0 \quad 0.00494694 \quad 0.0033909 \quad 0.0033909 \quad 0.00494694 \quad 0 \quad 0.00271213 \quad 0$$
$$0.00225656 \quad 0 \quad 0.00152523 \quad 0.00152523 \quad 0 \quad 0.00225656 \quad 0 \quad 0.00271213)$$

Element solution at $\{s, t\} = \{0, 0\} \Longrightarrow \{x, y\} = \{7.07107, 7.07107\}$

$$\boldsymbol{N}^T = \left\{-\tfrac{1}{4}, \tfrac{1}{2}, -\tfrac{1}{4}, \tfrac{1}{2}, -\tfrac{1}{4}, \tfrac{1}{2}, -\tfrac{1}{4}, \tfrac{1}{2}\right\}$$

Displacements $= \boldsymbol{N}^T\boldsymbol{d} = \{0.00201325, 0.00201325\}$

$$\boldsymbol{J} = \begin{pmatrix} 5. & 3.53553 \\ -5. & 3.53553 \end{pmatrix}; \qquad \det \boldsymbol{J} = 35.3553$$

$$\frac{\partial \boldsymbol{N}^T}{\partial s} = \begin{pmatrix} 0 & 0 & 0 & \frac{1}{2} & 0 & 0 & 0 & -\frac{1}{2} \end{pmatrix}$$

$$\frac{\partial \boldsymbol{N}^T}{\partial t} = \begin{pmatrix} 0 & -\frac{1}{2} & 0 & 0 & 0 & \frac{1}{2} & 0 & 0 \end{pmatrix}$$

$$\frac{\partial \boldsymbol{N}^T}{\partial x} = (0 \quad -0.0707107 \quad 0 \quad 0.05 \quad 0 \quad 0.0707107 \quad 0 \quad -0.05)$$

$$\frac{\partial \boldsymbol{N}^T}{\partial y} = (0 \quad -0.0707107 \quad 0 \quad -0.05 \quad 0 \quad 0.0707107 \quad 0 \quad 0.05)$$

$$B^T = \begin{pmatrix} 0 & 0 & -0.0707107 & 0 & 0 & 0 & 0.05 & 0 & 0 & 0 & 0.0707107 & 0 & 0 & 0 & -0.05 & 0 \\ 0 & 0 & 0 & -0.0707107 & 0 & 0 & 0 & -0.05 & 0 & 0 & 0 & 0.0707107 & 0 & 0 & 0 & 0.05 \\ 0 & 0 & -0.0707107 & -0.0707107 & 0 & 0 & -0.05 & 0.05 & 0 & 0 & 0.0707107 & 0.0707107 & 0 & 0 & 0.05 & -0.05 \end{pmatrix}$$

In-plane strain components, $\epsilon = B^T d = (3.6836 \times 10^{-6} \quad 3.6836 \times 10^{-6} \quad -0.000535058)$

In-plane stress components, $\sigma = C\epsilon = (0.212515 \quad 0.212515 \quad -6.17375)$

Computing out-of-plane strain and stress components using appropriate formulas, the complete strain and stress vectors are as follows:

$\epsilon^T = (3.6836 \times 10^{-6} \quad 3.6836 \times 10^{-6} \quad 0 \quad -0.000535058 \quad 0 \quad 0)$

$\sigma^T = (0.212515 \quad 0.212515 \quad 0.127509 \quad -6.17375 \quad 0 \quad 0)$

Substituting these stress components into appropriate formulas,

Principal stresses = $(6.38626 \quad 0.127509 \quad -5.96123)$

Effective stress (von Mises) = 10.6936

The exact solution corresponding to the element centroid is

$$\text{At } r = 5: \qquad \sigma_t = 8.125\,\text{ksi}; \qquad \sigma_r = 15\,\text{ksi}$$

We clearly need a finer mesh to get a better solution.

Example 7.10 ANSYS Solution The problem is solved using an eight-node quadrilateral element (Plane82) in ANSYS (AnsysFiles\Chap7\cylinderFine.inp on the book web site) with a 10 × 10 mapped mesh. In the resulting mesh the element size gradually increases with the radius, as shown in Figure 7.21. From the element solution the first principal stress plot is shown in the figure. The stress ranges from around 5 ksi on the outside to about 25 ksi on the inside. These values are almost identical to the exact solution.

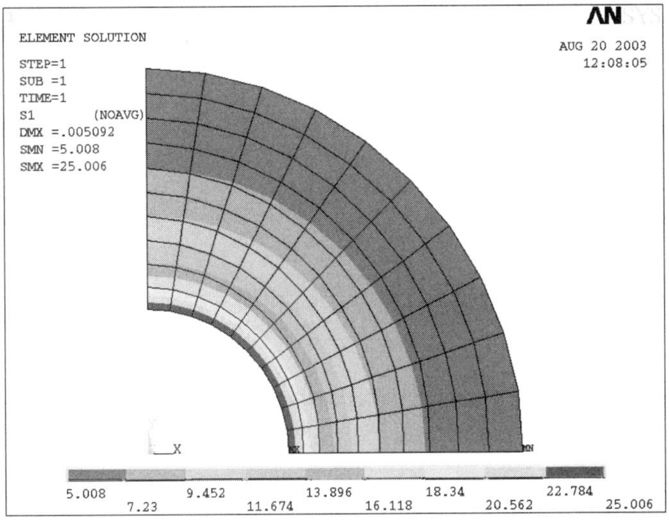

Figure 7.21. Plot of first principal stress

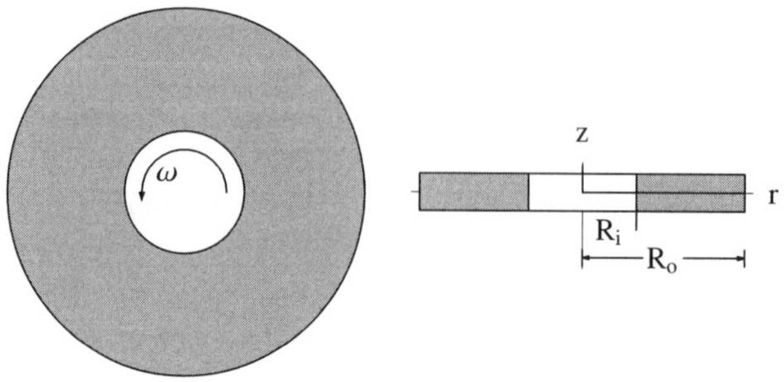

Figure 7.22. Rotating disk

7.5.2 Rotating Disks and Flywheels

Disks and flywheels are common components of rotating machines. Stresses are generated in these machine parts due to inertia forces that depend on the angular velocity and mass. Consider a disk of uniform thickness, shown in Figure 7.22, that is rotating at a constant angular velocity of ω rad/s. The inertia force is given by $mr\omega^2$, where m is the mass density per unit volume and r is the radial distance from the center of the disk. Treating this centrifugal force as the body force in the radial direction, a plane stress model of the disk can be used to determine stresses in the disk. The body force components in the x and y directions are as follows:

$$b_x = mx\omega^2 \qquad \text{and} \qquad b_y = my\omega^2$$

These forces are not constant over an element. For a simplified analysis we can assign constant values for each element by using the values of these forces at element centroids. For a more accurate analysis we can use these expressions directly and carry out integration to get the equivalent body force. This latter procedure is used in the numerical example presented in this section.

For thin disks, the following analytical solution is available:

$$u(r) = \frac{r}{E}\frac{\nu+3}{8}m\omega^2 R_o^2\left((\nu+1)\left(\frac{R_i}{r}\right)^2 - \frac{1-\nu^2}{\nu+3}\left(\frac{r}{R_o}\right)^2 + \left(\left(\frac{R_i}{R_o}\right)^2 + 1\right)(1-\nu)\right)$$

$$\sigma_r(r) = \frac{\nu+3}{8}m\omega^2 R_o^2\left(-\left(\frac{R_i}{r}\right)^2 - \left(\frac{r}{R_i}\right)^2 + \left(\frac{R_i}{R_o}\right)^2 + 1\right)$$

$$\sigma_t(r) = \frac{\nu+3}{8}m\omega^2 R_o^2\left(\left(\frac{R_i}{r}\right)^2 + \left(\frac{R_i}{R_o}\right)^2 - \frac{3\nu+1}{\nu+3}\left(\frac{r}{R_o}\right)^2 + 1\right)$$

where u is the radial displacement, σ_r is the radial stress, σ_t is the tangential stress, R_i is the inner radius (radius of the shaft), R_o is the outer radius, E is the modulus of elasticity, and ν is Poisson's ratio.

For disks with variable thickness and spoked flywheels, analytical solutions are not available. However, plane stress finite element models can be used effectively to determine stresses in these situations.

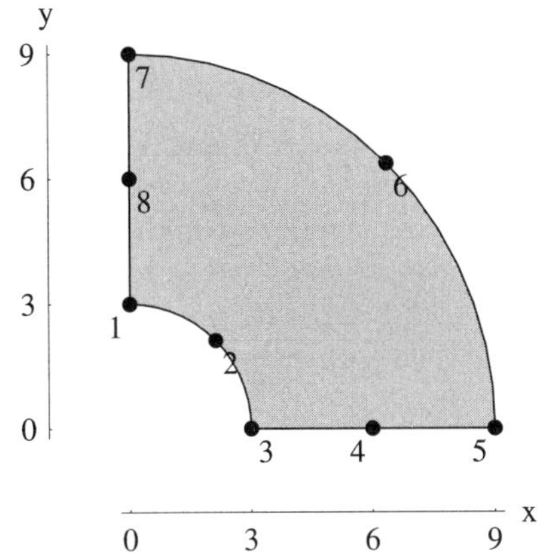

Figure 7.23. One-element plane stress model of rotating disk

Example 7.11 Analysis of a Rotating Disk As a numerical example, consider a 1-in-thick disk with inner radius 3 in and outer radius 9 in. The disk is rotating at 5000 rpm (revolutions per minute). The material properties are density = 0.283 lb/in^3, $E = 30 \times 10^6$ psi, and $\nu = 0.3$. The centrifugal force is

$$mr\omega^2 = \left(\frac{0.283\,\text{lb/in}^3}{386.4\,\text{in/s}^2}\right) r(5000 \times 2\pi/60\,\text{s})^2 = 200.792r\,\text{lb/in}^3$$

To show all calculations, only a single eight-node element is used, as shown in Figure 7.23. Because of symmetry, the following boundary conditions are imposed:

At nodes 3, 4, 5: y displacement = 0

At nodes 7, 8, 1: x displacement = 0

The following *Mathematica* functions are created to perform computations for the eight-node quadrilateral element. The inertia force components in the x and y directions are $b_x = mx\omega^2$ and $b_y = my\omega^2$. Since these terms are not constant, for establishing a body force vector, they are kept with the interpolation functions and are integrated by evaluating them at the Gauss points.

```
Needs["NumericalMath`GaussianQuadrature`"];
PlaneQuad8Element[type_, e_, nu_, h_, {α_, Δt_}, {m_, Ω_}, nodes_] :=
  Module[{c, e0, sloc, swts, tloc, twts, gpLocs, gpWts, n, s, t, dns, dnt, xst, yst,
      J, npt, Jpt, dnsPt, dntPt, detJ, k, r, dnx, dny, BT, B, Npt, xn, yn, bx, by,
      bxpt, bypt},
    {xn, yn} = Transpose[nodes];
    Switch[type,
      PlaneStress,
      e0 = {α Δt, α Δt, 0};
```

```
            c = e/(1 − nu^2){{1, nu, 0}, {nu, 1, 0}, {0, 0, (1 − nu)/2}},
            PlaneStrain,
            e0 = (1 + nu){α Δt, α Δt, 0};
            c = e/((1 + nu)(1 − 2nu)){{1 − nu, nu, 0}, {nu, 1 − nu, 0}, {0, 0, (1 − 2nu)/2}}
        ];
        {sloc, swts} = Transpose[GaussianQuadratureWeights[3, −1, 1]];
        {tloc, twts} = Transpose[GaussianQuadratureWeights[3, −1, 1]];
        gpLocs = Flatten[Outer[List, sloc, tloc], 1];
        gpWts = Flatten[Outer[Times, swts, twts], 1];
        n = {((1 − s) * (1 − t))/4 − ((1 − s^2) * (1 − t))/4 − ((1 − s) * (1 − t^2))/4,
              ((1 − s^2) * (1 − t))/2,
              ((1 + s) * (1 − t))/4 − ((1 − s^2) * (1 − t))/4 − ((1 + s) * (1 − t^2))/4,
              ((1 + s) * (1 − t^2))/2,
              ((1 + s) * (1 + t))/4 − ((1 − s^2) * (1 + t))/4 − ((1 + s) * (1 − t^2))/4,
              ((1 − s^2) * (1 + t))/2,
              ((1 − s) * (1 + t))/4 − ((1 − s^2) * (1 + t))/4 − ((1 − s) * (1 − t^2))/4,
              ((1 − s) * (1 − t^2))/2};
        {dns, dnt} = {D[n, s], D[n, t]};
        xst = n.xn; yst = n.yn;
        bx = m * xst * Ω^2; by = m * yst * Ω^2;
        J = {{D[xst, s], D[xst, t]}, {D[yst, s], D[yst, t]}};
        k = Table[0, {16}, {16}];
        r = Table[0, {16}];
        Do[pt = {s → gpLocs[[i, 1]], t → gpLocs[[i, 2]]};
            w = gpWts[[i]];
            {npt, Jpt, dnsPt, dntPt, bxpt, bypt} = {n, J, dns, dnt, bx, by}/.pt;
            Npt = Transpose[{Flatten[Table[{npt[[i]], 0}, {i, 1, 8}]],
                  Flatten[Table[{0, npt[[i]]}, {i, 1, 8}]]}];
            detJ = det[Jpt];
            dnx = (Jpt[[2, 2]] dnsPt − Jpt[[2, 1]] dntPt)/detJ;
            dny = (−Jpt[[1, 2]] dnsPt + Jpt[[1, 1]] dntPt)/detJ;
            BT = {Flatten[Table[{dnx[[i]], 0}, {i, 1, 8}]],
                  Flatten[Table[{0, dny[[i]]}, {i, 1, 8}]],
                  Flatten[Table[{dny[[i]], dnx[[i]]}, {i, 1, 8}]]};
        B = Transpose[BT];
        k+ = h detJ w B.c.BT;
        r+ = h detJ w Npt.{bxpt, bypt} + h detJ w B.c.e0,
        {i, 1, Length[gpWts]}
    ]; {k, r}
]

PlaneQuad8LoadTerm[side_, qn_, qt_, h_, nodes_] :=
    Module[{r, pts, wts, n, s, t, a, xa, ya, Jc, Jcpt, nx, ny, qx, qy, Npt, xn, yn,
                qxpt, qypt},
        {xn, yn} = Transpose[nodes];
```

```
        r = Table[0, {16}];
        {pts, wts} = Transpose[GaussianQuadratureWeights[3, −1, 1]];
        n = {((1 − s) ∗ (1 − t))/4 − ((1 − s^2) ∗ (1 − t))/4 − ((1 − s) ∗ (1 − t^2))/4,
            ((1 − s^2) ∗ (1 − t))/2,
            ((1 + s) ∗ (1 − t))/4 − ((1 − s^2) ∗ (1 − t))/4 − ((1 + s) ∗ (1 − t^2))/4,
            ((1 + s) ∗ (1 − t^2))/2,
            ((1 + s) ∗ (1 + t))/4 − ((1 − s^2) ∗ (1 + t))/4 − ((1 + s) ∗ (1 − t^2))/4,
            ((1 − s^2) ∗ (1 + t))/2,
            ((1 − s) ∗ (1 + t))/4 − ((1 − s^2) ∗ (1 + t))/4 − ((1 − s) ∗ (1 − t^2))/4,
            ((1 − s) ∗ (1 − t^2))/2};
        n = Switch[side, 1, n/.{s → a, t → −1},
                2, n/.{s → 1, t → a},
                3, n/.{s → −a, t → 1},
                4, n/.{s → −1, t → −a}
          ];
        xa = n.xn;
        ya = n.yn;
        Jc = Sqrt[D[xa, a]^2 + D[ya, a]^2];
        nx = D[ya, a]/Jc; ny = −D[xa, a]/Jc;
        qx = nx qn − ny qt; qy = nyqn + nxqt;
          Do[{npt, Jcpt, qxpt, qypt} = {n, Jc, qx, qy}/.a → pts[[i]];
          Npt = Transpose[{Flatten[Table[{npt[[i]], 0}, {i, 1, 8}]],
                  Flatten[Table[{0, npt[[i]]}, {i, 1, 8}]]}];
          r+ = h ∗ Jcpt ∗ wts[[i]]Npt.{qxpt, qypt},
          {i, 1, Length[pts]}
        ];
        r
];

ClearAll[PlaneQuad8Results];
PlaneQuad8Results[type_, e_, nu_, {α_, Δt_}, nodes_, d_] :=
  Module[{pt, n, dns, dnt, bx, by, xst, yst, J, npt, Jpt, dnsPt, dntPt, detJ, loc,
          c, s, t, xn, yn},
    {xn, yn} = Transpose[nodes];
    pt = {s → 0, t → 0};
    Switch[type,
      PlaneStress,
      e0 = {α Δt, α Δt, 0};
      c = e/(1 − nu^2){{1, nu, 0}, {nu, 1, 0}, {0, 0, (1 − nu)/2}},
      PlaneStrain,
      e0 = (1 + nu){α Δt, α Δt, 0};
      c = e/((1 + nu)(1 − 2nu)){{1 − nu, nu, 0}, {nu, 1 − nu, 0}, {0, 0, (1 − 2nu)/2}}
    ];
    n = {((1 − s) ∗ (1 − t))/4 − ((1 − s^2) ∗ (1 − t))/4 − ((1 − s) ∗ (1 − t^2))/4,
        ((1 − s^2) ∗ (1 − t))/2,
```

```
        ((1 + s) * (1 - t))/4 - ((1 - s^2) * (1 - t))/4 - ((1 + s) * (1 - t^2))/4,
        ((1 + s) * (1 - t^2))/2,
        ((1 + s) * (1 + t))/4 - ((1 - s^2) * (1 + t))/4 - ((1 + s) * (1 - t^2))/4,
        ((1 - s^2) * (1 + t))/2,
        ((1 - s) * (1 + t))/4 - ((1 - s^2) * (1 + t))/4 - ((1 - s) * (1 - t^2))/4,
        ((1 - s) * (1 - t^2))/2};
    {dns, dnt} = {D[n, s], D[n, t]};
    xst = n.xn;
    yst = n.yn;
    J = {{D[xst, s], D[xst, t]}, {D[yst, s], D[yst, t]}};
    {loc, npt, Jpt, dnsPt, dntPt} = {{xst, yst}, n, J, dns, dnt}/.pt;
    detJ = Det[Jpt];
    dnx = (Jpt[[2, 2]] dnsPt - Jpt[[2, 1]] dntPt)/detJ;
    dny = (-Jpt[[1, 2]] dnsPt + Jpt[[1, 1]] dntPt)/detJ;
    Npt = Transpose[{Flatten[Table[{npt[[i]], 0}, {i, 1, 8}]],
            Flatten[Table[{0, npt[[i]]}, {i, 1, 8}]]}];
    BT = {Flatten[Table[{dnx[[i]], 0}, {i, 1, 8}]],
        Flatten[Table[{0, dny[[i]]}, {i, 1, 8}]],
        Flatten[Table[{dny[[i]], dnx[[i]]}, {i, 1, 8}]]};
    eps = BT.d;
    {sx, sy, sxy} = c.(eps - e0);
    {loc, eps, {sx, sy, sxy},
      Eigenvalues[{{sx, sxy}, {sxy, sy}}],
      Sqrt[(sx - sy)^2 + sy^2 + sx^2 + 6sxy^2]/Sqrt[2]}
  ]
```

Using these functions the numerical solution (in k-inch units) is obtained as follows:

```
e = 30000; nu = 0.3; h = 1;
m = 0.283/386.4/1000; Ω = 5000 * 2 * π/60;
nodes = {{0, 3}, {3/Sqrt[2], 3/Sqrt[2]}, {3, 0}, {6, 0}, {9, 0},
        {9/Sqrt[2], 9/Sqrt[2]}, {0, 9}, {0, 6}};
{K, R} = PlaneQuad8Element[PlaneStress, e, ν, h, {0, 0}, {m, Ω}, nodes];
```

The essential boundary conditions are as follows:

```
debc{1, 6, 8, 10, 13, 15};
ebcVals = Table[0, {6}];
df = Complement[Range[16], debc];
Kf = K[[df, df]];
Rf = R[[df]] - K[[df, debc]].ebcVals;
```

The solution for nodal unknowns can now be obtained by using the LinearSolve function as follows:

```
dfVals = LinearSolve[Kf, Rf]
```

{0.00130927, 0.000933985, 0.000933985, 0.00130927, 0.00128101,
 0.00126707, 0.000901416, 0.000901416, 0.00126707, 0.00128101}

The complete vector of nodal values, in the original order established by the node numbering, is obtained by combining these values with those specified as essential boundary conditions as follows:

```
d = Table[0, {16}];
d[[debc]] = ebcVals;
d[[df]] = dfVals;
Partition[d, 2]
```

$$\begin{pmatrix} 0 & 0.00130927 \\ 0.000933985 & 0.000933985 \\ 0.00130927 & 0 \\ 0.00128101 & 0 \\ 0.00126707 & 0 \\ 0.000901416 & 0.000901416 \\ 0 & 0.00126707 \\ 0 & 0.00128101 \end{pmatrix}$$

Using these lists, the reactions can be calculated by the following expression:

```
reactions = K[[debc]].d − R[[debc]]
```

{−12.7742, −12.7742, −29.5583, −3.84672, −3.84672, −29.5583}

Finally, using the PlaneQuad8Results function, the strain and stress components, principal stresses, and equivalent von Mises stress for each element can be computed as follows:

```
PlaneQuad8Results[PlaneStress, e, nu, {0, 0}, nodes, d]
```

$\{\{3\sqrt{2}, 3\sqrt{2}\}, \{0.000102913, 0.000102913, -0.000221179\},$
$\{4.41054, 4.41054, -2.55206\}, \{6.96261, 1.85848\}, 6.24435\}$

Example 7.12 ANSYS Solution The problem is solved using an eight-node quadrilateral element (Plane82) in ANSYS (AnsysFiles\Chap7\diskFine.inp on the book web site) with a 10×10 mapped mesh. The model is developed in ANSYS using usual steps (see Appendix A). In the resulting mesh the element size gradually increases with the radius as shown in Figure 7.24. In order to compute the inertia forces due to rotation of the disk, mass density must be entered as a material property. The angular velocity is specified in terms of its components in the global coordinate system in the section where the loads are defined. (The complete menu path is Solution> Define loads> Apply> Structural> Other> Angular velocity). For this plane stress idealization the rotation vector points in the global z direction. Thus angular velocity is entered as the z component in rad/s. From the element solution the first principal stress plot is shown in the figure. Using the analytical expressions, we get the following maximum stress:

$$\text{Analytical solution at } r = 3 \text{ in:} \qquad \sigma_{t,\max} = 13.7342 \text{ ksi}$$

The finite element solution clearly compares very well with the analytical solution.

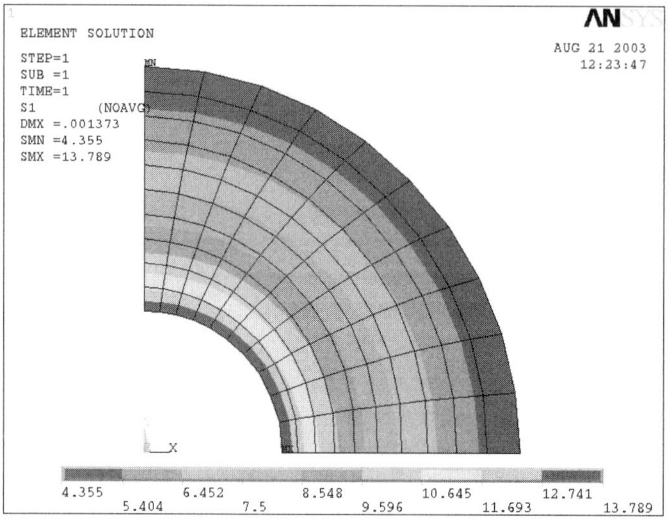

Figure 7.24. Plot of first principal stress in rotating disk

7.5.3 Residual Stresses due to Welding

During the welding process the weld material is melted at high temperature and is deposited at the location of a joint. As the *weld bead* cools, it generates large residual stresses in the parts that are welded together. Finite element analysis is an effective tool to quantify these stresses. As a simple illustration, consider a typical lap joint between two tension plates as shown in Figure 7.25. Assuming that the plates are wide in the plane perpendicular to the paper, a plane strain finite element model can be created to determine the residual stresses. We can also take advantage of the *skew symmetry* of the lap joint. This symmetry is recognized by the fact that, if the figure is rotated through 180° about the axis of symmetry, the original shape is recovered. All points along the axis of symmetry in this case have zero x and y displacements.

As a specific numerical example we consider the following data:

Steel plates: $\frac{1}{2} \times 10$ in; Overlap = 4 in; $E = 29 \times 10^3$ ksi; $\nu = 0.3$

Weld material: Weld size = $\frac{1}{2}$ in; $E = 25 \times 10^3$ ksi; $\nu = 0.3$

The weld bead is deposited at a temperature of 2000°F. Prior to welding the plates were heated to 120°F. We are interested in determining residual stresses when the assembly cools to the room temperature of 70°F. The coefficient of thermal expansion of both the steel plate and the weld material is 7×10^{-6}/°F.

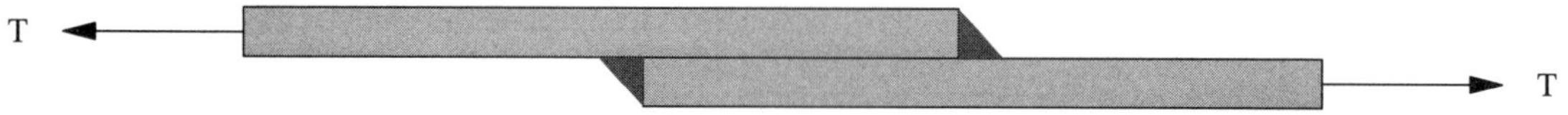

Figure 7.25. Welded lap joint

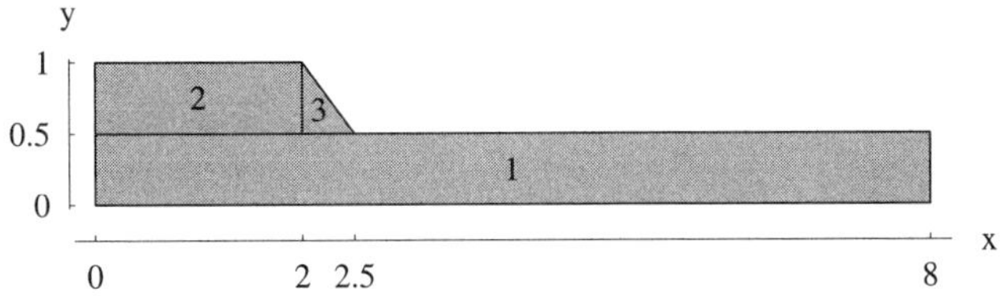

Figure 7.26. Finite element model of half of the assembly

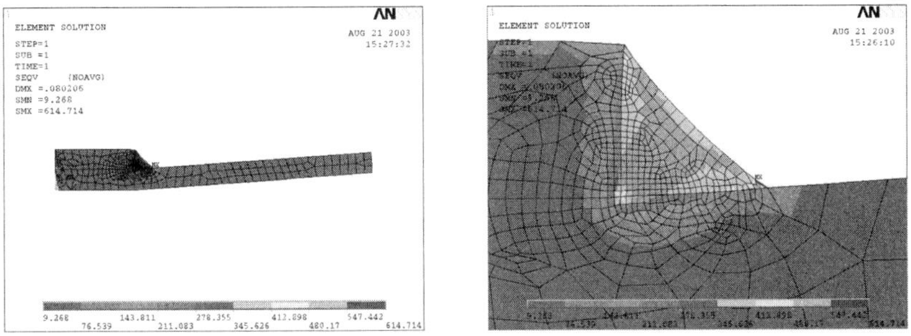

Figure 7.27. Computed von Mises stresses with a close up of the weld region

Half of the assembly is modeled in ANSYS by creating three areas as shown in Figure 7.26. All nodes on the left end are fixed. The areas 1 and 2 are assigned a reference temperature of 120°F and steel plate material properties. Area 3 is the weld material with a reference temperature of 2000°F. The *temperature loading* consists of the entire assembly having a temperature of 70°F.

A plane strain finite element model is constructed in ANSYS using Plane42 element (AnsysFiles\Chap7\weld.inp on the book web site). Fine mesh is created around the weld region to capture high stress gradients in this region. The resulting von Mises stresses are shown in Figure 7.27.

The analysis shows very high stresses in the weld region. Clearly, there will be some local yielding in the weld region which is not captured by this linear elastic analysis. This shows a need for a better model that takes into account material yielding. Another key limitation of the model is the assumed temperature distribution. The entire plate is assumed to be at 120°F at the time of welding. During the welding process the temperature of the plate in the vicinity of the weld is obviously much higher than this. Thus, for a more realistic analysis, first a thermal analysis should be carried out to determine the correct temperature distribution in the assembly just after welding. This temperature distribution becomes the reference temperature and the thermal loading is then cooling of various elements from the computed temperatures to room temperature.

7.5.4 Crack Tip Singularity

In fracture mechanics applications one is interested in accurately capturing high stress concentrations near a crack tip. Consider a typical situation of a thin plate with an edge crack as shown in Figure 7.28. Two sets of nodes are created along the crack length. To

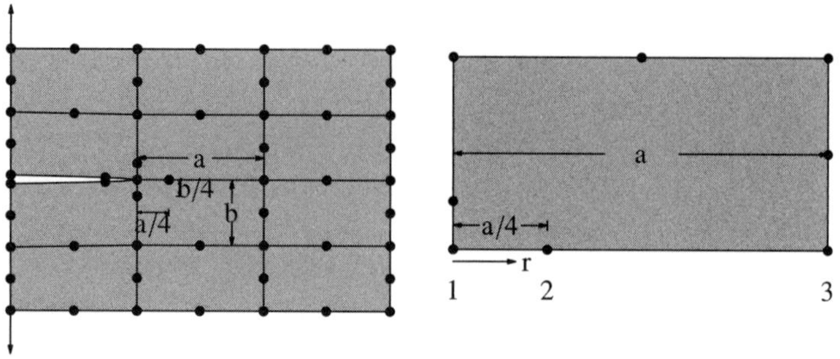

Figure 7.28. Finite element model of a thin plate with an edge crack and a typical *quarter-point* element

allow the crack to open, the elements on different sides of the crack must be connected to different nodes. It is well known that near the crack tip stresses are proportional to $1/\sqrt{r}$, where r is the radial distance from the tip of the crack. To capture this stress singularity, we must use higher order elements such as the eight-node quadrilateral or six-node triangles. A judicial placement of side nodes for the elements around the crack tip allows these elements to capture the stress singularity without having to use a large number of elements.

For normal applications it is best to place the side nodes at the middle points of the sides. However, it turns out that, if the side nodes are placed at quarter-points, then the excessive distortion introduced in the element as a result of mapping produces a beneficial result of giving a singular stress field that is proportional to $1/\sqrt{r}$ with r measured from the corner node closest to the quarter-point node. This is precisely the behavior of stresses around a crack tip.

To see this behavior clearly, consider the one-dimensional situation along side 1-2-3 of the quarter-point element shown in the figure. With r measured from node 1 the coordinates of the nodes are $(0, a/4, a)$. Multiplying these by quadratic Lagrange interpolations, the mapping for this side is

$$r = 0\left(\tfrac{1}{2}(s-1)s\right) + \tfrac{a}{4}(-(s-1)(s+1)) + a\left(\tfrac{1}{2}s(s+1)\right) = \tfrac{1}{4}a(s+1)^2$$

Using the same interpolation functions, the horizontal displacement for this side is

$$\begin{aligned} u &= u_1\left(\tfrac{1}{2}(s-1)s\right) + u_2(-(s-1)(s+1)) + u_3\left(\tfrac{1}{2}s(s+1)\right) \\ &= \tfrac{1}{2}((s-1)su_1 - 2(s-1)(s+1)u_2 + s(s+1)u_3) \end{aligned}$$

The axial strain is the derivative of this displacement with respect to r. Evaluating this derivative using the chain rule, we have

$$\frac{du}{ds} = \frac{du}{dr}\frac{dr}{ds} \Longrightarrow \frac{du}{dr} = \frac{du/ds}{dr/ds} = \frac{(2s-1)u_1 - 4su_2 + (2s+1)u_3}{a(s+1)}$$

By inverting the mapping, we can express this strain in terms of r as follows:

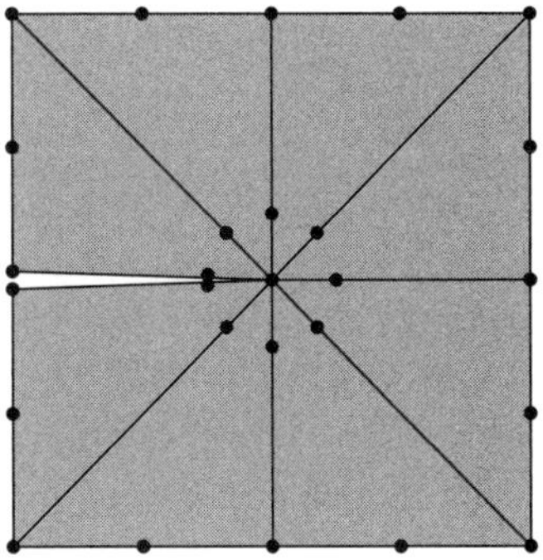

Figure 7.29. Finite element mesh around crack tip using triangular *quarter-point* elements

$$s = \frac{2\sqrt{r} - \sqrt{a}}{\sqrt{a}}$$

$$\frac{du}{dr} = \frac{(4\sqrt{r} - 3\sqrt{a})u_1 + 4(\sqrt{a} - 2\sqrt{r})u_2 - (\sqrt{a} - 4\sqrt{r})u_3}{2\sqrt{ra}}$$

This expression clearly shows that the strain is proportional to $1/\sqrt{r}$. Since the stress is proportional to strain, it follows that the stress distribution near the crack tip is proportional to the square root of the distance from the crack tip. Hence, defining the side node at the quarter point of the side captures the desired stress singularity. Thus, for analyzing fracture mechanics applications, we use the standard elements with the finite element mesh as shown in Figure 7.28. There is no need to create specialized elements or to refine the mesh excessively.

Elements with side nodes at quarter points are sometimes referred to as *singularity elements*. However, it should be clear from the discussion here that they are standard quadratic elements except that the side nodes are placed at quarter points instead of near the middle. Numerical experiments suggest even better performance with quarter-point triangular elements constructed by collapsing one side of an eight-node quadrilateral element. A typical mesh around a crack tip using these elements is as shown in Figure 7.29. It is recommended to have elements every 30° to 40° in the circumferential direction. The radius of the first row of elements around the crack tip should be approximately one-eighth of the crack length.

PROBLEMS

Fundamental Concepts in Elasticity

7.1 Stresses at a point on planes perpendicular to the coordinate directions are given as follows:

$$\sigma_x = 100; \quad \sigma_y = -30; \quad \sigma_z = -10;$$

$$\tau_{xy} = 0; \quad \tau_{yz} = 50; \quad \tau_{zx} = -50 \text{ MPa}$$

(a) Compute the stress vector, normal stress, and shear stress on a plane whose equation is

$$x + 2y - 2z = -4$$

(b) Compute principal stresses and unit normals for principal planes.

(c) If the yield stress of the material is 250 MPa, compute the factors of safety based on the maximum shear stress and the von Mises failure criteria.

7.2 Cartesian stress components at a point are given as follows:

$$\sigma_x = -100; \quad \sigma_y = 30; \quad \sigma_z = 10;$$
$$\tau_{xy} = 10; \quad \tau_{yz} = 50; \quad \tau_{zx} = -50 \text{ MPa}$$

(a) Compute the stress vector, normal stress, and shear stress on a plane whose unit normal is

$$\boldsymbol{n}^T = \{0.15199, -0.721676, -0.675339\}$$

(b) Compute principal stresses and unit normals for principal planes.

(c) If the yield stress of the material is 200 MPa, compute the factors of safety based on the maximum shear stress and the von Mises failure criteria.

7.3 A rectangular elastic solid with dimensions $a \times b \times c$ is simply supported along the three sides attached to the coordinate lines and is subjected to uniform pressure p along the remaining three sides, as shown in Figure 7.30. (Note that the loading is not concentrated but is applied uniformly over each face.) Thus the boundary conditions on all six sides of the solid are as follows:

On xy plane at $z = 0$:	$w(x, y, 0) = 0;$	$q_x(x, y, 0) = q_y(x, y, 0) = 0$
On xy plane at $z = c$:	$q_x(x, y, c) = q_y(x, y, c) = 0;$	$q_z(x, y, c) = -p$
On xz plane at $y = 0$:	$v(x, 0, z) = 0;$	$q_x(x, 0, z) = q_z(x, 0, z) = 0$

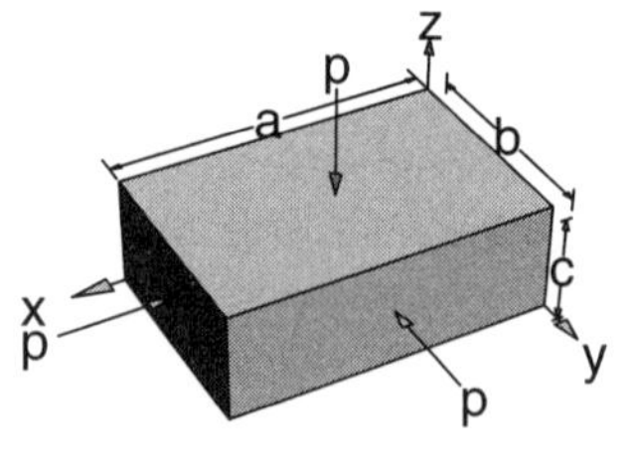

Figure 7.30.

On xz plane at $y = b$: $q_x(x, b, z) = q_z(x, b, z) = 0$; $q_y(x, b, z) = -p$

On yz plane at $x = 0$: $u(0, y, z) = 0$; $q_y(a, y, z) = q_z(a, y, z) = 0$

On yz plane at $x = a$: $q_y(a, y, z) = q_z(a, y, z) = 0$; $q_x(a, y, z) = -p$

Show that the following displacements satisfy the governing differential equations and all boundary conditions and hence represent the exact solution of the problem:

$$u(x, y, z) = -(1-2\nu)\frac{p}{E}x; \qquad v(x, y, z) = -(1-2\nu)\frac{p}{E}y; \qquad w(x, y, z) = -(1-2\nu)\frac{p}{E}z$$

where E = Young's modulus and ν = Poisson's ratio.

7.4 A cylindrical shaft of diameter 30 mm is subjected to an axial force of $F = 10\,\text{kN}$, a bending moment of $M = 170\ \text{N} \cdot \text{m}$, and a torque $T = 220\ \text{N} \cdot \text{m}$. The shaft is made of steel that has a yield stress $\sigma_f = 300\text{MPa}$. Determine the factor of safety against yielding using von Mises theory. Note the following relationships for stresses in shafts from the elementary mechanics of deformable bodies:

$$\text{Axial stress} = \frac{F}{A} + \frac{Mr}{I}; \qquad \text{Shear stress} = \frac{Tr}{J}$$

where r is the radius of the shaft, A is its area of cross section, I is the moment of inertia, and J is the torsional constant which is equal to the polar moment of inertia for a circular section.

Plane Stress and Plane Strain

7.5 A triangular cantilever plate 30 mm long, 20 mm high, and 10 mm thick is loaded as shown in Figure 7.31. Use only one triangular element to determine the factor of safety against stress failure (von Mises criterion). Use the following numerical data:

$E = 70$ GPa; $\nu = 0.3$; $p = 100$ MPa; Yield stress = 350 MPa

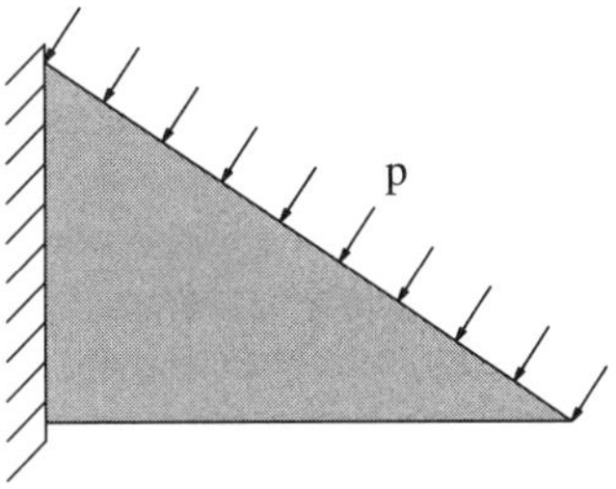

Figure 7.31.

7.6 You are asked to determine stresses in a thin triangular plate that is subjected to a temperature rise of 40°C. The plate is modeled with only *one triangular* element, as shown in Figure 7.32. The height of the plate is 50 mm and the base width is 30 mm. Other numerical data are as follows.

Thickness = 1 mm; $E = 30{,}000\ \text{N/mm}^2$; $\nu = \frac{1}{4}$; $\alpha = 11 \times 10^{-6}/°\text{C}$

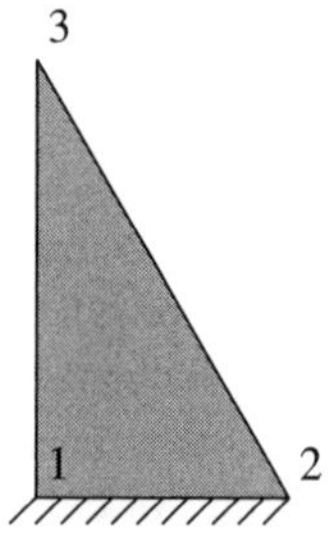

Figure 7.32.

7.7 Determine stresses in the cantilever plate shown in Figure 7.33 due to a temperature rise of 100°C. The plate is 5 mm thick, 50 mm long, 20 mm wide at the base, and 10 mm wide at the tip. Use the following material properties:

$E = 70\ \text{GPa}$; $\nu = 0.3$; $\alpha = 23 \times 10^{-6}/°\text{C}$

Consider a very coarse mesh with only two triangular elements as shown. Use N · mm units. Report displacements and effective stresses at element centroids.

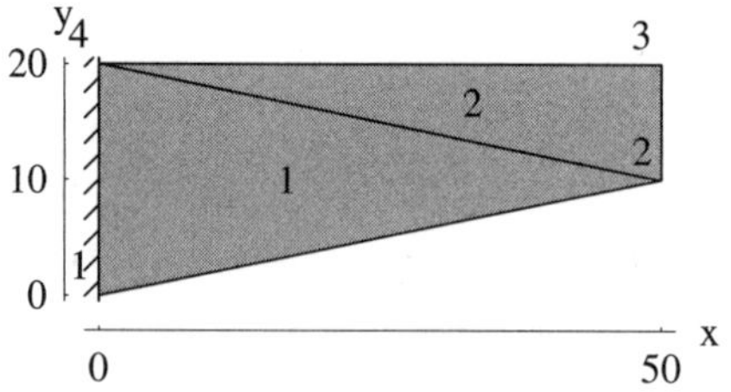

Figure 7.33.

7.8 The long diamond-shaped steel tube shown in Figure 7.34 is subjected to an internal pressure p. Compute displacements and principal stresses. Since the tube is long, we can take a unit slice and model it as a plane strain problem. Thus we need to create a finite element model of the planar cross section with thickness = 1. Use the following numerical values.

$$E = 200\,\text{GPa}; \quad \nu = 0.3; \quad \text{Wall thickness} = 15\,\text{mm};$$
$$d = 100\,\text{mm}; \quad p = 1\,\text{MPa}$$

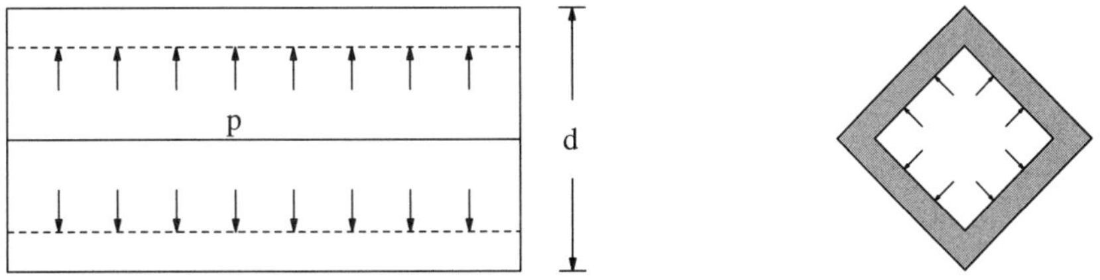

Figure 7.34.

Take advantage of symmetry and model one-quarter of the region using only a single four-node element, as shown in Figure 7.35.

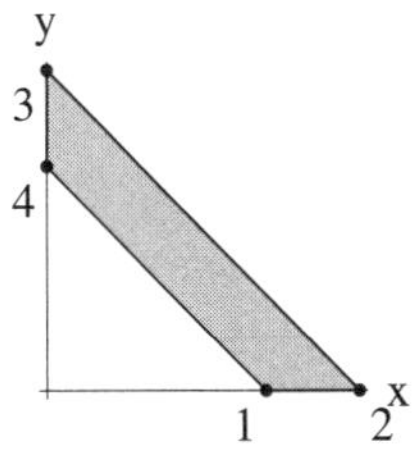

Figure 7.35.

7.9 A long steel pipe, shown in Figure 7.36, with outside diameter d and wall thickness t is subjected to an internal pressure p. Compute displacements and principal stresses. Use the following numerical values:

$$E = 30 \times 10^6\,\text{psi}; \quad \nu = 0.3; \quad t = 12\,\text{in}; \quad d = 120\,\text{in}; \quad p = 5000\,\text{psi}$$

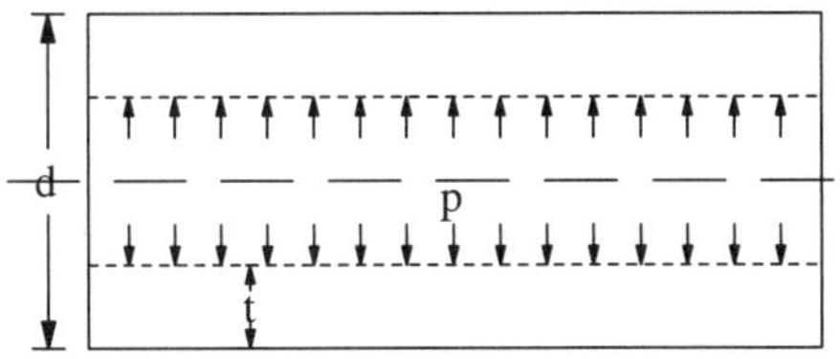

Figure 7.36.

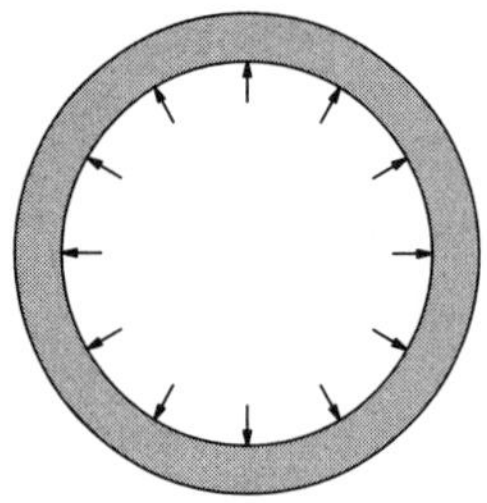

Figure 7.37.

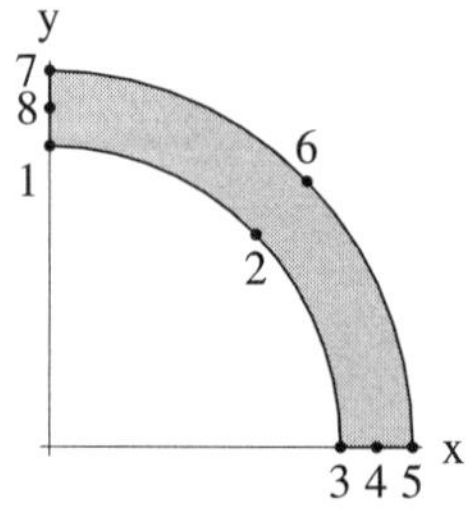

Figure 7.38.

Since the cylinder is long, we can take a unit slice and model it as a plane strain problem. Thus we need to create a finite element model of the planar region shown in Figure 7.37. We can further take advantage of symmetry and model one-quarter of the region. To show all calculations, only a single eight-node element is used, as shown in Figure 7.38.

Computational Projects

7.10 Determine stresses in the bolted bracket shown in Figure 7.39. Material properties are $E = 29000$ ksi and $\nu = 0.3$.

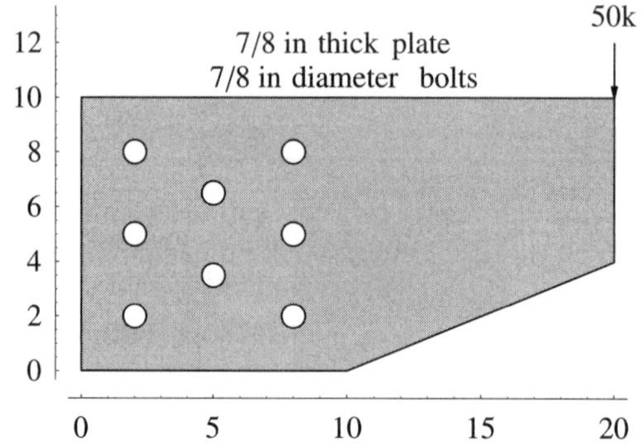

Figure 7.39. Bolted steel bracket

Modeling question: To include bolt holes in the model or not? The answer depends on the goal of the analysis. A reasonably accurate stress picture in the plate can be obtained by ignoring the bolt holes and simply constraining the bolt centers in the model. To get a more accurate stress distribution in the vicinity of the bolt holes, one must consider the bolt holes in the model.

7.11 The cross section of a concrete dam is shown in Figure 7.40. Using a planar finite element model, determine stresses in the dam due to self-weight of the dam and the water pressure. The depth of water behind the dam is 18 m. The density of concrete is 2400 kg/m^3 and that of water is 1000 kg/m^3. The modulus of elasticity of concrete is 30 GPa and Poisson's ratio is 0.15.

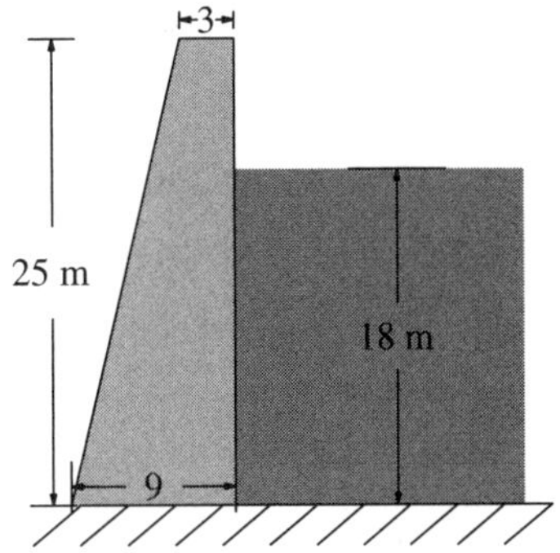

Figure 7.40. Concrete dam

7.12 The side view of a pry bar is shown in Figure 7.41. The cross section of the bar is rectangular with thickness $= \frac{1}{2}$ in and width (perpendicular to the plane of paper) $=$ 1.2 in. The other dimensions (in inches) are shown in the figure. The material properties are $E = 29 \times 10^6$ psi and $\nu = 0.3$. A load of $P = 200$ lb is applied at the center of the handle. Using a plane stress model, determine the maximum deflection and von Mises stress in the bar. To avoid stress concentration, assume the load to be uniformly distributed over a 3-in length.

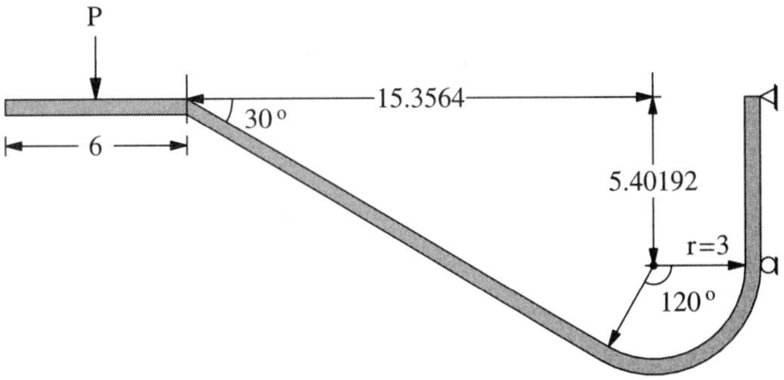

Figure 7.41. Pry bar

7.13 The shear wall shown in Figure 7.42 is subjected to a horizontal pressure p = 50 kN/m. The dimensions in meters are shown in the figure. The arches have a radius of 0.5 m. The thickness of the wall is 0.25 m and the material properties are $E = 21 \times 10^6$ kN/m^2 and $\nu = 0.15$. Using a plane stress model, determine the maximum deflection and von Mises stress in the wall.

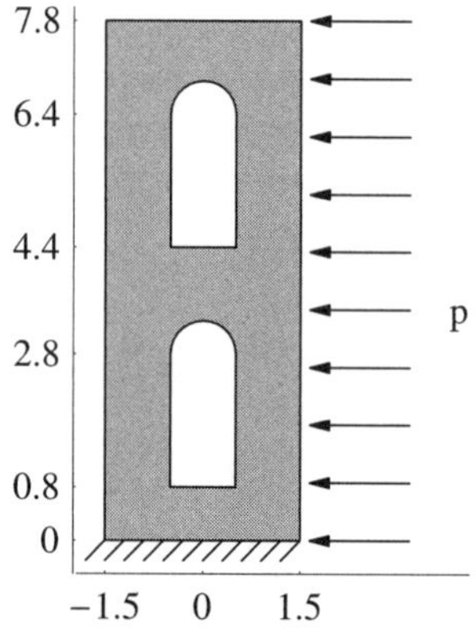

Figure 7.42. Shear wall

7.14 A concrete frame with in-filled masonry wall is shown in Figure 7.43. The dimensions in meters are shown in the figure. The beams and columns are 0.3 m thick and the wall is 0.15 m thick. The material properties of concrete are $E = 21 \times 10^6\ \mathrm{kN/m^2}$ and $\nu = 0.15$ and that of the masonry are $E = 10 \times 10^6\ \mathrm{kN/m^2}$ and $\nu = 0.15$. The horizontal load varies linearly from 0 at the top to 10 kN/m at the bottom. The distributed load on the beams is 2 kN/m. Using a plane stress model, determine the maximum deflection and von Mises stress.

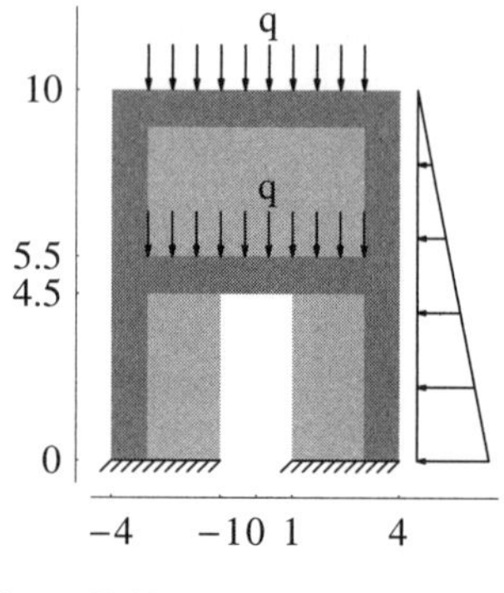

Figure 7.43. Frame with in-fill wall

7.15 Consider an aluminum machine part, as shown in Figure 7.44. A load of 1000 lb is applied near the tip. The other end can be considered fixed. The material properties are $E = 10.6 \times 10^6$ psi and $\nu = 0.35$.

(a) *Frame model—hand calculations:* From the geometry, loading, and support conditions, it is clear that the structure is basically a cantilever beam. Thus a very *simple frame model* of the part can be constructed by using the center line dimensions, as shown by the solid line in Figure 7.45. Determine the maximum stress using this model.

(b) Analyze the problem as a plane stress model. To avoid stress concentration due to point load, assume the load to be applied over a length of 1 in.

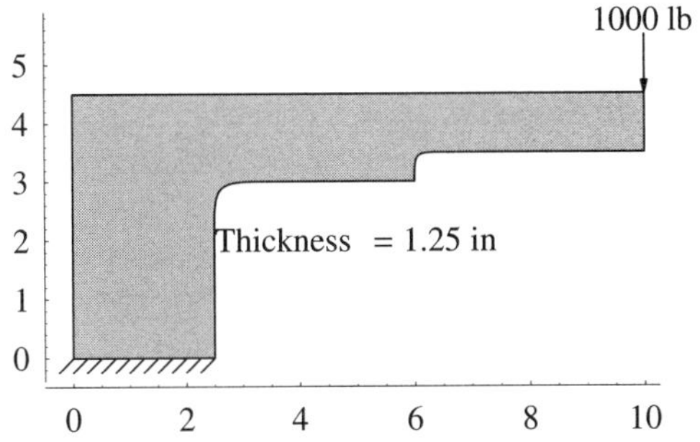

Figure 7.44. Machine part

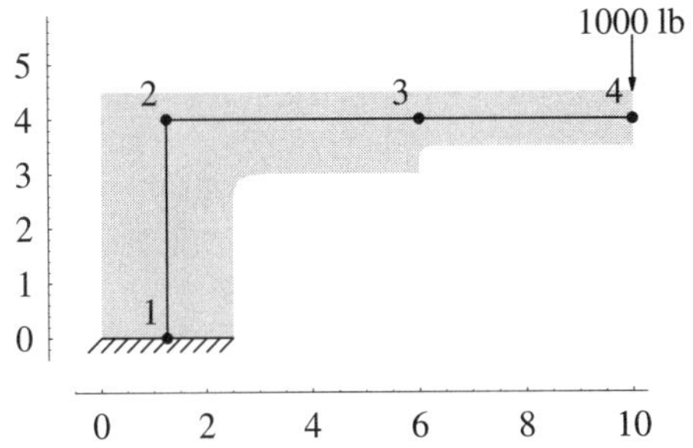

Figure 7.45.

7.16 A cast iron flywheel with spokes and rim is shown in Figure 7.46. The flywheel operates at speeds up to $\Omega = 4500$ rpm. The cast iron weighs 0.256 lb/in^3 and has an ultimate strength of 22,000 lb/in^2. Young's modulus is 30×10^6 lb/in^2 and Poisson's ratio is 0.3. The goal is to determine the factor of safety against stress failure. The dimensions of the flywheel can easily be seen from Figure 7.47, which shows the front view without fillets and a sectional view:

Shaft radius = 2 in	Inside radius = 3 in
Rim inner radius = 10 in	Rim outer radius = 12 in
Fillets near shaft = 1 in radius	Fillets at the rim = 0.5 in radius
$\alpha = 6.404°$	

The solution can be approached from a variety of different assumptions. Several models are suggested, from a simple analytical model to a fully three-dimensional finite element model. Your assignment is to use these models and compare solutions. Your report should contain a discussion on the appropriateness of different models. Comment on the cost/benefit of these models. Here the cost includes the time for model preparation, interpretation of results, and computer resources needed for the analysis. The benefit is the accuracy of the computed safety factor.

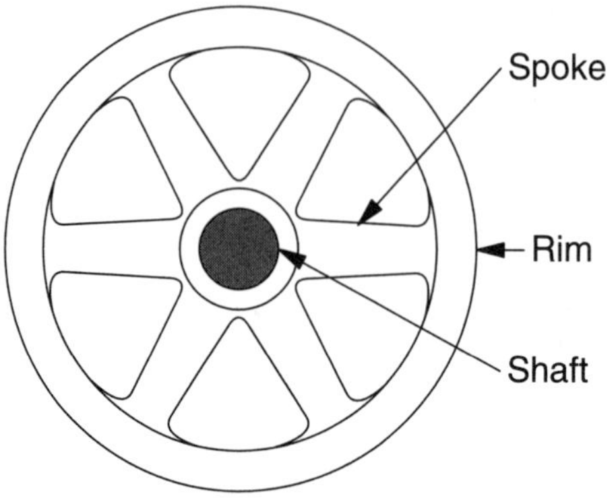

Figure 7.46.

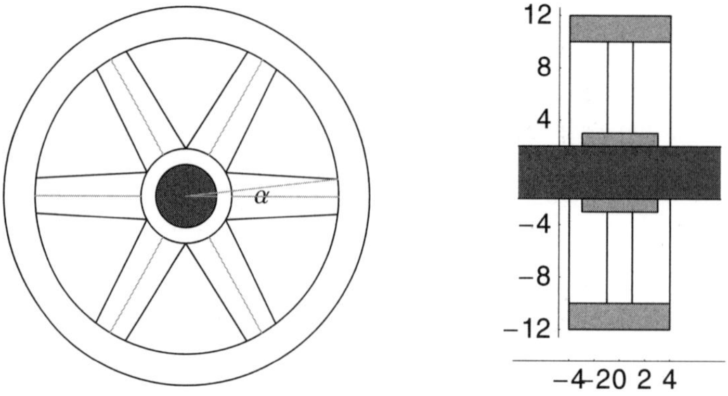

Figure 7.47.

(a) *Analytical solution based on spokes providing axial restraint to the rim:* A simple analytical solution is possible by considering the flywheel as essentially a ring with mean radius r. The rim expands under the influence of centrifugal force $F_c = mr\omega^2$ lb/in, where $m = (\rho/g)A_r$ is the mass per unit length of the rim, ρ lb/in^3 is the density, g in/s^2 is the acceleration due to gravity, A_r in^2 is the area of the cross section of the rim, and ω rad/s is rotational speed of the flywheel. The spokes provide axial restraint to the rim. The inertia forces in the spokes are neglected. By considering the rim as a circular beam, it is possible to derive analytical expressions for axial force (F_r) and moment (M_r) in the rim and axial force in the spokes (F_s). For a flywheel with six spokes, the following expressions are obtained:

$$F_s = F_c r H \text{ lb}$$

$$H = \frac{2/3}{(0.0203r^2/h^2) + 0.957 + (A_r/A_s)}$$

h = rim width in the front view

A_s = cross-sectional area of the spoke

$F_r = F_c r(1 - 0.866H)$ lb; $\quad M_r = 0.0889 F_s r$ in · lb

The stresses can then be computed as follows:

$$\text{Rim:} \quad \sigma = \frac{F_r}{A_r} + \frac{6M_r}{bh^2}; \qquad \text{Spoke:} \quad \sigma = \frac{F_s}{A_s}$$

where it is assumed that the rim is of rectangular cross section with width h (2 in) and thickness b (8 in).

(b) *Finite element model using beam elements:* A simple finite element model of the flywheel can be created by using beam elements for both the spoke and the rim. Create and analyze such a model and compare results with the previous solutions. Take advantage of symmetry and model only one-fourth of the flywheel using center line dimensions as shown by the dark lines in Figure 7.48. Use a reasonable number of beam elements to account for the circular geometry of the rim.

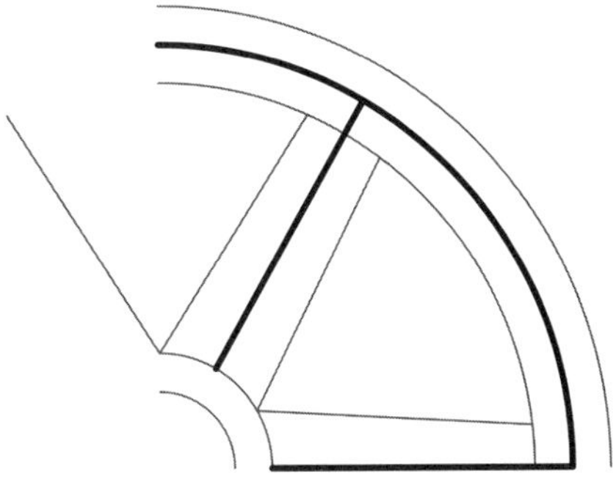

Figure 7.48.

(c) *Plane stress finite element model:* With the beam element model it is not possible to determine local stress concentrations at the junctions between the spokes and the rim. A two-dimensional plane stress model is necessary if one needs to determine a more complete stress distribution. Use a plane stress element to model the rim and the spokes. Make the model realistic by creating fillets between lines meeting at common points. Compare your results with those obtained from the previous simplified models. Take advantage of symmetry and model only one-fourth of the flywheel, as shown in Figure 7.49.

(d) *Three-dimensional finite element model:* The plane stress model assumes that the stresses in the thickness direction are negligible. A three-dimensional finite element model obviously does not need this assumption. Use an element equivalent to ANSYS Solid45 element to create a three-dimensional model

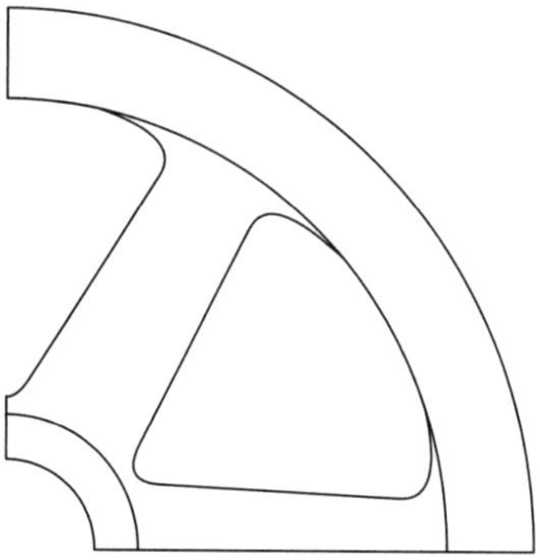

Figure 7.49.

of the flywheel. Compare your results with those obtained from the previous simplified models. Take advantage of symmetry and model only one-eighth of the flywheel (cut the wheel in half in the thickness direction also).

CHAPTER EIGHT

TRANSIENT PROBLEMS

8.1 TRANSIENT FIELD PROBLEMS

A variety of transient field problems, such as heat and fluid flow, are governed by a differential equation of the following form:

$$\frac{\partial}{\partial x}\left(k_x \frac{\partial u}{\partial x}\right) + \frac{\partial}{\partial y}\left(k_y \frac{\partial u}{\partial y}\right) + \frac{\partial}{\partial z}\left(k_z \frac{\partial u}{\partial z}\right) + pu + q = m\frac{\partial u}{\partial t}$$

where k_x, k_y, k_z, p, q, and m are known functions of (x, y, z). The solution variable u is a function both of space (x, y, z) and time t. Except for time dependence, this equation is a direct extension to three dimensions of the two-dimensional boundary value problem considered in Chapters 5 and 6. The possible boundary conditions are as follows:

(i) Essential boundary condition: u specified.

(ii) Natural boundary condition—specified normal derivative along a boundary:

$$k\frac{\partial u}{\partial n} \equiv \left(k_x \frac{\partial u}{\partial x} n_x + k_y \frac{\partial u}{\partial y} n_y + k_z \frac{\partial u}{\partial z} n_z\right) = \alpha u + \beta$$

where α and β are specified parameters along the boundary.

With respect to time, the differential equation is first order and therefore one initial condition is also needed in the following form:

$$u(x, y, z, t = 0) = u_0(x, y, z) \text{ specified}$$

8.1.1 Finite Element Equations

The general form of finite element equations for the problem can easily be written using exactly the same steps as those used for the steady-state problems in Chapters 5 and 6. In order to use Galerkin's method, we need to develop an appropriate weak form first. Assuming V refers to volume of an arbitrary element, moving all terms in the differential equation to the left-hand side, multiplying by the weighting functions N_i, and integrating over the volume, the Galerkin weighted residual for the problem is

$$\iiint_V \left(\frac{\partial}{\partial x}\left(k_x \frac{\partial u}{\partial x}\right) + \frac{\partial}{\partial y}\left(k_y \frac{\partial u}{\partial y}\right) + \frac{\partial}{\partial z}\left(k_z \frac{\partial u}{\partial z}\right) + pu + q - m\frac{\partial u}{\partial t} \right) N_i \, dV = 0; \qquad i = 1, 2, \ldots$$

where $dV = dx\,dy\,dz$ is the differential volume. Using the Green-Gauss theorem on the first three terms in the weighted residual, we have

$$\iint_S \left(k_x \frac{\partial u}{\partial x} n_x + k_y \frac{\partial u}{\partial y} n_y + k_z \frac{\partial u}{\partial z} n_z \right) N_i \, dS$$
$$+ \iiint_V \left(-k_x \frac{\partial u}{\partial x}\frac{\partial N_i}{\partial x} - k_y \frac{\partial u}{\partial y}\frac{\partial N_i}{\partial y} - k_z \frac{\partial u}{\partial z}\frac{\partial N_i}{\partial z} + puN_i + qN_i - m\frac{\partial u}{\partial t} N_i \right) dV = 0$$

where S is the surface of the element. On the surface there is a possibility of one of the two types of boundary conditions to be specified. Designating the surface over which u is specified as S_e and that over which its normal derivative is specified as S_n, the surface integral can be written as follows:

$$\iint_S \left(k_x \frac{\partial u}{\partial x} n_x + k_y \frac{\partial u}{\partial y} n_y + k_z \frac{\partial u}{\partial z} n_z \right) N_i \, dS$$
$$= \iint_{S_e} \left(k_x \frac{\partial u}{\partial x} n_x + k_y \frac{\partial u}{\partial y} n_y + k_z \frac{\partial u}{\partial z} n_z \right) N_i \, dS + \iint_{S_n} (\alpha u + \beta) N_i \, dS$$

Requiring the assumed solution to satisfy the essential boundary conditions results in weighting functions that are zero over S_e. Thus with admissible assumed solutions the weak-form equivalent to the given boundary value problem is as follows:

$$\iiint_V \left(-k_x \frac{\partial u}{\partial x}\frac{\partial N_i}{\partial x} - k_y \frac{\partial u}{\partial y}\frac{\partial N_i}{\partial y} - k_z \frac{\partial u}{\partial z}\frac{\partial N_i}{\partial z} + puN_i + qN_i - m\frac{\partial u}{\partial t} N_i \right) dV$$
$$+ \iint_{S_n} (\alpha u + \beta) N_i \, dS = 0; \qquad i = 1, 2, \ldots$$

Rearranging by keeping all terms involving unknown solution on the left-hand side, the weak form is

$$\iiint_V \left(k_x \frac{\partial u}{\partial x}\frac{\partial N_i}{\partial x} + k_y \frac{\partial u}{\partial y}\frac{\partial N_i}{\partial y} + k_z \frac{\partial u}{\partial z}\frac{\partial N_i}{\partial z} - puN_i + m\frac{\partial u}{\partial t}N_i\right) dV - \iint_{S_n} \alpha u N_i \, dS$$

$$= \iiint_V qN_i \, dV + \iint_{S_n} \beta N_i \, dS; \qquad i = 1, 2, \ldots$$

The assumed solution over an element is written as follows:

$$u(x, y, z, t) = (\, N_1(x, y, z) \quad N_2(x, y, z) \quad \cdots \quad N_n(x, y, z)) \begin{pmatrix} u_1(t) \\ u_2(t) \\ \vdots \\ u_n(t) \end{pmatrix} = \boldsymbol{N}^T \boldsymbol{d}$$

where $u_1, u_2, \ldots$ are the unknown solutions at the element nodes that are functions of time. Note the interpolation functions N_i are functions only of space variables and are exactly the same as those used in earlier chapters for steady-state problems. Differentiating the assumed solution, we get

$$\frac{\partial u}{\partial x} = \begin{pmatrix} \frac{\partial N_1}{\partial x} & \frac{\partial N_2}{\partial x} & \cdots & \frac{\partial N_n}{\partial x} \end{pmatrix} \begin{pmatrix} u_1 \\ u_2 \\ \vdots \\ u_n \end{pmatrix} = \boldsymbol{B}_x^T \boldsymbol{d}$$

$$\frac{\partial u}{\partial y} = \begin{pmatrix} \frac{\partial N_1}{\partial y} & \frac{\partial N_2}{\partial y} & \cdots & \frac{\partial N_n}{\partial y} \end{pmatrix} \begin{pmatrix} u_1 \\ u_2 \\ \vdots \\ u_n \end{pmatrix} = \boldsymbol{B}_y^T \boldsymbol{d}$$

$$\frac{\partial u}{\partial z} = \begin{pmatrix} \frac{\partial N_1}{\partial z} & \frac{\partial N_2}{\partial z} & \cdots & \frac{\partial N_n}{\partial z} \end{pmatrix} \begin{pmatrix} u_1 \\ u_2 \\ \vdots \\ u_n \end{pmatrix} = \boldsymbol{B}_z^T \boldsymbol{d}$$

$$\frac{\partial u}{\partial t} = (N_1 \quad N_2 \quad \cdots \quad N_n) \begin{pmatrix} \dot{u}_1 \\ \dot{u}_2 \\ \vdots \\ \dot{u}_n \end{pmatrix} = \boldsymbol{N}^T \dot{\boldsymbol{d}}$$

where an overdot indicates differentiation with respect to time. Substituting these into the weak form, we have

$$\iiint_V (mN_i\boldsymbol{N}^T)\,dV\,\dot{\boldsymbol{d}} + \iiint_V \left(k_x\frac{\partial N_i}{\partial x}\boldsymbol{B}_x^T + k_y\frac{\partial N_i}{\partial y}\boldsymbol{B}_y^T + k_z\frac{\partial N_i}{\partial z}\boldsymbol{B}_z^T\right)dV\,\boldsymbol{d}$$

$$-\iiint_V (pN_i\boldsymbol{N}^T)\,dV\,\boldsymbol{d} - \iint_{S_n} \alpha N_i\boldsymbol{N}^T\,dS\,\boldsymbol{d}$$

$$= \iiint_V qN_i\,dV + \iint_{S_h} \beta N_i\,dS; \qquad i = 1, 2, \ldots$$

This system represents n equations, one for each N_i, $i = 1, 2, \ldots, n$. Writing these equations explicitly and arranging them in matrix form, we have

$$\iiint_V (m\boldsymbol{N}\boldsymbol{N}^T)\,dV\,\dot{\boldsymbol{d}} + \iiint_V (k_x\boldsymbol{B}_x\boldsymbol{B}_x^T + k_y\boldsymbol{B}_y\boldsymbol{B}_y^T + k_z\boldsymbol{B}_z\boldsymbol{B}_z^T)\,dV\,\boldsymbol{d}$$

$$-\iiint_V (p\boldsymbol{N}\boldsymbol{N}^T)\,dV\,\boldsymbol{d} - \iint_{S_n} \alpha(\boldsymbol{N}\boldsymbol{N})^T\,dS\,\boldsymbol{d}$$

$$= \iiint_V q\boldsymbol{N}\,dV + \iint_{S_n} \beta\boldsymbol{N}\,dS$$

The three terms inside the second volume integral can be arranged in a more compact form by using matrices as follows:

$$k_x\boldsymbol{B}_x\boldsymbol{B}_x^T + k_y\boldsymbol{B}_y\boldsymbol{B}_y^T + k_z\boldsymbol{B}_z\boldsymbol{B}_z^T = (\boldsymbol{B}_x \quad \boldsymbol{B}_y \quad \boldsymbol{B}_z)\begin{pmatrix} k_x & 0 & 0 \\ 0 & k_y & 0 \\ 0 & 0 & k_z \end{pmatrix}\begin{pmatrix} \boldsymbol{B}_x^T \\ \boldsymbol{B}_y^T \\ \boldsymbol{B}_z^T \end{pmatrix} \equiv \boldsymbol{BCB}^T$$

where

$$\boldsymbol{B}^T = \begin{pmatrix} \frac{\partial N_1}{\partial x} & \frac{\partial N_2}{\partial x} & \cdots & \frac{\partial N_n}{\partial x} \\ \frac{\partial N_1}{\partial y} & \frac{\partial N_2}{\partial y} & \cdots & \frac{\partial N_n}{\partial y} \\ \frac{\partial N_1}{\partial z} & \frac{\partial N_2}{\partial z} & \cdots & \frac{\partial N_n}{\partial z} \end{pmatrix} \quad \text{and} \quad \boldsymbol{C} = \begin{pmatrix} k_x & 0 & 0 \\ 0 & k_y & 0 \\ 0 & 0 & k_z \end{pmatrix}$$

Thus the finite element equations are as follows:

$$\iiint_V (m\boldsymbol{N}\boldsymbol{N}^T)\,dV\,\dot{\boldsymbol{d}} + \left(\iiint_V \boldsymbol{BCB}^T\,dV - \iiint_V (p\boldsymbol{N}\boldsymbol{N}^T)\,dV - \iint_{S_n} \alpha(\boldsymbol{N}\boldsymbol{N})^T\,dS\right)\boldsymbol{d}$$

$$= \iiint_V q\boldsymbol{N}\,dV + \iint_{S_n} \beta\boldsymbol{N}\,dS$$

Define $n \times n$ matrices

$$\boldsymbol{k}_k = \iiint_V \boldsymbol{BCB}^T \, dV; \qquad \boldsymbol{k}_p = -\iiint_V (p\boldsymbol{NN}^T) \, dV$$

$$\boldsymbol{k}_\alpha = -\iint_{S_h} \alpha(\boldsymbol{NN})^T \, dS; \qquad \boldsymbol{m} = \iiint_V (m\boldsymbol{NN}^T) \, dV$$

Define $n \times 1$ vectors

$$\boldsymbol{r}_q = \iiint_V q\boldsymbol{N} \, dV; \qquad \boldsymbol{r}_\beta = \iint_A \beta\boldsymbol{N} \, dS$$

Thus the element equations are a system of first-order ordinary differential equations

$$\boldsymbol{m}\dot{\boldsymbol{d}} + (\boldsymbol{k}_k + \boldsymbol{k}_p + \boldsymbol{k}_\alpha)\boldsymbol{d} = \boldsymbol{r}_q + \boldsymbol{r}_\beta$$

Note that, except for the first term, these equations are direct generalizations to three dimensions of the corresponding two-dimensional steady-state problem considered in Chapters 5 and 6. The first term involves time derivatives of the nodal variables. Thus the element equations now represent a system of ordinary differential equations. The assembly and overall solution process remains the same. The global system of ordinary differential equations must be solved using appropriate numerical schemes such as Runge-Kutta or Gears methods.

In order to actually evaluate the integrals defining the element equations, we need to assume an element shape and develop appropriate interpolation functions. Any element shape can be created as long as it is useful in modeling practical shapes and it is possible to carry out required integrations and differentiations. Some typical elements are presented in the following sections.

8.1.2 Triangular Element

A triangular element is simple yet versatile for two-dimensional problems. Almost any two-dimensional shape can be discretized into triangular elements. A typical three-node triangular element is shown in Figure 8.1. The nodal coordinates are (x_1, y_1), (x_2, y_2), and (x_3, y_3). Assuming unknown solutions at the three corners as nodal degrees of freedom, we have a total of three degrees of freedom, u_1, u_2, and u_3. The following interpolation functions for the element were developed in Chapter 5:

$$u = (N_1 \quad N_2 \quad N_3) \begin{pmatrix} u_1 \\ u_2 \\ u_3 \end{pmatrix} \equiv \boldsymbol{N}^T \boldsymbol{d}$$

$$N_1 = \frac{1}{2A}(xb_1 + yc_1 + f_1); \qquad N_2 = \frac{1}{2A}(xb_2 + yc_2 + f_2); \qquad N_3 = \frac{1}{2A}(xb_3 + yc_3 + f_3)$$

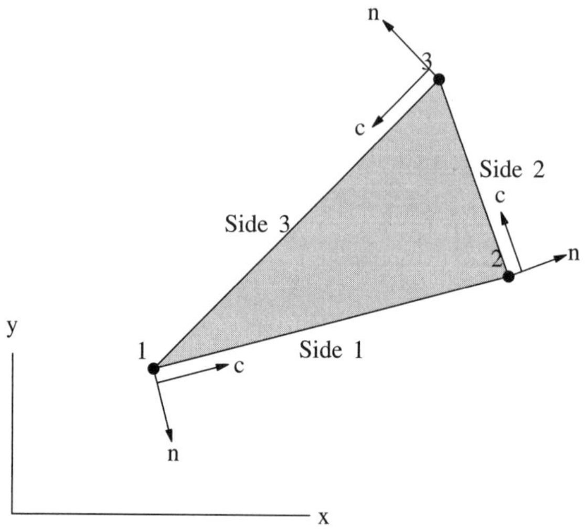

Figure 8.1. Three-node triangular element

$$f_1 = x_2y_3 - x_3y_2; \quad f_2 = x_3y_1 - x_1y_3; \quad f_3 = x_1y_2 - x_2y_1;$$
$$b_1 = y_2 - y_3; \quad b_2 = y_3 - y_1; \quad b_3 = y_1 - y_2;$$
$$c_1 = x_3 - x_2; \quad c_2 = x_1 - x_3; \quad c_3 = x_2 - x_1;$$
$$A = \text{area of triangle} = \tfrac{1}{2}(f_1 + f_2 + f_3)$$

Assuming a unit thickness, the volume integrals reduce to area integrals A and the surface integrals to line integrals C. Except for the $\boldsymbol{m}$ matrix, all other matrices and vectors are identical to those for the steady-state problem discussed in Chapter 5. To evaluate the $\boldsymbol{m}$ matrix, the integral over the triangle is evaluated using the procedure explained in Chapter 5. The final element equations are as follows:

$$\boldsymbol{m}\dot{\boldsymbol{d}} + (\boldsymbol{k}_k + \boldsymbol{k}_\alpha)\boldsymbol{d} = \boldsymbol{r}_q + \boldsymbol{r}_\beta$$

$$\boldsymbol{m} = \iiint_V (m\boldsymbol{N}\boldsymbol{N}^T)\,dV = \iint_A (m\boldsymbol{N}\boldsymbol{N}^T)\,dA = \frac{mA}{12}\begin{pmatrix} 2 & 1 & 1 \\ 1 & 2 & 1 \\ 1 & 1 & 2 \end{pmatrix}$$

$$\boldsymbol{B}^T = \begin{pmatrix} \frac{\partial N_1}{\partial x} & \frac{\partial N_2}{\partial x} & \frac{\partial N_3}{\partial x} \\ \frac{\partial N_1}{\partial y} & \frac{\partial N_2}{\partial y} & \frac{\partial N_3}{\partial y} \end{pmatrix} = \frac{1}{2A}\begin{pmatrix} b_1 & b_2 & b_3 \\ c_1 & c_2 & c_3 \end{pmatrix} \quad \text{and} \quad \boldsymbol{C} = \begin{pmatrix} k_x & 0 \\ 0 & k_y \end{pmatrix}$$

$$\boldsymbol{k}_k = \iint_A \boldsymbol{B}\boldsymbol{C}\boldsymbol{B}^T\,dA = A\boldsymbol{B}\boldsymbol{C}\boldsymbol{B}^T$$

$$= \frac{1}{4A}\begin{pmatrix} k_xb_1^2 + k_yc_1^2 & k_xb_1b_2 + k_yc_1c_2 & k_xb_1b_3 + k_yc_1c_3 \\ k_xb_1b_2 + k_yc_1c_2 & k_xb_2^2 + k_yc_2^2 & k_xb_2b_3 + k_yc_2c_3 \\ k_xb_1b_3 + k_yc_1c_3 & k_xb_2b_3 + k_yc_2c_3 & k_xb_3^2 + k_yc_3^2 \end{pmatrix}$$

$$k_p = -\frac{pA}{12}\begin{pmatrix} 2 & 1 & 1 \\ 1 & 2 & 1 \\ 1 & 1 & 2 \end{pmatrix}; \qquad r_q = \frac{qA}{3}\begin{pmatrix} 1 \\ 1 \\ 1 \end{pmatrix}$$

For a nonzero natural boundary condition on side 1,

Length of the side: $L_{12} = \sqrt{(x_2 - x_1)^2 + (y_2 - y_1)^2}$

$$k_\alpha = \iint_{S_n} \alpha(NN)^T \, dS = \int_C \alpha(NN)^T \, dc = -\frac{\alpha L_{12}}{6}\begin{pmatrix} 2 & 1 & 0 \\ 1 & 2 & 0 \\ 0 & 0 & 0 \end{pmatrix}$$

$$r_\beta = \iint_A \beta N \, dS = \int_c \beta N \, dc = \frac{\beta L_{12}}{2}\begin{pmatrix} 1 \\ 1 \\ 0 \end{pmatrix}$$

For a nonzero natural boundary condition on side 2,

Length of the side: $L_{23} = \sqrt{(x_3 - x_2)^2 + (y_3 - y_2)^2}$

$$k_\alpha = -\frac{\alpha L_{23}}{6}\begin{pmatrix} 0 & 0 & 0 \\ 0 & 2 & 1 \\ 0 & 1 & 2 \end{pmatrix}; \qquad r_\beta = \frac{\beta L_{23}}{2}\begin{pmatrix} 0 \\ 1 \\ 1 \end{pmatrix}$$

For a nonzero natural boundary condition on side 3,

Length of the side: $L_{31} = \sqrt{(x_1 - x_3)^2 + (y_1 - y_3)^2}$

$$k_\alpha = -\frac{\alpha L_{31}}{6}\begin{pmatrix} 2 & 0 & 1 \\ 0 & 0 & 0 \\ 1 & 0 & 2 \end{pmatrix}; \qquad r_\beta = \frac{\beta L_{31}}{2}\begin{pmatrix} 1 \\ 0 \\ 1 \end{pmatrix}$$

The equations are assembled in the usual manner to get a system of global first-order ordinary differential equations. After adjusting for essential boundary conditions, the first-order equations are then solved by one of the many numerical methods, such as Runge-Kutta and Gears methods, available for solving such equations. The *Mathematica* function NDSolve and MATLAB function ode use a variety of these methods to give solutions to these equations.

8.1.3 Transient Heat Flow

Consider the problem of finding temperature distribution T in an arbitrary three-dimensional body. Using conservation of energy on a differential volume, the following governing differential equation can easily be established:

$$\frac{\partial}{\partial x}\left(k_x \frac{\partial T}{\partial x}\right) + \frac{\partial}{\partial y}\left(k_y \frac{\partial T}{\partial y}\right) + \frac{\partial}{\partial z}\left(k_z \frac{\partial T}{\partial z}\right) + Q = \rho c_p \frac{\partial T}{\partial t}$$

where k_x, k_y, and k_z are thermal conductivities in the x, y, and z directions and $Q(x, y, z)$ is specified heat generation per unit volume, ρ is density of the material, and c_p is specific heat of the material. The temperature T is a function both of space (x, y, z) and of time t. The possible boundary conditions are as follows:

(i) Known temperature along a boundary: T specified.

(ii) Specified heat flux along a boundary:

$$-k\frac{\partial T}{\partial n} \equiv -\left(k_x\frac{\partial T}{\partial x}n_x + k_y\frac{\partial T}{\partial y}n_y + k_z\frac{\partial T}{\partial z}n_z\right) = q \text{ specified}$$

On an insulated boundary or across a line of symmetry there is no heat flow and thus $q = 0$. The sign convention for heat flow is that heat flowing into a body is positive and that out of the body is negative.

(iii) Heat loss due to convection along a boundary surface:

$$-k\frac{\partial T}{\partial n} \equiv -\left(k_x\frac{\partial T}{\partial x}n_x + k_y\frac{\partial T}{\partial y}n_y + k_z\frac{\partial T}{\partial z}n_z\right) = h(T - T_\infty)$$

where h is the convection coefficient, T is the unknown temperature on the boundary, and T_∞ is the known temperature of the surrounding fluid.

Both types of natural boundary conditions (ii) and (iii) can be handled by specifying the coefficients α and β in the following form:

$$k_x\frac{\partial T}{\partial x}n_x + k_y\frac{\partial T}{\partial y}n_y + k_z\frac{\partial T}{\partial z}n_z = \alpha T + \beta$$

If a heat flux is specified, then this boundary condition implies $\alpha = 0$ and $\beta = -q$. If convection is specified, then this boundary condition implies $\alpha = -h$ and $\beta = hT_\infty$.

The initial condition is of the following form:

$$T(x, y, z, t = 0) = T_0(x, y, z) \text{ specified}$$

Example 8.1 Transient Heat Flow The cross section of a long V-grooved ceramic strip is shown in Figure 8.2. The sides are insulated while convection heat loss takes place at the bottom with $h = 200\,\text{W/m}^2\cdot{}^\circ\text{C}$ and $T_\infty = 50^\circ$C. The grooved surface is maintained at a constant temperature of 300°C. For ceramic the thermal conductivity is $k = 3\,\text{W/m}\cdot{}^\circ\text{C}$, density $\rho = 1600\,\text{kg/m}^3$, and specific heat $c_p = 0.8\,\text{kJ/kg}\cdot{}^\circ\text{C}$. Assuming the entire strip to be at 50°C initially, determine the time history of temperature distribution. Estimate the time it will take to reach steady state conditions.

Taking advantage of symmetry, we need to model only half of the cross section. In order to show all calculations, a coarse mesh consisting of only two elements is used. Due to symmetry, there is no heat flow across side 1-2. Nodes 2 and 4 have specified temperature of 300°C. At time $t = 0$ the temperature at all other nodes is equal to 50°C.

The complete calculations are as follows:

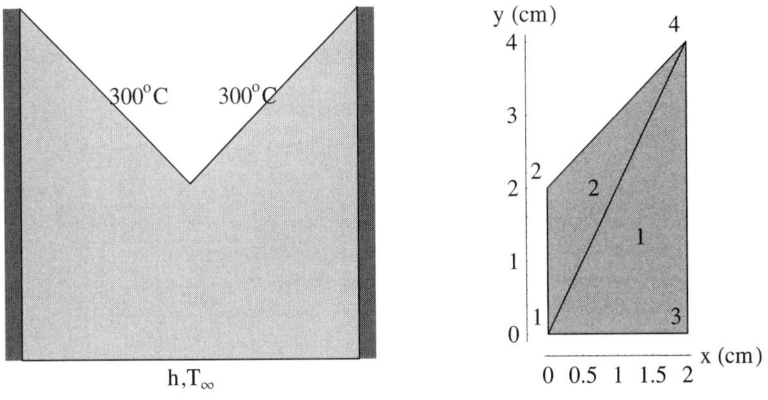

Figure 8.2. V-grooved strip and two-element model for half of the section

Global equations at the start of the assembly of element equations:

$$\begin{pmatrix} 0 & 0 & 0 & 0 \\ 0 & 0 & 0 & 0 \\ 0 & 0 & 0 & 0 \\ 0 & 0 & 0 & 0 \end{pmatrix}\begin{pmatrix} \dot{T}_1 \\ \dot{T}_2 \\ \dot{T}_3 \\ \dot{T}_4 \end{pmatrix} + \begin{pmatrix} 0 & 0 & 0 & 0 \\ 0 & 0 & 0 & 0 \\ 0 & 0 & 0 & 0 \\ 0 & 0 & 0 & 0 \end{pmatrix}\begin{pmatrix} T_1 \\ T_2 \\ T_3 \\ T_4 \end{pmatrix} = \begin{pmatrix} 0 \\ 0 \\ 0 \\ 0 \end{pmatrix}$$

Equations for element 1:

$$k_x = 3; \qquad k_y = 3; \qquad q = 0; \qquad m = \rho C_p = 1280000$$

$$\boldsymbol{C} = \begin{pmatrix} 3 & 0 \\ 0 & 3 \end{pmatrix}$$

Nodal Coordinates:

Element Node	Global Node Number	x	y
1	1	0	0
2	3	$\frac{1}{50}$	0
3	4	$\frac{1}{50}$	$\frac{1}{25}$

$$x_1 = 0; \qquad x_2 = \tfrac{1}{50}; \qquad x_3 = \tfrac{1}{50}$$

$$y_1 = 0; \qquad y_2 = 0; \qquad y_3 = \tfrac{1}{25}$$

Using these values, we get

$$b_1 = -\tfrac{1}{25}; \qquad b_2 = \tfrac{1}{25}; \qquad b_3 = 0$$

$$c_1 = 0; \qquad c_2 = -\tfrac{1}{50}; \qquad c_3 = \tfrac{1}{50}$$

$$f_1 = \tfrac{1}{1250}; \qquad f_2 = 0; \qquad f_3 = 0$$

$$\text{Element area } A = \frac{1}{2500}$$

$$\boldsymbol{B}^T = \begin{pmatrix} -\frac{1}{25} & \frac{1}{25} & 0 \\ 0 & -\frac{1}{50} & \frac{1}{50} \end{pmatrix}$$

$$\boldsymbol{m} = \begin{pmatrix} \frac{256}{3} & \frac{128}{3} & \frac{128}{3} \\ \frac{128}{3} & \frac{256}{3} & \frac{128}{3} \\ \frac{128}{3} & \frac{128}{3} & \frac{256}{3} \end{pmatrix}; \qquad \boldsymbol{k}_k = \begin{pmatrix} 3 & -3 & 0 \\ -3 & \frac{15}{4} & -\frac{3}{4} \\ 0 & -\frac{3}{4} & \frac{3}{4} \end{pmatrix}; \qquad \boldsymbol{r}_q = \begin{pmatrix} 0 \\ 0 \\ 0 \end{pmatrix}$$

NBC on side 1 (nodes {1, 3})—end nodal coordinates: $(\{0, 0\} \quad \{\frac{1}{50}, 0\})$, giving side length $L = \frac{1}{50}$.
Specified parameter values:

$$\alpha = -200; \qquad \beta = 10000$$

$$\boldsymbol{k}_\alpha = \begin{pmatrix} \frac{4}{3} & \frac{2}{3} & 0 \\ \frac{2}{3} & \frac{4}{3} & 0 \\ 0 & 0 & 0 \end{pmatrix}; \qquad \boldsymbol{r}_\beta = \begin{pmatrix} 100 \\ 100 \\ 0 \end{pmatrix}$$

Complete equations for element 1:

$$\begin{pmatrix} \frac{256}{3} & \frac{128}{3} & \frac{128}{3} \\ \frac{128}{3} & \frac{256}{3} & \frac{128}{3} \\ \frac{128}{3} & \frac{128}{3} & \frac{256}{3} \end{pmatrix} \begin{pmatrix} \dot{T}_1 \\ \dot{T}_2 \\ \dot{T}_3 \\ \dot{T}_4 \end{pmatrix} + \begin{pmatrix} \frac{13}{3} & -\frac{7}{3} & 0 \\ -\frac{7}{3} & \frac{61}{12} & -\frac{3}{4} \\ 0 & -\frac{3}{4} & \frac{3}{4} \end{pmatrix} \begin{pmatrix} T_1 \\ T_2 \\ T_3 \\ T_4 \end{pmatrix} = \begin{pmatrix} 100 \\ 100 \\ 0 \end{pmatrix}$$

The element contributes to {1, 3, 4} global degrees of freedom.

The global equations after assembly of this element are as follows:

$$\begin{pmatrix} \frac{256}{3} & 0 & \frac{128}{3} & \frac{128}{3} \\ 0 & 0 & 0 & 0 \\ \frac{128}{3} & 0 & \frac{256}{3} & \frac{128}{3} \\ \frac{128}{3} & 0 & \frac{128}{3} & \frac{256}{3} \end{pmatrix} \begin{pmatrix} \dot{T}_1 \\ \dot{T}_2 \\ \dot{T}_3 \\ \dot{T}_4 \end{pmatrix} + \begin{pmatrix} \frac{13}{3} & 0 & -\frac{7}{3} & 0 \\ 0 & 0 & 0 & 0 \\ -\frac{7}{3} & 0 & \frac{61}{12} & -\frac{3}{4} \\ 0 & 0 & -\frac{3}{4} & \frac{3}{4} \end{pmatrix} \begin{pmatrix} T_1 \\ T_2 \\ T_3 \\ T_4 \end{pmatrix} = \begin{pmatrix} 100 \\ 0 \\ 100 \\ 0 \end{pmatrix}$$

Equations for element 2:

$$k_x = 3; \qquad k_y = 3; \qquad q = 0; \qquad m = \rho C_p = 1280000$$

$$\boldsymbol{C} = \begin{pmatrix} 3 & 0 \\ 0 & 3 \end{pmatrix}$$

Nodal Coordinates:

Element Node	Global Node Number	x	y
1	1	0	0
2	4	$\frac{1}{50}$	$\frac{1}{25}$
3	2	0	$\frac{1}{50}$

$$x_1 = 0 \qquad x_2 = \tfrac{1}{50} \qquad x_3 = 0$$
$$y_1 = 0 \qquad y_2 = \tfrac{1}{25} \qquad y_3 = \tfrac{1}{50}$$

Using these values, we get

$$b_1 = \tfrac{1}{50}; \qquad b_2 = \tfrac{1}{50}; \qquad b_3 = -\tfrac{1}{25}$$
$$c_1 = -\tfrac{1}{50}; \qquad c_2 = 0; \qquad c_3 = \tfrac{1}{50}$$
$$f_1 = \tfrac{1}{2500}; \qquad f_2 = 0; \qquad f_3 = 0$$

$$\text{Element area } A = \frac{1}{5000}$$

$$\boldsymbol{B}^T = \begin{pmatrix} \frac{1}{50} & \frac{1}{50} & -\frac{1}{25} \\ -\frac{1}{50} & 0 & \frac{1}{50} \end{pmatrix}$$

$$\boldsymbol{m} = \begin{pmatrix} \frac{128}{3} & \frac{64}{3} & \frac{64}{3} \\ \frac{64}{3} & \frac{128}{3} & \frac{64}{3} \\ \frac{64}{3} & \frac{64}{3} & \frac{128}{3} \end{pmatrix}; \qquad \boldsymbol{k}_k = \begin{pmatrix} 3 & \frac{3}{2} & -\frac{9}{2} \\ \frac{3}{2} & \frac{3}{2} & -3 \\ -\frac{9}{2} & -3 & \frac{15}{2} \end{pmatrix}; \qquad \boldsymbol{r}_q = \begin{pmatrix} 0 \\ 0 \\ 0 \end{pmatrix}$$

Complete equations for element 2:

$$\begin{pmatrix} \frac{128}{3} & \frac{64}{3} & \frac{64}{3} \\ \frac{64}{3} & \frac{128}{3} & \frac{64}{3} \\ \frac{64}{3} & \frac{64}{3} & \frac{128}{3} \end{pmatrix} \begin{pmatrix} \dot{T}_1 \\ \dot{T}_2 \\ \dot{T}_3 \\ \dot{T}_4 \end{pmatrix} + \begin{pmatrix} 3 & \frac{3}{2} & -\frac{9}{2} \\ \frac{3}{2} & \frac{3}{2} & -3 \\ -\frac{9}{2} & -3 & \frac{15}{2} \end{pmatrix} \begin{pmatrix} T_1 \\ T_2 \\ T_3 \\ T_4 \end{pmatrix} = \begin{pmatrix} 0 \\ 0 \\ 0 \end{pmatrix}$$

The element contributes to {1, 4, 2} global degrees of freedom.

The global equations after assembly of this element are as follows:

$$\begin{pmatrix} 128 & \frac{64}{3} & \frac{128}{3} & 64 \\ \frac{64}{3} & \frac{128}{3} & 0 & \frac{64}{3} \\ \frac{128}{3} & 0 & \frac{256}{3} & \frac{128}{3} \\ 64 & \frac{64}{3} & \frac{128}{3} & 128 \end{pmatrix} \begin{pmatrix} \dot{T}_1 \\ \dot{T}_2 \\ \dot{T}_3 \\ \dot{T}_4 \end{pmatrix} + \begin{pmatrix} \frac{22}{3} & -\frac{9}{2} & -\frac{7}{3} & \frac{3}{2} \\ -\frac{9}{2} & \frac{15}{2} & 0 & -3 \\ -\frac{7}{3} & 0 & \frac{61}{12} & -\frac{3}{4} \\ \frac{3}{2} & -3 & -\frac{3}{4} & \frac{9}{4} \end{pmatrix} \begin{pmatrix} T_1 \\ T_2 \\ T_3 \\ T_4 \end{pmatrix} = \begin{pmatrix} 100 \\ 0 \\ 100 \\ 0 \end{pmatrix}$$

Essential boundary conditions:

Node	dof	Value
2	T_2	300
4	T_4	300

Delete equations {2, 4}.

$$\begin{pmatrix} 128 & \frac{64}{3} & \frac{128}{3} & 64 \\ \frac{128}{3} & 0 & \frac{256}{3} & \frac{128}{3} \end{pmatrix} \begin{pmatrix} \dot{T}_1 \\ 0 \\ \dot{T}_3 \\ 0 \end{pmatrix} + \begin{pmatrix} \frac{22}{3} & -\frac{9}{2} & -\frac{7}{3} & \frac{3}{2} \\ -\frac{7}{3} & 0 & \frac{61}{12} & -\frac{3}{4} \end{pmatrix} \begin{pmatrix} T_1 \\ 300 \\ T_3 \\ 300 \end{pmatrix} = \begin{pmatrix} 100 \\ 100 \end{pmatrix}$$

Extract columns {2, 4}, multiply by the corresponding known values, and move to the right-hand side.

The final global system of equations after adjusting for essential boundary conditions is as follows:

$$\begin{pmatrix} 128 & \frac{128}{3} \\ \frac{128}{3} & \frac{256}{3} \end{pmatrix} \begin{pmatrix} \dot{T}_1 \\ \dot{T}_3 \end{pmatrix} + \begin{pmatrix} \frac{22}{3} & -\frac{7}{3} \\ -\frac{7}{3} & \frac{61}{12} \end{pmatrix} \begin{pmatrix} T_1 \\ T_3 \end{pmatrix} = \begin{pmatrix} 1000 \\ 325 \end{pmatrix}$$

Solving the two ordinary differential equations using NDSolve, we get the following temperatures at the nodes. The results are computed for a time $t = 300$ s. The table lists values after every 30 s. The actual computations, however, are performed using a sophisticated automatic adaptive time integration scheme.

Transient Response Analysis

Initial values: $T \Longrightarrow$ for all nodes = 50

Time history of nodal solution:

Time	T_1	T_2	T_3	T_4
0	50.	300	50.	300
30	132.624	300	88.9721	300
60	160.102	300	120.496	300
90	172.66	300	135.335	300
120	178.479	300	142.218	300
150	181.176	300	145.409	300
180	182.427	300	146.889	300
210	183.007	300	147.575	300
240	183.276	300	147.893	300
270	183.4	300	148.04	300
300	183.458	300	148.108	300

The temperatures at nodes 1 and 3 are shown in Figure 8.3. After about 200 s the temperatures do not change much and thus a steady-state condition has been achieved.

For realistic results a finer mesh should be used. The problem is solved using ANSYS with element size of approximately 0.002 m (AnsysFiles\Chap8\VGrooveFine.inp on the book web site). The finite element mesh, showing the temperature distribution at $t = 300$ s, is as seen in Figure 8.4. The time history of temperatures at the bottom is shown in the time history plot.

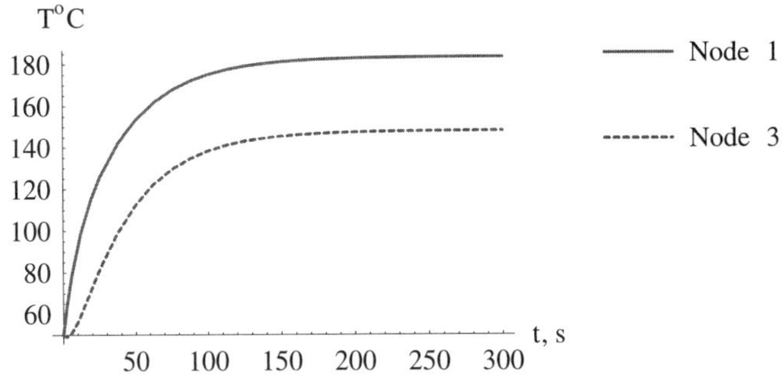

Figure 8.3. Temperature time history at nodes 1 and 3

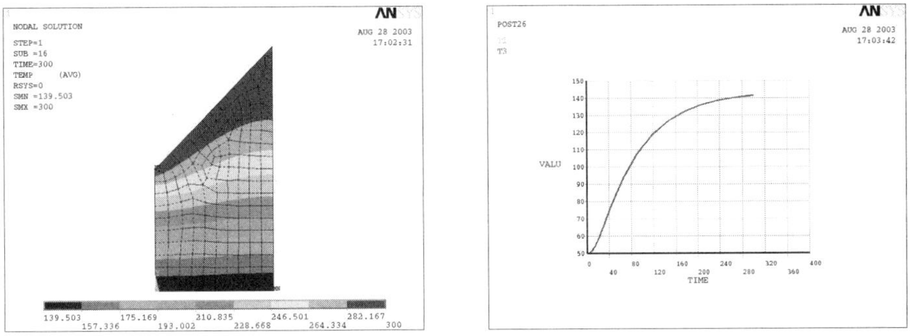

Figure 8.4. Solution using a finer mesh

◆ *Mathematica*/MATLAB Implementation 8.1 on the Book Web Site: *Triangular element for transient 2D BVP*

8.2 ELASTIC SOLIDS SUBJECTED TO DYNAMIC LOADS

The stress equilibrium equations presented in Chapter 7 considered a body in static equilibrium. If a body is in motion, then at any instant of time t Newton's law of motion implies that the sum of all forces must be equal to the inertia force. Denoting the mass density of the material by ρ and the accelerations in the coordinate directions by $\ddot{u} \equiv \partial^2 u/\partial t^2$, $\ddot{v}$, and $\ddot{w}$, the equations of motion can be written as follows:

$$\frac{\partial \sigma_x}{\partial x} + \frac{\partial \tau_{xy}}{\partial y} + \frac{\partial \tau_{xz}}{\partial z} + b_x = \rho \ddot{u}$$

$$\frac{\partial \tau_{yx}}{\partial x} + \frac{\partial \sigma_y}{\partial y} + \frac{\partial \tau_{yz}}{\partial z} + b_y = \rho \ddot{v}$$

$$\frac{\partial \tau_{zx}}{\partial x} + \frac{\partial \tau_{zy}}{\partial y} + \frac{\partial \sigma_z}{\partial z} + b_z = \rho \ddot{w}$$

The weak form corresponding to the these differential equations can be obtained following the same steps as those used in Chapter 7 for a static problem. Denoting the weighting functions by $\bar{u}$, $\bar{v}$, and $\bar{w}$, multiplying each equation by its weighting function, integrating over the volume, and adding all three terms, the total weighted residual is as follows:

$$\iiint_V \left(\frac{\partial \sigma_x}{\partial x} + \frac{\partial \tau_{xy}}{\partial y} + \frac{\partial \tau_{xz}}{\partial z} + b_x\right)\bar{u} + \left(\frac{\partial \tau_{yx}}{\partial x} + \frac{\partial \sigma_y}{\partial y} + \frac{\partial \tau_{yz}}{\partial z} + b_y\right)\bar{v}$$

$$+ \left(\frac{\partial \tau_{zx}}{\partial x} + \frac{\partial \tau_{zy}}{\partial y} + \frac{\partial \sigma_z}{\partial z} + b_z\right)\bar{w}\, dV - \iiint_V (\rho\ddot{u}\bar{u} + \rho\ddot{v}\bar{v} + \rho\ddot{w}\bar{w})\, dV = 0$$

Using the Green-Gauss theorem on each of the stress derivative terms, we get

$$\iint_S (\sigma_x n_x + \tau_{xy} n_y + \tau_{xz} n_z)\bar{u} + (\tau_{yx} n_x + \sigma_x n_y + \tau_{yz} n_z)\bar{v} + (\tau_{zx} n_x + \tau_{zy} n_y + \sigma_z n_z)\bar{w}\, dS$$

$$- \iiint_V \left(\sigma_x \frac{\partial \bar{u}}{\partial x} + \tau_{xy}\frac{\partial \bar{u}}{\partial y} + \tau_{xz}\frac{\partial \bar{u}}{\partial z}\right) + \left(\tau_{yx}\frac{\partial \bar{v}}{\partial x} + \sigma_y \frac{\partial \bar{v}}{\partial y} + \tau_{yz}\frac{\partial \bar{v}}{\partial z}\right)$$

$$+ \left(\tau_{zx}\frac{\partial \bar{w}}{\partial x} + \tau_{zy}\frac{\partial \bar{w}}{\partial y} + \sigma_z \frac{\partial \bar{w}}{\partial z}\right) dV$$

$$+ \iiint_V (b_x\bar{u} + b_y\bar{v} + b_z\bar{w})\, dV - \iiint_V (\rho\ddot{u}\bar{u} + \rho\ddot{v}\bar{v} + \rho\ddot{w}\bar{w})\, dV = 0$$

On the surface the applied forces are $q_x = \sigma_x n_x + \tau_{xy} n_y + \tau_{xz} n_z$, etc. Substituting these and rearranging terms, we get the following weak form:

$$\iiint_V \sigma_x \frac{\partial \bar{u}}{\partial x} + \sigma_y \frac{\partial \bar{v}}{\partial y} + \sigma_z \frac{\partial \bar{w}}{\partial z} + \tau_{xy}\left(\frac{\partial \bar{u}}{\partial y} + \frac{\partial \bar{v}}{\partial x}\right)$$

$$+ \tau_{yz}\left(\frac{\partial \bar{v}}{\partial z} + \frac{\partial \bar{w}}{\partial y}\right) + \tau_{xz}\left(\frac{\partial \bar{u}}{\partial z} + \frac{\partial \bar{w}}{\partial x}\right) dV + \iiint_V (\rho\ddot{u}\bar{u} + \rho\ddot{v}\bar{v} + \rho\ddot{w}\bar{w})\, dV$$

$$= \iint_S (q_x\bar{u} + q_y\bar{v} + q_z\bar{w})\, dS + \iiint_V (b_x\bar{u} + b_y\bar{v} + b_z\bar{w})\, dV$$

If we interpret the weighting functions as virtual displacements, then their derivatives are virtual strains,

$$\frac{\partial \bar{u}}{\partial x} = \bar{\epsilon}_x; \qquad \frac{\partial \bar{v}}{\partial y} = \bar{\epsilon}_y; \qquad \frac{\partial \bar{w}}{\partial z} = \bar{\epsilon}_z$$

$$\frac{\partial \bar{u}}{\partial y} + \frac{\partial \bar{v}}{\partial x} = \bar{\gamma}_{xy}; \qquad \frac{\partial \bar{v}}{\partial z} + \frac{\partial \bar{w}}{\partial y} = \bar{\gamma}_{yz}; \qquad \frac{\partial \bar{u}}{\partial z} + \frac{\partial \bar{w}}{\partial x} = \bar{\gamma}_{zx}$$

Substituting these into the weak form, we have

$$\iiint_V (\sigma_x \bar{\epsilon}_x + \sigma_y \bar{\epsilon}_y + \sigma_z \bar{\epsilon}_z + \tau_{xy}\bar{\gamma}_{xy} + \tau_{yz}\bar{\gamma}_{yz} + \tau_{xz}\bar{\gamma}_{zx})\, dV + \iiint_V (\rho\ddot{u}\bar{u} + \rho\ddot{v}\bar{v} + \rho\ddot{w}\bar{w})dV$$

$$= \iint_S (q_x\bar{u} + q_y\bar{v} + q_z\bar{w})\, dS + \iiint_V (b_x\bar{u} + b_y\bar{v} + b_z\bar{w})\, dV$$

Comparing this weak form to the one derived in Chapter 7 for static problems, we see that the inertia forces simply add one more body force term in the equations. By defining several vectors, the weak form can be written in a compact matrix notation as follows:

$$\iiint_V \rho\bar{\boldsymbol{u}}^T\ddot{\boldsymbol{u}},\, dV + \iiint_V \bar{\boldsymbol{\epsilon}}^T\boldsymbol{\sigma}\, dV = \iint_S \bar{u}^T q\, dS + \iiint_V \bar{\boldsymbol{u}}^T\boldsymbol{b}\, dV$$

where

$$\boldsymbol{\sigma} = (\sigma_x \quad \sigma_y \quad \sigma_z \quad \tau_{xy} \quad \tau_{yz} \quad \tau_{xz})^T$$

$$\bar{\boldsymbol{\epsilon}} = (\bar{\epsilon}_x \quad \bar{\epsilon}_y \quad \bar{\epsilon}_z \quad \bar{\gamma}_{xy} \quad \bar{\gamma}_{yz} \quad \bar{\gamma}_{xz})^T$$

$$\boldsymbol{q} = (q_x \quad q_y \quad q_z)^T; \qquad \boldsymbol{b} = (b_x \quad b_y \quad b_z)^T$$

$$\boldsymbol{u} = (u \quad v \quad w)^T; \qquad \ddot{\boldsymbol{u}} = (\ddot{u} \quad \ddot{v} \quad \ddot{w})^T; \qquad \bar{\boldsymbol{u}} = (\bar{u} \quad \bar{v} \quad \bar{w})^T$$

8.2.1 Finite Element Equations

The assumed solutions for displacements are as follows:

$$\boldsymbol{u} \equiv \begin{pmatrix} u \\ v \\ w \end{pmatrix} = \begin{pmatrix} N_1 & 0 & 0 & N_2 & 0 & \cdots \\ 0 & N_1 & 0 & 0 & N_2 & \cdots \\ 0 & 0 & N_1 & 0 & 0 & \cdots \end{pmatrix} \begin{pmatrix} u_1 \\ v_1 \\ w_1 \\ u_2 \\ \vdots \end{pmatrix} \equiv \boldsymbol{N}^T\boldsymbol{d}$$

where $u_1, v_1, w_1, u_2, \ldots$ are the nodal degrees of freedom. Here, $N_i(x, y, z)$ are suitable interpolation functions. The accelerations are written by differentiating the displacements. The interpolation functions do not change with time. The nodal displacements are the only quantities that are functions of time. Thus the accelerations are expressed as follows:

$$\ddot{u} \equiv \begin{pmatrix} \ddot{u} \\ \ddot{v} \\ \ddot{w} \end{pmatrix} = \begin{pmatrix} N_1 & 0 & 0 & N_2 & 0 & \cdots \\ 0 & N_1 & 0 & 0 & N_2 & \cdots \\ 0 & 0 & N_1 & 0 & 0 & \cdots \end{pmatrix} \begin{pmatrix} \ddot{u}_1 \\ \ddot{v}_1 \\ \ddot{w}_1 \\ \ddot{u}_2 \\ \vdots \end{pmatrix} \equiv \boldsymbol{N}^T\ddot{\boldsymbol{d}}$$

From the assumed solution the element strain vector can be computed by appropriate differentiation as follows:

$$\epsilon = \begin{pmatrix} \epsilon_x \\ \epsilon_y \\ \epsilon_z \\ \gamma_{xy} \\ \gamma_{yz} \\ \gamma_{zx} \end{pmatrix} = \begin{pmatrix} \frac{\partial u}{\partial x} \\ \frac{\partial v}{\partial y} \\ \frac{\partial w}{\partial z} \\ \frac{\partial u}{\partial y} + \frac{\partial v}{\partial x} \\ \frac{\partial v}{\partial z} + \frac{\partial w}{\partial y} \\ \frac{\partial u}{\partial z} + \frac{\partial w}{\partial x} \end{pmatrix} = \begin{pmatrix} \frac{\partial N_1}{\partial x} & 0 & 0 & \frac{\partial N_2}{\partial x} & 0 & 0 & \cdots \\ 0 & \frac{\partial N_1}{\partial y} & 0 & 0 & \frac{\partial N_2}{\partial y} & 0 & \cdots \\ 0 & 0 & \frac{\partial N_1}{\partial z} & 0 & 0 & \frac{\partial N_2}{\partial z} & \cdots \\ \frac{\partial N_1}{\partial y} & \frac{\partial N_1}{\partial x} & 0 & \frac{\partial N_2}{\partial y} & \frac{\partial N_2}{\partial x} & 0 & \cdots \\ 0 & \frac{\partial N_1}{\partial z} & \frac{\partial N_1}{\partial y} & 0 & \frac{\partial N_2}{\partial z} & \frac{\partial N_2}{\partial y} & \cdots \\ \frac{\partial N_1}{\partial z} & 0 & \frac{\partial N_1}{\partial x} & \frac{\partial N_2}{\partial z} & 0 & \frac{\partial N_2}{\partial x} & \cdots \end{pmatrix} \begin{pmatrix} u_1 \\ v_1 \\ w_1 \\ u_2 \\ \vdots \end{pmatrix} \equiv \boldsymbol{B}^T \boldsymbol{d}$$

Using the constitutive matrix appropriate for the material, the element stress vector can be written as follows:

$$\boldsymbol{\sigma} = \boldsymbol{C}(\boldsymbol{\epsilon} - \boldsymbol{\epsilon}_0) = \boldsymbol{C}\boldsymbol{B}^T\boldsymbol{d} - \boldsymbol{C}\boldsymbol{\epsilon}_0$$

where ϵ_0 is a vector of initial strains. The weighting functions in the weak form are derivatives of the assumed solution with respect to the nodal degrees of freedom. Thus

$$\bar{u} = \frac{\partial \boldsymbol{u}}{\partial \boldsymbol{d}} \equiv \boldsymbol{N}^T \qquad \text{and} \qquad \bar{\epsilon} = \boldsymbol{B}^T$$

Substituting these into the weak form, we have

$$\iiint_V \rho \boldsymbol{N}\boldsymbol{N}^T \ddot{\boldsymbol{d}}\, dV + \iiint_V \boldsymbol{B}(\boldsymbol{C}\boldsymbol{B}^T\boldsymbol{d} - \boldsymbol{C}\boldsymbol{\epsilon}_0)\, dV = \iint_S \boldsymbol{N}\boldsymbol{q}\, dS + \iint_V \boldsymbol{N}\boldsymbol{b}\, dV$$

or

$$\iiint_V \rho \boldsymbol{N}\boldsymbol{N}^T\, dV\, \ddot{\boldsymbol{d}} + \iiint_V \boldsymbol{B}\boldsymbol{C}\boldsymbol{B}^T\, dV\, \boldsymbol{d} = \iiint_V \boldsymbol{B}\boldsymbol{C}\boldsymbol{\epsilon}_0\, dV + \iint_S \boldsymbol{N}\boldsymbol{q}\, dS + \iiint_V \boldsymbol{N}\boldsymbol{b}\, dV$$

Thus we get the following element equations:

$$\boldsymbol{m}\ddot{\boldsymbol{d}} + \boldsymbol{k}\boldsymbol{d} = \boldsymbol{r}_\epsilon + \boldsymbol{r}_q + \boldsymbol{r}_b$$

where $\boldsymbol{m}$ is known as an element mass matrix

$$\boldsymbol{m} = \iiint_V \rho \boldsymbol{N}\boldsymbol{N}^T\, dV$$

and all other terms are exactly the same as those for the static case. That is, $\boldsymbol{k}$ is the element stiffness matrix, $\boldsymbol{r}_\epsilon$ is the equivalent nodal load vector due to initial strains, $\boldsymbol{r}_q$ is the equivalent nodal load vector due to surface forces, and $\boldsymbol{r}_b$ is the equivalent nodal load vector due to body forces:

$$\boldsymbol{k} = \iiint_V \boldsymbol{B}\boldsymbol{C}\boldsymbol{B}^T\, dV; \qquad \boldsymbol{r}_\epsilon = \iiint_V \boldsymbol{B}\boldsymbol{C}\epsilon_0\, dV$$

$$\boldsymbol{r}_q = \iint_S \boldsymbol{N}\boldsymbol{q}\, dS; \qquad \boldsymbol{r}_b = \iiint_V \boldsymbol{N}\boldsymbol{b}\, dV$$

The mass matrix term is the only new term in the element equations for dynamic problems. Explicit mass matrices for commonly used structural elements are derived in the following section. During assembly, the element mass matrices are assembled to form a global mass matrix in exactly the same manner as the stiffness matrix. The final equations after assembly are expressed as follows:

$$\boldsymbol{M}\ddot{\boldsymbol{d}} + \boldsymbol{K}\boldsymbol{d} = \boldsymbol{R}$$

This is a system of second-order ordinary differential equations. Any method for solving ordinary differential equations can be used to solve these equations. Newmark's method, generally discussed in courses on vibrations and structural dynamics, is one of the most popular methods for solving for a large system of second-order ordinary differential equations. For details of this method refer to any textbook on structural dynamics.

8.2.2 Mass Matrices for Common Structural Elements

The element mass matrix for a general three-dimensional elastic solid is given by the following integral:

$$\boldsymbol{m} = \iiint_V \rho \boldsymbol{N}\boldsymbol{N}^T\, dV$$

Explicit expressions for commonly used structural elements are derived in this section.

Axial Deformations A simple two-node axial deformation element, shown in Figure 8.5, was developed in Chapter 2. The element extends from x_1 to x_2 and has a length $L = x_2 - x_1$. For simplicity the element is assumed to have a uniform area of cross section A.

The element is based on the following interpolation functions:

$$N_1 = \frac{x - x_2}{x_1 - x_2} = -\frac{x - x_2}{L}; \qquad N_2 = \frac{x - x_1}{x_2 - x_1} = \frac{x - x_1}{L}$$

Using these interpolation functions, the mass matrix can be written as follows:

$$\boldsymbol{m} = \iiint_V \rho \boldsymbol{N}\boldsymbol{N}^T\, dV = \rho A \int_{x_1}^{x_2} \begin{pmatrix} -\frac{x-x_2}{L} \\ \frac{x-x_1}{L} \end{pmatrix} \begin{pmatrix} -\frac{x - x_2}{L} & \frac{x - x_1}{L} \end{pmatrix} dx$$

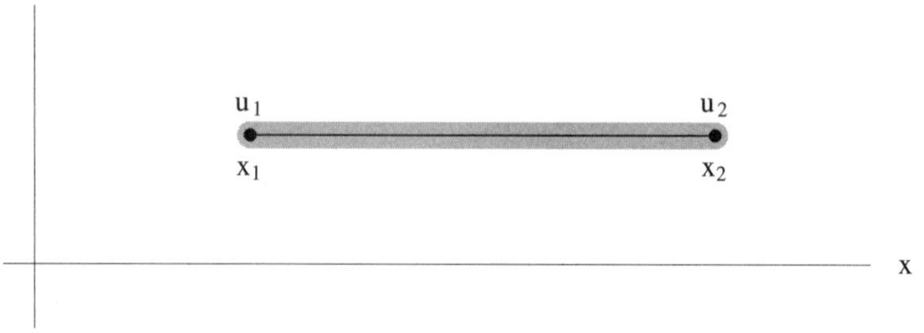

Figure 8.5. Axial deformation element

Carrying out integrations, we get the element mass matrix as follows:

$$\boldsymbol{m} = \frac{\rho AL}{6}\begin{pmatrix} 2 & 1 \\ 1 & 2 \end{pmatrix}$$

Plane Truss Element The two-node plane truss element shown in Figure 8.6 was developed in Chapter 4 by transforming an axial deformation element. The transformation works for the stiffness matrix since the strain energy expression for both the truss and the axial deformation elements is based on deformations along the axis of the element. For the mass matrix we must consider displacements both in the x and the y directions, since motion along both axes generates inertia forces. Therefore the mass matrix for a truss element is derived by writing the linear interpolation functions for x and y displacements.

Assuming a coordinate s along the axis of the element with $s = 0$ at node 1 and $s = L$, at node 2 the interpolation functions are as follows:

$$N_1 = -\frac{s-L}{L}; \qquad N_2 = \frac{s}{L}$$

$$\begin{pmatrix} u \\ v \end{pmatrix} = \begin{pmatrix} N_1 & 0 & N_2 & 0 \\ 0 & N_1 & 0 & N_2 \end{pmatrix}\begin{pmatrix} u_1 \\ v_1 \\ u_2 \\ v_2 \end{pmatrix} \equiv \boldsymbol{N}^T\boldsymbol{d}$$

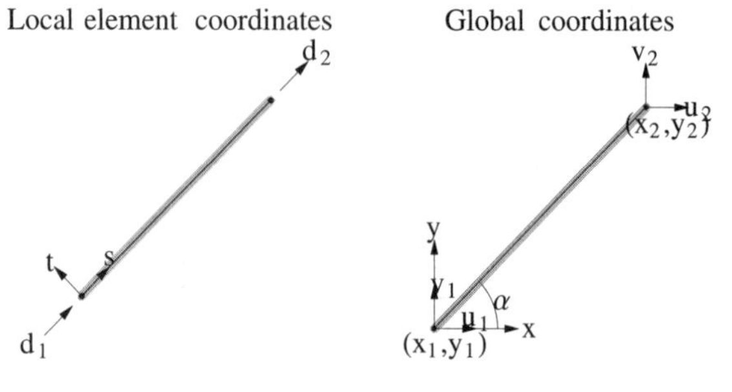

Figure 8.6. Plane truss element

Using these interpolation functions, the mass matrix can be written as follows:

$$\boldsymbol{m} = \iiint_V \rho N N^T \, dV = \rho A \int_0^L \begin{pmatrix} -\frac{s-L}{L} & 0 \\ 0 & -\frac{s-L}{L} \\ \frac{s}{L} & 0 \\ 0 & \frac{s}{L} \end{pmatrix} \begin{pmatrix} -\frac{s-L}{L} & 0 & \frac{s}{L} & 0 \\ 0 & -\frac{s-L}{L} & 0 & \frac{s}{L} \end{pmatrix} ds$$

Carrying out matrix multiplication and integrating each term, we get

$$\boldsymbol{m} = \frac{\rho A L}{6} \begin{pmatrix} 2 & 0 & 1 & 0 \\ 0 & 2 & 0 & 1 \\ 1 & 0 & 2 & 0 \\ 0 & 1 & 0 & 2 \end{pmatrix}$$

Space Truss Element The mass matrix for a three-dimensional space truss element, shown in Figure 8.7, can be written using exactly the same arguments as those for the plane truss.

The interpolation functions for x, y, and z displacements in terms of a coordinate s along the axis of the element with $s = 0$ at node 1 and $s = L$ at node 2 are as follows:

$$N_1 = -\frac{s-L}{L}; \qquad N_2 = \frac{s}{L}$$

$$\begin{pmatrix} u \\ v \\ w \end{pmatrix} = \begin{pmatrix} N_1 & 0 & 0 & N_2 & 0 & 0 \\ 0 & N_1 & 0 & 0 & N_2 & 0 \\ 0 & 0 & N_1 & 0 & 0 & N_2 \end{pmatrix} \begin{pmatrix} u_1 \\ v_1 \\ w_1 \\ u_2 \\ v_2 \\ w_2 \end{pmatrix} \equiv \boldsymbol{N}^T \boldsymbol{d}$$

Using these interpolation functions, the mass matrix can be written as follows:

$$\boldsymbol{m} = \iiint_V \rho N N^T dV = \rho A \int_0^L N N^T \, ds = \frac{\rho A L}{6} \begin{pmatrix} 2 & 0 & 0 & 1 & 0 & 0 \\ 0 & 2 & 0 & 0 & 1 & 0 \\ 0 & 0 & 2 & 0 & 0 & 1 \\ 1 & 0 & 0 & 2 & 0 & 0 \\ 0 & 1 & 0 & 0 & 2 & 0 \\ 0 & 0 & 1 & 0 & 0 & 2 \end{pmatrix}$$

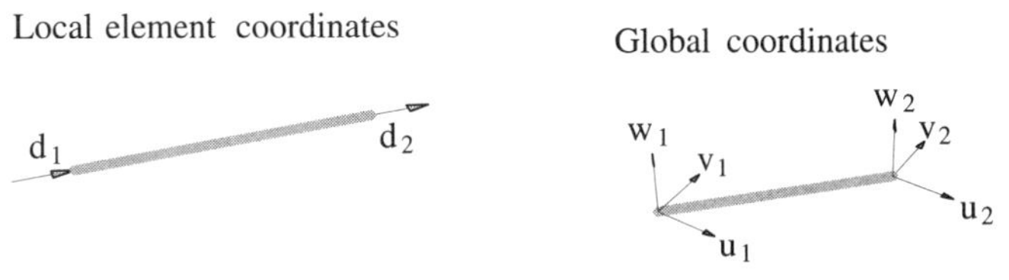

Figure 8.7. Space truss element

Beam Element The two-node beam element shown in Figure 8.8 was developed in Chapter 4. The interpolation functions are as follows:

$$v(s) = \begin{pmatrix} \frac{2s^3}{L^3} - \frac{3s^2}{L^2} + 1 & \frac{s^3}{L^2} - \frac{2s^2}{L} + s & \frac{3s^2}{L^2} - \frac{2s^3}{L^3} & \frac{s^3}{L^2} - \frac{s^2}{L} \end{pmatrix} \begin{pmatrix} v_1 \\ \theta_1 \\ v_2 \\ \theta_2 \end{pmatrix} \equiv \boldsymbol{N}^T \boldsymbol{d}$$

Assuming a uniform cross-sectional area A and using these interpolation functions, the mass matrix can be written as follows:

$$\boldsymbol{m} = \iiint_V \rho \boldsymbol{N}\boldsymbol{N}^T \, dV$$

$$= \rho A \int_0^L \begin{pmatrix} \frac{2s^3}{L^3} - \frac{3s^2}{L^2} + 1 \\ \frac{s^3}{L^2} - \frac{2s^2}{L} + s \\ \frac{3s^2}{L^2} - \frac{2s^3}{L^3} \\ \frac{s^3}{L^2} - \frac{s^2}{L} \end{pmatrix} \begin{pmatrix} \frac{2s^3}{L^3} - \frac{3s^2}{L^2} + 1 & \frac{s^3}{L^2} - \frac{2s^2}{L} + s & \frac{3s^2}{L^2} - \frac{2s^3}{L^3} & \frac{s^3}{L^2} - \frac{s^2}{L} \end{pmatrix} ds$$

Carrying out matrix multiplication and integrating each term, we get

$$\boldsymbol{m} = \frac{\rho A L}{420} \begin{pmatrix} 156 & 22L & 54 & -13L \\ 22L & 4L^2 & 13L & -3L^2 \\ 54 & 13L & 156 & -22L \\ -13L & -3L^2 & -22L & 4L^2 \end{pmatrix}$$

Plane Frame Element In terms of local coordinates, the plane frame element shown in Figure 8.9 is a combination of beam and axial deformation elements. Thus the interpolation

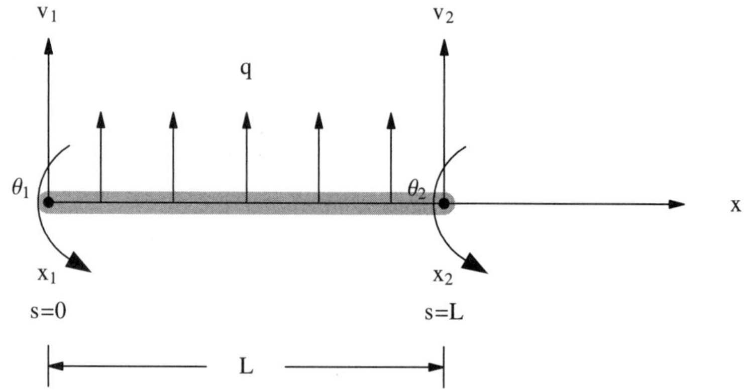

Figure 8.8. Beam element

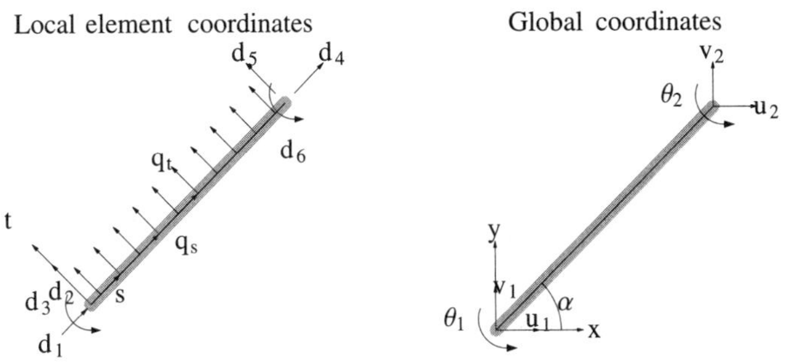

Figure 8.9. Plane frame element

functions are as follows:

$$\begin{pmatrix} u \\ v \end{pmatrix} = \begin{pmatrix} -\frac{s-L}{L} & 0 & 0 & \frac{s}{L} & 0 & 0 \\ 0 & \frac{2s^3}{L^3} - \frac{3s^2}{L^2} + 1 & \frac{s^3}{L^2} - \frac{2s^2}{L} + s & 0 & \frac{3s^2}{L^2} - \frac{2s^3}{L^3} & \frac{s^3}{L^2} - \frac{s^2}{L} \end{pmatrix} \begin{pmatrix} d_1 \\ d_2 \\ d_3 \\ d_4 \\ d_5 \\ d_6 \end{pmatrix} \equiv \boldsymbol{N}^T \boldsymbol{d}$$

Assuming uniform cross-sectional area A and using these interpolation functions, the mass matrix can be written as follows:

$$\boldsymbol{m} = \iiint_V \rho \boldsymbol{N}\boldsymbol{N}^T \, dV = \rho A \int_0^L \boldsymbol{N}\boldsymbol{N}^T \, ds$$

Carrying out matrix multiplication and integrating each term, we get the mass matrix in local coordinates as follows:

$$\boldsymbol{m}_l = \frac{\rho A L}{420} \begin{pmatrix} 140 & 0 & 0 & 70 & 0 & 0 \\ 0 & 156 & 22L & 0 & 54 & -13L \\ 0 & 22L & 4L^2 & 0 & 13L & -3L^2 \\ 70 & 0 & 0 & 140 & 0 & 0 \\ 0 & 54 & 13L & 0 & 156 & -22L \\ 0 & -13L & -3L^2 & 0 & -22L & 4L^2 \end{pmatrix}$$

The local-to-global transformation is given as

$$\begin{pmatrix} d_1 \\ d_2 \\ d_3 \\ d_4 \\ d_5 \\ d_6 \end{pmatrix} = \begin{pmatrix} l_s & m_s & 0 & 0 & 0 & 0 \\ -m_s & l_s & 0 & 0 & 0 & 0 \\ 0 & 0 & 1 & 0 & 0 & 0 \\ 0 & 0 & 0 & l_s & m_s & 0 \\ 0 & 0 & 0 & -m_s & l_s & 0 \\ 0 & 0 & 0 & 0 & 0 & 1 \end{pmatrix} \begin{pmatrix} u_1 \\ v_1 \\ \theta_1 \\ u_2 \\ v_2 \\ \theta_2 \end{pmatrix} \Rightarrow \boldsymbol{d}_l = \boldsymbol{T}\boldsymbol{d}$$

$$l_s = \cos\alpha = \frac{x_2 - x_1}{L}; \qquad m_s = \sin\alpha = \frac{y_2 - y_1}{L}$$

Using the transformation matrix $\boldsymbol{T}$, the mass matrix in the global coordinate system is obtained as follows:

$$\boldsymbol{m} = \boldsymbol{T}^T \boldsymbol{m}_l \boldsymbol{T}$$

Triangular Element for Plane Stress/Strain The three-node triangular element shown in Figure 8.10 was developed for plane stress and plane strain problems in Chapter 7. The interpolation functions are as follows:

$$\begin{pmatrix} u \\ v \end{pmatrix} = \begin{pmatrix} N_1 & 0 & N_2 & 0 & N_3 & 0 \\ 0 & N_1 & 0 & N_2 & 0 & N_3 \end{pmatrix} \begin{pmatrix} u_1 \\ v_1 \\ u_2 \\ v_2 \\ u_3 \\ v_3 \end{pmatrix} \equiv \boldsymbol{N}^T \boldsymbol{d}$$

$$N_1 = \frac{1}{2A}(xb_1 + yc_1 + f_1); \quad N_2 = \frac{1}{2A}(xb_2 + yc_2 + f_2); \quad N_3 = \frac{1}{2A}(xb_3 + yc_3 + f_3)$$

$$f_1 = x_2y_3 - x_3y_2; \quad f_2 = x_3y_1 - x_1y_3; \quad f_3 = x_1y_2 - x_2y_1$$

$$b_1 = y_2 - y_3; \quad b_2 = y_3 - y_1; \quad b_3 = y_1 - y_2$$

$$c_1 = x_3 - x_2; \quad c_2 = x_1 - x_3; \quad c_3 = x_2 - x_1$$

A = area of triangle = $\frac{1}{2}(f_1 + f_2 + f_3)$

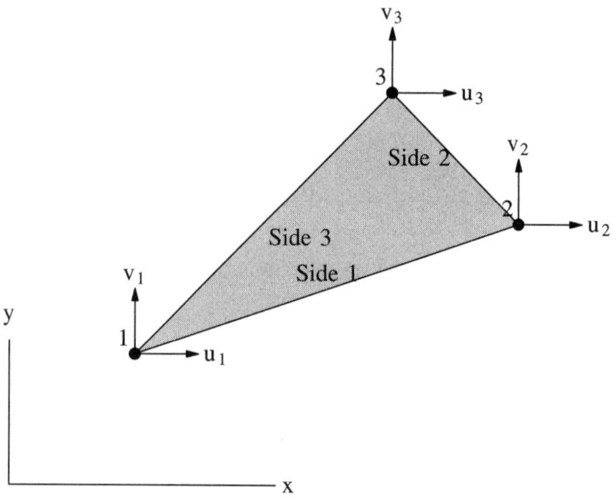

Figure 8.10. Three-node triangular element

With constant thickness h and using these interpolation functions, the mass matrix can be written as follows:

$$\boldsymbol{m} = \iiint_V \rho \boldsymbol{N}\boldsymbol{N}^T \, dV = \rho h \iint_A \boldsymbol{N}\boldsymbol{N}^T \, dA$$

Carrying out matrix multiplication and integrating each term over the triangle as explained in Chapter 5, we get

$$\boldsymbol{m} = \frac{\rho h A}{12} \begin{pmatrix} 2 & 0 & 1 & 0 & 1 & 0 \\ 0 & 2 & 0 & 1 & 0 & 1 \\ 1 & 0 & 2 & 0 & 1 & 0 \\ 0 & 1 & 0 & 2 & 0 & 1 \\ 1 & 0 & 1 & 0 & 2 & 0 \\ 0 & 1 & 0 & 1 & 0 & 2 \end{pmatrix}$$

8.2.3 Free-Vibration Analysis

Most structural dynamics and vibration problems typically start with the analysis of the free-vibration motion of the system. There is no externally applied load in this situation, and thus the global system of equations for the system is of the following form:

$$\boldsymbol{M}\ddot{\boldsymbol{d}} + \boldsymbol{K}\boldsymbol{d} = 0$$

where the global mass matrix $\boldsymbol{M}$ and the global stiffness matrix $\boldsymbol{K}$ are assembled from the corresponding element matrices using the usual assembly procedure. This system of equations is satisfied by a harmonic solution of the following form:

$$\boldsymbol{d} = \boldsymbol{\phi} \cos \omega t$$

where $\boldsymbol{\phi}$ and ω are parameters that will be later identified as the mode shape and the natural frequency. Substituting this into the differential equation, we have

$$-\boldsymbol{M}\boldsymbol{\phi}\omega^2 \cos \omega t + \boldsymbol{K}\boldsymbol{\phi} \cos \omega t = 0$$

Rearranging terms, we have what is known as the generalized eigenvalue problem

$$\boldsymbol{K}\boldsymbol{\phi} = \lambda \boldsymbol{M}\boldsymbol{\phi}$$

where $\lambda = \omega^2$ is introduced for convenience. From the solution of this system we get n natural frequencies $\omega_i = \sqrt{\lambda_i}$ and corresponding free-vibration mode shapes $\boldsymbol{\phi}_i$. Several methods are available for computing these frequencies and modes shapes and are discussed in books on vibrations and structural dynamics. Both MATLAB and *Mathematica* have built-in functions to solve these problems. In MATLAB the eigenvalues and corresponding eigenvectors are obtained by using the eig command as follows:

```
[V, Lam] = eig(K, M)
```

In the result Lam is a diagonal matrix of eigenvalues and V is a matrix whose columns are the eigenvectors (mode shapes). Through *Mathematica* version 4.2 the Eigensystem function can solve only the standard eigenvalue problem in which the $\boldsymbol{M}$ matrix is an identity matrix. When $\boldsymbol{M}$ is nonsingular, the generalized eigenvalue problem can easily be converted to a standard eigenvalue problem by multiplying both sides by $\boldsymbol{M}^{-1}$ as follows:

$$\boldsymbol{M}^{-1}\boldsymbol{K}\boldsymbol{\phi} = \lambda\boldsymbol{M}^{-1}\boldsymbol{M}\boldsymbol{\phi} \Longrightarrow \boldsymbol{M}^{-1}\boldsymbol{K}\boldsymbol{\phi} = \lambda\boldsymbol{\phi}$$

The function Eigensystem is then used to get the eigenvalues and the mode shapes:

```
{lam, V} = Eigensystem[Inverse[M].K]
```

In the result lam is a vector of eigenvalues and V is a matrix whose rows are the eigenvectors (mode shapes). The function returns the eigenvalues ordered in the descending order. In most structural dynamics applications one is interested in lower frequencies and thus preferred order is the ascending order. As demonstrated in the following example, using the Reverse function, the order of the results can easily be changed.

Example 8.2 Modal Analysis of a Plane Truss (MATLAB) The following TransientPlaneTrussElement function returns the mass and the stiffness matrix of a plane truss element. To generate the mass matrix the function needs the mass density (ρ).

MatlabFiles\Chap8\TransientPlaneTrussElement.m on the book web site

```
function [m, k] = TransientPlaneTrussElement(e, A,
                          rho, coord)
% [m, k] = TransientPlaneTrussElement(e, A, rho, coord)
% Generates mass & stiffness matrices for a plane truss
% element
% rho = mass density
% e = modulus of elasticity
% A = area of cross-section
% coord = coordinates at the element ends

x1=coord(1,1); y1=coord(1,2);
x2=coord(2,1); y2=coord(2,2);
L=sqrt((x2-x1)^2+(y2-y1)^2);
ls=(x2-x1)/L; ms=(y2-y1)/L;
k = e*A/L*[ls^2, ls*ms,-ls^2,-ls*ms;
     ls*ms, ms^2,-ls*ms,-ms^2;
     -ls^2,-ls*ms,ls^2,ls*ms;
     -ls*ms,-ms^2,ls*ms,ms^2];
m = ((rho*A*L)/6)*[2, 0, 1, 0; 0, 2, 0, 1;
     1, 0, 2, 0;0, 1, 0, 2];
```

Using this function, we consider the free-vibration analysis of the plane truss structure shown in Figure 8.11. All members have the same cross-sectional area and are made of the

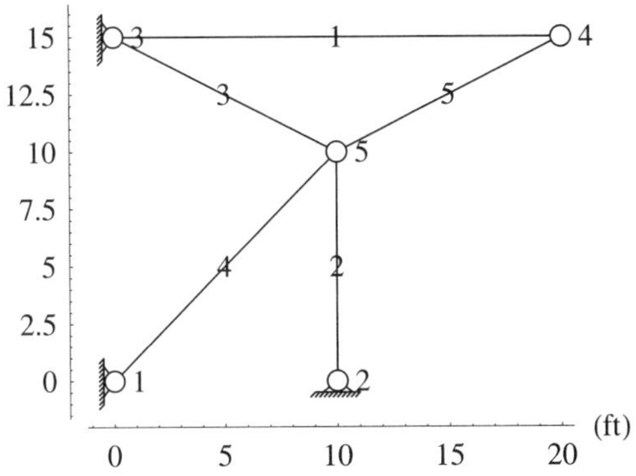

Figure 8.11. A plane truss

same material, $\rho = 490\,\text{lb/ft}^3$, $E = 30 \times 10^6\,\text{lb/in}^2$, and $A = 1.25\,\text{in}^2$. The truss supports a rotating machine weighing 2000 lb at its tip.

The weight of the machine is an added mass at node 4. This is defined as ma in the following code and is added directly to the global mass matrix corresponding to degrees of freedom 7 and 8 associated with node 4. The procedure for generating the global mass and stiffness matrix is exactly the same as that for the static analysis used in the previous chapters. Using in-lb units, the calculations are as follows:

MatlabFiles\Chap8\ModalTrussEx.m on the book web site

```
% Modal analysis of a plane truss
g = 386.4; e = 30*10^6; A = 1.25;
rho = (490/(12^3))/g; ma = 2000/g;
nodes = 12 * [0, 0; 10, 0; 0, 15; 20, 15; 10, 10];
conn = [3, 4; 2, 5; 5, 3; 1, 5; 5, 4];
elems = size(conn,1);
lmm=[];
for i=1:elems
    lmm = [lmm; [2*conn(i,1)-1, 2*conn(i,1),
                 2*conn(i,2)-1, 2*conn(i,2)]];
end
debc = [1:6]; ebcVals=zeros(length(debc),1);
dof=2*size(nodes,1);
M=zeros(dof); K=zeros(dof);
% Add nodal masses to the global M matrix.

M(7,7)=ma; M(8,8) = ma;
% Generate equations for each element and assemble them.
for i=1:elems
    con = conn(i,:);
    lm = lmm(i,:);
    [m, k] = TransientPlaneTrussElement(e, A, rho,
```

```
                    nodes(con,:));
    M(lm, lm) = M(lm, lm) + m;
    K(lm, lm) = K(lm, lm) + k;
end

% Adjust for essential boundary conditions
dof = length(R);
df = setdiff(1:dof, debc);
Mf = M(df, df);
Kf = K(df, df);

% Compute frequencies and mode shapes
[V, lam] = eig(Kf, Mf);
freq=sqrt(lam)
modeShapes = V

>> ModalTrussEx

freq =

      55.835           0           0           0
           0      235.46           0           0
           0           0      1598.4           0
           0           0           0      1971.6


modeShapes =

    0.099847    -0.42131    -0.01827   -0.036118
    -0.42305   -0.098538   0.0011071   -0.020645
   -0.041882    -0.18022      1.6041      1.8044
    -0.01476   -0.062462     -1.8123      1.6046
```

Example 8.3 Modal Analysis of a Plane Frame (*Mathematica*) The following TransientPlaneFrameElement function returns the mass and the stiffness matrix of a plane frame element. To generate the mass matrix, the function needs the mass density (ρ).

```
TransientPlaneFrameElement[e_, i_, a_, ρ_, {qs_, qt_}, {n1_, n2_}] :=
  Module[{EI = e * i, EA = e * a, L, ml, kl, rl, ls, ms},
    L = Sqrt[(n2 - n1).(n2 - n1)];
    {ls, ms} = (n2 - n1)/L;
    T = {{ls, ms, 0, 0, 0, 0}, {-ms, ls, 0, 0, 0, 0}, {0, 0, 1, 0, 0, 0}, {0, 0, 0, ls, ms, 0},
        {0, 0, 0, -ms, ls, 0}, {0, 0, 0, 0, 0, 1}};
    kl = {{EA/L, 0, 0, -(EA/L), 0, 0}, {0, (12 * EI)/L^3, (6 * EI)/L^2, 0, -((12 * EI)/L^3),
        (6 * EI)/L^2}, {0, (6 * EI)/L^2, (4 * EI)/L, 0, -((6 * EI)/L^2), (2 * EI)/L},
        {-(EA/L), 0, 0, EA/L, 0, 0}, {0, -((12 * EI)/L^3), -((6 * EI)/L^2), 0,
```

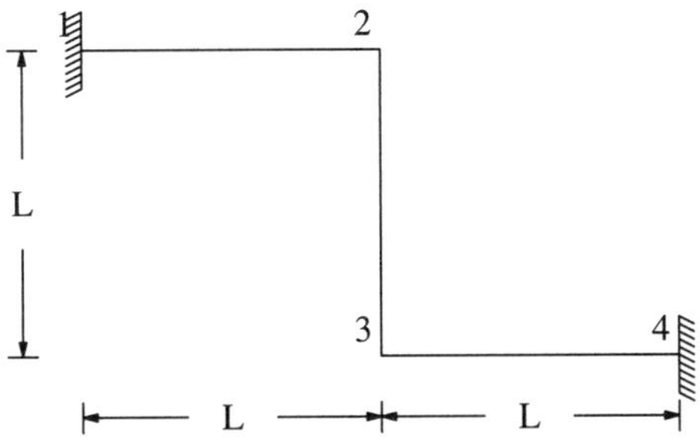

Figure 8.12. Plane frame

```
        (12 * EI)/L^3, -((6 * EI)/L^2)}, {0, (6 * EI)/L^2, (2 * EI)/L, 0,
        -((6 * EI)/L^2), (4 * EI)/L}};
    ml = ((ρ * a * L)/420) * {{140, 0, 0, 70, 0, 0}, {0, 156, 22 * L, 0, 54, -13 * L},
        {0, 22 * L, 4 * L^2, 0, 13 * L, -3 * L^2}, {70, 0, 0, 140, 0, 0},
        {0, 54, 13 * L, 0, 156, -22 * L}, {0, -13 * L, -3 * L^2, 0, -22 * L, 4 * L^2}};
    rl = {qs * (L/2), qt * (L/2), qt * (L^2/12),
    qs * (L/2), qt * (L/2), -qt * (L^2/12)};
    {Transpose[T].ml.T, Transpose[T].kl.T, Transpose[T].rl}
  ]
```

Using these functions, we consider the modal analysis of the plane frame shown in Figure 8.12. The horizontal members carry a weight of 200 N/m in addition to their own weight. The other numerical data are as follows (use N · mm units):

$$E = 200\,\text{GPa} = 200{,}000\,\text{N/mm}^2; \qquad \rho = 7840\,\text{kg/m}^3 = 7.84 \times 10^{-6}\,\text{kg/mm}^3;$$

$$L = 600\,\text{mm}; \qquad A = 240\,\text{mm}^2; \qquad I = 2000\,\text{mm}^4$$

Additional mass on horizontal members:

$$\frac{200}{9.81} = 20.3874\,\text{kg/m} = 0.0203874\,\text{kg/mm}$$

```
L = 1000; e = 200000; ρ = 7.84 * 10^(-6);
a = 240; i = 2000; ma = 0.0203874;
nodes = {{0, 0}, {L, 0}, {L, -L}, {2L, -L}};
lm[1] = {1, 2, 3, 4, 5, 6};
lm[2] = {4, 5, 6, 7, 8, 9};
lm[3] = {7, 8, 9, 10, 11, 12};
{m[1], k[1], r[1]} = TransientPlaneFrameElement[e, i, a, (ρ + ma/a), {0, 0},
                        nodes[[{1, 2}]]];
{m[2], k[2], r[2]} = TransientPlaneFrameElement[e, i, a, (ρ), {0, 0}, nodes[[{2, 3}]]];
{m[3], k[3], r[3]} = TransientPlaneFrameElement[e, i, a, (ρ + ma/a), {0, 0},
                        nodes[[{3, 4}]]];
M = K = Table[0, {12}, {12}];
Do[M[[lm[i], lm[i]]]+ = m[i];
K[[lm[i], lm[i]]]+ = k[i]; R[[lm[i]]]+ = r[i], {i, 1, 3}];
```

The essential boundary conditions are as follows:

```
debc = {1, 2, 3, 10, 11, 12};
ebcVals = Table[0, {6}];
df = Complement[Range[12], debc];
Mf = M[[df, df]]
```

$$\begin{pmatrix} 8.12188 & 0. & 98.56 & 0.24192 & 0. & -58.24 \\ 0. & 8.89854 & -1166.47 & 0. & 0.3136 & 0. \\ 98.56 & -1166.47 & 230006. & 58.24 & 0. & -13440. \\ 0.24192 & 0. & 58.24 & 8.12188 & 0. & -98.56 \\ 0. & 0.3136 & 0. & 0. & 8.89854 & 1166.47 \\ -58.24 & 0. & -13440. & -98.56 & 1166.47 & 230006. \end{pmatrix}$$

```
Kf = K[[df, df]]
```

$$\begin{pmatrix} \frac{240024}{5} & 0 & 2400 & -\frac{24}{5} & 0 & 2400 \\ 0 & \frac{240024}{5} & -2400 & 0 & -48000 & 0 \\ 2400 & -2400 & 3200000 & -2400 & 0 & 800000 \\ -\frac{24}{5} & 0 & -2400 & \frac{240024}{5} & 0 & -2400 \\ 0 & -48000 & 0 & 0 & \frac{240024}{5} & 2400 \\ 2400 & 0 & 800000 & -2400 & 2400 & 3200000 \end{pmatrix}$$

Computation of frequencies and mode shapes is as follows:

```
{lams, Vs} = Eigensystem[Inverse[Mf].Kf];
lam = Reverse[lams]
```

{0.331293, 18.4509, 19.7104, 5921.42, 6095.6, 41869.9}

The natural frequencies are the square roots of the eigenvalues:

```
Sqrt[lam]
```

{0.575581, 4.29546, 4.43964, 76.9508, 78.0743, 204.621}

The mode shapes are

```
V = Reverse[Vs]
```

$$\begin{pmatrix} 6.64393\times10^{-7} & 0.707107 & 0.000613879 & 6.64393\times10^{-7} & 0.707107 & -0.000613879 \\ -0.0585609 & -0.137874 & 0.691058 & 0.0585609 & 0.137874 & 0.691058 \\ 0.000392188 & 0.707081 & 0.00607015 & 0.000392188 & 0.707081 & -0.00607015 \\ -0.692288 & 0.143999 & 0.00113752 & -0.692288 & 0.143999 & -0.00113752 \\ 0.707018 & 0.0112127 & -0.000068896 & -0.707018 & -0.0112127 & -0.000068896 \\ -0.0227638 & 0.70673 & 0.00381233 & 0.0227638 & -0.70673 & 0.00381233 \end{pmatrix}$$

8.2.4 Transient Response Examples

For structures subjected to dynamic loads the global system of equations for the system is of the following form:

$$\boldsymbol{M}\ddot{\boldsymbol{d}} + \boldsymbol{K}\boldsymbol{d} = \boldsymbol{R}$$

where the global mass matrix $\boldsymbol{M}$, the global stiffness matrix $\boldsymbol{K}$, and the global load vector $\boldsymbol{R}$ are assembled from the corresponding element matrices using the usual assembly procedure. For a unique solution initial displacements and velocities at time 0 at all degrees of freedom must also be specified. Typically, these specified values are zero (the system is at rest before application of the dynamic load):

$$\boldsymbol{d}(0) = \boldsymbol{d}_0; \qquad \dot{\boldsymbol{d}}(0) = v_0$$

The second-order ordinary differential can be solved in *Mathematica* by using the NDSolve function. In MATLAB the solution can be obtained using one of the ode solvers, such as ode23. The following examples illustrate the use of these functions.

Example 8.4 Transient Analysis of a Plane Truss (*Mathematica*) Consider solution of the plane truss structure shown in Figure 8.13. All members have the same cross-sectional area and are of the same material, $\rho = 490\,\text{lb/ft}^3$, $E = 30 \times 10^6\,\text{lb/in}^2$, and $A = 1.25\,\text{in}^2$. The truss supports a rotating machine weighing 2000 lb at its tip. Due to unbalance in the machine, a harmonic force $P = 100\cos(7\pi t)$ lb is exerted on the truss.

The weight of the machine is an added mass at node 4. This is defined as ma in the following code and is added directly to the global mass matrix corresponding to degrees of freedom 7 and 8 associated with node 4. In defining the global load vector, a load of 1 unit is applied at degree of freedom 8. This global load vector is multiplied by $f(t) = 100\cos(7\pi t)$ before solving the differential equations of motion. The other procedure for generating the

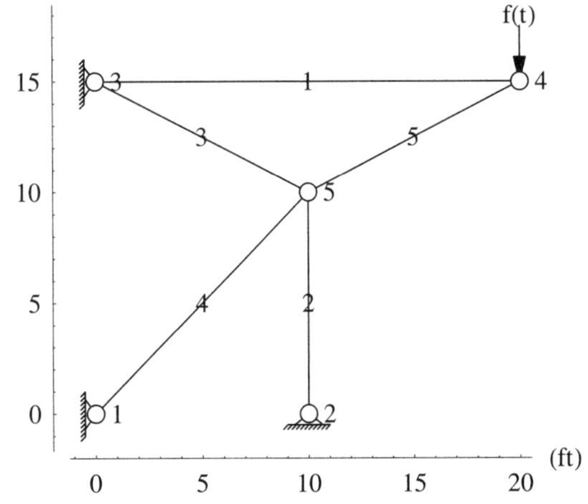

Figure 8.13. A plane truss

global mass, stiffness, and load vector is exactly the same as that for the static analysis. Using in-lb units the calculations are as follows:

```
f[t] = 100 Cos[7πt];
g = 386.4; e = 30 * 10^6; A = 1.25; ρ = 490/12^3/g; ma = 2000/g;
nodes = 12{{0, 0}, {10, 0}, {0, 15}, {20, 15}, {10, 10}};
en[1] = {3, 4}; en[2] = {2, 5}; en[3] = {5, 3}; en[4] = {1, 5}; en[5] = {5, 4};
lm[1] = {5, 6, 7, 8}; lm[2] = {3, 4, 9, 10}; lm[3] = {9, 10, 5, 6}; lm[4] = {1, 2, 9, 10};
lm[5] = {9, 10, 7, 8};
Do[{m[i], k[i]} = TransientPlaneTrussElement[e, A, ρ, nodes[[en[i]]]], {i, 1, 5}];
M = K = Table[0, {10}, {10}]; M[[8, 8]] = ma; M[[7, 7]] = ma;
R = Table[0, {10}]; R[[8]] = 1;
Do[M[[lm[i], lm[i]]]+ = m[i]; K[[lm[i], lm[i]]]+ = k[i], {i, 1, 5}];
```

The essential boundary conditions are as follows:

```
debc = Range[6];
ebcVals = Table[0, {6}];
df = Complement[Range[10], debc];
Mf = M[[df, df]]
```

$$\begin{pmatrix} 5.29039 & 0 & 0.0205121 & 0 \\ 0 & 5.29039 & 0 & 0.0205121 \\ 0.0205121 & 0 & 0.170634 & 0 \\ 0 & 0.0205121 & 0 & 0.170634 \end{pmatrix}$$

```
Kf = K[[df, df]]
```

$$\begin{pmatrix} 379857. & 111803. & -223607. & -111803. \\ 111803. & 55901.7 & -111803. & -55901.7 \\ -223607. & -111803. & 557699. & 110485. \\ -111803. & -55901.7 & 110485. & 534789. \end{pmatrix}$$

```
Rf = R[[df]] - K[[df, debc]].ebcVals
```

{0., 1., 0., 0.}

The equations can now be solved using the NDSolve function. To use this function, all equations and initial conditions must be expressed in the form of a list. This list of equations is generated as follows:

```
disp = {u4[t], v4[t], u5[t], v5[t]};
vel = D[disp, t];
acc = D[vel, t];
eqns = Thread[Mf.acc + Kf.disp == Rf * f[t]]
```

$\{379857.\, u4(t) - 223607.\, u5(t) + 111803.\, v4(t) - 111803.\, v5(t) + 5.29039\, u4''(t)$
$+ 0.0205121\, u5''(t) == 0.\cos(7\pi t),\ 111803.\, u4(t) - 111803.\, u5(t) + 55901.7\, v4(t)$
$- 55901.7\, v5(t) + 5.29039\, v4''(t) + 0.0205121\, v5''(t) == 100.\cos(7\pi t),$

$-223607.\,\mathrm{u4}(t) + 557699.\,\mathrm{u5}(t) - 111803.\,\mathrm{v4}(t) + 110485.\,\mathrm{v5}(t) + 0.0205121\,\mathrm{u4}''(t)$
$+ 0.170634\,\mathrm{u5}''(t) == 0.\cos(7\pi t), -111803.\,\mathrm{u4}(t) + 110485.\,\mathrm{u5}(t) - 55901.7\,\mathrm{v4}(t)$
$+ 534789.\,\mathrm{v5}(t) + 0.0205121\,\mathrm{v4}''(t) + 0.170634\,\mathrm{v5}''(t) == 0.\cos(7\pi t)\}$

```
initialVel = Thread[(vel/.t → 0) == 0]
```

$\{u4'(0) == 0, v4'(0) == 0, u5'(0) == 0, v5'(0) == 0\}$

```
initialDisp = Thread[(disp/.t → 0) == 0]
```

$\{\mathrm{u4}(0) == 0, \mathrm{v4}(0) == 0, \mathrm{u5}(0) == 0, \mathrm{v5}(0) == 0\}$

```
sol = NDSolve[Flatten[{eqns, initialVel, initialDisp}], disp, {t, 0, 1},
          MaxSteps → 10000]
```

{{u4(t) → InterpolatingFunction[(0. 1.), <>][t],
v4(t) → InterpolatingFunction[(0. 1.), <>][t],
u5(t) → InterpolatingFunction[(0. 1.), <>][t],
v5(t) → InterpolatingFunction[(0. 1.), <>][t]}}

The NDSolve function returns time histories of nodal displacements in the form of interpolation functions. Any of the values can be plotted in the usual manner using the Plot function, as shown in Figure 8.14. For example, the time history of vertical displacement at node 4 is as follows:

```
Plot[v4[t]/.sol, {t, 0, 1}, AxesLabel → {"t(s)", "disp(in)"},
    PlotLabel → "Verticaldisplacementatnode4"];
```

Once the nodal displacements are known, any other quantity of interest, say axial stresses in elements, can easily be computed using the same functions as those used in the static analysis case. Since the displacements are functions of time, obviously the element quantities are also functions of time. First we need the complete vector of nodal displacements, in the original order established by the node numbering. This requires combining the com-

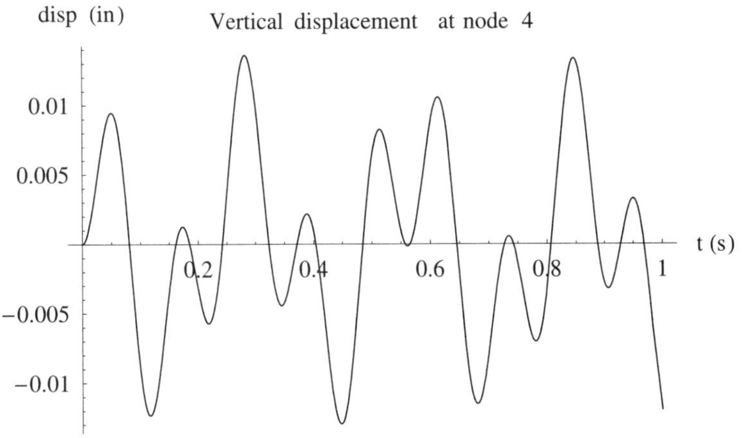

Figure 8.14. Displacement time history

puted solution with those specified as essential boundary conditions and is exactly the same process as used in the static analysis in earlier chapters.

```
d = Table[0, {10}];
d[[debc]] = ebcVals;
d[[df]] = disp/.sol[[1]];
d
```

{0, 0, 0, 0, 0, 0, InterpolatingFunction[(0. 1.), <>][t],
InterpolatingFunction[(0. 1.), <>][t], InterpolatingFunction[(0. 1.), <>][t],
InterpolatingFunction[(0. 1.), <>][t]}

Now, using the PlaneTrussResults function, the solution over any element can be computed. For example, the time history of strain, axial stress, and axial force for the first element is obtained as follows:

```
PlaneTrussResults[e_, A_, {{x1_, y1_}, {x2_, y2_}}, {u1_, v1_, u2_, v2_}] :=
Module[{ls, ms, L, T, d1, d2, eps},
L = Sqrt[(x2 - x1)^2 + (y2 - y1)^2];
ls = (x2 - x1)/L;
ms = (y2 - y1)/L;
T = {{ls, ms, 0, 0}, {0, 0, ls, ms}};
{d1, d2} = T.{u1, v1, u2, v2};
eps = (d2 - d1)/L;
{eps, e * eps, e * A * eps}]

elemSol = PlaneTrussResults[e, A, nodes[[en[1]]], d[[lm[1]]]]
```

$\left\{\frac{1}{240}\text{InterpolatingFunction}[(0.\quad 1.), <>][t],\right.$

125000 InterpolatingFunction[(0. 1.), <>][t],

$\left.156250.\ \text{InterpolatingFunction}[(0.\quad 1.), <>][t]\right\}$

Any of the response time histories can be plotted in a usual manner, as shown in Figure 8.15. For example, a plot of the axial stress time history for element 1 is as follows:

```
Plot[elemSol[[2]], {t, 0, 1}, AxesLabel → {"t(s)", "σ(lb/in²)"},
    PlotLabel → "Axialstressesinelement1"];
```

Note that the dynamic displacements, and hence the resulting stresses, are due only to the unbalance in the machine. For actual design of the truss, we must consider displacements and stresses due to static load of the machine (2000 lb) as well. These quantities can be calculated as usual (see Chapter 4).

Example 8.5 Transient Analysis of a Plane Frame (MATLAB) Consider solution of the plane frame in Figure 8.16. The load P varies with time, as shown in the load-time history. The horizontal members carry a weight of 200 N/m in addition to their own weight.

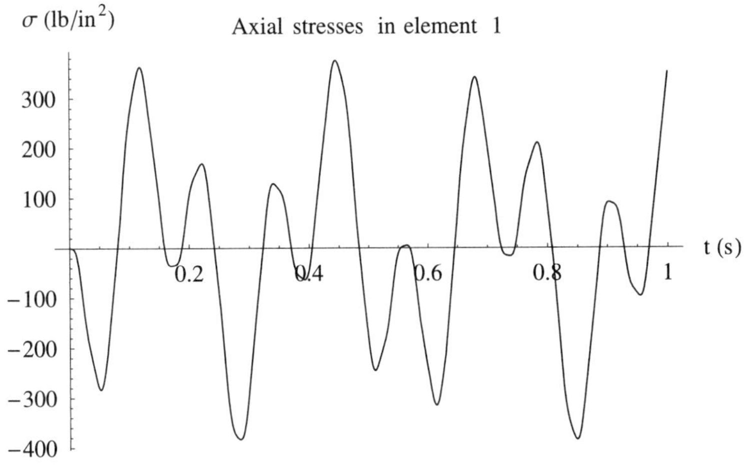

Figure 8.15. Axial stress time history

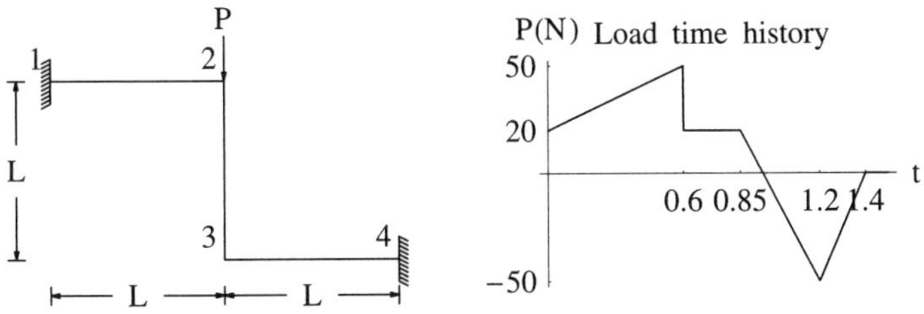

Figure 8.16. Plane frame

The other numerical data are as follows (use N · mm units):

$$E = 200\,\text{GPa} = 200{,}000\,\text{N/mm}^2; \qquad \rho = 7840\,\text{kg/m}^3 = 7.84\times10^{-6}\,\text{kg/mm}^3$$

$$L = 600\,\text{mm}; \qquad A = 240\,\text{mm}^2; \qquad I = 2000\,\text{mm}^4$$

Additional mass on horizontal members:

$$\frac{200}{9.81} = 20.3874\,\text{kg/m} = 0.0203874\,\text{kg/mm}$$

MatlabFiles\Chap8\TransientFrameEx.m on the book web site

```
% Transient analysis of a plane frame
global Mf Kf Rf
L = 1000; e = 200000; rho = 7.84*10^(-6); a = 240;
inertia = 2000; ma = 0.0203874;
nodes = [0, 0; L, 0; L, -L; 2*L, -L];
conn = [1,2; 2,3; 3,4];
elems = size(conn,1);
```

```
lmm=[];
for i=1:elems
    lmm = [lmm; [3*conn(i,1)-2, 3*conn(i,1)-1,
                 3*conn(i,1),...,3*conn(i,2)-2,
                 3*conn(i,2)-1, 3*conn(i,2)]];
end
debc = [1,2,3,10,11,12]; ebcVals=zeros(length(debc),1);
dof=3*size(nodes,1);
M=zeros(dof); K=zeros(dof);
R = zeros(dof,1); R(5)=1;

% Generate equations for each element and assemble them.
for i=1:2:elems
    con = conn(i,:);
    lm = lmm(i,:);
    [m, k, r] = TransientPlaneFrameElement(e, inertia,
        a, ..., rho+ma/a, 0, 0, nodes(con,:));
    M(lm, lm) = M(lm, lm) + m;
    K(lm, lm) = K(lm, lm) + k;
    R(lm) = R(lm) + r;
end
for i=2
    con = conn(i,:);
    lm = lmm(i,:);
    [m, k, r] = TransientPlaneFrameElement(e, inertia,
        a, ..., rho, 0, 0, nodes(con,:));
    M(lm, lm) = M(lm, lm) + m;
    K(lm, lm) = K(lm, lm) + k;
    R(lm) = R(lm) + r;
end

% Adjust for essential boundary conditions
dof = length(R);
df = setdiff(1:dof, debc);
Mf = M(df, df);
Kf = K(df, df);
Rf = R(df) - K(df, debc)*ebcVals;

% Setup and solve the resulting first order differential
% equations
u0 = zeros(length(Mf),1);
v0 = zeros(length(Mf),1);
[t,d] = ode23('FrameODE',[0,10],[u0; v0]);
plot(t,d(:,2)); xlabel('time'); ylabel('Disp');
title('Vertical displacement at node 2');
```

MatlabFiles\Chap8\FrameODE.m on the book web site

```
function ddot = FrameODE(t, d)
% ddot = FrameODE(t, d)
% function to set up equations for a transient frame problem
global Mf Kf Rf
if t <= 0.6
    ft = 50*t + 20;
elseif t <= .85
        ft = 20;
    elseif t <= 1.2
            ft = 190 - 200*t;
        elseif t <= 1.4
                ft = 250*t - 350;
            else
                ft = 0;
            end
        end
    end
end
n=length(d);
u = d(1:n/2);
v = d(n/2+1:n);
vdot = inv(Mf)*(Rf*ft - Kf*u);
udot = v;
% format short g
% soln=[t, ft, u(2), udot(2), vdot(2)]
ddot = [udot; vdot];
```

Executing the script file TransientFrameEx, the plot of vertical displacement at node 2 shown in Figure 8.17 is obtained. The script file can be modified to plot any other quantity of interest.

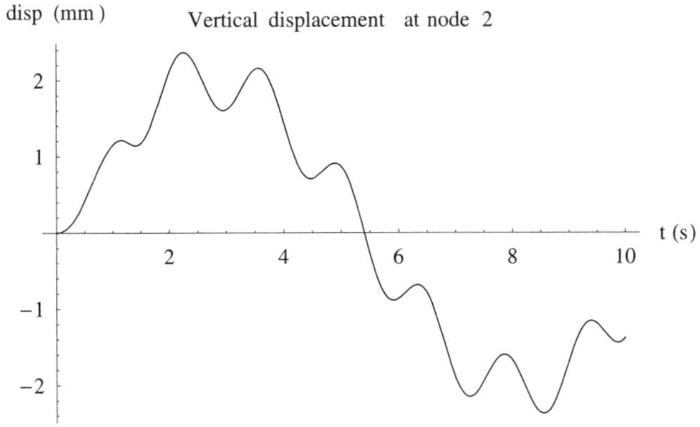

Figure 8.17. Displacement-time history

PROBLEMS

Transient Heat Flow

8.1 A square duct shown in Figure 8.18 is insulated by a layer of square fiberglass that has a thermal conductivity of 0.035 W/m · °C, density $\rho = 144\,\text{kg/m}^3$, and specific heat $C_p = 753\,\text{J/kg} \cdot °\text{C}$. The convection coefficient of the outside surface of the fiberglass is $h = 45\,\text{W/m}^2 \cdot °\text{C}$. Initially the duct and the insulation are at the room temperature of 25°C. A flow of hot gases suddenly raises the inside duct temperature to 1000°C. Determine what is the temperature distribution in the insulation 10 min after the flow of hot gases begins. Roughly how long will it take for the temperature to reach a steady-state condition. Take advantage of symmetry and model only one-eighth of the section, shown in dark shade. Assume that the temperature on the inside surface is the same as that of the hot gases.

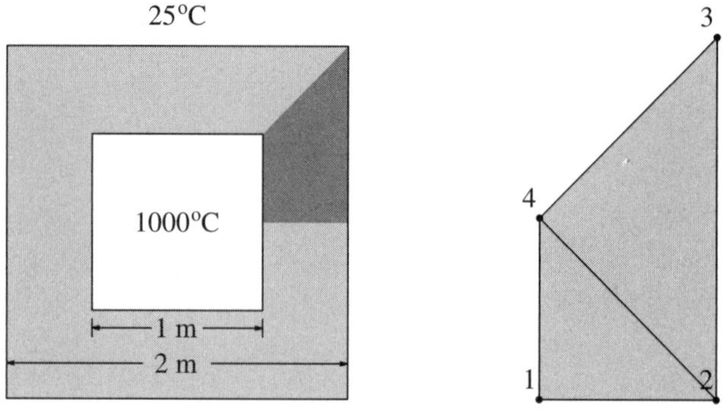

Figure 8.18.

8.2 Heating wires are embedded in a concrete slab, as shown in cross section in Figure 8.19, to keep snow from accumulating on the slab. In a proposed design wires are placed at 2 cm from the top and at 4-cm intervals. The heat generated by each wire is 10 W/m length. The bottom of the slab is insulated. The top is exposed to a convection heat loss with a convection coefficient of $h = 30\,\text{W/m}^2 \cdot °\text{C}$. For concrete the thermal conductivity is 1.046 W/m · °C, density $\rho = 2300\,\text{kg/m}^3$, and specific heat $C_p = 656.9\,\text{J/kg} \cdot °\text{C}$. Determine the slab surface temperature when the air temperature is −5°C. Assuming the entire slab is initially at the same temperature as the air, how long will it take to raise the temperature of the surface to above freezing

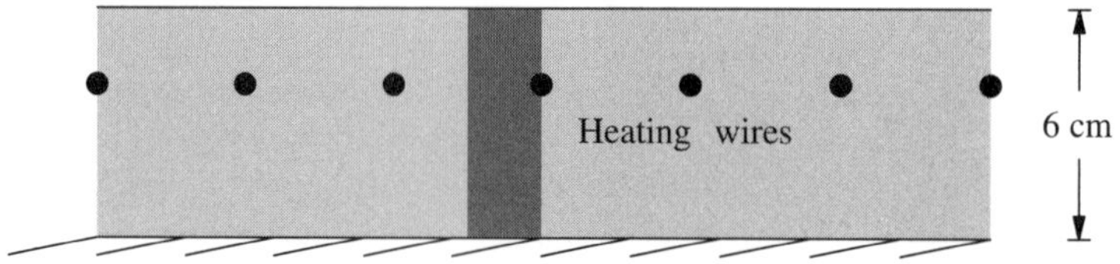

Figure 8.19. Heating wires embedded in concrete

after starting the heating process? Note that, because of symmetry, we need to model only the 2 cm × 6 cm dark shaded area. There will be no heat flow across the sides of this area. The heating wire represents a concentrated point heat source.

8.3 The cross section of a brick masonry chimney is shown in Figure 8.20. The thermal conductivity of brick is 0.711 W/m · °C, density $\rho = 2000\,\text{kg/m}^3$, and specific heat $C_p = 836.8\,\text{J/kg}\cdot{}^\circ\text{C}$. The convection coefficient of the outside surface of the chimney is $h = 15\,\text{W/m}^2\cdot{}^\circ\text{C}$. Initially the chimney is at the room temperature of 10°C. A flow of hot gases suddenly raises the inside chimney temperature to 250°C. Determine what is the temperature distribution in the insulation 5 min after the flow of hot gases begins. Roughly how long will it take for the temperature to reach a steady-state condition. Take advantage of symmetry and model only one-eighth of the section, shown in dark shade. Assume that the temperature on the inside surface is the same as that of the hot gases.

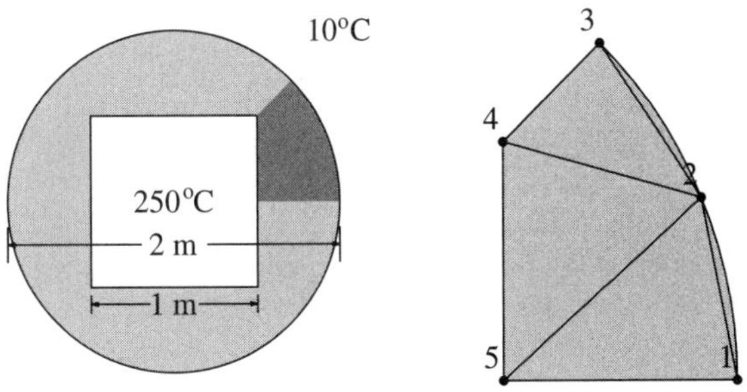

Figure 8.20.

Free Vibrations

8.4 Determine natural frequencies and mode shapes for an axially loaded bar fixed at the left end and restrained by a spring at the right end, as shown in Figure 8.21. Divide the bar into two equal-length elements and use the following numerical data:

$$E = 200 \times 10^9\,\text{N/m}^2; \qquad \rho = 7500\,\text{kg/m}^3; \qquad A = 1.6 \times 10^{-4}\,\text{m}^2$$
$$L = 2.5\,\text{m}; \qquad k = 1 \times 10^7\,\text{N/m}$$

Assume the spring is massless.

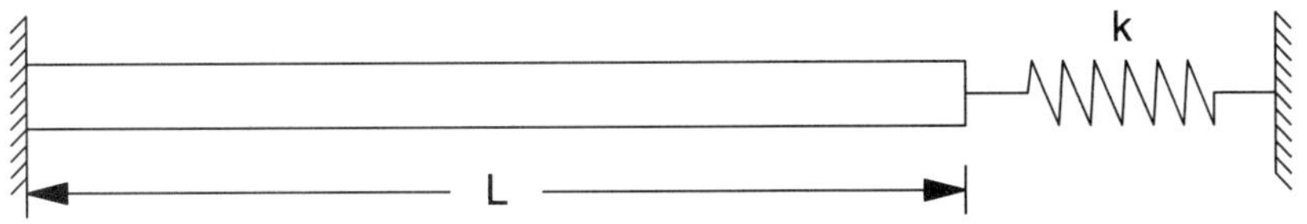

Figure 8.21.

8.5 Determine free-vibration frequencies and mode shapes for the three-bar pin-jointed structure shown in Figure 8.22. All members have the same cross-sectional area and are of the same material, $\rho = 490\,\text{lb/ft}^3$, $E = 30 \times 10^6\,\text{lb/in}^2$, and $A = 1\,\text{in}^2$. The dimensions in inches are shown in the figure.

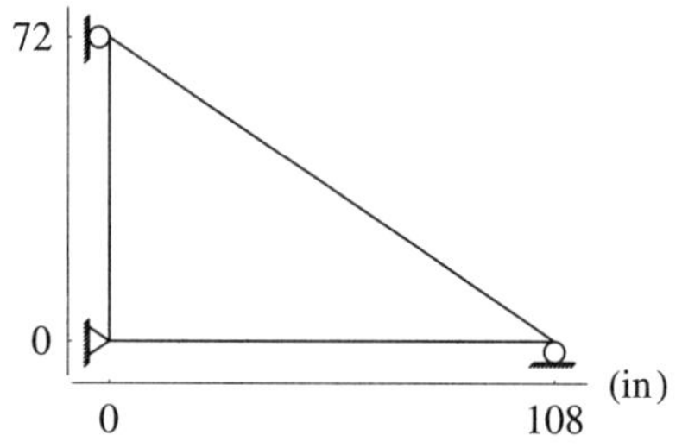

Figure 8.22.

8.6 Determine free-vibration frequencies and mode shapes for the three-bar pin-jointed structure shown in Figure 8.23. All members have the same cross-sectional area and are of the same material, $\rho = 490\,\text{lb/ft}^3$, $E = 30 \times 10^6\,\text{lb/in}^2$, and $A = 2\,\text{in}^2$. The attached mass is $m = 77.6\,\text{lb} \cdot \text{s}^2/\text{in}$. The dimensions in inches are shown in the figure.

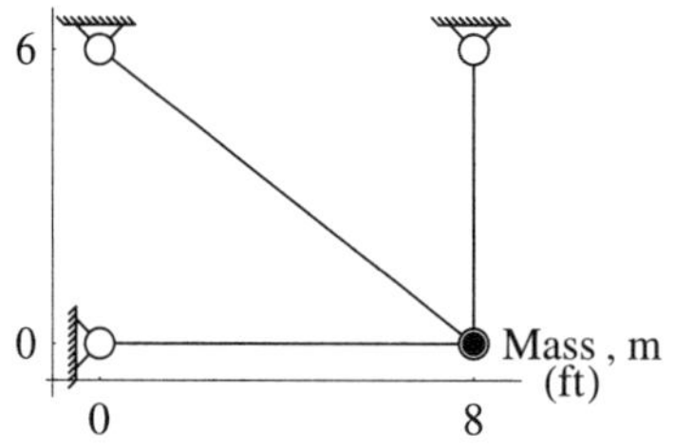

Figure 8.23.

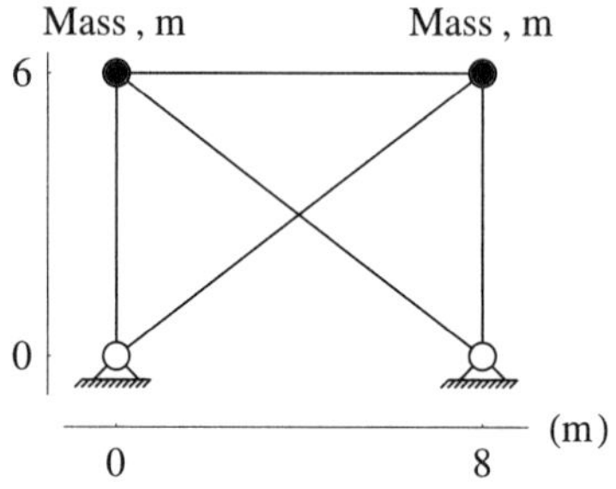

Figure 8.24.

8.7 Determine free-vibration frequencies and mode shapes for the truss shown in Figure 8.24. Note the diagonals are not connected to each other at their crossing. The cross-sectional area of vertical and horizontal members is $30 \times 10^{-4}\,\text{m}^2$ and that for the diagonals is $10 \times 10^{-4}\,\text{m}^2$. All members are made of steel with $\rho = 7850\,\text{kg}/m^3$ and $E = 210\,\text{GPa}$. The attached masses are $m = 500\,\text{kg}$. The dimensions in meters are shown in the figure.

8.8 Determine free-vibration frequencies and mode shapes for the truss shown in Figure 8.25. The cross-sectional area of members is $15 \times 10^{-4}\,\text{m}^2$. All members are made of

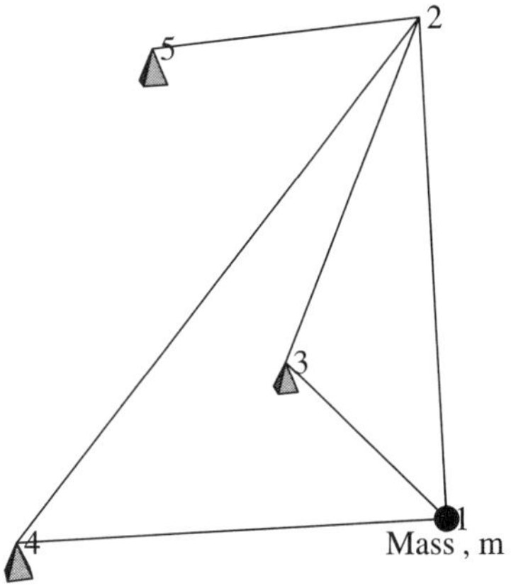

Figure 8.25.

aluminum with $E = 70\,\text{GPa}$ and $\rho = 2710\,\text{kg/m}^3$. The attached mass is $m = 500\,\text{kg}$. The nodal coordinates, in meters, are as follows:

	x	y	z
1	0.	4.	0.
2	−3.	2.	5.
3	−3.	0.	0.
4	3.	0.	0.
5	0.	0.	5.

8.9 Determine free-vibration frequencies and mode shapes for the uniform beam simply supported at ends and spring supported in the middle as shown in Figure 8.26. Assume the following numerical data:

$$L = 4\,\text{m}; \quad \rho = 2710\,\text{kg/m}^3; \quad E = 70\,\text{GPa}; \quad m = 350\,\text{kg}$$

$$I = 350.9 \times 10^{-6}\,\text{m}^4; \quad A = 8.39 \times 10^{-3}\,\text{m}^2; \quad k = 4 \times 10^5\,\text{N/m}$$

Assume the spring is massless.

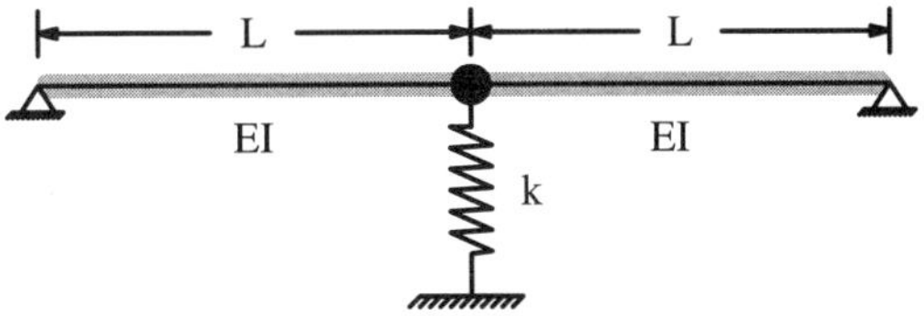

Figure 8.26. Beam supported by a spring

8.10 Determine free-vibration frequencies and mode shapes for the plane frame shown in Figure 8.27. Note the diagonal brace can carry axial forces only. The dimensions in meters are shown in the figure. Assume the following numerical data:

$\rho = 7850\,\text{kg/m}^3$; $E = 210\,\text{GPa}$; $m = 350\,\text{kg}$

Frame members: $I = 350.9 \times 10^{-6}\,\text{m}^4$; $A = 8.39 \times 10^{-3}\,\text{m}^2$;

Diagonal brace: $A = 1.252 \times 10^{-3}\,\text{m}^2$

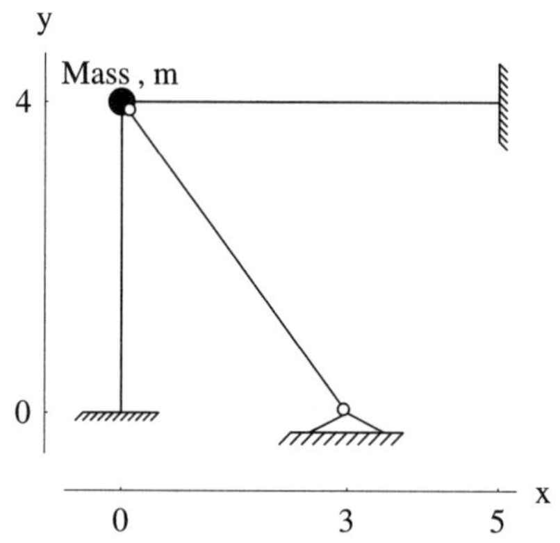

Figure 8.27.

Transient Vibrations

8.11 The left half of the beam shown in Figure 8.28 is supporting a rotating machine operating at 1500 rpm. Due to unbalance, the machine exerts a uniformly distributed load of 800 lb/in that varies harmonically. Using two beam elements, determine the time history of midspan displacement. Use the following numerical data:

$L = 150\,\text{in}$; $E = 30 \times 10^6\,\text{psi}$; $I = 10^3\,\text{in}^4$; $m = 0.15\,\text{lb} \cdot \text{s}^2/\text{in}^2$

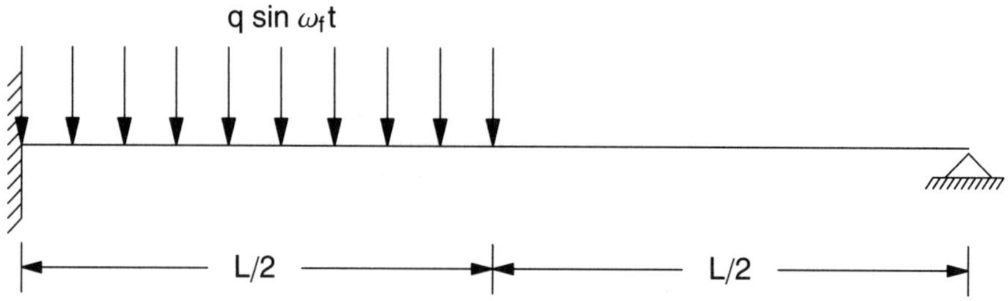

Figure 8.28.

8.12 The plane frame shown in Figure 8.29 is subjected to pressure q from a blast. A simple idealization of the blast loading is that of a triangular pulse as shown in the figure. Use the following numerical data:

$$q = 200\,\text{lb/in}; \qquad t_d = 0.02\,\text{s}; \qquad L = 180\,\text{in}; \qquad E = 30 \times 10^6\,\text{lb/in}^2$$

$$A = 100\,\text{in}^2; \qquad I = 1000\,\text{in}^4; \qquad m = 0.03\,\text{lb} \cdot \text{s}^2/\text{in}$$

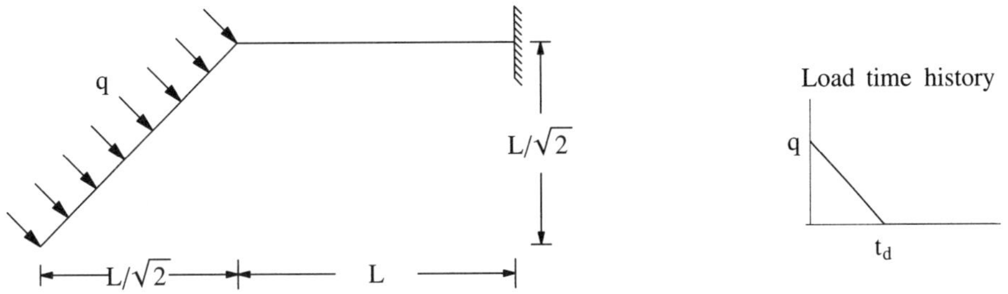

Figure 8.29. Plane frame with distributed load

CHAPTER NINE

p-FORMULATION

In conventional finite element formulation, each element is based on a specific set of interpolation functions. After choosing an element type, the only way to obtain a better solution is to refine the model. This formulation is called the *h-formulation*, where h indicates the generic size of an element. For one-dimensional problems h is the length of an element. For two- and three-dimensional problems it can be thought of as the diameter of the inscribing circle or sphere. In general, the finite element solution converges to the exact solution as $h \to 0$.

An alternative formulation, called the *p-formulation*, is presented in this chapter. In this formulation, the elements are based on interpolation functions that involve very high order terms. The initial finite element model is fairly coarse and is based primarily on geometric considerations. Refined solutions are obtained by increasing the order of the interpolation functions used in the formulation. Efficient interpolation functions have been developed so that higher order solutions can be obtained in a *hierarchical* manner from the lower order solutions.

9.1 p-FORMULATION FOR SECOND-ORDER 1D BVP

In this section p-finite element equations are derived for the one-dimensional boundary value problem of the following form:

$$\frac{d}{dx}\left(k(x)\frac{du(x)}{dx}\right) + p(x)u(x) + q(x) = 0; \qquad x_0 < x < x_L$$

where $k(x)$, $p(x)$, and $q(x)$ are given functions of x and $u(x)$ is the solution variable. The

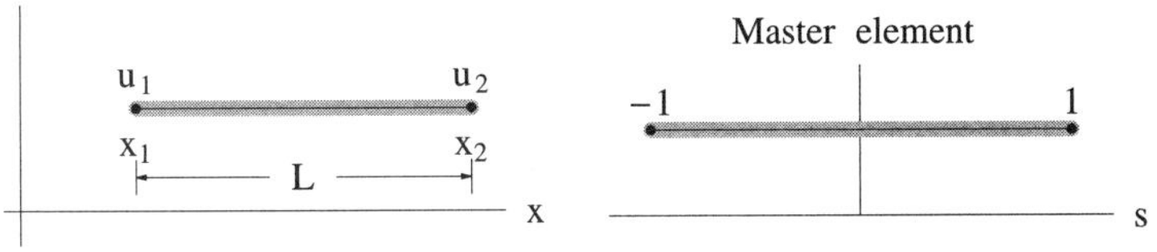

Figure 9.1. Two node *p*-element for second-order BVP

boundary conditions of the following form may be specified at one or more points:

$$\text{EBC:} \qquad u = \text{specified}$$

$$\text{NBC:} \qquad \frac{du}{dx} = \alpha u + \beta$$

where α and β are some specified constants.

The basic element is a two-node element shown in Figure 9.1. The mapping to the master element is based on linear Lagrangian interpolation functions for the two-node line element:

$$x(s) = N_1 x_1 + N_2 x_2 = \tfrac{1}{2}(1 - s)x_1 + \tfrac{1}{2}(s + 1)x_2$$

$$\frac{dx}{ds} = -\frac{x_1}{2} + \frac{x_2}{2} \equiv \frac{L}{2} \qquad \text{and} \qquad dx = \frac{L}{2}\,ds$$

Allowing for the possibility of natural boundary conditions at the element ends, over an element we must satisfy

$$\frac{d}{dx}\left(k(x)\frac{du(x)}{dx}\right) + p(x)u(x) + q(x) = 0; \qquad x_1 < x < x_2$$

$$-\frac{du(x_1)}{dx} = \alpha_1 u(x_1) + \beta_1; \qquad \frac{du(x_2)}{dx} = \alpha_2 u(x_2) + \beta_2$$

Note that in the natural boundary condition at x_1 the negative of the derivative is specified. As explained in Chapter 3, when applying this element to physical problems, the sign makes sense with the usual sign convention adopted for those problems. The weak form can be written using standard steps of writing the weighted residual, integration by parts, and incorporating the natural boundary conditions:

$$k(x_1)(\beta_1 + \alpha_1 u(x_1))w_i(x_1) + k(x_2)(\beta_2 + \alpha_2 u(x_2))w_i(x_2) + \int_{x_1}^{x_2} (qw_i + puw_i - ku'w_i')\,dx = 0$$

9.1.1 Assumed Solution Using Legendre Polynomials

Each element has two nodal degrees of freedom (u_1 and u_2) and any number of additional degrees of freedom identified as $\delta_1, \delta_2, \ldots$. The first two interpolation functions are the same as those used in the mapping and directly involve nodal unknowns. The remaining interpolation functions (called the *p-modes*) are not associated with nodal values. We can

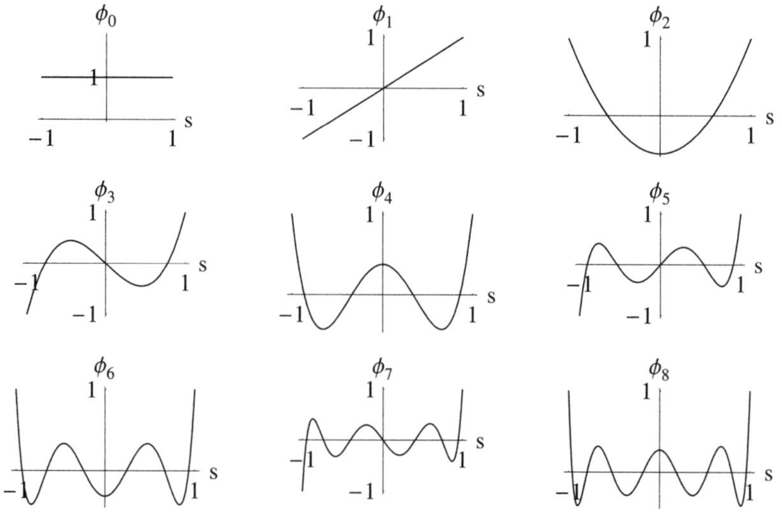

Figure 9.2. Plots of first nine Legendre polynomials

choose any set of functions for defining p-modes as long as they are easy to differentiate and integrate. Among several possibilities, p-modes written in terms of Legendre polynomials have been used quite successfully. The Legendre polynomials can be generated by the following recursive equation:

$$(i+1)\phi_{i+1}(s) = (2i+1)s\phi_i(s) - i\phi_{i-1}(s); \qquad i = 1, 2, \ldots$$
$$\phi_0(s) = 1; \qquad \phi_1(s) = s$$

where $\phi_i(s)$ is the ith Legendre polynomial. The first nine Legendre polynomials are given below explicitly. Figure 9.2 shows plots of these functions over the interval $(-1, 1)$. Note that the even Legendre polynomials have a value of 1 at the ends and the odd polynomials have values of -1 and 1 at the ends. The polynomial values fluctuate evenly over the entire interval. The Legendre polynomials also have the following orthogonality property that can easily be verified by direct computation:

$$\int_{-1}^{1} \phi_i(s)\phi_j(s)\,ds = \begin{cases} \dfrac{2}{2i+1} & \text{for } i = j \\ 0 & \text{for } i \neq j \end{cases}$$

$$\phi_0 = 1$$

$$\phi_1 = s$$

$$\phi_2 = \frac{3s^2}{2} - \frac{1}{2}$$

$$\phi_3 = \frac{5s^3}{2} - \frac{3s}{2}$$

$$\phi_4 = \frac{35s^4}{8} - \frac{15s^2}{4} + \frac{3}{8}$$

$$\phi_5 = \frac{63s^5}{8} - \frac{35s^3}{4} + \frac{15s}{8}$$

$$\phi_6 = \frac{231s^6}{16} - \frac{315s^4}{16} + \frac{105s^2}{16} - \frac{5}{16}$$

$$\phi_7 = \frac{429s^7}{16} - \frac{693s^5}{16} + \frac{315s^3}{16} - \frac{35s}{16}$$

$$\phi_8 = \frac{6435s^8}{128} - \frac{300s^6}{32} + \frac{3465s^4}{64} - \frac{315s^2}{32} + \frac{35}{128}$$

An important consideration in choosing appropriate functions for the p-modes is the requirement that the solution be continuous over the entire domain. In the standard finite element formulation the solution at a node common between two elements is made continuous by using the same nodal degree of freedom at the node. Since the p-modes are not associated with a node, they must be chosen carefully. The simplest approach is to define p-modes that are always zero at the element ends. Thus they do not influence the solution outside of the element and are unique to each element. Therefore the standard finite element assembly procedure can be used to get the global finite element equations. The numerical examples presented in the following section will further clarify this point.

From the plots of the Legendre polynomials, it is easy to see that we can get zero values at $s = \pm 1$ by subtracting two even or two odd polynomials from each other. Thus suitable p-modes can be defined by the following equation:

$$P_i(s) = \frac{1}{\sqrt{2(2i+1)}}(\phi_{i+1}(s) - \phi_{i-1}(s)); \qquad i = 1, 2, \ldots$$

Note the definition includes a factor that is chosen for convenience. The particular choice makes the integral for the orthogonality property come out to 1. By direct computation, it can be verified that the two p-modes satisfy the following orthogonality property:

$$\int_{-1}^{1} \frac{dP_i(s)}{ds}\frac{dP_j(s)}{ds}\,ds = \begin{cases} 1 & \text{for } i = j \\ 0 & \text{for } i \neq j \end{cases}$$

Because of this orthogonality property, the element matrices can be constructed in a very efficient manner. The first six p-modes are written explicity as follows and are also plotted in Figure 9.3. It can easily be seen that all the p-modes are zero at the ends. Being simple polynomials, they are easy to differentiate and integrate as well.

i	ϕ_{i-1}	ϕ_{i+1}	P_i
1	1	$\frac{3s^2}{2} - \frac{1}{2}$	$\frac{1}{2}\sqrt{\frac{3}{2}}(s^2 - 1)$
2	s	$\frac{5s^3}{2} - \frac{3s}{2}$	$\frac{1}{2}\sqrt{\frac{5}{2}}s(s^2 - 1)$

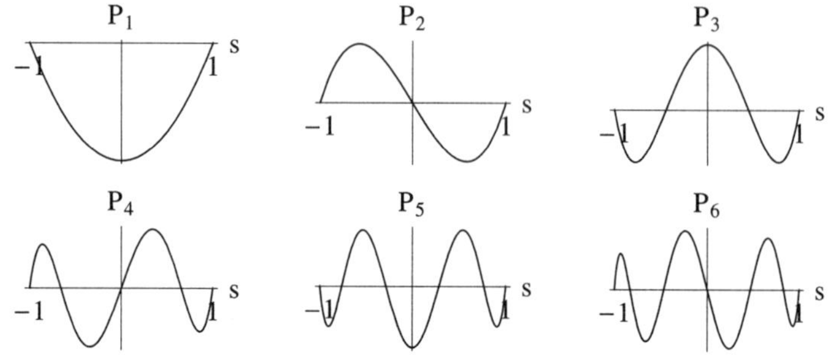

Figure 9.3. First six p-modes

i	ϕ_{i-1}	ϕ_{i+1}	P_i
3	$\frac{3s^2}{2} - \frac{1}{2}$	$\frac{35s^4}{8} - \frac{15s^2}{4} + \frac{3}{8}$	$\frac{1}{8}\sqrt{\frac{7}{2}}(5s^4 - 6s^2 + 1)$
4	$\frac{5s^3}{2} - \frac{3s}{2}$	$\frac{63s^5}{8} - \frac{35s^3}{4} + \frac{15s}{8}$	$\frac{3s(7s^4 - 10s^2 + 3)}{8\sqrt{2}}$
5	$\frac{35s^4}{8} - \frac{15s^2}{4} + \frac{3}{8}$	$\frac{231s^6}{16} - \frac{315s^4}{16} + \frac{105s^2}{16} - \frac{5}{16}$	$\frac{1}{16}\sqrt{\frac{11}{2}}(21s^6 - 35s^4 + 15s^2 - 1)$
6	$\frac{63s^5}{8} - \frac{35s^3}{4} + \frac{15s}{8}$	$\frac{429s^7}{16} - \frac{693s^5}{16} + \frac{315s^3}{16} - \frac{35s}{16}$	$\frac{1}{16}\sqrt{\frac{13}{2}}s(33s^6 - 63s^4 + 35s^2 - 5)$

It is interesting to contrast these p-modes with the high-order Lagrange interpolation functions. For example, the middle six interpolation functions for a seventh-order Lagrange interpolation are as shown in Figure 9.4. All these functions have zero values at the ends and thus could have been used as p-modes. However, the function values fluctuate widely over the interval, and thus the functions are not as nicely behaved as the p-modes based on Legendre polynomials.

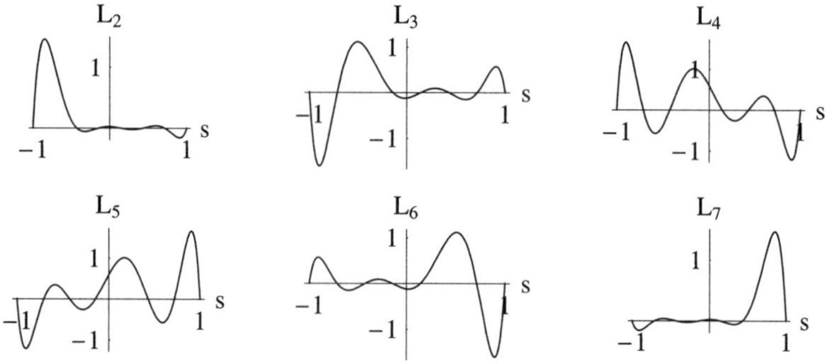

Figure 9.4. Second through seventh interpolation functions for a seventh-order Lagrange interpolation

9.1.2 Element Equations

The assumed solution consists of two linear Lagrange interpolation functions and any number of p-modes:

$$u(s) = N_1 u_1 + N_2 u_2 + P_1\delta_1 + P_2\delta_2 + P_3\delta_3 + \cdots = (N_1 \quad N_2 \quad P_1 \quad \cdots)\begin{pmatrix} u_1 \\ u_2 \\ \delta_1 \\ \vdots \end{pmatrix} \equiv \boldsymbol{N}^T\boldsymbol{d}$$

where u_1, u_2 are nodal degrees of freedom and $\delta_1, \delta_2, \ldots$ are unknown parameters associated with the p-modes. With the mapping the derivatives of the assumed solution with respect to actual element coordinate x can be written as follows:

$$\frac{du}{dx} = \frac{du}{ds}\frac{ds}{dx} = \frac{2}{L}\frac{du}{ds} = \frac{2}{L}\begin{pmatrix} \frac{dN_1}{ds} & \frac{dN_2}{ds} & \frac{dP_1}{ds} & \cdots \end{pmatrix}\begin{pmatrix} u_1 \\ u_2 \\ \delta_1 \\ \vdots \end{pmatrix} \equiv \boldsymbol{B}^T\boldsymbol{d}$$

The weak form over the element is

$$\int_{x_1}^{x_2} (q(x)w_i(x) + p(x)u(x)w_i(x) - k(x)u'(x)w_i'(x))\,dx$$
$$-k(x_1)(-\beta_1 - \alpha_1 u(x_1))w_i(x_1) + k(x_2)(\beta_2 + \alpha_2 u(x_2))w_i(x_2) = 0$$

Note the prime indicates differentiation with respect to x. Using the mapping $x = x(s)$, all terms inside the integral are expressed in terms of s, and the weak form is

$$\int_{-1}^{1}\left(q(s)w_i(s) + p(s)u(s)w_i(s) - k(s)\frac{2}{L}\frac{du}{ds}\frac{2}{L}\frac{dw_i}{ds}\right)\frac{L}{2}ds$$
$$- k(x_1)(-\beta_1 - \alpha_1 u(x_1))w_i(x_1) + k(x_2)(\beta_2 + \alpha_2 u(x_2))w_i(x_2) = 0$$

The weighting functions w_i are $N_1, N_2, P_1, \ldots$. We get one equation with each weighting function. Writing all equations together in a matrix form, we have

$$\int_{-1}^{1}\left(q(s)\begin{pmatrix} N_1 \\ N_2 \\ P_1 \\ \vdots \end{pmatrix} + p(s)\begin{pmatrix} N_1 \\ N_2 \\ P_1 \\ \vdots \end{pmatrix}u(s) - k(s)\frac{2}{L}\begin{pmatrix} dN_1/ds \\ dN_2/ds \\ dP_1/ds \\ \vdots \end{pmatrix}\frac{2}{L}\frac{du}{ds}\right)\frac{L}{2}ds$$
$$- k(x_1)(-\beta_1 - \alpha_1 u(x_1))\begin{pmatrix} \mathrm{N}_1(-1) \\ \mathrm{N}_2(-1) \\ P_1(-1) \\ \vdots \end{pmatrix} + k(x_2)(\beta_2 + \alpha_2 u(x_2))\begin{pmatrix} \mathrm{N}_1(1) \\ \mathrm{N}_2(1) \\ P_1(1) \\ \vdots \end{pmatrix} = \begin{pmatrix} 0 \\ 0 \\ 0 \\ \vdots \end{pmatrix}$$

Noting that $u(x_1) = u_1$, $u(x_2) = u_2$, $N_1(-1) = 1$, $N_1(1) = 0$, $N_2(-1) = 0$, $N_2(1) = 1$, and $P_i(\pm 1) = 0$, we have

$$\int_{-1}^{1}\left(q\boldsymbol{N} + p\boldsymbol{N}u(s) - k\boldsymbol{B}\frac{2}{L}\frac{du}{ds}\right)\frac{L}{2}\,ds - k(x_1)(-\beta_1 - \alpha_1 u_1)\begin{pmatrix}1\\0\\0\\\vdots\end{pmatrix}$$
$$+ k(x_2)(\beta_2 + \alpha_2 u_2)\begin{pmatrix}0\\1\\0\\\vdots\end{pmatrix} = \begin{pmatrix}0\\0\\0\\\vdots\end{pmatrix}$$

Substituting the assumed solution, we have

$$\int_{-1}^{1}\left(q\boldsymbol{N} + p\boldsymbol{N}\boldsymbol{N}^T\boldsymbol{d} - k\boldsymbol{B}\boldsymbol{B}^T\boldsymbol{d}\right)\frac{L}{2}\,ds - k(x_1)(-\beta_1 - \alpha_1 u_1)\begin{pmatrix}1\\0\\0\\\vdots\end{pmatrix}$$
$$+ k(x_2)(\beta_2 + \alpha_2 u_2)\begin{pmatrix}0\\1\\0\\\vdots\end{pmatrix} = \begin{pmatrix}0\\0\\0\\\vdots\end{pmatrix}$$

Rearranging the terms by keeping the terms involving unknown parameters on the left-hand side, we have

$$\int_{-1}^{1} k\boldsymbol{B}\boldsymbol{B}^T\boldsymbol{d}\frac{L}{2}\,ds - \int_{-1}^{1} p\boldsymbol{N}\boldsymbol{N}^T\boldsymbol{d}\frac{L}{2}\,ds - \begin{pmatrix}k(x_1)\alpha_1 u_1\\k(x_2)\alpha_2 u_2\\0\\\vdots\end{pmatrix} = \int_{-1}^{1} q\boldsymbol{N}\frac{L}{2}\,ds + \begin{pmatrix}k(x_1)\beta_1\\k(x_2)\beta_2\\0\\\vdots\end{pmatrix}$$

Define

$$k_k = \int_{-1}^{1} k\boldsymbol{B}\boldsymbol{B}^T\frac{L}{2}\,ds$$
$$k_p = -\int_{-1}^{1} p\boldsymbol{N}\boldsymbol{N}^T\frac{L}{2}\,ds$$
$$r_q = \int_{-1}^{1} q\boldsymbol{N}\frac{L}{2}\,ds$$
$$-\begin{pmatrix}k(x_1)\alpha_1 u_1\\k(x_2)\alpha_2 u_2\\0\\\vdots\end{pmatrix} = \begin{pmatrix}-\alpha_1 k(x_1) & 0 & 0 & \cdots\\0 & -\alpha_2 k(x_2) & 0 & \cdots\\0 & 0 & 0 & \cdots\\\vdots & \vdots & \vdots & \vdots\end{pmatrix}\begin{pmatrix}u_1\\u_2\\\vdots\end{pmatrix} = \boldsymbol{k}_\alpha\boldsymbol{d}$$
$$\begin{pmatrix}k(x_1)\beta_1\\k(x_2)\beta_2\\0\\\vdots\end{pmatrix} \equiv \boldsymbol{r}_\beta$$

Thus the element equations are

$$(\boldsymbol{k}_k + \boldsymbol{k}_p)\boldsymbol{d} + \boldsymbol{k}_\alpha \boldsymbol{d} = \boldsymbol{r}_q + \boldsymbol{r}_\beta$$

As explained in detail in Chapter 3, during assembly, the natural boundary condition terms are incorporated directly into the global equations as follows:

NBC at dof i:	Add to global $\boldsymbol{k}$	Add to global $\boldsymbol{r}$
	$k(i,i) = k(i,i) - \alpha_i k(x_i)$	$r(i) = r(i) + k(x_i)\beta_i$

Thus at the element level the equations are simply

$$(\boldsymbol{k}_k + \boldsymbol{k}_p)\boldsymbol{d} = \boldsymbol{r}_q \Longrightarrow \boldsymbol{kd} = \boldsymbol{r}$$

9.1.3 Numerical Examples

Example 9.1 Using only one element model, shown in Figure 9.5, obtain the finite element solution of the following boundary value problem. Consider 0, 1, and 2 p-modes.

$$\frac{d}{dx}(x^2 u'') + (x-1)u = 0; \qquad 0 < x < 1$$

$$u(0) = 1; \qquad u'(1) = 6$$

For this problem

$k(x) = x^2$; $p(x) = x - 1$; $q(x) = 0$

EBC at $x = 0$: $u = 1$

NBC at $x = 1$: $\alpha = 0, \beta = 6$

Element nodes: 0, 1; $L = 1$

Change of variables: $x(s) = \dfrac{s+1}{2}$

$$k(s) = \tfrac{1}{4}(s+1)^2; \qquad p(s) = \frac{s+1}{2} - 1; \qquad q(s) = 0$$

With no p-modes we get the standard linear element equations as follows:

Interpolation functions, $\boldsymbol{N}^T = \begin{pmatrix} \dfrac{1-s}{2} & \dfrac{s+1}{2} \end{pmatrix}$

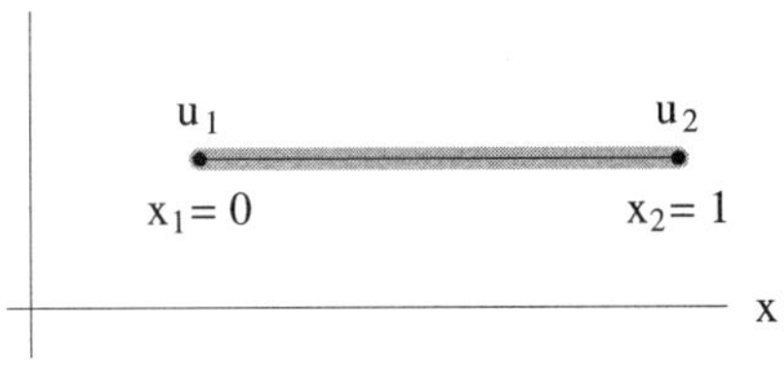

Figure 9.5.

$$B^T = dN^T/dx = (2/L)\,dN^T/ds = \begin{pmatrix} -1 & 1 \end{pmatrix}$$

$$k_k = \int_{-1}^{1} \left(\tfrac{1}{4}(s+1)^2\right) BB^T\, L/2\, ds = \begin{pmatrix} \frac{1}{3} & -\frac{1}{3} \\ -\frac{1}{3} & \frac{1}{3} \end{pmatrix}$$

$$k_p = \int_{-1}^{1} \tfrac{1}{2}\left(1 - \tfrac{1}{2}(s+1)\right) NN^T L\, ds = \begin{pmatrix} \frac{1}{4} & \frac{1}{12} \\ \frac{1}{12} & \frac{1}{12} \end{pmatrix}$$

The complete element equations are as follows:

$$\begin{pmatrix} \frac{7}{12} & -\frac{1}{4} \\ -\frac{1}{4} & \frac{5}{12} \end{pmatrix} \begin{pmatrix} u_1 \\ u_2 \end{pmatrix} = \begin{pmatrix} 0 \\ 0 \end{pmatrix}$$

The natural boundary conditions are

dof	α	β
u_2	0	6

dof	$k(x)$	$-k(x)\alpha$	$k(x)\beta$
u_2	1	0	6

The global equations after incorporating the NBC are

$$\begin{pmatrix} \frac{7}{12} & -\frac{1}{4} \\ -\frac{1}{4} & \frac{5}{12} \end{pmatrix} \begin{pmatrix} u_1 \\ u_2 \end{pmatrix} = \begin{pmatrix} 0 \\ 6 \end{pmatrix}$$

The essential boundary condition is

dof	Value
u_1	1

Incorporating the EBC, the final system of equations is as follows:

$$\left(\tfrac{5}{12}\right)(u_2) = \left(\tfrac{25}{4}\right)$$

Solution of global equations: $\{u_1 \to 1, u_2 \to 15\}$

The solution over elements is as follows:

Element 1

Nodes: $\{x_1 \to 0, x_2 \to 1\}$

Interpolation functions: $N^T = \left\{\tfrac{1}{2}(2 - 2x), x\right\}$

dof values: $d^T = \{1, 15\}$

Solution: $u(x) = N^T d = 14x + 1$

With one p-mode the element equations are as follows:

$$\text{Interpolation functions, } \boldsymbol{N}^T = \begin{pmatrix} \frac{1-s}{2} & \frac{s+1}{2} & \frac{1}{2}\sqrt{\frac{3}{2}}(s^2-1) \end{pmatrix}$$

$$\boldsymbol{B}^T = \frac{d\boldsymbol{N}^T}{dx} = (2/L)\frac{d\boldsymbol{N}^T}{ds} = \begin{pmatrix} -1 & 1 & \sqrt{6}s \end{pmatrix}$$

$$\boldsymbol{k}_k = \int_{-1}^{1} \tfrac{1}{2}\left(\tfrac{1}{4}(s+1)^2\right)\boldsymbol{B}\boldsymbol{B}^T L\, ds = \begin{pmatrix} \frac{1}{3} & -\frac{1}{3} & -\frac{1}{\sqrt{6}} \\ -\frac{1}{3} & \frac{1}{3} & \frac{1}{\sqrt{6}} \\ -\frac{1}{\sqrt{6}} & \frac{1}{\sqrt{6}} & \frac{4}{5} \end{pmatrix}$$

$$\boldsymbol{k}_p = \int_{-1}^{1} \tfrac{1}{2}\left(1-\tfrac{1}{2}(s+1)\right)\boldsymbol{N}\boldsymbol{N}^T L\, ds = \begin{pmatrix} \frac{1}{4} & \frac{1}{12} & -\frac{\sqrt{\frac{3}{2}}}{10} \\ \frac{1}{12} & \frac{1}{12} & -\frac{1}{5\sqrt{6}} \\ -\frac{\sqrt{\frac{3}{2}}}{10} & -\frac{1}{5\sqrt{6}} & \frac{1}{10} \end{pmatrix}$$

Note that the first 2×2 submatrix in these two matrices are exactly the same as those for the previous case of no p-modes. This is a direct result of orthogonality of the p-modes. In an actual implementation of p-formulation, we can take advantage of this fact and compute only the new terms that are added to the equations when considering higher p-modes. The complete element equations are as follows:

$$\begin{pmatrix} \frac{7}{12} & -\frac{1}{4} & -\frac{13}{10\sqrt{6}} \\ -\frac{1}{4} & \frac{5}{12} & \frac{2\sqrt{\frac{2}{3}}}{5} \\ -\frac{13}{10\sqrt{6}} & \frac{2\sqrt{\frac{2}{3}}}{5} & \frac{9}{10} \end{pmatrix} \begin{pmatrix} u_1 \\ u_2 \\ \delta_1^{(i)} \end{pmatrix} = \begin{pmatrix} 0 \\ 0 \\ 0 \end{pmatrix}$$

The superscript over δ indicates the element number while the subscript indicates the parameter number.

With only a single element the global equations before boundary conditions are the same as the equations for the first element:

$$\begin{pmatrix} 0.583333 & -0.25 & -0.530723 \\ -0.25 & 0.416667 & 0.326599 \\ -0.530723 & 0.326599 & 0.9 \end{pmatrix} \begin{pmatrix} u_1 \\ u_2 \\ \delta_1^{(1)} \end{pmatrix} = \begin{pmatrix} 0 \\ 0 \\ 0 \end{pmatrix}$$

The natural boundary conditions are

dof	$k(x)$	$-k(x)\alpha$	$k(x)\beta$
u_2	1.	0	6.

The global equations after incorporating NBC are

$$\begin{pmatrix} 0.583333 & -0.25 & -0.530723 \\ -0.25 & 0.416667 & 0.326599 \\ -0.530723 & 0.326599 & 0.9 \end{pmatrix} \begin{pmatrix} u_1 \\ u_2 \\ \delta_1^{(1)} \end{pmatrix} = \begin{pmatrix} 0 \\ 6. \\ 0 \end{pmatrix}$$

The essential boundary condition is

dof	Value
u_1	1.

Incorporating the EBC, the final system of equations is as follows:

$$\begin{pmatrix} 0.416667 & 0.326599 \\ 0.326599 & 0.9 \end{pmatrix} \begin{pmatrix} u_2 \\ \delta_1^{(1)} \end{pmatrix} = \begin{pmatrix} 6.25 \\ 0.530723 \end{pmatrix}$$

Solution of global equations: $\left\{u_1 \to 1., u_2 \to 20.3168, \delta_1^{(1)} \to -6.78301\right\}$

The solution over elements is as follows:

Element 1

Nodes: $\{x_1 \to 0, x_2 \to 1\}$

Interpolation functions: $N^T = \left\{\frac{1}{2}(2 - 2x), x, \frac{\frac{3}{2}(2x-1)^2 - \frac{3}{2}}{\sqrt{6}}\right\}$

dof values: $d^T = \{1., 20.3168, -6.78301\}$

Solution: $u(x) = N^T d = -16.6149x^2 + 35.9317x + 1.$

With two p-modes the element equations are as follows:

$$\text{Interpolation functions: } N^T = \begin{pmatrix} \frac{1-s}{2} & \frac{s+1}{2} & \frac{1}{2}\sqrt{\frac{3}{2}}(s^2-1) & \frac{1}{2}\sqrt{\frac{5}{2}}s(s^2-1) \end{pmatrix}$$

$$B^T = \frac{dN^T}{dx} = \left(\frac{2}{L}\right)\frac{dN^T}{ds} = \begin{pmatrix} -1 & 1 & \sqrt{6}s & \sqrt{\frac{5}{2}}(3s^2-1) \end{pmatrix}$$

$$k_k = \int_{-1}^{1} \frac{1}{2}\left(\frac{1}{4}(s+1)^2\right) BB^T L\, ds = \begin{pmatrix} \frac{1}{3} & -\frac{1}{3} & -\frac{1}{\sqrt{6}} & -\frac{1}{3\sqrt{10}} \\ -\frac{1}{3} & \frac{1}{3} & \frac{1}{\sqrt{6}} & \frac{1}{3\sqrt{10}} \\ -\frac{1}{\sqrt{6}} & \frac{1}{\sqrt{6}} & \frac{4}{5} & \frac{2}{\sqrt{15}} \\ -\frac{1}{3\sqrt{10}} & \frac{1}{3\sqrt{10}} & \frac{2}{\sqrt{15}} & \frac{16}{21} \end{pmatrix}$$

$$k_p = \int_{-1}^{1} \tfrac{1}{2}\left(1 - \tfrac{1}{2}(s+1)\right) NN^T L\, ds = \begin{pmatrix} \frac{1}{4} & \frac{1}{12} & -\frac{\sqrt{\frac{3}{2}}}{10} & \frac{1}{6\sqrt{10}} \\ \frac{1}{12} & \frac{1}{12} & -\frac{1}{5\sqrt{6}} & 0 \\ -\frac{\sqrt{\frac{3}{2}}}{10} & -\frac{1}{5\sqrt{6}} & \frac{1}{10} & -\frac{1}{14\sqrt{15}} \\ \frac{1}{6\sqrt{10}} & 0 & -\frac{1}{14\sqrt{15}} & \frac{1}{42} \end{pmatrix}$$

The complete element equations are as follows:

$$\begin{pmatrix} \frac{7}{12} & -\frac{1}{4} & -\frac{13}{10\sqrt{6}} & -\frac{1}{6\sqrt{10}} \\ -\frac{1}{4} & \frac{5}{12} & \frac{2\sqrt{\frac{2}{3}}}{5} & \frac{1}{3\sqrt{10}} \\ -\frac{13}{10\sqrt{6}} & \frac{2\sqrt{\frac{2}{3}}}{5} & \frac{9}{10} & \frac{9\sqrt{\frac{3}{5}}}{14} \\ -\frac{1}{6\sqrt{10}} & \frac{1}{3\sqrt{10}} & \frac{9\sqrt{\frac{3}{5}}}{14} & \frac{11}{14} \end{pmatrix} \begin{pmatrix} u_1 \\ u_2 \\ \delta_1^{(i)} \\ \delta_2^{(i)} \end{pmatrix} = \begin{pmatrix} 0 \\ 0 \\ 0 \\ 0 \end{pmatrix}$$

Once again, note that the lower p-mode equations are contained in the higher ones.

The global equations before the boundary conditions are

$$\begin{pmatrix} 0.583333 & -0.25 & -0.530723 & -0.0527046 \\ -0.25 & 0.416667 & 0.326599 & 0.105409 \\ -0.530723 & 0.326599 & 0.9 & 0.497955 \\ -0.0527046 & 0.105409 & 0.497955 & 0.785714 \end{pmatrix} \begin{pmatrix} u_1 \\ u_2 \\ \delta_1^{(1)} \\ \delta_2^{(1)} \end{pmatrix} = \begin{pmatrix} 0 \\ 0 \\ 0 \\ 0 \end{pmatrix}$$

The natural boundary conditions are

dof	$k(x)$	$-k(x)\alpha$	$k(x)\beta$
u_2	1.	0	6.

The global equations after incorporating the NBC are

$$\begin{pmatrix} 0.583333 & -0.25 & -0.530723 & -0.0527046 \\ -0.25 & 0.416667 & 0.326599 & 0.105409 \\ -0.530723 & 0.326599 & 0.9 & 0.497955 \\ -0.0527046 & 0.105409 & 0.497955 & 0.785714 \end{pmatrix} \begin{pmatrix} u_1 \\ u_2 \\ \delta_1^{(1)} \\ \delta_2^{(1)} \end{pmatrix} = \begin{pmatrix} 0 \\ 6. \\ 0 \\ 0 \end{pmatrix}$$

The essential boundary condition is

dof	Value
u_1	1.

Incorporating the EBC, the final system of equations is as follows:

$$\begin{pmatrix} 0.416667 & 0.326599 & 0.105409 \\ 0.326599 & 0.9 & 0.497955 \\ 0.105409 & 0.497955 & 0.785714 \end{pmatrix} \begin{pmatrix} u_2 \\ \delta_1^{(1)} \\ \delta_2^{(1)} \end{pmatrix} = \begin{pmatrix} 6.25 \\ 0.530723 \\ 0.0527046 \end{pmatrix}$$

Solution of global equations:

$$\left\{u_1 \to 1., u_2 \to 20.9794, \delta_1^{(1)} \to -8.47513, \delta_2^{(1)} \to 2.62375\right\}$$

Solution over elements is as follows:

Element 1

Nodes: $\{x_1 \to 0, x_2 \to 1\}$

Interpolation functions:

$$N^T = \left\{\frac{1}{2}(2 - 2x), x, \frac{\frac{3}{2}(2x-1)^2 - \frac{3}{2}}{\sqrt{6}}, \frac{\frac{5}{2}(2x-1)^3 - \frac{5}{2}(2x-1)}{\sqrt{10}}\right\}$$

dof values: $\boldsymbol{d}^T = \{1., 20.9794, -8.47513, 2.62375\}$

Solution: $u(x) = \boldsymbol{N}^T\boldsymbol{d} = 16.5941x^3 - 45.6508x^2 + 49.0361x + 1.$

These solutions, together with the one obtained by using three p-modes, are shown in Figure 9.6. The solutions are clearly converging. The plot of the first derivative shows that there is still room for further improvement. The natural boundary condition is not satisfied yet either but the solution is getting closer. The first derivative values at $x = 1$ are as follows:

	Computed $u'(1)$
Linear solution	14.
With 1 p-mode	2.70186
With 2 p-modes	7.51665
With 3 p-modes	5.17712

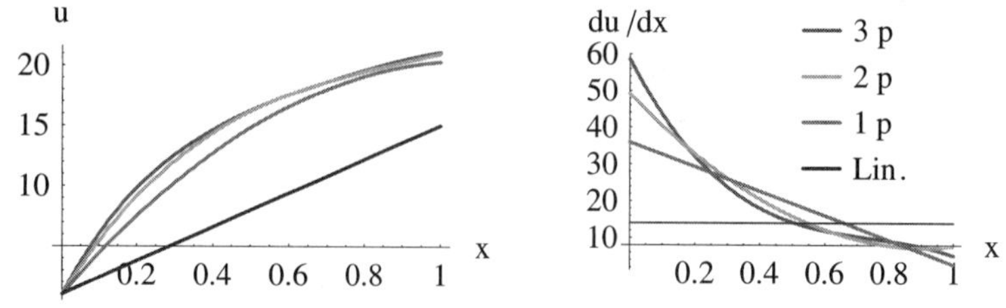

Figure 9.6.

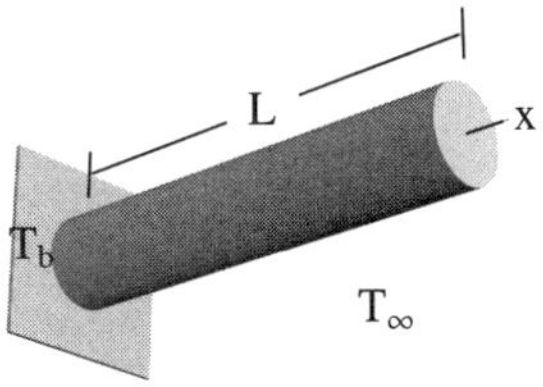

Figure 9.7. Heat transfer through a circular fin

Example 9.2 A 1 m × 1 m plate is at a temperature of 100°F. It is to be cooled by using 3-cm-long and 0.25-cm-diameter aluminum fins ($k = 237$ W/m·°C) with a center-to-center distance of 0.6 cm. A typical fin is as shown in Figure 9.7. The ambient air temperature is 30°C and average convection coefficient is 35 W/m^2 · °C. Determine the temperature distribution through the fins. What is the rate of heat transfer from the entire finned surface of the plate?

Assuming that the surrounding temperature T_∞ is uniform, the problem can be treated as one dimensional with the governing differential equation as follows:

$$\frac{d}{dx}\left(k_x A \frac{dT}{dx}\right) - hP(T - T_\infty) = 0; \qquad 0 < x < L$$

$$T = T_b \quad \text{at} \quad x = 0 \quad \text{and} \quad -k_x A \frac{dT}{dx} = hA(T - T_\infty) \quad \text{at} \quad x = L$$

where k_x = thermal conductivity, h = convection coefficient, $P = 2\pi r$ = surface area per unit length over which convection takes place, r = radius of the fin, $A = \pi r^2$ = area of cross section, and T_∞ = ambient air temperature. Once the temperature distribution is known, the total heat loss from a fin can be computed using the following equation:

$$\text{Heat loss} = \int_0^L hP(T(x) - T_\infty)\, dx$$

Comparing with the general form, we have

$$k_x = kA; \qquad p = -hP; \qquad q = hPT_\infty$$

$$\text{EBC at } x = 0: \quad T = T_b; \qquad \text{NBC at } x = L: \quad \alpha = -\frac{hA}{kA}; \qquad \beta = \frac{hAT_\infty}{kA}$$

The problem is solved using two elements with two p-modes each. The complete solution details are as follows:

Derivation of element equations:

Element nodes: 0, L; $\quad L = L$

Change of variables: $x(s) = \frac{1}{2}L(s + 1)$

$k(s) = kA$; $p(s) = -hP$; $q(s) = hPT_\infty$

$$\text{Interpolation functions, } N^T = \begin{pmatrix} \frac{1-s}{2} & \frac{s+1}{2} & \frac{1}{2}\sqrt{\frac{3}{2}}(s^2 - 1) & \frac{1}{2}\sqrt{\frac{5}{2}}s(s^2 - 1) \end{pmatrix}$$

$$\boldsymbol{B}^T = \frac{d\boldsymbol{N}^T}{dx} = \left(\frac{2}{L}\right)\frac{d\boldsymbol{N}^T}{ds} = \begin{pmatrix} -\frac{1}{L} & \frac{1}{L} & \frac{\sqrt{6}s}{L} & \frac{\sqrt{\frac{5}{2}}(3s^2-1)}{L} \end{pmatrix}$$

$$\boldsymbol{k}_k = \int_{-1}^{1} \tfrac{1}{2}(kA)\boldsymbol{B}\boldsymbol{B}^T L\,ds = \begin{pmatrix} \frac{kA}{L} & -\frac{kA}{L} & 0 & 0 \\ -\frac{kA}{L} & \frac{kA}{L} & 0 & 0 \\ 0 & 0 & \frac{2kA}{L} & 0 \\ 0 & 0 & 0 & \frac{2kA}{L} \end{pmatrix}$$

$$\boldsymbol{k}_p = \int_{-1}^{1} \tfrac{1}{2}(hP)\boldsymbol{N}\boldsymbol{N}^T L\,ds = \begin{pmatrix} \frac{hPL}{3} & \frac{hPL}{6} & -\frac{hPL}{2\sqrt{6}} & \frac{hPL}{6\sqrt{10}} \\ \frac{hPL}{6} & \frac{hPL}{3} & -\frac{hPL}{2\sqrt{6}} & -\frac{hPL}{6\sqrt{10}} \\ -\frac{hPL}{2\sqrt{6}} & -\frac{hPL}{2\sqrt{6}} & \frac{hPL}{5} & 0 \\ \frac{hPL}{6\sqrt{10}} & -\frac{hPL}{6\sqrt{10}} & 0 & \frac{hPL}{21} \end{pmatrix}$$

$$\boldsymbol{r}_q^T = \int_{-1}^{1} \tfrac{1}{2}(hPT_\infty)NL\,ds = \left\{\frac{hPT_\infty L}{2}, \frac{hPT_\infty L}{2}, -\frac{hPT_\infty L}{\sqrt{6}}, 0\right\}$$

The complete element equations are as follows:

$$\begin{pmatrix} \frac{kA}{L}+\frac{hPL}{3} & \frac{hPL}{6}-\frac{kA}{L} & -\frac{hPL}{2\sqrt{6}} & \frac{hPL}{6\sqrt{10}} \\ \frac{hPL}{6}-\frac{kA}{L} & \frac{kA}{L}+\frac{hPL}{3} & -\frac{hPL}{2\sqrt{6}} & -\frac{hPL}{6\sqrt{10}} \\ -\frac{hPL}{2\sqrt{6}} & -\frac{hPL}{2\sqrt{6}} & \frac{2kA}{L}+\frac{hPL}{5} & 0 \\ \frac{hPL}{6\sqrt{10}} & -\frac{hPL}{6\sqrt{10}} & 0 & \frac{2kA}{L}+\frac{hPL}{21} \end{pmatrix} \begin{pmatrix} T_1 \\ T_2 \\ \delta_1^{(i)} \\ \delta_2^{(i)} \end{pmatrix} = \begin{pmatrix} \frac{hPT_\infty L}{2} \\ \frac{hPT_\infty L}{2} \\ -\frac{hPT_\infty L}{\sqrt{6}} \\ 0 \end{pmatrix}$$

$k(x) = 0.00465348;\quad p(x) = -0.549779;\quad q(x) = 16.4934$

Two-element solution:

Nodal locations: {0, 0.015, 0.03}

Element 1:

Element nodes: $\{x_1 \to 0, x_2 \to 0.015\}$

$$\begin{pmatrix} 0.312981 & -0.308858 & -0.00168335 & 0.000434638 \\ -0.308858 & 0.312981 & -0.00168335 & -0.000434638 \\ -0.00168335 & -0.00168335 & 0.622114 & 0 \\ 0.000434638 & -0.000434638 & 0 & 0.620857 \end{pmatrix} \begin{pmatrix} T_1 \\ T_2 \\ \delta_1^{(1)} \\ \delta_2^{(1)} \end{pmatrix} = \begin{pmatrix} 0.1237 \\ 0.1237 \\ -0.101001 \\ 0 \end{pmatrix}$$

Element 2:

Element nodes: $\{x_2 \to 0.015, x_3 \to 0.03\}$

$$\begin{pmatrix} 0.312981 & -0.308858 & -0.00168335 & 0.000434638 \\ -0.308858 & 0.312981 & -0.00168335 & -0.000434638 \\ -0.00168335 & -0.00168335 & 0.622114 & 0 \\ 0.000434638 & -0.000434638 & 0 & 0.620857 \end{pmatrix} \begin{pmatrix} T_2 \\ T_3 \\ \delta_1^{(2)} \\ \delta_2^{(2)} \end{pmatrix} = \begin{pmatrix} 0.1237 \\ 0.1237 \\ -0.101001 \\ 0 \end{pmatrix}$$

Global equations before boundary conditions:

$$\begin{pmatrix} 0.312981 & -0.308858 & 0 & -0.00168335 & 0 & 0.000434638 & 0 \\ -0.308858 & 0.625962 & -0.308858 & -0.00168335 & -0.00168335 & -0.000434638 & 0.000434638 \\ 0 & -0.308858 & 0.312981 & 0 & -0.00168335 & 0 & -0.000434638 \\ -0.00168335 & -0.00168335 & 0 & 0.622114 & 0 & 0 & 0 \\ 0 & -0.00168335 & -0.00168335 & 0 & 0.622114 & 0 & 0 \\ 0.000434638 & -0.000434638 & 0 & 0 & 0 & 0.620857 & 0 \\ 0 & 0.000434638 & -0.000434638 & 0 & 0 & 0 & 0.620857 \end{pmatrix} \begin{pmatrix} T_1 \\ T_2 \\ T_3 \\ \delta_1^{(1)} \\ \delta_1^{(2)} \\ \delta_2^{(1)} \\ \delta_2^{(2)} \end{pmatrix} = \begin{pmatrix} 0.1237 \\ 0.2474 \\ 0.1237 \\ -0.101001 \\ -0.101001 \\ 0 \\ 0 \end{pmatrix}$$

Natural boundary conditions:

dof	α	β
T_3	−0.147679	4.43038

dof	$k(x)$	$-k(x)\alpha$	$k(x)\beta$
T_3	0.00465348	0.000687223	0.0206167

Global equations after incorporating NBC:

$$\begin{pmatrix} 0.312981 & -0.308858 & 0 & -0.00168335 & 0 & 0.000434638 & 0 \\ -0.308858 & 0.625962 & -0.308858 & -0.00168335 & -0.00168335 & -0.000434638 & 0.000434638 \\ 0 & -0.308858 & 0.313668 & 0 & -0.00168335 & 0 & -0.000434638 \\ -0.00168335 & -0.00168335 & 0 & 0.622114 & 0 & 0 & 0 \\ 0 & -0.00168335 & -0.00168335 & 0 & 0.622114 & 0 & 0 \\ 0.000434638 & -0.000434638 & 0 & 0 & 0 & 0.620857 & 0 \\ 0 & 0.000434638 & -0.000434638 & 0 & 0 & 0 & 0.620857 \end{pmatrix} \begin{pmatrix} T_1 \\ T_2 \\ T_3 \\ \delta_1^{(1)} \\ \delta_1^{(2)} \\ \delta_2^{(1)} \\ \delta_2^{(2)} \end{pmatrix} = \begin{pmatrix} 0.1237 \\ 0.2474 \\ 0.144317 \\ -0.101001 \\ -0.101001 \\ 0 \\ 0 \end{pmatrix}$$

Essential boundary condition:

dof	Value
T_1	100

Incorporating the EBC, the final system of equations is as follows:

$$\begin{pmatrix} 0.625962 & -0.308858 & -0.00168335 & -0.00168335 & -0.000434638 & 0.000434638 \\ -0.308858 & 0.313668 & 0 & -0.00168335 & 0 & -0.000434638 \\ -0.00168335 & 0 & 0.622114 & 0 & 0 & 0 \\ -0.00168335 & -0.00168335 & 0 & 0.622114 & 0 & 0 \\ -0.000434638 & 0 & 0 & 0 & 0.620857 & 0 \\ 0.000434638 & -0.000434638 & 0 & 0 & 0 & 0.620857 \end{pmatrix} \begin{pmatrix} T_2 \\ T_3 \\ \delta_1^{(1)} \\ \delta_1^{(2)} \\ \delta_2^{(1)} \\ \delta_2^{(2)} \end{pmatrix} = \begin{pmatrix} 31.1332 \\ 0.144317 \\ 0.0673339 \\ -0.101001 \\ -0.0434638 \\ 0. \end{pmatrix}$$

Solution of global equations:

$$\{T_1 \to 100, T_2 \to 97.1818, T_3 \to 96.1534, \delta_1^{(1)} \to 0.371193, \delta_1^{(2)} \to 0.360785,$$
$$\delta_2^{(1)} \to -0.00197291, \delta_2^{(2)} \to -0.00071994\}$$

Solution over element 1:

Nodes: $\{x_1 \to 0, x_2 \to 0.015\}$

Interpolation functions:

$$N^T = \left\{ \frac{1}{2}(1 - 66.6667(2x - 0.015)), \frac{1}{2}(66.6667(2x - 0.015) + 1), \frac{6666.67(2x - 0.015)^2 - \frac{3}{2}}{\sqrt{6}}, \frac{740741.(2x - 0.015)^3 - 166.667(2x - 0.015)}{\sqrt{10}} \right\}$$

dof values: $\boldsymbol{d}^T = \{100, 97.1818, 0.371193, -0.00197291\}$

Solution: $T(x) = \boldsymbol{N}^T\boldsymbol{d} = -3697.13x^3 + 4124.23x^2 - 248.912x + 100.$

Solution over element 2:

Nodes: $\{x_1 \to 0.015, x_2 \to 0.03\}$

Interpolation functions:

$$N^T = \left\{ \frac{1}{2}(1 - 66.6667(2x - 0.045)), \frac{1}{2}(66.6667(2x - 0.045) + 1), \frac{6666.67(2x - 0.045)^2 - \frac{3}{2}}{\sqrt{6}}, \frac{740741.(2x - 0.045)^3 - 166.667(2x - 0.045)}{\sqrt{10}} \right\}$$

dof values: $\boldsymbol{d}^T = \{97.1818, 96.1534, 0.360785, -0.00071994\}$

Solution: $T(x) = \boldsymbol{N}^T\boldsymbol{d} = -1349.13x^3 + 4018.8x^2 - 247.281x + 99.9913$

Solution summary:

	Range	Solution
1	$0 \leq x \leq 0.015$	$-3697.13x^3 + 4124.23x^2 - 248.912x + 100.$
2	$0.015 \leq x \leq 0.03$	$-1349.13x^3 + 4018.8x^2 - 247.281x + 99.9913$

A plot of the solution is shown in Figure 9.8.

$\{-3697.13x^3 + 4124.23x^2 - 248.912x + 100., -1349.13x^3 + 4018.8x^2 - 247.281x + 99.9913\}$

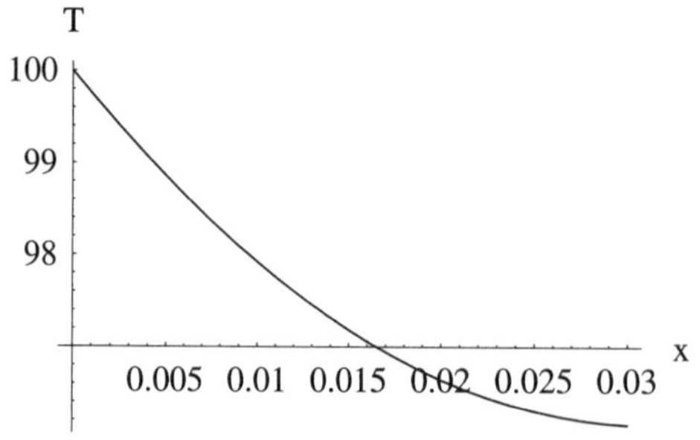

Figure 9.8. Temperature distribution in the fin

The heat loss over the length of a fin is determined as follows:

$$\int_0^L hP(T(x) - T_\infty)\,dx$$

$$= \int_0^{0.015} 0.549779(-3697.13x^3 + 4124.23x^2 - 248.912x + 100. - 30)\,dx$$

$$+ \int_{0.015}^{0.03} 0.549779(-1349.13x^3 + 4018.8x^2 - 247.281x + 99.9913 - 30)\,dx$$

$$= 1.11297$$

The following analytical solution for heat loss is available for the problem:

$$\mathtt{m} = \sqrt{\mathtt{hP/(kA)}};\ \sqrt{\mathtt{hPkA}}\,(\mathtt{Tb} - \mathtt{T}_\infty)\frac{\mathtt{Sinh[mL]} + \mathtt{h/(mk)\,Cosh[mL]}}{\mathtt{Cosh[mL]} + \mathtt{h/(mk)\,Sinh[mL]}}$$

1.07569

The finite element solution compares fairly well with the analytical solution. To compute the heat loss from the entire plate surface, we need to determine the total number of fins and multiply the heat loss per fin by this number:

$$\text{Number of fins} = \frac{1\,\text{m} \times 1\,\text{m}}{(0.006\,\text{m})(0.006\,\text{m})} = 27{,}778 \text{ fins}$$

$$\text{Heat loss through all fins} = 27778 \times 1.113 = 30{,}917\,\text{W}$$

As mentioned earlier, because of the orthogonality properties of p-modes, additional p-modes simply add new rows and columns to the element $\boldsymbol{k}$ matrix. The existing entries in the matrix do not change. By ordering the global equations as we have done in this example, the global $\boldsymbol{K}$ matrix also retains this property. Thus adding new p-modes simply requires computation of the new element rows and columns and their assembly into the existing global equations. This is clearly demonstrated in the following global equations obtained with two elements and zero, one, and two p-modes.

With no p-modes we get the following global equations:

$$\begin{pmatrix} 0.312981 & -0.308858 & 0 \\ -0.308858 & 0.625962 & -0.308858 \\ 0 & -0.308858 & 0.312981 \end{pmatrix} \begin{pmatrix} T_1 \\ T_2 \\ T_3 \end{pmatrix} = \begin{pmatrix} 0.1237 \\ 0.2474 \\ 0.1237 \end{pmatrix}$$

With one p-mode for each element two new rows and columns are added to the global equations. Note the first 3×3 system of equations is exactly the same as that with no p-modes:

$$\begin{pmatrix} 0.312981 & -0.308858 & 0 & -0.00168335 & 0 \\ -0.308858 & 0.625962 & -0.308858 & -0.00168335 & -0.00168335 \\ 0 & -0.308858 & 0.312981 & 0 & -0.00168335 \\ -0.00168335 & -0.00168335 & 0 & 0.622114 & 0 \\ 0 & -0.00168335 & -0.00168335 & 0 & 0.622114 \end{pmatrix} \begin{pmatrix} T_1 \\ T_2 \\ T_3 \\ \delta_1^{(1)} \\ \delta_1^{(2)} \end{pmatrix} = \begin{pmatrix} 0.1237 \\ 0.2474 \\ 0.1237 \\ -0.101001 \\ -0.101001 \end{pmatrix}$$

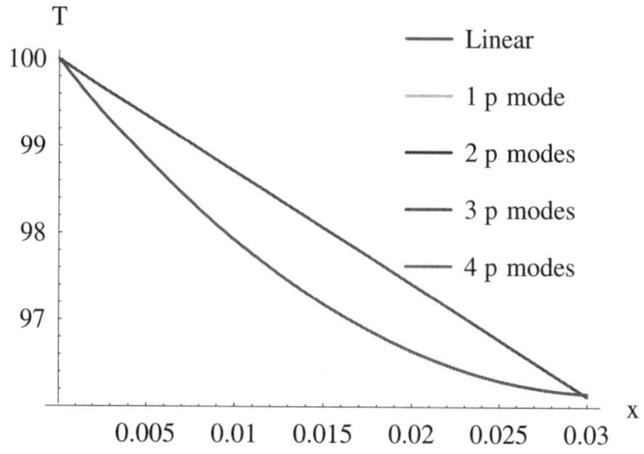

Figure 9.9. Comparison of solutions using one element with zero to four p-modes

With two p-modes for each element the first 5×5 system of equations is exactly the same as that with one p-mode:

$$\begin{pmatrix} 0.312981 & -0.308858 & 0 & -0.00168335 & 0 & 0.000434638 & 0 \\ -0.308858 & 0.625962 & -0.308858 & -0.00168335 & -0.00168335 & -0.000434638 & 0.000434638 \\ 0 & -0.308858 & 0.312981 & 0 & -0.00168335 & 0 & -0.000434638 \\ -0.00168335 & -0.00168335 & 0 & 0.622114 & 0 & 0 & 0 \\ 0 & -0.00168335 & -0.00168335 & 0 & 0.622114 & 0 & 0 \\ 0.000434638 & -0.000434638 & 0 & 0 & 0 & 0.620857 & 0 \\ 0 & 0.000434638 & -0.000434638 & 0 & 0 & 0 & 0.620857 \end{pmatrix} \begin{pmatrix} T_1 \\ T_2 \\ T_3 \\ \delta_1^{(1)} \\ \delta_1^{(2)} \\ \delta_2^{(1)} \\ \delta_2^{(2)} \end{pmatrix} = \begin{pmatrix} 0.1237 \\ 0.2474 \\ 0.1237 \\ -0.101001 \\ -0.101001 \\ 0 \\ 0 \end{pmatrix}$$

Thus, taking advantage of this hierarchical nature of element and global equations, it is possible to create very efficient computer implementations based on the p-formulation.

The solution convergence can be demonstrated by evaluating solutions with increasing number of p-modes. Figure 9.9 shows solutions obtained from using only one element with zero through four p-modes. Except for the linear solution, all others are practically indistinguishable from each other.

9.2 *p*-FORMULATION FOR SECOND-ORDER 2D BVP

In this section we consider p-formulation for the general two-dimensional boundary value problem first introduced in Chapter 5. The differential equation defined over an arbitrary two-dimensional area is as follows:

$$\frac{\partial}{\partial x}\left(k_x \frac{\partial u}{\partial x}\right) + \frac{\partial}{\partial y}\left(k_y \frac{\partial u}{\partial y}\right) + pu + q = 0$$

where $k_x(x, y)$, $k_y(x, y)$, $p(x, y)$, and $q(x, y)$ are known functions defined over the area. The solution variable is $u(x, y)$. The area of the solution domain is denoted by A and its boundary by C. The boundary is defined in terms of coordinate c and an outer unit normal to the boundary $\boldsymbol{n}$. The x and y components of the normal vector are denoted by n_x and n_y:

$$n = \begin{pmatrix} n_x \\ n_y \end{pmatrix}; |\boldsymbol{n}| = \sqrt{n_x^2 + n_y^2} = 1$$

The boundary conditions are of the following types:

(i) Essential boundary condition specified on portion of the boundary indicated by C_e:

$$u(c) \text{ specified on } C_e$$

(ii) Natural boundary condition specified on portion of the boundary indicated by C_n:

$$k\frac{\partial u}{\partial n} \equiv k_x\frac{\partial u}{\partial x}n_x + k_y\frac{\partial u}{\partial y}n_y = \alpha(c)u(c) + \beta(c) \text{ on } C_n$$

9.2.1 *p*-Mode Assumed Solution

The solution domain is discretized into arbitrary four-sided quadrilaterals. As discussed in Chapter 6, a quadrilateral element is mapped to a 2×2 square element called the master element. The mapping between the actual element coordinates and the master element is written as follows:

$$x(s,t) = N_1x_1 + N_2x_2 + \cdots \qquad y(s,t) = N_1y_1 + N_2y_2 + \cdots$$

where $N_1(s,t), N_2(s,t), \ldots$ are serendipity interpolation functions for the master element and $x_1, y_1, \ldots$ are the coordinates of the key points of the actual quadrilateral element. The interpolation functions for mapping a quadrilateral with straight sides and one with all sides curved, are as follows:

Interpolation functions for mapping when all four sides are straight lines: four key points (Figure 9.10):

$$N = \begin{pmatrix} \frac{1}{4}(-1+s)(-1+t) \\ -\frac{1}{4}(1+s)(-1+t) \\ \frac{1}{4}(1+s)(1+t) \\ -\frac{1}{4}(-1+s)(1+t) \end{pmatrix}$$

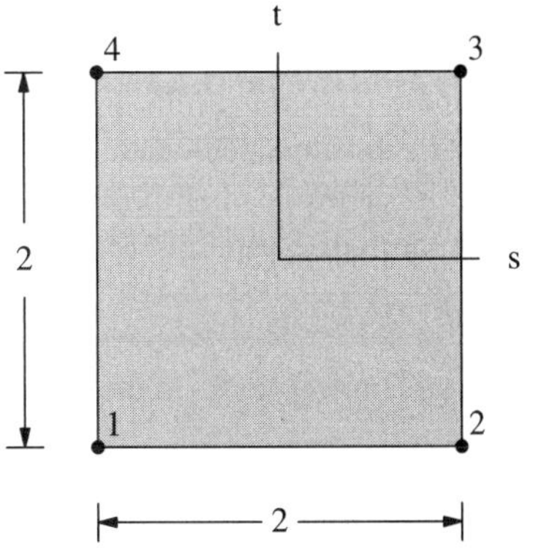

Figure 9.10.

Interpolation functions for mapping when all four sides are quadratic curves: eight key points (Figure 9.11):

$$N = \begin{pmatrix} -\frac{1}{4}(-1+s)(-1+t)(1+s+t) \\ \frac{1}{2}(-1+s^2)(-1+t) \\ \frac{1}{4}(-1+t)(1-s^2+t+st) \\ -\frac{1}{2}(1+s)(-1+t^2) \\ \frac{1}{4}(1+s)(1+t)(-1+s+t) \\ -\frac{1}{2}(-1+s^2)(1+t) \\ \frac{1}{4}(-1+s)(1+s-t)(1+t) \\ \frac{1}{2}(-1+s)(-1+t^2) \end{pmatrix}$$

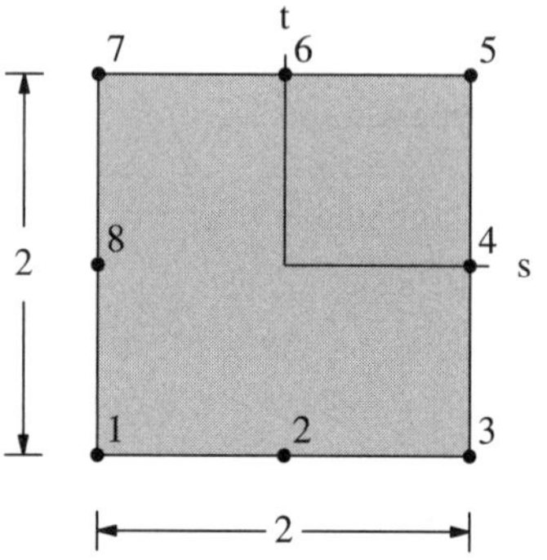

Figure 9.11.

The assumed solution for the p-formulation is developed in terms of s, t for the 2×2 master element. Each element has four corner nodal degrees of freedom u_1, u_2, u_3, and u_4 regardless of the number of key points used for mapping. Thus, even if an element has all sides that are curved and is defined with eight key points for mapping, it still has only four corner degrees of freedom. Linear Lagrange interpolation functions are used to interpolate between the corner nodal unknowns. Higher order solutions are written by including a number of p-modes identified with parameters $\delta_1, \delta_2, \ldots$. Thus the assumed solution is written as follows:

$$u(s,t) = N_1 u_1 + \cdots + N_4 u_4 + P_1 \delta_1 + P_2 \delta_2 + \cdots = \begin{pmatrix} N_1 & \cdots & N_4 & P_1 & \cdots \end{pmatrix} \begin{pmatrix} u_1 \\ \vdots \\ u_4 \\ \delta_1 \\ \vdots \end{pmatrix}$$

$$N_1 = \tfrac{1}{4}(1-s)(1-t); \qquad N_2 = \tfrac{1}{4}(s+1)(1-t); \qquad N_3 = \tfrac{1}{4}(s+1)(t+1);$$
$$N_4 = \tfrac{1}{4}(1-s)(t+1)$$

The p-modes must be chosen in such a way that during assembly it is possible to maintain continuity of the assumed solution over the entire solution domain. With the standard finite element formulation the continuity of the solution was maintained by simply matching appropriate nodal degrees of freedom at a common element interface during assembly. Since now the solution depends on the p-modes as well, we must account for nonzero p-modes across common element interfaces during assembly. To make this process computationally tractable, the p-modes are chosen carefully so that they are either zero on all four element sides or nonzero on one side of the element only. Thus the p-modes are classified as *side modes* and *internal modes*. The side modes are zero along three sides of the element and the internal modes are zero along all four sides of the element. The internal modes are independent for each element and do not require any special consideration during assembly.

The side-mode parameters along the common interface between the two elements must be matched during assembly to maintain continuity of the assumed solution. The process will become clear when we consider numerical examples.

The one-dimensional p-modes based on Legendre polynomials can easily be extended to define both the side and the internal modes. The one-dimensional p-modes in terms of s are as follows:

$$P_i(s) = \frac{1}{\sqrt{2(2i+1)}}(\phi_{i+1}(s) - \phi_{i-1}(s)); \qquad i = 1, 2, \ldots$$

where ϕ_i are the Legendre polynomials. The functions $P_i(s)$ are 0 at both $s = -1$ and $s = 1$. A suitable side mode for side 1-2 ($t = -1$; see Figure 9.10) of the master element must have a value of 1 on side 1-2 and zero on all other sides. This can be accomplished by multiplying $P_i(s)$ by $(1 - t)/2$. Thus suitable p-modes for side 1-2 of the element can be written as follows:

$$P_i^{(1-2)} = \frac{(1-t)/2}{\sqrt{2(2i+1)}}(\phi_{i+1}(s) - \phi_{i-1}(s)); \qquad i = 1, 2, \ldots$$

Following a similar reasoning, suitable p-modes for the other three sides of the element can be written as follows:

$$P_i^{(2-3)} = \frac{(1+s)/2}{\sqrt{2(2i+1)}}(\phi_{i+1}(t) - \phi_{i-1}(t)); \qquad i = 1, 2, \ldots$$

$$P_i^{(3-4)} = \frac{(1+t)/2}{\sqrt{2(2i+1)}}(\phi_{i+1}(s) - \phi_{i-1}(s)); \qquad i = 1, 2, \ldots$$

$$P_i^{(4-1)} = \frac{(1-s)/2}{\sqrt{2(2i+1)}}(\phi_{i+1}(t) - \phi_{i-1}(t)); \qquad i = 1, 2, \ldots$$

The internal modes can be defined by simply taking the product of one-dimensional p-modes in the s and the t directions:

$$P^{ij}(s, t) = P_i(s)P_j(t); \qquad i = 1, 2, \ldots; \qquad j = 1, 2, \ldots$$

Denoting the order of the assumed solution by n, with $n = 1$ for linear, $n = 2$ for quadratic, etc., an assumed solution of any order can be written systematically as follows:

Linear solution: $n = 1$. No p-modes:

$$N^T = \left(\tfrac{1}{4}(1-s)(1-t) \quad \tfrac{1}{4}(s+1)(1-t) \quad \tfrac{1}{4}(s+1)(t+1) \quad \tfrac{1}{4}(1-s)(t+1) \right)$$

Quadratic solution: $n = 2$. Four nodal modes + 4 side modes:

$$N^T = (N_1 \quad N_2 \quad N_3 \quad N_4 \quad P_1^{1-2} \quad P_1^{2-3} \quad P_1^{3-4} \quad P_1^{4-1})$$

Cubic solution: $n = 3$. Four nodal modes + 8 side modes:

$$N^T = (N_1 \quad N_2 \quad N_3 \quad N_4 \quad P_1^{1-2} \quad P_1^{2-3} \quad P_1^{3-4} \quad P_1^{4-1} \quad P_2^{1-2} \quad P_2^{2-3} \quad P_2^{3-4} \quad P_2^{4-1})$$

Higher order solutions: $n \geq 4$. Four nodal modes + $4(n-1)$ side modes + $(n-2)(n-3)/2$ internal modes:

$$\begin{aligned} N^T = (&N_1 \quad N_2 \quad N_3 \quad N_4 \quad P_1^{1-2} \quad P_1^{2-3} \quad P_1^{3-4} \quad P_1^{4-1} \\ &P_2^{1-2} \quad P_2^{2-3} \quad P_2^{3-4} \quad P_2^{4-1} \quad \cdots \quad P^{11} \quad P^{12} \quad \ldots) \end{aligned}$$

The total number of interpolation functions increases rapidly with n. For example, with $n = 5$ we have 4 nodal modes, 16 side modes, and 3 internal modes for a total of 27 interpolation functions for each element. However, the computational effort does not increase that rapidly because of orthogonality properties of the Legendre polynomials used in defining these modes.

Example 9.3 Write explicit expressions for a fourth-order interpolation for a 2×2 master element.

Here $n = 4$ and we have 4 nodal modes, 12 side modes, and 1 internal mode for a total of 17 interpolation functions. Using the above equations the expressions for these interpolation functions are as follows:

Nodal modes	$\frac{1}{4}(1-s)(1-t)$	$\frac{1}{4}(s+1)(1-t)$	$\frac{1}{4}(s+1)(t+1)$	$\frac{1}{4}(1-s)(t+1)$
Side modes: 1st set	$\dfrac{\left(\frac{3s^2}{2}-\frac{3}{2}\right)(1-t)}{2\sqrt{6}}$	$\dfrac{(s+1)\left(\frac{3t^2}{2}-\frac{3}{2}\right)}{2\sqrt{6}}$	$\dfrac{\left(\frac{3s^2}{2}-\frac{3}{2}\right)(t+1)}{2\sqrt{6}}$	$\dfrac{(1-s)\left(\frac{3t^2}{2}-\frac{3}{2}\right)}{2\sqrt{6}}$
Side modes: 2nd set	$\dfrac{\left(\frac{5s^3}{2}-\frac{5s}{2}\right)(1-t)}{2\sqrt{10}}$	$\dfrac{(s+1)\left(\frac{5t^3}{2}-\frac{5t}{2}\right)}{2\sqrt{10}}$	$\dfrac{\left(\frac{5s^3}{2}-\frac{5s}{2}\right)(t+1)}{2\sqrt{10}}$	$\dfrac{(1-s)\left(\frac{5t^3}{2}-\frac{5t}{2}\right)}{2\sqrt{10}}$
Side modes: 3rd set	$\dfrac{\left(\frac{35s^4}{8}-\frac{21s^2}{4}+\frac{7}{8}\right)(1-t)}{2\sqrt{14}}$	$\dfrac{(s+1)\left(\frac{35t^4}{8}-\frac{21t^2}{4}+\frac{7}{8}\right)}{2\sqrt{14}}$	$\dfrac{\left(\frac{35s^4}{8}-\frac{21s^2}{4}+\frac{7}{8}\right)}{2\sqrt{14}}$	$\dfrac{(1-s)\left(\frac{35t^4}{8}-\frac{21t^2}{4}+\frac{7}{8}\right)}{2\sqrt{14}}$
Internal mode	$\frac{1}{6}\left(\frac{3s^2}{2}-\frac{3}{2}\right)\left(\frac{3t^2}{2}-\frac{3}{2}\right)$			

9.2.2 Finite Element Equations

The finite element equations consist of exactly the same terms as those derived in Chapter 5. The only difference is that now we are using different interpolation functions. Thus the element equations are as follows:

$$(\boldsymbol{k}_k + \boldsymbol{k}_p + \boldsymbol{k}_\alpha)\boldsymbol{d} = \boldsymbol{r}_q + \boldsymbol{r}_\beta$$

where

$$\boldsymbol{d} = (u_1 \quad u_2 \quad u_3 \quad u_4 \quad \delta_1 \quad \delta_2 \quad \cdots)^T$$

$$\boldsymbol{k}_k = \iint_A \boldsymbol{BCB}^T\,dA; \qquad \boldsymbol{k}_p = -\iint_A p\boldsymbol{NN}^T\,dA; \qquad \boldsymbol{k}_\alpha = -\int_{C_n} \alpha \boldsymbol{N}_c \boldsymbol{N}_c^T\,dc$$

$$\boldsymbol{r}_q = \iint_A qN\,dA; \qquad \boldsymbol{r}_\beta = \int_{C_n} \beta N_c\,dc$$

N = vector of interpolation functions

N_c = vector of interpolation functions along a given element side

$$\boldsymbol{B}^T = \begin{pmatrix} \frac{\partial N_1}{\partial x} & \frac{\partial N_2}{\partial x} & \cdots \\ \frac{\partial N_1}{\partial y} & \frac{\partial N_2}{\partial y} & \cdots \end{pmatrix} \quad \text{and} \quad \boldsymbol{C} = \begin{pmatrix} k_x & 0 \\ 0 & k_y \end{pmatrix}$$

The quantities $\boldsymbol{k}_\alpha$ and $\boldsymbol{r}_\beta$ are a result of applied natural boundary conditions and will affect only those elements that share the boundary with the NBC. For the interior elements these terms will be zero. The interpolation functions are written over the master element in terms of s, t. However, the element equations need x and y derivatives of these interpolation functions. As shown in Chapter 6 using the mapping $x(s, t)$ and $y(s, t)$ and employing the chain rule of differentiation, the derivatives of the ith interpolation can be written as follows:

$$\boldsymbol{J} = \begin{pmatrix} \frac{\partial x}{\partial s} & \frac{\partial x}{\partial t} \\ \frac{\partial y}{\partial s} & \frac{\partial y}{\partial t} \end{pmatrix} = \begin{pmatrix} J_{11} & J_{12} \\ J_{21} & J_{22} \end{pmatrix}; \qquad \det \boldsymbol{J} = \left| \frac{\partial x}{\partial s}\frac{\partial y}{\partial t} - \frac{\partial x}{\partial t}\frac{\partial y}{\partial s} \right|$$

$$\begin{pmatrix} \frac{\partial N_i}{\partial x} \\ \frac{\partial N_i}{\partial y} \end{pmatrix} = \boldsymbol{J}^{-T} \begin{pmatrix} \frac{\partial N_i}{\partial s} \\ \frac{\partial N_i}{\partial t} \end{pmatrix} = \frac{1}{\det \boldsymbol{J}} \begin{pmatrix} \frac{\partial y}{\partial t} & -\frac{\partial y}{\partial s} \\ -\frac{\partial x}{\partial t} & \frac{\partial x}{\partial s} \end{pmatrix} \begin{pmatrix} \frac{\partial N_i}{\partial s} \\ \frac{\partial N_i}{\partial t} \end{pmatrix} \equiv \frac{1}{\det \boldsymbol{J}} \begin{pmatrix} J_{22} & -J_{21} \\ -J_{12} & J_{11} \end{pmatrix} \begin{pmatrix} \frac{\partial N_i}{\partial s} \\ \frac{\partial N_i}{\partial t} \end{pmatrix}$$

The matrix of derivatives of the interpolation functions $\boldsymbol{B}$ can be written as follows:

$$\boldsymbol{B}^T = \begin{pmatrix} \frac{\partial N_1}{\partial x} & \frac{\partial N_2}{\partial x} & \cdots & \frac{\partial N_n}{\partial x} \\ \frac{\partial N_1}{\partial y} & \frac{\partial N_2}{\partial y} & \cdots & \frac{\partial N_n}{\partial y} \end{pmatrix}$$

$$\boldsymbol{B}^T = \frac{1}{\det \boldsymbol{J}} \begin{pmatrix} J_{22}\frac{\partial N_1}{\partial s} - J_{21}\frac{\partial N_1}{\partial t} & J_{22}\frac{\partial N_2}{\partial s} - J_{21}\frac{\partial N_2}{\partial t} & \cdots & J_{22}\frac{\partial N_n}{\partial s} - J_{21}\frac{\partial N_n}{\partial t} \\ -J_{12}\frac{\partial N_1}{\partial s} + J_{11}\frac{\partial N_1}{\partial t} & -J_{12}\frac{\partial N_2}{\partial s} + J_{11}\frac{\partial N_2}{\partial t} & \cdots & -J_{12}\frac{\partial N_n}{\partial s} + J_{11}\frac{\partial N_n}{\partial t} \end{pmatrix}$$

In order to carry out integration, the given functions k_x, k_y, p, and q must be converted into the s, t form by using the mapping functions. Frequently it is assumed that these quantities are constant over an element, in which case they can simply be taken out of the integral sign. All integrands are expressed in terms of s, t and, therefore, the above element matrices can be established by evaluating integrals of the following form:

$$\boldsymbol{k}_k = \int_{-1}^{1}\int_{-1}^{1} \boldsymbol{BCB}^T \det \boldsymbol{J}\,ds\,dt$$

$$\boldsymbol{k}_p = -\int_{-1}^{1}\int_{-1}^{1} p\boldsymbol{NN}^T \det \boldsymbol{J}\,ds\,dt$$

$$\boldsymbol{r}_q = \int_{-1}^{1}\int_{-1}^{1} q\boldsymbol{N} \det \boldsymbol{J}\,ds\,dt$$

Using an $m \times n$ Gauss quadrature formula the element matrices are then evaluated numerically as follows:

$$\begin{aligned}
\boldsymbol{k}_k &= \int_{-1}^{1}\int_{-1}^{1} \boldsymbol{BCB}^T \det \boldsymbol{J}\, ds\, dt \\
&= \sum_{i=1}^{m}\sum_{j=1}^{n} w_i w_j \boldsymbol{B}(s_i, t_j)\boldsymbol{C}(s_i, t_j)\boldsymbol{B}^T(s_i, t_j) \det \boldsymbol{J}(s_i, t_j) \\
\boldsymbol{k}_p &= -\int_{-1}^{1}\int_{-1}^{1} p\boldsymbol{NN}^T \det \boldsymbol{J}\, ds\, dt \\
&= -\sum_{i=1}^{m}\sum_{j=1}^{n} w_i w_j p(s_i, t_j)\boldsymbol{N}(s_i, t_j)\boldsymbol{N}^T(s_i, t_j) \det \boldsymbol{J}(s_i, t_j) \\
\boldsymbol{r}_q &= \int_{-1}^{1}\int_{-1}^{1} q\boldsymbol{N} \det \boldsymbol{J}\, ds\, dt \\
&= \sum_{i=1}^{m}\sum_{j=1}^{n} w_i w_j q(s_i, t_j)\boldsymbol{N}(s_i, t_j) \det \boldsymbol{J}(s_i, t_j)
\end{aligned}$$

where (s_i, t_j) are the Gauss points and w_i and w_j are the corresponding weights.

Elements involving a boundary where a nonzero natural boundary condition is specified require computation of the $\boldsymbol{k}_\alpha$ matrix and the $\boldsymbol{r}_\beta$ vector:

$$\boldsymbol{k}_\alpha = -\int_{C_n} \alpha \boldsymbol{N}_c \boldsymbol{N}_c^T\, dc; \qquad \boldsymbol{r}_\beta = \int_{C_n} \beta \boldsymbol{N}_c\, dc$$

where c is a coordinate along the boundary and the integration is along the boundary of the element. In general, the element boundary is an arbitrary curve in the x, y coordinate system. After mapping each boundary curve, there is either a horizontal or a vertical line representing a side of the master square element. The boundary coordinate c for each side is mapped to a coordinate $-1 \le a \le 1$, as shown in Figure 9.12. The mapping for each side $x(a)$ and $y(a)$ can be obtained from the element mapping as follows:

For side 1: $a = s$; $t = -1$
For side 2: $s = 1$; $a = t$
For side 3: $a = -s$; $t = 1$
For side 4: $s = -1$; $a = -t$

The boundary integrals along the first side of the element can now be written as follows:

$$\boldsymbol{k}_\alpha = -\int_{-1}^{1} \alpha \boldsymbol{N}_c \boldsymbol{N}_c^T J_c\, da; \qquad \boldsymbol{r}_\beta = \int_{-1}^{1} \beta \boldsymbol{N}_c J_c\, da$$

where J_c is the Jacobian of the side:

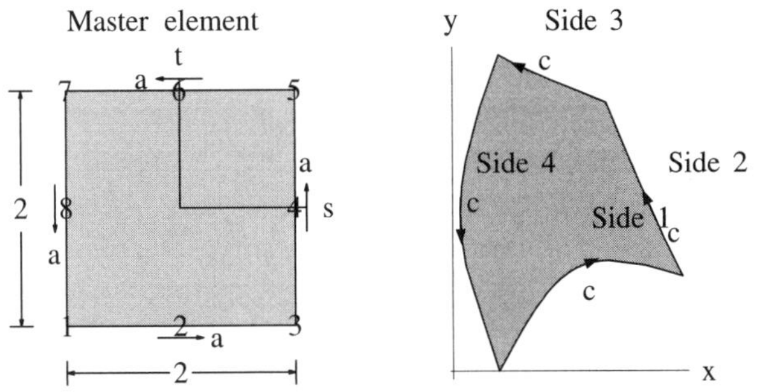

Figure 9.12. Boundary coordinates on the master and the actual element

$$J_c = \sqrt{\left(\frac{dx}{da}\right)^2 + \left(\frac{dy}{da}\right)^2}$$

Of course, the known quantities α and β, if not constant, must be expressed in terms of a using the mapping for the side. Now these integrals can easily be evaluated using one-dimensional Gauss quadrature formulas:

$$\boldsymbol{k}_\alpha = -\int_{-1}^{1} \alpha \boldsymbol{N}_c \boldsymbol{N}_c^T J_c\, da = -\sum_{i=1}^{n} w_i \alpha(a_i) \boldsymbol{N}_c(a_i) \boldsymbol{N}_c^T(a_i) J_c(a_i)$$

$$\boldsymbol{r}_\beta = \int_{-1}^{1} \beta \boldsymbol{N}_c J_c\, da = \sum_{i=1}^{n} w_i \beta(a_i) \boldsymbol{N}_c(a_i) J_c(a_i)$$

where a_i are the Gauss points and w_i are the corresponding weights.

Notation for the p-Mode Parameters The p-mode parameters are denoted by δ with subscripts and superscripts. For the side modes the subscripts indicate the mode number. The superscripts consist of two numbers indicating the node numbers of the line segment on which that side mode is nonzero. Thus $\delta_2^{\{3-5\}}$ indicates the second side mode with nonzero value on line 3-5 of the element. For the internal modes the subscript indicates the mode number while the superscript indicates the element number. Thus δ_2^3 indicates the second internal mode for the third element.

Example 9.4 Eight-Node Quadrilateral Element Consider solution of the following equation over a quarter-circle annular domain using only one element as shown in Figure 9.13:

$$2\frac{\partial^2 u}{\partial x^2} + 2\frac{\partial^2 u}{\partial y^2} + 3u + 4 = 0$$

$$2\frac{\partial u}{\partial n} = 5u + 6 \text{ on side 3-4-5}$$

$$u = 100 \text{ on side 1-2-3}$$

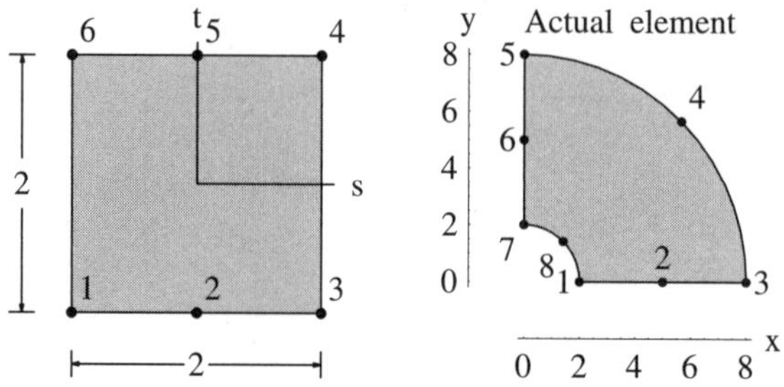

Figure 9.13. Eight-node master element and actual element

Comparing the given equation to the general form, we have $k_x = k_y = 2$, $p = 3$, and $q = 4$. On side 3-4-5 we have a natural boundary condition with $\alpha = 5$ and $\beta = 6$. On side 1-2-3 we have an essential boundary condition with $u = 100$.

The nodal coordinates are as follows:

	1	2	3	4	5	6	7	8
x	2	5.	8	$4\sqrt{2}$	0	0.	0	$\sqrt{2}$
y	0	0.	0	$4\sqrt{2}$	8	5.	2	$\sqrt{2}$

Use second-order interpolation ($n = 2$) to evaluate element equations:

Interpolation functions for mapping:

$$\Big\{\tfrac{1}{4}(1-s)(1-t) - \tfrac{1}{4}(1-s^2)(1-t) - \tfrac{1}{4}(1-s)(1-t^2),\ \tfrac{1}{2}(1-s^2)(1-t),$$
$$\tfrac{1}{4}(s+1)(1-t) - \tfrac{1}{4}(1-s^2)(1-t) - \tfrac{1}{4}(s+1)(1-t^2),\ \tfrac{1}{2}(s+1)(1-t^2),$$
$$\tfrac{1}{4}(s+1)(t+1) - \tfrac{1}{4}(1-s^2)(t+1) - \tfrac{1}{4}(s+1)(1-t^2),\ \tfrac{1}{2}(1-s^2)(t+1),$$
$$\tfrac{1}{4}(1-s)(t+1) - \tfrac{1}{4}(1-s^2)(t+1) - \tfrac{1}{4}(1-s)(1-t^2),\ \tfrac{1}{2}(1-s)(1-t^2)\Big\}$$

Interpolation functions for assumed solution:

$$N^T = \Big\{\tfrac{1}{4}(1-s)(1-t),\ \tfrac{1}{4}(s+1)(1-t),\ \tfrac{1}{4}(s+1)(t+1),\ \tfrac{1}{4}(1-s)(t+1),$$
$$\frac{(3s^2/2 - 3/2)(1-t)}{2\sqrt{6}},\ \frac{(s+1)(3t^2/2 - 3/2)}{2\sqrt{6}},$$
$$\frac{(3s^2/2 - 3/2)(t+1)}{2\sqrt{6}},\ \frac{(1-s)(3t^2/2 - 3/2)}{2\sqrt{6}}\Big\}$$

$$\frac{\partial N^T}{\partial s} = \left\{\frac{t-1}{4}, \frac{1-t}{4}, \frac{t+1}{4}, \frac{1}{4}(-t-1),\right.$$

$$\left.\frac{1}{2}\sqrt{\frac{3}{2}}s(1-t), \frac{3t^2/2-3/2}{2\sqrt{6}}, \frac{1}{2}\sqrt{\frac{3}{2}}s(t+1), -\frac{3t^2/2-3/2}{2\sqrt{6}}\right\}$$

$$\frac{\partial N^T}{\partial t} = \left\{\frac{s-1}{4}, \frac{1}{4}(-s-1), \frac{s+1}{4}, \frac{1-s}{4},\right.$$

$$\left.-\frac{3s^2/2-3/2}{2\sqrt{6}}, \frac{1}{2}\sqrt{\frac{3}{2}}(s+1)t, \frac{3s^2/2-3/2}{2\sqrt{6}}, \frac{1}{2}\sqrt{\frac{3}{2}}(1-s)t\right\}$$

The element equations are as follows:

Element coordinates:

$$\left(\{2,0\} \quad \{5.,0.\} \quad \{8,0\} \quad \{4\sqrt{2},4\sqrt{2}\} \quad \{0,8\} \quad \{0.,5.\} \quad \{0,2\} \quad \{\sqrt{2},\sqrt{2}\}\right)$$

$$x(s,t) = -\frac{3st^2}{\sqrt{2}} + \frac{3st^2}{2} - \frac{5t^2}{\sqrt{2}} + \frac{5t^2}{2} - \frac{3st}{2} - 2.5t + \frac{3s}{\sqrt{2}} + 3.53553$$

$$y(s,t) = -\frac{3st^2}{\sqrt{2}} + \frac{3st^2}{2} - \frac{5t^2}{\sqrt{2}} + \frac{5t^2}{2} + \frac{3st}{2} + 2.5t + \frac{3s}{\sqrt{2}} + 3.53553$$

$$\boldsymbol{J} = \begin{pmatrix} -\frac{3t^2}{\sqrt{2}} + \frac{3t^2}{2} - \frac{3t}{2} + \frac{3}{\sqrt{2}} & -3\sqrt{2}ts + 3ts - \frac{3s}{2} - 5\sqrt{2}t + 5t - 2.5 \\ -\frac{3t^2}{\sqrt{2}} + \frac{3t^2}{2} + \frac{3t}{2} + \frac{3}{\sqrt{2}} & -3\sqrt{2}ts + 3ts + \frac{3s}{2} - 5\sqrt{2}t + 5t + 2.5 \end{pmatrix}$$

$$\det \boldsymbol{J} = 1.86396st^2 + 3.1066t^2 + \frac{9s}{\sqrt{2}} + 10.6066$$

Given element data: $k_x = 2$; $k_y = 2$; $p = 3$; $q = 4$

Use 2×2 Gauss quadrature for integration.

Gauss point = $\{s \to -0.57735, t \to -0.57735\}$; Weight = 1.

$$N^T = \{0.622008, 0.166667, 0.0446582, 0.166667, -0.321975, -0.086273, -0.086273, -0.321975\}$$

$$\frac{\partial N^T}{\partial s} = (-0.394338 \quad 0.394338 \quad 0.105662 \quad -0.105662 \quad -0.557678 \quad -0.204124 \quad -0.149429 \quad 0.204124)$$

$$\frac{\partial N^T}{\partial t} = (-0.394338 \quad -0.105662 \quad 0.105662 \quad 0.394338 \quad 0.204124 \quad -0.149429 \quad -0.204124 \quad -0.557678)$$

$$J^{-T} = \begin{pmatrix} 0.317444 & -0.137753 \\ 0.11203 & 0.36538 \end{pmatrix}; \det J = 7.60918$$

$$B^T = \begin{pmatrix} -0.0708591 & 0.139736 & 0.0189866 & -0.0878632 & -0.20515 & -0.0442137 & -0.0193168 & 0.14162 \\ -0.188261 & 0.00557075 & 0.0504442 & 0.132246 & 0.0121062 & -0.0774664 & -0.0913233 & -0.180896 \end{pmatrix}$$

$$k_k = \begin{pmatrix} 0.615781 & -0.166645 & -0.164998 & -0.284138 & 0.186541 & 0.269621 & 0.282473 & 0.365553 \\ -0.166645 & 0.297627 & 0.0446525 & -0.175634 & -0.435236 & -0.10059 & -0.0488202 & 0.285825 \\ -0.164998 & 0.0446525 & 0.0442111 & 0.0761344 & -0.0499836 & -0.0722447 & -0.0756885 & -0.0979496 \\ -0.284138 & -0.175634 & 0.0761344 & 0.383637 & 0.298678 & -0.0967861 & -0.157965 & -0.553429 \\ 0.186541 & -0.435236 & -0.0499836 & 0.298678 & 0.642721 & 0.123766 & 0.0434829 & -0.475472 \\ 0.269621 & -0.10059 & -0.0722447 & -0.0967861 & 0.123766 & 0.121076 & 0.12066 & 0.11797 \\ 0.282473 & -0.0488202 & -0.0756885 & -0.157965 & 0.0434829 & 0.12066 & 0.132599 & 0.209776 \\ 0.365553 & 0.285825 & -0.0979496 & -0.553429 & -0.475472 & 0.11797 & 0.209776 & 0.803218 \end{pmatrix}$$

$$k_p = \begin{pmatrix} -8.83185 & -2.36649 & -0.634098 & -2.36649 & 4.5717 & 1.22498 & 1.22498 & 4.5717 \\ -2.36649 & -0.634098 & -0.169906 & -0.634098 & 1.22498 & 0.328234 & 0.328234 & 1.22498 \\ -0.634098 & -0.169906 & -0.0455262 & -0.169906 & 0.328234 & 0.0879499 & 0.0879499 & 0.328234 \\ -2.36649 & -0.634098 & -0.169906 & -0.634098 & 1.22498 & 0.328234 & 0.328234 & 1.22498 \\ 4.5717 & 1.22498 & 0.328234 & 1.22498 & -2.36649 & -0.634098 & -0.634098 & -2.36649 \\ 1.22498 & 0.328234 & 0.0879499 & 0.328234 & -0.634098 & -0.169906 & -0.169906 & -0.634098 \\ 1.22498 & 0.328234 & 0.0879499 & 0.328234 & -0.634098 & -0.169906 & -0.169906 & -0.634098 \\ 4.5717 & 1.22498 & 0.328234 & 1.22498 & -2.36649 & -0.634098 & -0.634098 & -2.36649 \end{pmatrix}$$

$$r_q^T = (18.9319 \quad 5.07279 \quad 1.35925 \quad 5.07279 \quad -9.79987 \quad -2.62587 \quad -2.62587 \quad -9.79987)$$

Similarly, evaluating at the remaining Gauss points and adding contributions,

$$k_k = \begin{pmatrix} 1.61408 & -0.592297 & -0.72701 & -0.294777 & -0.85304 & 0.523564 & 0.206715 & -0.553646 \\ -0.592297 & 1.29396 & 0.0253502 & -0.72701 & 0.149446 & -0.523564 & 0.49688 & 0.553646 \\ -0.72701 & 0.0253502 & 1.29396 & -0.592297 & 0.49688 & -0.523564 & 0.149446 & 0.553646 \\ -0.294777 & -0.72701 & -0.592297 & 1.61408 & 0.206715 & 0.523564 & -0.85304 & -0.553646 \\ -0.85304 & 0.149446 & 0.49688 & 0.206715 & 2.04632 & -0.25158 & 0.592297 & 0.25158 \\ 0.523564 & -0.523564 & -0.523564 & 0.523564 & -0.25158 & 1.21948 & -0.25158 & 0.134714 \\ 0.206715 & 0.49688 & 0.149446 & -0.85304 & 0.592297 & -0.25158 & 2.04632 & 0.25158 \\ -0.553646 & 0.553646 & 0.553646 & -0.553646 & 0.25158 & 0.134714 & 0.25158 & 1.95799 \end{pmatrix}$$

$$k_p = \begin{pmatrix} -10.866 & -7.76142 & -3.88071 & -5.433 & 7.60461 & 4.75288 & 3.80231 & 6.65403 \\ -7.76142 & -20.1797 & -10.0899 & -3.88071 & 11.4069 & 12.3575 & 5.70346 & 4.75288 \\ -3.88071 & -10.0899 & -20.1797 & -7.76142 & 5.70346 & 12.3575 & 11.4069 & 4.75288 \\ -5.433 & -3.88071 & -7.76142 & -10.866 & 3.80231 & 4.75288 & 7.60461 & 6.65403 \\ 7.60461 & 11.4069 & 5.70346 & 3.80231 & -7.76142 & -6.98528 & -3.88071 & -4.65685 \\ 4.75288 & 12.3575 & 12.3575 & 4.75288 & -6.98528 & -10.0899 & -6.98528 & -3.88071 \\ 3.80231 & 5.70346 & 11.4069 & 7.60461 & -3.88071 & -6.98528 & -7.76142 & -4.65685 \\ 6.65403 & 4.75288 & 4.75288 & 6.65403 & -4.65685 & -3.88071 & -4.65685 & -5.433 \end{pmatrix}$$

$$r_q^T = (37.2548 \quad 55.8823 \quad 55.8823 \quad 37.2548 \quad -38.0231 \quad -45.6277 \quad -38.0231 \quad -30.4184)$$

The natural boundary conditions are as follows:

Specified NBC values for side 2: $\alpha = 5; \beta = 6$

Interpolation functions for mapping:

$$\left\{0, 0, \frac{1-a}{2} + \frac{1}{2}(a^2 - 1), 1 - a^2, \frac{a+1}{2} + \frac{1}{2}(a^2 - 1), 0, 0, 0\right\}$$

Interpolation functions for solution:

$$\left\{0, \frac{1-a}{2}, \frac{a+1}{2}, 0, 0, \frac{3a^2/2 - 3/2}{\sqrt{6}}, 0, 0\right\}$$

$x(a) = -4\sqrt{2}a^2 + 4a^2 - 4a + 4\sqrt{2}$; $y(a) = -4\sqrt{2}a^2 + 4a^2 + 4a + 4\sqrt{2}$

$$J_c = \sqrt{\left(-8\sqrt{2}a + 8a - 4\right)^2 + \left(-8\sqrt{2}a + 8a + 4\right)^2}$$

Value in mapped coordinate: $\alpha(a) = 5$; $\beta(a) = 6$

Gauss point = −0.57735; Weight = 1.

$N_c^T = \{0., 0.788675, 0.211325, 0., 0., -0.408248, 0., 0.\}$

$J_c = 6.2706$; $\alpha = 5.$; $\beta = 6.$

$$\boldsymbol{k}_\alpha = \begin{pmatrix} 0. & 0. & 0. & 0. & 0. & 0. & 0. & 0. \\ 0. & 19.5018 & 5.2255 & 0. & 0. & -10.0949 & 0. & 0. \\ 0. & 5.2255 & 1.40017 & 0. & 0. & -2.70492 & 0. & 0. \\ 0. & 0. & 0. & 0. & 0. & 0. & 0. & 0. \\ 0. & 0. & 0. & 0. & 0. & 0. & 0. & 0. \\ 0. & -10.0949 & -2.70492 & 0. & 0. & 5.2255 & 0. & 0. \\ 0. & 0. & 0. & 0. & 0. & 0. & 0. & 0. \\ 0. & 0. & 0. & 0. & 0. & 0. & 0. & 0. \end{pmatrix}; \ \boldsymbol{r}_\beta = \begin{pmatrix} 0. \\ -29.6728 \\ -7.9508 \\ 0. \\ 0. \\ 15.3598 \\ 0. \\ 0. \end{pmatrix}$$

Gauss point = 0.57735; Weight = 1.

$N_c^T = \{0., 0.211325, 0.788675, 0., 0., -0.408248, 0., 0.\}$

$J_c = 6.2706$; $\alpha = 5.$; $\beta = 6.$

$$\boldsymbol{k}_\alpha = \begin{pmatrix} 0. & 0. & 0. & 0. & 0. & 0. & 0. & 0. \\ 0. & 1.40017 & 5.2255 & 0. & 0. & -2.70492 & 0. & 0. \\ 0. & 5.2255 & 19.5018 & 0. & 0. & -10.0949 & 0. & 0. \\ 0. & 0. & 0. & 0. & 0. & 0. & 0. & 0. \\ 0. & 0. & 0. & 0. & 0. & 0. & 0. & 0. \\ 0. & -2.70492 & -10.0949 & 0. & 0. & 5.2255 & 0. & 0. \\ 0. & 0. & 0. & 0. & 0. & 0. & 0. & 0. \\ 0. & 0. & 0. & 0. & 0. & 0. & 0. & 0. \end{pmatrix}; \ \boldsymbol{r}_\beta = \begin{pmatrix} 0. \\ -7.9508 \\ -29.6728 \\ 0. \\ 0. \\ 15.3598 \\ 0. \\ 0. \end{pmatrix}$$

Adding contributions from all Gauss points,

$$\boldsymbol{k}_\alpha = \begin{pmatrix} 0. & 0. & 0. & 0. & 0. & 0. & 0. & 0. \\ 0. & 20.902 & 10.451 & 0. & 0. & -12.7998 & 0. & 0. \\ 0. & 10.451 & 20.902 & 0. & 0. & -12.7998 & 0. & 0. \\ 0. & 0. & 0. & 0. & 0. & 0. & 0. & 0. \\ 0. & 0. & 0. & 0. & 0. & 0. & 0. & 0. \\ 0. & -12.7998 & -12.7998 & 0. & 0. & 10.451 & 0. & 0. \\ 0. & 0. & 0. & 0. & 0. & 0. & 0. & 0. \\ 0. & 0. & 0. & 0. & 0. & 0. & 0. & 0. \end{pmatrix}; \ \boldsymbol{r}_\beta = \begin{pmatrix} 0. \\ -37.6236 \\ -37.6236 \\ 0. \\ 0. \\ 30.7195 \\ 0. \\ 0. \end{pmatrix}$$

Thus the complete element equations are as follows:

$$\begin{pmatrix} -9.25191 & -8.35372 & -4.60772 & -5.72777 & 6.75157 & 5.27645 & 4.00902 & 6.10039 \\ -8.35372 & 2.01626 & 0.386502 & -4.60772 & 11.5564 & -0.965882 & 6.20034 & 5.30653 \\ -4.60772 & 0.386502 & 2.01626 & -8.35372 & 6.20034 & -0.965882 & 11.5564 & 5.30653 \\ -5.72777 & -4.60772 & -8.35372 & -9.25191 & 4.00902 & 5.27645 & 6.75157 & 6.10039 \\ 6.75157 & 11.5564 & 6.20034 & 4.00902 & -5.71511 & -7.23686 & -3.28842 & -4.40527 \\ 5.27645 & -0.965882 & -0.965882 & 5.27645 & -7.23686 & 1.58063 & -7.23686 & -3.746 \\ 4.00902 & 6.20034 & 11.5564 & 6.75157 & -3.28842 & -7.23686 & -5.71511 & -4.40527 \\ 6.10039 & 5.30653 & 5.30653 & 6.10039 & -4.40527 & -3.746 & -4.40527 & -3.47501 \end{pmatrix} \begin{pmatrix} u_1 \\ u_3 \\ u_5 \\ u_7 \\ \delta_1^{\{1,3\}} \\ \delta_1^{\{3,5\}} \\ \delta_1^{\{5,7\}} \\ \delta_1^{\{1,7\}} \end{pmatrix} = \begin{pmatrix} 37.2548 \\ 18.2586 \\ 18.2586 \\ 37.2548 \\ -38.0231 \\ -14.9081 \\ -38.0231 \\ -30.4184 \end{pmatrix}$$

Since there are no other elements, these are the global equations. The essential boundary conditions are as follows:

On element 1, side 1, specified value = 100

$\{u_1 = 100, u_3 = 100, \delta_1^{\{1,3\}} = 0\}$

(See the following section for details on setting parameter values corresponding to a given EBC.)

Global equations after the EBC:

$$\begin{pmatrix} 2.01626 & -8.35372 & 5.30653 & -0.965882 & 11.5564 \\ -8.35372 & -9.25191 & 6.10039 & 5.27645 & 6.75157 \\ 5.30653 & 6.10039 & -3.47501 & -3.746 & -4.40527 \\ -0.965882 & 5.27645 & -3.746 & 1.58063 & -7.23686 \\ 11.5564 & 6.75157 & -4.40527 & -7.23686 & -5.71511 \end{pmatrix} \begin{pmatrix} u_5 \\ u_7 \\ \delta_1^{\{1,7\}} \\ \delta_1^{\{3,5\}} \\ \delta_1^{\{5,7\}} \end{pmatrix} = \begin{pmatrix} 440.381 \\ 1070.8 \\ -1171.11 \\ -445.964 \\ -1058.96 \end{pmatrix}$$

Solving the final system of global equations, we get

$$\{u_5 = -392.717, u_7 = -943.4, \delta_1^{\{1,7\}} = -1263.19, \delta_1^{\{3,5\}} = -558.828, \delta_1^{\{5,7\}} = -41.996\}$$

The element solution is as follows:

The dof values for the element:

$$\{u_1 = 100, u_3 = 100, u_5 = -392.717, u_7 = -943.4, \delta_1^{\{1,3\}} = 0,$$
$$\delta_1^{\{3,5\}} = -558.828, \delta_1^{\{5,7\}} = -41.996, \delta_1^{\{1,7\}} = -1263.19\}$$

$$\boldsymbol{d}^T = (100 \quad 100 \quad -392.717 \quad -943.4 \quad 0 \quad -558.828 \quad -41.996 \quad -1263.19)$$

Element solution at $\{s \to -0.57735, t \to -0.57735\}$

Location: $\{3.02856, 1.14181\}$

$$\boldsymbol{N}^T = (0.622008 \quad 0.166667 \quad 0.0446582 \quad 0.166667 \quad -0.321975 \quad -0.086273 \quad -0.086273 \quad -0.321975)$$

$$\frac{\partial N^T}{\partial s} = (-0.394338 \quad 0.394338 \quad 0.105662 \quad -0.105662 \quad -0.557678 \quad -0.204124 \quad -0.149429 \quad 0.204124)$$

$$\frac{\partial N^T}{\partial t} = (-0.394338 \quad -0.105662 \quad 0.105662 \quad 0.394338 \quad 0.204124 \quad -0.149429 \quad -0.204124 \quad -0.557678)$$

$$J^{-T} = \begin{pmatrix} 0.317444 & -0.137753 \\ 0.11203 & 0.36538 \end{pmatrix}$$

$$B_x^T = \frac{\partial N^T}{\partial x} = (-0.0708591 \quad 0.139736 \quad 0.0189866 \quad -0.0878632 \quad -0.20515 \quad -0.0442137 \quad -0.0193168 \quad 0.14162)$$

$$B_y^T = \frac{\partial N^T}{\partial y} = (-0.188261 \quad 0.00557075 \quad 0.0504442 \quad 0.132246 \quad 0.0121062 \quad -0.0774664 \quad -0.0913233 \quad -0.180896)$$

$$u = N^T d = 362.646$$

$$\frac{\partial u}{\partial x} = B_x^T d = -71.052$$

$$\frac{\partial u}{\partial y} = B_y^T d = 112.791$$

Solution at any other location over the element can be computed in a similar manner.

9.2.3 Assembly of Element Equations

As seen from the previous section, in the p-formulation there are three types of degrees of freedom: the nodal degrees of freedom, side-mode parameters, and internal-mode parameters. The internal modes do not influence solution outside of the element because they are zero along the element sides. However, side modes are nonzero on one of the sides of an element. Therefore, to maintain solution continuity, during assembly we must assign the same global parameter to the common sides between elements. This will add contributions of elements to the appropriate parameters in a manner similar to that for the nodal degrees of freedom. The following example numerically illustrates the procedure.

Example 9.5 Consider solution of the Laplace equation over a rectangular domain using two elements as shown in Figure 9.14.

$$\frac{\partial^2 u}{\partial x^2} + \frac{\partial^2 u}{\partial y^2} = 0; \qquad 0 < x < 3; \qquad 0 < y < 1$$

Write element equations with order of interpolation $n = 2$. Assemble these equations to form the global system of equations.

For this problem $k_x = k_y = 1$, $p = q = 0$. There are no natural boundary conditions. The element equations can be obtained easily using the procedure discussed in the previous section. With the second-order interpolation, there will be four side modes for each element, no internal modes, and the element equations will be 8×8. The complete model has

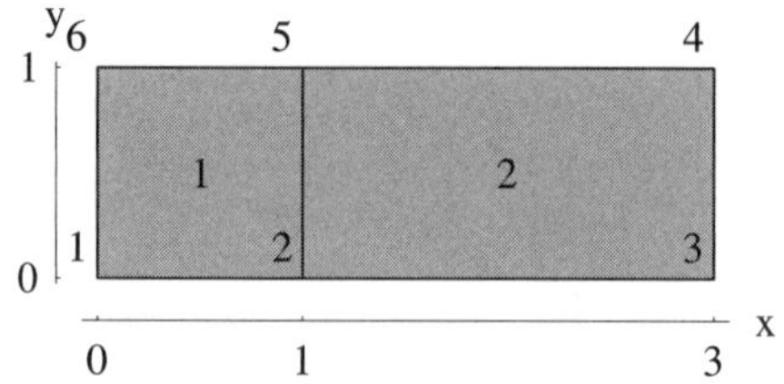

Figure 9.14. Two-element model

six nodes and seven unique element sides. Thus the global system will be 13 × 13. Using the convention for numbering the nodal and the p-mode parameters as described earlier, the global degrees of freedom are as follows:

$$\text{dof} = \left\{u_1, u_2, u_3, u_4, u_5, u_6, \delta_1^{\{1,2\}}, \delta_1^{\{1,6\}}, \delta_1^{\{2,3\}}, \delta_1^{\{2,5\}}, \delta_1^{\{3,4\}}, \delta_1^{\{4,5\}}, \delta_1^{\{5,6\}}\right\}$$

The subscript 1 on the δ terms indicates that all these parameters are associated with the first set of side modes. As mentioned earlier, the superscripts indicate node numbers of lines over which the mode has a nonzero value.

Global equations at the start of the element assembly process are

$$\begin{pmatrix} 0&0&0&0&0&0&0&0&0&0&0&0&0\\ 0&0&0&0&0&0&0&0&0&0&0&0&0\\ 0&0&0&0&0&0&0&0&0&0&0&0&0\\ 0&0&0&0&0&0&0&0&0&0&0&0&0\\ 0&0&0&0&0&0&0&0&0&0&0&0&0\\ 0&0&0&0&0&0&0&0&0&0&0&0&0\\ 0&0&0&0&0&0&0&0&0&0&0&0&0\\ 0&0&0&0&0&0&0&0&0&0&0&0&0\\ 0&0&0&0&0&0&0&0&0&0&0&0&0\\ 0&0&0&0&0&0&0&0&0&0&0&0&0\\ 0&0&0&0&0&0&0&0&0&0&0&0&0\\ 0&0&0&0&0&0&0&0&0&0&0&0&0\\ 0&0&0&0&0&0&0&0&0&0&0&0&0 \end{pmatrix} \begin{pmatrix} u_1\\ u_2\\ u_3\\ u_4\\ u_5\\ u_6\\ \delta_1^{\{1,2\}}\\ \delta_1^{\{1,6\}}\\ \delta_1^{\{2,3\}}\\ \delta_1^{\{2,5\}}\\ \delta_1^{\{3,4\}}\\ \delta_1^{\{4,5\}}\\ \delta_1^{\{5,6\}} \end{pmatrix} = \begin{pmatrix} 0\\0\\0\\0\\0\\0\\0\\0\\0\\0\\0\\0\\0 \end{pmatrix}$$

The equations for element 1 are as follows:

$$\begin{pmatrix} 0.666667 & -0.166667 & -0.333333 & -0.166667 & -0.204124 & 0.204124 & 0.204124 & -0.204124\\ -0.166667 & 0.666667 & -0.166667 & -0.333333 & -0.204124 & -0.204124 & 0.204124 & 0.204124\\ -0.333333 & -0.166667 & 0.666667 & -0.166667 & 0.204124 & -0.204124 & -0.204124 & 0.204124\\ -0.166667 & -0.333333 & -0.166667 & 0.666667 & 0.204124 & 0.204124 & -0.204124 & -0.204124\\ -0.204124 & -0.204124 & 0.204124 & 0.204124 & 0.833333 & 0 & 0.166667 & 0\\ 0.204124 & -0.204124 & -0.204124 & 0.204124 & 0 & 0.833333 & 0 & 0.166667\\ 0.204124 & 0.204124 & -0.204124 & -0.204124 & 0.166667 & 0 & 0.833333 & 0\\ -0.204124 & 0.204124 & 0.204124 & -0.204124 & 0 & 0.166667 & 0 & 0.833333 \end{pmatrix} \begin{pmatrix} u_1\\ u_2\\ u_5\\ u_6\\ \delta_1^{\{1,2\}}\\ \delta_1^{\{2,5\}}\\ \delta_1^{\{5,6\}}\\ \delta_1^{\{1,6\}} \end{pmatrix} = \begin{pmatrix} 0\\0\\0\\0\\0\\0\\0\\0 \end{pmatrix}$$

Global locations to which this element contributes: {1, 2, 5, 6, 7, 10, 13, 8}
Assembly locations for

$$
k \Longrightarrow \begin{pmatrix}
[1,1] & [1,2] & [1,5] & [1,6] & [1,7] & [1,10] & [1,13] & [1,8] \\
[2,1] & [2,2] & [2,5] & [2,6] & [2,7] & [2,10] & [2,13] & [2,8] \\
[5,1] & [5,2] & [5,5] & [5,6] & [5,7] & [5,10] & [5,13] & [5,8] \\
[6,1] & [6,2] & [6,5] & [6,6] & [6,7] & [6,10] & [6,13] & [6,8] \\
[7,1] & [7,2] & [7,5] & [7,6] & [7,7] & [7,10] & [7,13] & [7,8] \\
[10,1] & [10,2] & [10,5] & [10,6] & [10,7] & [10,10] & [10,13] & [10,8] \\
[13,1] & [13,2] & [13,5] & [13,6] & [13,7] & [13,10] & [13,13] & [13,8] \\
[8,1] & [8,2] & [8,5] & [8,6] & [8,7] & [8,10] & [8,13] & [8,8]
\end{pmatrix}
$$

and for

$$
r \Longrightarrow \begin{pmatrix} 1 \\ 2 \\ 5 \\ 6 \\ 7 \\ 10 \\ 13 \\ 8 \end{pmatrix}
$$

Global equations after assembling this element:

$$
\begin{pmatrix}
0.666667 & -0.166667 & 0 & 0 & -0.333333 & -0.166667 & -0.204124 & -0.204124 & 0 & 0.204124 & 0 & 0 & 0.204124 \\
-0.166667 & 0.666667 & 0 & 0 & -0.166667 & -0.333333 & -0.204124 & 0.204124 & 0 & -0.204124 & 0 & 0 & 0.204124 \\
0 & 0 & 0 & 0 & 0 & 0 & 0 & 0 & 0 & 0 & 0 & 0 & 0 \\
0 & 0 & 0 & 0 & 0 & 0 & 0 & 0 & 0 & 0 & 0 & 0 & 0 \\
-0.333333 & -0.166667 & 0 & 0 & 0.666667 & -0.166667 & 0.204124 & 0.204124 & 0 & -0.204124 & 0 & 0 & -0.204124 \\
-0.166667 & -0.333333 & 0 & 0 & -0.166667 & 0.666667 & 0.204124 & -0.204124 & 0 & 0.204124 & 0 & 0 & -0.204124 \\
-0.204124 & -0.204124 & 0 & 0 & 0.204124 & 0.204124 & 0.833333 & 0 & 0 & 0 & 0 & 0 & 0.166667 \\
-0.204124 & 0.204124 & 0 & 0 & 0.204124 & -0.204124 & 0 & 0.833333 & 0 & 0.166667 & 0 & 0 & 0 \\
0 & 0 & 0 & 0 & 0 & 0 & 0 & 0 & 0 & 0 & 0 & 0 & 0 \\
0.204124 & -0.204124 & 0 & 0 & -0.204124 & 0.204124 & 0 & 0.166667 & 0 & 0.833333 & 0 & 0 & 0 \\
0 & 0 & 0 & 0 & 0 & 0 & 0 & 0 & 0 & 0 & 0 & 0 & 0 \\
0 & 0 & 0 & 0 & 0 & 0 & 0 & 0 & 0 & 0 & 0 & 0 & 0 \\
0.204124 & 0.204124 & 0 & 0 & -0.204124 & -0.204124 & 0.166667 & 0 & 0 & 0 & 0 & 0 & 0.833333
\end{pmatrix}
\begin{pmatrix} u_1 \\ u_2 \\ u_3 \\ u_4 \\ u_5 \\ u_6 \\ \delta_1^{\{1,2\}} \\ \delta_1^{\{1,6\}} \\ \delta_1^{\{2,3\}} \\ \delta_1^{\{2,5\}} \\ \delta_1^{\{3,4\}} \\ \delta_1^{\{4,5\}} \\ \delta_1^{\{5,6\}} \end{pmatrix}
=
\begin{pmatrix} 0 \\ 0 \\ 0 \\ 0 \\ 0 \\ 0 \\ 0 \\ 0 \\ 0 \\ 0 \\ 0 \\ 0 \\ 0 \end{pmatrix}
$$

The equations for element 2 are as follows:

$$
\begin{pmatrix}
0.833333 & 0.166667 & -0.416667 & -0.583333 & -0.408248 & 0.102062 & 0.408248 & -0.102062 \\
0.166667 & 0.833333 & -0.583333 & -0.416667 & -0.408248 & -0.102062 & 0.408248 & 0.102062 \\
-0.416667 & -0.583333 & 0.833333 & 0.166667 & 0.408248 & -0.102062 & -0.408248 & 0.102062 \\
-0.583333 & -0.416667 & 0.166667 & 0.833333 & 0.408248 & 0.102062 & -0.408248 & -0.102062 \\
-0.408248 & -0.408248 & 0.408248 & 0.408248 & 0.666667 & 0 & -0.166667 & 0 \\
0.102062 & -0.102062 & -0.102062 & 0.102062 & 0 & 1.41667 & 0 & 0.583333 \\
0.408248 & 0.408248 & -0.408248 & -0.408248 & -0.166667 & 0 & 0.666667 & 0 \\
-0.102062 & 0.102062 & 0.102062 & -0.102062 & 0 & 0.583333 & 0 & 1.41667
\end{pmatrix}
\begin{pmatrix} u_2 \\ u_3 \\ u_4 \\ u_5 \\ \delta_1^{\{2,3\}} \\ \delta_1^{\{3,4\}} \\ \delta_1^{\{4,5\}} \\ \delta_1^{\{2,5\}} \end{pmatrix}
=
\begin{pmatrix} 0 \\ 0 \\ 0 \\ 0 \\ 0 \\ 0 \\ 0 \\ 0 \end{pmatrix}
$$

Global locations to which this element contributes: {2, 3, 4, 5, 9, 11, 12, 10}
Assembly locations for

$$
k \Longrightarrow \begin{pmatrix}
[2,2] & [2,3] & [2,4] & [2,5] & [2,9] & [2,11] & [2,12] & [2,10] \\
[3,2] & [3,3] & [3,4] & [3,5] & [3,9] & [3,11] & [3,12] & [3,10] \\
[4,2] & [4,3] & [4,4] & [4,5] & [4,9] & [4,11] & [4,12] & [4,10] \\
[5,2] & [5,3] & [5,4] & [5,5] & [5,9] & [5,11] & [5,12] & [5,10] \\
[9,2] & [9,3] & [9,4] & [9,5] & [9,9] & [9,11] & [9,12] & [9,10] \\
[11,2] & [11,3] & [11,4] & [11,5] & [11,9] & [11,11] & [11,12] & [11,10] \\
[12,2] & [12,3] & [12,4] & [12,5] & [12,9] & [12,11] & [12,12] & [12,10] \\
[10,2] & [10,3] & [10,4] & [10,5] & [10,9] & [10,11] & [10,12] & [10,10]
\end{pmatrix}
$$

and for

$$
r \Longrightarrow \begin{pmatrix} 2 \\ 3 \\ 4 \\ 5 \\ 9 \\ 11 \\ 12 \\ 10 \end{pmatrix}
$$

Global equations after assembling this element:

$$
\begin{pmatrix}
0.666667 & -0.166667 & 0 & 0 & -0.333333 & -0.166667 & -0.204124 & -0.204124 & 0 & 0.204124 & 0 & 0 & 0.204124 \\
-0.166667 & 1.5 & 0.166667 & -0.416667 & -0.75 & -0.333333 & -0.204124 & 0.204124 & -0.408248 & -0.306186 & 0.102062 & 0.408248 & 0.204124 \\
0 & 0.166667 & 0.833333 & -0.583333 & -0.416667 & 0 & 0 & 0 & -0.408248 & 0.102062 & -0.102062 & 0.408248 & 0 \\
0 & -0.416667 & -0.583333 & 0.833333 & 0.166667 & 0 & 0 & 0 & 0.408248 & 0.102062 & -0.102062 & -0.408248 & 0 \\
-0.333333 & -0.75 & -0.416667 & 0.166667 & 1.5 & -0.166667 & 0.204124 & 0.204124 & 0.408248 & -0.306186 & 0.102062 & -0.408248 & -0.204124 \\
-0.166667 & -0.333333 & 0 & 0 & -0.166667 & 0.666667 & 0.204124 & -0.204124 & 0 & 0.204124 & 0 & 0 & -0.204124 \\
-0.204124 & -0.204124 & 0 & 0 & 0.204124 & 0.204124 & 0.833333 & 0 & 0 & 0 & 0 & 0 & 0.166667 \\
-0.204124 & 0.204124 & 0 & 0 & 0.204124 & -0.204124 & 0 & 0.833333 & 0 & 0.166667 & 0 & 0 & 0 \\
0 & -0.408248 & -0.408248 & 0.408248 & 0.408248 & 0 & 0 & 0 & 0.666667 & 0 & 0 & -0.166667 & 0 \\
0.204124 & -0.306186 & 0.102062 & 0.102062 & -0.306186 & 0.204124 & 0 & 0.166667 & 0 & 2.25 & 0.583333 & 0 & 0 \\
0 & 0.102062 & -0.102062 & -0.102062 & 0.102062 & 0 & 0 & 0 & 0 & 0.583333 & 1.41667 & 0 & 0 \\
0 & 0.408248 & 0.408248 & -0.408248 & -0.408248 & 0 & 0 & 0 & -0.166667 & 0 & 0 & 0.666667 & 0 \\
0.204124 & 0.204124 & 0 & 0 & -0.204124 & -0.204124 & 0.166667 & 0 & 0 & 0 & 0 & 0 & 0.833333
\end{pmatrix}
\begin{pmatrix} u_1 \\ u_2 \\ u_3 \\ u_4 \\ u_5 \\ u_6 \\ \delta_1^{\{1,2\}} \\ \delta_1^{\{1,6\}} \\ \delta_1^{\{2,3\}} \\ \delta_1^{\{2,5\}} \\ \delta_1^{\{3,4\}} \\ \delta_1^{\{4,5\}} \\ \delta_1^{\{5,6\}} \end{pmatrix}
=
\begin{pmatrix} 0 \\ 0 \\ 0 \\ 0 \\ 0 \\ 0 \\ 0 \\ 0 \\ 0 \\ 0 \\ 0 \\ 0 \\ 0 \end{pmatrix}
$$

Since there are no other elements left, these are the global equations for the system.

9.2.4 Incorporating Essential Boundary Conditions

For the second-order boundary value problem being considered in this chapter, the essential boundary condition is $u =$ specified along portion of the boundary. In the conventional h-formulation, the specified value is incorporated into the global equations by assigning appropriate values to the nodal degrees of freedom. If the specified value is constant along a line, the nodes at the ends of the line are simply assigned this value. The interpolation functions ensure that the solution along the entire line is equal to the specified value. If the specified value is not constant but is a function of x, y, then the essential boundary condition is only approximated. The quality of the approximation depends on the order of interpolation and can be improved by placing more nodes on the line.

In the p-formulation the situation is a little more complicated. In addition to the nodal degrees of freedom, we must assign values to the side-mode parameters that are nonzero for the side with the essential boundary condition. If the specified value is constant, then the node at the ends of the line are assigned the specified value and all side-mode parameters for that line are set to zero. This will ensure that the solution is equal to the specified value along the entire line. The was the situation in Example 9.4 and hence we assigned $u = 100$ to the end nodes and set the associated side-mode parameter to 0. However, if the specified value is a function, then we must find suitable values for the side-mode parameters as well.

As an example consider a situation where $u = u(c)$ is specified along line 1-2 of an element, where c is a coordinate along the line. Over the mapped element each line is represented in terms of a coordinate $-1 \le a \le 1$, as illustrated in Figure 9.12. The specified value in terms of the mapped coordinate is $u(a)$. The assumed solution along this line is written as $u = \boldsymbol{N}_c^T \boldsymbol{d}$, where $\boldsymbol{N}_c$ is a vector of interpolation functions and $\boldsymbol{d} = \{u_1, u_2, \delta_1^{1-2}, \delta_2^{1-2}, \ldots\}$ is a vector of unknown parameters for this line. The values of nodal parameters are simply the specified function values at the ends. Thus

$$u_1 = u(-1) \qquad \text{and} \qquad u_2 = u(1)$$

Our next task is to determine the parameter values $\delta_1^{1-2}, \delta_2^{1-2}, \ldots$ such that the solution along the line is as close to $u(a)$ as possible. To determine the values of these parameters, we define the total squared error between the interpolated solution and the specified value as follows:

$$e = \int_{-1}^{1} (u(a) - \boldsymbol{N}_c^T \boldsymbol{d})^2 J_c \, da$$

where J_c is the Jacobian for the side introduced as a result of mapping. Writing $\boldsymbol{N}_c^T = (N_1, N_2, P_1, P_2, \ldots)$, the error can be written as follows:

$$e = \int_{-1}^{1} (u(a) - N_1 u_1 - N_2 u_2 - P_1 \delta_1 - P_2 \delta_2 - \ldots)^2 J_c \, da$$

where superscripts on δ terms are dropped for convenience. The necessary condition for the error to be a minimum is that the partial derivative of e with respect to the unknown parameters δ_i be zero. Thus we get the following equations for determining these parameters:

$$\frac{\partial e}{\partial \delta_i} = 2 \int_{-1}^{1} (u(a) - N_1 u_1 - N_2 u_2 - P_1 \delta_1 - P_2 \delta_2 - \cdots)(-P_i) J_c da = 0; \qquad i = 1, 2, \ldots$$

Writing all equations together in a matrix form and rearranging terms, we get the following system of equations:

$$\begin{pmatrix} \int_{-1}^{1} P_1^2 J_c da & \int_{-1}^{1} P_1P_2J_c da & \int_{-1}^{1} P_1P_3J_c da & \cdots \\ \int_{-1}^{1} P_2P_1J_c da & \int_{-1}^{1} P_2^2 J_c da & \int_{-1}^{1} P_2P_3J_c da & \cdots \\ \vdots & \vdots & \vdots & \vdots \end{pmatrix} \begin{pmatrix} \delta_1 \\ \delta_2 \\ \vdots \end{pmatrix} = \begin{pmatrix} \int_{-1}^{1} (u(a) - N_1u_1 - N_2u_2)P_1J_c\, da \\ \int_{-1}^{1} (u(a) - N_1u_1 - N_2u_2)P_2J_c\, da \\ \vdots \end{pmatrix}$$

In compact form these equations can be written as follows:

$$\int_{-1}^{1} \boldsymbol{P}\boldsymbol{P}^T J_c\, da\, \boldsymbol{\delta} = \int_{-1}^{1} \bar{u}\boldsymbol{P}J_c\, da \Longrightarrow \boldsymbol{A}\boldsymbol{\delta} = \boldsymbol{b}$$

where

$$\boldsymbol{P}^T = (P_1 \quad P_2 \quad \ldots), \text{ vector of } p\text{-modes that are nonzero for the side}$$

$$\boldsymbol{\delta}^T = (\delta_1 \quad \delta_2 \quad \ldots), \text{ vector of parameters corresponding to modes in vector } \boldsymbol{P}$$

$$\bar{u} = u(a) - N_1u_1 - N_2u_2$$

$$\boldsymbol{A} = \int_{-1}^{1} \boldsymbol{P}\boldsymbol{P}^T J_c\, da$$

$$\boldsymbol{b} = \int_{-1}^{1} (u(a) - N_1u_1 - N_2u_2)\boldsymbol{P}J_c\, da = \int_{-1}^{1} \bar{u}\boldsymbol{P}J_c\, da$$

The integrals in matrix $\boldsymbol{A}$ and vector $\boldsymbol{b}$ are evaluated numerically using the one-dimensional Gaussian quadrature. Solution of these equations give appropriate side-mode parameter values to enforce the essential boundary condition along a given side of an element. The procedure is further illustrated through the following numerical example.

Example 9.6 Consider the two-element model shown in Figure 9.14. The following essential boundary condition is specified along side 3-4 of the domain:

$$u = y\sin(\pi y/6) \text{ along } x = 3; \qquad 0 < y < 1$$

Determine the appropriate parameter values to incorporate this boundary condition if a third-order interpolation is to be used for developing element equations.

With the third-order interpolation, there will be two sets of side modes for a total of eight side modes for each element. The unknown parameters along side 3-4 are as follows:

$\{u_3, u_4, \delta_1^{3-4}, \delta_2^{3-4}\}$

On element 2, side 2, specified value $= y\sin\left(\frac{y\pi}{6}\right)$

Mapping functions along the side: $\left\{\frac{1-a}{2}, \frac{a+1}{2}\right\}$

Mapped coordinates: $x(a) = 3;\ y(a) = \frac{a}{2} + \frac{1}{2};\ J_c = \frac{1}{2}$

Specified value in mapped coordinates: $u(a) = \left(\frac{a}{2} + \frac{1}{2}\right)\sin\left(\frac{1}{6}\left(\frac{a}{2} + \frac{1}{2}\right)\pi\right)$

The specified values at the two end nodes are as follows:

Left end $u(-1) = 0$
Right end $u(1) = \frac{1}{2}$.

Thus

Values at ends $\{u_3, u_4\} = \left\{0, \frac{1}{2}\right\}$

Nodal interpolation functions: $\left\{\dfrac{1-a}{2}, \dfrac{a+1}{2}\right\}$

$$\bar{u} = \frac{1}{4}(-a-1) + \left(\frac{a}{2} + \frac{1}{2}\right)\sin\left(\frac{1}{6}\left(\frac{a}{2} + \frac{1}{2}\right)\pi\right)$$

p-Modes for the side: $\boldsymbol{P}^T = \left\{\dfrac{\frac{3a^2}{2} - \frac{3}{2}}{\sqrt{6}}, \dfrac{\frac{5a^3}{2} - \frac{5a}{2}}{\sqrt{10}}\right\}$

Gauss point = −0.774597; Weight = 0.555556
$J_c = 0.5$; $\bar{u} = -0.0497041$; $\boldsymbol{P}^T = \{-0.244949, 0.244949\}$

$$\boldsymbol{A} = \begin{pmatrix} 0.0166667 & -0.0166667 \\ -0.0166667 & 0.0166667 \end{pmatrix}; \boldsymbol{b} = \begin{pmatrix} 0.00338194 \\ -0.00338194 \end{pmatrix}$$

Gauss point = 0.; Weight = 0.888889
$J_c = 0.5$; $\bar{u} = -0.12059$; $\boldsymbol{P}^T = \{-0.612372, 0.\}$

$$\boldsymbol{A} = \begin{pmatrix} 0.166667 & 0. \\ 0. & 0. \end{pmatrix}; \boldsymbol{b} = \begin{pmatrix} 0.0328206 \\ 0. \end{pmatrix}$$

Gauss point = 0.774597; Weight = 0.555556
$J_c = 0.5$; $\bar{u} = -0.0460909$; $\boldsymbol{P}^T = \{-0.244949, -0.244949\}$

$$\boldsymbol{A} = \begin{pmatrix} 0.0166667 & 0.0166667 \\ 0.0166667 & 0.0166667 \end{pmatrix}; \boldsymbol{b} = \begin{pmatrix} 0.00313609 \\ 0.00313609 \end{pmatrix}$$

Adding contributions from all Gauss points:

$$\begin{pmatrix} 0.2 & 0. \\ 0. & 0.0333333 \end{pmatrix}\begin{pmatrix} \delta_1^{\{3,4\}} \\ \delta_2^{\{3,4\}} \end{pmatrix} = \begin{pmatrix} 0.0393386 \\ -0.000245849 \end{pmatrix}$$

Solution gives $\left\{\delta_1^{\{3,4\}}, \delta_2^{\{3,4\}}\right\} = \{0.196693, -0.00737546\}$

$$\left\{u_3, u_4, \delta_1^{\{3,4\}}, \delta_2^{\{3,4\}}\right\} = \left\{0, \tfrac{1}{2}, 0.196693, -0.00737546\right\}$$

Known values from EBC:

$$\left\{u_3 = 0, u_4 = \tfrac{1}{2}, \delta_1^{\{3,4\}} = 0.196693, \delta_2^{\{3,4\}} = -0.00737546\right\}$$

9.2.5 Applications

Example 9.7 Heat Flow in an L-Shaped Body Consider two-dimensional heat flow over an L-shaped body with thermal conductivity $k = 45\ \mathrm{W/m \cdot {}^\circ C}$ shown in Figure 9.15. The bottom is maintained at $T_0 = 110^\circ\mathrm{C}$. Convection heat loss takes place on the top where the ambient air temperature is 20°C and the convection heat transfer coefficient is $h = 55\ \mathrm{W/m^2 \cdot {}^\circ C}$. The right side is insulated. The left side is subjected to heat flux at a uniform rate of $q_L = 8000\ \mathrm{W/m^2}$. Heat is generated in the body at a rate of $Q = 5 \times 10^6\ \mathrm{W/m^3}$. Determine the temperature distribution in the body.

The simplest finite element model using quadrilateral elements is the two-element model shown in Figure 9.16.

With $n = 2$ the finite element solution is as follows.

Equations for element 1 are as follows:

Natural boundary conditions

Specified NBC values for side 3: $\alpha = -55; \beta = 1100$

$$\begin{pmatrix} 25.0962 & 19.9038 & -6.05769 & -38.9423 & -29.6765 & -5.65267 & 4.2395 & 5.65267 \\ 19.9038 & 70.0962 & -38.9423 & -51.0577 & -43.8082 & 5.65267 & 32.5028 & 31.0897 \\ -6.05769 & -38.9423 & 43.5846 & 0.590385 & 14.1317 & -20.4909 & -27.9265 & -16.2514 \\ -38.9423 & -51.0577 & 0.590385 & 88.5846 & 59.353 & 20.4909 & -8.1422 & -20.4909 \\ -29.6765 & -43.8082 & 14.1317 & 59.353 & 57.6923 & 4.61538 & -10.3846 & -34.6154 \\ -5.65267 & 5.65267 & -20.4909 & 20.4909 & 4.61538 & 96.0577 & 5.76923 & 38.9423 \\ 4.2395 & 32.5028 & -27.9265 & -8.1422 & -10.3846 & 5.76923 & 35.4942 & 24.2308 \\ 5.65267 & 31.0897 & -16.2514 & -20.4909 & -34.6154 & 38.9423 & 24.2308 & 96.0577 \end{pmatrix} \begin{pmatrix} T_1 \\ T_2 \\ T_3 \\ T_4 \\ \delta_1^{\{1,2\}} \\ \delta_1^{\{2,3\}} \\ \delta_1^{\{3,4\}} \\ \delta_1^{\{1,4\}} \end{pmatrix} = \begin{pmatrix} 937.5 \\ 937.5 \\ 733.5 \\ 733.5 \\ -765.466 \\ -688.919 \\ -598.9 \\ -688.919 \end{pmatrix}$$

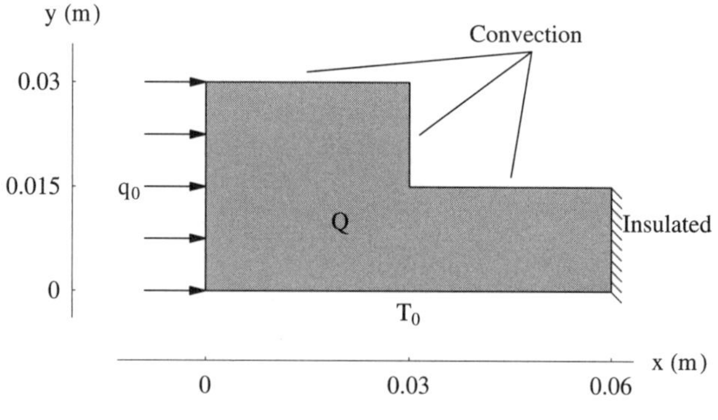

Figure 9.15. L-shaped body

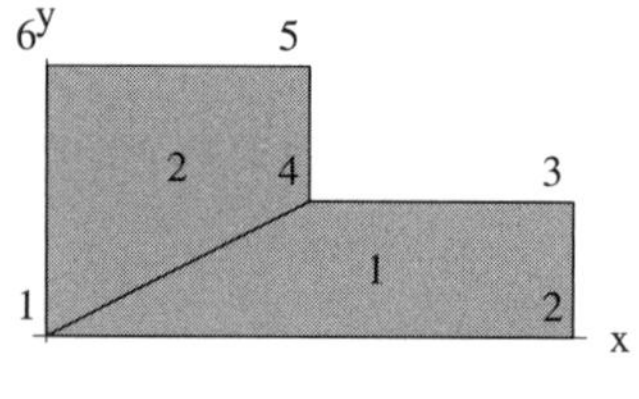

Figure 9.16. Two element model

Equations for element 2 are as follows:

Natural boundary conditions

Specified NBC values for side 2: $\alpha = -55; \beta = 1100$

Specified NBC values for side 3: $\alpha = -55; \beta = 1100$

Specified NBC values for side 4: $\alpha = 0; \beta = 8000$

$$\begin{pmatrix} 22.5 & 0 & -11.25 & -11.25 & -9.18559 & 1.05988 & 9.18559 & -7.41913 \\ 0 & 44.725 & -22.6375 & -22.5 & -18.3712 & -1.95135 & 18.3712 & 14.8383 \\ -11.25 & -22.6375 & 32.925 & -0.275 & 9.18559 & -6.89743 & -18.0344 & 3.53292 \\ -11.25 & -22.5 & -0.275 & 33.2 & 18.3712 & 8.12571 & -8.84878 & -10.952 \\ -9.18559 & -18.3712 & 9.18559 & 18.3712 & 33.75 & 6.05769 & 0 & -8.65385 \\ 1.05988 & -1.95135 & -6.89743 & 8.12571 & 6.05769 & 54.2375 & 1.44231 & 16.875 \\ 9.18559 & 18.3712 & -18.0344 & -8.84878 & 0 & 1.44231 & 33.475 & 1.15385 \\ -7.41913 & 14.8383 & 3.53292 & -10.952 & -8.65385 & 16.875 & 1.15385 & 46.875 \end{pmatrix} \begin{pmatrix} T_1 \\ T_4 \\ T_5 \\ T_6 \\ \delta_1^{\{1,4\}} \\ \delta_1^{\{4,5\}} \\ \delta_1^{\{5,6\}} \\ \delta_1^{\{1,6\}} \end{pmatrix} = \begin{pmatrix} 817.5 \\ 741.75 \\ 725.25 \\ 801. \\ -688.919 \\ -605.636 \\ -675.447 \\ -667.486 \end{pmatrix}$$

The global equations after assembling all elements are

$$\begin{pmatrix} 47.5962 & 19.9038 & -6.05769 & -38.9423 & -11.25 & -11.25 & -29.6765 & -3.53292 & -7.41913 & -5.65267 & 4.2395 & 1.05988 & 9.18559 \\ 19.9038 & 70.0962 & -38.9423 & -51.0577 & 0 & 0 & -43.8082 & 31.0897 & 0 & 5.65267 & 32.5028 & 0 & 0 \\ -6.05769 & -38.9423 & 43.5846 & 0.590385 & 0 & 0 & 14.1317 & -16.2514 & 0 & -20.4909 & -27.9265 & 0 & 0 \\ -38.9423 & -51.0577 & 0.590385 & 133.31 & -22.6375 & -22.5 & 59.353 & -38.8621 & 14.8383 & 20.4909 & -8.1422 & -1.95135 & 18.3712 \\ -11.25 & 0 & 0 & -22.6375 & 32.925 & -0.275 & 0 & 9.18559 & 3.53292 & 0 & 0 & -6.89743 & -18.0344 \\ -11.25 & 0 & 0 & -22.5 & -0.275 & 33.2 & 0 & 18.3712 & -10.952 & 0 & 0 & 8.12571 & -8.84878 \\ -29.6765 & -43.8082 & 14.1317 & 59.353 & 0 & 0 & 57.6923 & -34.6154 & 0 & 4.61538 & -10.3846 & 0 & 0 \\ -3.53292 & 31.0897 & -16.2514 & -38.8621 & 9.18559 & 18.3712 & -34.6154 & 129.808 & -8.65385 & 38.9423 & 24.2308 & 6.05769 & 0 \\ -7.41913 & 0 & 0 & 14.8383 & 3.53292 & -10.952 & 0 & -8.65385 & 46.875 & 0 & 0 & 16.875 & 1.15385 \\ -5.65267 & 5.65267 & -20.4909 & 20.4909 & 0 & 0 & 4.61538 & 38.9423 & 0 & 96.0577 & 5.76923 & 0 & 0 \\ 4.2395 & 32.5028 & -27.9265 & -8.1422 & 0 & 0 & -10.3846 & 24.2308 & 0 & 5.76923 & 35.4942 & 0 & 0 \\ 1.05988 & 0 & 0 & -1.95135 & -6.89743 & 8.12571 & 0 & 6.05769 & 16.875 & 0 & 0 & 54.2375 & 1.44231 \\ 9.18559 & 0 & 0 & 18.3712 & -18.0344 & -8.84878 & 0 & 0 & 1.15385 & 0 & 0 & 1.44231 & 33.475 \end{pmatrix} \begin{pmatrix} T_1 \\ T_2 \\ T_3 \\ T_4 \\ T_5 \\ T_6 \\ \delta_1^{\{1,2\}} \\ \delta_1^{\{1,4\}} \\ \delta_1^{\{1,6\}} \\ \delta_1^{\{2,3\}} \\ \delta_1^{\{3,4\}} \\ \delta_1^{\{4,5\}} \\ \delta_1^{\{5,6\}} \end{pmatrix} = \begin{pmatrix} 1755. \\ 937.5 \\ 733.5 \\ 1475.25 \\ 725.25 \\ 801. \\ -765.466 \\ -1377.84 \\ -667.486 \\ -688.919 \\ -598.9 \\ -605.636 \\ -675.447 \end{pmatrix}$$

The essential boundary conditions are as follows:

On element 1, side 1, specified value = 110

$$\left\{T_1, T_2, \delta_1^{\{1,2\}}\right\} = \{110, 110, 0\}$$

Known values from EBC:

$$\left\{T_1 = 110, T_2 = 110, \delta_1^{\{1,2\}} = 0\right\}$$

Global equations after EBC:

$$\begin{pmatrix} 43.5846 & 0.590385 & 0 & 0 & -16.2514 & 0 & -20.4909 & -27.9265 & 0 & 0 \\ 0.590385 & 133.31 & -22.6375 & -22.5 & -38.8621 & 14.8383 & 20.4909 & -8.1422 & -1.95135 & 18.3712 \\ 0 & -22.6375 & 32.925 & -0.275 & 9.18559 & 3.53292 & 0 & 0 & -6.89743 & -18.0344 \\ 0 & -22.5 & -0.275 & 33.2 & 18.3712 & -10.952 & 0 & 0 & 8.12571 & -8.84878 \\ -16.2514 & -38.8621 & 9.18559 & 18.3712 & 129.808 & -8.65385 & 38.9423 & 24.2308 & 6.05769 & 0 \\ 0 & 14.8383 & 3.53292 & -10.952 & -8.65385 & 46.875 & 0 & 0 & 16.875 & 1.15385 \\ -20.4909 & 20.4909 & 0 & 0 & 38.9423 & 0 & 96.0577 & 5.76923 & 0 & 0 \\ -27.9265 & -8.1422 & 0 & 0 & 24.2308 & 0 & 5.76923 & 35.4942 & 0 & 0 \\ 0 & -1.95135 & -6.89743 & 8.12571 & 6.05769 & 16.875 & 0 & 0 & 54.2375 & 1.44231 \\ 0 & 18.3712 & -18.0344 & -8.84878 & 0 & 1.15385 & 0 & 0 & 1.44231 & 33.475 \end{pmatrix} \begin{pmatrix} T_3 \\ T_4 \\ T_5 \\ T_6 \\ \delta_1^{\{1,4\}} \\ \delta_1^{\{1,6\}} \\ \delta_1^{\{2,3\}} \\ \delta_1^{\{3,4\}} \\ \delta_1^{\{4,5\}} \\ \delta_1^{\{5,6\}} \end{pmatrix} = \begin{pmatrix} 5683.5 \\ 11375.3 \\ 1962.75 \\ 2038.5 \\ -4409.08 \\ 148.618 \\ -688.919 \\ -4640.56 \\ -722.223 \\ -1685.86 \end{pmatrix}$$

Solving the final system of global equations, we get

$$\{T_3 = 129.195, T_4 = 138.48, T_5 = 162.431, T_6 = 160.634, \delta_1^{\{1,4\}} = -12.0338,$$
$$\delta_1^{\{1,6\}} = -15.6763, \delta_1^{\{2,3\}} = -4.97674, \delta_1^{\{3,4\}} = 11.6987, \delta_1^{\{4,5\}} = -5.63838, \delta_1^{\{5,6\}} = 4.39332\}$$

The solution for element 1 is as follows:

dof values for the element

$$\{T_1 = 110, T_2 = 110, T_3 = 129.195, T_4 = 138.48, \delta_1^{\{1,2\}} = 0,$$
$$\delta_1^{\{2,3\}} = -4.97674, \delta_1^{\{3,4\}} = 11.6987, \delta_1^{\{1,4\}} = -12.0338\}$$

$$\boldsymbol{d}^T = (110 \quad 110 \quad 129.195 \quad 138.48 \quad 0 \quad -4.97674 \quad 11.6987 \quad -12.0338)$$

	x	y	T	$\partial T/\partial x$	$\partial T/\partial y$
1	0.0375	0.0075	123.545	−199.205	1310.77

The solution for element 2 is as follows:

dof values for the element

$$\{T_1 = 110, T_4 = 138.48, T_5 = 162.431, T_6 = 160.634,$$
$$\delta_1^{\{1,4\}} = -12.0338, \delta_1^{\{4,5\}} = -5.63838, \delta_1^{\{5,6\}} = 4.39332, \delta_1^{\{1,6\}} = -15.6763\}$$

$$\boldsymbol{d}^T = (110 \quad 138.48 \quad 162.431 \quad 160.634 \quad -12.0338 \quad -5.63838 \quad 4.39332 \quad -15.6763)$$

	x	y	T	$\partial T/\partial x$	$\partial T/\partial y$
1	0.015	0.01875	151.752	−2.87239	1210.35

Nodal Solution Summary:

dof	x	y	Value
T_1	0	0	110
T_2	0.06	0	110
T_3	0.06	0.015	129.195
T_4	0.03	0.015	138.48
T_5	0.03	0.03	162.431
T_6	0	0.03	160.634

Element Solution Summary:

	x	y	T	$\partial T/\partial x$	$\partial T/\partial y$
1	0.0375	0.0075	123.545	−199.205	1310.77
2	0.015	0.01875	151.752	−2.87239	1210.35

With $n = 3$ the following finite element solution is obtained:

Nodal Solution Summary:

dof	x	y	Value
T_1	0	0	110
T_2	0.06	0	110
T_3	0.06	0.015	125.645
T_4	0.03	0.015	138.336
T_5	0.03	0.03	161.689
T_6	0	0.03	159.31

Element Solution Summary:

	x	y	T	$\partial T/\partial x$	$\partial T/\partial y$
1	0.0375	0.0075	124.347	−220.205	1502.78
2	0.015	0.01875	151.263	−52.3684	1258.08

With $n = 4$ the following finite element solution is obtained:

Nodal Solution Summary:

dof	x	y	Value
T_1	0	0	110
T_2	0.06	0	110
T_3	0.06	0.015	126.672
T_4	0.03	0.015	139.179
T_5	0.03	0.03	161.022
T_6	0	0.03	157.838

Element Solution Summary:

	x	y	T	$\partial T/\partial x$	$\partial T/\partial y$
1	0.0375	0.0075	124.811	−260.161	1481.78
2	0.015	0.01875	151.195	−23.981	1318.8

Example 9.8 Seepage through Soil The problem of determining the amount of water that seeps through dams or from underneath sheet piles can be formulated in terms of the following equation:

$$\frac{\partial}{\partial x}\left(k_x \frac{\partial \phi}{\partial x}\right) + \frac{\partial}{\partial y}\left(k_y \frac{\partial \phi}{\partial y}\right) = 0$$

where $\phi(x, y)$ is the hydraulic head (or hydraulic potential) and k_x and k_y are coefficients of permeability in the x and y directions. Typical units for ϕ are meters and those for k_x and k_y are m/day. The fluid velocity components in the x and y directions are related to the hydraulic head as follows:

$$v_x = -k_x \frac{\partial \phi}{\partial x} \qquad \text{and} \qquad v_y = -k_y \frac{\partial \phi}{\partial y}$$

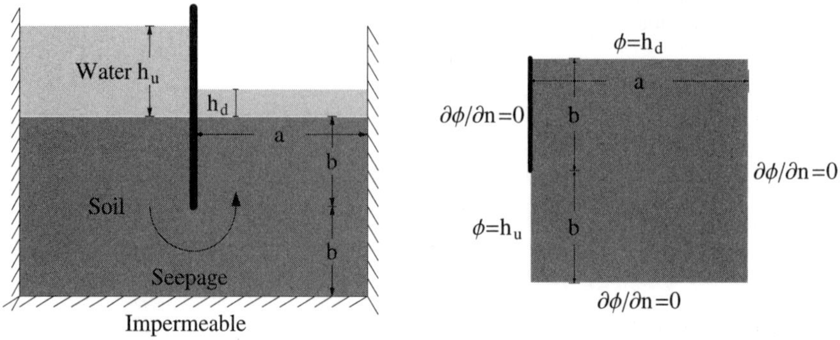

Figure 9.17. Seepage through soil underneath a sheet pile wall

A typical situation is illustrated in Figure 9.17. The analytical model consists of the soil on the downstream side. On the impermeable sides the no-flow condition is expressed in terms of the normal derivative of ϕ being zero. On the top $\phi = h_d$, the hydraulic head on the downstream side. On the left side the boundary condition on the soil below the pile is $\phi = h_u$, the hydraulic head on the upstream side.

As a numerical example, consider the following numerical values:

$$a = b = 10\,\text{m}; \qquad h_u = 10\,\text{m}; \qquad h_d = 1\,\text{m}; \qquad k_x = k_y = 1\,\text{m/s}$$

The solution domain is divided into four elements, as shown in Figure 9.18. A solution using a second-order ($n = 2$) p-formulation is as follows:

Equations for element 1:

$$\begin{pmatrix} 0.666667 & -0.166667 & -0.333333 & -0.166667 & -0.204124 & 0.204124 & 0.204124 & -0.204124 \\ -0.166667 & 0.666667 & -0.166667 & -0.333333 & -0.204124 & -0.204124 & 0.204124 & 0.204124 \\ -0.333333 & -0.166667 & 0.666667 & -0.166667 & 0.204124 & -0.204124 & -0.204124 & 0.204124 \\ -0.166667 & -0.333333 & -0.166667 & 0.666667 & 0.204124 & 0.204124 & -0.204124 & -0.204124 \\ -0.204124 & -0.204124 & 0.204124 & 0.204124 & 0.833333 & 0 & 0.166667 & 0 \\ 0.204124 & -0.204124 & -0.204124 & 0.204124 & 0 & 0.833333 & 0 & 0.166667 \\ 0.204124 & 0.204124 & -0.204124 & -0.204124 & 0.166667 & 0 & 0.833333 & 0 \\ -0.204124 & 0.204124 & 0.204124 & -0.204124 & 0 & 0.166667 & 0 & 0.833333 \end{pmatrix} \begin{pmatrix} \phi_1 \\ \phi_4 \\ \phi_5 \\ \phi_2 \\ \delta_1^{\{1,4\}} \\ \delta_1^{\{4,5\}} \\ \delta_1^{\{2,5\}} \\ \delta_1^{\{1,2\}} \end{pmatrix} = \begin{pmatrix} 0 \\ 0 \\ 0 \\ 0 \\ 0 \\ 0 \\ 0 \\ 0 \end{pmatrix}$$

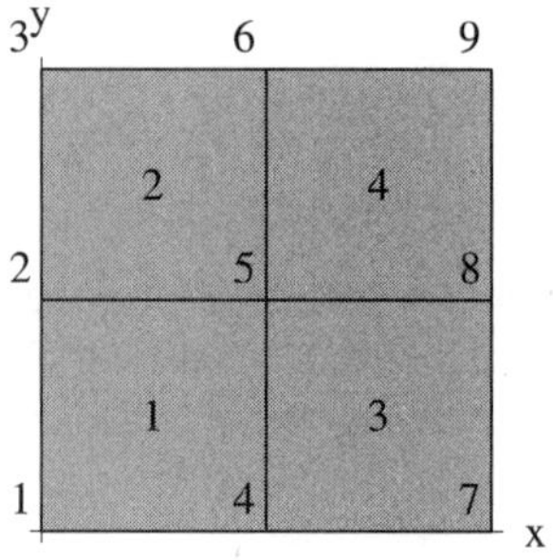

Figure 9.18. Four-element model for seepage under a sheet pile wall

Equations for element 2:

$$\begin{pmatrix} 0.666667 & -0.166667 & -0.333333 & -0.166667 & -0.204124 & 0.204124 & 0.204124 & -0.204124 \\ -0.166667 & 0.666667 & -0.166667 & -0.333333 & -0.204124 & -0.204124 & 0.204124 & 0.204124 \\ -0.333333 & -0.166667 & 0.666667 & -0.166667 & 0.204124 & -0.204124 & -0.204124 & 0.204124 \\ -0.166667 & -0.333333 & -0.166667 & 0.666667 & 0.204124 & 0.204124 & -0.204124 & -0.204124 \\ -0.204124 & -0.204124 & 0.204124 & 0.204124 & 0.833333 & 0 & 0.166667 & 0 \\ 0.204124 & -0.204124 & -0.204124 & 0.204124 & 0 & 0.833333 & 0 & 0.166667 \\ 0.204124 & 0.204124 & -0.204124 & -0.204124 & 0.166667 & 0 & 0.833333 & 0 \\ -0.204124 & 0.204124 & 0.204124 & -0.204124 & 0 & 0.166667 & 0 & 0.833333 \end{pmatrix} \begin{pmatrix} \phi_2 \\ \phi_5 \\ \phi_6 \\ \phi_3 \\ \delta_1^{\{2,5\}} \\ \delta_1^{\{5,6\}} \\ \delta_1^{\{3,6\}} \\ \delta_1^{\{2,3\}} \end{pmatrix} = \begin{pmatrix} 0 \\ 0 \\ 0 \\ 0 \\ 0 \\ 0 \\ 0 \\ 0 \end{pmatrix}$$

Equations for element 3:

$$\begin{pmatrix} 0.666667 & -0.166667 & -0.333333 & -0.166667 & -0.204124 & 0.204124 & 0.204124 & -0.204124 \\ -0.166667 & 0.666667 & -0.166667 & -0.333333 & -0.204124 & -0.204124 & 0.204124 & 0.204124 \\ -0.333333 & -0.166667 & 0.666667 & -0.166667 & 0.204124 & -0.204124 & -0.204124 & 0.204124 \\ -0.166667 & -0.333333 & -0.166667 & 0.666667 & 0.204124 & 0.204124 & -0.204124 & -0.204124 \\ -0.204124 & -0.204124 & 0.204124 & 0.204124 & 0.833333 & 0 & 0.166667 & 0 \\ 0.204124 & -0.204124 & -0.204124 & 0.204124 & 0 & 0.833333 & 0 & 0.166667 \\ 0.204124 & 0.204124 & -0.204124 & -0.204124 & 0.166667 & 0 & 0.833333 & 0 \\ -0.204124 & 0.204124 & 0.204124 & -0.204124 & 0 & 0.166667 & 0 & 0.833333 \end{pmatrix} \begin{pmatrix} \phi_4 \\ \phi_7 \\ \phi_8 \\ \phi_5 \\ \delta_1^{\{4,7\}} \\ \delta_1^{\{7,8\}} \\ \delta_1^{\{5,8\}} \\ \delta_1^{\{4,5\}} \end{pmatrix} = \begin{pmatrix} 0 \\ 0 \\ 0 \\ 0 \\ 0 \\ 0 \\ 0 \\ 0 \end{pmatrix}$$

Equations for element 4:

$$\begin{pmatrix} 0.666667 & -0.166667 & -0.333333 & -0.166667 & -0.204124 & 0.204124 & 0.204124 & -0.204124 \\ -0.166667 & 0.666667 & -0.166667 & -0.333333 & -0.204124 & -0.204124 & 0.204124 & 0.204124 \\ -0.333333 & -0.166667 & 0.666667 & -0.166667 & 0.204124 & -0.204124 & -0.204124 & 0.204124 \\ -0.166667 & -0.333333 & -0.166667 & 0.666667 & 0.204124 & 0.204124 & -0.204124 & -0.204124 \\ -0.204124 & -0.204124 & 0.204124 & 0.204124 & 0.833333 & 0 & 0.166667 & 0 \\ 0.204124 & -0.204124 & -0.204124 & 0.204124 & 0 & 0.833333 & 0 & 0.166667 \\ 0.204124 & 0.204124 & -0.204124 & -0.204124 & 0.166667 & 0 & 0.833333 & 0 \\ -0.204124 & 0.204124 & 0.204124 & -0.204124 & 0 & 0.166667 & 0 & 0.833333 \end{pmatrix} \begin{pmatrix} \phi_5 \\ \phi_8 \\ \phi_9 \\ \phi_6 \\ \delta_1^{\{5,8\}} \\ \delta_1^{\{8,9\}} \\ \delta_1^{\{6,9\}} \\ \delta_1^{\{5,6\}} \end{pmatrix} = \begin{pmatrix} 0 \\ 0 \\ 0 \\ 0 \\ 0 \\ 0 \\ 0 \\ 0 \end{pmatrix}$$

Essential boundary conditions:

On element 1, side 4, specified value = 10:

$$\left\{\phi_2, \phi_1, \delta_1^{\{1,2\}}\right\} = \{10, 10, 0\}$$

On element 2, side 3, specified value = 1:

$$\left\{\phi_6, \phi_3, \delta_1^{\{3,6\}}\right\} = \{1, 1, 0\}$$

On element 4, side 3, specified value = 1:

$$\left\{\phi_9, \phi_6, \delta_1^{\{6,9\}}\right\} = \{1, 1, 0\}$$

Known values from EBC:

$$\left\{\phi_1 = 10, \phi_2 = 10, \phi_3 = 1, \phi_6 = 1, \phi_9 = 1, \delta_1^{\{1,2\}} = 0, \delta_1^{\{3,6\}} = 0, \delta_1^{\{6,9\}} = 0\right\}$$

Global equations after EBC:

$$\begin{pmatrix} 1.33333 & -0.333333 & -0.166667 & -0.333333 & -0.204124 & 0 & 0.204124 & -0.408248 & -0.204124 & 0 & 0.204124 & 0.204124 & 0 \\ -0.333333 & 2.66667 & -0.333333 & -0.333333 & 0.204124 & 0.204124 & -0.408248 & -0.408248 & 0.204124 & -0.408248 & -0.408248 & 0.204124 & 0.204124 \\ -0.166667 & -0.333333 & 0.666667 & -0.166667 & 0 & 0 & 0 & 0.204124 & -0.204124 & 0 & 0.204124 & -0.204124 & 0 \\ -0.333333 & -0.333333 & -0.166667 & 1.33333 & 0 & 0 & 0 & 0.204124 & 0.204124 & 0.204124 & -0.408248 & -0.204124 & -0.204124 \\ -0.204124 & 0.204124 & 0 & 0 & 0.833333 & 0 & 0.166667 & 0 & 0 & 0 & 0 & 0 & 0 \\ 0 & 0.204124 & 0 & 0 & 0 & 0.833333 & 0 & 0 & 0 & 0.166667 & 0 & 0 & 0 \\ 0.204124 & -0.408248 & 0 & 0 & 0.166667 & 0 & 1.66667 & 0 & 0 & 0 & 0 & 0 & 0 \\ -0.408248 & -0.408248 & 0.204124 & 0.204124 & 0 & 0 & 0 & 1.66667 & 0 & 0 & 0 & 0.166667 & 0 \\ -0.204124 & 0.204124 & -0.204124 & 0.204124 & 0 & 0 & 0 & 0 & 0.833333 & 0 & 0.166667 & 0 & 0 \\ 0 & -0.408248 & 0 & 0.204124 & 0 & 0.166667 & 0 & 0 & 0 & 1.66667 & 0 & 0 & 0.166667 \\ 0.204124 & -0.408248 & 0.204124 & -0.408248 & 0 & 0 & 0 & 0 & 0.166667 & 0 & 1.66667 & 0 & 0 \\ 0.204124 & 0.204124 & -0.204124 & -0.204124 & 0 & 0 & 0 & 0.166667 & 0 & 0 & 0 & 0.833333 & 0 \\ 0 & 0.204124 & 0 & -0.204124 & 0 & 0 & 0 & 0 & 0 & 0.166667 & 0 & 0 & 0.833333 \end{pmatrix} \begin{pmatrix} \phi_4 \\ \phi_5 \\ \phi_7 \\ \phi_8 \\ \delta_1^{\{1,4\}} \\ \delta_1^{\{2,3\}} \\ \delta_1^{\{2,5\}} \\ \delta_1^{\{4,5\}} \\ \delta_1^{\{4,7\}} \\ \delta_1^{\{5,6\}} \\ \delta_1^{\{5,8\}} \\ \delta_1^{\{7,8\}} \\ \delta_1^{\{8,9\}} \end{pmatrix} = \begin{pmatrix} 5. \\ 7.66667 \\ 0 \\ 0.5 \\ 0 \\ 2.04124 \\ 1.63299 \\ -4.08248 \\ 0 \\ -2.04124 \\ -0.408248 \\ 0 \\ 0 \end{pmatrix}$$

Solving the final system of global equations, we get

$$\{\phi_4 = 6.12921, \phi_5 = 4.66596, \phi_7 = 5.08248, \phi_8 = 3.96195,$$
$$\delta_1^{\{1,4\}} = 0.0857292, \delta_1^{\{2,3\}} = 1.44833, \delta_1^{\{2,5\}} = 1.36347, \delta_1^{\{4,5\}} = -0.887807, \delta_1^{\{4,7\}} = 0.54473,$$
$$\delta_1^{\{5,6\}} = -0.708827, \delta_1^{\{5,8\}} = 0.440829, \delta_1^{\{7,8\}} = -0.251281, \delta_1^{\{8,9\}} = -0.0306799\}$$

Solution for element 1:

dof values for the element

$$\{\phi_1 = 10, \phi_4 = 6.12921, \phi_5 = 4.66596, \phi_2 = 10, \delta_1^{\{1,4\}} = 0.0857292,$$
$$\delta_1^{\{4,5\}} = -0.887807, \delta_1^{\{2,5\}} = 1.36347, \delta_1^{\{1,2\}} = 0\}$$

$$\boldsymbol{d}^T = (10 \quad 6.12921 \quad 4.66596 \quad 10 \quad 0.0857292 \quad -0.887807 \quad 1.36347 \quad 0)$$

	x	y	ϕ	$\partial\phi/\partial x$	$\partial\phi/\partial y$
1	2.5	2.5	7.5269	−0.811749	−0.302817

Solution for element 2:

dof values for the element

$$\{\phi_2 = 10, \phi_5 = 4.66596, \phi_6 = 1, \phi_3 = 1, \delta_1^{\{2,5\}} = 1.36347, \delta_1^{\{5,6\}} = -0.708827,$$
$$\delta_1^{\{3,6\}} = 0, \delta_1^{\{2,3\}} = 1.44833\}$$

$$\boldsymbol{d}^T = (10 \quad 4.66596 \quad 1 \quad 1 \quad 1.36347 \quad -0.708827 \quad 0 \quad 1.44833)$$

	x	y	ϕ	$\partial\phi/\partial x$	$\partial\phi/\partial y$
1	2.5	7.5	3.52259	−0.269207	−1.09961

Solution for element 3:

dof values for the element

$\{\phi_4 = 6.12921, \phi_7 = 5.08248, \phi_8 = 3.96195, \phi_5 = 4.66596, \delta_1^{\{4,7\}} = 0.54473,$
$\delta_1^{\{7,8\}} = -0.251281, \delta_1^{\{5,8\}} = 0.440829, \delta_1^{\{4,5\}} = -0.887807\}$

$$d^T = (6.12921 \quad 5.08248 \quad 3.96195 \quad 4.66596 \quad 0.54473 \quad -0.251281 \quad 0.440829 \quad -0.887807)$$

	x	y	ϕ	$\partial\phi/\partial x$	$\partial\phi/\partial y$
1	7.5	2.5	5.00691	−0.253032	−0.245653

Solution for element 4:

dof values for the element

$\{\phi_5 = 4.66596, \phi_8 = 3.96195, \phi_9 = 1, \phi_6 = 1, \delta_1^{\{5,8\}} = 0.440829, \delta_1^{\{8,9\}} = -0.0306799,$
$\delta_1^{\{6,9\}} = 0, \delta_1^{\{5,6\}} = -0.708827\}$

$$d^T = (4.66596 \quad 3.96195 \quad 1 \quad 1 \quad 0.440829 \quad -0.0306799 \quad 0 \quad -0.708827)$$

	x	y	ϕ	$\partial\phi/\partial x$	$\partial\phi/\partial y$
1	7.5	7.5	2.74843	−0.153456	−0.608801

Computing $\partial\phi/\partial x$ and $\partial\phi/\partial y$ at several points within each element, the velocity field shown in Figure 9.19 is obtained.

The following tables summarize solutions obtained at the element centers by increasing the order of interpolation from 1 to 6. Complete calculations of these solutions can be seen on the book web site. The $n = 1$ solution corresponds to the conventional rectangular

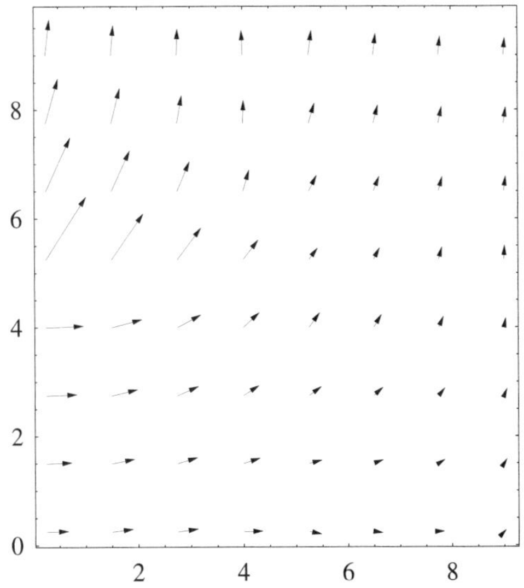

Figure 9.19. Velocity field for seepage under a sheet pile wall

element with bilinear solution and no p-modes.

	x	y	ϕ	$\partial\phi/\partial x$	$\partial\phi/\partial y$
1	2.5	2.5	7.83036	−0.867857	−0.173571
2	2.5	7.5	4.19821	−0.520714	−1.27929
3	7.5	2.5	5.03393	−0.250714	−0.289286
4	7.5	7.5	2.65536	−0.0964286	−0.662143

Solution summary with $n = 1$:

	x	y	ϕ	$\partial\phi/\partial x$	$\partial\phi/\partial y$
1	2.5	2.5	7.83036	−0.867857	−0.173571
2	2.5	7.5	4.19821	−0.520714	−1.27929
3	7.5	2.5	5.03393	−0.250714	−0.289286
4	7.5	7.5	2.65536	−0.0964286	−0.662143

Solution summary with $n = 2$:

	x	y	ϕ	$\partial\phi/\partial x$	$\partial\phi/\partial y$
1	2.5	2.5	7.5269	−0.811749	−0.302817
2	2.5	7.5	3.52259	−0.269207	−1.09961
3	7.5	2.5	5.00691	−0.253032	−0.245653
4	7.5	7.5	2.74843	−0.153456	−0.608801

Solution summary with $n = 3$:

	x	y	ϕ	$\partial\phi/\partial x$	$\partial\phi/\partial y$
1	2.5	2.5	7.51643	−0.804721	−0.279215
2	2.5	7.5	3.54361	−0.240051	−1.06869
3	7.5	2.5	5.00409	−0.243475	−0.245966
4	7.5	7.5	2.73426	−0.140519	−0.642457

Solution summary with $n = 4$:

	x	y	ϕ	$\partial\phi/\partial x$	$\partial\phi/\partial y$
1	2.5	2.5	7.52003	−0.825071	−0.312494
2	2.5	7.5	3.73481	−0.189595	−0.996245
3	7.5	2.5	4.94684	−0.226552	−0.242614
4	7.5	7.5	2.63491	−0.0857831	−0.641989

Solution summary with $n = 5$:

	x	y	ϕ	$\partial\phi/\partial x$	$\partial\phi/\partial y$
1	2.5	2.5	7.48739	−0.863053	−0.276887
2	2.5	7.5	3.68376	−0.316262	−1.10231
3	7.5	2.5	4.9193	−0.228476	−0.247952
4	7.5	7.5	2.62783	−0.0930979	−0.617761

Solution summary with $n = 6$:

	x	y	ϕ	$\partial\phi/\partial x$	$\partial\phi/\partial y$
1	2.5	2.5	7.44143	−0.8635	−0.280942
2	2.5	7.5	3.57625	−0.322093	−1.10679
3	7.5	2.5	4.90475	−0.23055	−0.25038
4	7.5	7.5	2.62078	−0.0989251	−0.613286

Example 9.9 Groundwater Flow—Free-Surface Problem The flow of water through porous media gives rise to the so-called free-surface problem. A typical situation, illustrated in Figure 9.20, considers flow of groundwater toward a well. At a sufficient distance away from the well the groundwater level is equal to the established water table in the area and is unaffected by the well. Closer to the well the groundwater level is lower because of the water flowing into the well. The exact shape and depth below ground of the top of the groundwater surface depend on the coefficient of permeability of the soil and are not known a priori. An iterative procedure is used in the analysis to establish this free surface.

The governing differential equation for the problem is as follows:

$$\frac{\partial}{\partial x}\left(k_x \frac{\partial \phi}{\partial x}\right) + \frac{\partial}{\partial y}\left(k_y \frac{\partial \phi}{\partial y}\right) = 0$$

where $\phi(x, y)$ is the hydraulic head (or hydraulic potential) and k_x and k_y are coefficients of permeability in the x and y directions. Typical units for ϕ are meters and those for k_x and k_y are m/day. The fluid velocity components in the x and y directions are related to the hydraulic head as follows:

$$v_x = -k_x \frac{\partial \phi}{\partial x} \qquad \text{and} \qquad v_y = -k_y \frac{\partial \phi}{\partial y}$$

The computational domain consists of the region bounded by the *unknown* top of the groundwater suface, side of the well, line extending from the water line in the well, and a

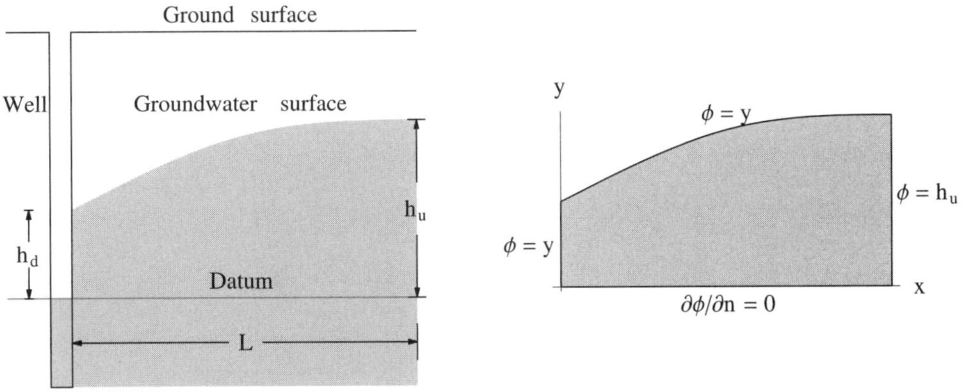

Figure 9.20. Groundwater flow problem and the computational domain

vertical line at a sufficient distance away from the well. The boundary conditions on this computational domain are as illustrated in Figure 9.20. Considering the water line in the well as datum for the hydraulic head, ϕ on the side away from the well is equal to the water depth h_u and that on the well side is equal to the elevation of the groundwater surface. On the bottom there is no flow and therefore the boundary condition is $\partial\phi/\partial n = 0$. On the top surface $\phi = y$, the height of the surface above the datum, and also $\partial\phi/\partial n = 0$. It is unusual to have two specified boundary conditions for one boundary. However, since the boundary itself is not known, there is no inconsistency. One of the conditions is used to establish the boundary while the other is used as a usual boundary condition. In finite elements a zero natural boundary condition does not require any work. Thus computationally we simply have to establish the free surface.

In an iterative procdure we arbitrarily choose the height of the free surface, say, equal to the water level h_u. The finite model is now established with zero natural boundary conditions at the top and bottom, an essential boundary condition on the far side, and essential and natural boundary conditions on the well side. The problem is solved in the usual manner. In general, the computed ϕ values on the top suface will not be equal to the assumed height. The top surface is now reestablished by setting y equal to the computed ϕ values and the process is repeated until the difference between the computed ϕ values for the nodes on the top surface are equal to the y coordinate for these nodes. The procedure is illustrated using the following numerical data:

$$k_x = k_y = 1\ \text{m/s}; \qquad h_u = 1\ \text{m}; \qquad L = 2\ \text{m}$$

Initially the groundwater surface is assumed to be horizontal. Thus the computational domain is an $L \times h_u$ rectangle as shown in Figure 9.21. It is discretized into 15 elements. The essential boundary condition of $\phi = h_u$ is specified for element 13, side 23-24; element 14, side 22-23; and element 15, side 21-22. For element 5, side 2-3, and element 6, side 1-2, the specified boundary condition is $\phi = y$. It is important not to specify ϕ value at node 4; otherwise its elevation will not change and we will not be able to adjust the groundwater surface.

A linear ($n = 1$) element is used to solve the problem. The computed nodal ϕ values are as follows:

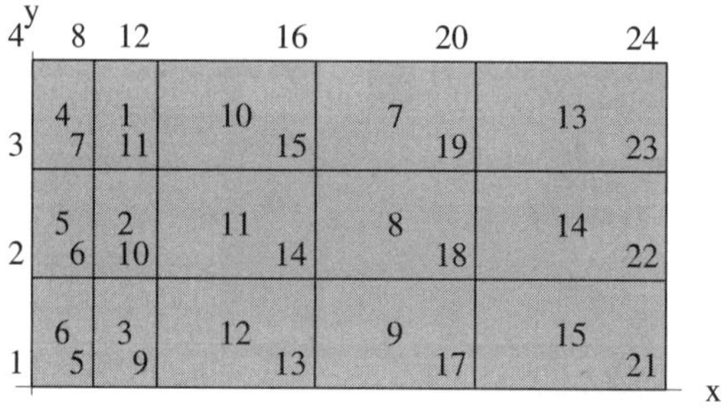

Figure 9.21. Initial finite element model for groundwater flow

Nodal Solution Summary

dof	x	y	Value
ϕ_1	0	0	0
ϕ_2	0	$\frac{1}{3}$	$\frac{1}{3}$
ϕ_3	0	$\frac{2}{3}$	$\frac{2}{3}$
ϕ_4	0	1	0.596157
ϕ_5	0.2	0	0.306957
ϕ_6	0.2	$\frac{1}{3}$	0.423017
ϕ_7	0.2	$\frac{2}{3}$	0.581671
ϕ_8	0.2	1	0.620208
ϕ_9	0.4	0	0.463903
ϕ_{10}	0.4	$\frac{1}{3}$	0.509264
ϕ_{11}	0.4	$\frac{2}{3}$	0.587326
ϕ_{12}	0.4	1	0.619842
ϕ_{13}	0.9	0	0.676938
ϕ_{14}	0.9	$\frac{1}{3}$	0.681975
ϕ_{15}	0.9	$\frac{2}{3}$	0.693494
ϕ_{16}	0.9	1	0.700009
ϕ_{17}	1.4	0	0.828093
ϕ_{18}	1.4	$\frac{1}{3}$	0.829002
ϕ_{19}	1.4	$\frac{2}{3}$	0.830666
ϕ_{20}	1.4	1	0.831418
ϕ_{21}	2	0	1
ϕ_{22}	2	$\frac{1}{3}$	1
ϕ_{23}	2	$\frac{2}{3}$	1
ϕ_{24}	2	1	1

For the next iteration, the y coordinates of the nodes on the top surface are set equal to the computed ϕ values at nodes 4, 8, 12, 16, 20, and 24:

$$\text{New } y \text{ coordinates of top surface}$$
$$= \{0.596157, 0.620208, 0.619842, 0.700009, 0.831418, 1\}$$

To avoid badly shaped elements, a completely new finite element mesh, shown in Figure 9.22, is created.

Using this new finite element model, we get the following nodal solution:

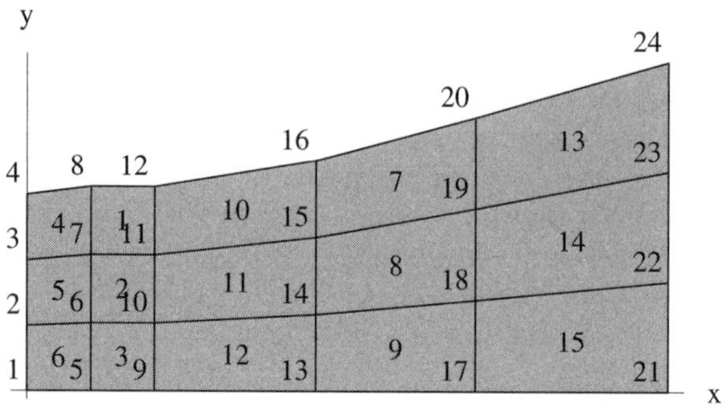

Figure 9.22. Second finite element model for groundwater flow

Nodal Solution Summary

dof	x	y	Value
ϕ_1	0	0	0.
ϕ_2	0	0.198719	0.198719
ϕ_3	0	0.397438	0.397438
ϕ_4	0	0.596157	0.398856
ϕ_5	0.2	0	0.289648
ϕ_6	0.2	0.206736	0.32478
ϕ_7	0.2	0.413472	0.389915
ϕ_8	0.2	0.620208	0.420311
ϕ_9	0.4	0	0.419679
ϕ_{10}	0.4	0.206614	0.430365
ϕ_{11}	0.4	0.413228	0.452748
ϕ_{12}	0.4	0.619842	0.46802
ϕ_{13}	0.9	0	0.637184
ϕ_{14}	0.9	0.233336	0.640101
ϕ_{15}	0.9	0.466673	0.648756
ϕ_{16}	0.9	0.700009	0.663723
ϕ_{17}	1.4	0	0.812016
ϕ_{18}	1.4	0.277139	0.815298
ϕ_{19}	1.4	0.554278	0.825998
ϕ_{20}	1.4	0.831418	0.846926
ϕ_{21}	2	0	1
ϕ_{22}	2	$\frac{1}{3}$	1
ϕ_{23}	2	$\frac{2}{3}$	1
ϕ_{24}	2	1	1

A new finite element mesh is created using the new coordinate values from this solution:

New y coordinates of top surface

$$= \{0.398856, 0.420311, 0.46802, 0.663723, 0.846926, 1\}$$

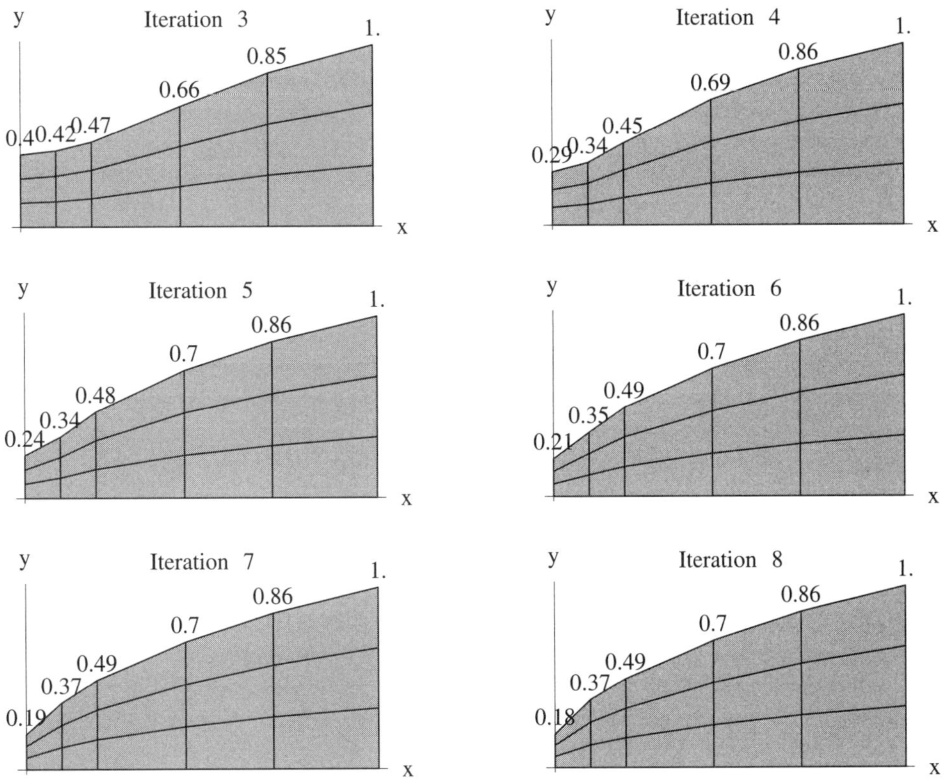

Figure 9.23. Computed profiles for groundwater flow

The process is repeated several more times. The finite element meshes and the coordinates of the groundwater surface are shown in Figure 9.23. Complete solution details can be seen on the book web site.

The coordinates of the top surface have converged now. We can use the solution from the final mesh to compute the flow rate and fluid velocity.

PROBLEMS

9.1 By direct computation, verify the following orthogonality property for first three p-modes:

$$\int_{-1}^{1} \frac{dP_i(s)}{ds}\frac{dP_j(s)}{ds} ds = \begin{cases} 1 & \text{for } i = j \\ 0 & \text{for } i \neq j \end{cases}$$

9.2 Determine the axial stress in a bar that is rotating at 500 rpm, as shown in Figure 9.24. The problem can be treated as one dimensional with the governing differential

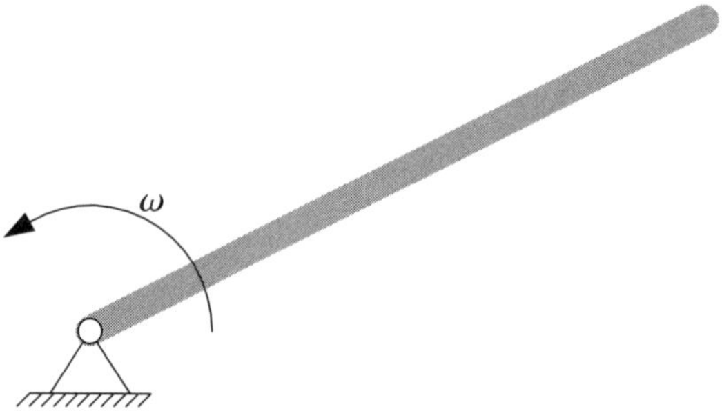

Figure 9.24. Rotating bar

equation as follows:

$$\frac{d}{dx}\left(EA\frac{du}{dx}\right) + \rho A x \omega^2 = 0; \qquad 0 < x < L$$

$$u(0) = 0\, EA\frac{du(L)}{dx}$$

where x is the coordinate along the axis of the bar, $u(x)$ is the axial displacement, L = length of the bar, E = Young's modulus, A = area of cross section, ρ = mass density, and ω = angular velocity in rad/s. The axial stress is $\sigma_x = E\, du/dx$. An exact analytical solution of the problem is

$$\sigma_{x,\text{Exact}} = \frac{\rho\omega^2}{2}(L^2 - x^2)$$

Use one finite element. Compare solutions with zero, one, and two p-modes. Use the following numerical data:

$$L = 80\,\text{cm}; \qquad E = 200\,\text{GPa}; \qquad A = 250\,\text{mm}^2; \qquad \rho = 7850\,\text{kg/m}^3$$

9.3 The square duct shown in Figure 9.25 carries hot gases at a temperature of 300°C. The duct is insulated by a layer of circular fiberglass that has a thermal conductivity of 0.04 W/m · °C. The outside surface of the fiberglass is subjected to convection with h = 15 W/m^2 · °C and T_∞ = 30°C. Taking advantage of symmetry, a one-element model of one-eighth of the section is shown in figure. Use one element with p-formulation of order 2 to obtain a solution of the problem.

	1	2	3	4	5
x	1.	0.92388	0.707107	0.5	0.5
y	0.	0.382683	0.707107	0.5	0.

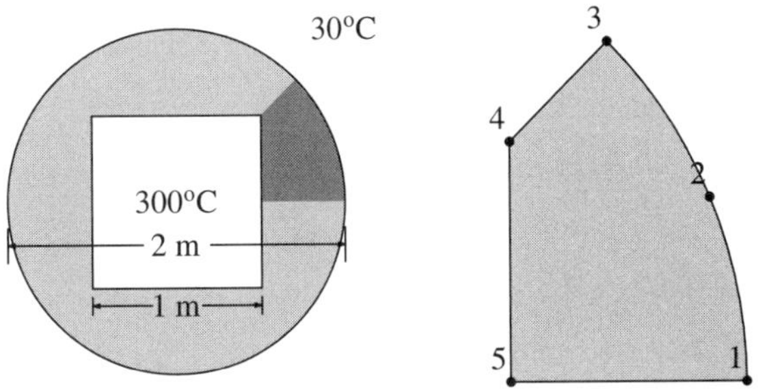

Figure 9.25.

9.4 Consider solution of the Laplace equation over a rectangular domain using two elements as shown in Figure 9.26.

$$\frac{\partial^2 u}{\partial x^2} + \frac{\partial^2 u}{\partial y^2} + 2(x + y) - 4 = 0; \qquad 0 < x < 1; \qquad 0 < y < 2$$

$$\frac{\partial u}{\partial x} = 2 - 2y - y^2 \text{ along } x = 0; \qquad \frac{\partial u}{\partial y} = 2 - 2x - x^2 \text{ along } y = 0$$

$$u = y^2 \text{ along } x = 1; \qquad u = x^2 \text{ along } y = 2$$

Use two square elements with the order of interpolation $n = 2$ to obtain an approximate solution of the problem. Report the solution and its x and y derivatives at the element centers.

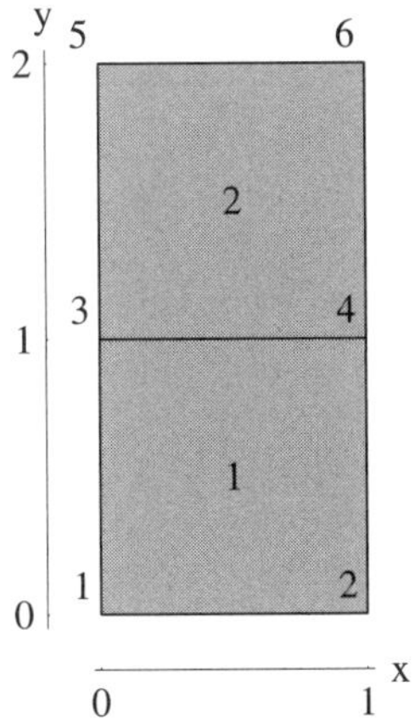

Figure 9.26.

9.5 Consider solution of the Laplace equation over a rectangular domain using two elements as shown in Figure 9.27:

$$\frac{\partial^2 u}{\partial x^2} + \frac{\partial^2 u}{\partial y^2} + (\pi^2 y^2 - 2)\sin(\pi x) = 0; \qquad 0 < x < 1; \qquad 0 < y < 1$$

$$\frac{\partial u}{\partial y} = 0 \text{ along } y = 0$$

$$u = 0 \text{ along } x = 0; \qquad u = 0 \text{ along } x = 1; \qquad u = \sin(\pi x) \text{ along } y = 1$$

Use two rectangular elements with the order of interpolation $n = 2$ to obtain an approximate solution of the problem. Report the solution and its x and y derivatives at the element centers.

It can easily be verified that the exact solution of the problem is

$$u(x, y) = y^2 \sin(\pi x)$$

Compare the finite element solution and its derivatives at the element centers with the exact values.

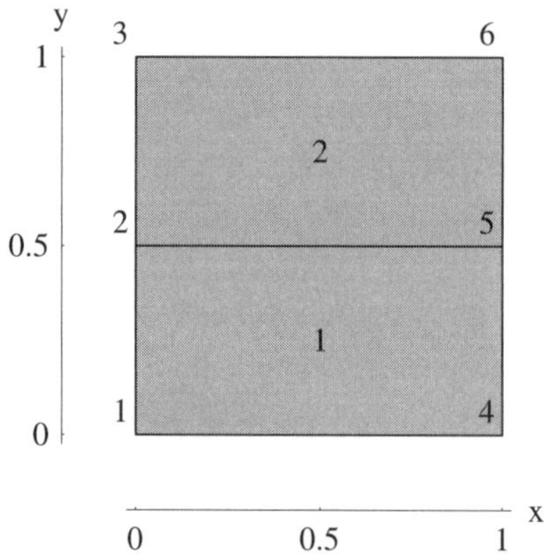

Figure 9.27.

APPENDIX A

USE OF COMMERCIAL FEA SOFTWARE

A large number of finite element programs are commercially available around the world. Some of the more widely used programs in the United States are ABAQUS, ALGOR, ANSYS, and NASTRAN. Each program requires its own specific sequence of commands and menu options to build and solve a finite element model. However, the general procedures are very similar among different programs. Thus, if one is familiar with one program, it is relatively easy to learn a different package. With this in mind this Appendix presents some examples of finite element analyses performed using ANSYS and ABAQUS. The following are the main reasons for choosing these programs.

1. They are widely available in the universities, in part because of generous academic discounts offered to universities by their developers. The academic versions of these programs are exactly the same as the commercial versions except for some restrictions on the size of the models that can be processed. In most cases the limits on the model size are generous enough that the program can be used effectively in an academic setting and one is not restricted to playing with simple toy problems.
2. Both programs consist of a large number of integrated modules that include pre- and postprocessing and analysis modules for linear and nonlinear static and dynamic analysis, buckling, fluid flow, optimization, and fatigue. Thus with these packages students can be exposed to a variety of finite element applications.
3. There are several translators available for interchange of data between these programs and several popular modeling and design packages and other finite element software. Thus, if one is already familiar with another modeling package, ABAQUS or ANSYS can be employed as analysis engines.
4. Both programs are used widely in the industry and for research and are available for Windows and Unix operating systems. ABAQUS was designed right from the be-

ginning for nonlinear analysis and supports a variety of nonlinear material models. Early versions of ANSYS had fewer material models and nonlinear analysis capabilities. However, the current versions of the two programs have comparable capabilities in this regard. Also, until fairly recently ABAQUS had a more primitive text-based user interface while ANSYS has had powerful integrated graphical pre- and post-processing capabilities for quite some time. The latest release of ABAQUS includes an integrated cae (computer aided engineering) environment for creating complex finite element models and viewing analysis results.

A.1 ANSYS APPLICATIONS

In the ANSYS release 6.1, used for the examples in this section, the user interface consists of six main windows. The windows can be closed, resized, and moved around on the screen. Their default locations are indicated in the following description.

1. ANSYS Utility Menu: This menu is usually near the top of the screen. It contains 10 pull-down menus: File, Select, List, Plot, PlotCtrls, WorkPlane, Parameters, Macro, MenuCtrls, and Help. The File menu contains usual commands for interacting with the file system. The Select menu contains commands that are used to select various finite element entities. The List and Plot menus have commands that give similar information about the finite element model; the list commands display results as tables or lists while the plot commands show this information graphically. The PlotCtrls and Workplane menus are used to set various graphics options. The Parameters and Macro menus are used primarily for more advanced ANSYS applications. The MenuCtrls menu can be used to turn various menus off or on. The Help menu gives access to comprehensive on-line help.
2. ANSYS Input: This small window is usually directly below the Utility menu. This can be used for typing commands or entering numerical data. When creating models graphically, it is often more convenient to enter some data numerically (for example, nodal coordinates) rather than to use a mouse to pick a precise location on the screen.
3. ANSYS Toolbar: This small window is usually to the right of the Input window. It contains buttons for commonly used operations such as saving database to disk and quitting the program. To safeguard against loss of data in case of a program crash, the Save-DB button should be used frequently to save the current values to the disk.
4. ANSYS Main Menu: This menu is generally located on the left side of the screen directly below the Input window. It provides access to the primary ANSYS functions through a series of cascading menus.
5. ANSYS Graphics: This window, covering most of the remaining screen, is used for displaying graphics.
6. ANSYS Output: Any text messages generated by ANSYS are placed in this window. It usually is hidden behind the other windows. However, it is strongly encouraged to periodically review the information in this window to see exactly how ANSYS is interpreting the information supplied through the user interface.

Each graphical user interface action (selection of a menu item, clicking in a dialog box, or picking an item from the graphical display) has a corresponding text-based command associated with it. This feature makes ANSYS convenient for situations that require several repetitions of a series of steps. You can prepare a text file containing these commands using any text editor (such as pico under Unix) and then use "Read Input From" option from the File menu to execute these commands.

During an interactive session, ANSYS automatically writes commands equivalent to a user's graphical actions to a file called the log file. It is a text file and is saved in the working directory with a name *jobname*.log. (The jobname refers to the name a user has assigned to the current analysis job. The default is 'File'.) The log file is very useful if you want to repeat the same analysis with perhaps slightly modified values for some of the parameters. If you are interested in this use, make sure to change the name of the file from *jobname*.log to any other convenient name before relaunching ANSYS. The file is overwritten when you enter ANSYS the next time. It is also important to remember that the log file contains everything that you did in an interactive session, including your mistakes and Exit command that you used to get out of the program. Therefore some editing is always necessary before you can use it as an input file.

ANSYS uses a combination of cascading menus, input windows, and dialog boxes. In the examples presented in this section, the required selections to perform a specific task are shown through a series of key words separated by '>' symbol. For example,

ANSYS Utility Menu > PlotCtrls > Capture Image > File > Print > Print to: [lp -dlaser] > OK

This line means that we start with the ANSYS Utility Menu and choose PlotCtrls. This choice leads to a submenu from which we select Capture Image. This menu opens up a window with a File menu. From this menu we select Print option, which leads to the Print to: [lp -dlaser] dialog box, which is closed by clicking on the OK button.

A.1.1 General Steps

The ANSYS program is organized into several processors. A typical analysis involves using three processors:

1. Preprocessing (PREP7 processor), where you provide data such as the geometry, materials, and element types to the program
2. Solution (SOLUTION processor), where you define analysis type and select associated options, apply loads, and initiate the finite element solution
3. Postprocessing (POST1 or POST26 processors), where you review the results of the analysis through graphics display and tabular listings

Choose One or More Element Types for Analysis A proper choice of element types is crucial to the success of the analysis effort. ANSYS has over 100 different element types. The ANSYS elements corresponding to those discussed in this Appendix are as follows:

LINK1-2D Spar: Plane truss element

PLANE55-2D Thermal Solid: Plane element for heat flow

PLANE42-2D Structural Solid: Element for stress analysis of plane stress, plane strain, and axisymmetric problems

If needed, several different element types can be used in the same analysis. Elements chosen for an analysis are identified as Type1, Type2,. . . , in the order that they are selected. A typical menu path to define a truss element type is as follows:

Preprocessor> Element Type> Add/Edit/Delete> Add> Link> 2D spar 1> Element type reference number [enter 1, 2, etc.]

The equivalent text commands for defining this element type are as follows:

```
/PREP7       !* Enters the Preprocessor
ET,1,LINK1   !* Define element type 1 as LINK1
```

Any text following "!*" characters is treated as a comment.

Assign Appropriate Physical Properties to Each Type For each element type appropriate physical properties, such as area of cross section and thickness, must be defined. ANSYS calls these properties real constants. Required real constants are given for each element type in the ANSYS element documentation. For the elements identified above, some of the common required real constants are as follows:

Real constant for LINK1: A (areas of cross section)

No real constant needed for PLANE55

Real constant for PLANE42: Thickness for plane stress model

Each element type can have several sets of real constants associated with it. They are identified as Real1, Real2,. . . , in the order that they are defined. A typical menu path to define a real constant for a truss element is as follows:

Preprocessor> Real Constants> Add/Edit/Delete> Add> Type 1 LINK1> AREA [enter value]

The equivalent text command for defining this real constant ($A = 20$) is as follows:

```
R,1,20, ,
```

Define One or More Sets of Materials Used in the Model Material properties, such as modulus of elasticity E, Poisson's ratio ν, and thermal conductivity, must be defined for each material that is to be used in the analysis. Required material properties are given for each element type in the ANSYS element documentation. For the elements identified above, some of the common required material properties are as follows:

Material properties for LINK1: E (modulus of elasticity)

Material properties for PLANE55: Thermal conductivities in the x and y directions

Material properties for PLANE42: For an isotropic material E (modulus of elasticity) and ν (Poisson's ratio)

Several different materials can be used in an analysis, in which case they are identified as Mat1, Mat2,..., in the order that they are created. A typical menu path to define material properties is as follows.

Preprocessor> Material Props> Material Models> Material Model Number 1> Structural> Linear> Elastic> Isotropic> EX [enter value] (double click on icons that look like folders)

The equivalent text commands for defining this material property ($E = 29000000$) are as follows:

```
MPTEMP,,,,,,,,
MPTEMP,1,0
MPDATA,EX,1,,29000000
```

Create Nodes The finite element model consists of elements connected together at the nodes. For structural frameworks we create first the nodes and then the elements. The nodes can be created manually by entering x, y, z coordinates for each node. The origin of the coordinate system can be any conveniently chosen point and does not have to be a node.

For two- and three-dimensional solid elements, there are a variety of tools available for automatically creating finite element meshes. A typical process consists of defining key points, joining key points to create lines, joining lines to create areas, and for three-dimensional solids using areas to create volumes. Plane meshes can be created through areas by defining the target size of elements or choosing a number of subdivisions along selected lines. Three-dimensional meshes are created in a similar manner through volumes.

Typical menu paths to define nodes directly are as follows:

Preprocessor> Modeling> Create> Nodes> On working plane [Pick locations if you have created a suitable grid]

Preprocessor> Modeling> Create> Nodes> In the active CS [Enter node # and x, y, z coordinates]

The equivalent text command for direct creation of nodal coordinates is N. For example,

```
N,1,-90,,,,,,      !* Creates node 1 with coordinates (−90, 0, 0)
N,2,90,,,,,,       !* Creates node 2 with coordinates (90, 0, 0)
N,3,-90,180,,,,,   !* Creates node 3 with coordinates (−90, 180, 0)
```

Nodes that are equally spaced along a line can be created more easily by defining a line and then *meshing* this line. There are several different ways of creating lines as well. The simplest is to create several key points and then define a line passing through these key points.

It is important to understand the distinction between the key points and the nodes. Both are just points defined at specified *x, y, z* locations. However, nodes are finite element entities and can be used to create elements. On the other hand, a key point is simply a geometric entity and an element cannot be defined between a set of key points.

Create Elements For structural frameworks elements can be created by choosing the nodes at the element ends. As the elements are created, ANSYS automatically numbers them as E1, E2, etc. When creating elements, it is important to choose appropriate element type, real constant type, and material type. By default, these attributes are set as type1, real1, and mat1. Appropriate attributes must be selected before creating elements with those attributes.

Typical menu paths to create elements are as follows:

Preprocessor> Modeling> Create> Elements> Elem Attributes [Select appropriate element type, material number, and real constant]

Preprocessor> Modeling> Create> Elements> Auto numbered> Thru nodes [pick nodes associated with the element]

The equivalent text command for direct creation of elements is E. For example, the following lines create an element between nodes 1 and 2 with attributes TYPE 1, MAT 1, REAL 1.

```
TYPE, 1
MAT,1
REAL,1
E,1,2
```

The following line creates an element between nodes 1 and 3 with attributes TYPE 1, MAT 1, REAL 2.

```
REAL,2
E,1,3
```

Apply Loads and Specify Boundary Conditions By default each node has six degrees of freedom (three translations and three rotations). However, some of these degrees of freedom may be automatically suppressed depending on the type of element used. For example, nodes in a model consisting only of the plane truss elements have only two translational degrees of freedom.

Concentrated loads are specified at the nodes in the global directions. Distributed loads are defined along element sides. Specified boundary conditions in the global directions at selected nodes can easily be defined. Typical menu paths to specify nodal displacements and to create nodal loads are as follows:

Preprocessor> Loads> Define Loads> Apply> Structural> Displacement> On Nodes [Pick all nodes with the same specified values]

Preprocessor> Loads> Define Loads> Apply> Structural> Force/Moment> On Nodes [Pick all nodes with the same specified values]

In the resulting dialog box select appropriate degrees of freedom and specify known values (typically zero for displacements). The equivalent text commands for specifying displacements and nodal loads are D and F, respectively. For example, the following lines specify zero values for x and y displacements at node 2 and a load of 1000 in the negative y direction at node 9.

```
D,2, ,0, , , ,UX,UY, , , ,
F,9,FY,-1000
```

Boundary conditions that are not along the global directions (for example, an inclined support) must be simulated through appropriate use of spring elements or by defining constraint equations. Time-dependent boundary conditions and loads are created as steps. Between any two time steps the loads are assumed to vary linearly with time. The initial time is set to 0. Loads specified without changing this time are suddenly applied at time 0. At each time when there is a change in load magnitude or position, the process of defining loads is repeated.

Choose Appropriate Analysis Type and Initiate Solution The default is the linear static analysis, assuming small displacements. Other analysis types include static analysis considering large displacement effects, modal analysis, transient response analysis, etc. Within each analysis type several options are available to select an appropriate solution algorithm or to set key solution parameters.

Typical menu paths to specify a solution type and initiate a solution are as follows:

Solution> Analysis Type> New Analysis> [Pick desired analysis type, e.g., Static (default)]

Solution> Solve> Current LS (starts the finite element solution process for the current load case)

The equivalent text commands are as follows:

```
/SOLU
ANTYPE,0
SOLVE
FINISH
```

This step may take a few seconds or several hours depending on the complexity of the model and the type of solution required. ANSYS goes through the entire finite element process of writing element equations, assembling to form global equations, imposing boundary conditions, and then solving for the nodal unknowns. The results are saved in a database that is retrieved in the next step for postprocessing.

View Results The Post1 module can display results at a specified time step. The default is the first step. It is possible to draw deformed shapes and stress contours (for solid models). For time-dependent problems, the entire time history of a chosen solution quantity, say displacement at a selected node, can be plotted by using commands given in the Post26 module.

Typical menu paths to plot and list results are as follows:

General postproc> Plot Results> Contour Plot> Element solu> By sequence num (Last entry. Use scroll buttons)> SMISC, 1

General postproc> List Results> Nodal solution> DOF solution> All DOFs

General postproc> List Results> Element solution> By sequence num> SMISC, 1

In these examples the "By sequence num" option is used. This is the last entry in the dialog box that results from the above menu paths. Selecting this results in a display of commands such as SMISC, NMISC. These commands, and with the appropriate number associated with them, give access to a large number of quantities computed for each element. For example, to get axial forces at the first node of each truss element, we use the sequence SMISC, 1. The information on the available options must be obtained from the documentation of the element being used in the analysis. This information is accessible from the help menu. For example, for the LINK1 element use the following menu path:

Help> Help Topics> ANSYS Documentation> ANSYS Element Reference> Element Library> LINK1

In the element descriptions the specific information related to these commands is given in the Element Output sections in Tables entitled "Item and Sequence Numbers for the ETABLE and ESOL commands."

The equivalent text commands are as follows:

```
/POST1                !* Enter the postprocessor
SET,FIRST             !* Process results of the first load step (default)
PRNSOL,DOF,           !* Create a list of nodal displacements
PRESOL,SMISC, 1       !* Create a list of element axial forces
PLESOL,SMISC,1,1,1    !* Show axial forces on a plot with deformed shape
```

A.1.2 Truss Analysis

Determine nodal displacements and axial forces in the transmission tower shown in Figure A.1. The tower is made of steel ($E = 29 \times 10^6\ \text{lb/in}^2$) angle sections. The main vertical members (elements 2, 4, 6, 8, 10, and 12) have a cross-sectional area of 20 in^2. All other members have a cross-sectional area of 10 in^2.

The best way to create frame and truss models in ANSYS is to enter nodal coordinates and element definitions directly into a text file and then use the following menu path to run the commands:

File> Read Input from ...> [Select the appropriate file]

For this example the model can be created by using the input as follows:

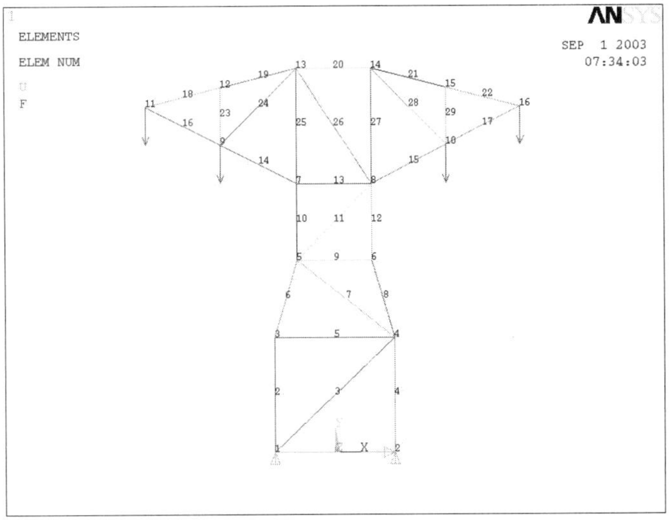

Figure A.1. Tower model in ANSYS

```
!* Model creation
/PREP7
!* Element type
ET,1,LINK1
!* Real constants
R,1,20,,
R,2,10,0,
!* Material property
MPTEMP,,,,,,,,
MPTEMP,1,0
MPDATA,EX,1,,29000000
MPDATA,PRXY,1,,
!* Create nodes
N,1,-96,,,,,,
N,2,96,,,,,,
N,3,-96,180,,,,,
N,4,96,180,,,,,
N,5,-60,300,,,,,
N,6,60,300,,,,,
N,7,-60,420,,,,,
N,8,60,420,,,,,
N,9,-180,480,,,,,
N,10,180,480,,,,,
N,11,-300,540,,,,,
N,12,-180,570,,,,,
N,13,-60,600,,,,,
N,14,60,600,,,,,
N,15,180,570,,,,,
N,16,300,540,,,,,
!* Create elements
!* with appropriate attributes
MAT,1
REAL,2
E,1,2
REAL,1
E,1,3
REAL,2
E,1,4
REAL,1
E,2,4
REAL,2
E,3,4
REAL,1
E,3,5
REAL,2
E,4,5
REAL,1
E,4,6
REAL,2
E,5,6
REAL,1
E,5,7
REAL,2
E,5,8
REAL,1
E,6,8
REAL,2
E,7,8
E,7,9
E,8,10
E,9,11
E,10,16
E,11,12
E,12,13
E,13,14
E,14,15
E,15,16
E,9, 12
E,9,13
E,7, 13,
E,8, 13
E,8, 14
E,10, 14
E,10, 15
!*
```

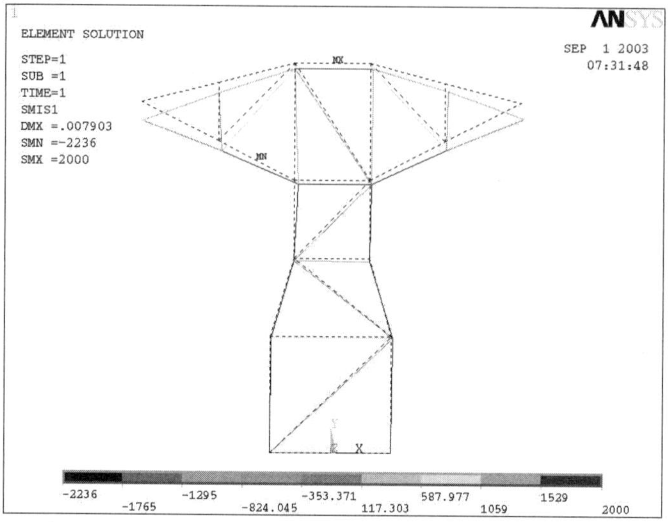

Figure A.2. Axial forces shown on a deformed shape of the model

The remaining steps of specifying boundary conditions, loads, initiating solution, and extracting results are accomplished through the following commands:

```
!* Specify displacement boundary conditions
D,2, ,0, , , ,UX,UY, , , ,
D,1, ,0, , , ,UY, , , , ,
!* Specify applied forces
F,9,FY,-1000
F,10,FY,-1000
F,11,FY,-1000
F,16,FY,-1000
!* Solution module
/SOLU
SOLVE
FINISH
```

```
!* Postprocessing
/POST1
SET,FIRST
!* List nodal displacements
PRNSOL,DOF,
!* List element results
PRESOL,ELEM
!* List element axial forces
PRESOL,SMISC, 1
!* Show axial forces on a plot with deformed shape
PLESOL,SMISC,1,1,1
```

A plot of the axial forces in the elements superimposed on a highly exaggerated deformed shape of the tower is shown in Figure A.2. The numerical values of the maximum and minimum displacements and axial forces obtained from the ANSYS output are as follows:

	Nodal displacements		Axial Compression	Axial Tension
Node	7	11	Elem 14	Elem 20
Value:	1.63×10^{-3}	-7.89×10^{-3}	Value: −2236.1	Value: 2000.

A.1.3 Steady-State Heat Flow

Consider two-dimensional heat flow over the L-shaped body shown in Figure A.3. To demonstrate how to handle situations involving different material properties, here we consider the solid to be made of two different materials. The thermal conductivity in both directions for the larger left area is 45 W/m · °C and that for the smaller right area is 25 W/m · °C. Heat is generated only in the left area at a rate of $Q = 5 \times 10^6$ W/m^3. The bottom is maintained at a temperature of $T_0 = 110$°C. Convection heat loss takes place on the top where the ambient air temperature is 20°C and the convection heat transfer coefficient is $h = 55$ W/m^2 · °C. The right side is insulated. The left side is subjected to heat flux at a uniform rate of $q_0 = 8000$ W/m^2.

For this analysis the most suitable element type is PLANE55. There are no real constants for this element. We must define two different materials. These are defined by using the following menu paths:

Preprocessor> Element Type> Add/Edit/Delete> Add> Thermal Solid> Quad 4node 55 [1 (default)]

Preprocessor> Material Props> Material Models>

Material> New Model> Material Model Number 1> Thermal> Conductivity> Isotropic> k [45]

Material> New Model> Material Model Number 2> Thermal> Conductivity> Isotropic> k [25]

Material> Exit

The next step is to create a suitable model. Instead of creating nodes and elements directly, as was done for the truss example, it is much simpler to first create the geometry of the model and then use automatic meshing to divide it into a suitable number of elements. The simplest way to create the geometry is to first define seven key points at the corners of the two areas as shown in Figure A.3. The key points can be created by using the following menu path:

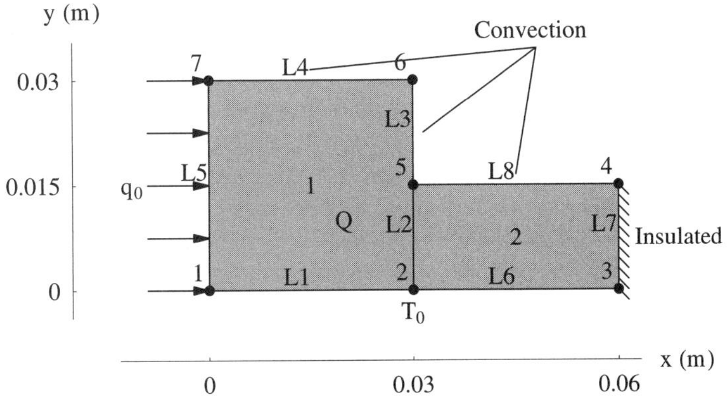

Figure A.3. L-shaped region for heat flow

Preprocessor> Modeling> Create> Keypoints> In the active CS (Enter keypoint # and x, y, z coordinates)

After defining all key points, the two areas can be created by using the following menu path and picking the appropriate key points for the areas:

Preprocessor> Modeling> Create> Areas> Arbitrary> Through KPs (Pick key points)

For assigning boundary conditions using text commands, such as SFL and DL, it is necessary to know the line numbers. When creating areas through key points, ANSYS automatically creates lines around the boundary of the areas. The lines are sequentially numbered as one moves around the area. Assuming the order of key points picked for defining area 1 is 1-2-5-6-7 and that for area 2 is 2-3-4-5, the line numbers are as shown in Figure A.3 with labels L1 through L8.

After creating the two areas and before actually creating a mesh, we must assign appropriate material properties to each area created. This is done by the following menu path:

Preprocessor> Meshing> Meshing Attributes> Picked Areas> [Pick area 1]

From the resulting dialog box associate material 1 with this area (it is actually the default and thus not really necessary for this area).

Preprocessor> Meshing> Meshing Attributes> Picked Areas> [Pick area 2]

From the resulting dialog box associate material 2 with this area.

Next comes an important operation called Glue in ANSYS. Without this operation ANSYS will create meshes in the two areas independently, and most likely along the common line between key points 2 and 5 the nodes from the two areas will not match up. During the gluing operation, ANSYS goes through the two areas and determines the common lines between the areas and keeps track of them during meshing. The gluing operation does not cause any physical change to the areas. The menu path for this operation is as follows:

Preprocessor> Modeling> Operate> Booleans> Glue> Areas [Pick the two areas]

Note that ANSYS also provides an operation by which two areas can be "added" together. As might be expected, this operation creates an entirely new area by combining the two areas together and the common line is completely eliminated. This will not be a useful operation here because then we will not be able to assign different material attributes to two areas.

The next step is to actually create a finite element mesh. We must decide on an approximate size of each element, which will obviously determine the total number of elements and nodes. One can specify a global element size for the entire model. To provide further control over the mesh, it is possible to override the global size by specifying different element sizes for a given area. One can also specify the target number of subdivisions of a given line or a specific element size near a key point (for example, to capture a solution near a corner where the solution may change rapidly). For this example, looking at the physical dimensions of the model, we choose a global element size of 0.002 m. The length of the model will then be divided roughly into 30 segments and the width into 15, resulting in a mesh of the order of $30 \times 15 = 450$ elements, which should give us reasonable results. Near the corner at key point 5 we expect rapid solution change and thus we specify a smaller element size of 0.0005 m there. The following menu paths are used to accomplish these tasks:

Preprocessor> Meshing> Size Cntrls> ManualSize> Global> Size [0.002]

Preprocessor> Meshing> Size Cntrls> ManualSize> Keypoints> Picked KPs [Pick key point 5 and enter size as 0.0005]

Preprocessor> Meshing> Mesh> Areas> Free [Pick both areas]

The final task before a solution can be initiated is to specify boundary conditions. This can be done either in the Preprocessor or the Solution modules. Here we will use the Solution module. Even though it is the default, we first explicitly specify that we are conducting a steady-state thermal analysis as follows:

Solution> Analysis type> New Analysis> Steady-State

The heat flow on the line between key points 1 and 7 is specified as follows:

Solution> Define loads> Apply> Thermal> Heat Flux> On lines [Pick line and enter Load FLUX value = 8000]

The heat generation over the left area is specified as follows:

Solution> Define loads> Apply> Thermal> Heat Generat> On areas [Pick area and enter Load HGEN value = 5000000]

The constant temperature on the bottom is specified as follows:

Solution> Define loads> Apply> Thermal> Temperature> On lines [Pick the two bottom lines and enter Load TEMP value = 110]

The convection on the top surface is specified as follows:

Solution> Define loads> Apply> Thermal> Convection> On lines [Pick the three lines defining top of the solid and enter Film coefficient = 55 and Bulk temperature = 20]

In a heat flow problem, if no boundary condition is specified along a boundary, it automatically means there is no heat flow across that face, which is equivalent to the insulated boundary condition. Thus there is no need to explicitly specify a zero heat flux on the right end of the solid. The model is shown in Figure A.4.

We are now ready to actually perform the finite element analysis, which is done by using the following menus:

Solution> Analysis Type> New Analysis> [Pick Steady-State (default)]

Solution> Solve> Current LS

After the solution is done, the results are viewed using the general postprocessor. First the results are read from the database:

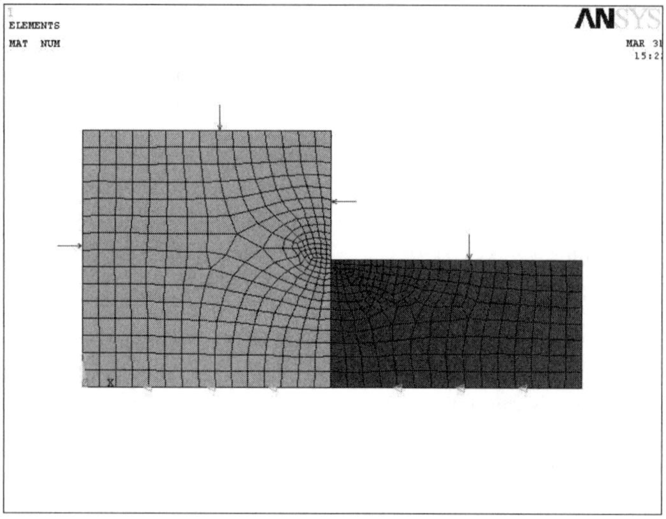

Figure A.4. Heat flow model in ANSYS

General postproc> Postprocessor> Read results> First set

The results can now be viewed as numerical lists or plotted in various forms. For example, Figure A.5 shows a contour plot of nodal temperatures and a vector plot of element heat flux obtained using the following menus:

General postproc> Plot Results> Contour Plot> Nodal Soln> DOF Solution> Temperature TEMP

General postproc> Plot Results> Vector Plot> Predefined> Flux and Gradient> Thermal Flux TF

Thermal flux can be plotted using either Nodal Soln or Element Soln. The Nodal Soln values are average values from all elements connected to that node. The Elem Soln uses

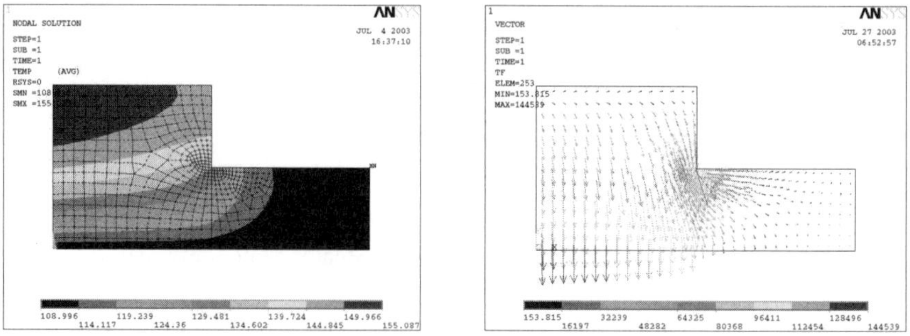

Figure A.5. Temperature contours and heat flux vectors

values only from one element. To assess the accuracy of the finite element model, you should always use the element solutions. If the flux or the gradients change drastically from one element to the next, then the mesh needs further refinement. Since the nodal values are averaged, this information is lost in the nodal solution plots.

The complete set of equivalent text commands to solve this example is as follows:

```
!* Model creation
/PREP7
!* Element type
ET,1,PLANE55
!* Material properties
MPTEMP,,,,,,,,
MPTEMP,1,0
MPDATA,KXX,1,,45
MPTEMP,,,,,,,,
MPTEMP,1,0
MPDATA,KXX,2,,25
!* Key points
K,1,0, 0,
K,2,0.03, 0
K,3,0.06, 0
K,4,0.06, 0.015
K,5,0.03, 0.015
K,6,0.03, 0.03
K,7,0, 0.03
!* Create areas
A,1,2,5,6,7
A,2,3,4,5
ASEL,,,,   1
!* Set mat 1 and type 1
AATT,   1,, 1,   0,
!* Select area 2
ASEL,,,,   2
!* Set mat 2 and type 1
AATT,   2,, 1,   0,
!* Define global element size
ESIZE,0.002,0,
!* Select key point 5
KSEL,,,,5
!* Size around this point
KESIZE,ALL,0.0005
ASEL,ALL
KSEL,ALL
!* Specify free meshing
MSHKEY,0
!* Mesh areas
AMESH,1,2
FINISH
!* Solution module
/SOLU
!* Specify analysis type
ANTYPE,0
!* Heat generation for area 1
BFA, 1, HGEN, 5000000
!* Heat flux along line 5
SFL,5,HFLUX,8000,
!* Temperature on lines
DL,1, ,TEMP,110,0
DL,6, ,TEMP,110,0
!* Convection on lines
SFL,8,CONV,55, ,20,
SFL,3,CONV,55, ,20,
SFL,4,CONV,55, ,20,
SOLVE
FINISH
!* Postprocessing
/POST1
SET,FIRST
!* Nodal temperature contours
PLNSOL,TEMP, ,1,
!* Thermal gradient vectors
PLVECT,TG, , , ,VECT,ELEM,ON,0
```

A.1.4 Plane Stress Analysis

As a final example, consider the problem of finding stresses in a notched beam of rectangular cross section. Taking advantage of symmetry, we model just the right half of the beam, which is shown in Figure A.6. The beam is 4 in thick in the direction perpendicular to the plane of the paper and is made of concrete with modulus of elasticity $E = 3 \times 10^6$ lb/in^2 and Poisson's ratio $\nu = 0.2$. The beam is loaded by a uniform pressure of 50 lb/in^2 on the top surface.

For this analysis the most suitable element type is PLANE42. By default, this element assumes a plane stress problem with a unit thickness. For any other thickness we must set the element option to Plane strs w/thk (plane stress analysis with specified thickness). The actual thickness value is entered as a real constant for the element. The menu paths for this setup are as follows:

> Preprocessor> Element Type> Add/Edit/Delete> Add> Structural Solid> Quad 4node 42 [1 (default)]

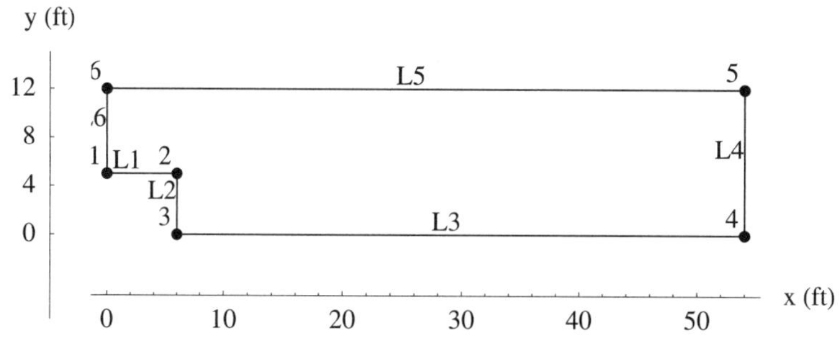

Figure A.6. Key point locations for notched beam

In the Element types dialog box click on the Options button and set Element behavior K3 to Plane strs w/thk.

> Preprocessor> Real Constants> Add/Edit/Delete> Add> Type 1 PLANE42> Thickness [4]
>
> Preprocessor> Material Props> Material Models> Material Model Number 1> Structural> Linear> Elastic> Isotropic> EX [3000000] and PRXY [0.2]
>
> Material> Exit

The geometry of the model can easily be created from the six key points at the corners of the area, as shown in Figure A.6. For later reference the line numbers are also shown in the figure. The key points can be created by using the following menu path:

> Preprocessor> Modeling> Create> Keypoints> In the active CS (Enter keypoint # and x, y, z coordinates)

After defining all key points, the area can be created by using the following menu path and picking the key points:

> Preprocessor> Modeling> Create> Areas> Arbitrary> Through KPs [Pick key points]

The next step is to actually create a finite element mesh. We must decide on an approximate size of each element. Looking at the physical dimensions of the model, we choose a global element size of 2 in. Thus the length of the model will be divided roughly into 27 segments and the width into 6, resulting in a mesh of the order of $27 \times 6 = 162$ element mesh. We expect high stresses in the notch area and near the fixed end. To capture these stresses, we use a finer mesh in these parts by specifying element size of 0.5 in along lines L1 and L6 and around key point 2. An element size of 1 is specified around key points 4 and 5. The following menu paths are used to accomplish these tasks:

> Preprocessor> Meshing> Size Cntrls> ManualSize> Global> Size [2]
>
> Preprocessor> Meshing> Size Cntrls> ManualSize> Lines> Picked lines [Pick lines L1 and L6 and enter size as 0.5]
>
> Preprocessor> Meshing> Size Cntrls> ManualSize> Keypoints> Picked KPs [Pick key point 2 and enter size as 0.5]

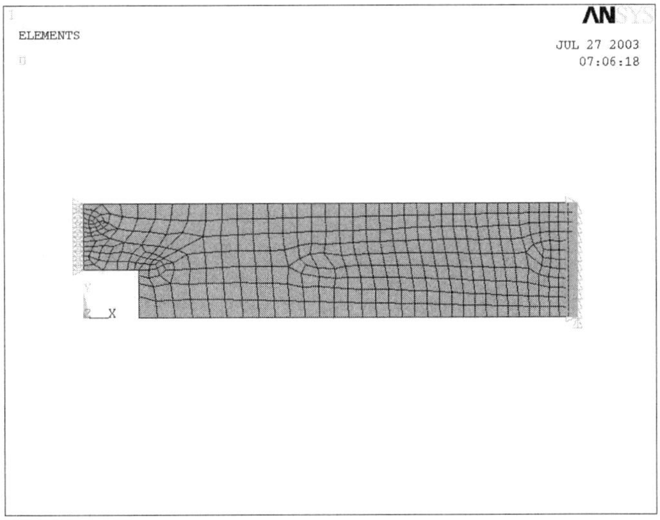

Figure A.7. Plane stress model of notched beam in ANSYS

Preprocessor> Meshing> Size Cntrls> ManualSize> Keypoints> Picked KPs [Pick key points 4 & 5 and enter size as 1]

Preprocessor> Meshing> Mesh> Areas> Free [Pick area]

The final task before a solution can be initiated is to specify boundary conditions. The fixed-end condition on the right end is specified by constraining all nodes on line L4 as follows:

Solution> Loads> Define Loads> Apply> Structural> Displacement> On lines [Pick line L4 and set all dof to 0]

The symmetry boundary condition along line L6 is defined as follows:

Solution> Loads> Define Loads> Apply> Structural> Displacement> Symmetry B.C.> On lines [Pick line L6]

The pressure along line L5 is defined as follows:

Solution> Loads> Define Loads> Apply> Structural> Pressure> On lines [Pick line L5 and specify Pressure = 50]

The model is now complete and is as shown in Figure A.7.

We are now ready to actually perform the finite element analysis, which is done by using the following menus:

Solution> Analysis Type> New Analysis> [Pick Static analysis (default)]

Solution> Solve> Current LS

After the solution is done, the results are viewed using the general postprocessor. First the results are read from the database:

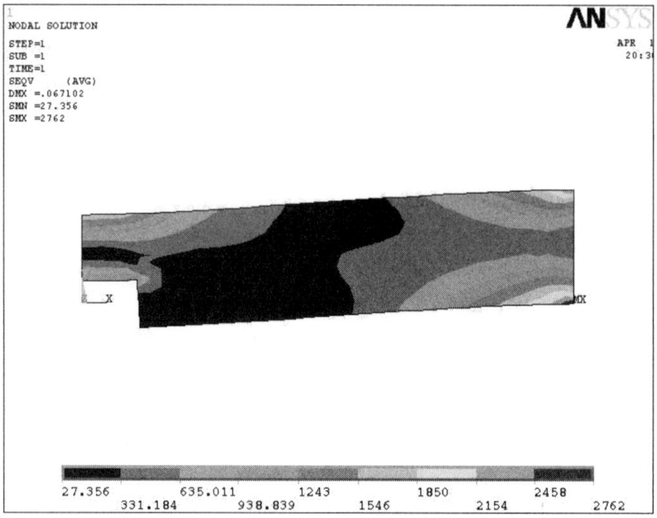

Figure A.8. von Mises equivalent stress contours

General postproc> Postprocessor> Read results> First set

The results can now be viewed as numerical lists or plotted in various forms. For example, Figure A.8 shows a contour plot of equivalent von Mises stresses obtained using the following menu path:

General postproc> Plot Results> Contour Plot> Nodal Soln> Stress> von Mises SEQV

The complete set of equivalent text commands to solve this example is as follows:

```
!* Model creation
/PREP7
!* Element type
ET,1,PLANE42
KEYOPT,1,1,0
KEYOPT,1,2,0
KEYOPT,1,3,3
KEYOPT,1,5,0
KEYOPT,1,6,0
!* Real constant
R,1,4,
!* Material properties
MPTEMP,,,,,,,,
MPTEMP,1,0
MPDATA,EX,1,,3000000
MPDATA,PRXY,1,,.2
!* Key points
K,1,0, 5
K,2,6, 5
K,3,6, 0
K,4,54, 0
K,5,54, 12
K,6,0, 12
!* Create areas
A,1,2,3,4,5,6
!* Select area 1
ASEL,,,, 1
!* Set mat 1 and type 1
AATT, 1,, 1, 0,
!* Define global element size
ESIZE,2,0,
!* Select key points
KSEL,,,,2
!* Define element size
KESIZE,ALL,.5
KSEL,,,,4
KSEL,A,,,5
KESIZE,ALL,1
!* Select lines
LSEL,,,,1
LSEL,,,,6
!* Define element size
LESIZE,ALL,.5
ASEL,ALL
KSEL,ALL
LSEL,ALL
```

```
!* Specify free meshing
MSHKEY,0
!* Mesh area
AMESH,1
FINISH
!* Solution module
/SOLU
!* Specify analysis type
ANTYPE,0
!* Zero disp along line 4
DL,4, ,ALL,
!* Symmetry along line 6
DL,    6, ,SYMM
!* Pressure on lines 5
SFL,5,PRES,50,
SOLVE
FINISH
!* Postprocessing
/POST1
SET,FIRST
!* Equivalent stress plot
PLNSOL,S,EQV,0,0
```

A.2 OPTIMIZING DESIGN USING ANSYS

Most finite element analyses are carried out with an ultimate goal of making improvements in a given design. Several iterations, each requiring a new finite element analysis, may be necessary to achieve a satisfactory design. If an analysis is carried out using an input file, it is generally easy to make the design changes and reanalyze the model. The process can be streamlined even further by using the optimization capability built into ANSYS. By defining certain parameters as design variables and creating suitable objective and constraint functions, it is possible to let ANSYS perform iterations to automatically create optimum designs.

A.2.1 General Steps

1. *Prepare an input file for analysis using variables for key parameters.* In order to use optimization capability, all analysis steps must be defined in an input file. Furthermore, the file must use variables, instead of fixed numerical values, for the key design parameters that are to be changed. Even if one is not interested in optimization, it generally is a good idea to use variables in the input file to improve readability and to facilitate making manual changes if necessary. The input file is used repeatedly during optimization iterations, and thus it should not contain unnecessary commands (for example, plot commands). These commands slow down the execution and the overall process will take much longer.
2. *Run an analysis using the input file and define additional parameters.* From the analysis results define additional parameters that are based on computed results. These parameters will be used to define constraints and the objective function in the optimization module. Add the equivalent commands to the input file. A convenient way to find the equivalent commands to menu selections is to use the Session Editor window. It is accessed by selecting the Session Editor command, which is the second to the last option in the ANSYS Main Menu. For all menu selections, ANSYS records equivalent commands in this window. These commands can simply be copied and pasted into the input file.
3. *Enter the Optimization module interactively and define an optimization problem by using the following menu paths:*

   ANSYS Main Menu> Design Opt>

Analysis File> Assign. Assign Analysis file [Enter input file name]

Design variables> Add. Define a Design Variable [Select name, enter min and maximum values (Convergence tolerance can be left blank)] OK

State variables> Add (State variables are really the constraints.)

Define a State Variable [Select name, enter lower limit and upper limit (Feasibility tolerance can be left blank)] OK

Objective> Define Objective Function [Select parameter name (Convergence tolerance can be left blank)] OK

Method/Tool> Specify Optimization Method/Tool [Sub-Problem] OK.

Controls for Sub-problem optimization [Maximum iterations, Maximum infeasible sets (ANSYS will stop iterations if this limit on infeasible designs is reached), Print frequency [1 - will save results for each optimization iteration] OK

Run. Starts the optimization iterations

There are equivalent commands for these menu selections as well. However, these commands cannot be placed in the file used for analysis. One can create a separate file containing these commands, if desired.

Once the optimization iterations stop, the results can be viewed by using Design Opt> Design sets> List. For each iteration ANSYS lists values of design and state variables (constraints) and the objective function. The solutions are labeled Feasible or Infeasible based on whether the design satisfies all constraints or not. The best feasible design is indicated by an asterisk. Changes in any of the optimization variables as a function of design iterations can be plotted on a graph using Design Opt> Design sets> Graphs/Tables.

A.2.2 Heat Flow Example

Consider design of an insulation system for a square duct as shown in Figure A.9. The inside surface of the duct is at a temperature of 300°C due to hot gases flowing from a furnace. We would like to determine the suitable thickness of the insulation so that the duct is not too hot to touch, say the outside surface temperature be less than 30°C. It has been decided that the insulation geometry will be square to match that of the duct. The heat conduction coefficient of insulation is 1.4 W/m · °C. The average convection heat transfer coefficient on the outside of the duct is 27 W/m · °C. The ambient air temperature is 20°C.

Because of symmetry, we create model of one-eighth of the domain as shown. The insulation dimension is variable. We must choose some starting value based on prior experience. Here we start with $w = 0.1$ m. In the analysis file we create a variable parameter using the following command:

```
*set, w, 0.1
```

Note the asterisk is part of the command and cannot be omitted.

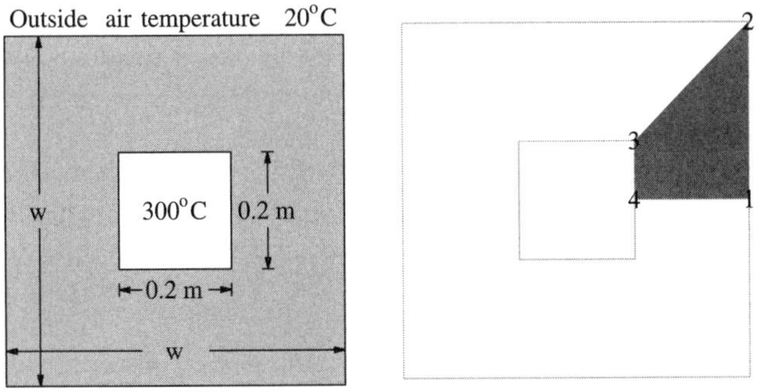

Figure A.9. Square duct

The area must be divided into a reasonable number of elements. The automatic mesh generation capability of ANSYS is a powerful tool to accomplish this. To use this, we have to create the area first. There are several ways of creating areas in ANSYS. For this simple example the area can be defined in terms of four key points. With the origin at keypoint 4, the coordinates of these points are as follows:

	x	y
1	w	0
2	w	$w + 0.1$
3	0	0.1
4	0	0

In ANSYS, the key points are defined by using the k command, which takes the key point number and its coordinates as arguments. From the key points one can create an area using the *a* command. The arguments to the *a* command are the key point numbers, defined by moving counterclockwise around the area. In the process ANSYS automatically creates lines between the pairs of key points. Each line is numbered sequentially, starting with line 1 defined between the first two key points used to define the area. The lines are used for defining boundary conditions.

After creating an area, we need to tell ANSYS how to create a mesh for the area. There are several controls to accomplish this. Here we simply define a global element size of say w/10 with the ESIZE command. This will give us a mesh of roughly 100 elements, and as you will see from the plots of the element solution later, it produces reasonable results. We also need to specify the type of finite element mesh to be generated using the MSHKEY, 0 command. The argument 0 specifies the free-meshing method as opposed to the mapped meshing (specified with 1). The free meshing is more flexible and can handle essentially any geometry. The mapped meshing creates a more uniform mesh but is restricted to quadrilateral areas. For this simple example we can use either method. The actual finite element mesh is generated using the AMESH command (area mesh). The argument to the command is area number.

We can specify the boundary conditions on nodes and elements directly. However, in order to have flexibility in changing the mesh size, we should specify the boundary conditions on lines and areas. This way, if a new mesh is created, ANSYS will automatically apply these boundary conditions to the nodes and elements defined along these lines. The temperature along a line is specified using the DL command. The convection is specified using the SFL command.

The model is now complete and we can use the Solve command to perform the analysis. To prepare for optimization, in the postprocessor we must select quantities that we need to define constraints and objective function. For a satisfactory design we have to make sure that the outside surface temperature is less than 30°C. This requires that we monitor temperature at a node on the outside. With automatic mesh generation we do not have a prior knowledge of the node numbers. However, after some experience with ANSYS, you will discover some of the patterns that ANSYS employs in numbering nodes and elements. One such observation is that ANSYS typically assigns the same node number as key point number at the locations where a key point exists. With this observation, we expect node 1 to be at the location of key point 1. (The numbering used for the key points was chosen with this in mind.) Since this point is on the outside surface and is closest geometrically to the hot duct, it is a good candidate to monitor temperature to satisfy the stated design goal. The actual definition of this as a parameter is accomplished by the following command:

```
*GET,nt,NODE,1,TEMP
```

where nt is a label of the user's choice and 1 refers to node 1. The command defines a parameter nt that is equal to the computed temperature at node 1.

The temperature goal can be met by making the insulation very thick. However, this will be uneconomical. Thus we need to keep the total volume of insulation used to a minimum. In order to obtain the total volume, we must compute the volume of each element and sum them together. This is done by the following commands:

```
ETABLE,volu,VOLU,
SSUM
*GET,totalvol,SSUM, ,ITEM,VOLU
```

ETABLE is a general command that is used to extract a variety of data from elements. Here we are asking for VOLU (for volume) data. The second argument is any label of user's choice. The second line tells ANSYS to create a sum over all elements of the ETABLE quantity. The last line actually defines a parameter called totalvol (the name is the user's choice) that is equal to the sum of element volumes.

The complete input data file for the example is as follows:

```
/PREP7
!* Element type
ET,1,PLANE55
!* Material properties
MPTEMP,,,,,,,,
MPTEMP,1,0
MPDATA,KXX,1,,1.4
set,w,0.1
!* Create key points
k,1,w,0
k,2,w,w+0.1
k,3,0.0,0.1
k,4,0,0,
!* Create area
a,1,2,3,4
```

```
!* Define global element size
ESIZE,w/10,0,
!* Specify free meshing
MSHKEY,0
!* Mesh areas
AMESH,1
FINISH
!* Solution module
/SOLU
!* Specify analysis type
ANTYPE,0
!* Specified temperature along line
DL,3,,TEMP,300,
!* Convection on line
SFL,1,CONV,27, ,20,
SOLVE
FINISH
!* Postprocessing
/POST1
GET,nt,NODE,1,TEMP,
ETABLE,volu,VOLU,
SSUM
GET,totalvol,SSUM, ,ITEM,VOLU
```

ANSYS Results

```
LIST OPTIMIZATION SETS FROM SET   1 TO SET  5 AND SHOW
ONLY OPTIMIZATION PARAMETERS. (A "*" SYMBOL IS USED TO
INDICATE THE BEST LISTED SET)

                 SET  1          SET  2        SET  3          SET  4
              (INFEASIBLE)  (FEASIBLE)    (INFEASIBLE)  (INFEASIBLE)
NT       (SV)  >  109.72       28.584     >  36.232    >  31.981
W        (DV)    0.10000      0.86246       0.51147      0.65667
TOTALVOL(OBJ)    0.15000E-01   0.45816      0.18195       0.28127

               *SET  5*
              (FEASIBLE)
NT       (SV)     29.053
W        (DV)    0.82655
TOTALVOL(OBJ)    0.42425
```

The plot in Figure A.10 shows how the outside surface temperature changed with each iteration. Figure A.11 shows the temperature distribution in the initial and the optimum designs.

A.3 ABAQUS APPLICATIONS

A.3.1 Execution Procedure

To perform a finite element analysis using ABAQUS, one must prepare a text file and save it with an extension .inp. Assuming a file named feajob.inp exists, the program can be executed by typing the following command at the shell prompt:

```
ABAQUS job=feajob
```

The analysis proceeds entirely in the background without any user interaction. ABAQUS writes analysis results in several files, all starting with the same job name but with dif-

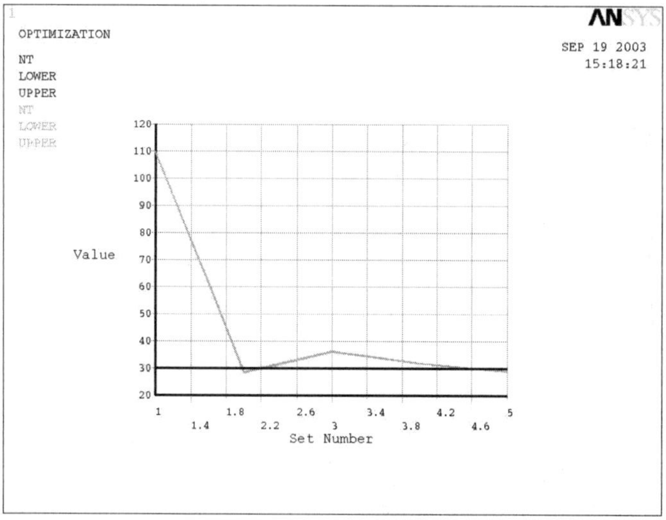

Figure A.10. Outside surface temperature during optimization iterations

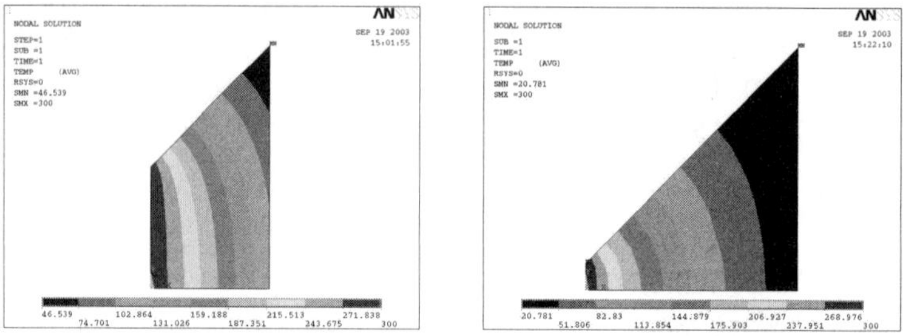

Figure A.11. Comparison of temperature in the initial and optimum designs

ferent extensions. The file with a .dat extension is a text file containing a summary of the execution steps and any other requested output. Any warnings or errors produced during execution are saved in this file as well. The file with extension .odb contains the entire analysis database in a binary form. This file is used for graphical viewing of results using the ABAQUS viewer as follows:

```
ABAQUS viewer database=feajob
```

The viewer has menus for standard operations of plotting contours of nodal and element solutions, deformed shape, etc.

The input file can be created either manually by typing appropriate commands in a text file or by using a graphical preprocessor for ABAQUS. To use the preprocessor, issue the following command at the shell prompt:

```
ABAQUS cae
```

The main screen of the cae environment contains the usual file manipulation, visualization, and help menus. Near the top left-hand side, under the buttons for File Save, Print, etc., is a drop-down list for accessing different modules for creating a finite element model. The order in which these modules appear in the list reflects the intended sequence in which the modules are to be used.

The following sections define typical commands and procedures from ABAQUS cae to create an input file. For trusses, structural frameworks, and other situations involving simple structured finite element meshes, the direct creation of the input file is generally the most convenient option. For modeling situations involving generation of complex unstructured finite element mesh, the use of ABAQUS cae is recommended. The following sections show examples of both approaches.

A.3.2 Truss Analysis

Determine nodal displacements and axial forces in the transmission tower shown in Figure A.1. The tower is made of steel ($E = 29 \times 10^6$ lb/in^2) angle sections. The main vertical members (elements 2, 4, 6, 8, 10, and 12) have a cross sectional area of 20 in^2. All other members have a cross-sectional area of 10 in^2.

The simplest way to create a model for structural frameworks is to manually create an input data file. For this example the model can be created by using the input as follows:

```
*Heading
Transmission tower
Node
1,-96,0,0
2,96,0,0
3,-96,180,0
4,96,180,0
5,-60,300,0
6,60,300,0
7,-60,420,0
8,60,420,0
9,-180,480,0
10,180,480,0
11,-300,540,0
12,-180,570,0
13,-60,600,0
14,60,600,0
15,180,570,0
16,300,540,0
* Node sets for bc
Nset, nset=pin
2
Nset, nset=roller
1
*Nset, nset=loaded
9,10,11,16
boundary
pin,1,2
boundary
roller,2
Element, type=T2D2
1,1,2
2,1,3
3,1,4
4,2,4
5,3,4
6,3,5
7,4,5
8,4,6
9,5,6
10,5,7
11,5,8
12,6,8
13,7,8
14,7,9
15,8,10
16,9,11
17,10,16
18,11,12
19,12,13
20,13,14
21,14,15
22,15,16
23,9, 12
24,9,13
25,7, 13,
26,8, 13
27,8, 14
28,10, 14
29,10, 15
* Element sets
Elset, elset=set1
2,4,6,8,10,12
Elset, elset=set2, generate
1,11,2
13,29,1
solidsection, elset=set1, material=steel
20
solidsection, elset=set2, material=steel
10
material, name=steel
elastic
```

```
29000000, 0.3
Step, name=Step-1
Vertical load 1000 lb
static
cload
loaded, 2, -1000
output, field, variable=preselect
output, history, variable=preselect
elprint, frequency=1
S,
*nodeprint, frequency=1
U,
RF,
endstep
```

The keywords starting with an asterisk are ABAQUS commands. The commands are not case sensitive. Lines starting with two asterisks are comment lines. The line following the *Heading command is used on output and plots to identify the analysis. The nodal coordinates are entered as node number and x, y, and z coordinates following the *Node command. In preparation for defining boundary conditions and loads, three nodal sets are created using the *Nset command. Nodes with pin supports are identified as a set called "pin," those with roller support are called "roller," and those with vertical load are called "loaded." Using the *Boundary command, the degrees of freedom 1 and 2 (x and y displacements) are suppressed for node set "pin." The roller support is represented by suppressing only the second degree of freedom (y displacement). The command *Element starts the definition of elements. The plane truss element in ABAQUS is identified as T2D2. The following lines define elements as element number, node at one end, and node at the other end. For assigning material and section properties the elements are classified into two sets identified as set1 and set2. Using the *Elset command, the elements in set1 are assigned material as "steel" and section as *solidsection with area = 20 and those in set2 are assigned the same material but an area = 10. The elastic material properties for steel are defined using *Material and *Elastic commands. A load of 1000 is applied in the $-y$ direction to the "loaded" nodes using the *Cload command.

The actual analysis commands are between the *Step and *Endstep commands. *Static specifies a static analysis. *Output, *Elprint, and *Nodeprint commands control the results information written to the database, element results written to the .dat file, and nodal results written to .dat file, respectively. The label S refers to stresses, U to displacements, and RF to reaction forces.

Using this input file the following command is used to perform the analysis:

```
ABAQUS job=tower
```

The ABAQUS output consisting of results and any error or warning messages are saved in a text file called tower.dat. The results can be viewed using the viewer as follows:

```
ABAQUS viewer database=tower
```

A plot of the axial stresses in the elements on a highly exaggerated deformed shape of the tower is shown in Figure A.12. The numerical results are exactly the same as those obtained from ANSYS.

A.3.3 Steady-State Heat Flow

Consider two-dimensional heat flow over an L-shaped body shown in Figure A.3 using ABAQUS. The thermal conductivity in both directions for the larger left area is 45 W/m·°C and that for the smaller right area is 25 W/m · °C. Heat is generated only in the left area at

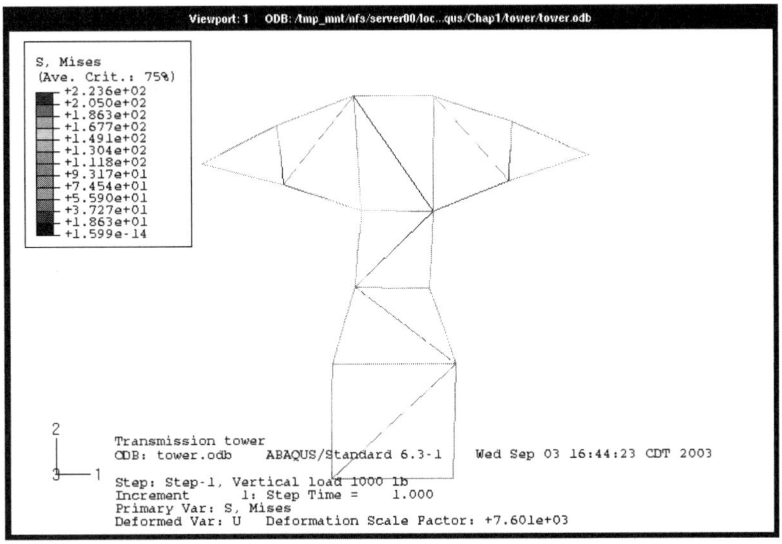

Figure A.12. Axial stresses shown on a deformed shape of the model

a rate of $Q = 5 \times 10^6$ W/m^3. The bottom is maintained at a temperature of $T_0 = 110$°C. Convection heat loss takes place on the top where the ambient air temperature is 20°C and the convection heat transfer coefficient is $h = 55$ W/m$^2\cdot$°C. The right side is insulated. The left side is subjected to heat flux at a uniform rate of $q_0 = 8000$ W/m^2.

For this example it is convenient to use the ABAQUS cae to prepare the input file.

Model Creation Using ABAQUS cae

1. *Part Module.* The overall geometry is created using the Part module in the ABAQUS cae system. A model may consist of several different parts. Interactions among these parts are specified in a later module. For this heat flow example we create a single part consisting of the L-shaped region:

Part> Create [Name: Part-1, Modeling space: 2D Planar, Type: Deformable, Base feature: Shell, Approximate size: 0.1]. Continue

Add> Line> Connected lines [Enter x,y coordinates of corners of the L-shape]. Enter return after entering each pair: 0,0; 0.06,0; 0.06,0.015; 0.03,0.015; 0.03,0.03; 0,0.03; 0,0. Click on any other icon to get out of line define mode. Click on Done next to "Sketch the section for planar shell."

In order to assign different properties, we must partition the L-shaped region into two as follows:

Tools> Partition> Face> Shortest path between two points. Select a point (0.03,0) at the bottom edge and the inside corner of L-shape (0.03,0.015) and click on Create partition. The L-shaped region should now be divided into the left square area and the right rectangular area.

2. *Property Module.* Different material and section properties are created in this module. For this example two materials are defined with appropriate thermal conductivity values:

Material> Create> Material-1> Continue>Thermal> Conductivity [Enter Conductivity (k) = 45] OK

Material> Create> Material-2> Continue>Thermal> Conductivity [Enter Conductivity (k) = 25] OK

In order to use these materials later, they must be associated with appropriate sections. Two sections are defined and associated with these materials as follows:

Section> Create> Section-1> Solid> Homogeneous> Continue>[Set Material to Material-1 and plane stress/strain thickness to 1] OK

Section> Create> Section-2> Solid> Homogeneous> Continue>[Set Material to Material-2 and plane stress/strain thickness to 1] OK

The next task is to actually associate these sections to different areas of the part as follows:

Assign> Section> [Select left square area and click Done on the button next to 'Select the regions to be assigned a section']> Section-1> OK

Assign> Section> [Select right rectangular area and click Done]> Section-2> OK

3. *Assembly Module.* Different parts in a model are assembled in this module. Different contact and interface conditions can be created between these parts. For this example, we only have a single part and thus our task is very simple. The assembly is just the part and is defined as follows:

Instance> Create> Part-1> OK

4. *Step Module.* This module defines the analysis type and sets desired output solution parameters:

Step> Create> [Name: Step-1, Procedure type: General]> Heat transfer> Continue> [Response: Steady state]

The desired output solution parameters are defined as follows:

Output> Field output requests> Create> [Name: F-Output-2, Step-1]> Continue> Thermal [Select NT, Nodal temperature and HFL, Heat flux vector]

5. *Interaction Module.* For structural problems contact and sliding-type boundary conditions are defined in this module. For heat transfer problems, the main purpose of this module is to define the convection boundary condition:

Interaction> Create> [Name: Int-1, Step: Step-1]> Film condition> Continue. Select the lines on the top of the region by clicking. Keep the shift key pressed to be able to select multiple lines. Click on Done next to "Select the surface." In the resulting dialog box enter h value (55) in the Film coefficient and surrounding fluid temperature (20) in the Sink temperature fields and click OK.

6. *Load Module.* For structural problems loads and boundary conditions are created in this module. For heat transfer problems, this module is used to specify known temperature and heat flux boundary conditions.

Load> Create> [Name: Load-1, Step: Step-1, Category: Thermal]> Surface heat flux> Continue. Select the line on the left of the region by clicking. Click on Done next to the "Select surfaces for the load." In the resulting dialog box enter Magnitude = 8000 for the specified heat flux on this side and click OK.

Load> Create> [Name: Load-2, Step: Step-1, Category: Thermal]> Body heat flux> Continue. Select the left area by clicking. Click on Done next to the "Select bodies for the load." In the resulting dialog box enter Magnitude = 5000000 for the specified body heat flux on this area and click OK.

BC> Create> [Name: BC-1, Step: Step-1, Category: Other]> Temperature> Continue. Select the bottom lines by clicking. Click on Done next to the "Select regions for the boundary condition." In the resulting dialog box enter Magnitude = 110 for the specified temperature on this bottom and click OK.

If no boundary condition is specified along a boundary, it automatically means there is no heat flow across that face, which is equivalent to the insulated boundary condition. Thus there is no need to explicitly specify a zero heat flux on the right end of the solid.

7. *Mesh Module.* The next step is to actually create a finite element mesh. We must decide on an approximate size of each element, which will obviously determine the total number of elements and nodes. One can specify a global element size for the entire model. To provide further control over the mesh, it is possible to specify the target number of subdivisions of a given line. The subdivisions can be biased toward a chosen end of the line (for example, to capture a rapidly changing solution near a corner). For this example, looking at the physical dimensions of the model, we choose a global element size of 0.002 m. Thus the length of the model will be divided roughly into 30 segments and the width into 15, resulting in a mesh of the order of $30 \times 15 = 450$ elements, which should give us reasonable results. Near the corner at key point 5 we expect a rapid solution change and thus we specify 20 subdivisions for each of the three lines meeting at the corner. A bias toward the corner is created by clicking near the corner end of each line (the bias direction is indicated by an arrow by ABAQUS graphics). The following steps are used to accomplish these tasks:

Seed> Edge biased> Click on the three lines meeting at the interior corner (pick near the end where the mesh must be denser). Click on Done near "Enter edges to be assigned local seeds." Enter Bias ratio = 2 (larger value implies more bias).

Seed> Instance> Enter global element size (approximate): 0.002.

Mesh> Controls> Select both areas (with shift key pressed to be able to select multiple areas) and then click on Done next to "Select the regions to be assigned mesh controls." In the resulting dialog box choose Element shape: Quad-dominated, Technique: Free, and Algorithm: Advancing front and then click OK and Done.

Mesh> Controls> Select both areas and then click on Done next to "Select the regions to be assigned element types." In the resulting dialog box choose Element Library: Standard, Geometric Order: Linear, and Family: Heat transfer and then click OK and Done.

Mesh> Instance> Click on Yes button next to "OK to mesh part instance?"

The resulting finite model is shown in Figure A.13. The model generation is now complete. Additional information, such as boundary conditions and load symbols, can be displayed on the model by selecting View> Assembly display options. The graphics can be saved to a file by using File> Print menu and setting the Destination to File, giving a file name and specifying the desired format.

8. *Job Module.* We are now ready to actually create the finite element analysis input file as follows:

Job> Create> [Name: LShapeHeat, Model: Model-1]> Continue. In the resulting dialog box accept the default options [Job Type: Full analysis, Run Mode: Background, Submit Time: Immediately]

This step will create an input file named LShapeHeat.inp.

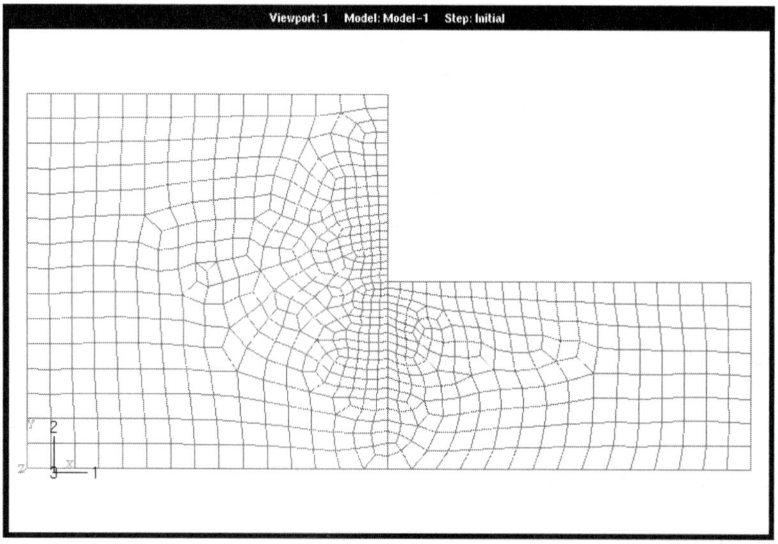

Figure A.13. Heat flow model in ABAQUS

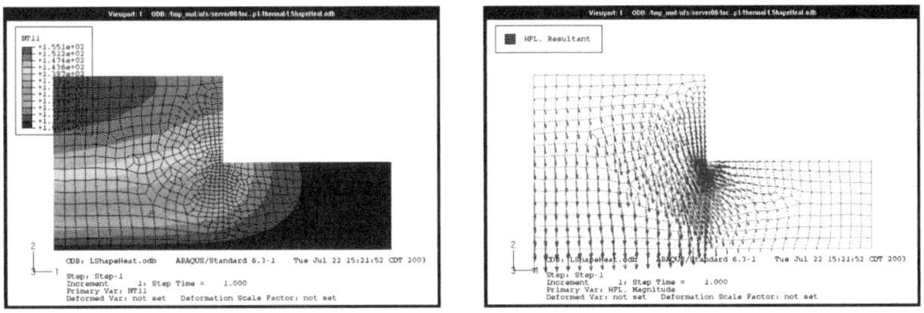

Figure A.14. Temperature contours and heat flux vectors

Solution The actual analysis can be performed in the batch mode using the input file by issuing the command abaqus job=LShapeHeat at the shell prompt. Alternatively we can use the menu Job> Submit from within the job module of ABAQUS cae. With this option the progress of the solution can be monitored using the Job> Monitor option. Once the analysis is complete, the results will be saved in a database LShapeHeat.odb and summary of the execution steps and requested output will be written to a text file LShapeHeat.dat.

Postprocessing Using ABAQUS Viewer The postprocessing of the results is done using the viewer module. If the ABAQUS cae environment is already open, one can simply switch to the viewer module and open the results database. Otherwise the viewer can be launched separately by issuing the abaqus viewer database=LShapeHeat command at the shell prompt.

The results can now be viewed as numerical lists or plotted in various forms. For example, Figure A.14 shows a contour plot of nodal temperatures and vector plot of element heat flux obtained using the following steps:

Results> Field output> [Select Name: NT11, Nodal temperatures at nodes]

Plot> Contours

Results> Field output> [Select Name: HFL, Heat flux vector at integration points, Invariant: Magnitude]

Plot> Contours

A.3.4 Plane Stress Analysis

As a final example, consider the problem of finding stresses in a notched beam of rectangular cross section shown in Figure A.6. The beam is 4 in thick in the direction perpendicular to the plane of paper and is made of concrete with modulus of elasticity $E = 3 \times 10^6$ lb/in^2 and Poisson's ratio $\nu = 0.2$. The beam is loaded by a uniform pressure of 50 lb/in^2 on the top surface.

Model Creation Using ABAQUS cae

1. *Part Module.* The overall geometry is created using the Part module in the ABAQUS cae system. A model may consist of several different parts. Interactions among these parts are specified in a later module. For this example we create a single part as follows:

Part> Create [Name: Part-1, Modeling space: 2D Planar, Type: Deformable, Base feature: Shell, Approximate size: 60] Continue

Add> Line> Connected lines [Enter *x, y* coordinates of the corners (Figure A.6)]. Enter return after entering each pair: 0,5; 6,5; 6,0; 54,0; 54,12; 0,12; 0,5. Click on any other icon to get out of line define mode. Click on Done next to "Sketch the section for planar shell."

2. *Property Module.* Different material and section properties are created in this module. For this example one material is defined with appropriate modulus of elasticity and Poisson's ratio values:

Material> Create> Material-1> Continue>Mechanical> Elasticity> Elastic [Enter $E = 30000000$ and $\nu = 0.2$] OK

In order to use this material later, it must be associated with appropriate sections. For this example one section is defined and associated with the material as follows:

Section> Create> Section-1> Solid> Homogeneous> Continue>[Set Material to Material-1 and plane stress/strain thickness to 4] OK

The next task is to actually associate the sections to different areas of the part. In this example there is only one section and one area:

Assign> Section> [Select the area and click on Done next to "Select the regions to be assigned a section"]> Section-1> OK

3. *Assembly Module.* Different contact and interface conditions can be created between parts using this module. For this example, we only have a single part and thus the assembly is just the part and is defined as follows:

Instance> Create> Part-1> OK

4. *Step Module.* This module defines the analysis type and sets desired output solution parameters:

Step> Create> [Name: Step-1, Procedure type: General]> Static General> Continue> [Nlgeom: Off]

Nonlinear large displacement effects can be included in the analysis by setting Nlgeom: On. For a conventional small-displacement linear analysis Nlgeom must be set to off.

The desired output solution parameters are defined as follows:

Output> Field output requests> Create> [Name: F-Output-2, Step-1]> Continue> [Select Stresses: S, Stress components and invariants, Displacement/Velocity/Acceleration: U, Translations and rotations, and

Forces/Reactions: Loads, Uniform distributed loads and RF, Reaction forces and moments]

5. *Interaction Module.* Contact and sliding-type boundary conditions are defined in this module. These conditions do not exist in this example and thus there is no need to use this module.

6. *Load Module.* Loads and boundary conditions are created in this module:

Load> Create> [Name: Load-1, Step: Step-1, Category: Mechanical]> Pressure> Continue. Select the line on the top of the region by clicking. Click on Done next to "Select surfaces for the load." In the resulting dialog box enter Magnitude = 50 for the specified load on this side and click OK.

BC> Create> [Name: BC-1, Step: Step-1, Category: Mechanical]> Symmetry/Antisymmetry/Encastre> Continue. Select the left side symmetry line by clicking. Click on Done next to "Select regions for the boundary condition." In the resulting dialog box click XSYMM for symmetry in the x direction and click OK.

BC> Create> [Name: BC-2, Step: Step-1, Category: Mechanical]> Symmetry/Antisymmetry/Encastre> Continue. Select the right line by clicking. Click on Done next to "Select regions for the boundary condition." In the resulting dialog box click ENCASTRE for fixed-end condition and click OK.

7. *Mesh Module.* The next step is to actually create a finite element mesh. We must decide on an approximate size of each element. Looking at the physical dimensions of the model, we choose a global element size of 2 in. Thus the length of the model will be divided roughly into 27 segments and the width into 6, resulting in a mesh of the order of $27 \times 6 = 162$ element mesh. We expect high stresses in the notch area and near the fixed end. To capture these stresses, we use a finer mesh by specifying element size of 0.5 in the notch region. An element size of 1 is used for the line on the fixed end. The following menu paths are used to accomplish these tasks:

Seed> Edge by size> Click on the three lines in the notch region. Click on Done near "Enter edges to be assigned local seeds." Enter element size along edges (approximate) = 0.5.

Seed> Edge by size> Click on the line on the fixed end. Click on Done near "Enter edges to be assigned local seeds." Enter element size along edges (approximate) = 1.

Seed> Instance> Enter global element size (approximate): 2.

Mesh> Controls. In the resulting dialog box choose Element shape: Quad-dominated, Technique: Free, and Algorithm: Advancing front and then click OK and Done.

Mesh> Controls. In the resulting dialog box choose Element Library: Standard, Geometric Order: Linear, and Family: Plane stress and then click OK and Done. Make sure Reduced integration and Incompatible modes are not selected. (These options could be useful but one must understand associated theoretical details before relying on these ad hoc measures at improving solution accuracy.)

Mesh> Instance> Click on Yes button next to "OK to mesh part instance?"

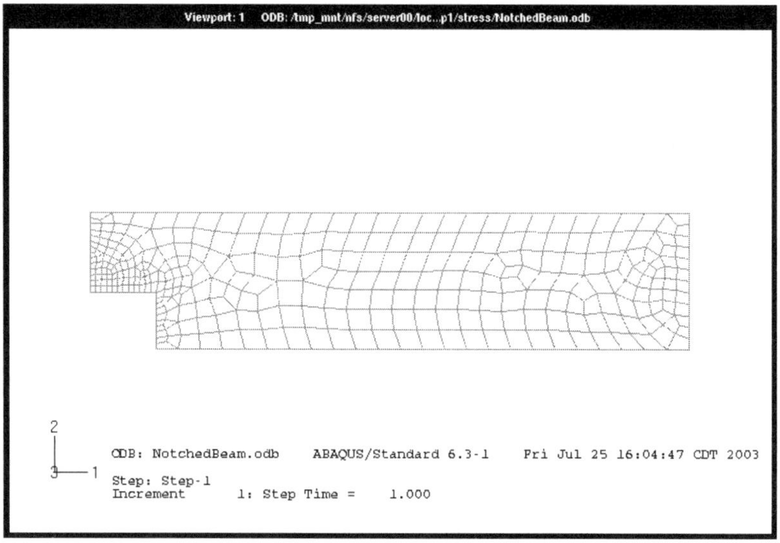

Figure A.15. Plane stress model of notched beam in ABAQUS

The resulting finite model is shown in Figure A.15. The model generation is now complete. Additional information, such as boundary conditions and load symbols, can be displayed on the model by selecting View> Assembly display options. The graphics can be saved to a file by using File> Print menu and setting the Destination to File, giving a file name and specifying the desired format.

8. *Job Module.* We are now ready to actually create the finite element analysis input file as follows:

> Job> Create> [Name: NotchedBeam, Model: Model-1]> Continue. In the resulting dialog box accept the default options [Job Type: Full analysis, Run Mode: Background, Submit Time: Immediately]

This step will create an input file named NotchedBeam.inp.

Solution The actual analysis can be performed in the batch mode using the input file by issuing the command abaqus job=NotchedBeam at the shell prompt. Alternatively, we can use the menu Job> Submit from within the job module of ABAQUS cae. With this option the progress of the solution can be monitored using the Job> Monitor option. Once the analysis is complete, the results will be saved in a database NotchedBeam.odb and summary of the execution steps and requested output will be written to a text file Notched-Beam.dat.

Postprocessing Using ABAQUS Viewer The postprocessing of the results is done using the viewer module. If the ABAQUS cae environment is already open, one can simply switch to the viewer module and open the results database. Otherwise the viewer can be

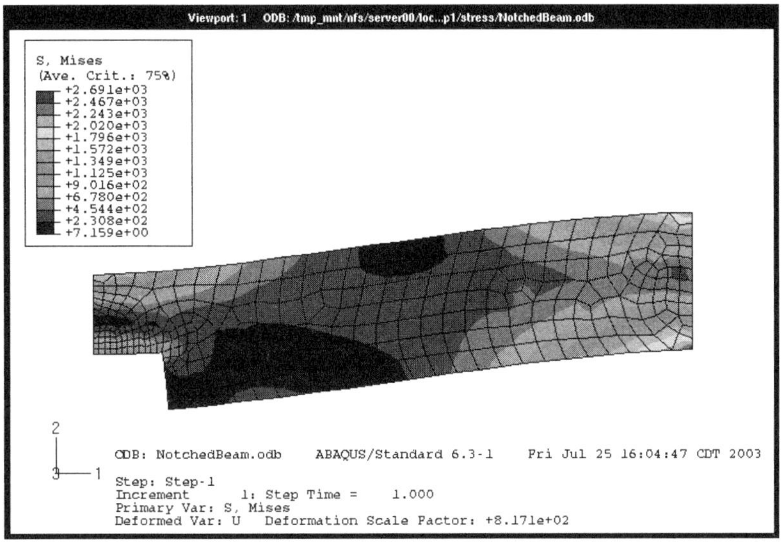

Figure A.16. von Mises equivalent stress contours

launched separately by issuing the abaqus viewer database=NotchedBeam command at the shell prompt.

The results can now be viewed as numerical lists or plotted in various forms. For example Figure A.16 shows a contour plot of equivalent von Mises stresses obtained using the following menu path:

Results> Field output> [Select Name: S, Stress components at integration points, Invariant: Mises]

Plot> Contours

APPENDIX B

VARIATIONAL FORM FOR BOUNDARY VALUE PROBLEMS

For some classes of boundary value problems it is possible to derive an equivalent variational or energy form. The variational form is written as an integral functional (function of a function) whose necessary conditions for a minimum imply that the boundary value problem is satisfied. Thus, instead of deriving the weak form using the Galerkin criteria, we can use this functional to obtain an approximate solution of the problem.

The derivation of functionals employs concepts from calculus of variations. The basic concept of variation of a function is presented in the first section. The second section presents a procedure to develop a suitable functional for a given boundary value problem. The last section presents a procedure to do the opposite, i.e., to find the equivalent boundary value problem corresponding to a given functional.

B.1 BASIC CONCEPT OF VARIATION OF A FUNCTION

In ordinary calculus the derivative of a function is defined by using the limiting process on the difference of values of the function at two neighboring points. In a similar manner we can define the *variation of a function* that is related to the difference between two closely related functions. As an example, consider the following function of three parameters a_0, a_1, a_2:

$$u(x) = a_0 + a_1 x + a_2 x^2$$

Denoting infinitesimal changes in the coefficients as $\delta a_0, \delta a_1, \delta a_2$, a closely related function is written as follows:

$$\tilde{u}(x) = (a_0 + \delta a_0) + (a_1 + \delta a_1)x + (a_2 + \delta a_2)x^2$$

The *variation* of $u(x)$, denoted by δu is defined as the difference between these two functions:

$$\delta u(x) \equiv \tilde{u}(x) - u(x) = \delta a_0 + \delta a_1 x + \delta a_2 x^2$$

Using the same concept, we can talk about the *variation of functionals*. For example, the variation of a square of function $u(x)$ is written as follows:

$$F \equiv u(x)^2$$

$$\delta(F) \equiv \tilde{u}(x)^2 - u(x)^2 = ((a_0 + \delta a_0) + (a_1 + \delta a_1)x + (a_2 + \delta a_2)x^2)^2 - (a_0 + a_1 x + a_2 x^2)^2$$

Expanding

$$\begin{aligned}\delta(u(x)^2) = (&\delta a_2^2 x^4 + 2a_2\delta a_2 x^4 + 2a_2\delta a_1 x^3 + 2a_1\delta a_2 x^3 + 2\delta a_1\delta a_2 x^3 + \delta a_1^2 x^2 + 2a_2\delta a_0 x^2 \\ &+ 2a_1\delta a_1 x^2 + 2a_0\delta a_2 x^2 + 2\delta a_0\delta a_2 x^2 + 2a_1\delta a_0 x + 2a_0\delta a_1 x + 2\delta a_0\,\delta a_1 x \\ &+ \delta a_0^2 + 2a_0\delta a_0)\end{aligned}$$

Since the changes in parameters are infinitesimal, the higher order terms are neglected, giving

$$\begin{aligned}\delta(u(x)^2) = (&2a_2\delta a_2 x^4 + 2a_2\delta a_1 x^3 + 2a_1\delta a_2 x^3 + 2a_2\delta a_0 x^2 \\ &+ 2a_1\delta a_1 x^2 + 2a_0\delta a_2 x^2 + 2a_1\delta a_0 x + 2a_0\delta a_1 x + 2a_0\delta a_0)\end{aligned}$$

As another example, consider the variation of the first derivative of function $u(x)$:

$$\delta\left(\frac{du}{dx}\right) \equiv \frac{d\tilde{u}}{dx} - \frac{du}{dx} = (a_1 + \delta a_1 + 2(a_2 + \delta a_2)x) - (a_1 + 2a_2 x)$$

Simplifying

$$\delta\left(\frac{du}{dx}\right) = \delta a_1 + 2x\delta a_2$$

Using this basic concept, we can demonstrate that the order of differentiation and variation can be interchanged:

$$\delta\left(\frac{du}{dx}\right) \equiv \frac{d(\delta u)}{dx} = \frac{d}{dx}(\delta a_0 + \delta a_1 x + \delta a_2 x^2) = \delta a_1 + 2x\delta a_2$$

which is same as before. Thus the variation of the derivative of a function is the same as the derivative of its variation.

From a computational point of view, it is convenient to relate variation to the total derivative. To accomplish this, we note that the coefficients of the changes in parameters are simply the derivatives of the function $u(x)$ with respect to the parameters:

$$\frac{\partial u}{\partial a_0} = 1; \qquad \frac{\partial u}{\partial a_1} = x; \qquad \frac{\partial u}{\partial a_2} = x^2$$

Using this, the definition of a variation of function can be expressed as a total differential, as follows:

$$\delta u(x) = \frac{\partial u}{\partial a_0}\delta a_0 + \frac{\partial u}{\partial a_1}\delta a_1 + \frac{\partial u}{\partial a_2}\delta a_2$$

Using this concept of variation, it is easy to compute the variation of functionals. For example,

$$F \equiv u(x)^2$$

$$\delta(F) \equiv 2u\delta u = 2(a_0 + a_1 x + a_2 x^2)(\delta a_0 + \delta a_1 x + \delta a_2 x^2)$$

Expanding

$$\delta(u(x)^2) = (2a_2\delta a_2 x^4 + 2a_2\delta a_1 x^3 + 2a_1\delta a_2 x^3 + 2a_2\delta a_0 x^2 + 2a_1\delta a_1 x^2 + 2a_0\delta a_2 x^2 + 2a_1\delta a_0 x + 2a_0\delta a_1 x + 2a_0\delta a_0)$$

which is same as the one obtained from the basic definition. The following example shows the computation of variations of more complicated functionals.

Example B.1 Here are some more examples of the computation of the variation of functionals:

(i) $F(u(x), x) = u^3 + x^3 u + 4$

$$\delta F = 3u^2\delta u + x^3\delta u$$

(ii) $F(u(x), x) = \left(\frac{du}{dx}\right)^3 + u^2\sin(x) + 4u$

$$\delta F = 3\left(\frac{du}{dx}\right)^2 \delta\left(\frac{du}{dx}\right) + 2u\sin(x)\delta u + 4\delta u \equiv 3\left(\frac{du}{dx}\right)^2 \frac{d(\delta u)}{dx} + 2u\sin(x)\delta u + 4\delta u$$

(iii) $F(u) = \int_0^1 \left(\frac{1}{2}\left(\frac{du}{dx}\right)^2 + u\sin(x)\right) dx$

$$\delta F = \int_0^1 \left(\frac{du}{dx}\delta\left(\frac{du}{dx}\right) + u\sin(x)\delta u\right) dx \equiv \int_0^1 \left(\frac{du}{dx}\frac{d(\delta u)}{dx} + \delta(u\sin(x))\right) dx$$

Useful Properties of Variation With the interpretation of variation as a total derivative, it is easy to derive the following properties of the variation of functionals:

(i) Given two functional F and G, we have

$$\delta(F + G) = \delta F + \delta G$$

$$\delta(FG) = (\delta F)G + (F)\delta G$$

$$\delta(aF) = a\delta F; \qquad a \text{ is a given constant}$$

$$\delta(F^n) = nF^{n-1}\delta F; \qquad n \text{ is a given integer}$$

(ii) We can interchange the order in which integration or differentiation and variation is carried out. That is,

$$\delta\left(\int F dx\right) = \int \delta F dx; \qquad \delta\left(\frac{dF}{dx}\right) = \frac{d}{dx}(\delta F)$$

(iii) The necessary condition for the minimum of a functional is that its first variation must be equal to zero:

$$\text{Necessary condition for minimum of } F \Longrightarrow \delta F = 0$$

(iv) The following identities, used frequently in the derivation of equivalent variational forms, follow immediately from these properties:

$$f(x)\delta u = \delta(f(x)u(x)); \qquad f(x) \text{ a given function of } x$$

$$u(x)\delta u = \frac{1}{2}\delta(u^2)$$

$$\frac{du}{dx}\frac{d(\delta u)}{dx} = \frac{1}{2}\delta\left[\left(\frac{du}{dx}\right)^2\right]$$

B.2 DERIVATION OF EQUIVALENT VARIATIONAL FORM

For certain types of boundary value problems it is possible to find equivalent variational forms by a series of mathematical manipulations. The procedure is illustrated through the following examples.

Example B.2 Find an equivalent variational functional for the following boundary value problem:

$$-u'' + \sin(\pi x) = 0; \qquad a < x < b$$

with the boundary conditions $u'(a) = c$ and $u(b) = d$, where a, b, c, and d are given constants.

The process starts by multiplying the differential equation by the variation in the solution and integrating over the solution domain:

$$\int_a^b (-u'' + \sin(\pi x))\delta u \, d\, x = 0$$

Our goal now is to manipulate this expression so that we end up with the following form:

$$\delta(\cdots) = 0$$

The reason for doing this is because this form indicates that the variation of the quantity inside the parentheses is equal to zero, which is a necessary condition for the minimum of a functional. Thus the quantity inside the parentheses is the required functional and its minimum corresponds to the solution of the given boundary value problem.

In order to create the desired form, we use integration by parts to make each term appear with the same order of derivative for the solution and its variation. In this example, the second term inside the integral is fine because it involves a function of x and δu. The first term, however, involves u'' and δu. Using integration by parts, we need to move one of the derivatives from u'' to δu, which will result in both parts involving u'. Thus integrating the first term by parts, we get

$$(-\delta u u')_{x=b} - (-\delta u u')_{x=a} + \int_a^b (\delta u' u' + \sin(\pi x)\delta u)\, dx = 0$$

Using properties of variations, the terms inside the integral can be written as follows:

$$-\delta u(b)u'(b) + \delta u(a)u'(a) + \int_a^b \left(\delta\left(\tfrac{1}{2}u'^2\right) + \delta(\sin(\pi x)u)\right) dx = 0$$

Using the boundary condition $u'(a) = c$, combining the terms inside the integral, and changing the order of integration and variation, we get

$$-\delta u(b)u'(b) + \delta u(a)c + \delta\left(\int_a^b \left(\tfrac{1}{2}u'^2 + \sin(\pi x)u\right) dx\right) = 0$$

The boundary term at $x = b$ still involves δu and u'. We cannot simplify this term like the others. The only way to proceed is to require that the admissible trial solutions satisfy the boundary condition at $x = b$. Then for all such functions $\delta u(b) = 0$, and we have

$$\delta(cu(a)) + \delta\left(\int_a^b \left(\tfrac{1}{2}u'^2 + \sin(\pi x)u\right) dx\right) = 0$$

We now have the sum of two terms involving variations. These can be combined together to give

$$\delta\left[cu(a)) + \int_a^b \left(\tfrac{1}{2}u'^2 + \sin(\pi x)u\right) dx\right] = 0$$

This is the form that we were seeking, and thus the equivalent functional for the given boundary value problem is

$$F(x, u, u') = cu(a) + \int_a^b \left(\tfrac{1}{2}u'^2 + \sin(\pi x)u\right) dx$$

with assumed solutions satisfying $u(b) = d$.

Example B.3 Find an equivalent variational functional for the following boundary value problem:

$$x^2u'' + 2xu' + u + 1 = 0 \qquad \text{or} \qquad \frac{d(x^2u')}{dx} + u + 1 = 0$$

with the boundary conditions $u(1) = 1$, $u'(2) = 1$.

Multiply the differential equation by the variation in the solution and integrate over the solution domain:

$$\int_1^2 \left(\frac{d(x^2u')}{dx} + u + 1 \right) \delta u \, dx = 0 \qquad \text{or} \qquad \int_1^2 \left(\frac{d(x^2u')}{dx} \delta u + u\,\delta u + \delta u \right) dx = 0$$

Use integration by parts to make each term have the same order of derivative for the solution and its variation. In this example, the second and the third terms inside the integral are fine because they involve u and δu. The first term, however, involves u'' and δu. Using integration by parts, we need to move one of the derivatives from u'' to δu, which will result in both involving u'. Thus integrating the first term by parts, we get

$$(\delta u\, x^2 u')_{x=2} - (\delta u\, x^2 u')_{x=1} + \int_1^2 (-\delta u'\, x^2 u' + \delta u\, u + \delta u) dx = 0$$

or

$$4\,\delta u(2)\, u'(2) - \delta u(1)\, u'(1) + \int_1^2 \left(-\delta \left(\tfrac{1}{2} x^2 u'^2 \right) + \delta \left(\frac{1}{2} u^2 \right) + \delta u \right) dx = 0$$

Using the boundary condition $u'(2) = 1$, we get

$$4\,\delta u(2) - \delta u(1)\, u'(1) + \delta \left[\int_1^2 \left(-\tfrac{1}{2} x^2 u'^2 + \tfrac{1}{2} u^2 + u \right) dx \right] = 0$$

The boundary term at $x = 1$ still involves δu and u'. We cannot simplify this term like the others. The only way to proceed is to require that the admissible trial solutions satisfy the boundary condition at $x = 1$. Then for all such functions $\delta u(1) = 0$, and we have

$$\delta(4u(2)) + \delta \left[\int_1^2 \left(-\tfrac{1}{2} x^2 u'^2 + \tfrac{1}{2} u^2 + u \right) dx \right] = 0$$

$$\delta \left[4u(2) + \int_1^2 \left(-\tfrac{1}{2} x^2 u'^2 + \tfrac{1}{2} u^2 + u \right) dx \right] = 0$$

Then the equivalent functional is

$$F(x, u, u') = 4u(2) + \int_1^2 \left(-\tfrac{1}{2} x^2 u'^2 + \tfrac{1}{2} u^2 + u \right) dx$$

Example B.4 Two-Dimensional BVP Find an equivalent variational functional for the following boundary value problem:

$$\frac{\partial u^2}{\partial x^2} + \frac{\partial u^2}{\partial y^2} = \sin(x); \qquad 0 < x < \pi; \qquad -\pi/2 < y < \pi/2$$

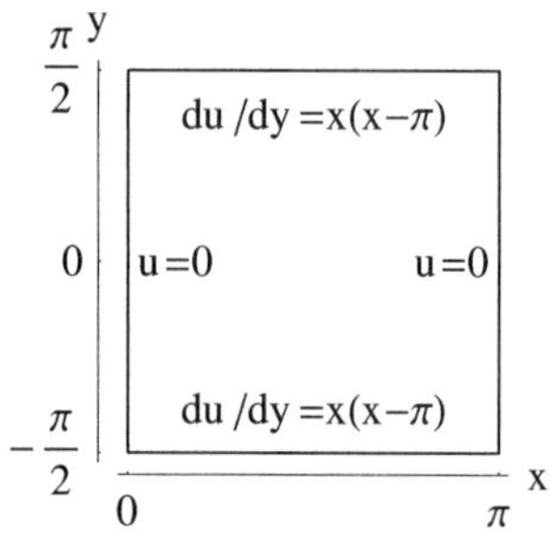

Figure B.1.

with the boundary conditions (see Figure B.1)

$$u = 0 \quad \text{on} \quad (x = 0, y) \quad \text{and} \quad (x = \pi, y)$$

$$\frac{\partial u}{\partial y} = x(x - \pi) \quad \text{on} \quad (x, y = -\pi/2) \quad \text{and} \quad (x, y = \pi/2)$$

The variational functional can be obtained as follows:

$$\iint_A \left(\frac{\partial u^2}{\partial x^2} + \frac{\partial u^2}{\partial y^2} - \sin(x) \right) \delta u \, dx \, dy = 0$$

Using Green's theorem

$$\int_c \left(\frac{\partial u}{\partial x} \delta u \, n_x + \frac{\partial u}{\partial y} \delta u \, n_y \right) dc - \iint_A \left(\frac{\partial u}{\partial x} \frac{\partial \delta u}{\partial x} + \frac{\partial u}{\partial y} \frac{\partial \delta u}{\partial y} + \sin(x) \delta u \right) dx \, dy = 0$$

Using admissible trial solutions, δu will be zero over the boundaries on which u is specified. Thus

$$\int_{x=0}^{\pi} \left(\frac{\partial u}{\partial x} \delta u \, n_x + \frac{\partial u}{\partial y} \delta u \, n_y \right)_{y=-\pi/2} dx + \int_{x=0}^{\pi} \left(\frac{\partial u}{\partial x} \delta u \, n_x + \frac{\partial u}{\partial y} \delta u \, n_y \right)_{y=\pi/2} dx$$
$$- \iint_A \left(\frac{\partial u}{\partial x} \frac{\partial \delta u}{\partial x} + \frac{\partial u}{\partial y} \frac{\partial \delta u}{\partial y} + \sin(x) \delta u \right) dx \, dy = 0$$

Using the specified natural boundary condition and noting direction cosines for the outer normals for top side ($n_x = 0$ and $n_y = 1$) and for the bottom side ($n_x = 0$ and $n_y = -1$), we have

$$\int_{x=0}^{\pi} x(x - \pi) \delta u \left(x, \frac{\pi}{2} \right) dx - \int_{x=0}^{\pi} x(x - \pi) \delta u \left(x, -\frac{\pi}{2} \right) dx$$
$$- \iint_A \delta \left(\frac{1}{2} \left(\frac{\partial u}{\partial x} \right)^2 + \frac{1}{2} \left(\frac{\partial u}{\partial y} \right)^2 + \sin(x) u \right) dx \, dy = 0$$

This can be written as follows:

$$\delta\left[\int_{x=0}^{\pi} x(x-\pi)u\left(x,\frac{\pi}{2}\right)dx - \int_{x=0}^{\pi} x(x-\pi)u\left(x,-\frac{\pi}{2}\right)dx \right.$$
$$\left. - \iint_A \left(\frac{1}{2}\left(\frac{\partial u}{\partial x}\right)^2 + \frac{1}{2}\left(\frac{\partial u}{\partial y}\right)^2 + \sin(x)u\right)dx\,dy\right] = 0$$

Thus the variational functional is as follows:

$$I(u) = \int_{x=0}^{\pi} x(x-\pi)u\left(x,\frac{\pi}{2}\right)dx - \int_{x=0}^{\pi} x(x-\pi)u\left(x,-\frac{\pi}{2}\right)dx$$
$$- \iint_A \left(\frac{1}{2}\left(\frac{\partial u}{\partial x}\right)^2 + \frac{1}{2}\left(\frac{\partial u}{\partial y}\right)^2 + \sin(x)u\right)dx\,dy$$

B.3 BOUNDARY VALUE PROBLEM CORRESPONDING TO A GIVEN FUNCTIONAL

In this section we consider the situation opposite to that of Section B.2. Here we assume that a functional is known. Our goal then is to find the corresponding boundary value problem. The procedure is useful in verifying that a given functional indeed is the correct functional for the problem.

Since the minimum of the functional corresponds to the equivalent differential equation, we start the process by setting the first variation of the functional to zero. Mathematical manipulations involving integration by parts are then carried out to bring boundary conditions into the picture. The process is illustrated through the following example.

Example B.5 Axial Deformations In Chapter 2 we used the following potential energy functional for the nonuniform bar shown in Figure B.2:

$$\Pi = U - W = \frac{1}{2}\int_0^L EA(x)\left(\frac{du}{dx}\right)^2 dx - \int_0^L q(x)u(x)dx - Pu(L)$$

To show that this is a suitable functional for the governing equation for the axial deforma-

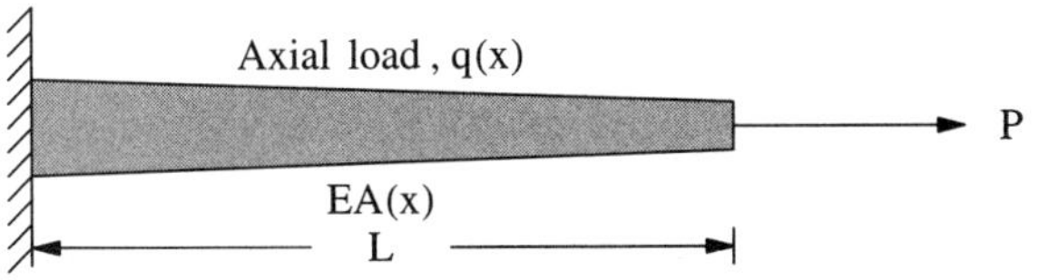

Figure B.2. Tapered axially loaded bar

tions, we take its first variation as follows:

$$\delta\Pi = \frac{1}{2}\int_0^L \delta\left[EA(x)\left(\frac{du}{dx}\right)^2\right]dx - \int_0^L \delta[q(x)u(x)]dx - P\delta u(L)$$

From the definition of variation

$$\delta\Pi = \int_0^L EA(x)\frac{du}{dx}\delta\left(\frac{du}{dx}\right)dx - \int_0^L q(x)\delta u dx - P\delta u(L)$$

Noting that order of differentiation and variation can be interchanged, we have

$$\delta\Pi = \int_0^L EA(x)\frac{du}{dx}\frac{d(\delta u)}{dx}dx - \int_0^L q(x)\,\delta u\,dx - P\,\delta u(L)$$

Integrating the first term by parts, we have

$$\delta\Pi = \left[EA(x)\frac{du}{dx}\delta u\right]_0^L - \int_0^L \left[\frac{d}{dx}\left(EA(x)\frac{du}{dx}\right)\right]\delta u\,dx - \int_0^L q(x)\,\delta u\,dx - P\,\delta u(L)$$

At $x = L$, $EA\,du/dx = P$, the applied load, and thus at the this point the boundary term cancels with the last term, and we have

$$\delta\Pi = -EA(0)\frac{du(0)}{dx}\,\delta u(0) - \int_0^L \left[\frac{d}{dx}\left(EA(x)\frac{du}{dx}\right)\right]\delta u\,dx - \int_0^L q(x)\,\delta u\,dx$$

At $x = 0$ we have a specified displacement. Requiring all assumed solutions to satisfy this boundary condition means that $\delta u(0) = 0$. Thus, with this restriction, the variation of the potential energy functional can be written as follows:

$$\delta\Pi = -\int_0^L \left[\frac{d}{dx}\left(EA(x)\frac{du}{dx}\right) + q(x)\right]\delta u\,dx$$

For $\delta\Pi$ to be zero (a necessary condition for the minimum), we have

$$\int_0^L \left[\frac{d}{dx}\left(EA(x)\frac{du}{dx}\right) + q(x)\right]\delta u\,dx = 0$$

Since the variation δu is arbitrary, the only way this integral can always be zero is if the term inside the square brackets is zero. Thus

$$\frac{d}{dx}\left(EA(x)\frac{du}{dx}\right) + q(x) = 0$$

This is exactly the differential equation governing the axial deformation of a nonuniform bar. Thus, as long as we restrict ourselves to solutions that satisfy the displacement boundary condition, the given potential energy is the appropriate functional for the problem.

PROBLEMS

B.1 Compute the variation of the following functionals:

(a) $F[u, x] = \dfrac{u^2}{x^3} + e^x$

(b) $F[u', u, x] = \displaystyle\int_0^1 (x^2 u'^2 + u^3 + x)\,dx$

(c) $F[u, x] = u^2 u'^2 + x^3$

(d) $F[u'', u', u, x] = \displaystyle\int_{-1}^1 \left(\frac{u''}{u} + u'x\right) dx$

B.2 Determine an equivalent variational form for the following boundary value problem:

$$\frac{d^2u}{dx^2} + x^2 = 0; \qquad 0 < x < 1$$

with the boundary conditions

$$u(0) = 1; \qquad u'(1) + 2u(1) = 1$$

B.3 Consider the following boundary value problem:

$$u'' \sin(x) + u' \cos(x) + u \sin(x) = 0; \qquad \pi/4 < x < \pi/2$$

$$u(\pi/4) = 1 \qquad \text{and} \qquad u'(\pi/2) = 2$$

Derive an equivalent variational functional for the problem. Note that the differential equation can be written as follows:

$$\frac{d}{dx}[u' \sin(x)] + u \sin(x) = 0$$

B.4 Consider the finite element solution of the following boundary value problem:

$$-u'' + x = 0; \qquad \pi/4 < x < \pi/2$$

$$u(\pi/4) = 1 \qquad \text{and} \qquad u'(\pi/2) + 2 = 0$$

Verify that the following is an appropriate functional for the problem:

$$I(u) = 2u\left(\frac{\pi}{2}\right) + \int_{\pi/4}^{\pi/2} \left(\frac{1}{2}u'^2 + xu\right) dx$$

B.5 A functional is given as follows:

$$I(u) = \int_0^l \left(\frac{1}{2}EAu'^2 - u \sin\left(\frac{\pi x}{l}\right)\right) dx - \frac{1}{2}ku(l)^2 - \frac{1}{\pi}u(0)$$

Determine the corresponding boundary value problem.

B.6 The potential energy for the problem of the torsion of a thin-walled section with warping restraint can be written as follows:

$$\Pi = \int_0^L \left(\frac{EJ_w}{2} \phi''^2 + \frac{GJ_0}{2} \phi'^2 - t(x)\phi \right) dx$$

where E is the modulus of elasticity, G is the shear modulus, J_w is the warping constant, J_0 is the torsional constant, t is the thickness of the section, and ϕ is the angle through which a cross section rotates. Determine the governing differential equation and appropriate boundary conditions.

BIBLIOGRAPHY

Basic Finite Element Books

1. Baker, A.J. and Pepper, D.W., Finite Elements 1-2-3, McGraw-Hill, New York, 1991.
2. Bickford, W.B., A First Course in the Finite Element Method, Irwin, Homewood, IL, 1990.
3. Buchanan, G.R., Theory and Problems of Finite Element Analysis, Schaum's Outline Series, McGraw-Hill, New York, 1995.
4. Burnett, D.S., Finite Element Analysis: From Concepts to Applications, Addison-Wesley, Reading, MA, 1987.
5. Chandrupatla, T.R. and Belegundu, A.D., Introduction to the Finite Elements in Engineering, Third Edition, Prentice-Hall, Englewood Cliffs, NJ, 2002.
6. Desai, C.S. and Kundu, T., Introductory Finite Element Method, CRC Press, Boca Raton, FL, 2001.
7. Fagan, M.J., Finite Element Analysis, Theory and Practice, Longman Group, Essex, U.K., 1992.
8. Hutton, D.V., Fundamentals of Finite Element Analysis, McGraw-Hill, Boston, MA, 2004.
9. Knight, C.E., The Finite Element Method in Mechanical Design, PWS-KENT Publishing Co., Boston, MA, 1993.
10. Kwon, Y.W. and Bang, H., The Finite Element Method Using Matlab, Second Edition, CRC Press, Boca Raton, FL, 2000.
11. Liu, G.R. and Quek, S.S., Finite Element Method, A Practical Course, Butterworth-Heinemann, Oxford, UK, 2003.
12. Logan, D.L., A First Course in the Finite Element Method, PWS Engineering, Boston, MA, 1986.

13. Moaveni, S., Finite Element Analysis: Theory and Application with Ansys, Second Edition, Prentice-Hall, Upper Saddle River, NJ, 2003.
14. Ottosen, N.S. and Petersson, H., Introduction to the Finite Element Method, Prentice-Hall, Englewood Cliffs, NJ, 1992.
15. Pepper, D.W. and Heinrich, J.C., The Finite Element Method: Basic Concepts and Applications, Taylor & Francis, Bristol, PA, 1992.
16. Rao, S.S., The Finite Element Method in Engineering, Third Edition, Butterworth-Heinemann, Boston, MA, 1999.
17. Reddy, J.N., An Intoduction to the Finite Element Method, Second Edition, McGraw-Hill, New York, 1993.
18. Segerlind, L.J., Applied Finite Element Analysis, Second Edition, John Wiley, New York, 1984.
19. Stasa, F.L., Applied Finite Element Analysis for Engineers, Holt, Rinehart, and Winston, New York, 1985.

Advanced Finite Element Books

20. Bathe, K.J., Finite Element Procedures, Prentice-Hall, Englewood Cliffs, NJ, 1996.
21. Cook, R.D., Malkus, D.S., Plesha, M.E., and Witt, R.J., Concepts and Applications of Finite Element Analysis, Fourth Edition, John Wiley, New York, 2002.
22. Gallagher, R.H., Finite Element Analysis Fundamentals, Prentice-Hall, Englewood Cliffs, NJ, 1975.
23. Gupta, K.K. and Meek, J.L., Finite Element Multidisciplinary Analysis, Second Edition, American Institute of Aeronautics and Astronautics, Reston, VA, 2003.
24. Huebner, K.H., Thornton, E.A., and Byrom, T.G., The Finite Element Method for Engineers, Third Edition, John Wiley, New York, 1995.
25. Kardestuncer, H., Editor-in-Chief, Finite Element Handbook, McGraw-Hill, New York, 1987.
26. Tong, P. and Rossettos, J.N., Finite Element Method: Basic Technique and Implementation, MIT Press, Cambridge, MA, 1977.
27. Zienkiewicz, O.C. and Taylor, R.L., The Finite Element Method, Fifth Edition, Volume 1, The Basics, Butterworth-Heinemann, Oxford, U.K., 2000.
28. Zienkiewicz, O.C. and Taylor, R.L., The Finite Element Method, Fifth Edition, Volume 2, Solid and Fluid Mechanics, Butterworth-Heinemann, Oxford, U.K., 2000.

Mathematically Oriented Books

29. Ainsworth, M. and Oden, J.T., A Posteriori Error Estimation in Finite Element Analysis, John Wiley, New York, 2000.
30. Jiang, B., The Least-Squares Finite Element Method, Theory and Applications in Computational Fluid Dynamics and Electromagnetics, Springer-Verlag, Berlin, 1998.
31. Braess, D., Finite Elements: Theory, Fast Solvers, and Applications in Solid Mechanics, Cambridge University Press, Cambridge, England, 1997.

32. Celia, M.A. and Gray, W.G., Numerical Methods for Differential Equations: Fundamental Concepts for Scientific and Engineering Applications, Prentice-Hall, Englewood Cliffs, NJ, 1992.
33. Davies, A.J., The Finite Element Method: A First Approach, Oxford University Press, Oxford, England, 1980.
34. Dow, J.O., A Unified Approach to the Finite Element Method and Error Analysis Procedures, Academic Press, San Diego, CA, 1999.
35. Dwoyer, D.L., Hussaini, M.Y., and Voigt, R.G., Editors, Finite Elements: Theory and Application, Springer-Verlag, New York, 1988.
36. Kleiber, M. and Hien, T.D., The Stochastic Finite Element Method: Basic Perturbation Technique and Computer Implementation, John Wiley, Chichester, England, 1992.
37. Lapidus, L. and G.F. Pinder, Numerical Solution of Partial Differential Equations in Science and Engineering, John Wiley, New York, 1982.
38. Lewis, P.E. and Ward J.P., The Finite Element Method: Principles and Applications, Addison-Wesley, Wokingham, England, 1991.
39. Martin, H.C. and Carey, G.F., Intoduction to Finite Element Analysis, McGraw-Hill, New York, 1973.
40. Mori, M., The Finite Element Method and Its Applications, Macmillan Publishing Company, New York, 1986.
41. Szabo, B. and Babuska, I., Finite Element Analysis, John Wiley, New York, 1991.
42. Taigbenu, A.E., The Green Element Method, Kluwer Academic Publishers, Boston, MA, 1999.
43. White, R.E., An Intoduction to the Finite Element Method with Applications to Nonlinear Problems, Wiley-Interscience, New York, 1985.
44. Zenisek, A., Nonlinear Elliptic and Evolution Problems and Their Finite Element Approximations, Academic Press, San Diego, CA, 1990.
45. Zienkiewicz, O.C. and Morgan, K., Finite Elements and Approximation, Wiley-Interscience, New York, 1982.

Finite Element Programming

46. ABAQUS User's Manual, Volumes I, II, and III, Version 6.3, Hibbet, Karlsson & Sorensen, Pawtucket, RI, 2000.
47. ABAQUS Theory Manual, Version 6.3, Hibbet, Karlsson & Sorensen, Pawtucket, RI, 2000.
48. ANSYS User's Manual, Volumes I, II, and III, Version 7.1, Swanson Analysis Systems, Houston, PA, 2003.
49. ANSYS Theory Manual, Version 7.1, Swanson Analysis Systems, Houston, PA, 2003.
50. Cheung, Y.K., Lo, S.H., and Leung, A.Y.T., Finite Element Implementation, Blackwell Science, Cambridge, MA, 1996.

51. George, P.L., Automatic Mesh Generation: Application to Finite Element Methods, John Wiley, New York, 1991.
52. Sabonnadiere, J.C. and Coulomb, J.L., Finite Element Methods in CAD, North Oxford Academic, London, 1987.
53. Smith, I.M. and Griffiths, D.V., Programming the Finite Element Method, Second Edition, John Wiley, New York, 1988.

Structural and Geotechnical Applications

54. Adams, V. and Askenazi, A., Building Better Products with Finite Element Analysis, OnWorld Press, Santa Fe, NM, 1999.
55. Boast, D. and Coveney, V.A., Finite Element Analysis of Elastomers, Professional Engineering Publishing, London, 1999.
56. Cook, R.D., Finite Element Modeling for Stress Analysis, John Wiley, New York, 1995.
57. Crisfield, M.A., Nonlinear Finite Element Analysis of Solids and Structures, Volume 1, Essentials, John Wiley, Chichester, England, 1991.
58. Eisley, J.G., Mechanics of Elastic Structures: Classical and Finite Element Methods, Prentice-Hall, Englewood Cliffs, NJ, 1989.
59. Hendricks, M.A.N. and Rots, J.G., Finite Elements in Civil Engineering Applications, A.A. Balkema Publishers, Lisse, The Netherlands, 2002.
60. Hinton, E., Numerical Methods and Software for Dynamic Analysis of Plates and Shells, Pineridge Press, Swansea, U.K., 1988.
61. Hoa, S.V. and Feng, W., Hybrid Finite Element Method for Stress Analysis of Laminated Composites, Kluwer Academic Publishers, Boston, MA, 1998.
62. Hoit, M., Computer-Assisted Structural Analysis and Modeling, Prentice-Hall, Englewood Cliffs, NJ, 1995.
63. Hsu, T.R., The Finite Element Method in Thermomechanics, Allen & Unwin, Boston, MA, 1986.
64. Hsieh, Y., Elementary Theory of Structures, Third Edition, Prentice-Hall, Englewood Cliffs, NJ, 1988.
65. Kleiber, M., Incremental Finite Element Modelling in Nonlinear Solid Mechanics, Ellis Horwood, John Wiley, New York, 1989.
66. Lewis, R.W. and Schrefler, B.A., The Finite Element Method in the Deformation and Consolidation of Porous Media, John Wiley, New York, 1987.
67. Lewis, R.W. and Schrefler, B.A., The Finite Element Method in the Static and Dynamic Deformation and Consolidation of Porous Media, Second Edition, John Wiley, New York, 1998.
68. Meyer, C., Finite Element Idealization, American Society of Civil Engineers, New York, 1987.
69. Owen, D.R.J. and Hinton, E. Finite Elements in Plasticity: Theory and Practice, Pineridge Press, Swansea, England, 1980.

70. Potts, D.M. and Zdravkovic, L., Finite Element Analysis in Geotechnical Engineering, Theory, Thomas Telford, London, 1999.
71. Przemieniecki, J.S., Theory of Matrix Structual Analysis, McGraw-Hill, New York, 1968.
72. Reddy, J.N., Energy Principles and Variational Methods in Applied Mechanics, Second Edition, John Wiley, New York, 2002.
73. Ross, C.T.F., Advanced Applied Finite Element Methods, Horwood Publishing, Chichester, England, 1998.
74. Shames, I.H. and Dym, C.L., Energy and Finite Element Methods in Structural Mechanics, Taylor & Francis, Bristol, PA, 1985.
75. Spotts, M.F. and Shoup, T.E., Design of Machine Elements, Seventh Edition, Prentice-Hall, Englewood Cliffs, NJ, 1998.
76. Tenek, L.T. and Argyris, J., Finite Element Analysis for Composite Structures, Kluwer Academic Publishers, Boston, MA, 1998.
77. Weaver, W. Jr. and Johnston, P.R., Finite Elements for Structural Analysis, Prentice-Hall, Englewood Cliffs, NJ, 1984.
78. Yang, T.Y., Finite Element Structural Analysis, Prentice-Hall, Englewood Cliffs, NJ, 1986.

ThermoFluid Applications

79. Comini, G., Giudice, S.D., and Nonino, C., Finite Element Analysis in Heat Transfer: Basic Formulation and Linear Problems, Taylor & Francis, Washington, DC, 1994.
80. Donea, J. and Huerta, A., Finite Element Methods for Flow Problems, John Wiley, New York, 2003.
81. Gresho, P.M. and Sani, R.L., Incompressible Flow and the Finite Element Method, Volume 2, Isothermal Laminar Flow, John Wiley, Chichester, England, 1998.
82. Jaluria, Y. and Torrance, K.E., Computational Heat Transfer, Second Edition, Taylor and Francis, New York, 2003.
83. Noda, N., Hetnarski, R.B., and Tanigawa, Y., Thermal Stresses, Lastran, Rochester, 2000.
84. Rohsenow, W.M., Hartnett, J.P., and Ganic, E.N., Handbook of Heat Transfer Applications, Second Edition, McGraw-Hill, New York, 1985.
85. Zienkiewicz, O.C. and Taylor, R.L., The Finite Element Method, Fifth Edition, Volume 3, Fluid Dynamics, Butterworth-Heinemann, Oxford, U.K., 2000.

Electromagnetics Applications

86. Chari, M.V.K. and Salon, S.J., Numerical Methods in Electromagnetism, Academic Press, San Diego, CA, 2000.
87. Pelosi, G., Coccioli, R., and Selleri, S., Quick Finite Elements for Electromagnetic Waves, Artech House, Boston, 1998.

88. Sadiku, M.N.O., Numerical Techniques in Electromagnetics, Second Edition, CRC Press, Boca Raton, FL, 2001.
89. Salazar-Palma, M., Sarkar, T.K., Garcia-Castillo, L., Roy, T., and Djordjevic, A., Iterative and Self-Adaptive Finite-Elements in Electromagnetic Modeling, Artech House, Boston, MA, 1998.
90. Silvester, P.S. and Ferrari, R.L., Finite Elements for Electrical Engineers, Cambridge University Press, Cambridge, MA, 1983.
91. Swanson, D.G. Jr., and Hoefer, W.J.R., Microwave Circuit Modeling Using Electromagnetic Field Simulation, Artech House, Boston, MA, 2003.
92. Volakis, J.L., Chatterjee, A., and Kempel, L.C., Finite Element Method for Electromagnetics, IEEE Press, New York, 1998.

Related Topics

93. Anderson, T.L., Fracture Mechanics, Fundamentals and Applications, Second Edition, CRC Press, Boca Raton, FL, 1995.
94. Basar, Y. and Weichert, D., Nonlinear Continuum Mechanics of Solids: Fundamental Mathematical and Physical Concepts, Springer-Verlag, New York, NY, 2000.
95. Bhatti, M.A., Practical Optimization Methods, Springer-Verlag, New York, 2000.
96. Bickford, W.B., Advanced Mechanics of Materials, Addison-Wesley, Menlo Park, CA, 1998.
97. Bonet, J. and Wood, R.D., Nonlinear Continuum Mechanics for Finite Element Analysis, Cambridge University Press, Cambridge, England, 1997.
98. Budynas, R.G., Advanced Strength and Applied Stress Analysis, Second Edition, WCB McGraw-Hill, Boston, MA, 1999.
99. Carroll, M.M. and Hayes, M.A., Nonlinear Effects in Fluids and Solids, Plenum Press, New York, 1996.
100. Chen, W.F., Plasticity in Reinforced Concrete, McGraw-Hill, New York, 1982.
101. Chen, W.F. and Han, D.J., Plasticity for Structural Engineers, Springer-Verlag, New York, 1988.
102. Cook, R.D. and Young, W.C., Advanced Mechanics of Materials, Second Edition, Prentice-Hall, Englewood Cliffs, NJ, 1999.
103. Desai, C.S. and Siriwardane, H.J., Constitutive Laws for Engineering Materials with Emphasis on Geologic Materials, Prentice-Hall, Englewood Cliffs, NJ, 1984.
104. Fischer-Cripps, A.C., Introduction to Contact Mechanics, Springer-Verlag, New York, 2000.
105. Fung, Y.C., Foundations of Solid Mechanics, Prentice-Hall, Englewood Cliffs, NJ, 1965.
106. Fung, Y.C., A First Course in Continuum Mechanics, Third Edition, Prentice-Hall, Englewood Cliffs, NJ, 1994.
107. Harman, T.L., Dabney, J.B., and Richert, N.J., Advanced Engineering Mathematics with Matlab, Second Edition, Brooks/Cole, Pacific Grove, CA, 2000.

108. Haslinger, J. and Neitaanmaki, P., Finite Element Approximation for Optimal Shape, Material, and Topology Design, Second Edition, John Wiley, New York, 1996.

109. Hjelmstad, K.D., Fundamentals of Structural Mechanics, Prentice-Hall, Englewood Cliffs, NJ, 1997.

110. Hyer, M.W., Stress Analysis of Fiber-Reinforced Composite Materials, WCB McGraw-Hill, Boston, MA, 1998.

111. Kikuchi, N., Finite Element Methods in Mechanics, Cambridge University Press, Cambridge, England, 1986.

112. Mura, T. and Koya, T., Variational Methods in Mechanics, Oxford University Press, Oxford, England, 1992.

113. Narasimhan, M.N.L., Principles of Continuum Mechanics, John Wiley, New York, 1993.

114. Pilkey, W.D. and Wunderlich, W., Mechanics of Structures: Variational and Computational Methods, CRC Press, Boca Raton, FL, 1994.

115. Wilde, A., A Dual Boundary Element Formulation for Three-Dimensional Fracture Analysis, WIT Press, Southampton, U.K., 2000.

INDEX